Handbook of
Australasian Biogeography

CRC Biogeography Series

Series Editor

Malte C. Ebach

School of Biological, Earth and Environmental Sciences, Australia, University of New South Wales

Neotropical Biogeography: Regionalization and Evolution, *Juan J. Morrone*
Handbook of Australasian Biogeography, *Malte C. Ebach*
Biogeography and Evolution in New Zealand, *Michael Heads*

Handbook of Australasian Biogeography

Edited by

Malte C. Ebach

School of Biological, Earth and Environmental Sciences, Australia,
University of New South Wales

CRC Press
Taylor & Francis Group
Boca Raton London New York

CRC Press is an imprint of the
Taylor & Francis Group, an **informa** business

Front Cover: Kings Tableland and Jamison Valley, Wentworth Falls, Greater Blue Mountains Area UNESCO World Heritage Site, New South Wales, Australia 2016. Photo copyright Melina L. Tursky. Used with permission. All Rights Reserved.

CRC Press
Taylor & Francis Group
6000 Broken Sound Parkway NW, Suite 300
Boca Raton, FL 33487-2742

Printed on acid-free paper
Version Date: 20161013

International Standard Book Number-13: 978-1-4822-3636-1 (hardback)

Library of Congress Cataloging-in-Publication Data

Names: Ebach, Malte C.
Title: Handbook of Australasian biogeography / [edited by] Malte C. Ebach.
Description: Boca Raton : CRC Press, 2017. | Includes bibliographical references.
Identifiers: LCCN 2016033795| ISBN 9781482236361 (hardback : alk. paper) | ISBN 9781315373096 (ebook) | ISBN 9781482236378 (ebook) | ISBN 9781315355771 (ebook) | ISBN 9781315336718 (ebook)
Subjects: LCSH: Biogeography--Australasia.
Classification: LCC QH84.3 .H36 2017 | DDC 578.099--dc23
LC record available at https://lccn.loc.gov/2016033795

Visit the Taylor & Francis Web site at
http://www.taylorandfrancis.com

and the CRC Press Web site at
http://www.crcpress.com

Contents

Contents

Preface

The present work is borne out of a frustration at the lack of a single reference work that covers the entire Australasian biogeography taxon by taxon. The last major attempt was the *Monographiae Biologicae* edited by Illes for Dr. W. Junk Publishers. Volumes 25, 27, 41 and 42 cover a total of six tomes: *Biogeography and Ecology in Tasmania*, edited by Williams (1974, 1 volume), *Biogeography and Ecology in New Zealand* (Kuschel 1975, 1 volume), *Ecological Biogeography of Australia*, edited by Keast (1981, 3 volumes) and *Biogeography and Ecology of New Guinea*, edited by Gressitt (1982, 2 volumes). These works included biogeographic and ecological revisions of taxa and vegetation. Succeeding volumes were method specific, such as the 'Panbiogeography Special Issue' of the *New Zealand Journal of Zoology* (Matthews 1989) and 'Austral Biogeography' issue of *Australian Systematic Botany* (republished as Ladiges *et al.* 1991), or they were taxon specific, such as *Fauna of Australia, Volume 1* (Dyne and Walton 1987), and both editions of *Flora of Australia, Volume 1* (George 1981; Orchard and Thompson 1999) also contain important chapters on the biogeography of Australian fauna and flora (Heatwole 1987; Barlow 1981; Crisp *et al.* 1999). Other notable taxon-specific works include *Vertebrate Zoogeography and Evolution in Australasia* (Archer and Clayton 1984), *Evolution and Biogeography of Australasian Vertebrates* (Merrick *et al.* 2006), *Ecology* of *Australian Freshwater Fishes* (Humphries and Walker 2013) and *New Zealand Freshwater Fishes* (McDowall 2010), or geographic/taxon syntheses such as *Biogeography of Australasia* (Heads 2014) and *Biogeography and Evolution of New Zealand* (Heads 2016).

The chapters in this book are biogeographic revisions/syntheses of significant taxonomic groups, including algae, plants, fungi, insects, arachnids, marine invertebrates, marine fishes, reptiles, amphibians, and mammals, including a chapter on our current understanding of Australasian biodiversity. Recent biogeographic revisions, however, are not included in this book, such as freshwater fishes. While not covering all organisms (e.g. bacteria, freshwater planarians, freshwater crustaceans), this volume is part of the CRC Biogeography Series, and elements that may be missing from New Zealand biogeography, for example, are covered in Volume 1 of this series, *Biogeography and Evolution of New Zealand* (Heads 2016). I have also decided not to include the customary introductory palaeogeography/geology chapter, as these date quickly and rarely highlight the many disagreements in palaeogeographical reconstruction, such as neotectonics, fission tracking and traditional geomorphological approaches (Quigley *et al.* 2010). I would rather point researchers in Australasian biogeography towards the current literature.

The majority of authors focus on the recent biogeography literature, which for some taxonomic groups is greater in size than for others (hence the various sizes of the chapters). As this book is an account of the recent literature, I direct the reader to Keast (1981), Barlow (1981), Dyne and Walton (1987) and Ebach (2012, 2017) for the early history of Australasian biogeography.

I am indebted to the authors and reviewers for helping to produce a solid text that will be a reference for Australasian biogeographers for years to come. I thank my editor John Sulzycki of CRC/Taylor & Francis for endorsing the idea of a book series and a book on Australasian biogeography. Thanks also to Jill Jurgensen and Jennifer Blaise for their help with preparing the final manuscript.

Malte C. Ebach
Kensington, New South Wales, Australia

References

Archer, M., and Fox, B. (1984) Background to vertebrate zoogeography in Australia. In Archer, M., and Clayton, G. (Eds.), *Vertebrate Zoogeography and Evolution in Australasia*, pp. 1–15. Hesperian Press, Perth, Australia.

Barlow, B.A. (1981) The Australian flora: Its origin and evolution. In George, A.S. (Ed.), *Flora of Australia, Volume 1, Introduction*, pp. 25–75. Australian Government Publishing Service for Bureau of Flora and Fauna, Canberra, Australia.

Crisp, M.D., West, J.G., and Linder, H.P. (1999) Biogeography of the terrestrial flora. In Orchard, A.E., and Thompson, H.S. (Eds.) *Flora of Australia, Volume 1*, Second Edition, pp. 321–367. CSIRO Publishing, Melbourne, Australia.

Dyne, G.R., and Walton, D.W. (1987) *Fauna of Australia, Volume 1A, General Articles*. Australian Government Publishing Service, Canberra, Australia.

Ebach, M.C. (2012) A history of biogeographical regionalisation in Australia. *Zootaxa* 3392: 1–34.

Ebach, M.C. (2017) *Reform, Revolt and Revival: The Impact of Biogeography in Australasia*. CSIRO Publishing, Melbourne, Australia.

George, A.S. (Ed.) (1981) *Flora of Australia, Volume 1, Introduction*. Australian Government Publishing Service for Bureau of Flora and Fauna, Canberra, Australia.

Gressitt, J.L. (1982) *Biogeography and Ecology of New Guinea*. Dr. W. Junk Publishers, The Hague, The Netherlands.

Heads, M. (2014) *Biogeography of Australasia: A Molecular Analysis*. Cambridge University Press, Cambridge, UK.

Heads, M. (2016) *Biogeography and Evolution in New Zealand*. CRC Press, Boca Raton, FL.

Heatwole, H. (1987) Major components and distributions of the terrestrial fauna. In Dyne, G.R., and Walton, D.W. (Eds.), *Fauna of Australia, Volume 1A, General Articles*, pp. 101–135. Australian Government Publishing Service, Canberra, Australia.

Humphries, P., and Walker, K. (2013) *Ecology of Australian Freshwater Fishes*. CSIRO Publishing, Collingwood, Australia.

Keast, A. (1981) *Ecological Biogeography of Australia*. Dr. W. Junk, The Hague, the Netherlands.

Kuschel, G. (1975) *Biogeography and Ecology in New Zealand*. Dr. W. Junk, The Hague, the Netherlands.

Ladiges, P.Y., Humphries, C.J., and Martinelli, L.W. (1991) *Austral Biogeography*. CSIRO, Canberra, Australia.

Matthews, C. (Ed.) (1989) Panbiogeography special issue. *New Zealand Journal of Zoology* 16: 471–815.

McDowall, R.M. (2010) *New Zealand Freshwater Fishes: An Historical and Ecological Biogeography*. Springer Science & Business Media, New York.

Merrick, J.R., Archer, M., Hickey, G.M., and Lee, M.S.Y. (2006) *Evolution and Biogeography of Australasian Vertebrates*. Auscipub, Sydney, Australia.

Orchard, A.E., and Thompson, H.S. (1999) *Flora of Australia, Volume 1*, Second Edition. CSIRO Publishing, Melbourne, Australia.

Quigley, M.C., Clark, D., and Sandiford, M. (2010) Tectonic geomorphology of Australia. In Bishop, P., and Pillans, B. (Eds.), *Australian Landscapes*, Special Publications 346, pp. 243–265. Geological Society, London.

Williams, W.D. (1974) Biogeography and Ecology in Tasmania. Dr. W. Junk, The Hague, The Netherlands.

Contributors

Shane T. Ahyong
Australian Museum, Sydney, Australia and School of Biological, Earth and Environmental Sciences
University of New South Wales
Sydney, Australia

Robin M.D. Beck
School of Environment and Life Sciences
University of Salford
Salford, United Kingdom

and

School of Biological, Earth and Environmental Sciences
University of New South Wales
Sydney, Australia

Gerasimos Cassis
School of Biological, Earth and Environmental Sciences
University of New South Wales
Sydney, Australia

David G. Chapple
School of Biological Sciences
Monash University
Melbourne, Australia

Alex Coughlan
CSIRO Wealth from Oceans Flagship
Brisbane, Australia

Roberta A. Cowan
School of Veterinary and Life Sciences
Murdoch University
Murdoch, Australia

and

Western Australian Herbarium
Department of Parks and Wildlife
Bentley, Australia

Darren M. Crayn
Australian Tropical Herbarium
James Cook University
Cairns, Australia

Olivier De Clerck
Department of Biology
Ghent University
Ghent, Belgium

Malte C. Ebach
Palaeontology, Geobiology and Earth Archives Research Centre (PANGEA)
School of Biological, Earth and Environmental Sciences
University of New South Wales
Sydney, Australia

Anthony C. Gill
Macleay Museum and School of Biological Sciences
The University of Sydney
Sydney, Australia

and

Australian Museum
Sydney, Australia

Gonzalo Giribet
Museum of Comparative Zoology
Department of Organismic and Evolutionary Biology
Harvard University
Cambridge, Massachusetts, USA

Gustaaf M. Hallegraeff
Institute for Marine and Antarctic Studies
University of Tasmania
Hobart, Australia

Danilo Harms
Center of Natural History Zoological Museum
University of Hamburg
Hamburg, Germany

Mark S. Harvey
Department of Terrestrial Zoology
Western Australian Museum
Welshpool, Australia

John M. Huisman
School of Veterinary and Life Sciences
Murdoch University
Perth, Australia

and

Western Australian Herbarium
Department of Parks and Wildlife
Bentley, Australia

William F. Humphreys
Collections and Research Centre
Western Australian Museum
Welshpool, Australia

and

School of Animal Biology
University of Western Australia
Nedlands, Australia

and

Australian Centre for Evolutionary Biology and Biodiversity
School of Earth and Environmental Sciences
The University of Adelaide
Adelaide, Australia

J. Scott Keogh
Research School of Biology
The Australian National University
Canberra, Australia

J. Pat Kociolek
Museum of Natural History and Department of Ecology and Evolutionary Biology
University of Colorado
Boulder, Colorado

Shawn W. Laffan
School of Biological, Earth and Environmental Sciences
University of New South Wales
Sydney, Australia

Tom W. May
Royal Botanic Gardens Victoria
Melbourne, Australia

Randall D. Mooi
The Manitoba Museum
Winnipeg, Canada

and

Department of Biological Sciences
University of Manitoba
Winnipeg, Canada

Daniel J. Murphy
Royal Botanic Gardens Victoria
Melbourne, Australia

Mitzy Pepper
Research School of Biology
The Australian National University
Canberra, Australia

Anthony J. Richardson
CSIRO Wealth from Oceans Flagship
Division of Marine and Atmospheric Research
Ecosciences Precinct
Brisbane, Australia

and

Centre for Applications in Natural Resource Mathematics
School of Mathematics and Physics
University of Queensland
St. Lucia, Australia

Michael G. Rix
Australian Centre for Evolutionary Biology and Biodiversity and School of Earth and Environmental Sciences
The University of Adelaide
Adelaide, Australia and Biodiversity and Geosciences Program Queensland Museum
Brisbane, Australia

Cor J. Vink
Canterbury Museum
Christchurch, New Zealand

David E. Walter
University of the Sunshine Coast
Maroochydore, Australia

and

Department of Biological Sciences University of Alberta
Edmonton, Canada

David M. Williams
Department of Life Sciences
Natural History Museum
London, United Kingdom

David K. Yeates
Australian National Insect Collection
CSIRO National Research Collections Australia
Canberra, Australia

1

Biodiversity and Bioregionalisation Perspectives on the Historical Biogeography of Australia

Gerasimos Cassis, Shawn W. Laffan and Malte C. Ebach

CONTENTS

Introduction

Australia's biota has long fascinated scientists and nonscientists alike for its uniqueness and taxon richness (Keast, 1981; Cranston and Naumann, 1991; Crisp *et al.*, 1999; Yeates *et al.*, 2003; Austin *et al.*, 2004; Chapman, 2009; Cranston, 2010). In reference to these biodiversity values, much has been made of Australia's separation and isolation, and its high endemism is diagnostic for the continent (Crisp *et al.*, 1999). This is in large part an outcome of intracontinental drivers during the Late Palaeogene (Byrne *et al.*, 2008, 2011; Rix *et al.*, 2015), which implicitly characterise Australia as a biogeographic unit in itself, with many lineage radiations attributed to the aridification of Australia (Clayton, 1984; Schodde, 1989; Greenwood and Christophel, 2005). As a consequence, Australia is stamped with arid and semi-arid biogeographic regions, particularly the southwest corner of Western Australia (Hopper, 1979; Rix *et al.*, 2015; Cassis and Symonds, 2016) and the interior deserts (Cracraft, 1991; Crisp *et al.*, 1995; Byrne *et al.*, 2011), as well as mesic refugia (e.g. the Wet Tropics; Williams and Pearson, 1997; Boyer *et al.*, 2016). The counterpoint to this is that Australia is a biogeographic composite (Giribet and Edgecombe, 2006), particularly for supraspecific taxa, with multiple and in most cases older histories, including a replicated east Gondwanan signature, which couples Australia's biota with New Zealand (Stow *et al.*, 2015), the rises (Lord Howe and Norfolk Island), Melanesia (New Guinea, New Caledonia, the southwest Pacific archipelagos) (Burbidge, 1960), and cool, temperate South America (Brundin, 1966; Swenson *et al.*, 2001). In contrast, other components of Australia's biota have a palaeotropical signature (Herbert, 1932; Webb and Tracey, 1981), with some taxon–area relationships explained by an Indo-Pacific model, which connects monsoonal Australia to West Africa through the Indian subcontinent (Schuh and Stonedahl, 1986), or with more restricted area relationships, such as mammal taxa east of Lydekker's Line (Simpson, 1977). More recently, there has been characterisation of the Australian Monsoon Tropics, highlighting its biogeographic complexity, intermixing palaeotropical and Gondwanan elements as well as local endemics (Bowman *et al.*, 2010).

For 150 years, biogeographers have searched for overarching theories to explain the origins and diversification of Australia's biota (e.g. Cracraft, 1991; Crisp *et al.*, 1995, 2009). Explanations have vacillated between invasion (Hedley, 1893; Burbidge, 1960; Heatwole, 1987; Byrne *et al.*, 2011) and vicariant models (Cracraft, 1991; Unmack, 2001), and combinations thereof (Sanmartín and Ronquist, 2004). These alternative models mirror in part paradigmatic shifts in biogeographic theory and practice, such as the incorporation of continental drift theory (Nelson and Rosen, 1981), integrated methodologies (viz. patterns of distribution and phylogenetics; Nelson and Platnick, 1981; Humphries and Parenti, 1999; Parenti and Ebach, 2009; Morrone, 2013), and bioregional classifications (e.g. elements, biomes, areas of endemism, hotspots) (Burbidge, 1960; Byrne *et al.*, 2008, 2011; Crisp *et al.*, 2009; Ebach *et al.*, 2013, 2015), and, more recently, routinely include spatial analysis (Crisp *et al.*, 2001; Laffan and Crisp, 2003; González-Orozco *et al.*, 2013, 2014a,b).

Historical biogeography is now in a transformational period and is demonstrably more hypothesis driven (Crisp *et al.*, 2011). Having said this, the field is not conceptually united; there are disputes about methods (Morrone, 2013) and modes of diversification (Ladiges *et al.*, 2012; Heads, 2014; Crisp *et al.*, 2004; e.g. centres of origin, progression rule), the vicariance and dispersalist divide persists (cf. Heads, 2015 and McGlone, 2016), and there are data impediments, such as sampling inadequacy and a lack of fossil data for most lineages. In cases where fossil calibrations are available, the use of divergence dating (Crisp *et al.*, 2004; Rix *et al.*, 2015) has become a favoured line of evidence, providing biogeographers with a means to differentiate between dispersal and earth history events. This has resulted in rewritings of Australian biogeography, such as the overturning of the iconic Gondwanan vicariant hypothesis for *Nothofagus* (Cook and Crisp, 2005) and a hypothesis that cycad distributions in Central Australia is not an ancient relic (Ingham *et al.*, 2013).

Regardless of conceptual divides, historical biogeography has at its base a requirement for taxonomic knowledge, phylogenetic reconstruction and bioregional classifications. Despite the sizeable taxonomic impediment in Australia (Taylor, 1983; Cassis *et al.*, 2007; Chapman, 2009), there has been a concerted effort of late to compile and distribute information on Australia's biota, through big data and online exercises such as the *Atlas of Living Australia*, *Australia's Virtual Herbarium*, the *Online Zoological Collections of Australian Museums* and the Bush Blitz (Preece *et al.*, 2015) species discovery program. Online catalogues and authoritative nomenclatorial lists in turn backstop these portals, with the *Australian Faunal Directory*, the *Australian Plant Census* and the *Australian Plant Name Index* representing the world's best practice.

With the purpose of providing a biodiversity baseline for historical biogeographic analysis, this chapter provides a brief overview of Australia's biodiversity from a taxon perspective. This includes the documentation of species richness and endemism by higher taxonomic categories, revising numbers given by Chapman (2009), particularly for terrestrial animals and vascular plants. It also touches on the ongoing integration of ecological and historical approaches to biogeography, through species modelling (Nix and Switzer, 1991; Franklin, 2010; Hallgren *et al.*, 2016) and other spatial analysis tools (Laffan *et al.*, 2010). Such techniques result in the agglomeration of species distributions and the derivation of biodiversity patterns such as species richness, endemism, phylogenetic endemism and hotspots (Morrone, 2013). Lastly, we present an assessment of the distribution patterns for a set of biogeographic surrogates, including invertebrate, vertebrate, vascular plant and fungal lineages, based on a harvesting of data from the *Atlas of Living Australia* and using the *Biodiverse* software (Laffan *et al.*, 2010).

Biodiversity Perspective: How Many Australian and Planetary Species Are There?

Australia is notable for its high species richness and endemism, and is recognised as one of 17 megadiverse countries (World Conservation Monitoring Centre, 2000), with the southwest of Western Australia categorised as one of only 34 global biodiversity hotspots (Myers *et al.*, 2000; Mittermeier *et al.*, 2011). Hyperdiverse lineages such as emu bush (Chinnock, 2007; *Eremophila*) and bulldog ants (Ogata and Taylor, 1991; *Myrmecia*) also highlight the uniqueness of Australia's biota, with significant genus-level endemicity and intracontinental diversifying processes.

General knowledge about the biogeography of Australia is impeded by a lack of taxonomic knowledge of both Australian and planetary species. The question about how many species exist is a perennial question in biology (May, 1988; Stork, 1988; Erwin, 1991; Stork *et al.*, 2015), with ca. two million formally described worldwide (Chapman, 2009). Since the breakthrough canopy knockdown study on beetles in Panama (Erwin,

1982) this question came into sharper focus. Erwin's study spawned planetary species richness estimates that ranged wildly between 2 and 80 million species (summary in Cassis *et al.*, 2007). More recent estimates have been more conservative, with one highly cited paper predicting 8.7 million planetary species (Mora *et al.*, 2011). Although there is tacit agreement among many taxonomists that there are less than 10 million planetary species, Caley *et al.* (2014) rightly argues that there is no convergence in estimates based on independent lines of evidence (e.g. species interactions, body size and rates of description/synonymy).

Chapman (2009), in his seminal taxon-based biodiversity study of Australia, compared the described and estimated number of species of Australia with those assembled for the world. In comparison with Mora *et al.* (2011), Chapman tabulated almost 1.9 million described and over 11.3 million estimated planetary species (Table 1.1). In comparison, he estimated that Australia carries 566,398 species, of which 147,579 are described. From 2009 to the present, an additional 14,838 species have been either described and/or catalogued, for animals and vascular plants (Table 1.1; note all planetary and estimated Australian species have not been reassessed beyond Chapman, 2009), based on records derived from the *Australian Plant Census*, the *Australian Plant Name Index* and the *Australian Faunal Directory*. On this basis, Australia comprises 8.6% of described and 5% of estimated planetary species (Table 1.1). As expected, invertebrates comprise almost 70% of the described species in Australia, which is comparable to that found on a worldwide basis (i.e. 72%). Plants are the next most speciose clade, comprising 15.7% of the Australian biota, which is commensurate with the ubiquity of insect–plant interactions in terrestrial ecosystems (Grimaldi and Engel, 2005; Janz, 2011).

In terms of estimated species, the percentage of unknown species in Australia is 71%, which is a little less than unknown planetary species estimates (84%), but significantly greater than most Northern Hemisphere countries, especially for insects (Taylor, 1983; Cassis *et al.*, 2007). The percentage of unknown Australian invertebrates decreases relative to described invertebrates, and this reflects the enhanced taxonomic activity on terrestrial arthropods due to two interdisciplinary programs: the

TABLE 1.1

Numbers and Percentages of Described and Estimated Planetary and Australian Species by Major Clade and in Total

Taxon	Planetary Described	Planetary Estimated	Australia Described	Australia Estimated	Australia: Planetary Described (%)	Australia: Planetary Estimated (%)	Australia Described (%)	Australia Estimated (%)
Chordates	64,788	~80,500	9,991[a]	~9,088	15.4%	11.3%	6.2%	1.6%
Invertebrates	1,359,365	~6,755,830	110832[a]	~320,465	8.2%	4.7%	68%	56.6%
Plants	310,129	~390,800	25,562[b]	26,845	8.2%	6.9%	15.7%	4.7%
Fungi[c]	98,998	1,500,000	11,846	50,000	12%	3.3%	7.3%	8.8%
Others[d]	~66,307	2,600,500	>4,186	~160,000	6.3%	6.2%	2.6%	28.3%
Total	1,899,587	~11,327,630	162,417	~566,398	8.6%	5%	—	—

Note: Percentages are given for the number of Australian species relative to planetary species for described and estimated species, and for described and estimated species by higher taxon for Australia. The data are chiefly redrawn from Chapman (2009), with updates for the number of described species from the *Australian Faunal Directory* for 'Chordates' and 'Invertebrates' (from 1758–present), and for 'Plants' from the *Australian Plant Census* and the *Australian Plant Name Index* (1758–2014). The described and estimated number of planetary species, and the estimated number of Australian species, are identical to those of Chapman (2009).

[a] Updates to Chapman (2009) for 'Chordates'. Chapman lists 8128 species. The updated number given in the table of 9991 is based on 9226 for all vertebrates from the *Australian Faunal Directory* (accessed 2 April 2016) plus the Cephalochordata and Tunicata numbers from Chapman (2009) of 8 and 757 described species, respectively, and ~8 and ~850 estimated species, respectively. For 'Invertebrates' Chapman (2009) lists 98,703 species; the updated number given in the table is derived from the *Australian Faunal Directory* (accessed 2 April 2016).

[b] For 'Plants' Chapman gives 24,716 species inclusive of vascular plants, bryophytes and algae. The updated number given in the table is based on 20,170 for vascular plants derived from the *Australian Plant Census* and the *Australian Plant Index* (accessed 20 April 2016), plus Chapman's (2009) numbers for Bryophyta and Algae of 1847 and ~3545 described species, respectively.

[c] Data for 'Fungi' include lichens and are redrawn from Chapman (2009).

[d] Data for 'Others' are redrawn from Chapman (2009) and refer mostly to single-celled organisms, including Prokaryota, Chromista, Protoctista and Cyanophyta.

Planetary Biodiversity Inventory (e.g. Baehr *et al.*, 2010; Valerio *et al.*, 2013) and Bush Blitz programs (e.g. Lambkin and Bartlett, 2011; Cassis and Symonds, 2016). Chapman (2009; also repeated in Table 1.1) estimated that, after the invertebrates, prokaryotes and their allies represent the greatest unknown species diversity in Australia in comparison with the other higher taxa tabulated. It is also well accepted that the fungi have a significant taxonomic impediment, particularly for the microfungi, and Chapman's numbers are a likely underestimate.

Based on updated described species numbers for vascular plants (Table 1.1), there is apparent near taxonomic saturation (95%), although the curve of accumulated described species was elevated between 1980 and 2008 (Figure 1.1). In comparison, the taxonomic accumulation curve for animals has never approached a horizontal asymptote since the *Systema Naturae* (Linnaeus, 1758), with elevated species descriptions since the 1970s (Figure 1.2). In Table 1.1, described species counterintuitively exceed the estimated species, which is undoubtedly impacted by the recent discovery of cryptic species and the enhanced cataloguing of marine and freshwater fishes.

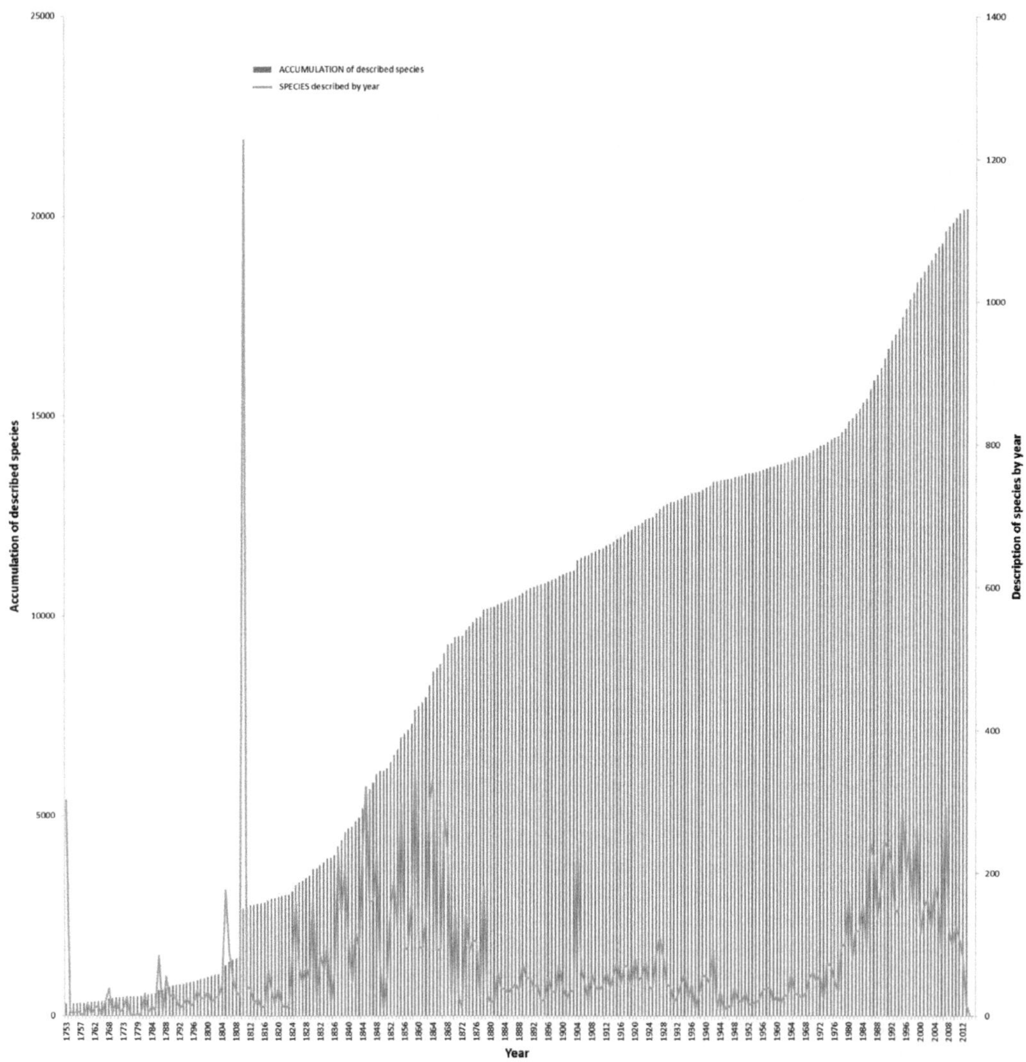

FIGURE 1.1 (See colour insert.) Taxonomic accumulation of Australian plant species (1758–2014). (Data from *Australian Plant Census*, IBIS database, Centre for Australian National Biodiversity Research, Council of Heads of Australasian Herbaria, 2016, http://www.chah.gov.au/apc/index.html; *Australian Plant Name Index*, IBIS database, Centre for Australian National Biodiversity Research, Australian Government, Canberra, Australia, 2016, http://www.cpbr.gov.au/cgi-bin/apni.)

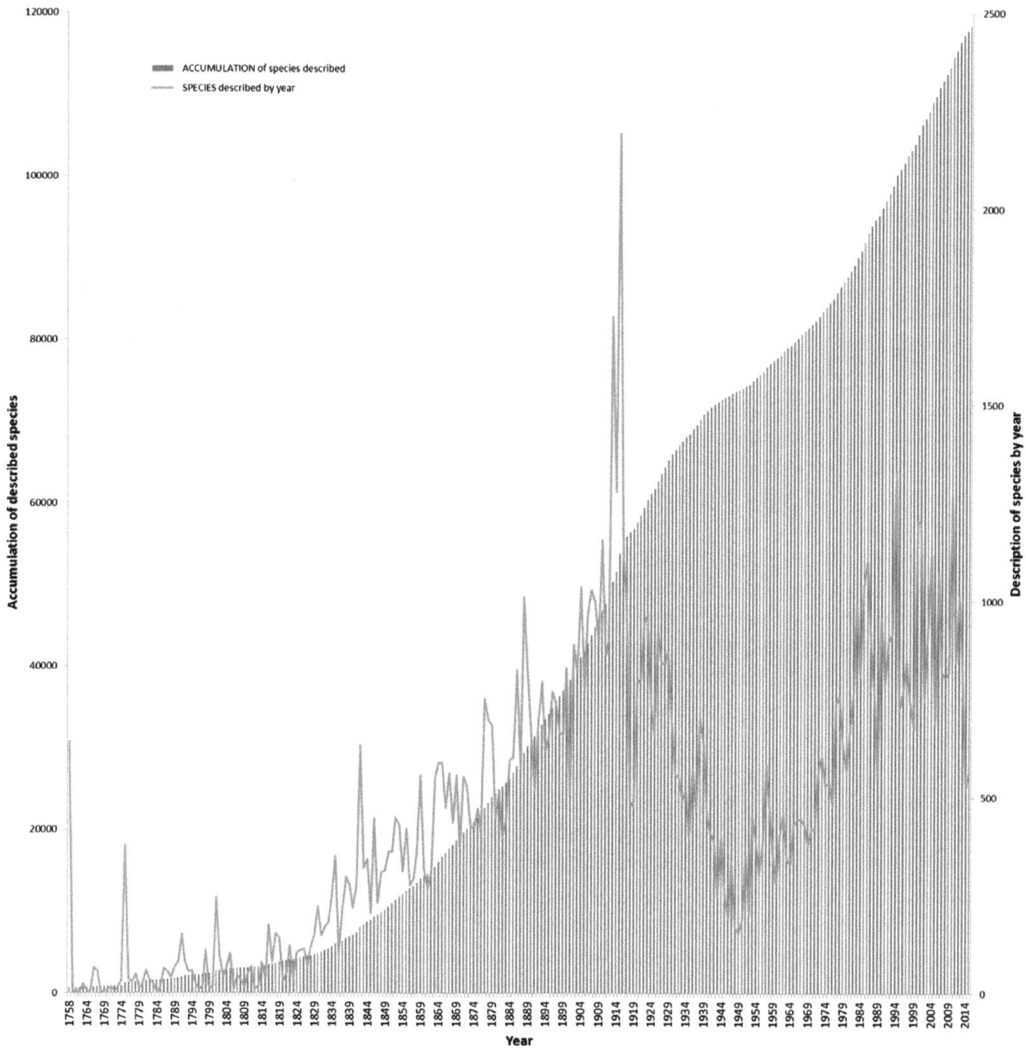

FIGURE 1.2 (See colour insert.) Taxonomic accumulation of Australian animal species (1758–2015). (Data from *Australian Faunal Directory*, Australian Biological Resources Study, Canberra, Australia, 2009, http://www.environment.gov.au/biodiversity/abrs/online-resources/fauna/afd/index.html.)

Atlas of Living Australia and Bioregional Analysis: Richness, Endemicity and Sampling Adequacy

The *Atlas of Living Australia* provides a freely available portal to the taxonomy of Australia's species and their distribution records. It currently stores more than 50 million occurrence records, something inconceivable 20 years ago (Belbin, 2011). Mapping the distribution of Australia's biodiversity is now routinely achieved on a continental scale and enables biogeographic analyses of multiple taxa (e.g. species richness and endemism), as well as assessing sampling adequacy and identifying new areas for biodiversity surveys.

In a novel analysis, we acquired species distribution data from the *Atlas of Living Australia* for a set of 19 exemplar taxon groups across four major groups (vascular plants, fungi, vertebrates and invertebrates; Table 1.2). The selection of these groups was subjective and thus these results serve as indicators of sampling and broad patterns rather than being definitive. Birds were not analysed because these comprise

BOX 1.1 BIOREGIONALISATION PERSPECTIVE: ELEMENTS, PROVINCES, ENDEMIC AREAS, BIOMES

The division of Australia into biogeographic regions has a long history, having the common aim to determine areas of biotic overlap (Keast, 1981; Cranston and Naumann, 1991). The first phytogeographical regions were proposed by von Mueller (1858), which he based on vegetation types (e.g. plants of the dense coast forests, plants of the desert). Tate's (1889) phytogeographical areas were based on rainfall, the assumed driver of distribution; he also foresaw them as 'species [drawn] into physiographic and regional complexes'. Drude (1890) and Diels (1906) used physiographic and climatic factors in their vegetation classification of Australia. This combination of vegetation and climate came to be known as *biomes*, which sometimes included *elements*. Unlike an *area*, elements were defined as species that had a designated origin. Tate (1889) recognised two 'immigrant' elements, the Oriental and Andean, and an Australian element. Cockayne (1921) recognised seven elements: endemic, palaeozelandic, Australian, subantarctic, palaeotropical, cosmopolitan, Lord Howe and Norfolk Island.

Burbidge (1960) replaced Tate's (1889) Australian elements and erected the Australian, Melanesian and Indo-Malayan elements. Elements underpinned Hedley's (1893) *invasion* hypothesis, which proposed that the biota of Australian and New Zealand arrived as two northern invasions via the Melanesian Archipelago (with southern elements arriving via an Antarctic land bridge). This hypothesis dispensed with land bridges between Australia and New Zealand. Wallace (1880) and Hutton (1896) proposed geological scenarios such as sunken Pacific continents and the flooding of the interior of Australia to explain biotic similarities between Australia and New Zealand. This gave rise to the east–west divide of the Australian biota, in which the eastern half was connected to New Zealand and the western half, particularly the southwest, was isolated. Hedley (1893) critiqued the east–west hypothesis, stating, 'Most European writers who have touched on the zoogeography of Australia have described the fauna and flora as falling into a temperate and a tropical division, which again subdivide into eastern and western sections. A little real experience proves these divisions to be quite artificial.' Like land bridges, the invasion hypothesis was abandoned.

Tenison-Woods (1878, 1882) proposed the first zoogeographical areas, which Hedley (1904) rejected as being 'neither natural nor well defined'. Hedley's use of the term *natural* is indicative of how early biogeographers sought to discover natural areas, even resulting in a debate between Hedley and Wallace in 1900 about the role of New Zealand within a natural classification (i.e. Australasia) – a debate that lasted throughout the twentieth century and into the twenty-first century (Udvardy, 1987; Fleming, 1987; Pole, 1994; Campbell, 2013). The drowning of New Zealand is perhaps the most recent attempt to address this question. Had New Zealand emerged from the sea, its surface barren of terrestrial life, 30 million years ago, then there would be good cause to suggest that it is an oceanic island within Australasia (or even an overlap zone of Pacific and Australasian regions). These debates re-emerge and are abandoned as biogeographers switch between classifications of biomes and endemic areas.

By the 1980s and the development of cladistics and further discoveries of fossil pollen had shown that most Australian and New Zealand taxa were endemic (Archer and Fox, 1984; Webb and Tracey, 1981; Schodde, 1989). Cladistic biogeographers adopted endemic areas during the late 1980s and 2000s (e.g. Schodde, 1989; Cracraft, 1991; Crisp *et al.*, 1995, 1999; Cassis and Symonds, 2011; Ladiges *et al.*, 2005, 2001; Unmack, 2001). The move from a division of north–south, east–west biomes to endemics areas and elements and the switch from an invasion hypothesis to an endemic Australasian biota was short-lived. By the turn of the twenty-first century, there is a return to notions of east–west, north–south biomes, with tropical elements invading from the north (Crisp *et al.*, 2004, 2009; Byrne *et al.*, 2008; Bowman *et al.*, 2010).

TABLE 1.2

Taxonomic Exemplars of Plants, Fungi, Vertebrates and Invertebrates Used in the Biodiversity Analyses

Group	Taxon	Taxonomic Rank and Name
Plants	*Acacia*	Genus: *Acacia*
Plants	*Daviesia*	Genus: *Daviesia*
Plants	*Eremophila*	Genus: *Eremophila*
Plants	Eucalypts	Genus: *Corymbia*, *Eucalyptus*, *Symphyomyrtus*
Fungi	Fungi	Kingdom: Fungi
Vertebrates	Amphibians	Family: Hylidae, Microhylidae, Myobatrachidae
Vertebrates	Geckos	Suborder: Gekkota
Vertebrates	Mammals	Class: Mammalia (excluding Order: Artiodactyla, Carnivora, Cetacea, Insectivora, Lagomorpha, Perissodactyla, Sirenia; Species: *Mus musculus*, *Rattus rattus*, *R. norvegicus*)
Vertebrates	Skinks	Suborder: Lacertilia
Vertebrates	Snakes	Suborder: Serpentes
Invertebrates	Acari	Suprageneric: Acari
Invertebrates	Araneae	Order: Araneae
Invertebrates	Camaenidae	Family: Camaenidae
Invertebrates	Carabidae	Family: Carabidae
Invertebrates	Diptera	Order: Diptera
Invertebrates	Formicidae	Family: Formicidae
Invertebrates	Hemiptera	Order: Hemiptera
Invertebrates	Hymenoptera	Order: Hymenoptera
Invertebrates	Lepidoptera	Order: Lepidoptera

more than 30 million records, greatly exceeding the *Atlas of Living Australia* download limits in place at the time of this analysis.

Records were downloaded for each group and then filtered to exclude those without names at the species level, as well as all crosses and unnamed species (e.g. '*Acacia* sp.'). Variety and subspecies records were used at the species level. Any records flagged as suspicious were removed, as were those with coordinate uncertainty values greater than 10 km. Data for each group were then projected into an Albers equal area coordinate system (EPSG:3577) and aggregated to 50 × 50 km cells. Any cells that did not overlap with Australia and its proximal offshore islands were then excluded, resulting in a maximum of 3437 possible cells for any group.

For each major group we calculated estimates of species richness (Figure 1.3), weighted endemism (WE) (Figure 1.4) and sample redundancy (Figure 1.5) for each 50 km cell using *Biodiverse* version 1.99_002 (Laffan *et al.*, 2010). Species richness is simply the count of unique species in each cell. WE is a measure of relative, as opposed to absolute, endemism (Crisp *et al.*, 2001) and is interpreted as a range-weighted richness score. It is calculated as a weighted sum of species in a location (each cell in this analysis), where the weight of each species is the fraction of its geographic range found in that location. A species range in this analysis is the number of cells containing it. Thus, for the single-cell analyses used here, a species found in two cells contributes a weight of 0.5, a species found in 10 cells contributes 0.1 and a species found in 500 cells contributes 0.002. Sample redundancy (Garcillán *et al.*, 2003) is calculated as the ones' complement of the ratio of the number of species in a cell to the number of samples. It has a value of 0 when there is only one sample per species and approaches 1 as the average number of samples approaches infinity. There are no clear guidelines for what constitutes a 'good' level of sample redundancy, but a value of 0.3 could be considered a useful benchmark, as in that case each species should have, on average, close to 1.4 samples.

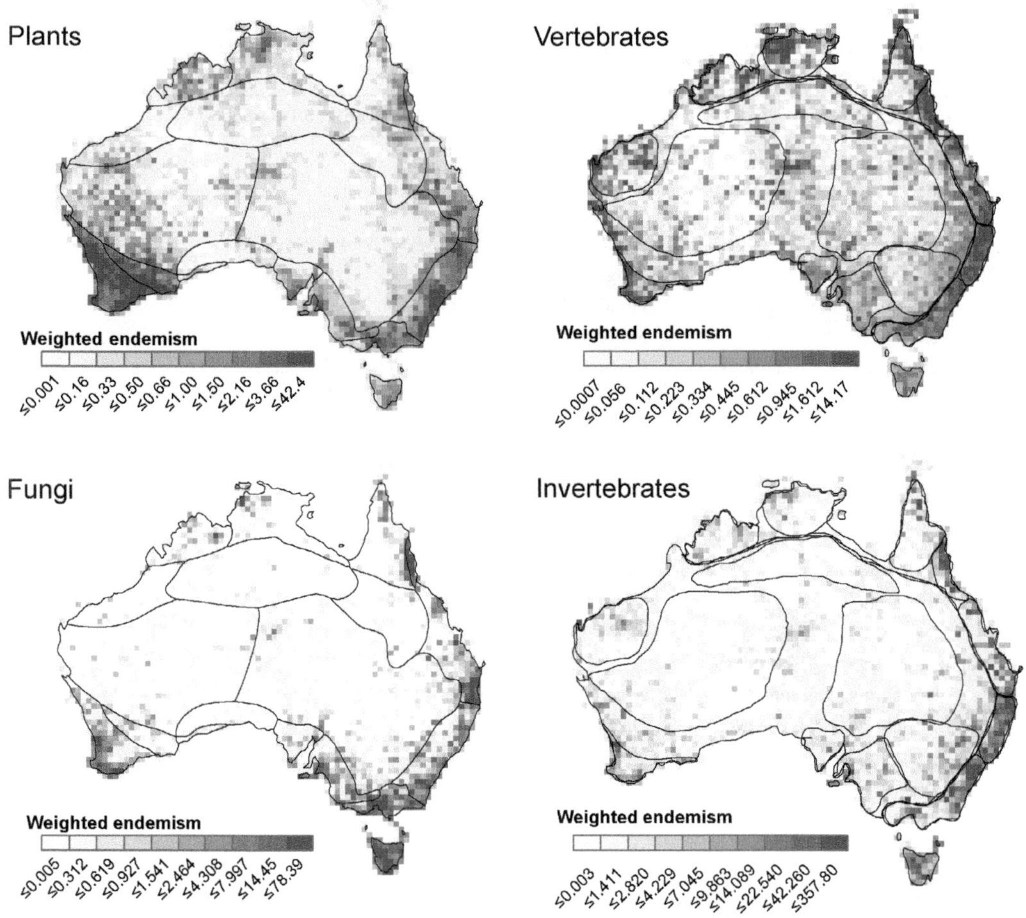

FIGURE 1.3 (See colour insert.) Species richness for exemplars of plants, fungi, vertebrates and invertebrates (Table 1.2). Polygon outlines are the phytogeographical or zoogeographical subregions defined in the *Australian Bioregionalisation Atlas*. Colours are assigned using deciles. (From Ebach, M.C., *et al.*, *Zootaxa*, 3619(3), 315–342, 2013.)

We also calculated an estimate of potential species richness for each group using the Chao1 estimator (Chao, 1984, 1987; Table 1.3), collating sample counts from across all 50 km cells. These analyses were also done using *Biodiverse* version 1.99_002 (Laffan *et al.*, 2010).

The spatial analysis results (Figures 1.3–1.5) show a clear pattern, where the arid areas of Australia have lower species richness and WE across the groups, consistent with the results of Crisp *et al.* (2001) and Laffan and Crisp (2003) for a sample of the vascular flora. It is well known that WE is correlated with species richness (Crisp *et al.*, 2001), but here there are marked differences in the relative values of species richness and WE in the arid zone. Many of the higher WE values across the groups are more tightly constrained to the coastal regions than the richness values, generally aligning well with the mapped bioregions. It is clearly evident from all maps that many areas of the arid zone have not been sampled at all.

Sample redundancy for vascular plants and vertebrates is well represented overall in cells that have been sampled (Figure 1.5). For example, invertebrates are best sampled near Canberra, whereas the fungi are best sampled in Tasmania. Nonetheless, there are major gaps for the exemplar taxa that were investigated. For example, 1028 *Acacia* species are known from Australia, but either not all are represented in the *Atlas of Living Australia* or their records did not pass the data quality process. Regardless, it is evident that each group analysed is likely to have a number of undiscovered species.

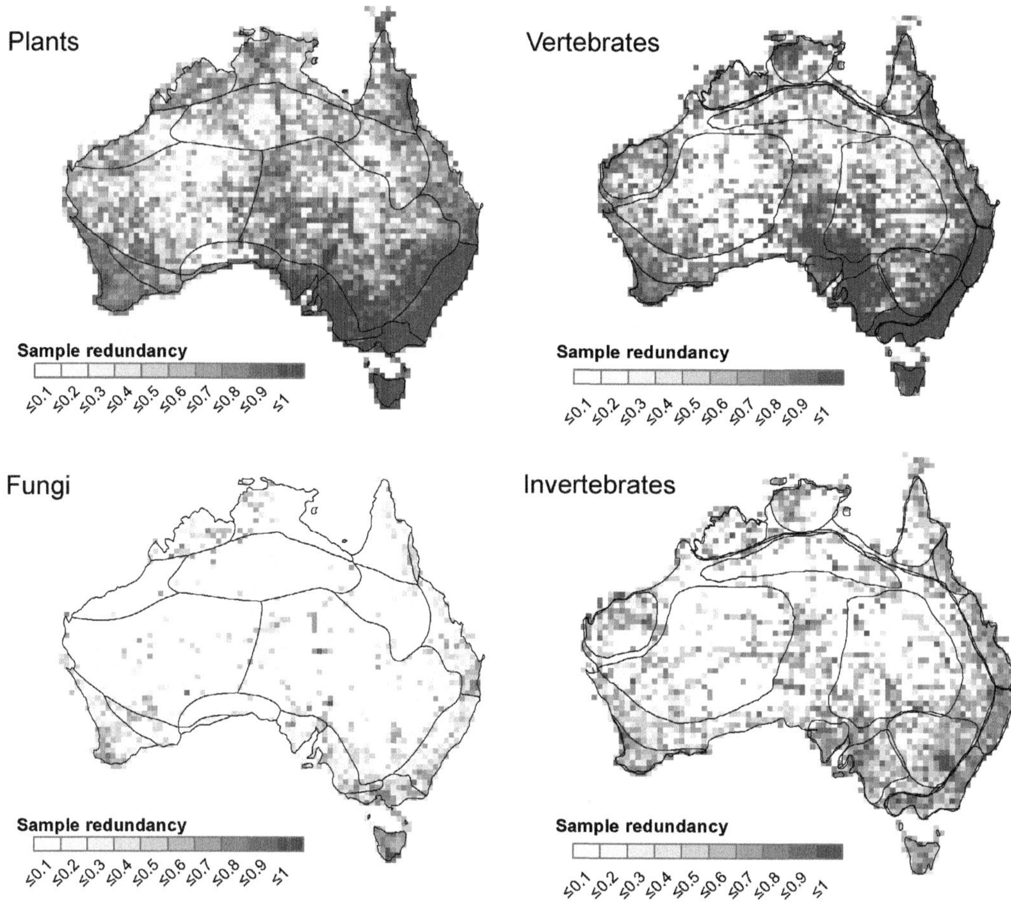

FIGURE 1.4 (See colour insert.) WE scores for exemplars of plants, fungi, vertebrates and invertebrates (Table 1.2). Polygon outlines are the phytogeographical or zoogeographical subregions defined in the *Australian Bioregionalisation Atlas*. Colours are assigned using deciles. (From Ebach, M.C., *et al.*, *Zootaxa*, 3619(3), 315–342, 2013.)

As with the spatial analyses, the fungi and invertebrates are likely to have many more unsampled species, with as many as 15%–35% of their total numbers of species being undetected. The wide confidence intervals for these groups suggest this could be substantially higher or lower. Vertebrates and plants are generally well described (Table 1.3). Amphibians are estimated to have no undiscovered species, but the upper confidence interval estimate indicates there could be 10 such species. Some of the low undiscovered species numbers could also be due to undescribed species not being considered in the analyses. Such species are more likely to have low sample counts, and their inclusion would thus influence the Chao1 estimator since it is a function of the number of species found only once and twice in the sample.

Filling in the Gaps

The discovery of new species will help facilitate a better understanding of Australia's biodiversity and biogeography. Collecting in undersampled areas, across all taxa, will further help fill in taxonomic and distribution gaps. Over the past 5 years the species discovery program Bush Blitz has reinforced the notion of Australia's taxonomic impediment, with the discovery of 1137 new animal and 57 new plant

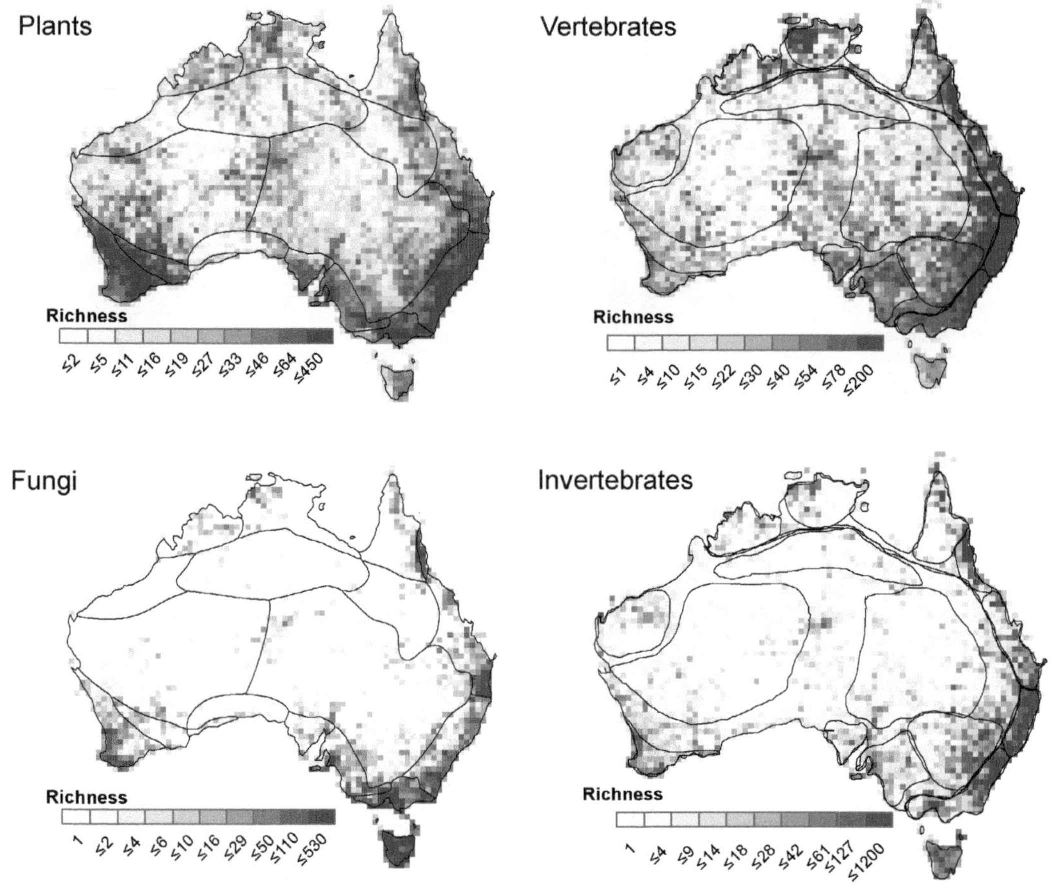

FIGURE 1.5 (See colour insert.) Sample redundancy for exemplars of plants, fungi, vertebrates and invertebrates (Table 1.2). Polygon outlines are the phytogeographical or zoogeographical subregions defined in the *Australian Bioregionalisation Atlas*. Colours are assigned using equal interval classes. (From Ebach, M.C., *et al.*, *Zootaxa*, 3619(3), 315–342, 2013.)

species from 23 surveys between 2010 and 2015 (www.bushblitz.org.au/statistics). Despite a decline in national taxonomic capacity (FASTS, 2008), the rate of species description is still high (Figures 1.1 and 1.2); for example, more than 850 new animal species have been described each year on average over the past 10 years (Figures 1.1 and 1.2).

Bush Blitz surveys have also documented major range extensions for designated core taxa, which exemplify limitations on the distributional range of many taxa (Preece *et al.*, 2015). Such shortfalls can be overcome to some extent by species distribution modelling (Nix and Switzer, 1991; Franklin, 2010; Hallgren *et al.*, 2016), although there are caveats concerning the use of ad hoc collection data and overestimating distributional ranges. Recently, the remarkable consolidation of collection event data at the national scale, largely driven by the taxonomic community, has resulted in benchmark biodiversity portals that provide open access to tens of millions of species records (Belbin, 2011; *Atlas of Living Australia*, 2016). Access to new and integrated databases has helped us establish more accurate and representative bioregional classifications (e.g. González-Orozco *et al.*, 2013; González-Orozco *et al.*, 2014a,b) and rates of endemism (Crisp *et al.*, 2001; Laffan and Crisp, 2003; Laffan *et al.*, 2013). Notwithstanding current national survey and bioinformatics efforts and monographic and cybertaxonomy (Wheeler, 2009) efforts, it is critical that survey data collection is ongoing to refine and test bioregionalisation and biogeographic theories.

TABLE 1.3

Estimates of the Number of Species for Exemplars of Plants, Fungi, Vertebrates and Invertebrates

Group	Taxon	Observed	Estimated	95% CI	Undetected (%)	Undetected (95% CI)
Plants	*Acacia*	1,015	1,054	(1,034, 1,096)	3.7	(1.8, 7.4)
Plants	*Daviesia*	122	123	(122, 130)	0.5	(0.0, 6.0)
Plants	*Eremophila*	204	209	(205, 224)	2.5	(0.7, 8.7)
Plants	Eucalypts	829	835	(830, 854)	0.7	(0.2, 3.0)
Fungi	Fungi	3,647	4,775	(4,602, 4,978)	23.6	(20.8, 26.7)
Vertebrates	Amphibians	234	234	(234, 242)	0.2	(0.0, 3.5)
Vertebrates	Geckos	175	178	(175, 194)	1.5	(0.2, 9.7)
Vertebrates	Mammals	225	236	(227, 277)	4.5	(0.9, 18.6)
Vertebrates	Skinks	501	524	(508, 574)	4.4	(1.4, 12.7)
Vertebrates	Snakes	178	181	(179, 192)	1.6	(0.3, 7.4)
Invertebrates	Acari	458	545	(511, 601)	16.0	(10.4, 23.8)
Invertebrates	Araneae	1,752	2,157	(2,073, 2,263)	18.8	(15.5, 22.6)
Invertebrates	Camaenidae	341	366	(352, 402)	6.9	(3.0, 15.1)
Invertebrates	Carabidae	741	1,143	(1,036, 1,288)	35.2	(28.5, 42.5)
Invertebrates	Diptera	2,198	3,068	(2,915, 3,253)	28.4	(24.6, 32.4)
Invertebrates	Formicidae	896	1,020	(980, 1,078)	12.1	(8.6, 16.9)
Invertebrates	Hemiptera	1,339	1,805	(1,705, 1,933)	25.8	(21.5, 30.7)
Invertebrates	Hymenoptera	1,947	2,355	(2,267, 2,467)	17.3	(14.1, 21.1)
Invertebrates	Lepidoptera	2,177	2,588	(2,503, 2,695)	15.9	(13.0, 19.2)

Source: Values were generated using the Chao1 index (from Chao, A., *Scandinavian Journal of Statistics*, 11, 265–270, 1984; Chao, A., *Biometrics*, 43, 783–791, 1987) and implemented in the *Biodiverse* software (from Laffan, S.W., *et al.*, *Ecography*, 33, 643–647, 2010). Values are indicative as, for example, 1028 *Acacia* species are known from Australia, but clearly not all are represented in the *Atlas of Living Australia* or their records were removed through a data quality filtering process.

As stated in the introduction, there are three baseline requirements in historical biogeography – namely, taxonomic information, distribution information and phylogenetic reconstruction. This chapter has dealt with the former, under the guise of biodiversity and bioregional assessments. The chapters following address phylogenies at higher taxonomic categories and are not repeated here, but we make note that the routine inclusion of molecular data has resulted in divergence dating becoming a fourth pillar in a biogeographer's toolkit. The integration of these information sources is without question heralding an exciting phase in Australian biogeography.

Acknowledgements

The authors thank Christy Geromboux, Anna Monro and Sandra Knapp from the Australian Biological Resources Study for the data used to compile Figures 1.1 and 1.2 and Table 1.1. Marina Cheng of the University of New South Wales is thanked for her assistance in preparing Figures 1.1 and 1.2.

References

Archer, M., and Clayton, G. (1984) *Vertebrate Zoogeography and Evolution in Australasia*. Hesperian Press, Perth, Australia.

Archer, M., and Fox, B. (1984) Background to vertebrate zoogeography in Australia. In *Vertebrate Zoogeography and Evolution in Australasia: Animals in Space and Time* (eds. M. Archer and G. Clayton), pp. 1–15. Hesperian Press, Perth, Australia.

Atlas of Living Australia. http://www.ala.org.au (accessed 1 June 2016).

Austin, A.D., Yeates, D.K., Cassis, G., Fletcher, M.J., La Salle, J., Lawrence, J.F., McQuillan, P.B., *et al.* (2004) Insects 'Down Under': Diversity, endemism and evolution of the Australian insect fauna; Examples from select orders. *Australian Journal of Entomology*, 43, 216–234.

Australian Faunal Directory. Australian Biological Resources Study, Canberra. http://www.environment.gov.au/biodiversity/abrs/online-resources/fauna/afd/index.html (accessed 1 June 2016).

Australian Plant Census. IBIS database. Centre for Australian National Biodiversity Research, Council of Heads of Australasian Herbaria. http://www.chah.gov.au/apc/index.html (accessed 20 April 2016).

Australian Plant Name Index. IBIS database. Centre for Australian National Biodiversity Research, Australian Government, Canberra, Australia. http://www.cpbr.gov.au/cgi-bin/apni (accessed 20 April 2016).

Australia's Virtual Herbarium. Council of Heads of Australasian Herbaria. http://avh.ala.org.au (accessed 1 June 2016).

Baehr, B.C., Harvey, M.S., and Smith, H.M. (2010) The goblin spiders of the new endemic Australian genus *Cavisternum* (Araneae: Oonopidae). *American Museum Novitates*, 3684, 1–40.

Belbin, L. (2011) The Atlas of Living Australia's Spatial Portal. In *Proceedings of the 2011 Environmental Information Management Conference (EIM2011)*. Santa Barbara, CA (pp. 28–29).

Bowman, D.M.J.S., Brown, G.K., Braby, M.F., Brown, J.R., Cook, L.G., Crisp, M.D., Ford, F., *et al.* (2010) Biogeography of the Australian Monsoon Tropics. *Journal of Biogeography*, 37, 201–216.

Boyer, S.L., Markle, T.M., Baker, C.M., Luxbacher, A.M., and Kozak, K.H. (2016) Historical refugia have shaped biogeographical patterns of species richness and phylogenetic diversity in mite harvestmen (Arachnida, Opiliones, Cyphophthalmi) endemic to the Australian Wet Tropics. *Journal of Biogeography*, 43(7), 1400–1411.

Brundin, L. (1966) Transantarctic relationships and their significance, as evidenced by chironomid midges. *Kungliga Svenska vetenskapsakademiens handlinger*, 11, 1–472.

Burbidge, N.T. (1960) The phytogeography of the Australian region. *Australian Journal of Botany*, 8, 75–211.

Bush Blitz. http://bushblitz.org.au (accessed 1 June 2016).

Byrne, M., Steane, D.A., Joseph, L., Yeates, D.K., Jordan, G.J., Crayn, D., Aplin, K., *et al.* (2011) Decline of a biome: Evolution, contraction, fragmentation, extinction and invasion of the Australian mesic zone biota. *Journal of Biogeography*, 38, 1635–1656.

Byrne, M., Yeates, D.K., Joseph, L., Kearney, M., Bowler, J., Williams, M.A.J., Cooper, S., *et al.* (2008) Birth of a biome: Insights into the assembly and maintenance of the Australian arid zone biota. *Molecular Ecology*, 17, 4398–4417.

Caley, M.J., Fisher, R., and Mengersen, K. (2014) Global species richness estimates have not converged. *Trends in Ecology and Evolution*, 29, 187–188.

Campbell H. (2013) *The Zealandia Drowning Debate: Did New Zealand Sink Beneath the Waves?* Bridget Williams Books, Wellington, New Zealand.

Cassis, G., and Symonds, C. (2011) Systematics, biogeography and host plant associations of the lacebug genus *Lasiacantha* Stal (Insecta: Heteroptera: Tingidae). *Zootaxa*, 2818, 1–63.

Cassis, G., and Symonds, C. (2016) Plant bugs, plant interactions and the radiation of a species rich clade in Southwest Australia: *Naranjakotta* nov. gen. and eighteen new species (Insecta: Heteroptera: Miridae: Orthotylinae). *Invertebrate Systematics*, 30(2), 95–186.

Cassis, G., Wall, M., and Schuh, R.T. (2007) Insect biodiversity and industrializing the taxonomic process: The plant bug case study (Insecta: Heteroptera: Miridae). In *Taxonomy and Systematics of Species Rich Taxa: Towards the Tree of Life* (eds. T.R. Hodkinson, J. Parnell and S. Waldren), pp. 193–212. CRC Press, Boca Raton, FL.

Chao, A. (1984) Non-parametric estimation of the number of classes in a population. *Scandinavian Journal of Statistics*, 11, 265–270.

Chao, A. (1987) Estimating the population size for capture-recapture data with unequal catchability. *Biometrics*, 43, 783–791.

Chapman, A.D. (2009) *Numbers of Living Species in Australia and the World, 2nd edn: A Report for the Australian Biological Resources Study September 2009*. Australian Biodiversity Information Services, Toowoomba, Australia.

Chinnock, R.J. (2007) *Eremophila and Allied Genera: A Monograph of the Plant Family Myoporaceae*. Rosenberg Publishing, Adelaide, Australia.

Cockayne, L. (1921) The vegetation of New Zealand. In Die Vegetation der Erde XIV. (Eds. A. Engler and O. Drude) pp. 1–364. Wilhelm Engelmann, Leipzig, Germany.

Cook, L.G., and Crisp, M.D., 2005. Not so ancient: The extant crown group of *Nothofagus* represents a post-Gondwanan radiation. *Proceedings of the Royal Society of London B: Biological Sciences*, 272, 2535–2544.

Cracraft, J. (1991) Patterns of diversification within continental biotas: Hierarchical congruence among the areas of endemism of Australian vertebrates. *Australian Systematic Botany*, 4, 211–227.

Cranston, P.S. (2010) Insect biodiversity and conservation in Australasia. *Annual Review of Entomology*, 55(1), 55–75.

Cranston, P.S., and Naumann, I.D. (1991) The insects of Australia: A textbook for students and research workers. In *Biogeography* (ed. I.D. Naumann), pp. 180–197. Cornell University Press, Ithaca, NY.

Crisp, M., Cook, L., and Steane, D. (2004) Radiation of the Australian flora: What can comparisons of molecular phylogenies across multiple taxa tell us about the evolution of diversity in present-day communities? *Philosophical Transactions of the Royal Society of London B: Biological Sciences*, 359(1450), 1551–1571.

Crisp, M.D., Arroyo, M.T.K., Cook, L.G., Gandolfo, M.A., Jordan, G.J., McGlone, M.S., Weston, P.H., Westoby, M., Wilf, P., and Linder, H.P. (2009) Phylogenetic biome conservatism on a global scale. *Nature*, 458, 754–756.

Crisp, M.D., Laffan, S.W., Linder, H.P., and Monro, A. (2001) Endemism in the Australian flora. *Journal of Biogeography*, 28, 183–198.

Crisp, M.D., Linder, H.P., and Weston, P.H. (1995) Cladistic biogeography of plants in Australia and New Guinea: Congruent pattern reveals two endemic tropical tracks. *Systematic Biology*, 44(4), 457–473.

Crisp, M.D., Trewick, S.A., and Cook, L.G. (2011) Hypothesis testing in biogeography. *Trends in Ecology and Evolution*, 26(2), 66–72.

Crisp, M.D., West, J.G., and Linder, H.P. (1999) Biogeography of the terrestrial flora. In *Flora of Australia*, 2nd edn., vol. 1 (eds. A.E. Orchard and H.S. Thompson), pp. 321–367. Australian Biological Resources Study, Canberra, Australia.

Diels, L. (1906) Die Pflanzenwelt von West-Australien südlich des Wendekreises: Mit einer einleitung über die Pflanzenwelt Gesamt-Australiens in Grundzügen. In *Vegetation der Erde VII* (eds. O. Drude and A. Engler), pp. 1–143. Wilhelm Engelmann, Leipzig, Germany.

Drude, O. (1890) *Handbuch der Pflanzengeographie*. J. Engelhorn, Stuttgart, Germany.

Ebach, M.C. (2012) A history of bioregionalisation in Australia. *Zootaxa*, 3392, 1–34.

Ebach, M.C., Gill, A.C., Kwan, A., Ahyong, S.T., Murphy, D.J., and Cassis, G. (2013) Towards an Australian Bioregionalisation Atlas: A provisional area taxonomy of Australia's biogeographical regions. *Zootaxa*, 3619(3), 315–342.

Ebach, M.C., González-Orozoco, C.E., Miller, J.T., and Murphy, D.J. (2015) A revised area taxonomy of phytogeographical regions within the Australian Bioregionalisation Atlas. *Phytotaxa*, 208(4), 261–277.

Erwin, T.L. (1982) Tropical forests: Their richness in Coleoptera and other arthropod species. *Coleopterists Bulletin*, 36, 74–75.

Erwin, T.L. (1991) How many species are there? Revisited. *Conservation Biology*, 5, 330–333.

FASTS. (2008) Federation of Australian Scientific and Technological Societies 2008. In *Proceedings of the National Taxonomy Forum*. Australian Museum, Sydney, Australia. 3–4 October 2007.

Fleming, C.A. (1987) Comments on Udvardy's biogeographical realm Antarctica. *Journal of the Royal Society of New Zealand*, 17(2), 195–200.

Franklin, J. (2010) *Mapping Species Distributions: Spatial Inference and Prediction*. Cambridge University Press, Cambridge, UK.

Garcillán, P.P., Ezcurra, E., and Riemann, H. (2003) Distribution and species richness of woody dryland legumes in Baja California, Mexico. *Journal of Vegetation Science*, 14, 475–486.

Giribet, G., and Edgecombe, G.D. (2006) The importance of looking at small-scale patterns when inferring Gondwanan biogeography: A case study of the centipede *Paralamyctes* (Chilopoda, Lithobiomorpha, Henicopidae). *Biological Journal of the Linnean Society*, 89, 65–78.

González-Orozco, C.E., Ebach, M.C., Laffan, S.W., Thornhill, A.H., Knerr, N., Schmidt-Lebuhn, A., Cargill, C.C., *et al.* (2014b) Quantifying phytogeographical regions of Australia using geospatial turnover in species composition. *PLoS One*, 9, e92558.

González-Orozco, C.E., Laffan, S.W., Knerr, N., and Miller, J.T. (2013) A biogeographic regionalisation of Australian *Acacia* species. *Journal of Biogeography*, 40, 2156–2166.

González-Orozco, C.E., Thornhill, A.H., Knerr, N., Laffan, S.W., and Miller, J.T. (2014a) Biogeographical regions and phytogeography of the eucalypts. *Diversity and Distributions*, 20, 46–58.

Greenwood, D.R., and Christophel, D.C. (2005) The origins and tertiary history of Australian 'tropical' rainforests. In *Tropical Rainforests: Past, Present, and Future* (eds. E. Bermingham, C.W. Dick and C. Moritz), pp. 336–373. University of Chicago Press, Chicago, IL.

Grimaldi, D., and Engel, M.S. (2005) *Evolution of the Insects*. Cambridge University Press, Cambridge, UK.

Hallgren, W., Beaumont, L., Bowness, A., Chambers, L., Graham, E., Holewa, H., Laffan, S., *et al.* (2016) The biodiversity and climate change virtual laboratory: Where ecology meets big data. *Environmental Modelling and Software*, 76, 182–186.

Heads, M. (2014) *Biogeography of Australasia: A Molecular Analysis*. Cambridge University Press, Cambridge, UK.

Heads, M. (2015) Panbiogeography, its critics, and the case of the ratite birds. *Australian Systematic Botany*, 27(4), 241–256.

Heatwole, H. (1987) Major components and distributions of the terrestrial fauna. In *Fauna of Australia. General Articles*, 1A (eds. G.R. Dyne and D.W. Walton), pp. 101–135. Australian Government Publishing Service, Canberra, Australia.

Hedley, C. (1893) The faunal regions of Australia. In *Report of the Fourth Meeting of the Australasian Association for the Advancement of Science*. January 1892, Hobart (ed. A. Morton), pp. 444–446. Australasian Association for the Advancement of Science, Sydney, Australia.

Hedley, C. (1904) The effect of the Bassian Isthmus upon the existing marine fauna: A study in ancient geography. *Proceedings of the Linnean Society of New South Wales*, 28, 876–883.

Herbert, D.A. (1932) The relationships of the Queensland flora. *Proceedings of the Royal Society of Queensland*, 44, 2–22.

Hopper, S.D. (1979) Biogeographical aspects of speciation in the Southwest Australian flora. *Annual Review of Ecology and Systematics*, 10, 399–422.

Humphries, C.J., and Parenti, L.R. (1999) *Cladistic biogeography*. Oxford University Press, Oxford, UK.

Hutton, F.W. (1896) Theoretical explanations of the distribution of southern faunas. *Proceedings of the Linnean Society of New South Wales*, 21, 36–47.

Ingham, J.A., Forster, P.I., Crisp, M.D., and Cook, L.G. (2013) Ancient relicts or recent dispersal: How long have cycads been in central Australia? *Diversity and Distributions*, 19(3), 307–316.

Janz, N. (2011) Ehrlich and Raven revisited: Mechanisms underlying codiversification of plants and enemies. *Annual Review of Ecology, Evolution, and Systematics*, 42(1), 71–89.

Keast, A. (ed.) (1981) *Ecological Biodiversity of Australia*. Dr. W. Junk, The Hague, the Netherlands.

Ladiges, P., Kellermann, J., Nelson, G., Humphries, C., and Udovicic, F. (2005) Historical biogeography of Australian Rhamnaceae, tribe Pomaderreae. *Journal of Biogeography*, 32, 1909–1919.

Ladiges, P., Parra, O.C., Gibbs, A., Udovicic, F., Nelson, G., and Bayly, M. (2011) Historical biogeographical patterns in continental Australia: Congruence among areas of endemism of two major clades of eucalypts. *Cladistics*, 27, 29–41.

Ladiges, P.Y., Bayly, M.J., and Nelson, G. (2012) Searching for ancestral areas and artifactual centers of origin in biogeography: With comment on east–west patterns across southern Australia. *Systematic biology, 61*(4), pp. 703–708.

Laffan, S.W., and Crisp, M.D. (2003) Assessing endemism at multiple spatial scales, with an example from the Australian vascular flora. *Journal of Biogeography*, 30, 511–520.

Laffan, S.W., Lubarsky, E., and Rosauer, D.F. (2010) *Biodiverse*, a tool for the spatial analysis of biological and related diversity. *Ecography*, 33, 643–647.

Laffan, S.W., Ramp, D., and Roger, E. (2013) Using endemism to assess representation of protected areas: The family myrtaceae in the greater blue mountains world heritage area. *Journal of Biogeography*, 40, 570–578.

Lambkin, C.L. and Bartlett, I.S. (2011) Bush blitz aids description of three new species and a new genus of Australian beeflies (*Diptera, Bombyliidae, Exoprosopini*). *ZooKeys*, 150, 231–280.

Linnaeus, C. (1758) *Systema naturae per regna tria naturae, secundum classes, ordines, genera, species, cum characteribus, differentis, synonymis, locis*. 10th edn., revised. Laurentii Salvii, Stockholm, Sweden.

McGlone, M.S. (2016) Once more into the wilderness of panbiogeography: A reply to Heads (2014). *Australian Systematic Botany*, 28(6), 388–393.

May, R.H. (1988) How many species are there on Earth? *Science*, 241, 1441–1449.

Mittermeier, R.A., Turner, W.R., Larsen, F.W., Brooks, T.M., and Gascon, C. (2011) Global biodiversity conservation: The critical role of hotspots. In *Biodiversity Hotspots*: Distribution and Protection of Conservation Priority Areas (eds. F.E. Zachos and J.C. Habel), pp. 3–22. Springer, Heidelberg, Germany.

Mora, C., Tittensor, D.P., Adl, S., Simpson, A.G., and Worm, B. (2011) How many species are there on Earth and in the ocean? *PLoS Biology*, 9, e1001127.

Morrone, J.J. (2013) *Evolutionary Biogeography: An Integrative Approach with Case Studies*. Columbia University Press, New York.

Myers, N., Mittermeier, R.A., Mittermeier, C.G., Da Fonseca, G.A.B., and Kent, J. (2000) Biodiversity hotspots for conservation priorities. *Nature*, 403, 853–858.

Nelson, G., and Platnick, N.I. (1981) *Systematics and Biogeography: Cladistics and Vicariance*. Columbia University Press, New York.

Nelson, G., and Rosen, D.E. (1981) Vicariance biogeography: A critique. *Symposium of the Systematics Discussion Group of the American Museum of Natural History*. Columbia University Press, New York.

Nix, H.A., and Switzer, M.A. (1991) *Rainforest Animals: Atlas of Vertebrates Endemic to the Wet Tropics*. Australian National Parks and Wildlife Service, Canberra, Australia.

Ogata, K., and Taylor, R.W. (1991) Ants of the genus *Myrmecia* Fabricius: A preliminary review and key to the named species (Hymenoptera: Formicidae: Myrmeciinae). *Journal of Natural History*, 25(6), 1623–1673.

Online Zoological Collections of Australian Museums. http://ozcam.org.au (accessed 1 June 2016).

Parenti, L.R., and Ebach, M.C. (2009) *Comparative Biogeography: Discovering and Classifying Biogeographical Patterns of a Dynamic Earth*. University of California Press, Berkeley, CA.

Pole, M. (1994) The New Zealand flora: Entirely long-distance dispersal? *Journal of Biogeography*, 21, 625–635.

Preece, M., Harding, J., and West, J.G. (2015) Bush Blitz: Journeys of discovery in the Australian outback. *Australian Systematic Botany*, 27(6), 325–332.

Rix, M.G., Edwards, D.L., Byrne, M., Harvey, M.S., Joseph, L., and Roberts, J.D. (2015) Biogeography and speciation of terrestrial fauna in the south-western Australian biodiversity hotspot. *Biological Reviews*, 90, 762–793.

Sanmartín, I., and Ronquist, F. (2004) Southern hemisphere biogeography inferred by event-based models: Plant versus animal patterns. *Systematic Biology*, 53, 216–243.

Schodde, R. (1989) Origins, radiations and sifting in the Australasian biota: Changing concepts from new data and old. *Australian Systematic Botanical Society Newsletter*, 60, 2–11.

Schuh, R.T., and Stonedahl, G.M. (1986) Historical biogeography of the Indo-Pacific: A cladistic approach. *Cladistics*, 2, 337–335.

Simpson, G.G. (1977) Too many lines: The limits of the Oriental and Australian zoogeographic regions. *Proceedings of the American Philosophical Society*, 121(2), 107–120.

Stork, N.E. (1988) Insect diversity: Facts, fiction, and speculation. *Biological Journal of the Linnean Society*, 35, 321–337.

Stork, N.E., McBroom, J., Gely, C., and Hamilton, A.J. (2015) New approaches narrow global species estimates for beetles, insects, and terrestrial arthropods. *Proceedings of the National Academy of Sciences of the United States of America*, 112, 7519–7523.

Stow, A., Maclean, N., and Holwell, G.I. (2015) *Austral Ark: The State of Wildlife in Australia and New Zealand*. Cambridge University Press, Cambridge, UK.

Swenson, U., Hill, R.S., and McLoughlin, S. (2001) Biogeography of *Nothofagus* supports the sequence of Gondwana break-up. *Taxon*, 50, 1025–1041.

Tate, R. (1889) On the influence of physiological changes in the distribution of life in Australia. In *Report of the First Meeting of the Australasian Association for the Advancement of Science*, August and September 1888, Sydney (eds. A. Liversidge and R. Etheridge), pp. 312–326. Australasian Association for the Advancement of Science, Sydney, Australia.

Taylor, R.W. (1983) Descriptive taxonomy: Past, present, and future. In: *Australian Systematic Entomology: A Bicentenary Perspective* (eds. E. Highley and R.W. Taylor), pp. 93–134. CSIRO, Canberra, Australia.

Tenison-Woods, J.E. (1878) The echini of Australia. *Proceedings of the Linnean Society of New South Wales*, 2, 145–176.

Tenison-Woods, J.E. (1882) On the natural history of New South Wales: An essay. Thomas Richards, Government Printers, Sydney, Australia.

Udvardy, M.D. (1987) The biogeographical realm Antarctica: A proposal. *Journal of the Royal Society of New Zealand*, 17, 187–194.

Unmack, P.J. (2001) Biogeography of Australian freshwater fishes. *Journal of Biogeography*, 28, 1053–1089.

Valerio, A.A., Austin, A.D., Masner, L., and Johnson, N.F. (2013) Systematics of Old World *Odontacolus* Kieffer sl (Hymenoptera, Platygastridae sl): Parasitoids of spider eggs. *ZooKeys*, 314, 1–151.

von Mueller, F. (1858) Botanical report on the North-Australian Expedition, under the command of A.C. Gregory, Esq. *Journal of the Proceedings of the Linnean Society*, 2, 137–163.

Wallace, A.R. (1880) *Island Life*. Macmillan, London.

Webb, L.J, and Tracey, J.G. (1981) Australian rainforest: Patterns and change. In *Ecological Biogeography of Australia* (ed. A. Keast), pp. 605–694. Dr. W. Junk, The Hague, the Netherlands.

Wheeler, Q.D. (2009) The science of insect taxonomy: Prospects and needs. In *Insect Biodiversity: Science and Society* (eds. R.G. Foottit and P.H. Adler), pp. 357–380. Wiley-Blackwell, Chichester, UK.

Williams, S.E., and Pearson, R.G. (1997) Historical rainforest contractions, localized extinctions and patterns of vertebrate endemism in the rainforests of Australia's Wet Tropics. *Proceedings of the Royal Society B: Biological Sciences*, 264, 709–716.

World Conservation Monitoring Centre. (2000) *Annual Report*. UNEP-WCMC, Cambridge, UK.

Yeates, D.K., Harvey, M.S., and Austin, A.D. (2003) New estimates for terrestrial species-richness in Australia. *Records of the South Australian Museum Monograph Series*, 7, 231–241.

2

Historical Biogeography of Diatoms in Australasia: A Preliminary Assessment

David M. Williams and J. Pat Kociolek

CONTENTS

Introduction

Several decades ago, almost no diatomist would have been interested in historical biogeography, as the notion that there were any reasonable levels of species endemism to be worthy of study would have been disputed (for a review see Kociolek and Spaulding 2000). This viewpoint, with very few exceptions, would have applied to diatom species occurring on almost any and every part of the globe, not just Australia or Australasia. Any historical perspective on geographical distribution is intimately tied to endemism, for without endemic taxa there would be no unique patterns of distribution to explain. When focusing on any particular region, whatever size that region might be – a pond, an ancient lake, even a continent as physically large as Australia – the issue of diatom endemism arises, but usually in the context of a lack of expectation, rather than its extent (Williams 2011).

The idea that all microbial ('small') organisms have barely any level of endemism was recently revived and captured in the *Everything is everywhere* (EIE) hypothesis, a hypothesis to account for the apparent widespread occurrence of many microbial groups, including diatoms (Finlay 2002; Fenchel and Finlay 2004; reviewed in Fontaneto 2011). That viewpoint, or at least approaches to its study, is succinctly summed up in this recent title: 'Does the cosmopolitan diatom *Gomphonema parvulum* (Kützing) Kützing have a biogeography?' (Abarca *et al.* 2014); lacking a *biogeography* implies taxon ubiquity, hence the alternative name the *ubiquity hypothesis* for the EIE proposal (Finlay 2002).

The EIE hypothesis can be thought of as having two parts. The first is that microorganisms are indeed ubiquitous.

> According to advocates of the ubiquity hypothesis, the vast population sizes of micro-organisms drive ubiquitous dispersal and make local extinction virtually impossible. Geographic isolation is therefore absent and as a result, allopatric speciation should be rare or nonexistent, which would explain the perceived low global morphospecies diversity of microbial eukaryotes. (Vanormelingen *et al.* 2008: 394)

The notion that there are 'vast population sizes of micro-organisms' is derived from another idea – the second part of the EIE hypothesis – that endemicity is scarce. Discovering levels of endemicity is an empirical enterprise, largely the domain of taxonomic endeavour. To study historical biogeography as applied to diatoms, then, is to first establish whether there are indeed any appreciable levels of endemism, and if so, whether those endemic taxa are distributed in such a way as to reflect some common cause relative to other organisms having the same distribution (Williams 2011). Establishing endemic taxa has been hampered in the past by the use of European floras to identify non-European biotas (Kociolek and Spaulding 2000). That approach to new and understudied floras has now largely ceased, as evidenced by the *Iconographia Diatomologica* series, for example, which was established in 1995 and now (in 2016) has 26 volumes published, many dedicated to diatom floras of unknown areas of the world.

In the last two decades it has been established that diatom endemism is accepted as a fact, with several summaries documenting that view from a general perspective (Kociolek and Spaulding 2000; Flower 2005: 111; Williams and Reid 2006a; Williams 2011; Kulikovskiy and Kuznetsova 2014; for a more general summary of microbial distribution see Logares 2006; for the alternative view of the *Everything is endemic* thesis see Williams 2011) and for particular areas or particular taxa (e.g. Moser *et al.* 1998; Metzeltin and Lange-Bertalot 1998, 2007; Rumrich *et al.* 2000; Kulikovskiy *et al.* 2010; Vishnyakova *et al.* 2014; Kociolek *et al.* in press).

This chapter will attempt to set out some discoveries relevant to the distribution of diatoms in Australasia and relate those discoveries to distributions elsewhere on the globe. We concentrate primarily on inland waters (freshwater), but, when seeking historical interpretations, brackish and marine environments cannot be ignored as relationships between taxa from both environments have intertwined historical dimensions (Williams and Reid 2006c). We briefly review what is known of freshwater diatom diversity in Australasia and illustrate in more detail how one might tackle the problem of diatom historical biogeography by focusing on one particular diatom order, the Eunotiales, a monophyletic group of genera and species. This group has already shed light on explaining diatom diversity elsewhere in the Southern Hemisphere, and its further application to Australia and the surrounding areas should be productive.

For the purposes of this chapter, we consider Australia, New Zealand, New Caledonia and New Guinea to be Australasia, which largely follows Michael Heads's recent book *Biogeography of Australasia* (Heads 2014).

Diatom Distribution and Endemism in Australasia

First, it is worth considering the extent of diatom endemism currently recognised with respect to recently published Australian floras and identification guides. Typical comments are found in *An Illustrated Key to Common Diatom Genera from Southern Australia* (Gell *et al.* 1999) and its companion volume *An Illustrated Guide to Common Stream Diatom Species from Temperate Australia* (Sonneman *et al.* [1999] 2000).

> The recent expansion of diatom research in Australia … has suggested that a great majority of diatom species found in Australia are indeed cosmopolitan if the present taxonomic schemes are used. (Gell *et al.* 1999: 1)
>
> The vast majority of common species encountered in southern Australia are accommodated by taxonomic systems and keys developed outside Australia. (Sonneman *et al.* [1999] 2000)

The perception that the freshwater diatom flora of Australia is comprised mostly of cosmopolitan species may simply be due to a lack of study or ready acceptance of existing non-Australian determinations ('if the present taxonomic schemes are used'). As John (2007) noted in an introductory account to the diatoms of Australia, 'Apart from occasional early records..., study of diatoms systematics (particularly of inland taxa) was sparse in Australia until the 1970s' (304). Peter Tyler, who has written significant essays on endemism in Australian (and Tasmanian) freshwater algae (Tyler and Wickham 1988; Tyler 1996), contributed the foreword to *An Illustrated Guide to Common Stream Diatom Species from Temperate Australia*, in which he noted that in the 'highly professional series *flora of Australia* ... the higher plants and animals ... have had the lion's share of endeavour' (Tyler [1999] 2000: iii). John's (2007) summary appeared in the introductory volume of the Australian Biological Resources Study (ABRS)'s *Algae of Australia* series (an ongoing project). In the same volume there is a somewhat more perceptive essay by Vyverman *et al.* (2007a) titled 'Biogeography of Freshwater Microalgae', in which they discuss 'the major causes of the gaps in our current knowledge [of microalgal diversity]' (580). Vyverman *et al.* chart the beginning of a modern understanding of diatom diversity and note,

> Information on microalgal diversity (i.e. species numbers) is generally sparse and reflects the research efforts exerted in a certain group or geographical region rather than realistic estimates.... Therefore it would be inappropriate to use currently available data as the basis for a discussion of microalgal diversity. (Vyverman *et al.* 2007a: 580)

These are sensible words and ones we heed, but past studies may well illuminate contemporary problems. In the following, we briefly review current knowledge, which obviously has a historical component.

Australia

There are a number of early nineteenth-century accounts that include diatoms from Australia, such as Ehrenberg's *Mikrogeologie* (1854), Schmidt's *Atlas der Diatomaceen-kunde* (1874–1937) and Cleve's two-volume *Synopsis of the Naviculoid Diatoms* (1894–5). Alongside these are two widely available exsiccatae sets: Cleve and Möller's *Diatoms* (1877–1882), which includes samples collected by Sven Berggren in the 1870s (Bagnall 1970a,b), and Tempère and Peragallo's *Diatomées du Monde Entier* (1907–1915), specimens that would repay reexamination. But the first focused studies on freshwater diatom diversity in Australia were made by George Israel Playfair (1871–1922), who took up phycology as a hobby (Anonymous 1923). During his studies he described just under 60 new diatom taxa, most of which, as far as we are aware, have never been recorded again since his original descriptions of them (see Cowan and Ducker 2007: 31, Table 20, for a summary of all algal species described by Playfair). Playfair's illustrations are, by today's standards, not so easy to interpret and may account for the neglect of his taxa and their (unjust, perhaps) passing into obscurity. Yet they are not entirely devoid of informative content. For example, his written description of *Synedra lismorensis* is reasonably well detailed and the illustrations (seven in all) are sufficient to indicate an alternative placement in the revised genus *Fragilaria** (Williams and Round 1987); the illustrations of *Achnanthes woodlawnensis* and *Amphora lagerheimii* var. *minuta* are less clear. Playfair's material is available in the Academy of Natural Sciences (ANSP, Marina Potapova pers. comm.; Mahoney and Reimer 1987, 1997), but there is enough published information to suggest that Playfair was not entirely mistaken in his taxon judgements and his material is well worth reexamination.

Following shortly after Playfair, Giovanni Battista de Toni (1864–1924) and Achilli Forti (1878–1937), in their *Alghe di Australia, Tasmania e Nuova Zelanda*, described a few taxa that appear to be endemic (De Toni and Forti 1922), but nearly all were marine. The next most significant contribution was from L.H.J. Crosby and E.J. Ferguson-Wood (1904–1972), who, in a series of papers they published on diatom diversity in Australia and New Zealand, described a number of new taxa (Crosby and Ferguson-Wood 1958, 1959; Wood 1961a,b, 1963; the pair were later joined by Vivienne Cassie Cooper: Wood *et al.* 1959). Although these were also primarily marine, the total number of diatom species recorded by

* It should be noted that *Fragilaria*, as it currently stands, is probably not monophyletic.

them exceeded 800, of which roughly one-quarter were freshwater. But only a handful of species were described as new for Australia and New Zealand, among them a species they named *Amphora hendeyi* (probably a *nom. invalid*, but now recognised as a synonym of *Eunophora tasmanica*; Levkov 2009: 280), a species described from Lake Dobson, Tasmania, and one which we will return to later in the chapter (Figure 2.1f is a reproduction of the original illustration for *A. hendeyi* by Crosby and Wood 1959; Figure 2.1g is a specimen of '*Amphora* sp.', but not type material).

The contributions by Crosby, Ferguson-Wood and Cassie were followed in the late 1970s by a series of regional floras (Foged 1978; Brady 1979; John 1983; Thomas 1983). Most are restricted to one area, varying in size and differing from study to study. For example, Foged's *Diatoms in Eastern Australia* was based on '212 samples ... from 152 localities in the eastern part of Australia' (Foged 1978: 3, abstract); Brady's *The Diatom flora of Australia* was from 'the Northern Territory, especially in the Magela Creek System' (Brady 1979); and John's first study was *The Diatom flora of the Swan River Estuary, Western Australia* (John 1983, revised and updated in John 2012c). All these works, valuable in their own right, cover quite different areas and ecologies. But taken in total, very few endemic taxa were recognised and described until Vyverman and colleagues' studies in Tasmania (Vyverman *et al.* 1995; Hodgson *et al.* 1996). These are largely floristic accounts, however; most of the new taxa were described elsewhere (Vyverman *et al.* 1997; Vyverman *et al.* 1998a; Sabbe *et al.* 2000; Sabbe *et al.* 2001; Kilroy *et al.* 2003; Kociolek *et al.* 2004; Vanhoutte *et al.* 2004).

Results from the Australian floristic studies are summarised in Table 2.1, which reveals that, roughly speaking, the Australian freshwater diatom flora has, at present, a total of between 900 and 1000 species, of which no more than ca. 50 have been recognised as endemic to Australia (including Tasmania). This is a surprisingly low number given the conventional expectation of the diversity of the Australian biota (Heads 2014).

New Zealand

Cassie Cooper and Harper (2012) published a comprehensive and useful checklist for freshwater New Zealand diatoms (marine species and fossils are also included as separate entries: Chang and Harper 2012 and Harper 2012), developed and extended from an earlier version (Cassie 1984). In total, ca. 800 taxa (freshwater) are recorded for New Zealand, of which only nine are identified as endemic (eight of those are indicated to be questionable): *Frankophila biggsii*, *Gomphoneis minuta* var. *cassieae*, *Frustulia aotearoa*,* *Frustulia maoriana*, *Pinnularia segariana*, *Actinella aotearoaia* (this species is not found in Tasmania; see '*Actinella*' in this chapter), *Tabularia variostriata*, *Hantzschia doigiana* and *Skeletomastus coelatus* (some of these species overlap with those listed in Kilroy *et al.* 2007; see 'Comparisons of Endemism' in this chapter); this estimate may be too low as the checklist does not include all of Foged's (1979) new species. For example, *Achnanthes crosbyana* is identified only as a brackish water species but not as an endemic, and *Navicula barberiana* Foged, *N. cassieana* Foged, *N. tairuaensis* Foged and so on all go unrecognised as endemics (it may be that these have been found elsewhere; if so, it is not clear where those data are recorded).

For the new species described by Lowe *et al.* (2006), Harper *et al.* remark, "A curious new species is *Frankophila biggsii* [marked as endemic in their checklist] a fragilarioid with a raphe-like slit (Lowe *et al.* 2006). This genus and *Veigaludwigia* were first recognised in South America, and the latter has since been reported from New Zealand mountain tarns (Van Houtte *et al.* 2006)" (Harper *et al.* 2012: 133), thus suggesting a South America –New Zealand distribution for some taxa. Likewise, *Frustulia cassieae** and *Pleurosigma inscriptura* both have a South American–New Zealand distribution (Beier and Lange-Bertalot 2007; Sar *et al.* 2013).

* *Frustulia aotearoa* was one of four recently described by Beier and Lange-Bertalot (2007) and was said to occur 'exclusively in ombrotrophic peat bogs'. Of the other three, *F. cassieae* had been previously recorded from eastern Australia and New Zealand (but under a different name), *F. gondwana* was considered to 'resemble' specimens from 'other regions of the southern hemisphere and the Neotropics' and *F. maoriana* was, like *F. aotearoa*, a New Zealand endemic.

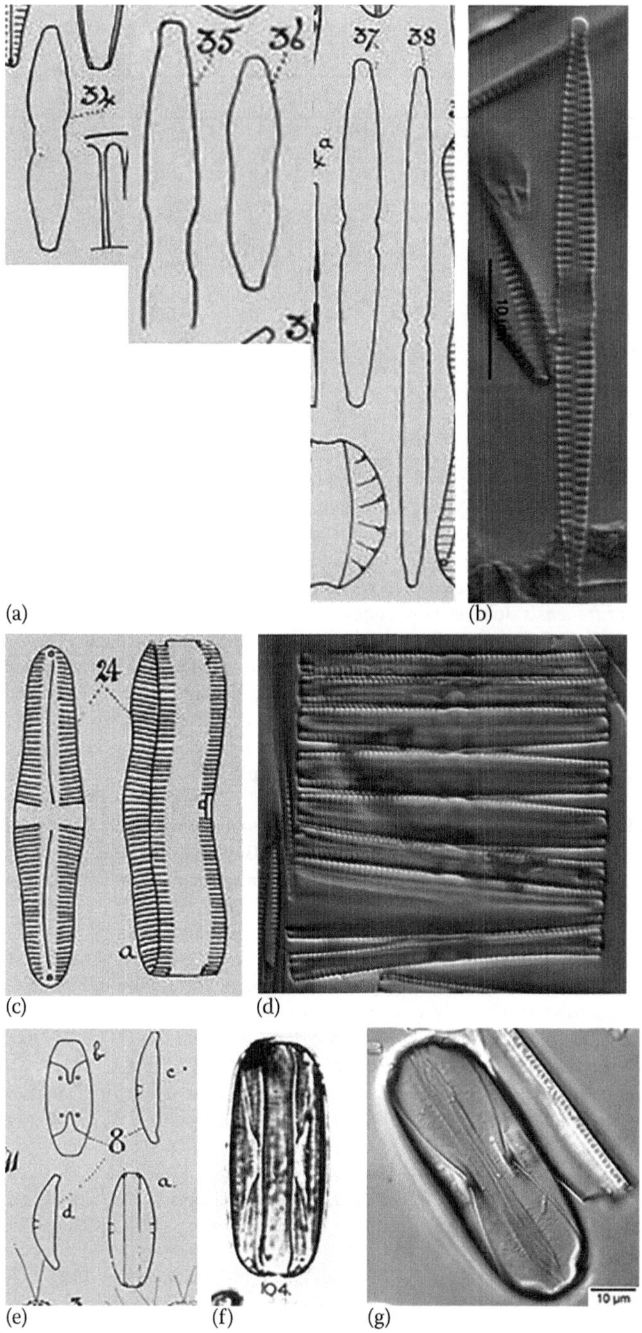

FIGURE 2.1 (a) Reproduction of *Synedra lismorensis* (from Playfair, G.I., *Proceedings of the Linnean Society of New South Wales* 39, 93–151, 1914, 118, pl. 4); Playfair included seven figures in all, 33–38, but 33 and 34a are possibly other diatoms (pers. comm. Marina Potapova, who examined the type material); reproduced here are Playfair's Figures 34–38; (b and d) *S. lismorensis* from type material ANSP AGC 400238a, image courtesy of Marina Potapova; (c) *Achnanthes woodlawnensis* (from Playfair, G.I., *Proceedings of the Linnean Society of New South Wales* 40, 310–362, 1915: 347, pl. 45, Figure 24); (e) *Amphora lagerheimii* var. *minuta* (from Playfair, G.I., *Proceedings of the Linnean Society of New South Wales* 40, 310–362, 1915, 344, pl. 45, Figure 8); (f) *Amphora hendeyi* (from Crosby, L.H., and Wood, E.J.F., *Transactions of the Royal Society of New Zealand* 86, 1–58, 1959, 34, pl. 7, Figure 104); (g) '*Amphora* sp.' = *A. hendeyi*, Lake Dobson, Tasmania, specimen from the Hustedt collection, BRM 345/44a, image courtesy of Bank Beszteri and Sarah Olischläger.

TABLE 2.1
Summary of Diatom Diversity for Australia Derived from floras/Identification Guides

Author	Date	Total	New
Playfair	1913–1915	300+	57
Crosby/Ferguson-Wood/Cassie	1958–1963		1
Foged	1978	ca. 860	10
Brady	1979	ca. 80	3[a]
John	1983	360	6[b]
Thomas	1983	160	2
Holland and Clark[c]	1989[c]		
Vyverman *et al.*	1995	ca. 280	0[d]
Hodgson *et al.*	1996	ca. 400	0[e]
Gell *et al.*; Sonneman *et al.*[e]	1999; 2000	ca. 150–200	0
John	2012c	365	0[f]

[a] All these names are *nom. nud.*; a few were identified differently by Thomas, D.P., *A Limnological Survey of the Alligator River System: I. The Diatoms (Bacillariophyceae) of the Region*, Australia Government Publishing Service, Canberra, 1983 (see later).

[b] From 1980–1991 John described 25 new endemic taxa, most either brackish or marine (John 2012c). Some of these have since been revised or are clearly in need of revision. For example, *Amphora australiensis* John is now considered to be a synonym of *A. pleniluna* Hohn and Hellerman and *A. polita* Krasske (Levkov, Z., *Amphora* sensu lato, in *Diatoms of Europe: Diatoms of the European Inland Waters and Comparable Habitats Elsewhere*, A.R.G. Gantner, Ruggell, Liechtenstein, 2009; Desianti, N., *et al.*, Examination of the type materials of diatoms described by Hohn and Hellerman from the Atlantic Coast of the USA, *Diatom Research* 30, 93–116, 2015) and hence should now be regarded as a widespread species; on the other hand, *Synedra acus* var. *varipunctata* John would be better placed in a different genus, but which one is not clearly evident (Williams pers. obs.), but it has remained under this name in the revised second edition of the Swan River Flora (John, J., *Diatoms in the Swan River Estuary, Western Australia: Taxonomy and Ecology*, Koeltz Scientific Books, Königstein, Germany, 2012c, 89). Some of John's new taxa are included in the 1983 flora but many were described separately.

[c] We have had no access to this publication.

[d] The floras published here included no new species descriptions, although new taxa were identified but not named. Vyverman and his coworkers went on to describe a number of new taxa from Tasmania (ca. 15 endemics), including the new genus *Eunophora* (Vyverman, W., *et al.*, *Phycologia* 36, 91–102, 1997; Vyverman, W., *et al.*, *European Journal of Phycology* 33, 95–111, 1998a; Sabbe, K., *et al.*, *Systematics and Geography of Plants* 70, 237–243, 2000; Sabbe, K., *et al.*, *European Journal of Phycology* 36, 321–340, 2001.).

[e] Gell, P.A., *et al.*, *An Illustrated Key to Common Diatom Genera from Southern Australia*, Cooperative Research Centre for Freshwater Ecology, Thurgoona, NSW, 1999, is a guide to common diatom genera; and Sonneman, J.A., *et al.*, *An Illustrated Guide to Common Stream Diatom Species from Temperate Australia*, Cooperative Research Centre for Freshwater Ecology, Thurgoona, Australia, [1999] 2000, is a guide to common species; it is a companion volume to Gell *et al.*, hence the two are taken together. It would not be expected to find new descriptions in these kinds of identification guides.

[f] This is the second edition of John, J., *Bibliotheca Phycologia* 64, 1–358, 1983.

A revised estimate of the number of endemic taxa should be in the region of ca. 30 but that is still a relatively small number.

New Guinea and New Caledonia

For New Guinea, there are over 500 known species but only eight endemic taxa have been recorded. The flora includes the remarkable genus *Nupela* (Vyverman and Compére 1991), remarkable as at present it has some 45 or more species and, although distributed worldwide, many are restricted to specific regions, including Australia and New Caledonia (Monnier *et al.* 2003). In almost total contrast, three recent publications have documented an exceptional quantity of endemic taxa found in New Caledonia (christened the 'Galapagos of diatoms' in Moser *et al.* 1998: 14), the primary flora entitled *Insel der Endemiten: Geobotanisches Phänomem Nuekaledonien* (Island of endemics: New Caledonia, a geobotanical phenomenon) (Moser *et al.* 1998). Of the 643 taxa recorded, a massive 257 are considered endemic (Moser *et al.* 1998; also Moser *et al.* 1995; Moser 1999).

Comparisons of Endemism

The calculation of percentage endemism yields an average of only ca. 2% for New Guinea, Australia and New Zealand; New Caledonia is the exception with 40% endemism (Table 2.2). New Caledonia aside, and even using revised figures for Australia and New Zealand, one might still conclude that there is barely any endemism to the freshwater diatoms of Australasia.

The concept of endemism primarily concerns taxa restricted to a particular land mass, usually defined geopolitically: Australia, New Zealand and so on. For example, Australia might be considered an area of endemism, and the search is for species endemic to it. Of interest, then, is the study of Kilroy *et al.* (2007), who provided a different definition of endemism for application to Australian diatoms. Their definition is 'species described from New Zealand and/or Tasmania/Eastern Australia, and, to date, not recorded elsewhere to our knowledge', which they contrasted with a *Southern Hemisphere* category, 'stated in the literature to occur in the S. Hemisphere only; not recorded in the N. Hemisphere' (Kilroy *et al.* 2007: Table 2.1). These are more useful categories (a similar approach was taken by Williams and Reid 2004, 2006a). Yet Kilroy *et al.* (2007) still found only one Southern Hemisphere species, *Cymbella kappii*, and nine endemic taxa: *Gomphoneis minuta* var. *cassieae*, *Rhopalodia novae-zealandiae*, *Pinnularia segariana*, *Encyonema tasmaniense*, *Actinella aotearoaia*, *Fragilariforma cassieae*, *Fragilariforma rakiuriensis*, *Eunophora oberonica* and *Eunophora (Amphora) bergrennii*, the last six being described relatively recently (Kilroy *et al.* 2007: 562, Appendix B; also Vyverman *et al.* 2007b; note the overlap with the Cassie Cooper and Harper checklist in Harper *et al.* 2012), and the two species of *Eunophora* have since been recorded from Australia (John 2009a,b, 2012a–c, 2015).

Nevertheless, a definition of endemism defined by taxa found in particular areas is one we wish to pursue. As we have noted, endemism need not be limited to any particular size of area or even landmass. Our view also focuses on taxa that, supposedly at least, are historically bound via monophyly. For historical biogeography the monophyly and relationships of the species found endemic are of great significance. We have chosen one order to illustrate these points.

TABLE 2.2

Comparison between Number of Diatom Species and Numbers of Endemics Found in New Guinea, Australia, New Zealand and New Caledonia

	Publication	Total	New	%
Australia	Data in this chapter	900	50	0.5
[Eastern] Australia	Foged (1978)	477	10	2.1
New Zealand	Foged (1979)	525	12	2.3
New Zealand	Cassie and Harper (2012)	800	9 [30]	[4]
Papua New Guinea	Vyverman (1991)	525	8	1.5
New Caledonia	Moser *et al.* (1998)	643 [750]	257	40 [34]

Source: Moser, G., *et al.*, *Bibliotheca Diatomologica* 38, 464, 1998, 10, Table; alternative numbers for New Caledonia are based on commentary that suggests the total number of diatom taxa may be nearer 750, although it is not certain how much of that as yet unidentified additional number will be new species: Moser, G., *et al.*, *Bibliotheca Diatomologica* 38, 464, 1998, 11).

Eunotiales and Australasian Diversity

Approaching an understanding of the biogeographic meaning of diatom distribution in Australia, as elsewhere, greatly depends on the taxonomy of the species found, or recorded in, various floras and checklists. It is also greatly dependent on the relationships of the species themselves. As with other organisms' relationships, these are subject to change; however, the more primitive the taxonomy, the more primitive the geographical conclusions. It is possible to focus more clearly on taxa that might display something of significance to the understanding of the spatial relationships of Australian diatoms. While there are a number of potential candidates, we have chosen to elaborate on the order Eunotiales, conventionally partitioned into two families, Eunotiaceae and Peroniaceae. Much has been learnt in the past few decades about this group, but, relevant to this discussion of Australasian biogeography, three initial points need to be made.

1. Establishing monophyletic groups within Eunotiales is progressing, although given the vast number of species that have been, and are still being, placed under the default genus name *Eunotia*, much effort will be needed to establish not just the monophyly of the genus but its proper composition and the interrelationships of the various included taxa. There is at present, however, compelling morphological and molecular evidence to suggest that the Eunotiales are monophyletic.
2. Only after can the respective geography of each monophyletic unit be established and then related to Australia and, in the wider context, Australasia and the Southern Hemisphere.
3. Most genera currently in the family Eunotiaceae do have known endemics, most from South America. For example, the first volume of *Tropical Diatoms of South America* (Metzeltin and Lange-Bertalot 1998) yielded over 221 new taxa, of which 46 (20%) belong to the Eunotiales; the second volume (Metzeltin and Lange-Bertalot 2007) yielded a further 196 new taxa, of which 63 (30%) belong to the Eunotiales. In contrast, an account of the diatoms from Uruguay (Metzeltin *et al.* 2005) yielded 95 taxa, of which only 8 (9%) belong to the Eunotiales. Besides these recent floras, there are many studies for specific geographical areas, especially Brazil (e.g. Cavalcante *et al.* 2014; Ludwig and Flôres 1995; Oliveira *et al.* 2012; Tremarin *et al.* 2008).
4. The discovery of the genus *Eunophora*, nearly 20 years ago, led to the realisation that Eunotiales would now have to include taxa that had previously been placed in the distantly related genus *Amphora* (Vyverman *et al.* 1998a; Levkov 2009; Williams and Reid 2006a, 2006b, Williams and Reid 2009) – indeed, *Amphora*, as originally or even currently conceived, is anything but monophyletic (Stepanek and Kociolek 2014). It is of significance that amphoroid genera described in the context of the Eunotiales since the discovery of *Eunophora*, such as *Amphorotia* and *Colliculoamphora*, have species that are not only geographically constrained but cross broad ecological barriers (having both freshwater and marine members). It is likely, then, that other species currently residing in *Amphora* sensu lato will be members of Eunotiales (Levkov 2009). We comment briefly on a few of these 'non-*Amphora*' species.

For the purposes of this chapter, with its focus on Australian/Australasian biogeography, we include notes on the following generic names in Eunotiaceae: *Actinella*, *Amphicampa*, *Amphorotia*, *Bicudoa*, *Climacidium*, *Colliculoamphora*, *Desmogonium*, *Eunophora*, *Eunotia*, *Eunotioforma*, *Heterocampa*, *Ophidocampa*, *Perinotia* and *Semiorbis*. It is necessary to mention some members of Peroniaceae – *Actinellopsis* in particular – as it appears likely that the two families Eunotiaceae and Peroniaceae that comprise Eunotiales as currently recognised may not themselves both be monophyletic. The use of these generic names should not be taken to represent known monophyletic groups; thus, we first need to briefly digress and discuss the approach we have taken to taxon monophyly in this chapter.

Many diatom genera currently recognised are not demonstrably monophyletic. By 'demonstrably' we mean that many currently recognised generic names in common use (as well as many that have been prematurely neglected rather than refuted) have had little, if any, work on them in any formal way, and thereby no clear synapomorphies or suites of synapomorphies that define them precisely have been identified. We do not mean that genera not so defined are not monophyletic; we mean that they have yet to be demonstrated as such. This is, we believe, not unique to diatom taxonomy, as an examination of any organisms' most recent classification, especially those with large numbers of species, will show that many of the names are composed of groups that have yet to be demonstrated as monophyletic; many taxon names in use today are of unknown status relative to their *-phyly* (in the sense of being mono-, para- or polyphyletic). In a paper published a few years ago it was suggested that groups of unknown -phyly might better be referred to as *aphyletic*, 'meaning that they require taxonomic revision' (Ebach and Williams 2010: 124; see Wilkins and Ebach 2013; Minelli 2014: 845; Williams and Ebach in press) and have no status with respect to phylogenetic interpretation. The application of this terminology avoids the often-made assumption that when a monophyletic group of species is discovered (diagnosed, recognised, defined by synapomorphies) and removed from a larger group of species in an already known and named genus, the species that remain are thereby rendered paraphyletic. It would be more accurate to state that the group of species left behind is *aphyletic*, as we do not know what its status is relative to the data at hand; the group by necessity requires revision. This is common taxonomic practice. As it happens, Willi Hennig created the term *restkörper*, meaning 'remainder', prior to introducing the term *paraphyletic*, to represent those species left behind after a monophyletic group had been discovered (Hennig 1948, 1953). As an example, the genus *Eunotia*, central to our discussions here, is an extremely large genus by any standards (Williams and Reid 2006d), with over 2000 names and probably just as many, if not more, species, as many await description (from 2005 to 2010 over 80 species were described, most from South America). The genus *Eunotia* was more or less defined by the presence of a very short, apical raphe on the valve, which extends along the valve mantle rather than the valve face as it does in most raphid diatoms. Since its initial description (back in 1837, Table 2.3), many subgroups have been removed from it, such as

TABLE 2.3

Generic Names Considered here for Inclusion in Eunotiales with Respect to Australian Biogeography (Figures 2.2 and 2.3)

Taxon Name		Date	Estimated Number of Species Recognised	-phyly
Eunotiales				
Eunotiaceae	*Eunotia*	1837	2300+	a
	Desmogonium	1848	15	a
	Amphicampa	1861	25	a
	Actinella	1864	53	a
	Heterocampa	1870	3	a
	Ophidocampa	1870	15	a
	Climacidiuma	1870	15	a
	Semiorbis	1966	3	m
	Eunophora	1998	5	m
	Amphorotia	2006	19	m
	Colliculoamphora	2006	10	m
	Perinotia	2009	2	a
	Bicudoa	2012	1	a
	Eunotioforma	2013	8	m
Peroniaceae	*Peronia*	1868	14	a
	Peroniopsis	1952	2	a
	Actinellopsis	2014	2	m

Note: Organised first by family, genera are listed chronologically (*Eunotia* being described first in 1837, *Eunotioforma*, the most recent, in 2013), each followed by the number of estimated species at present and a statement of their -phyly; m: monophyletic, a: aphyletic; generic names in bold have species present in Australia/Australasia.

[a] Not to be confused with the diatom genus *Climacodium*, which is a genus of marine phytoplankton species.

Amphorotia (19 species), *Bicudoa* (1 species), *Eunotioforma* (8 species) and *Semiorbis* (3 species). This reduces the number of species in *Eunotia*, but only by a mere handful (less than 30 from more than 2000). Significantly, though, it removes species that had the defining character(s) of the genus. Thus, following one particular viewpoint, all species now remaining in the genus *Eunotia*, after the removal of some with its defining characters, will leave a paraphyletic group as it now does not include all its members. This is a premature judgement, of course, as the status of the remaining group is, at the very least, uncertain: it may be mono-, para- or even polyphyletic. Without evidence to support any of these options, the group remaining is thus literally aphyletic: it is without -phyly (Ebach and Williams 2010 and submitted). It should be noted that in almost all cases the relationships of monotypic taxa are problematic (Williams and Ebach 2009, 2013) as their -phyly will be deemed unknown and they are therefore, by default, aphyletic. In the case of Eunotiaceae, this applies only to the recently described genus *Bicudoa*.

Table 2.3 summarises some relevant data for Eunotiales. The table is composed of genera organised by their currently recognised family (Eunotiaceae and Peroniaceae). Within each family, genera are placed in chronological sequence with current estimated numbers of included species and their -phyly, as discussed previously. It is pointless to formally, or even informally, recognise para- or polyphyletic groups as they have no existence beyond their artificial definition with respect to a particular monophyletic group (Kociolek and Williams 2015).

The cladogram in Figure 2.2 is based on the analysis of Williams and Reid (2006a). In this scheme of relationships there are three significant branches: branch 1 relates *Actinella* and *Peronia* most closely, branch 2 relates *Semiorbis* and *Eunotia biseriatoides* most closely and branch 3 relates *Eunotia*, *Amphorotia* and *Eunophora* most closely. Branches 2 and 3 together relate most closely relative to branch 1 (Figure 2.2, node a). An additional box, basal to the entire tree, enclosed in hashed lines and attached via the branch marked '?', represents taxa not included in the original analysis, many of which have Australian representatives.

Two further relationships require comment (Figure 2.2, branch 4). In the original analysis *Amphorotia* was placed as most closely related to *Eunotia lapponica*, a species currently assumed to be widespread but which deserves further attention (here it is added with a short-dashed line). An alternative, represented in Figure 2.2 with a short-dashed paired line, is the close relationship of *Eunotia arcuoides* with *Amphorotia*:

> Notably, the genus *Amphorotia* shares a number of characters with *E. arcuoides*.... Based on the presence of spines, Williams and Reid (2006a) suggested that some species related to *Eunotia*

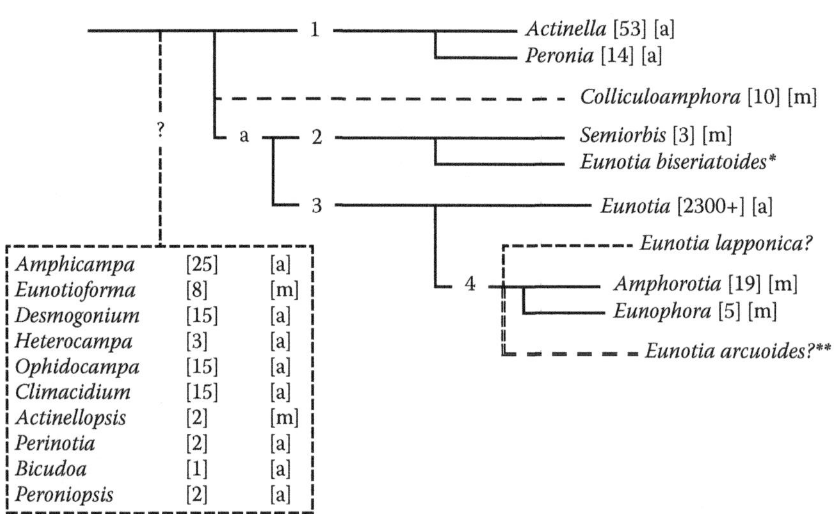

FIGURE 2.2 Cladogram of relationships in Eunotiales; genera are marked with respect to their status either as mono- [m] or aphyletic [a] groups (summarised in Table 2.3). (Based on the analysis of Williams, D.M., and Reid, G., *Diatom Monographs* 6, 1–153, 2006a.)

> *lapponica* Grunow may be sister taxa to *Amphorotia*.... *E. arcuoides* may represent an alternative tentative sister taxon to *Amphorotia*. (Beals and Popatova 2013: 31)

Other alternative close relatives to *Amphorotia* that might be considered are *Eunotia relicta* ('Chapada Diamantina region, Northeast Brazil'), *E. charliereimeri* (United States) and *E. sarraceniae* (United States; Brazil).

In Figure 2.2, *Semiorbis* (a largely Northern Hemisphere genus) is most closely related to *Eunotia biseriatoides*, a species initially found in Japan. (Relevant, too, is a second species from Japan, *E. sparsistriata* Mayama in Mayama 1993; Kulikovskiy *et al.* 2015). Cavalcante *et al.*, on the other hand, have drawn attention to a number of species still in *Eunotia* that might be better placed in the genus *Perinotia* should evidence (synapomorphies) be found to support such a view (2014: 3–4). With that in mind, they propose that *E. odebrechtiana* Metzeltin and Lange-Bertalot (1998: 71; Brazil) is a synonym of *E. biseriatoides* Kobayasi *et al.* (1981: 98; Izu Peninsula, Japan) and transferred *E. odebrechtiana* var. *essequiboensis* Metzeltin and Lange-Bertalot (1998: 72) to *E. biseriatoides* as *E. biseriatoides* var. *essequiboensis* (Guyana). *E. monodon* var. *constricta* (Demerara River, Guyana) is also a synonym, according to Metzeltin and Lange-Bertalot (1998: 71), and *E. biseriata* Hustedt (1952: 143; Brazil) may well be too. While this branch does not directly impinge on Australasian taxa, it points to yet another trans-Pacific relationship within this group, and some species described from New Caledonia (e.g. *E. melanogaster* and *E. sparsiornata*) and Lake Baikal (e.g. *E. dorsicrenata*, *E. leptocrenata*, *E. pseudincisa* and *E. immanis*) may be worth exploring in terms of a potential *Semiorbis–E. biseriatoides–Peronotia* relationship, which represents a possible bipolar distribution (Moser *et al.* 1998; Kulikovskiy *et al.* 2015).

The second cladogram (Figure 2.3) is based on the analysis of Burliga *et al.* (2013), a somewhat more conservative representation proposing that *Eunotioforma* is basal to all Eunotiales genera with the exception of *Peronia* (*Eunophora* has been added to Figure 2.3). The significant difference between Figures 2.2 and 2.3 is that *Actinella* is included in the more derived group in Figure 2.3, rather than related to the basal *Peronia* in Figure 2.2 (this conflict was briefly discussed in Williams and Reid 2008). Thus, from the point of view of Australian distributional data, Figure 2.3 includes all genera with Australian taxa grouped together in a more derived, if unresolved, position, whereas in Figure 2.2 they are scattered among the various clades.

Eunotia

As previously noted, *Eunotia* is a large genus and is extremely diverse in its structural complexity and its many growth forms. Perhaps rather more so than other large diatom genera, it is significant not just because it is morphologically diverse but also because it has clear geographical components. There has been, for example, an acknowledged diversity in South America for some while (e.g. Hustedt in Schmidt 1913; Hustedt 1952) and elsewhere in the Southern Hemisphere, a diversity that is still being

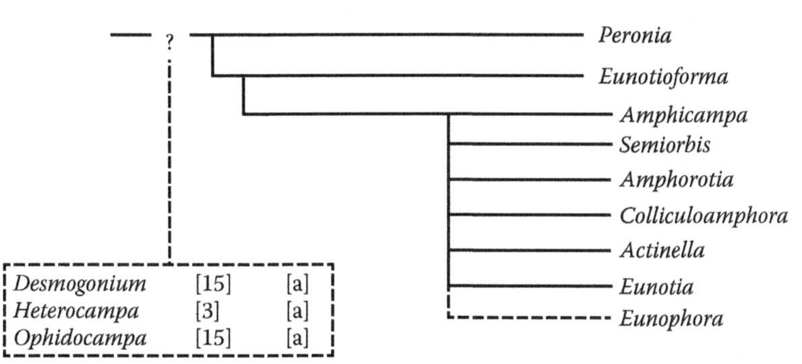

FIGURE 2.3 Cladogram of relationships in Eunotiales; genera are marked with respect to their status either as mono- [m] or aphyletic [a] groups (summarised in Table 2.3). (Based on the analysis of Burliga, A.L., *et al.*, *Phytotaxa* 79, 47–57, 2013. *Eunophora* has been added.)

uncovered, described and explained today (Metzeltin and Lange-Bertalot 1998, 2007; Cavalcante *et al.* 2014; Ludwig and Flôres 1995; Oliveira *et al.* 2012; Tremarin *et al.* 2008: Wetzel *et al.* 2011). Yet from the perspective of Australasian diatom diversity, the many species of *Eunotia* recorded in various books and papers mostly have European names, with the conclusion that they relate to Northern Hemisphere species. It remains unclear, then, as to whether those species found in Australasia are really the 'same' as those found in Europe or if they are distinct in ways yet to be explored. For example, DNA studies on clones of the widespread '*E. bilunaris*' from New Zealand and Tasmania were found to be similar enough to each other but less so to the European species (Vanormelingen *et al.* 2007, 2008: 399). It is evident that specimens captured by the binomial *E. bilunaris* are doubtfully one single widespread species.

In the introduction to this chapter, two identification guides were mentioned: *An Illustrated Key to Common Diatom Genera from Southern Australia* and *An Illustrated Guide to Common Stream Diatom Species from Temperate Australia*. Neither conveys the extent of diversity in the genus *Eunotia* in Australia. Diversity data happen to accumulate piecemeal and, at this time, it is almost impossible to summarise in the absence of a detailed and documented taxon checklist. Reliance on identification guides may suggest, misleadingly we believe, that the preponderance of diatom diversity in the genus *Eunotia* in Australia is European (not just in this genus, as it happens). That is, the majority of the species found were first discovered in Europe (or elsewhere). The total numbers of species in the genus *Eunotia* that have appeared in floras published since 1978 (see Table 2.1) are summarised in Table 2.4. Most suggest a low level of endemism, with Foged (23%) and Thomas (29%) much higher than the others.

In the absence of any useful checklist, however, one has to rely on various published sources, which include diatom species lists but where the study was designed for a purpose other than determining the geographical distribution of Australian diatom species. For example, a recent paper by Chessman *et al.*, 'A diatom species index for bio-assessment of Australian rivers', includes an appendix, 'Average sensitivity values (SVs) for diatom species'. The appendix lists a number of diatom species, including 22 from *Eunotia* (and one species of *Actinella*) (Chessman *et al.* 2007). Among the 22 species, three, *E. pseudoserra*, *E. silvae*, and *E. tecta*, are all found in South America (as is the one species of *Actinella*, *A. eunotioides*). Thus, Chessman *et al.*'s study is also useful for diversity assessments.

In addition to the floras (and ecological studies), Thomas (1987) described *E. didyma* var. *maxima* f. *tumida*, another Australian endemic, and while the distinguishing feature and taxonomic rank suggests a relatively trivial difference (a 'consistently smaller length to breadth ratio' of the valves; Thomas 1987: 53), *E. didyma* is considered to be a highly variable taxon occurring in the Southern Hemisphere, mainly in South America (Metzeltin and Lange-Bertalot 1998) but also in China (Huang Chengyan and Liu Shicheng 1998), Australia (Foged 1978) and New Zealand (Harper *et al.* 2012). Hustedt described (and named) many varieties of *E. didyma* from the Demerara River, Brazil, and Banka, Indonesia, the latter locality noted for the original description of *Desmogonium rabenhorstianum* (Hustedt in Schmidt 1913;

TABLE 2.4

Data from Several Australian Floras for Numbers of Species in the Genus *Eunotia*: Total Numbers and Numbers of Those of Restricted Distribution

		***Eunotia* sp.**	
Author	**Date**	**Total Number Recorded**	**Number with Restricted Distribution**[a]
Foged	1978	42	ca. 10
Brady	1979	20	ca. 3
John	1983	8	–
Thomas	1983	34[b]	ca. 10
Vyverman *et al.*	1995	30+	ca. 2
Hodgson *et al.*	1996	22	ca. 3
John	2012[c]	9	ca. 2

[a] By restricted, we mean not found in the Northern Hemisphere.
[b] This number includes two taxa referable to *Desmogonium*.

see 'Desmogonium' in this chapter). This example suggests that at least some Australian diversity in the genus *Eunotia* relates directly to that in South America (Brazil: *E. didyma* f. *obtusa*; Peru: *E. didyma* var. *jugulata*, *E. didyma* var. *pileus*; Argentina: *E. didyma* var. *papilio*).

Three other endemic species of *Eunotia* have been named, two described by Brady (1979, both are *nom. nud.*, as they lacked a Latin description at the time of publication): *E. ossa* (Brady 1979; Thomas [1983] believes this to be *E. zasuminensis*) and *E. diodon* var. *inflata* (Brady 1979*; Thomas [1983] believes this to be *E. camelus*), neither being illustrated adequately enough to make further judgement. *E. ossa* does, however, resemble *Actinella siolii* (Metzeltin and Lange-Bertalot 2007: taf. 39, Figures 1–10, from Brazil) – without electron microscope images that resemblance may be entirely superficial – and (possibly) *E. zasuminensis* (Thomas 1983: 23; a full account of *E. zasuminensis* can be found in Wetzel *et al.* 2010), a widespread, cosmopolitan species. The third endemic species, *E. ballinaensis*, was described by Foged (1978); it too requires further examination.

As opposed to Australia, there are useful checklists for the diatoms of New Zealand. Cassie Cooper and Harper (in Harper *et al.* 2012, the most recent checklist for New Zealand), list over 50 species of *Eunotia*, but only 2 are recorded as endemic for Australia: *E. sudetica* var. *hamuranaensis*† and *E. cassia*. *E. ophidocampa* is included in the checklist and should also be considered endemic as it was first described from Rotorua Lake, New Zealand (Cleve and Möller 1882 in Cleve and Möller 1877–1882); a few taxa are shared between New Zealand and South American countries. Harper *et al.* (2012: 133) refer to the *New Zealand–Tasmania* link, a geographical relationship first suggested by Foged, who included Australia, Tasmania and New Zealand (Foged 1979: 113).

Finally, we must mention here the occurrence of a few marine fossil species of *Eunotia* from the Oamaru deposit in New Zealand, specimens that require more detailed interpretation and analysis but which, nevertheless, appear to be genuinely marine and do belong to *Eunotia* (synapomorphies in Novitski and Kociolek 2005: 141; Berg 1948). Their relationships are not yet evident, but they provide some dates as the Oamaru deposit is Late Eocene or Early Oligocene (Edwards 1991).

Actinella

The monophyly of *Actinella* has yet to be unequivocally established, but current data suggest that, with the presently recognised species composition, it will be demonstrated as such (Spaulding and Kociolek 2003). In all, there are around 70 species: 21 from Africa and Madagascar (10 from Madagascar alone; Spaulding and Kociolek 2003; Metzeltin and Lange-Bertalot 2002), 31 from South America, mostly from Brazil (at least 20), Guyana (7), Venezuela (3) and 2 species shared between various localities: *A. guianensis* Grunow in Van Heurck occurs in Brazil (Oliveira *et al.* 2012) and Guyana, and *A. disjuncta* Metzeltin and Lange-Bertalot occurs in Brazil, Venezuela and Borneo.

For Australasia, there are a number of relevant occurrences: one species, *A. aotearoaia*, occurs in New Zealand and Australia (Sydney); two species, *A. comperei* and *A. muylaertii*, occur in Australia (Tasmania); three species, *A. indistincta*, *A. parva* and *A. pulchella*, are shared between New Zealand and Tasmania (all described in Sabbe *et al.* 2001); one species, *A. tasmaniensis*, is shared between Tasmania, New Zealand and Brazil‡; and one species, *A. brasiliensis*, is shared between New Zealand and Brazil, but has been recorded as widespread and deserves closer attention.

Two species, *A. cuneiformis* and *A. modesta*,§ occur in New Caledonia (Metzeltin and Lange-Bertalot 2007; Moser *et al.* 1998), and one, *A. giluwensis* (Sabbe *et al.* 2001), occurs in Papua New Guinea.

The occurrence of *A. punctata* in the United States and Europe is considered to be a widespread distribution. This species is related to European fossils (such as *A. pliocenica*, *A. pliocenica* var. *serpentina* and *A. pliocenica* var. *tenuistriata*, all described some while ago by Héribaud 1902, 1903).

---

* Brady's specimens are apparently available at the Royal Botanic Gardens, Sydney, Australia.

† This may be a synonym of *E. sudetica* var. *australis*. In which case, it occurs in New Zealand and Tierra del Fuego (Cleve-Euler 1948; Thomas 1983).

‡ See Uherkovich and Franken (1980).

§ According to Tremarin *et al.*, *A. leontopithecus-rosalia* (and its synonym *A. lange-bertaloti*), a southeast Brazilian species, may be most closely related to *A. modesta* (Tremarin *et al.* 2016: 264).

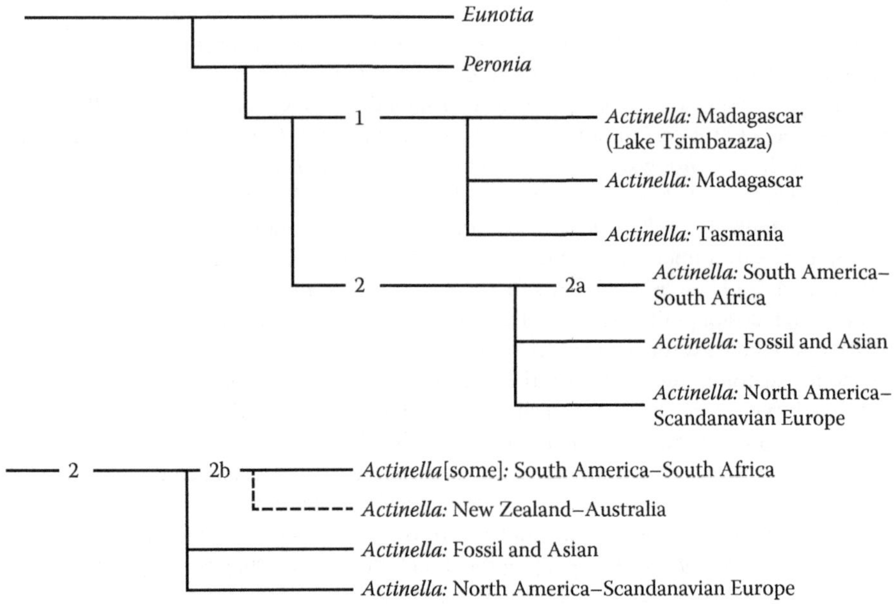

FIGURE 2.4 Cladogram of relationships for a number of species in *Actinella* (Spaulding, S.A., and Kociolek, J.P., *Natural History of Madagascar*, University of Chicago Press, Chicago, 2003, 281, Figure 7.11, redrawn and modified).

Consideration of these fossils does not make it clear as to their status in relation to the many taxa in the Southern Hemisphere. Finally, the relationship of the genus *Actinellopsis* – created for one living species from Zambia, *A. murphyi*, and for an Eocene fossil from the Canadian Arctic, *Actinella giraffensis* – to *Actinella* is unclear, as are the implications of their geographical spread (Taylor *et al.* 2014; Siver *et al.* 2010; Siver *et al.* 2015).

Spaulding and Kociolek offered a cladogram for a number of species in *Actinella* (Spaulding and Kociolek 2003: 281, Figure 7.11; redrawn and modified here as Figure 2.4), where they showed that the Tasmanian species of *Actinella* are related most closely to the Madagascan species. Developments since include the description of *Cultria*, first described as a subgenus of *Eunotia* for the species *E. falcifera* (Metzeltin and Lange-Bertalot 1998: 46; Brazil, Rio Tapajós), later transferred to *Actinella* as *A. falcifera* (Metzeltin and Lange-Bertalot 2007: 29; Guyana, Essequibo River). Species that may be most closely related to *A. falcifera* (*Cultria*) are *A. aotearoaia*, which, as previously noted, occurs in New Zealand and Australia but not Tasmania, and *A. eunotioides*, which occurs in South America (Brazil) and Australia. Thus, some South American species of *Actinella* (Figure 2.4, node 2) may be more closely related to some New Zealand–Australian species of *Actinella* (compare Figure 2.4, node 2a and the inset cladogram node 2b) if species such as *A. falcifera* (*Cultria*) are established as their closest relatives. Canani and Torgan (2013), developing ideas first proposed by Metzeltin and Lange-Bertalot (2007), suggested that *A. sabbei* and *A. lyonae* (Metzeltin and Lange-Bertalot 2007, both from Guyana, Essequibo River) may also be most closely related to *A. falcifera*, but at present relevant data are lacking. Canani and Torgan go on to suggest a number of problematic species in *Actinella* with respect to current generic definitions and suggest that 'cladistic analysis of morphological characters and characters states and/or molecular analyses could provide additional information to help [solve] these taxonomic puzzles, clarifying the natural relationships among these taxa' (Canani and Torgan 2013: 11), a sentiment we wholly concur with, and we suggest that the geography of these species would be clarified as well (see also Ripple and Kociolek 2013: 618, for comments on the geographical relationships of species of *Actinella* recently described from Hawaii, which may have some significance to the more global picture of this genus*).

* There are three endemic species of *Actinella* in Hawaii (Ripple and Kociolek 2013; Main 2004).

In summary, Australasian species of *Actinella* occur in Tasmania, New Zealand and New Caledonia, suggesting relationships among these areas.

Amphicampa, Ophidocampa, Heterocampa, Climacidium

Amphicampa, a genus name barely used today, was introduced as an 'Unterabteilung von *Eunotia* [subdivision of *Eunotia*]' (Ehrenberg 1854: 373) and at first included just two species, *A. eruca* and *A. mirabilis*, both from 'Tisar [Tizar], Mexico' (Ehrenberg 1854: pl. 33, 7/1, Figures 1–2, both figures reproduced here as our Figure 2.5a,b). Early reports of *A. (mirifica) mirabilis* (Ehrenberg 1854: pl. 33, 7/1, Figure 2, reproduced here as Figure 5b; Ralfs in Pritchard 1861: 765, pl. 4, Figure 5, reproduced here as

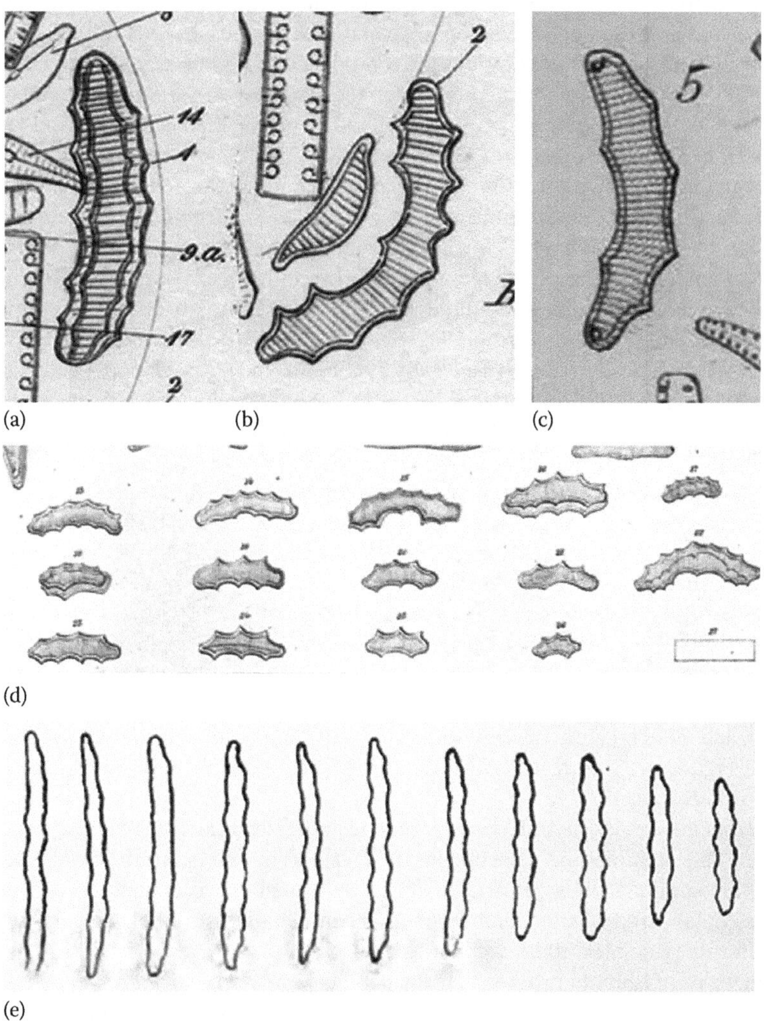

FIGURE 2.5 (a,b) Reproduction of *Amphicampa eruca* and *A. mirabilis* (from Ehrenberg, G.C., *Mikrogeologie: Das Erden und Felsen schaffende Wirken der unsichtbar kleinen selbständigen Lebens auf der Erde*, L. Voss, Leipzig, 1854, pl. 33, 7/1, Figures 1–2); (c) Reproduction of *A. (mirifica) mirabilis* (from Ralfs in Pritchard, A., *A History of Infusoria, Living and Fossil: Arranged according to 'Die Infusionsthierchen' of C.G. Ehrenberg…*, Whittaker, London, 1861, 765, pl. 4, Figure 5); (d) Ehrenberg's species of *Amphicampa* (13 in total) (reproduced from Ehrenberg, G.C., *Abhandlungen der Königliche Akademie der Wissenschaften zu Berlin* (1869), 43(2), 1–66, 1870, 44); (e) *Eunotia serpentina* (from de Toni, G.B., and Forti, A., *Memorie del Reale Institute Veneto di Scienze, Lettere ed Arti* 29(3), 1922, 118, an account of Australian algae).

Figure 2.5c; Ehrenberg 1870: 46, 62, taf. 1, B, Figure 14) mention only Mexico ('Tisar [Tizar], Mexico'), while more contemporary accounts document specimens from other areas but with greater variation (e.g. Metzeltin and Lange-Bertalot 1998: taf. 4, Figures 1–5, from 'Brasilein'; Metzeltin and Lange-Bertalot 2007: taf. 24, Figures 1–6, from Brazil; both under the name *Actinella mirabilis*). Ehrenberg later named a number of species in *Amphicampa* (13 in total), all from Mexican material; their individuality almost certainly requires detailed examination (Ehrenberg 1870: 44, reproduced here as Figure 2.5d).

Since Ehrenberg's descriptions, Australian occurrences of species seemingly like *Amphicampa* have consistently emerged, *A.* (*Eunotia*) *serpentina* being the most common named used; its synonymy with *A. eruca* and/or *A. mirabilis*, if any, is a matter yet to be resolved (John 1983: 44, pl. 17, Figures 2 and 3 [Swan River]; Sonneman *et al.* [1999] 2000: 58; for occurrences in New Zealand and a fuller explanation of the nomenclatural complexity, see Raeside 1970; Foged 1979; Cassie 1989). Ehrenberg first recorded *Eunotia serpentina* from various Australian rivers (Ehrenberg 1854: 6, 12, but only described later in Ehrenberg 1870), and an extended discussion of the species appears in de Toni and Forti's (1922: 118) account of Australian algae (their illustrations are reproduced in Figure 2.5e) and in Foged's *Diatoms in Eastern Australia* (1978: 60). Ehrenberg complicated the taxonomy by separating some fossil specimens into another genus, *Ophidocampa*, others into yet another genus, *Heterocampa*, and living specimens into the species *Ophidocampa septenaria* (Ehrenberg 1870: 50).

Other taxonomic problems with *A.* (*Eunotia*) *serpentina* remain, as, for example, Tsumura suggested that it is simply a form of *A. eruca*, and named it *A. eruca* f. *serpentina* (Ehrenb.) Tsumura (1967: 232; it is not clear where Tsumura's specimens are from). Hustedt suggested that *Eunotia transylvanica* (*transsilvanica*) (Pantocsek 1892: pl. 3, Figure 36, fossil from Köpecz, Romania) was either a variety of *E. serpentina* or *E. eruca* (Hustedt in Schmidt 1911, pl. 274, Figures 6–8, pl. 290, Figure 8, corrected names given in the legend to pl. 293; Burbury [1902: 4] records *E. transylvanica* from Hobart, Tasmania; and John [2012c: 44] includes *E. transylvanica* as a synonym), and Cassie Cooper and Harper considered it to be a variety of *A. mirabilis* (Harper *et al.* 2012: 152). Some valve outline images, of limited use, for '*Amphicampa* sp.?' have been published by Foged from Rennell Island in the Solomon Islands (Foged 1957: Figures 9–14).

Doubts over the validity of *Amphicampa* as a genus separate from *Eunotia* was expressed early on by Thomas Brightwell, who placed *A. eruca* in *Eunotia* as *E. eruca* (Ehrenberg) Brightwell (1859: 179); his specimens were from a 'fresh-water lagoon, near Melbourne, New South Wales; Mackie' (Brightwell 1859: 179, reproduced as Figure 2.6c).

Further names worth a mention are *E. crispula* G.S. West (1909: 78, pl. 3, Figure 14, reproduced as Figure 2.6b), recorded from Australia (Yarra Yarra Reservoir and Heidelberg, Victoria), *Himantidium maskellii* Inglis (1883: 358, pl. 22, Figure 2, reproduced as Figure 2.6a), recorded from New Zealand ('River Avon, Canterbury ... Aukland ... North Canterbury (fossil)' [*E. maskellii* (Inglis)]; Chapman *et al.* 1957: 733), and more recently the new variety *E. serpentina* var. *eskdalensis* as described by Foged (1978: 61, pl. 12, Figure 14, 'Australia'). With the exception of Foged's new variety, these specimens are more like Hustedt's *E. eruca*.

Suffice to say that all species in *Amphicampa* that occur in Australasia (and elsewhere, for that matter) require attention. The examination of specimens of *A. eruca* (in The Natural History Museum, London, most specimens are similar to those illustrated at first by Ehrenberg) is expanding its range to Australia, Tasmania, New Zealand and Mexico, and the most recent detailed study of the genus examined specimens from California (Kociolek 2000: 15, Figures 21–35).

At some point, three of Ehrenberg's neglected genera, *Ophidocampa*, *Heterocampa* and *Climacidium*, will deserve further attention, if for no other reason than *Ophidocampa* has numerous names, none of which are currently recognised species (13 names in all; Ehrenberg 1870: 44). All but two are from New Zealand fossil material (Ehrenberg 1870: 50–51, pl. 2/2, Figures 1–13, reproduced as Figure 2.6d). To complicate matters, *O. crocodilus* is possibly based on the earlier name *E. crocodilus* Ehrenberg, described from 'Viva in Guinea anglica. Eadem in Senegambia' and *O. tapacumae* Ehrenberg is also based on an earlier name, *E. tapacumae* from 'Barima-flusse in Englischen Guiana', thus suggesting a possible African–South American–New Zealand connection.

Heterocampa has three species, these too being fossils from New Zealand (Ehrenberg 1870: 49–50, pl. 2/2, Figures 14 through 16, reproduced as Figure 2.6e), and *Climacidium* now has ca. 15 species

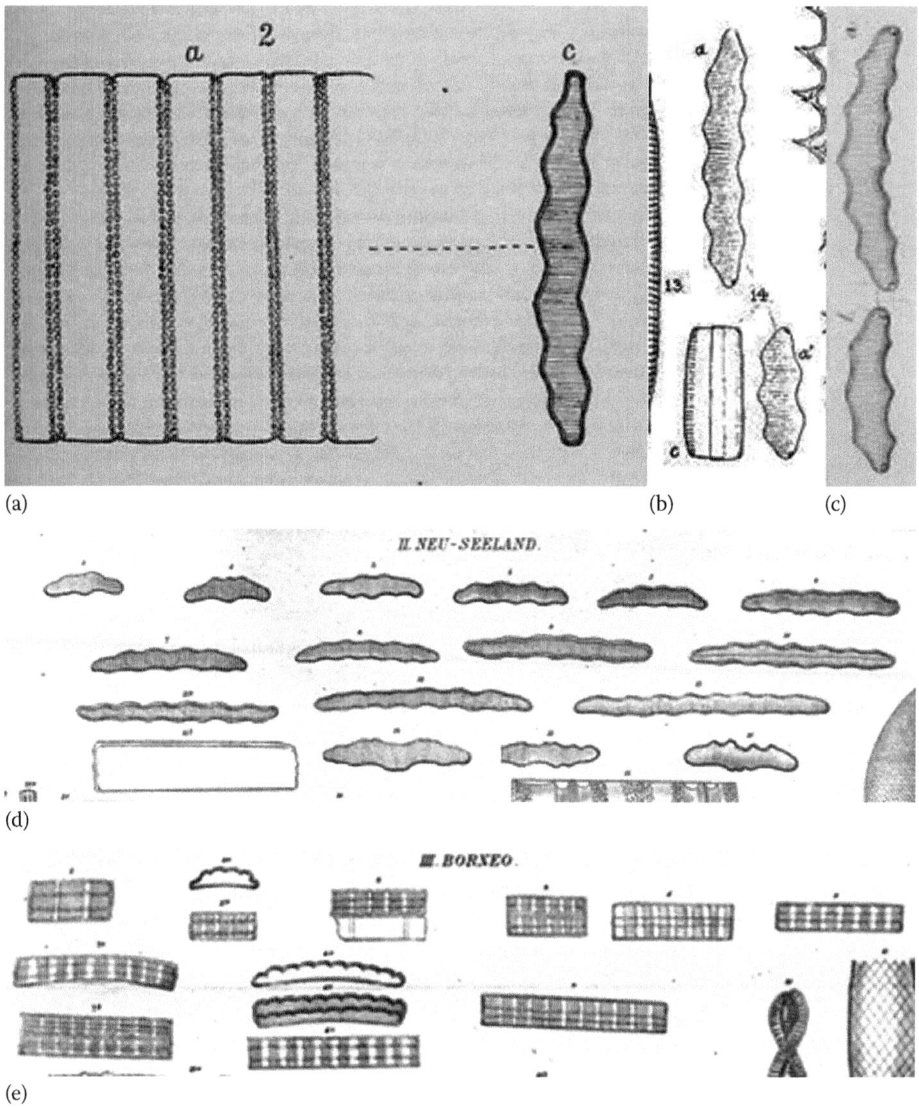

FIGURE 2.6 (a) *Himantidium maskellii* (reproduced from Inglis, J., *Transactions and Proceedings of the New Zealand Institute* 15, 1883, 358, pl. 22, Figure 2), recorded from New Zealand, 'River Avon, Canterbury ... Aukland ... North Canterbury (fossil)...'; (b) *Eunotia crispula*, Yarra Yarra Reservoir and Heidelberg, Victoria (reproduced from West, G.S., *Journal of the Linnean Society Botany* 39, 1–88, 1909, 78, pl. 3, Figure 14); (c) *E. eruca*, 'Fresh-water lagoon, near Melbourne, New South Wales; Mackie' (reproduced from Brightwell, T., *Quarterly Journal of Microscopical Science* 7, 179–181, 1859, 179); (d) Ehrenberg's *Ophidocampa*, New Zealand fossil material (reproduced from Ehrenberg, G.C., *Abhandlungen der Königliche Akademie der Wissenschaften zu Berlin* [1869], 43[2], 1–66, 1870, 50–51, pl. 2/2, Figures 1–13); (e) Ehrenberg's *Climacidium*, Borneo (reproduced from Ehrenberg, G.C., *Abhandlungen der Königliche Akademie der Wissenschaften zu Berlin* [1869], 43[2], 1–66, 1870, 47, pl. 2/3, Figures 4 and 8, 1–10).

(many since moved to *Eunotia*), of which the original 9 described by Ehrenberg were all from Borneo (Ehrenberg 1870: 47, pl. 2/3, Figures 4 and 8, 1–10, reproduced as Figure 2.6e).

In summary, in spite of the complex taxonomic problems and difficulty in differentiating species, those that have been assigned to *Amphicampa*, *Ophidocampa* and *Heterocampa* occur in certain parts of Australia, Tasmania, New Zealand, Mexico and California, with the preponderance of (apparent) species diversity divided between New Zealand and Mexico. It should be stressed that these conclusions depend entirely on current taxonomy, which, as previously documented in this chapter and in other publications, is in dire need of attention.

Desmogonium

There are currently ca. 10–12 species in this genus (if it is recognised at all), distributed largely (but not exclusively) in the Southern Hemisphere (excluding fossils, which are distributed more extensively, but all require reevaluation): South America (Brazil, Guyana, Venezuela), Sumatra, Thailand, Guadeloupe, Africa (*Eunotia theronii* and var. *capensis*, for example, which may be better placed in *Desmogonium*) and Australia. Of note for Australia are two species: *D. rabenhorstianum* Grunow (1865: 6) and its varieties, and *D. guianense* Ehrenberg (1848: 539).

For *D. rabenhorstianum*, the nominate species was first described from Insela Banka (Bangka), an island off Sumatra, Indonesia (Grunow 1865: 1) and has since been found in a number of places, including Sri Lanka (Ceylon, West and West 1901: 209), Malaysia (Prowse and Ratnasabapathy 1970), Brazil (Patrick 1940; Metzeltin and Lange-Bertalot 1998) and, more significantly for Australasia, Rennell Island in the Solomon Islands (Foged 1957); *D. rabenhorstianum* var. *crassa* Østrup (1902: 37, Figure 15) was first described from 'Koh Chang: Lem Dan … Siam [Thailand]'; and *D. rabenhorstianum* var. *elongatum* (Patrick 1940: 3, Figures 1–3) was first described from Brazil but later found in Florida (Patrick in Patrick and Reimer 1966: 221). *D. rabenhorstianum* var. *elongatum* was subsequently transferred to *Eunotia* by Metzeltin and Lange-Bertalot as *E. rabenhorstiana* var. *elongata* (Metzeltin and Lange-Bertalot 1998: 76; Brazil, Rio Negro), but an earlier transfer, effected by Brady, for *E. rabenhorstiana* var. *elongatum* (a name change missed by a number of diatom nomenclature lists), was made from specimens found from a number of localities in the Northern Territory (Brady 1979).

As Hustedt did not recognise the genus *Desmogonium*, he transferred *D. guianense* to *Eunotia*, a change that required a new name, *E. lineolata*. Recorded under that name, it has been noted for Australia (Foged 1978; Brady 1979) and New Zealand (Foged 1979; Harper *et al.* 2012). It has been noted from Brazil (de Azevedo Barros *et al.* 2013) and the Colombian Amazon (Upper Solimões, Iça and Japurá Rivers; Sala *et al.* 2002: 597), but these specimens may belong to *D. transfugum* (*E. transfuga* Metzeltin and Lange-Bertalot [1998]; Tremarin *et al.* 2008: 858; Ludwig and Flôres 1995). It has also been noted from Borneo (Hirano 1974), but these specimens may be from yet another species. A form, *E. lineolata* f. *minor* (Woodhead and Tweed 1960: 127), has been recorded from Sierra Leone, but the published illustrations and description are unhelpful in determining what this taxon might actually be.

In summary, Australian species of *Desmogonium* (given the genus will at some stage be found monophyletic) are shared geographically with Brazil, Guyana, Venezuela, Sumatra, New Zealand and Thailand (and possibly, but doubtfully, Africa and Borneo; specimens in these areas require further investigation).

Eunotioforma, Bicudoa, Perinotia

Several recently described genera appear endemic to Brazil (with a few exceptions). *Eunotioforma* has eight species, of which five are endemic to Brazil; of the remaining three, one occurs in the United States (Columbia River, Oregon), another occurs in Finland and the third is widespread (Burliga *et al.* 2013). *Bicudoa* is a monotypic genus, endemic to Brazil (Wetzel *et al.* 2012). *Perinotia* has, at present, two species, both from Brazil (Ferrari *et al.* 2009; Metzeltin and Lange-Bertalot 2007). In the second of the two cladograms published here (Figure 2.3), *Eunotioforma* is primitive to all other members of the Eunotiales.

Cavalcante *et al.* (2014) have suggested some further species as possible members of *Perinota*, as discussed previously. A search might reveal further relevant species, such as *Eunotia luna* var. *globosa* (United States and Brazil).

No representatives have been encountered in Australasia.

Eunophora

First described in 1998, now with five species, *E. berggrenii* (Figure 2.7a–d) and *E. novaezealandiae* occur in New Zealand (Vyverman *et al.* 1998a; Levkov 2009; Harper *et al.* 2009); the remaining three species, *E. indistincta*, *E. oberonica* and *E. tasmanica* occur in Tasmania (Vyverman *et al.* 1998a;

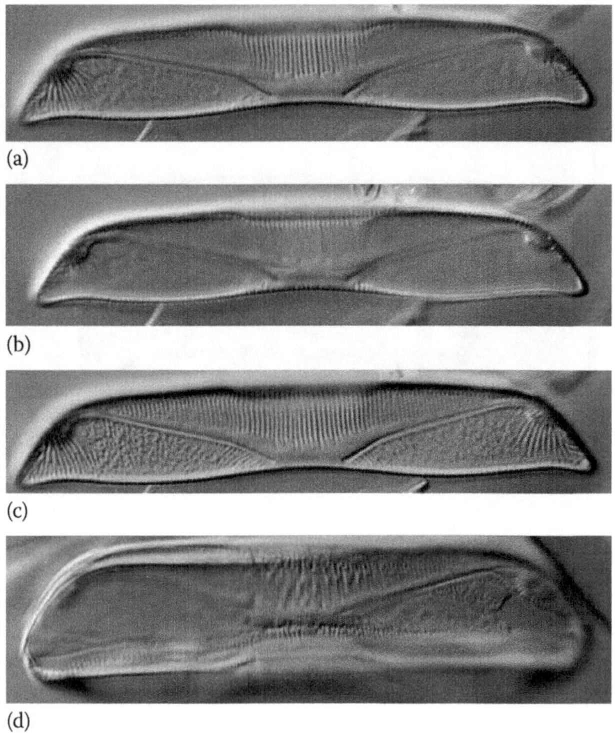

FIGURE 2.7 (a–d) *Eunophora berggrenii*, from type material, BM.

Levkov 2009) and Australia (John 2009a,b, 2012a–c), the latter species probably synonymous with *Amphora hendeyi* (Lake Dobson, Tasmania; Levkov 2009: 280; Figure 2.8b,c).

It is worth a short digression to comment briefly on the genus *Amphora* as, in the paper that presented the original description of *Eunophora*, the authors noted that

> *Eunophora* most closely resembles species of Cleve's subgenera *Psammamphora* and *Amblyamphora*, [such as] the marine species *A. obtusa* Gregory, *A. arenaria* Donkin, *A. spectabilis* Gregory, *A. cingulata* Cleve and *A. latecingulata* M. Peragallo and the freshwater taxa *A. delphinea* Bailey and *Amphora* sp. 1 of Vyverman *et al.* (1995). (Vyverman *et al.* 1998a; see Levkov 2009 for illustrations of some of these species of *Amphora* and Figure 2.8b,c for examples).

Rather than dwell on the details of these proposed relationships, it is of note that most of the species of *Amphora* listed in the quotation are marine, possibly some with restricted distributions in the Southern Hemisphere, with the few freshwater species also exhibiting distributions relevant to a wider more global understanding of Australasian geography. (We noted previously the demonstrated polyphyly of *Amphora* and emphasise here the immensely valuable flora/monograph of Levkov [2009], which documents the genus in its broadest sense). For example, the freshwater species *A. obtusa* var. *baikalensis* (Figure 2.8a, reproduced from Skvortzow 1937: Figures 20 and 28; Figure 2.8f, reproduced from Jansnitsky 1936: Figure 8; Figure 2.8d, specimen from Lake Baikal), endemic to Lake Baikal, was initially thought to be a possible member of *Eunophora* (Vyverman *et al.* 1998a), a judgement in need of revision (Vyverman *et al.* 1998b). Skvortzow mentions and illustrates a specimen of *A. delphinea* (Skvortzow 1937: Figure 4), a species found in South America, Japan and the United States (see Levkov 2009: pl. 116 for illustrations of specimens from Demerara River, Guyana, 'distributed in South America.... Some records of this species in freshwater fossil deposits are probably misidentifications' [260]). The Lake Baikal specimens are probably *A. delphineiformis* Levkov (from Lake Baikal, Russia, and Lake Hövsgöl, Mongolia; Levkov 2009: 261; Levkov also described the related *A. nipponica* from Japan). It remains to be seen how these

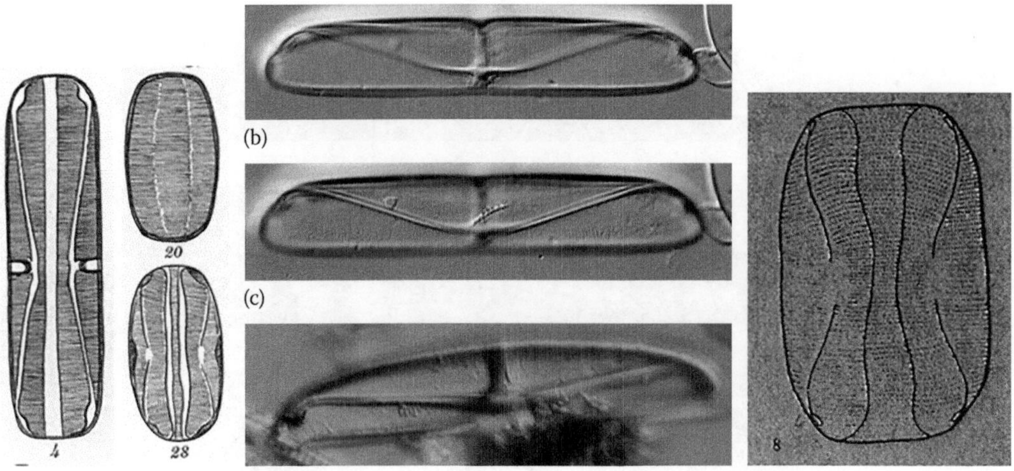

(a) (b) (c) (d) (e)

FIGURE 2.8 (a) *Amphora obtusa* var. *baikalensis* (reproduced from Skvortzow, B.W., *The Phillipine Journal of Science* 62, 293–377, 1937, Figures 20 and 28); (b,c) '*Amphora*' spp., BM; (d) *A. obtusa* var. *baikalensis* (*A. delphineiformis* Levkov?) specimen from Lake Baikal, BM; (f) *A. obtusa* var. *baikalensis* (reproduced from Jansnitsky, V., *Journal Botanique de l'URSS* 21(6), 689–703, 1936, Figure 8).

'*Amphora*' species are related to *Eunophora* and how those relationships impinge on the understanding of their biogeography with respect to Australasia.

Amphorotia

This genus was first described for a number of endemic species from Lake Baikal, Siberia, Russia. It was initially based on an unusual species of *Eunotia*, *E. clevei* (Williams and Reid 2006a). In all, 19 species (including *A. clevei*) were described (fossil and recent) from various places. For Recent species: *Amphorotia A. baicalensis*, *A. linearis*, *A. lacusbaikali*, *A. lunata*, *A. hispida* (all Lake Baikal endemics), *A. clevei* (Scandinavia, Russia, fossil and Recent), *A. stoermeri* (Mongolia), *A. asiatica* (Indochina), *A. sinica* (China, Japan), *A. mekonensis* (Mekong River, Vietnam) and *A. reimeri* (China). For fossils: *A. spinusnullosi* (United States, Finland), *A. voigtii*, *A. curvata*, *A. miocenica* (all China), *A. americana* (United States), *A. aculeata*, *A. maculata* and *A. penzhica* (all Russia). Broadly speaking, *Amphorotia* has two basic patterns: one in the cold boreal (Lake Baikal, Scandinavia, Russia, Mongolia), the other trans-Pacific (Indochina, China, Japan, Vietnam, United States). Several species remain undescribed (from China, Mongolia and Russia; Figure 9a,b, undescribed specimens from Lake Baikal, Russia).

Colliculoamphora

This genus, another based on species previously placed in *Amphora* and *Eunotia*, was first described for just two species: *C. reedii* and *C. reichardtiana*. *C. reedii* is found in the well-known Oamaru fossil deposit in New Zealand ('Jackson's Neu-Seeland, marin fossil'; Schrader 1969), and *C. reichardtiana* is a living species found in various places (Campeche Bay, Mexico, Cuba and the Adriatic Sea*; Figure 9c,d; specimens from 'Sansego', type material) (Williams and Reid 2009; Levkov 2009). Since its original description nine more species have been described (Williams and Reid 2009; Lobban 2015): *C. minima* (Beaufort, NC, United States), *C. fabiformis* and *C. panduriformis* (Galapagos), *C. edgarensis* (Detroit de Foveaux, New Zealand), *C. reniformis* (Colon, Panama), *C. stoermeri* and *C. palawanensis*

* When *C. reichardtiana* was transferred from *Amphora*, its type material was erroneously assumed to have been from 'Sargassum von Honduras'. This is incorrect. Grunow's type specimens, from the Sansego, Adriatic Sea, and from Campeche Bay, Mexico, have since been examined and the results published (Williams 2016).

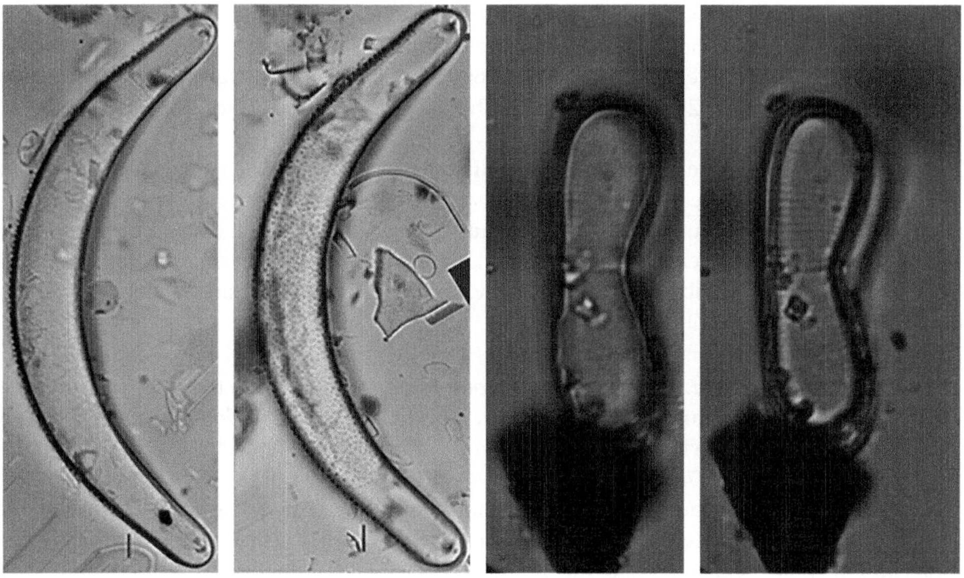

(a) (b) (c) (d)

FIGURE 2.9 (a) Undescribed species of *Amphorotia* (from Mongolia); (b) *Colliculoamphora reichardtiana*, from Sansego, Adriatic Sea (type specimens; for details see Williams, D.M., *Diatom Research* 31, 77–83, 2016).

(Sulu Sea, Palawan, Philippines), *C. gibba* (Campeche Bay, Mexico) and *C. gabgabensis* from Guam (Lobban 2015), with as yet undescribed species from other localities.

Its distribution, not straightforward at present, includes Central America (including the Galapagos and Cuba), New Zealand, the southern part of the United States, the Philippines and Guam, but not Australia.

Discussion

Broadly speaking, to explore historical biogeography there are two requirements: endemic species and a scheme of their relationships. At present, historical biogeography of diatoms sits uneasily within the broader scope of the subject, partly because schemes of relationships are few and partly because the majority of practitioners of diatom biogeography treat the subject as a branch of applied ecology. Historical factors in determining diatom distributions, however, are now recognised to be of some significance (Vanormelingen *et al.* 2007; Williams and Reid 2006c; Williams 2011), even if most explanations of present-day diatom distributions defer to the primacy of ecological processes (Vanormelingen *et al.* 2007; Vyverman *et al.* 2007b; Verleyen *et al.* 2009) rather than incorporating historical (evolutionary) considerations (Ripple and Kociolek 2013; Spaulding and Kociolek 2003; Williams and Reid 2006a,c; Williams 2011). Nevertheless, general explanations of distributions are moving away from human (Harper 1994) or other forms of random dispersal (e.g. Potapova and Charles 2002) to explain 'odd' and 'unexpected' distributions (Kristiansen 1996). With respect to diatom endemism, it is now clear that the simple EIE hypothesis is largely redundant; it is worth repeating, however, that at present reported endemism in Australasia is surprisingly low, with only ca. 2% for New Guinea, Australia and New Zealand, and with New Caledonia an exception at 40% (Table 2.2). Explanations for these figures could be varied, but at present the fact that a good deal of Australasia remains unexplored with respect to diatom diversity may indeed be the main factor.

As noted previously, a more useful definition of endemism is where the area under consideration is not necessarily a physically isolated land mass or a politically drawn boundary, but is defined, more or less, by the species found there, such as that proposed by Kilroy *et al.* (2007). While their definitions are ultimately geographical, their relevance for diatom studies was that they were based on empirical factors – the distribution of the diatoms themselves – rather than having a predefined area and then searching for its endemics. In relation to defining areas of endemism in Australasia, Heads's recent book *Biogeography of Australasia* (2014) outlines a number of geographical relationships that are relevant to diatom studies.

With respect to schemes of relationships among diatoms, we have reviewed a number of taxa from the order Eunotiales as an example of a higher taxon's species distribution relevant to Australasia. In short, a Southern Hemisphere distribution is evident in many taxa, a Gondwanan distribution evident in others. Within the Gondwanan areas (excluding Antarctica, not discussed here, but see Kociolek *et al.* in press), there are more localised distributions. For example, the following area relationships occur frequently when considering diatom distributions.

- *Tasmania–New Zealand*: As noted for '*Eunotia bilunaris*' (Vanormelingen *et al.* 2008: 399) and species belonging to *Actinella*, *Biremis*, *Fragilariforma*, and *Gomphonema*. This area relationship is well known for many organisms; for diatoms it is documented and summarised in Vyverman *et al.* (2007a: 586) and Harper *et al.* (2012), who refer to the earlier comment of Foged (1979: 113): '[Eastern] Australia, Tasmania and New Zealand'. A general consideration of the *Tasmania–New Zealand* relationship is discussed in Heads (2014: 165–166). The relationship must now exclude *Eunophora* as it has since been found in Australia (John 2009b, 2012a–c).
- *Tasmania–New Zealand–New Caledonia*: Species belonging to *Actinella*.
- *South America–Australia*: As seen with *Eunotia pseudoserra*, *E. silvae*, *E. tecta* and *Actinella eunotioides*.
- *African–South American–New Zealand*: Various species.

Other permutations involving areas in the Southern Hemisphere exist (including various places on the African continent) and have been touched on. The point we wish to make is that such area relationships need to be identified and examined prior to exploring historical explanations for current distribution patterns of diatoms within Australasia.

Three general points emerge.

- Diatom taxonomy is relatively unsophisticated in Australasia and any generalities or explanations of taxon distributions, historical or otherwise, are hampered by the lack of focused taxonomic studies. There is an imperative to conduct and complete a comprehensive freshwater diatom *flora of Australia*, as far as that is possible, given the size of the landmass involved. A more practical suggestion would be to approach a flora on a region-by-region basis following what might be considered to be natural regions (such as those outlined in Ebach *et al.* 2013, 2015). It is also worth noting that some of the areas that have already been collected (e.g. those in Foged 1978) may be degraded systems where any endemic taxa that may have once existed are now extinct. Thus, alongside collecting in natural regions, natural systems (as opposed to degraded systems) are required too. Studies such as Levkov (2009), which focused on *Amphora* sensu stricto, and Kulikovskiy *et al.* (2012, 2105), which focused on the diatoms of Lake Baikal, are indispensable for biogeographic studies; both kinds of study are lacking for Australasia. It would seem crucial to have a functional and detailed checklist (similar to that now available for Lake Ohrid and Prespa in Macedonia, [Levkov and Williams 2012] or for the United States [Kociolek 2005]).
- Establishing phylogenetic relationships among species is crucial prior to the judgement of any historically relevant distribution pattern(s), studies which are lacking in diatom systematics.
- Regardless of our lack of knowledge of diatom interrelationships, there is emerging evidence for both a Southern Hemisphere relationship and a North–South (bipolar) relationship, at least in the Eunotiales, that transcends the freshwater–marine boundaries (surprisingly most of these data for a North–South [bipolar] relationship emerged from revisions of the Lake Baikal

diatom flora; Williams and Reid 2006a,b), relationships that extend into the marine waters of Australasia (Williams and Reid 2006a,b). As a consequence, in some areas it may be useful, if not necessary, to focus on both the freshwater and marine components simultaneously.

In the final analysis, we expect diatom studies in Australasia to yield insights into global biogeography, insights that may go beyond our current understanding of the ecological processes that maintain existing biotas.

Acknowledgements

We would like to thank Bank Beszteri (Alfred Wegener Institute) for permission to use the images of '*Amphora* sp.' (*A. hendeyi*) from the Hustedt collection (BRM 345/44a) and Marina Popatova (ANSP) for the image of *Synedra lismorensis* from ANSP (AGC 400238a). We also wish to thank the following for their information, guidance and references: Jennifer Beals, Bank Beszteri, Vivienne Cassie Cooper, Margaret Harper, Jacob John, Sarah Olischläger, Marina Popatova, Sarah Spaulding, Frithjof Sterrenberg and Stuart Stidolph.

References

Abarca, N., Jahn, R., Zimmermann, J., and Enke, N. 2014. Does the cosmopolitan diatom *Gomphonema parvulum* (Kützing) Kützing have a biogeography? *PLoS ONE* 9(1): 1–18.

Anonymous. 1923. George Israel Playfair. *Proceedings of the Linnean Society of New South Wales* 48: vi–vii.

Bagnall, A.G. 1970a. Sven Berggren in New Zealand. *The Turnbull Library Record* 3: 29–12.

Bagnall, A.G. 1970b. Sven Berggren in New Zealand, Section 2. *The Turnbull Library Record* 3: 143–151.

de Azevedo Barros, C.F., dos Santos, A.M.M., and Barbosa, F.A.R. 2013. Phytoplankton diversity in the middle Rio Doce lake system of southeastern Brazil. *Acta Botanica Brasilica* 27: 327–346.

Beals, J., and Popatova, M. 2013. Type material of the diatom *Eunotia arcuoides* Foged. *Proceedings of the Academy of Natural Sciences of Philadelphia* 162: 25–32.

Beier, T., and Lange-Bertalot, H. 2007. A synopsis of cosmopolitan, rare and new *Frustulia* species (Bacillariophyceae) from ombrotrophic peat bogs and minerotrophic swamps in New Zealand. *Nova Hedwigia* 85: 73–91.

Berg, Å. 1948. Observations on the development of the *Eunotia* raphe. *Arkiv für Botanik* 33A(15): 1–10.

Brady, H.T. 1979. *The Diatom flora of Australia: Report 1. Freshwater Diatoms of the Northern Territory, Especially in the Magela Creek System*. North Ryde, Australia: Macquarie University School of Biological Sciences, 57 p.

Brightwell, T. 1859. On some of the rarer or undescribed species of Diatomaceae, Part I. *Quarterly Journal of Microscopical Science* 7: 179–181.

Burbury, F.E. 1902. Tasmanian Diatomaeae. *Papers and Proceedings of the Royal Society of Tasmania* 1900–1901: 4–8.

Burliga, A.L., Kociolek, J.P., Salomoni, S.E., and Figueiredo, D. 2013. A new genus and species in the diatom family Eunotiaceae Kützing (Bacillariophyceae) from the Amazonian hydrographic region, Brazil. *Phytotaxa* 79: 47–57.

Cassie, V. 1984. Checklist of the freshwater diatoms of New Zealand. *Bibliotheca Diatomologica* 4: 1–129.

Cassie, V. 1989. A contribution to the study of New Zealand diatoms. *Bibliotheca Diatomologica* 17: 1–266.

Cassie Cooper, V., and Harper, M.J. 2012. Freshwater diatoms. In Harper, M.A., Cassie Cooper, V., Chang, F.H., Nelson, W.A., and Broady, P.A., Phylum Ochrophyta: Brown and golden-brown algae, diatoms, silicoflagellates, and kin. In *New Zealand Inventory of Biodiversity*, Vol. 3, *Kingdoms Bacteria, Protozoa, Chromista, Plantae, Fungi* (Gordon, D.P., ed.), pp. 147–153. Christchurch, New Zealand: Canterbury University Press.

Cavalcante, K.P., Tremarin, P.I., de Castro, E.C., Junqueira de Azevedo Tibiric, C.E.J., Wojciechowski, J., and Ludwig, T.A.V. 2014. Epiphytic *Eunotia* (Bacillariophyceae) on *Podostemum* from Santa Catarina, southern Brazil, including new observations on morphology and taxonomy of some rare recorded species. *Biota Neotropica* 14(3): 1–12.

Chang, H., and Harper, M. 2012. Marine diatoms. In Harper, M.A., Cassie Cooper, V., Chang, F.H., Nelson, W.A., and Broady, P.A., Phylum Ochrophyta: Brown and golden-brown algae, diatoms, silicoflagellates, and kin. In *New Zealand Inventory of Biodiversity*, Vol. 3, *Kingdoms Bacteria, Protozoa, Chromista, Plantae, Fungi* (Gordon, D.P., ed.), pp. 153–157. Christchurch, New Zealand: Canterbury University Press.

Chapman, V.J., Thompson, R.H., and Segar, E.C.M. 1957. Check list of the fresh-water algae of New Zealand. *Transactions of the Royal Society of New Zealand* 84: 695–747.

Chessman, B.C., Bate, N., Gell, P.A., and Newall, P. 2007. A diatom species index for bioassessment of Australian rivers. *Marine and Freshwater Research* 58: 542–557.

Cleve-Euler, A. 1948. Süsswasserdiatomeen aus dem Feuerland von Prof. Väinö Auer Gesammelt. *Acta Geographica* 10: 3–6.

Cleve, P.T. 1894–1895. Synopsis of the Naviculoid diatoms. *Kongliga Svenska-Vetenskaps Akademiens Handlingar* 26 (2): 1–194 (Part I); *Kongliga Svenska-Vetenskaps Akademiens Handlingar* 27 (3): 1–219 (Part II).

Cleve, P.T., and Möller, J.D. 1877–1882. *Diatoms*, Parts I–VI, Nos. 1–324. Uppsala, Sweden: Edquist.

Cowan, R.A., and Ducker, S.C. 2007. A history of systematic phycology in Australia. In *Algae of Australia* (McCarthy, P.M., and Orchard, A.E., eds.), pp. 1–65. Melbourne: CSIRO Publishing.

Crosby, L.H., and Wood, E.J.F. (Ferguson-Wood, E.J.) 1958. Studies on Australian and New Zealand diatoms: I. Planktonic and allied species. *Transactions of the Royal Society of New Zealand* 85: 483–530.

Crosby, L.H., and Wood, E.J.F. (Ferguson-Wood, E.J.) 1959. Studies on Australian and New Zealand diatoms: II. Normally epontic and benthic genera. *Transactions of the Royal Society of New Zealand* 86: 1–58.

de Castro Canani, L.G., and Torgan, L.C. 2013. Two new *Eunotia* species with subapical costae, an uncommon structure for the genus. *Diatom Research* 28: 395–406.

de Toni, G.B., and Forti, A. 1922. Alghe di Australia, Tasmania e Nuova Zelanda racolte del rev. dott. Giuseppe Capra nel 1908–1909. *Memorie del Reale Institute Veneto di Scienze, Lettere ed Arti* 29(3): 1–183.

Desianti, N., Potapova, M., and Beals, J. 2015. Examination of the type materials of diatoms described by Hohn and Hellerman from the Atlantic Coast of the USA. *Diatom Research* 30: 93–116.

Ebach, M.C., Gill, A.C., Kwan, A., Ahyong, S.T., Murphy, D.J., and Cassis, G. 2013. Towards an Australian Bioregionalisation Atlas: A provisional area taxonomy of Australia's biogeographical regions. *Zootaxa* 3619: 315–342.

Ebach, M.C., Murphy, D.J., González-Orozco, C.E., and Miller, J.T. (2015). A revised area taxonomy of phytogeographical regions within the Australian Bioregionalisation Atlas. *Phytotaxa* 208: 261–277.

Ebach, M.C., and Williams, D.M. 2010. Aphyly: A systematic designation for a taxonomic problem. *Evolutionary Biology* 37: 123–127.

Edwards, A.R. (comp). 1991. The Oamaru diatomite. *New Zealand Geological Survey Paleontological Bulletin* 64: 260.

Ehrenberg, C.G. 1848. Die Mikroskopischen Lebenformen. *In Versuch einer Fauna und flora von Britisch-Guiana*, Vol. 3 (Schomburgk, R.H., ed.), pp. 537–544. Leipzig: J.J. Weber.

Ehrenberg, G.C. 1854. *Mikrogeologie: Das Erden und Felsen schaffende Wirken der unsichtbar kleinen selbständigen Lebens auf der Erde*, Vol. 1, pp. i–xxviii, 1–374, Leipzig, Germany: L. Voss.

Ehrenberg, G.C. 1870. Über mächtige Gebirgs-Schichten vorherrschend aus mikroskopischen Bacillarien unter und bei der Stadt Mexiko. *Abhandlungen der Königliche Akademie der Wissenschaften zu Berlin* (1869) 43(2): 1–66, pls. 1–2.

Ferrari, F., Wetzel, C.E., Ector, L., Blanco, S., Viana, J.C.C., Silva, E.M., and Bicudo, D.C. 2009. *Perinotia diamantina* sp. nov., a new diatom species from the Chapada Diamantina, northeastern Brazil. *Diatom Research* 24: 79–100.

Fenchel, T., and Finlay, B. J. 2004. The ubiquity of small species: Patterns of local and global diversity. *Bioscience* 54: 777–784.

Finlay, B.J. 2002. Global dispersal of free-living microbial eukaryote species. *Science* 296: 1061–1063.

Flower, R.J. 2005. A review of diversification trends in diatom research with special references to taxonomy and environmental applications using examples from Lake Baikal and elsewhere. *Proceedings of the California Academy of Sciences* 56: 107–128.

Foged, N. 1957. Diatoms from Rennell Island. In *The Natural History of Rennell Island, British Solomon Islands*, Vol. 3, 117 p. Copenhagen, Denmark: Danish Science Press.

Foged, N. 1978. Diatoms in eastern Australia. *Bibliotheca Phycologia* 41: 1–243.

Foged, N. 1979. Diatoms in New Zealand, the North Island. *Bibliotheca Phycologia* 47: 1–130.

Fontaneto, D. (ed.) 2011. *Biogeography of Microscopic Organisms: Is Everything Small Everywhere?* Cambridge, UK: Cambridge University Press.

Gell, P.A., Sonneman, J.A., Reid, M.A., Illman, M.A., and Sincock, A.J. 1999. *An Illustrated Key to Common Diatom Genera from Southern Australia*. Cooperative Research Centre for Freshwater Ecology Identification Guide No. 26, 63 p. Thurgoona, Australia: Cooperative Research Centre for Freshwater Ecology.

Grunow, A. 1865. Uber die von Herrn Gerstenberger in Rabenhorst's Decaden ausgegeben Süsswasser Diatomaceen und Desmidiaceen von der Insel Banka, nebst Untersuchungen über die Gattungen *Ceratoneis* und *Frustulia*. In *Beiträge zur näheren Kenntniss und Verbreitung der Algen*, Bk. 2, 1–16. Herausgegeben von Dr. L. Rabenhorst. Leipzig, Germany: Eduard Kummer.

Harper, M. 2012. Fossil marine diatoms. In Harper, M.A., Cassie Cooper, V., Chang, F.H., Nelson, W.A., and Broady, P.A. Phylum Ochrophyta: Brown and golden-brown algae, diatoms, silicoflagellates, and kin. In *New Zealand Inventory of Biodiversity*, Vol. 3, *Kingdoms Bacteria, Protozoa, Chromista, Plantae, Fungi* (Gordon, D.P., ed.), pp. 157–162. Christchurch, New Zealand: Canterbury University Press.

Harper, M.A. 1994. Did Europeans introduce *Asterionella* formosa Hassall into New Zealand? *Memoirs of the Californian Academy of Sciences (Proceedings of the 11th International Diatom Symposium)* 17: 479–484.

Harper, M.A., Cassie Cooper, V., Chang, F.H., Nelson, W.A., and Broady, P.A. 2012. Phylum Ochrophyta: Brown and golden-brown algae, diatoms, silicoflagellates, and kin. In *New Zealand Inventory of Biodiversity*, Vol. 3, *Kingdoms Bacteria, Protozoa, Chromista, Plantae, Fungi* (Gordon, D.P., ed.), pp. 114–163. Christchurch, New Zealand: Canterbury University Press.

Harper, M.A., Mann, D.G., and Patterson, J.E. 2009. Two unusual diatoms from New Zealand: *Tabularia variostriata* a new species and *Eunophora berggrenii. Diatom Research* 24: 291–306.

Heads, M. 2014. *Biogeography of Australasia: A Molecular Analysis*. Cambridge, UK: Cambridge University Press, 493 p.

Hennig, W. 1948. *Die Larvenformen der Dipteren*, Vol. 1. Berlin: Akademie-Verlag, 628 p.

Hennig, W. 1953. Kritische Bemerkungen zum phylogenetischen System der Insekten. *Beiträge zur Entomologie* 3 (Sonderh.): 1–85.

Héribaud, J. 1902. *Les Diatomées Fossiles d'Auvergne*. Paris: Librairie des Sciences Naturelles.

Héribaud, J. 1903. *Les Diatomées Fossiles d'Auvergne* (Second Mémoire). Paris: Librairie des Sciences Naturelles.

Hirano, M. 1974. Freshwater Algae from North Borneo. *Contributions from the Biological Laboratory, Kyoto University* 24: 121–144.

Hodgson, D.A., Vyverman, W., and Tyler, P.A. 1996. Diatoms of meromictic lakes adjacent to the Gordon River, and of the Gordon River estuary in south-west Tasmania. *Bibliotheca Diatomologica* 35: 174.

Holland, J., and Clark, R.L. 1989. *Diatoms of Burrinjuck Reservoir, New South Wales, Australia*. CSIRO Divisional Report 89/1. Canberra, Australia: Institute of Natural Resources and Environment, Division of Water Resources.

Hustedt, F. 1952. Neue und wenig bekannte Diatomeen: III. Philogenetische Variationen bei den rhaphidioiden Diatomeen. *Berichte der Deutschen Botanischen Gesellschaft* 65: 133–144.

Huang, C., Mao, Y., Liu, S., and Cheng, Z. 1998. *Atlas of Limnetic Fossil Diatoms from China* [In Chinese]. Beijing: China Ocean Press.

Inglis, J. 1883. Notes on some of the diatomaceous deposits in New Zealand. *Transactions and Proceedings of the New Zealand Institute* 15: 340–346.

Jansnitsky, V. 1936. Neue und interessante Arten der Diatomeen aus dem Baikalsee. *Journal Botanique de l'URSS* 21(6): 689–703.

John, J. 1983. The diatom flora of the Swan River Estuary, Western Australia. *Bibliotheca Phycologia* 64: 1–358.

John, J. 2007. Heterokontophyta: Bacillariophyceae. In *Algae of Australia*. (McCarthy, P.M., and Orchard, A.E., eds.), pp. 288–310. Melbourne, Australia: CSIRO Publishing.

John, J. 2009a. Diatom flora of tropical Australia: High lights and salient features. *Scripta Botanica Belgica* 45: 40.

John, J. 2009b. *Eunophora* in Western Australia and Queensland. *Scripta Botanica Belgica* 45: 41.

John, J. 2012a. *A Beginner's Guide to Diatoms*. Ruggell, Liechtenstein: A.R.G. Ganter Verlag K.G.

John, J. 2012b. Sand, dingos and diatoms. In *22nd International Diatom Symposium, Aula Academica, Ghent, 26–31 August 2012, Abstracts* (Sabbe, K., Van de Vijver, B., and Vyverman, W., eds.), p. 176. Ostend, Belgium: VLIZ Special Publication 58.

John, J. 2012c. *Diatoms in the Swan River Estuary, Western Australia: Taxonomy and Ecology*. Königstein, Germany: Koeltz Scientific Books, 456 p.

John, J. 2015. *A Beginner's Guide to Diatoms*, 2nd Edn. Ruggell, Liechtenstein: A.R.G. Ganter Verlag K.G.

Kilroy, C., Biggs, B.J.F., Vyvermann, W. 2007. Rules for macro-organisms applied to microorganisms: Patterns of endemism in benthic freshwater diatoms. *Oikos* 116: 550–564.

Kilroy, C., Sabbe, K., Bergey, E.A., Vyverman, W., and Lowe, R. 2003. New species of *Fragilariforma* (Bacillariophyceae) from New Zealand and Australia. *New Zealand Journal of Botany* 41: 535–554.

Kobayasi, H., Ando, K., and Nagumo, T. 1981. On some endemic species of the genus *Eunotia* in Japan. In *Proceedings of the Sixth Symposium on Recent and Fossil Diatoms* (Ross, R., ed.), pp. 93–114. Königstein, Germany: Otto Koetz, Science Publishers.

Kociolek, J.P. 2000. Valve ultrastructure of some Eunotiaceae (Bacillariophyceae), with comments on the evolution of the raphe system. *Proceedings of the California Academy of Sciences* 52: 11–21.

Kociolek, J.P. 2005. A checklist and preliminary bibliography of the recent, freshwater diatoms of inland environments of the continental United States. *Proceedings of the California Academy of Sciences* (4th Series) 56(27): 395–525.

Kociolek, J.P., Kopalová, K., Hamsher, S., Kohler, T., Van de Vijver, B., Convey, P. and McKnight, D. In Press. Freshwater diatom biogeography and and the genus *Luticola*: An extreme case of endemism in Antarctica. *Polar Biology.*

Kociolek, J.P., and Spaulding, S.A. 2000. Freshwater diatom biogeography. *Nova Hedwigia* 71: 223–241.

Kociolek, J.P., Spaulding, S.A., Sabbe, K., and Vyverman, W. 2004. *New Gomphonema* (Bacillariophyta) species from Tasmania. *Phycologia* 43: 427–444.

Kociolek, J.P., and Williams, D.M. 2015. How to define a diatom genus? Notes on the creation and recognition of taxa, and a call for revisionary studies of diatoms. *Acta Botanica Croatia* 74(2): 195–210.

Kristiansen, J. 1996. Dispersal of freshwater algae: A review. *Hydrobiologia* 336: 151–157.

Kulikovskiy, M., Lange-Bertalot, H., Genkal, S., and Witkowski, A. 2010. *Eunotia* (Bacillariophyta) in the Holarctic: New species from the Russian Arctic. *Polish Botanical Journal* 55: 93–107.

Kulikovskiy, M., Lange-Bertalot, H., Witkowski, A., Khursevich, G.K., and Kociolek, J.P. 2015. New species of *Eunotia* (Bacillariophyta) from Lake Baikal with comments on morphology and biogeography of the genus. *Phycologia* 54: 248–260.

Kulikovskiy, M.S., and Kuznetsova, I.V. 2014. Biogeography of Bacillariophyta: Main concepts and approaches. *International Journal on Algae* 16: 207–228.

Levkov, Z. 2009. *Amphora* sensu lato. In *Diatoms of Europe: Diatoms of the European Inland Waters and Comparable Habitats Elsewhere*, Vol. 5. Ruggell, Liechtenstein: A.R.G. Gantner Verlag, 916 p.

Levkov, Z., and Williams, D.M. 2012. Checklist of diatom (Bacillariophyta) from Lake Ohrid and Lake Prespa (Macedonia), and their watersheds. *Phytotaxa* 45: 1–76.

Lobban, C.S. 2015. Benthic marine diatom flora of Guam: New records, redescription of *Psammodictyon pustulatum* n. comb., n. stat., and three new species (*Colliculoamphora gabgabensis*, *Lauderia excentrica*, and *Rhoiconeis pagoensis*). *Micronesica* 2015-02, 49 p. http://micronesica.org/volumes/2015 (published online 22 July 2015).

Logares, R.E. 2006. Does the global microbiota consist of a few cosmopolitan species? *Ecología Austral* 16: 85–90.

Lowe, R., Morales, E., and Kilroy, C. 2006. *Frankophila biggsii* (Bacillariophyceae), a new diatom species from New Zealand. *New Zealand Journal of Botany* 44: 41–46.

Ludwig, T.A.V., and Flôres, T.L. 1995. Diatomoflórula dos rios da região a ser inundada para a construção da usina hidrelétrica de Segredo, PR: I. Coscinodiscophyceae, Bacillariophyceae (Achnanthales e Eunotiales) e Fragilariophyceae (Meridion e Asterionella). *Arquivos de Biologia e Tecnologia* 38: 31–65.

Mahoney, R.K., and Reimer, C.W. 1987. Current status of the type collection (Bacillariophyceae) in the Diatom herbarium, the Academy of Natural Sciences of Philadelphia. *Proceedings of the Academy of Natural Sciences of Philadelphia* 139: 261–305.

Mahoney, R.K., and Reimer, C.W. 1997. Updated status of the type collection (Bacillariophyceae) in the Diatom herbarium, the Academy of Natural Sciences of Philadelphia. *Proceedings of the Academy of Natural Sciences of Philadelphia* 147: 125–192.

Main, S.P. 2004. Observations of diatoms from Hawai'i: Morphological features of infrequently reported taxa and description of new taxa. In *Proceedings of the Seventeenth International Diatom Symposium* (Poulin, M., ed.), pp. 203–221. Bristol, UK: BioPress.

Mayama, S. 1993. *Eunotia sparsistriata* sp. nov., a moss diatom from Mikura Island, Japan. *Nova Hedwigia, Beiheft* 106: 143–150.

Metzeltin, D., and Lange-Bertalot, H. 1998. Tropical diatoms of South America: I. About 700 predominantly rarely known or new taxa representative of the neotropical flora. *Iconographia Diatomologica* 5: 1–695.

Metzeltin, D., and Lange-Bertalot, H. 2002. Diatoms from the 'Island Continent' Madagascar. *Iconographia Diatomologica* 11: 1–286.

Metzeltin, D., and Lange-Bertalot, H. 2007. Tropical diatoms of South America: II. *Iconographia Diatomologica* 18: 1–877.

Metzeltin, D., Lange-Bertalot, H., and García-Rodríguez, F. 2005. Diatoms of Uruguay: Compared with other taxa from South America and elsewhere. *Iconographia Diatomologica* 15: 1–736.

Minelli, A. 2014. Review of *The Nature of Classification: Relationships and Kinds in the Natural Sciences* by John S. Wilkins and Malte C. Ebach. *Systematic Biology* 63: 844–846.

Monnier, O., Lange-Bertalot, H., and Bertrand, J. 2003. *Nupela exotica* species nova: Une diatomée d'un aquarium tropical d'eau douce; Avec des remarques sur la biogéographie du genre [*Nupela exotica* sp. nov.: A diatom from a tropical freshwater aquarium; With comments on the genus biogeograph]. *Diatom Research* 18: 273–291.

Moser, G. 1999. Die Diatomeenflora von Neukaledonien. *Bibliotheca Diatomologica* 43: 205.

Moser, G., Lange-Bertalot, H., and Metzeltin, D. 1998. Insel der Endemiten: Geobotanisches Phänomen Neukaledonien [Island of endemics: New Caledonia, a geobotanical phenomenon]. *Bibliotheca Diatomologica* 38, 464 p.

Moser, G., Steindorf, A., and Lange-Bertalot, H. 1995. Neukaledonien Diatomeenflora einer Tropeninsel: Revision der collection Maillard und Untersuchung neuen materials. *Bibliotheca Diatomologica* 32: 340.

Novitski, L., and Kociolek, J.P. 2005. Preliminary light and scanning electron microscope observations of marine fossil *Eunotia* species with comments on the evolution of the genus *Eunotia*. *Diatom Research* 1: 137–143.

Oliveira, B.D., de Souza Nogueira, I., and da Graça Machado de Souza, M. 2012. Eunotiaceae Kützing (Bacillariophyceae) planctônicas do Sistema Lago dos Tigres, Britânia, GO, Brasil. *Hoehnea* 39(2): 297–313.

Patrick, R. 1940. Some new diatoms from Brazil. *Notulae Naturae* 59: 2–7.

Patrick, R.M., and Reimer, C.W. 1966. The diatoms of the United States exclusive of Alaska and Hawaii, Vol. 1: Fragilariaceae, Eunotoniaceae, Achnanthaceae, Naviculaceae. *Monographs of the Academy of Natural Sciences of Philadelphia* 13: 688, 64 pls.

Pantocsek, J. 1892. *Beiträge zur Kenntnis der Fossilen Bacillarien Ungarns*, Vol. 3, *Susswasser Bacillarien: Anhang-analysen 15 neuer Depots von Bulgarien, Japan, Mahern, Russland und Ungarn*. Nagy-Tapolcsány, Slovakia: Buchdrucherei von Julius Platzko, 42 pls.

Playfair, G.I. 1913. Plankton of the Sydney water-supply. *Proceedings of the Linnean Society of New South Wales* 37: 512–552.

Playfair, G.I. 1914. Contributions to a knowledge of the biology of the Richmond river. *Proceedings of the Linnean Society of New South Wales* 39: 93–151.

Playfair, G.I. 1915. Freshwater algae of the Lismore district: With an appendix on the algal fungi and Schizomycetes. *Proceedings of the Linnean Society of New South Wales* 40: 310–362.

Potapova, M.G., and Charles, D.F. 2002. Benthic diatoms in USA rivers: Distributions along spatial and environmental gradients. *Journal of Biogeography* 29: 167–187.

Pritchard, A. 1861. *A History of Infusoria, Living and Fossil: Arranged According to Die Infusionsthierchen of C.G. Ehrenberg....* 4th Edn. London: Whittaker, xiii + 968 pp., 40 pls.

Prowse, G.A., and Ratnasabapathy, M. 1970. A species list of freshwater algae from the Taiping Lakes, Perak, Malaysia. *Gardens' Bulletin Singapore* 25: 179–187.

Raeside, J.D. 1970. Note on the Occurrence of the Diatoms *Cymbella jordani* Grun. ex Cleve, and *Eunotia serpentina* Ehr., in New Zealand. *Nova Hedwigia* 31: 537–541.

Ripple, H., and Kociolek, J.P. 2013. The diatom (Bacillariophyceae) genus *Actinella* Lewis in Hawai'i. *Pacific Science* 67: 609–621.

Rumrich, U., Lange-Bertalot, H., and Rumrich, M. 2000. Diatomeen der Anden: Von Venezuela bis Patagonien/Feuerland. *Iconographia Diatomologica* 9: 1–649.

Sabbe, K., Vanhoutte, K., and Vyverman, W. 2000. *Actinella comperei* (Eunotiophycidae, Bacillariophyta): A new endemic freshwater diatom from Tasmania (Australia). *Systematics and Geography of Plants* 70: 237–243.

Sabbe, K., Vanhoutte, K., Lowe, R.L., Bergey, E.A., Biggs, B.J.F., Francoeur, S., Hodgson, D., and Vyverman, W. 2001. Six new *Actinella* (Bacillariophyta) species from Papua New Guinea, Australia and New Zealand: Further evidence for widespread diatom endemism in the Australasian region. *European Journal of Phycology* 36: 321–340.

Sala, S.E., Duque, S.R., Núñex-Avellaneda, M., and Lamaro, A.A. 2002. Diatoms from the Colombian Amazon: Some species of the genus *Eunotia* (Bacillariophyceae). *Acta Amazonica* 32: 589–603.

Sar, E.A., Sterrenburg, F.A.S., Lavignem, A.S., and Sunesen, I. 2013. Diatomeas de ambientes marinos costeros de Argentina: Especies del género *Pleurosigma* (Pleurosigmataceae). *Boletín de la Sociedad Argentina de Botánica*. 48(1): 17–51.

Schmidt, A. 1874–1937. *Atlas der Diatomaceen-Kunde*. Leipzig, Germany: O.R. Reisland.

Schmidt, A. 1911. *Atlas der Diatomaceen-Kunde*, Bk. 68, pls. 269–272. Leipzig, Germany: O.R. Reisland.

Schmidt, A. 1913. *Atlas der Diatomaceen-Kunde*, Bk. 72–73, pls 285–292. Leipzig, Germany: O.R. Reisland.

Schrader, H.J. 1969. Die Pennaten Diatomeen aus dem Obereozän von Oamaru, Neuseeland. *Nova Hedwigia, Beiheft* 28: 1–163.

Siver, P.A., Bishop, J., Lott, A., and Wolfe, A.P. 2015. Heteropolar eunotioid diatoms (Bacillariophyceae) were common in the North American Arctic during the Middle Eocene. *Journal of Micropalaeontology* 34: 151–163.

Siver, P.A., Wolfe, A.P., and Edlund, M.B. 2010. Taxonomic descriptions and evolutionary implications of Middle Eocene pinnate diatoms representing the extant genera *Oxyneis, Actinella* and *Nupela* (Bacillariophyceae). *Plant Ecology and Evolution* 143: 340–351.

Skvortzow, B.W. 1937. Bottom diatoms from Olchon gate of Baikal Lake, Siberia. *The Phillipine Journal of Science* 62: 293–377.

Sonneman, J.A., Sincok, A., Fluin, J., Reid, M., Newall, P., Tibby, J., and Gell, P. (eds.) [1999] 2000. *An Illustrated Guide to Common Stream Diatom Species from Temperate Australia*. Cooperative Research Centre for Freshwater Ecology Identification Guide No. 33. Thurgoona, Australia: Cooperative Research Centre for Freshwater Ecology.

Spaulding, S.A., and J.P. Kociolek. 2003. Freshwater diatoms (Bacillariophyceae). In *Natural History of Madagascar* (Goodman, S., and Benstead, J., eds.), pp. 276–282. Chicago, IL: University of Chicago Press.

Stepanek, J.G, and Kociolek, J.P. 2014. Molecular phylogeny of *Amphora* sensu lato (Bacillariophyta): An investigation into the monophyly and classification of the amphoroid diatoms. *Protist* 165: 177–195.

Taylor, J.C., Karthick, B., Kociolek, J.P., Wetzel, C.E., and Cocquyt, C. 2014. *Actinellopsis murphyi* gen. et spec. nov.: A new small celled freshwater diatom (Bacillariophyta, Eunotiales) from Zambia. *Phytotaxa* 178(2): 128–137.

Tempère, J., and Peragallo, H. 1907–1915. *Diatomées du Monde Entier*, 2nd Edn. 30 fascicules. Arcachon: Gironde.

Thomas, D.P. 1983. *A Limnological Survey of the Alligator River system: I. The Diatoms (Bacillariophyceae) of the Region*. Supervising Scientist of the Alligator Rivers Region, Research Report 3. Canberra, Australia: Australia Government Publishing Service.

Thomas, D.P. 1987. New freshwater diatom taxa from tropical Northern Australia. *Transactions of The Royal Society of South Australia* 111: 53–58.

Tremarin, P.I., Ludwig, T.A.V., and Moreira Filho, H. 2008. *Eunotia* Ehrenberg (Bacillariophyceae) do rio Guaraguaçu, litoral do Paraná, Brasil. *Acta botanica bras.* 22: 845–862.

Tremarin, P.I., Kwang, K.I.M.B., Marra, R.C., and Ludwig, T.A.V. 2016. Additional data on morphology of *Actinella leontopithecus-rosalia* Costa (Bacillariophyta, Eunotiaceae). *Phytotaxa* 247: 259–266.

Tsumura, K. 1967. Notice of the preserved specimens of some curious diatoms. *Journal of Technological Researches* 12(2): 231–245.

Tyler, P.A. 1996. Endemism in freshwater algae with special references to Australian region. *Hydrobiologia* 336: 127–135.

Tyler, P.A. [1999] 2000. Foreword. In *An Illustrated Guide to Common Stream Diatom Species from Temperate Australia*. Cooperative Research Centre for Freshwater Ecology Identification Guide No. 33 (Sonneman, J.A., Sincok, A., Fluin, J., Reid, M., Newall, P., Tibby, J., and Gell, P., eds.), p. iii. Thurgoona, Australia: Cooperative Research Centre for Freshwater Ecology.

Tyler, P.A., and Wickham, R.P. 1988. Yan Yean revisited: A Bicentennial window on Australian freshwater algae. *British Phycological Journal* 23: 105–114.

Uherkovich, G., and Franken, M. 1980. Periphytic algae from central Amazonian rain forest streams. *Amazoniana* 7: 49–79.

Vanhoutte, K., Verleyen, E., Sabee, K., Kilroy, C., Sterken, M., and Vyverman, W. 2006. Congruence and disparity in benthic community structure of small lakes in New Zealand and Tasmania. *Marine and Freshwater Research* 57: 789–801.

Vanhoutte, K., Verleyen, E., Vyverman, W., Chepurnov, V., and Sabbe, K. 2004. The Freshwater diatom genus *Kobayasiella* (Bacillariophyta) in Tasmania, Australia. *Australian Systematic Botany* 17: 483–496.

Vanormelingen, P., Cherpurnov, V.A., Mann, D.G., Cousin, S., Vyverman, W. 2007. Congruence of morphological, reproductive and ITS-rDNA sequence data in some Australasian *Eunotia bilunaris* (Bacillariophyta). *European Journal of Phycology* 42: 61–79.

Vanormelingen, P., Verleyen, E., and Vyverman, W. 2008. The diversity and distribution of diatoms: From cosmopolitanism to narrow endemism. *Biodiversity and Conservation* 17: 393–405.

Verleyen, E., Vyverman, W., Sterken, M., Hodgson, D.A., de Wever, A., Juggins, S., Van de Vijver, B., *et al.* 2009. The importance of dispersal related and local factors in shaping the taxonomic structure of diatom metacommunities. *Oikos* 118: 1239–1249.

Vishnyakov, V.A., Kulikovskiy, M.S., Genkal, S.I, and Dorofeyuk, N.I., Lange-Bertalot, H., and Kuznetsova, I.V. 2014. Taxonomy and geographical distribution of the diatom genus *Epithemia* Kützing in water bodies of Central Asia. *Inland Water Biology* 7: 318–330.

Vyverman, W. 1991. Diatoms from Papua New Guinea. *Bibliotheca Diatomologica* 22: 223 p., 208 pls.

Vyverman, W., and Compére, P. 1991. *Nupela giluwensis* gen. & spec. nov. A new genus of naviculoid diatoms. *Diatom Research* 6: 175–179.

Vyverman, W., Sabbe, K., Mann, D.G., Kilroy, C., Vyverman, R., Vanhoutte, K., and Hodgson, D. 1998a. *Eunophora* gen. nov. (Bacillariophyta) from Tasmania and New Zealand: Description and comparison with *Eunotia* and amphoroid diatoms. *European Journal of Phycology* 33: 95–111.

Vyverman, W., Sabbe, K., Mann, D.G., Vyverman, R., Hodgson, D.A., Muylaert, K., and Vanhoutte, K. 1998b. *Clepsydra truganiniae* gen. nov., spec. nov. prov. from Tasmanian highland lakes and its relationships with other amphoroid diatoms. In *Populations: Natural and Manipulated; Symposium Organized by the Royal Society of Natural Sciences, Dodonaea, University of Gent, 29 October 1997* (Beeckman, T., *et al.*, ed.). *Biologisch Jaarboek (Dodonaea)* 65: 205.

Vyverman, W., Sabbe, K., and Vyverman, R. 1997. Five new freshwater species of *Biremis* (Bacillariophyta) from Tasmania. *Phycologia* 36: 91–102.

Vyverman, W., Verleyen, E., and Sabbe, K. 2007a. Biogeography of freshwater microalgae. In *Algae of Australia* (McCarthy, P.M., and Orchard, A.E., eds.), pp. 580–593. Melbourne, Australia: CSIRO Publishing.

Vyverman, W., Verleyen, E., Sabbe, K., Vanhoutte, K., Sterken, M., Hodgson, D.A., Mann, D.G., *et al.* 2007b. Historical processes constrain patterns in global diatom diversity. *Ecology* 88: 1924–1931.

Vyverman, W., Vyverman, R., Hodgson, D., and Tyler, P.A. 1995. Diatoms from Tasmanian mountain lakes: A reference data-set (TASDIAT) for environmental reconstruction and a systematic and autecological study. *Bibliotheca Diatomologica* 33: 1–193.

West, G.S. 1909. The algae of the Yan Yean Reservoir, Victoria. *Journal of the Linnean Society, Botany* 39: 1–88.

West, W., and West, G.S. 1901. A contribution to the freshwater algae of Ceylon. *Transactions of the Linnean Society of London*, 2nd ser. *Botany* 6: 123–215.

Wetzel, C.E., Ector, L., Hoffmann, L., and Bicudo, D.C. 2010. Colonial planktonic *Eunotia* (Bacillariophyceae) from Brazilian Amazon: Taxonomy and biogeographical considerations on the *E. asterionelloides* species complex. *Nova Hedwigia* 91: 49–86.

Wetzel, C.E., Ector, L., Hoffmann, L., Lange-Bertalot, H., and Bicudo, D.C. 2011. Two new periphytic *Eunotia* species from the neotropical Amazonian 'black waters', with a type analysis of *E. braunii. Diatom Research* 26: 135–146.

Wetzel, C.E., Lange-Bertalot, H., Morales, E.A., Bicudo, D.C., and Ector, L. 2012. *Bicudoa amazonica* gen. nov. et sp. nov. (Bacillariophyta): A new freshwater diatom from the Amazon basin with a complete raphe loss in the Eunotioid lineage. *Phytotaxa* 75: 1–18.

Wood, E.J.F. (Ferguson-Wood, E.J.) 1961a. Studies on Australian and New Zealand Diatoms: IV. Descriptions of further sedentary species. *Transactions of the Royal Society of New Zealand* 88: 669–698.

Wood, E.J.F. (Ferguson-Wood, E.J.) 1961b. Studies on Australian and New Zealand diatoms: V. The Rawson collection of recent diatoms. *Transactions of the Royal Society of New Zealand* 88: 699–712.

Wood, E.J.F. (Ferguson-Wood, E.J.) 1963. Studies on Australian and New Zealand Diatoms: VI. Tropical and subtropical species. *Transactions of the Royal Society of New Zealand* 2: 189–218.

Wood, E.J.F. (Ferguson-Wood, E.J.), Crosby, L.H., and Cassie, V. 1959. Studies on Australian and New Zealand diatoms: III. Descriptions of further discoid species. *Transactions of the Royal Society of New Zealand* 87: 211–219.

Woodhead, N., and Tweed, R.D. 1960. Freshwater algae of Sierra Leone: 3. The algae of Rokupr and the Great Scarcies River. *Revue Algologique* 5: 116–150.

Wilkins, J.S., and Ebach, M.C. 2014. *The Nature of Classification: Relationships and Kinds in the Natural Sciences*. Basingstoke, UK: Palgrave Macmillan.

Williams, D.M. 2009. 'Araphid' diatom classification and the 'absolute standard'. *Acta Botanica Croatia* 68: 455–463.

Williams, D.M. 2011. Historical biogeography, microbial endemism and the role of classification: Everything is endemic. In *Biogeography of Microorganisms: Is Everything Small Everywhere?* (Fontaneto, D., ed.), pp. 11–32. Cambridge, UK: Cambridge University Press.

Williams, D.M. 2013. Why is *Synedra berolinensis* so hard to classify? More on monotypic taxa. *Phytotaxa* 127: 113–127.

Williams, D.M. 2016. Examination of type specimens for *Colliculoamphora reichardtiana* (Grunow) Williams and Reid, with a description of a new species, *Colliculoamphora johnwrightii* nov. sp. *Diatom Research* 31: 77–83.

Williams, D.M., and Ebach, M.C. In press. Aphyly: Identifying the flotsam and jetsam of systematics. *Cladistics*. (In press)

Williams, D.M., and Reid, G. 2004. Origin and diversity of the diatom genus *Eunotia* in Lake Baikal: Some preliminary considerations. In *Long Continental Records from Lake Baikal* (Kashiwaya, K., ed.), pp. 259–269. Tokyo: Kluwer Academic Publishers.

Williams, D.M., and Reid, G. 2006a. *Amphorotia* nov. gen., a new genus in the family Eunotiaceae (Bacillariophyceae), based on *Eunotia clevei* Grunow in Cleve and Grunow 1880. *Diatom Monographs* 6: 1–153.

Williams, D.M., and Reid, G. 2006b. Fossils and the tropics, the Eunotiaceae (Bacillariophyta) expanded: The Upper Eocene fossil diatom *Eunotia reedi* and the recent marine diatom *Amphora reichardtiana* from the tropics. *European Journal of Phycology* 41: 147–154.

Williams, D.M., and Reid, G. 2006c. Diatom biogeography and water babies: The search for meaning among the protists. *Diatom Research* 21: 457–462.

Williams, D.M., and Reid, G. 2006d. Dealing with large taxonomic groups: Diatoms species and geography. In *Towards the Tree of Life: The Taxonomy and Systematics of Large and Species Rich Taxa* (Hodkinson, T.R., and Parnell, J.A.N., eds.), pp. 305–322. Boca Raton, FL: CRC Press.

Williams, D.M., and Reid, G. 2008. The type material of *Peronia fibula*: Morphology, systematics and relationships. In *Proceedings of the 19th International Diatom Symposium* (Likoshway, Y., ed.). Pages 141–150 Bristol, UK: Biopress.

Williams, D.M., and Reid, G. 2009. New species in the genus *Colliculoamphora* Williams & Reid with commentary on species concepts in diatom taxonomy. *Beihefte zur Nova Hedwigia* 135: 185–200.

Williams, D.M., and Round, FE. 1987. Revision of the genus *Fragilaria*. *Diatom Research* 2: 267–288.

3

Marine Phytoplankton Bioregions in Australian Seas

Gustaaf M. Hallegraeff, Anthony J. Richardson and Alex Coughlan

CONTENTS

Introduction

Knowledge on the biogeography of pelagic phytoplankton serves (1) as a baseline for climate change and ocean acidification, (2) to assess the impact from human impacts such as eutrophication and ship ballast water introductions, (3) as a guide for natural resource planning and management, and (4) to provide insight into palaeobiogeography and understand drivers of biodiversity. The very first known microalgal examinations from Australian waters date back to the world voyage by HMS *Challenger* in 1874, collecting in the Southern Ocean, the Bass Strait and the Tasman, Coral and Arafura Seas (Castracane 1886). A limited number of collections in Australian waters were also made on the German *Valdivia* expedition (1898–1899), which resulted in accounts of the phytoplankton from the Antarctic and Indian Oceans (Karsten 1905, 1907). More comprehensive early investigations on the taxonomy and distribution of Australian marine microalgae include those of the Great Barrier Reef region (Marshall 1933), the neighbouring Java Sea (Allen and Cupp 1935), Sydney coastal waters (Dakin and Colefax 1933, 1940) and the Australia-wide netphytoplankton surveys by Wood (1954, 1964a,b,c,d) and Crosby and Wood (1958, 1959). These early studies compared Pacific, Indian and Antarctic Ocean basins (Karsten 1905, 1907) or tried to compare Australian phytoplankton communities with those of better-studied European coastal waters (Dakin and Colefax 1933, 1940). The first effort to produce an Australian map of marine phytoplankton provinces (based on dinoflagellate associations) was provided by Wood (1954, 1964a,b; reproduced here as Figure 3.1). Although hampered by limited taxonomic discrimination, this work is impressive in its geographic coverage, but unfortunately, precise sample coordinates cannot be readily retrieved. These latter studies produced the first conclusive biological evidence for the transport of Indian Ocean dinoflagellates southwards along the west coast of Australia and all the way to the west coast of Tasmania (Wood 1954); this flow is now known as the Leeuwin Current. A subsequent bioregionalisation by Markina (1972, 1974, 1976) was limited to netphytoplankton from the west coast of Australia. In the period 1978–1984, as part of a series of CSIRO Division of Fisheries and Oceanography cruises, significant advances in plankton collection methodologies (finer mesh plankton nets, water bottle sampling, more sophisticated preservatives, enrichment cultures), and notably the use of electron microscopy for species identification, considerably improved our knowledge of marine microalgal species communities in the Australian region. Definitive studies were published on the phytoplankton of New South Wales coastal waters (Hallegraeff and Reid 1986; Hallegraeff and Jeffrey 1993), East Australian Current eddies (Jeffrey and Hallegraeff 1980, 1987), the Coral Sea (Revelante and Gilmartin 1982; Revelante

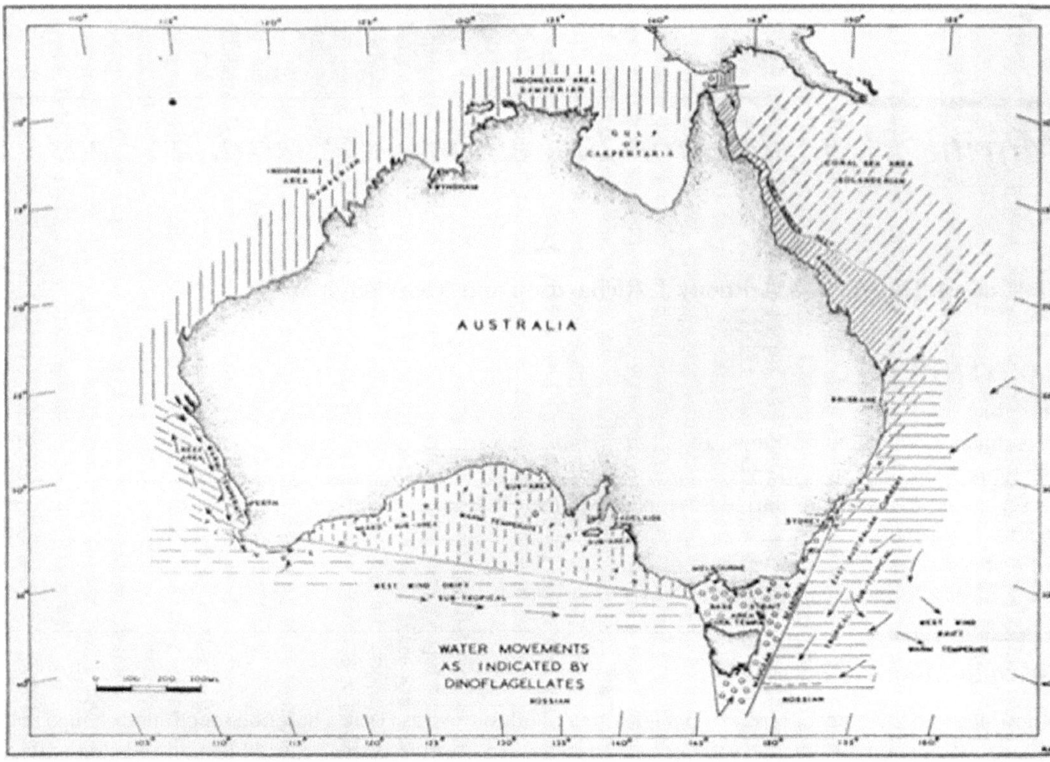

FIGURE 3.1 Dinoflagellate bioregions recognised by Wood (1954). (Reproduced with permission from Wood, E. J. F., *Australian Journal of Marine and Freshwater Research*, 5, 171–351, 1954).

et al. 1982) and the North West Shelf and Gulf of Carpentaria (Hallegraeff and Jeffrey 1984). A new recognition at that time was the importance of delicate nanoplankton flagellates (2–20 μm size) and minute coccoid picoplankton (0.2–2.0 μm size). These small size fractions accounted for up to 90% of the total phytoplankton chlorophyll biomass, except during episodic large diatom or dinoflagellate blooms (Hallegraeff 1981). Few of these species would have been recognised in earlier studies that used harsh preservatives such as formaldehyde or Lugol's iodine, or used coarse (>20 μm mesh) plankton nets for collection. As elsewhere in the world, Australian nanoplankton is primarily composed of haptophytes (notably *Chrysochromulina* spp. and calcareous coccolithophorids; Hallegraeff 1983, 1984a; LeRoi and Hallegraeff 2004, 2006; Moestrup 1979), prasinophytes, small unarmoured dinoflagellates, chrysophytes, cryptomonads and small diatoms (e.g. *Minidiscus*, *Thalassiosira*; Hallegraeff 1984b), while the picoplankton fraction is composed predominantly of prokaryote cyanobacteria (e.g. *Synechococcus*) and prochlorophytes (e.g. *Prochlorococcus*) (Jeffrey and Hallegraeff 1987). Surprisingly, the nanoplankton species of Australian tropical and temperate inshore and offshore waters appeared morphologically remarkably similar. In contrast, the netphytoplankton diatoms and dinoflagellates (20–200 μm size) could be differentiated into a temperate neritic community of the coastal waters of New South Wales, Victoria and Tasmania, a tropical neritic community confined to the Gulf of Carpentaria and North West Australia, and a tropical oceanic community in the offshore waters of the Coral Sea and Indian Ocean (first summarised by Jeffrey and Hallegraeff 1990 and also covered by Hallegraeff 2007). Further phytoplankton species data that have become available since 1990 include: the Gulf of Carpentaria work by Rothlisberg *et al.* (1994), Burford *et al.* (1995) and Burford and Rothlisberg (1999); the Great Barrier Reef surveys by Furnas and Mitchell (1996, 1999); the New South Wales coastal studies by Lee *et al.* (2001), Ajani *et al.* (2001, 2013a,b, 2014), Farrell *et al.* (2013) and Armbrecht *et al.* (2014); the Western Australian cruises by Hanson *et al.* (2005) and Thompson and Bonham (2011); the South Australian Shellfish Quality Assurance Program (SASQAP) algal monitoring by Wilkinson (2005); the Gulf of

Saint Vincent surveys by Leterme *et al.* (2014); the Bass Strait phytoplankton surveys by the University of Melbourne (e.g. Huisman 1989; McFadden *et al.* 1986); and the Tasmanian coastal phytoplankton surveys by Hallegraeff *et al.* (2010).

Australian Phytoplankton Database

In 2007, the Australian government initiated the Integrated Marine Observing System (IMOS) to monitor the marine environment around Australia. It introduced two major initiatives to monitor phytoplankton over large areas of the Australian coast: the National Reference Station network (NRS, established in 2007) and the Australian Continuous Plankton Recorder program (AusCPR, established in 2009). The NRS consists of a network of nine reference stations around Australia, collecting information on biophysical properties in addition to monthly plankton sampling. The AusCPR program uses commercial ships of opportunity to tow a torpedo-like device that collects plankton samples (Richardson *et al.* 2006). These AusCPR tows collect plankton offshore in oligotrophic waters over large spatial and temporal scales on eight routes around Australia. In 2012, the CSIRO and the University of Tasmania initiated the development of a historical phytoplankton database, with the objective to digitise and centralise known Australian phytoplankton species presence and abundance records. The Australian Phytoplankton Database (Davies *et al.* 2016) has 225,429 presence records for phytoplankton taxa, including a significant part of the early Wood (1954) data. In addition to this historic literature, results from the AusCPR tows and NRS stations have been compiled to create the most comprehensive Australian phytoplankton data set, spanning over 70 years. While in the past we relied on single-snapshot samples, for the first time we can now construct a synoptic view of phytoplankton distributions. Locations of the total phytoplankton data set available for the present work are summarised in Table 3.1 and Figure 3.2. Only data covering full taxonomic phytoplankton species resolution in either netplankton and/or water bottle samples were included, with emphasis on the presence or implied absence of key phytoplankton indicator species (Figure 3.3). Species presence–only data obtained from the NRS, AusCPR and Australian Phytoplankton databases were compiled and used to produce the distribution maps provided. The creation of the Australian phytoplankton database will increasingly allow for more rigorous analysis of numerical multispecies census data and underpin efforts to quantitatively relate community composition to environmental properties.

Phytoplankton Indicator Species

While the dominant phytoplankton species account for most of the chlorophyll biomass and primary productivity, it is the rare species surviving at the borderline of environmental tolerances that best define the boundaries of bioregions (Smayda 1978). Latitudinal ranges imply species preferences to temperature, while onshore versus offshore habitats relate to requirements for nutrients and salinity tolerance. Rare species are better collected by water column net hauls or continuous plankton recorder tows for subsurface dwelling species because of the greater volume of water sampled (cubic metres rather than millilitres). Figure 3.3 illustrates examples of distribution maps for the key phytoplankton indicator species used to discriminate between tropical and temperate, neritic and oceanic, and subantarctic communities. The tropical neritic dinoflagellates *Ceratium dens* (Figure 3.3a) and *Dinophysis miles* (Figure 3.3b) are essentially Indo-West Pacific species. The distribution of *C. dens* occasionally can traverse the Torres Strait and the species has only once been seen as far south as Sydney (Wood 1954). The colonial tropical cyanobacterium *Trichodesmium* (Figure 3.3c) causes widespread blooms off North West Australia and in the Coral Sea, from where the Leeuwin Current and East Australian Current carry it southwards to Perth and Sydney (Jervis Bay) beaches. The tropical neritic diatom *Bacteriastrum* (Figure 3.3d) has a broadly similar distribution, being found throughout the year in the Gulf of Carpentaria (Burford *et al.* 1995), but only in warmer months off Sydney (Ajani *et al.* 2014) and more rarely as far south as Tasmania (Hallegraeff *et al.* 2010). The warm-water dinoflagellates *Ornithocercus* (Figure 3.3e) and *Histioneis* (Figure 3.3f) have broader neritic and oceanic distributions. *Histioneis* is easily overlooked, never abundant and more confined to oceanic waters than the more widespread *Ornithocercus*. The large

TABLE 3.1

Summary of Published Marine Phytoplankton Species Data Used in the Present Work, Arranged by Region and in Chronological Order

Australia-Wide
Crosby and Wood 1958, Wood 1954
Hallegraeff 1984a,b
Jeffrey and Hallegraeff 1990
North West Shelf
Allen and Cupp 1935
Desrosieres 1965
Hallegraeff and Jeffrey 1984
Thompson and Bonham 2011
Gulf of Carpentaria
Markina 1972, 1974, 1976
Hallegraeff and Jeffrey 1984
Rothlisberg *et al.* 1994
Burford *et al.* 1995, Burford and Rothlisberg 1999
West Australia
Hanson 2004
John 2012
South Australia
Wilkinson 2005
Leterme *et al.* 2014
Coral Sea
Marshall 1933
Hallegraeff and Jeffrey 1980, unpublished data
Revelante *et al.* 1982
Revelante and Gilmartin 1982
Furnas and Mitchell 1989
Gottschalk *et al.* 2007
East Australian Current
Dakin and Colefax 1933, 1940
Hallegraeff 1981, 1983
Hallegraeff and Reid 1986
Hallegraeff and Jeffrey 1993
Farrell *et al.* 2013
Ajani *et al.* 2001, 2013a,b, 2014
Armbrecht *et al.* 2014
East Australian Current Eddies
Jeffrey and Hallegraeff 1980, 1987
Bass Strait and Tasmania
McFadden *et al.* 1986
Von Stosch 1986
Huisman 1989
Bolch and Hallegraeff 1990
Leroi and Hallegraeff 2004, 2006
Hallegraeff *et al.* 2010
Subantarctic
Findlay and Flores 2000
Scott and Marchant 2005
Cubillos *et al.* 2007

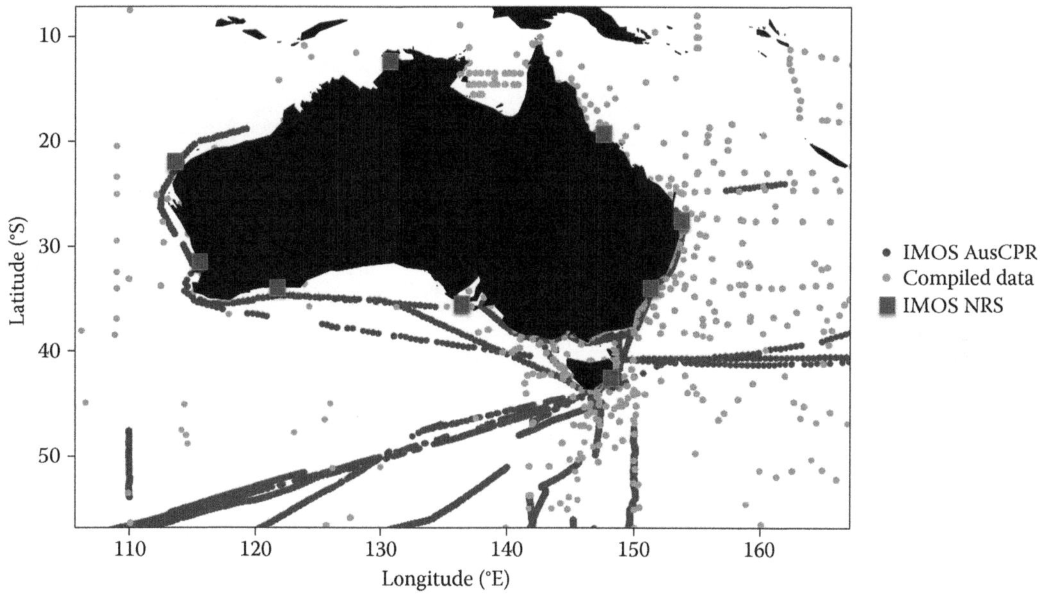

FIGURE 3.2 (See colour insert.) Sampling locations of the total phytoplankton data archive available for the present work. The IMOS national reference stations (NRS; red squares), continuous plankton recorder (CPR) transects (grey lines) have been combined with literature data. Note the sparsity of phytoplankton information for the west coast of Australia and notably the region between Cape Leeuwin and Port Lincoln.

warm-water coccolithophorid *Scyphosphaera apsteinii* (Figure 3.3g) can be carried by the Leeuwin Current from the Great Australian Bight down to the west coast of Tasmania, and by the East Australian Current down the east coast of Australia to the east coast of Tasmania. The subantarctic coccolithophorid *Coccolithus pelagicus* subsp. *braarudii* (Figure 3.3h) is mainly confined to the Southern Ocean and Tasmanian and New Zealand waters. Other subantarctic species delineating that cold-water community are the dinoflagellate *Dinophysis truncata* (occasionally identified in the Great Australian Bight and off Tasmania) and the diatom *Fragilariopsis kerguelensis*. Often, attempts to define biogeographic regions are qualitative at best (Wood 1964a,b) and merely based on presence or absence data.

Description of Australian Marine Phytoplankton Provinces

A recent effort towards generating a map of Australian phytoplankton bioregions was initiated in 2005 as part of a National Oceans Office project to collate spatial, physical and biological information to summarise ecological patterns for natural resource planning and management (Commonwealth of Australia 2005). Figure 3.4 depicts the Australian marine phytoplankton provinces recognised. The productive phytoplankton province (1) found in the shelf waters of North West Australia, the Gulf of Carpentaria, the Arafura Sea and the Timor Sea is basically a tropical diatom flora (keystone species are the diatoms *Bacteriastrum*, *Chaetoceros*, *Coscinodiscus*, *Rhizosolenia sensu lato*, *Thalassionema*, and *Thalassiothrix*). Subtle differences in species dominance and phytoplankton chlorophyll biomass are discriminants of the biomes 1a, b, c, while the floristically distinct shallow waters of the Great Barrier Reef lagoon (3) are dominated by fast-growing nanoplankton diatoms. These tropical neritic communities are distinct from the tropical oceanic, predominantly dinoflagellate flora (*Gonyaulax birostris*, *Histioneis*, *Ornithocercus*) of the Indian Ocean (2a) and Coral Sea (2c), which are carried southwards by the Leeuwin Current (2b) and East Australian Current (2d), respectively. The productive temperate neritic province (4) includes the coastal waters of New South Wales, Tasmania, Victoria and South Australia and has predictable phytoplankton species succession patterns from small diatoms (*Asterionellopsis*, *Pseudo-nitzschia*, *Skeletonema*, *Thalassiosira*) to large diatoms (*Detonula*, *Ditylum*, *Eucampia*) to

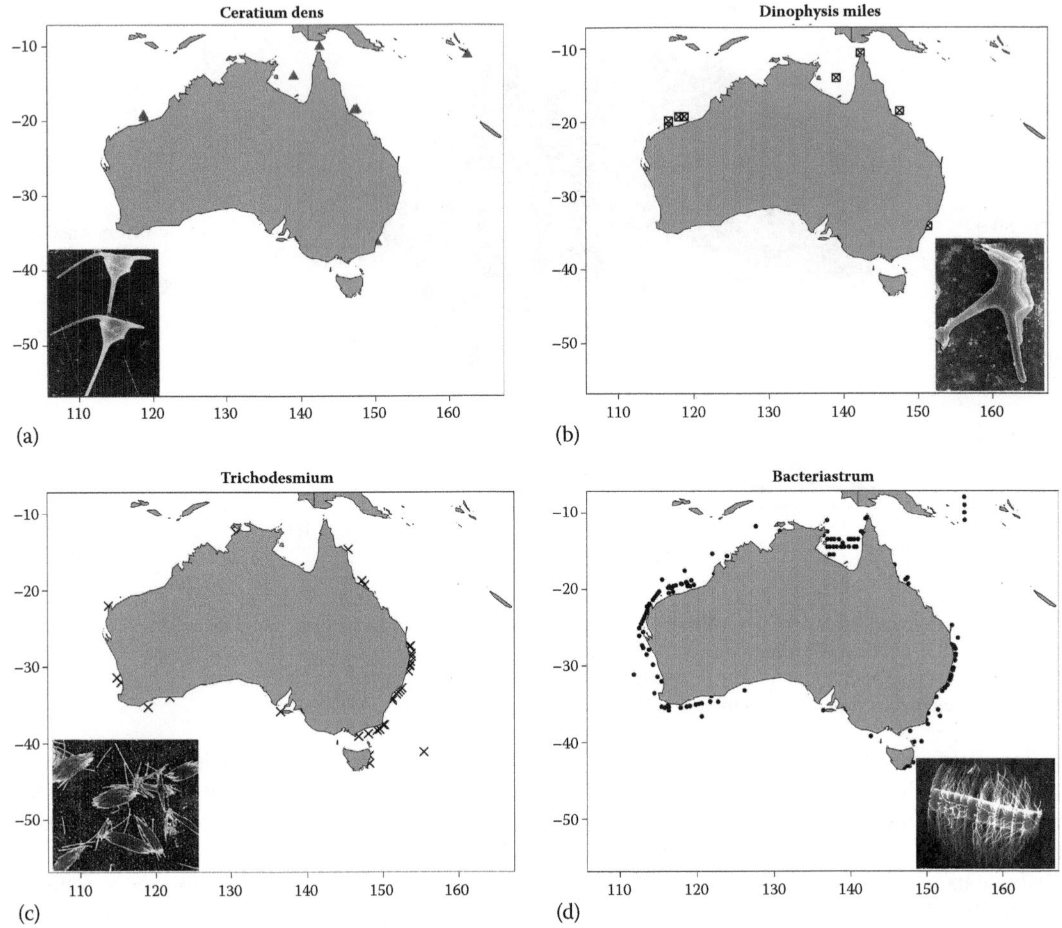

FIGURE 3.3 Semiquantitative Australian distribution maps for key phytoplankton indicator species used to discriminate between tropical and temperate, neritic and oceanic, and cosmopolitan and subantarctic communities, derived from the data archive in Figure 3.2. (a) Tropical neritic dinoflagellate *Ceratium dens* (80 μm long); (b) Tropical neritic dinoflagellate *Dinophysis miles* (100 μm long); (c) Colonial tropical cyanobacterium *Trichodesmium erythraeum* (filaments up to 750 μm long); (d) Tropical neritic diatom *Bacteriastrum furcatum* (30 μm diameter).

larger dinoflagellates (*Ceratium*, *Protoperidinium*), in response to nutrient enrichment from current and/or wind-induced upwelling phenomena. A highly variable oceanic transition zone (5), distinct from inshore phytoplankton communities (4) and embedded tropical flora (2b,d), is bordered to the south by a subantarctic phytoplankton province (6), including the coccolithophorid *Coccolithus pelagicus* subsp. *braarudii* as an indicator organism (Findlay and Flores 2000; Hiramatsu and De Deckker 1996; Scott and Marchant 2005). Tasmania is subject to episodic incursions from the south of cold-water subantarctic species (diatom *Chaetoceros criophilum*, dinoflagellate *Dinophysis truncata*) and from the north of warm-water East Australian Current species (diatom *Bacteriastrum furcatum*), with the red-tide dinoflagellate *Noctiluca* (since 1994) recently extending its range from New South Wales (McLeod *et al.* 2012). Wood (1964d) reported *Noctiluca* as an insignificant species in Australian waters, with sparse records only from the Sydney region and Moreton Bay. By 2008–2013, there were major bloom events from tropical Cairns down to temperate Tasmania, as well as from Port Esperance and Port Lincoln to Port Phillip Bay. A combination of anthropogenic nutrient enrichment (Sydney region), climate change (Tasmania) and possibly even ship ballast water transport (Cairns, Port Esperance) could explain such a

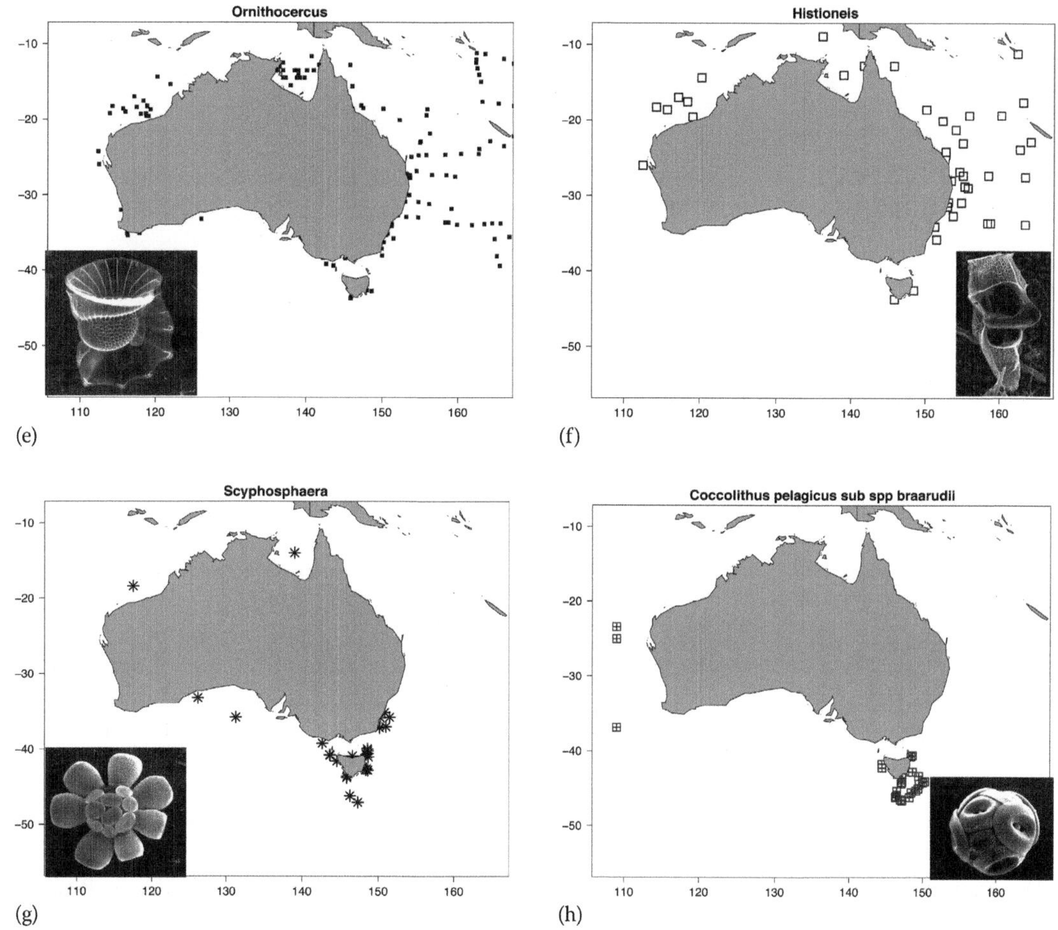

FIGURE 3.3 (CONTINUED) Semiquantitative Australian distribution maps for key phytoplankton indicator species used to discriminate between tropical and temperate, neritic and oceanic, and cosmopolitan and subantarctic communities, derived from the data archive in Figure 3.2. (e) Warm-water dinoflagellate *Ornithocercus steinii* (80 µm long); (f) Tropical oceanic dinoflagellate *Histioneis mitchellana* (60 µm long); (g) Large warm-water coccolithophorid *Scyphosphaera apsteinii* (40 µm diameter); (h) Subantarctic coccolithophorid *Coccolithus pelagicus* subsp. *braarudii* (20 µm diameter).

dramatic shift in abundance and distribution (Hallegraeff 2010). In the Southern Ocean since the 1990s, the coccolithophorid *Emiliania huxleyi* has expanded its range south of 60°S (Cubillos *et al.* 2007; Winter *et al.* 2014).

There exists broad agreement between the regions defined by qualitative phytoplankton data and the regions defined on water masses from physical oceanographic data (temperature, salinity, oxygen; Commonwealth of Australia 2005, Condie and Dunn 2006). For example, phytoplankton regions 1a and 1b reflect the different water masses in the Gulf of Carpentaria. Detailed features of the phytoplankton map, such as those associated with the Leeuwin Current (2b) offshore of the shelf break in the western Great Australian Bight, also correspond with finer-scale circulation regimes and oceanographic features used in the bioregionalisation. Notwithstanding, there are some minor discrepancies between regions based on physical and biological data, indicating that water mass data are not a perfect surrogate for biological data, with pelagic communities also responding differently compared with benthic communities (Commonwealth of Australia 2005). The pelagic phytoplankton map presented here provides a baseline against which to assess future ship ballast water–mediated microalgal species

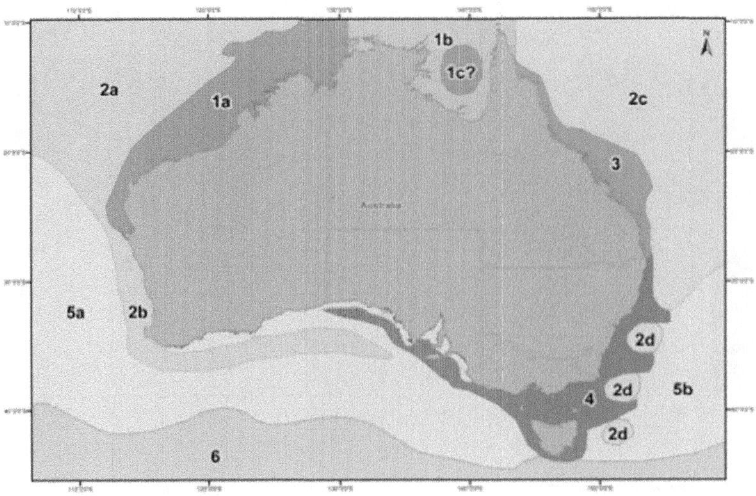

FIGURE 3.4 (See colour insert.) Summary of the Australian marine phytoplankton provinces recognised as described in the text. (Reproduced with permission from Commonwealth of Australia, *National Marine Bioregionalisation of Australia: Summary*, Department of Environment and Heritage, Canberra, Australia, 2005.)

introductions (Hallegraeff and Bolch 1992), climate-driven range extensions (McLeod *et al.* 2012), coastal eutrophication or even the future impact of processes such as ocean acidification (Hallegraeff 2010).

Acknowledgements

The first author is indebted to the late Dr. Shirley Jeffrey of CSIRO Marine and Atmospheric Research for organising his participation in the CSIRO cruises by RV *Sprightly* SP11/81, SP 3/82, SP16/80 and SP6/82, and by FRV *Soela* SO4/80, SO6/82 and SO3/83. The Department of the Environment and Heritage and Dr. Vincent Lyne from CSIRO Marine and Atmospheric Research provided the stimulus for first collating the present material as part of a 2005 effort by the National Oceans Office towards pelagic bioregionalisation of Australia's marine jurisdiction for natural resource planning and management. The maps presented here form part of an effort to generate an Australian phytoplankton database supported by the IMOS, which is funded by the Australian government through the National Collaborative Research Infrastructure Strategy and the Super Science Initiative.

References

Ajani, P., S. Brett, M. Krogh, P. Scanes, G. Webster and L. Armand. 2013a. The risk of harmful algal blooms (HABs) in the oyster-growing estuaries of New South Wales, Australia. *Environmental Monitoring Assessment* 185: 5295–5316.

Ajani, P., R. Lee, T. Pritchard and M. Krogh. 2001. Phytoplankton dynamics at a long-term coastal station off Sydney, Australia. *Journal of Coastal Research* 34: 60–73.

Ajani, P., S. Murray, G. Hallegraeff, N. Lundholm, M. Gillings, S. Brett and L. Armand. 2013b. The diatom genus *Pseudo-nitzschia* (Bacillariophyceae) in New South Wales, Australia: Morphotaxonomy, molecular phylogeny, toxicity and distribution. *Journal of Phycology* 49: 765–785.

Ajani, P. A., A. P. Allen, T. Ingleton and L. Armand. 2014. A decadal decline in relative abundance and a shift in microphytoplankton composition at a long-term coastal station off southeast Australia. *Limnology and Oceanography* 59: 519–531.

Allen, W. E., and E. E. Cupp. 1935. Plankton diatoms of the Java Seas. *Annales Jardin Botanique Buitenzorg* 44: 101–174.

Armbrecht, L. H., M. Roughan, V. Rossi, A. Schaeffer, P. L. Davies, A. M. Waite and L. K. Armand. 2014. Phytoplankton composition under contrasting oceanographic conditions: Upwelling and downwelling (Eastern Australia). *Continental Shelf Research* 75: 54–67.

Bolch, C. J., and G. M. Hallegraeff. 1990. Dinoflagellate cysts from recent marine sediments of Tasmania, Australia. *Botanica Marina* 33: 173–192.

Burford, M. A., and P. C. Rothlisberg. 1999. Factors limiting phytoplankton production in a tropical continental shelf ecosystem. *Estuarine, Coastal and Shelf Science* 48: 541–549.

Burford, M. A., P. C. Rothlisberg and Y. Wang. 1995. Spatial and temporal distribution of tropical phytoplankton species and biomass in the Gulf of Carpentaria, Australia. *Marine Ecology Progress Series* 118: 255–266.

Castracane degli Antelminelli, F. 1886. Report on the diatomaceae collected by H.M.S. *Challenger* during the Years 1873–1876; Report of the scientific results of the voyage of the H.M.S. *Challenger. Botany* 2(4), i–iii, 1–178.

Commonwealth of Australia. 2005. *National Marine Bioregionalisation of Australia: Summary*. Department of Environment and Heritage, Canberra, Australia, p. 142 (accessible via: http://www.oceans.gov.au) (accessed 23 September 2016).

Condie, S. A., and J. R. Dunn. 2006. Seasonal characteristics of the surface mixed layer in the Australasian region: Implications for primary production regimes and biogeography. *Marine and Freshwater Research* 57: 569–590.

Crosby, L. H., and E. J. F. Wood. 1958. Studies on Australian and New Zealand diatoms: 1. Planktonic and allied species. *Transactions of the Royal Society of New Zealand* 85: 483–530.

Cubillos, J. C., S. W. Wright, G. Nash, M. F. de Salas, B. Griffiths, B. Tilbrook, A. Poisson and G. M. Hallegraeff. 2007. Calcification morphotypes of the coccolithophorid *Emiliania huxleyi* in the Southern Ocean: Changes in 2001 to 2006 compared to historical data. *Marine Ecology Progress Series* 348: 47–54.

Dakin, W. J., and A. Colefax. 1933. The marine plankton of the coastal waters of New South Wales: 1. The chief planktonic forms and their seasonal distribution. *Proceedings Linnean Society New South Wales* 58: 186–222.

Dakin, W. J., and A. Colefax. 1940. The plankton of Australian coastal waters of New South Wales: Part 1. *University of Sydney, Department of Zoology Monographs* 1: 221.

Davies, C. H. *et al.* 2016. A database of marine phytoplankton abundance, biomass and species composition in Australian waters. *Nature Scientific Data* 3:160043. doi:10.1038/sdata.2016.43.

Desrosieres, R. 1965. Observations sur le phytoplankton superficial de l'Ocean Indien Oriental. *Cahiers ORSTOM, Oceanographie* 3: 31–37.

Farrell, H., S. Brett, P. Ajani and S. Murray. 2013. Distribution of the genus *Alexandrium* (Halim) along the coastline of New South Wales, Australia. *Marine Pollution Bulletin* 72: 133–145.

Findlay, C. S., and J. A. Flores. 2000. Subtropical front fluctuations south of Australia (45°09′S, 146°17′E) for the last 130 ka years based on calcareous nannoplankton. *Marine Micropaleontology* 40: 403–416.

Furnas, M. J., and A. W. Mitchell. 1996. Nutrient inputs into the Great Barrier Reef (Australia) from subsurface intrusions of Coral Sea Waters: A two-dimensional displacement model. *Continental Shelf Research* 16: 1127–1148.

Furnas, M. J., and A. W. Mitchell. 1999. Wintertime carbon and nitrogen fluxes on Australia's Northwest Shelf. *Estuarine, Coastal and Shelf Science* 49: 165–175.

Gottschalk, S., S. Uthicke and K. Heimann. 2007. Benthic diatom community composition in three regions of the Great Barrier Reef, Australia. *Coral Reefs* 26: 345–357.

Hallegraeff, G. M. 1981. Seasonal study of phytoplankton pigments and species at a coastal station off Sydney: Importance of diatoms and the nanoplankton. *Marine Biology* 61: 107–118.

Hallegraeff, G. M. 1983. Scale-bearing and loricate nanoplankton from the East Australian Current. *Botanica Marina* 26: 493–515.

Hallegraeff, G. M. 1984a. Coccolithophorids (calcareous nanoplankton) from Australian waters. *Botanica Marina* 27: 229–247.

Hallegraeff, G. M. 1984b. Species of the diatom genus *Thalassiosira* in Australian waters. *Botanica Marina* 27: 495–513.

Hallegraeff, G. M. 2007. Biogeography of marine microalgae. In: P. M. McCarthy and A. E. Orchard (eds.), *Algae of Australia: Introduction*, pp. 560–565. ABRS, Canberra, Australia.

Hallegraeff, G. M. 2010. Ocean climate change, phytoplankton community responses and harmful algal blooms: A formidable predictive challenge. *Journal of Phycology* 46: 220–235.

Hallegraeff, G. M., and C. J. Bolch. 1992. Transport of dinoflagellate cysts in ship's ballast water: Implications for plankton biogeography and aquaculture. *Journal of Plankton Research* 14: 1067–1084.

Hallegraeff, G. M., C. J. S. Bolch, D. R. A. Hill, I. Jameson, J. M. LeRoi, A. McMinn, S. Murray, M. F. De Salas and K. Saunders. 2010. In: *Algae of Australia: Phytoplankton of Temperate Coastal Waters*, p. 421. ABRS, Canberra, Australia.

Hallegraeff, G. M., and S. W. Jeffrey. 1984. Tropical phytoplankton species and pigments in continental shelf waters of North and North-West Australia. *Marine Ecology Progress Series* 20: 59–74.

Hallegraeff, G. M., and S. W. Jeffrey. 1993. Annually recurrent diatom blooms in spring along the New South Wales coast of Australia. *Australian Journal of Marine and Freshwater Research* 44: 325–334.

Hallegraeff, G. M., and D. D. Reid. 1986. Phytoplankton species successions and their hydrological environment at a coastal station off Sydney. *Australian Journal of Marine and Freshwater Research* 37: 361–377.

Hanson, C. E., C. B. Pattiaratchi and A. M. Waite. 2005. Seasonal production regimes off south-western Australia: Influence of the Capes and Leeuwin Currents on phytoplankton dynamics. *Marine and Freshwater Research* 56: 1011–1026.

Hiramatsu, C., and P. De Deckker. 1996. Distribution of calcareous nanoplankton near the subtropical convergence, south of Tasmania, Australia. *Marine and Freshwater Research* 47: 707–713.

Huisman, J. M. 1989. The genus *Ceratium* Schrank (Dinophyceae) in Bass Strait and adjoining waters, southern Australia. *Australian Systematic Botany* 2: 425–454.

Jeffrey, S. W., and G. M. Hallegraeff. 1980. Studies of phytoplankton species and photosynthetic pigments in a warm-core eddy of the East Australian Current: I. Summer populations. *Marine Ecology Progress Series* 3: 285–294.

Jeffrey, S. W., and G. M. Hallegraeff. 1987. Phytoplankton pigments, species and light climate in a complex warm-core eddy (Mario) of the East Australian Current. *Deep-Sea Research* 34: 649–673.

Jeffrey, S. W., and G. M. Hallegraeff. 1990. Phytoplankton ecology of Australasian waters. In: M. N. Clayton and R. J. King (eds.), *Biology of Marine Plants*, pp. 310–348. Longman Cheshire, Melbourne, Austraslia.

John, J. 2012. *Diatom Flora of the Swan River Estuary, Western Australia*, 2nd edn., p. 456. Koeltz Scientific Books, Konigstein, Germany.

Karsten, G. H. H. 1905. *Das Phytoplankton des Antarktischen Meeres nach dem Material der Deutschen Tiefsee-Expedition 1898–1899.* Wissenschafliche Ergebnisse Deutsche Tiefsee-Expedition 'Valdivia' 2(2), pp. 1–136. Gustaf Fischer, Jena, Germany.

Karsten, G. H. H. 1907. *Das Indische Phytoplankton nach dem Material der Deutschen Tiefsee-Expedition 1898–1899.* Wissenschafliche Ergebnisse Deutsche Tiefsee-Expedition 'Valdivia' 2(2), pp. 223–548. Gustaf Fischer, Jena, Germany.

Lee, R., P. Ajani, M. Krogh and T. R. Pritchard. 2001. Resolving climatic variance in the context of retrospective phytoplankton pattern investigations off the east coast of Australia. *Journal of Coastal Research* 34: 74–86.

LeRoi, J. M., and G. M. Hallegraeff. 2004. Scale-bearing nanoplankton flagellates from Southern Tasmanian coastal waters, Australia: 1. Species of the genus *Chrysochromulina* (Haptophyta). *Botanica Marina* 47: 73–102.

LeRoi, J. M., and G. M. Hallegraeff. 2006. Scale-bearing nanoplankton flagellates from Southern Tasmanian coastal waters, Australia: 2. Species of Chrysophyceae (Chrysophyta), Prymnesiophyceae (Haptophyta, excluding *Chrysochromulina*) and Prasinophyceae (Chlorophyta). *Botanica Marina* 49: 216–235.

Leterme, S. C., J. G. Jendyk, A. V. Ellis, M. H. Brown and T. Kildea. 2014. Annual phytoplankton dynamics in the Gulf Saint Vincent, South Australia, in 2011. *Oceanologia* 56: 757–778.

Markina, N. P. 1972. Special features of plankton distribution around northern coasts of Australia during different seasons of 1968–1969 *Izv. Tikhvokean nauchno-issled. Inst. ryb. Khoz. Okeanogr.* 81: 57–68 (in Russian).

Markina, N. P. 1974. Biogeographic regionalization of Australian waters of the Indian Ocean. *Oceanology* 15: 602–604.

Markina, N. P. 1976. Ecological diversity of plankton in the Australian region of the Indian Ocean. *Biologiya Morya* 3: 49–57 (in Russian).

Marshall, S. M. 1933. The production of microplankton in the Great Barrier Reef region. *Scientific Reports Great Barrier Reef Expedition* 2: 112–157.

McFadden, G. I., D. R. A. Hill and R. Wetherbee. 1986. A study of the genus *Pyramimonas* (Prasinophyceae) from south-eastern Australia. *Nordic Journal of Botany* 6: 209–234.

McLeod, D. J., G. M. Hallegraeff, G. W. Hosie and A. J. Richardson. 2012. Climate-driven range expansion of the red-tide dinoflagellate *Noctiluca scintillans* into the Southern Ocean. *Journal of Plankton Research* 34: 332–337.

Moestrup, Ø. 1979. Identification by electron microscopy of marine nanoplankton from New Zealand, including the description of four new species. *New Zealand Journal of Botany* 17: 61–95.

Revelante, N., and M. Gilmartin. 1982. Dynamics of phytoplankton in the Great Barrier Reef Lagoon. *Journal of Plankton Research* 4: 47–76.

Revelante, N., W. T. Williams and J. S. Bunt. 1982. Temporal and spatial distribution of diatoms, dinoflagellates and *Trichodesmium* in waters of the Great Barrier Reef. *Journal of Experimental Marine Biology and Ecology* 63: 27–45.

Richardson, A. J., A. W. Walne, A. W. G. John, T. D. Jonas, J. A. Lindley, D. W. Sims, D. Stevens and M. Witt. 2006. Using continuous plankton recorder data. *Progress in Oceanography* 68: 27–74.

Rothlisberg, P. C., P. C. Pollard, P. D. Nichols, D. J. W. Moriarty, A. M. G. Forbes, C. J. Jackson and D. Vaudrey. 1994. Phytoplankton community structure and productivity in relation to the hydrological regime of the Gulf of Carpentaria, Australia, in summer. *Australian Journal of Marine and Freshwater Research* 45: 265–282.

Scott, F. J., and H. J. Marchant (eds.). 2005. *Antarctic Marine Protists*. ABRS and Australian Antarctic Division, Canberra, Australia.

Smayda, T. J. 1978. Biogeographical meaning: Indicators. In: A. Sournia (ed.), *Phytoplankton Manual*, pp. 225–229. Monographs on oceanographic methodology 6. UNESCO, Paris, France.

Thompson, P. A., and P. Bonham. 2011. New insights into the Kimberley phytoplankton and their ecology. *Journal of the Royal Society of Western Australia* 94: 161–169.

von Stosch, H. A. 1986. Some marine diatoms from the Australian region, especially from Port Phillip Bay and tropical north-eastern Australia. *Brunonia* 8: 293–348.

Wilkinson, C. 2005. Common phytoplankton of South Australia. South Australian Shellfish Quality Assurance Program, Port Lincoln Marine Science Centre, unpublished report.

Winter, A., J. Henderiks, L. Beaufort, R. E. M. Rickaby and C. W. Brown. 2014. Poleward expansion of the coccolithophore *Emiliania* huxleyi. *Journal of Plankton Research* 36: 316–325.

Wood, E. J. F. 1954. Dinoflagellates in the Australian region. *Australian Journal of Marine and Freshwater Research* 5: 171–351.

Wood, E. J. F. 1964a. Studies in microbial ecology of the Australasian region: 1. Relation of oceanic species of diatoms and dinoflagellates to hydrology. *Nova Hedwigia* 8: 5–20.

Wood, E. J. F. 1964b. Studies in microbial ecology of the Australasian region: 2. Ecological relations of oceanic and neritic diatom species. *Nova Hedwigia* 8: 5–34.

Wood, E. J. F. 1964c. Studies in microbial ecology of the Australasian region: 3. Ecological relations of some oceanic dinoflagellates. *Nova Hedwigia* 8: 35–54.

Wood, E. J. F. 1964d. Studies in microbial ecology of the Australasian region: 7. Ecological relations of Australian estuarine dinoflagellates. *Nova Hedwigia* 8: 548–568.

4

Biogeography of Australian Seaweeds

John M. Huisman, Roberta A. Cowan and Olivier De Clerck

CONTENTS

Introduction

Marine benthic algae, or seaweeds, form a conspicuous element on most Australian coasts and are of considerable ecological importance, providing habitat, food and substratum stabilisation. Along with the seagrasses, the only angiosperms that have truly recolonised marine habitats, the seaweeds are the benthic primary producers that sustain most shallow coastal ecosystems. Australia is well known as a centre of biodiversity for seaweeds, with numerous species recorded and a particularly high percentage of endemic taxa, especially along the temperate southern coast (Phillips 2001; Kerswell 2006; Hommersand 2007). This chapter will review the biogeography of Australian seaweeds, including historical studies that have aimed to place the Australian flora in local and global scenarios, examine current thinking about the Australian biogeographic provinces and conclude with some of the exciting developments that will enhance seaweed taxonomic and biogeographic studies but at the same time introduce an entirely new suite of problems.

Historical Studies of Australian Seaweed Biogeography

> Biogeography is dependent on good taxonomy and extensive well-documented collections: without these, conclusions are dubious and often misleading.
>
> **– H. B. S. Womersley (1984)**

Here we will briefly mention the history of Australian seaweed studies, primarily to set the scene for our restrained approach to biogeographical speculation inherent in the these cautionary words of Bryan Womersley, the leading Australian floristic phycologist of recent times. Several detailed accounts of Australian phycological history are available, notably those of Womersley (1959), Ducker (1979, 1981a–c, 1983, 1986, 1988) and Cowan and Ducker (2007), the latter providing an excellent overview and incorporating the earlier publications.

Knowledge of the diversity of Australian seaweeds began in August 1699, when the English adventurer and privateer William Dampier initiated the European collection of marine algae from 'New Holland' with specimens gathered in or near Shark Bay on the west coast. Dampier (1703) described his specimens as '*Fucus foliis capillaceis brevissimis, vesiculis minimis donatis*' but we now call plants of this taxon *Sirophysalis trinodis* (Forssk.) Kütz., a widespread Indian Ocean species. Since this early inception, progress documenting the Australian seaweed flora has been erratic (Figure 4.1), with major

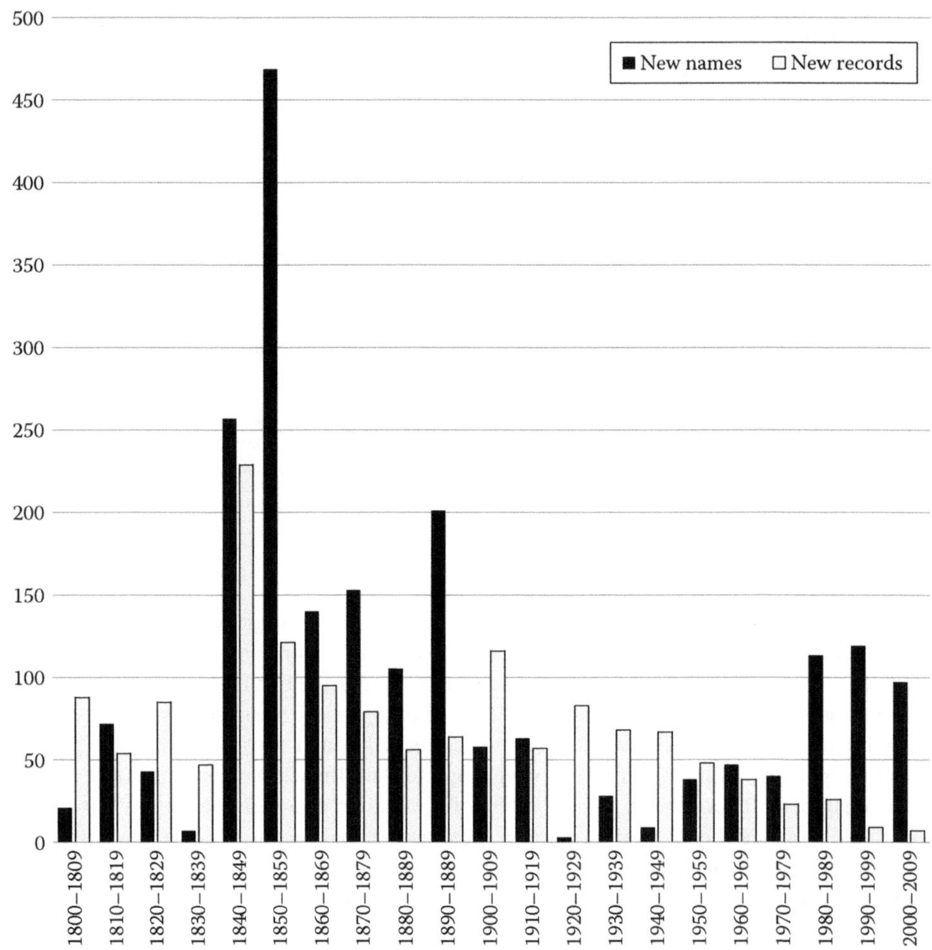

FIGURE 4.1 Historical accumulation of seaweed names for Australia. Note the peaks around the mid-nineteenth century and the late twentieth century, these due primarily to W. H. Harvey and H. B. S. Womersley, respectively.

advances resulting from the English and French 'voyages of discovery' (1800–1839) and, following those, typically aligned to the presence of enthused collectors and botanists. Herein we will mention only two of the latter group, the first being William Henry Harvey (1811–1866), the Irish botanist who visited Australia in the mid-nineteenth century and wrote the first great Australian seaweed Flora, the five-volume *Phycologia Australica* (1858, 1859, 1860, 1862, 1863). Specimens collected by Harvey were subsequently used by many other botanists to describe new Australian taxa. The second is Bryan Womersley (1922–2011), the South Australian phycologist who dedicated his working life to documenting Australia's temperate seaweed flora. In mentioning only these two, we do not wish to undervalue the contributions of others, but as can be seen in Figure 4.1, the documentation of Australia's seaweeds experienced significant peaks due to Harvey and Womersley. Notably, and of relevance to this chapter, both attempted to place the Australian flora in a global context.

William Henry Harvey visited Australia in 1854, at which time he collected numerous specimens and described a large number of new taxa. As noted by Womersley (1981a), this was the first occasion when 'an expert was seeing, collecting, and describing the living plants'. Harvey was also seemingly the first to attempt a biogeographic analysis; his 1855 paper 'Some Account of the Marine Botany of the Colony of Western Australia' compared his collection with species recorded from elsewhere. Harvey listed 14 species he described as 'pelagic, or those which inhabit many very distant places and dissimilar climates', 18 from the Mediterranean, 27 'natives of the coasts of the British Islands', six species showing affinities with the Red Sea and Indian Ocean, and smaller numbers showing affinities with the South Pacific, the North Pacific, the Cape of Good Hope, and Antarctica. While many of Harvey's names have since proved to be misidentifications, it is interesting to note that the essence of his methodology, that of comparing species composition from various localities, has persisted until the present time. Similar exercises with Australian seaweeds were undertaken by May (1940), Womersley and Edmonds (1958), Womersley (1959, 1981a), Millar (1990), Kraft (2000), Phillips (2001), Hommersand (2007), Huisman (2007), Huisman *et al.* (2009), Phillips and Huisman (2009) and Waters *et al.* (2010).

In more recent times, the contributions of Bryan Womersley stand out. As described by Kraft (2011), Womersley was 'surely the finest macroalgal regional monographer of the twentieth century'. His working career spanned some 60 years, during which time he and his students advanced knowledge of the southern Australian macroalgae to the point where the region's marine flora is today recognised as one the most diverse and taxonomically rich in the world. The contributions of Womersley are numerous, culminating in his six-volume *Marine Benthic Flora of Southern Australia* (1984, 1987, 1994, 1996, 1998, 2003), but equally he should be remembered for his focal role in the training of students, several of whom (Gerry Kraft, Bill Woelkerling, Murray Parsons) went on to similarly illustrious careers.

As previously mentioned, the recognition of Australia's seaweed flora has progressed erratically, with considerable geographic variation in effort, despite the major contributions of Harvey, Womersley and others (Figure 4.1). Appreciation of the intermittent and inconsistent nature of this accumulation of knowledge is particularly important if one is to make a realistic appraisal of Australia's seaweed biogeography, which, as Womersley so aptly stated, 'is dependent on good taxonomy and extensive well-documented collections'.

Global Biogeography of Australian Seaweeds

Womersley (1981b) commented on the paucity of discussions concerning marine algal biogeography on a world basis, suggesting this was probably due to taxonomic uncertainties. The present molecular-based taxonomy, while removing many of these uncertainties, has also led to an entirely new suite of problems that further obfuscate our understanding of world seaweed biogeography. As noted by Gary Saunders in an address to the Australasian Society for Phycology and Aquatic Botany (2010), there are 'far more species on the planet than currently recognized' and 'levels of introduced seaweeds likely far exceed current knowledge'. It is clear that attempting to place Australian seaweeds in a global scenario is likely to be somewhat fraught, as the level of taxonomic uncertainty must increase progressively with the distance from Australian shores. Nevertheless, Hommersand (2007), while acknowledging these uncertainties, presented an excellent overview of the Australian seaweed flora and the likely pathways that led to its current composition. He suggested (2007, 511) that the richness and diversity of the flora were the result of an unusual sequence of vicariance events in the geological history of Australasia under conditions of relative isolation, and the maintenance of a temperate climate for extended periods and on long stretches of coastline. Hommersand (2007) concluded with the prescient statement, that while 'much can be learned ... from comparative morphological studies but more and more our understanding of the global relationships of this remarkable continent will rely on evidence from molecular investigations' (2007, 535).

Kerswell (2006) considered the Australian seaweed biodiversity in her analysis of global diversity patterns, confirming long-held views that southern Australia constitutes a hotspot for seaweed diversity and that, globally, diversity at the genus level is the highest in temperate regions, with richness decreasing towards the tropics and polar regions. At the species level a different pattern can occur, as shown in Kerswell's analysis of the green algal order Bryopsidales. This group shows the greatest species diversity in the tropics, suggesting that groups that radiated in the modern tropics are relatively few but often incredibly speciose.

Australian Marine Biogeographical Provinces

Inception and Twentieth Century

Hedley (1904) was the first to propose the subdivision of the Australian coast into biogeographical provinces, based on the distribution of intertidal molluscs. His initial focus was the southeastern Australian coast, proposing that the region west of approximately Wilson's Promontory in Victoria supported an *Adelaidean* fauna, whereas the region to the east supported a very different species composition that he termed the *Peronian* fauna. Hedley also introduced the tropical *Dampierian* and *Solanderian* faunas, which he did only as an aside and seemingly without justification, stating, 'To these names [Adelaidean and Peronian] I might take this opportunity of adding the Dampierian for the marine fauna which extends from the Torres Straits to Houtman's Abrolhos; and the Solanderian for the marine fauna of the Queensland coast from Moreton Bay to Torres Strait.' The Adelaidean Province was subsequently renamed the Flindersian by Cotton (1930: 219), who felt the name was too 'localized in meaning for a faunal area extending over nearly all the southern and half the western coasts of Australia', but for the most part Hedley's provinces were accepted by later workers (Knox 1963).

Bennett and Pope (1953, 1960) undertook a detailed analysis of the southern Australian provinces, focusing particularly on the Maugean Province (adopting the name from Iredale and May 1916, who proposed this province for the east coast of Tasmania based on chiton distributions), which they applied to the cool temperate region of Tasmania and southern Victoria. They did not discuss Hedley's tropical provinces, which were reproduced in their map (1953: 139; Figure 4.5) 'without prejudice'. Again, their work was based primarily on the composition of the intertidal fauna, although they also incorporated several marine algae in their analysis, including the bull kelp *Durvillaea potatorum* (Labill.) Aresch. and *Macrocystis pyrifera* (L.) C.Agardh (as *M. angustifolia* Bory), which they regarded as characteristic of colder waters. Bennett and Pope (1953) provided a map of the provinces, which is reproduced in Figure 4.2. As these provinces were proposed primarily on faunal distributions, their legitimacy for the algae required further analysis, and several studies have attempted to provide a numerical basis for at least some of these provinces. Womersley and Edmonds (1958) were the first to test the southern provinces based on a suite of seaweed distributions and questioned the status of the Maugean, which they felt should be relegated to a subprovince of the Flindersian. They presented evidence that a greater distinction exists between the eastern warm temperate (Peronian) and the cool temperate (Maugean), than between the Flindersian and the Maugean, concluding that 'a large proportion of algal species ranged along southern Australia, with a very considerable eastern element and smaller western element. Of the eastern element, a high proportion is common to South Australia, Victoria, and Tasmania, while a smaller proportion is limited to Victoria or Tasmania, or both.' Thus, while not denying Bennett and Pope's acceptance of *Durvillaea* and *Macrocystis* as indicators of cool-temperate conditions ('fully justified'), they argued that with the large proportion of species common to the regions, the Maugean should

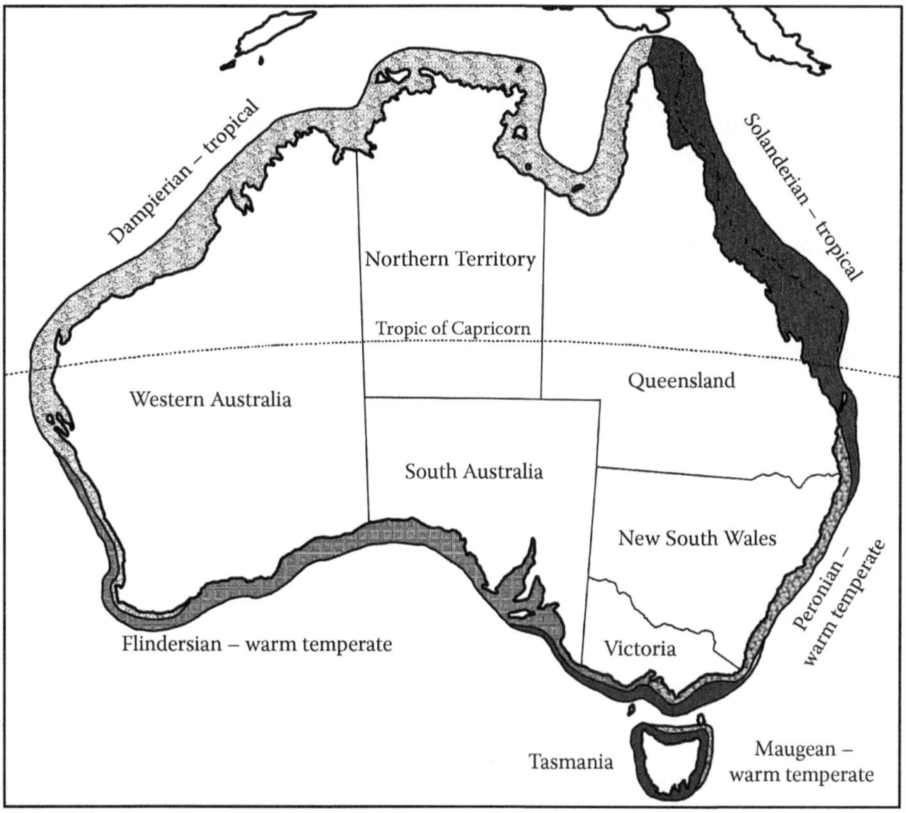

FIGURE 4.2 Biogeographic provinces of Australian marine macroalgae (redrawn and revised from I. Bennett and E. C. Pope, *Aust. J. Mar. Freshw. Res.* 4, 105–159, 1953).

be a subprovince of the Flindersian. Subsequently, Bennett and Pope (1960) disputed this, suggesting that a greater difference is to be expected between the warm- and cool-temperate biotas in the east in Australia because of the additional presence, in the Peronian, of a large Pacific element. Womersley, however, maintained his stance (1981a,b), although later (1984) was not decisive regarding the status of the Maugean.

Hedley's (1904) scheme proposed only the Solanderian Province for tropical eastern Australia, in which he initially included only the Queensland coast from Moreton Bay to the Torres Strait, although later he expanded this to encompass the 'waters of the Coral Sea from Torres Straits to Wide Bay, including the Great Barrier Reef' (Hedley 1926). Whitley (1932) recognised the offshore Great Barrier Reef as a province distinct from that of the Queensland mainland. He adopted Hedley's Solanderian for the Great Barrier Reef and proposed the name *Banksian* for the mainland province, and his scheme was followed by Endean *et al.* (1956) and Endean (1957). According to Endean (1957) the echinoderm fauna of the mainland has strong affinities to that of the East Indies (the Indo-Malay Archipelago), whereas that of the Great Barrier Reef was closer to the West Pacific. Despite Hedley's (1926) later inclusion of the Great Barrier Reef in his Solanderian Province, Womersley (1959) argued that the Solanderian was originally applied to the mainland, and if two provinces are to be recognised for tropical Queensland the inshore should retain the name. He used the eponymous *Great Barrier Reef* for the offshore province. Womersley (1959: 593) also subsumed the Solanderian into the Dampierian, although acknowledging that it might be recognised as a subprovince. Later, based on the belief of a widespread, tropical Indo-Pacific algal flora, Womersley (1981b) suggested that the 'whole of the tropical Australian coasts is probably best regarded as one province, with possible subprovincial regions if these can be justified by more comprehensive faunistic and floristic studies'.

Recent Insights

In the recent *Algae of Australia: Introduction* (McCarthy and Orchard, 2007), the Australian provinces were discussed individually (Dampierian by Huisman [2007]; Solanderian by Phillips [2007], and Peronian and Flindersian [incorporating the Maugean as a subprovince] by Millar [2007]). Their analyses and more recent studies are reviewed in the following sections.

The Dampierian Province

The Dampierian biogeographical province includes much of northern and northwestern Australia, from the Torres Strait and the tip of Cape York Peninsula westwards to approximately the Houtman Abrolhos and Geraldton on the coast of Western Australia (Womersley 1981b; Huisman 2007; see Figure 4.1). Despite being the largest in terms of coastline length, it was until relatively recently the most poorly documented of the Australian provinces (Knox 1963; Huisman *et al.* 1998), due primarily to its relative inaccessibility and remoteness from population centres, plus the difficulties in dealing with large tidal amplitudes, turbid water and life-threatening fauna. However, the northwest coast has been the focus of considerable recent taxonomic interest, with numerous surveys collecting and documenting the macroalgal flora (e.g. Huisman 2001, 2002, 2004; Huisman and Borowitzka 2003; Huisman *et al.* 2009, 2011; Leliaert *et al.* 2007; Belton *et al.* 2014), these culminating in two volumes of the *Algae of Australia* series (Huisman 2015, in prep.). The progress can be gauged by the manyfold increase in recorded species, from a previous total of 28 (Huisman *et al.* 1998) to the currently known total approaching 500 (Huisman 2015, in prep), equalling the estimates by Womersley (1981a; '400–500') and Huisman (2007; 'around 500') and far exceeding the earlier 'over 300' of Huisman *et al.* (1998).

The eastern boundary of the province is the Torres Strait and the tip of Cape York, but if based on algal species this is essentially arbitrary as there is little evidence for a common disjunction in distributions (Huisman 2007). Endean (1957), based on echinoderm distributions, doubted the separation of the Solanderian and Dampierian Provinces. Knox (1963) noted that to the echinoderms of the Queensland mainland (i.e. the Solanderian *sensu stricto*), the Torres Strait does not present a biogeographical barrier, but that very few Great Barrier Reef species have extended their range through the strait.

With recent surveys undertaken in the northwest, leading to a marked increase in knowledge regarding the seaweed diversity of the region, the central north coast (particularly the Northern Territory coast and Gulf of Carpentaria) remains the only area about which very little is known. Sonder (1871) recorded 162 species from *tropischen Australiens*, which included 43 brown algae, 84 red algae and 35 green algae. The first significant gatherings from the Northern Territory were by Raymond L. Specht during an American–Australian expedition to Arnhem Land in 1948. Specht collected 59 species of macroalgae, with good locality and ecological data, and these were included in a report by Womersley (1958), where he described the algae as 'largely typically tropical and subtropical species which are widely distributed in warm waters of the world' (139). The macroalgal flora of Darwin is now known to include 76 taxa (Wynne and Luong-van Thinh 1997). Phillips *et al.* (1999) described 64 species from the Gulf of Carpentaria, but noted that this was 'an underestimate of macroalgal biodiversity for the region'. This is undoubtedly true, but the gulf is also likely to support an impoverished algal flora due to the lack of extensive hard substratum.

At the western boundary of the Dampierian, a broad overlap zone exists between the Dampierian Province and the southern Australian Flindersian Province (primarily between the Houtman Abrolhos and the southwest tip of Western Australia, extending into but less prominent along the south coast). Tropical species can often be found in this zone (Huisman and Walker 1990; Huisman 1997), occasionally in unexpected locations, but their occurrence is clearly sporadic. For example, Harvey (1855) recorded the green alga *Penicillus nodulosus* (J.V.Lamour.) Blainv. (as *P. arbuscula* Mont.) as abundant on the reef flats at Rottnest Island. This species is common in tropical northwest Australia but modern records do not extend any farther south than the Jurien Bay/Geraldton region (Huisman and Walker 1990). Several unique species of otherwise tropical genera also occur in the southwest region – for example, the red alga *Gibsmithia womersleyi* Kraft, recorded from scattered locations east of Cape Leeuwin, Western Australia, to the Adelaide region. Species such as these are presumably relictual from earlier warmer periods and evolved vicariantly. Similarly, the ranges of typically Flindersian species can often extend into the tropics – for example, the green alga *Struvea plumosa* Sonder, which was recently recorded from Barrow Island, some 1000 km north of the previous most northerly location. This record was confirmed by DNA analysis (Huisman and Leliaert 2015). The region from Fremantle to Albany was designated the *Baudinian Province* by Kott (1952), based on the distributions of ascidians. Womersley (1981a) noted that the algae of that region were not well known at the time and he was unable to judge whether there was any support for the Baudinian Province. The distributions of fish and algae in the southwest Capes region were examined by Westera *et al.* (2009). This region experiences the southerly flowing, warm-water Leeuwin Current (LC), as well as the northerly flowing, cool-water Capes Current. Westera *et al.* (2009) found that virtually all algal species were of temperate rather than tropical affinities, suggesting that the LC played only a minor role in influencing seaweed distributions in this region.

This west coast overlap zone is of particular interest as it essentially has a north–south alignment (thus a latitudinal gradient) and includes the southern range limits of principally tropical species and the northern range limit of principally temperate species. Of particular interest is the flora of the offshore Houtman Abrolhos, an archipelago of 122 mainly coral islands lying some 65–90 km offshore from Geraldton, Western Australia, at a latitude of 28° S–29° S. They include the southernmost examples of major coral reefs in the eastern Indian Ocean and are one of the highest latitude reef systems in the world (Wells 1997). Phillips and Huisman (2009) analysed the Houtman Abrolhos marine flora and compared it with that of the coastal Jurien Bay, using the tropical and temperate Western Australian floras as benchmarks. Their results demonstrated that, in terms of assemblage structure and taxonomic distinctness, the marine flora of the Houtman Abrolhos clearly represent a transitional zone between tropical and temperate regions, with the strong tropical influence a direct result of the LC. In contrast, the nearby inshore flora of the Jurien Bay region exhibited a much lower, almost negligible tropical influence.

Understanding the distribution of seaweeds in the western overlap zone is particularly important as it represents a region where the floristic composition might be acutely susceptible to variations in sea temperature. On ecological timescales, the ranges of seaweeds are primarily dependent on sea temperature (Breeman 1988) and might be expected to alter as a result of climate change. A recent extreme warming event on the west coast resulted in the decline of the temperate fucalean brown alga *Scytothalia dorycarpa* (Turner) Grev. from the Jurien Bay region (Smale and Wernberg 2013), although it is unknown whether this is a temporary change. In a more controversial study, Wernberg *et al.* (2011) analysed historical specimen records in *Australia's Virtual Herbarium* (AVH) and suggested that the ranges of numerous temperate species were contracting south due to ocean warming. The methods used in this study were criticised by Huisman and Millar (2013), however, and there is ongoing debate as to their validity (Wernberg *et al.* 2014; Huisman and Millar 2015).

The Solanderian and Great Barrier Reef Provinces

Based on the concept of the widespread distribution of most tropical Indo-Pacific algae, Womersley (1981b) suggested that the recognition of distinct tropical Australian provinces was unlikely, although subprovinces might be justified by more comprehensive studies. In part following this suggestion, Huisman *et al.* (1998), Huisman (2007) and Phillips (2007) combined the Solanderian and Great Barrier Reef Provinces and questioned the distinction between the Dampierian and Solanderian. Huisman (2007) referred to an unpublished honours study by Telford (1998) that utilised recently collected (mostly by Huisman) herbarium specimens and published records of marine benthic algae, available in the *Australian Algal Name Index* (Cowan 2006), to test the distinction between the northern Dampierian (western) and Solanderian (eastern) tropical provinces. Records were divided across the 17 northern Australian ecological regions recognised in the Interim Marine and Coastal Regionalisation for Australia (an early version of the Integrated Marine and Coastal Regionalisation of Australia [IMCRA]) (Thackway and Cresswell 1997). About half of the species recorded for the Dampierian Province also occurred in the Queensland Provinces (the Solanderian and Great Barrier Reef). However, in many cases the species richness and affinities of an area were invariably due to the level of research effort, and until a reasonably uniform knowledge is achieved over the entire northern region, proposals regarding biogeographic provinces must remain speculative. As with all such studies, however, Huisman (2007) highlighted numerous concerns regarding the available data and suggested a 'clear need for further studies of the tropical marine algae'. Phillips (2007) also pointed to the fragmentary knowledge of the macroalgae in the Solanderian Province, although this has markedly improved following the publication of Kraft's treatments of the green and brown algae (2007, 2009).

Many species were found to be common to both regions, but there were numerous exceptions. In most cases this can be explained by the collecting effort, which highlights the caution that must accompany these exercises. The case of *Corynocystis prostrata* Kraft is an interesting example. When first described by Kraft *et al.* (1999) it had a reported distribution of the Philippines, Fiji, Tahiti, Mauritius and the Great Barrier Reef, so would not have been considered part of the Dampierian flora. The species was later reported by Huisman *et al.* (2009) from the Rowley Shoals, so is now regarded as widespread in the Indo-Pacific. Perhaps the only common, easily recognised and conspicuous element that distinguishes the two regions is the green alga *Chlorodesmis fastigiata* (C.Agardh) Ducker, the bright green 'turtle weed' that is a familiar sight on the Great Barrier Reef (Figure 4.7). This species has apparently never been collected in northern or northwestern Australia, and was not found during surveys of the offshore atolls (Huisman *et al.* 2009) or of Ashmore Reef and Hibernia Reef. Ducker (1967) reported numerous records from the Pacific, but very few from the Indian Ocean and none from Western Australia. Numerous records are found in AVH, but these are all on the east coast of Queensland and essentially map the extent of the Great Barrier Reef Province, and to a lesser extent Hedley's coastal Solanderian Province. The reasons for the absence of *Chlorodesmis* from northwestern Australia are unclear, but again point to the need for further study. Its distribution might serve to distinguish the Great Barrier Reef at subprovincial level.

Since the review by Phillips (2007), there have been two major publications describing the marine flora of the southern Great Barrier Reef (Solanderian) and Lord Howe Island (Peronian) (Kraft 2007, 2009). These publications recorded a perhaps typical mix of wide-ranging species, those with limited distribution and short-range endemics. These treatments did not specifically discuss the biogeographic affinities of the floras, but the predominance of tropical Indo-Pacific species would suggest strong links with that region.

The Flindersian and Maugean Provinces

The most recent analysis of the southern provinces was that of Waters *et al.* (2010), who examined the southern Australian provinces based on species richness as identified from herbarium records. Relative to the tropical north coast, the algal flora of southern Australia is considerably better known, primarily due to the pioneering work of Bryan Womersley. In addition to his published works, Womersley and his associates established an extensive herbarium collection, initially at Adelaide University, later transferred to and extended at the Adelaide Herbarium. The data relating to these specimens (and others) have only recently become available through AVH and were mined by Waters *et al.* (2010) to provide numerical support for the three traditionally recognised broad marine biogeographic provinces in southern Australia (Bennett and Pope 1953, 1960): the Peronian (east), Flindersian (west) and Maugean (southeast) Provinces. As originally proposed, these zones had considerable overlap at their boundaries (Waters *et al.* 2010; Figure 4.1). Waters *et al.* (2010) suggested slight modifications to the province's boundaries, extending the Maugean Province into southern New South Wales (NSW) and placing the Maugean–Flindersian boundary in western Victoria. As proposed by Bennett and Pope, however, the northern boundary of the Maugean was placed at Bermagui, at the northernmost location of *D. potatorum* and well into the NSW coast. Millar (2007) conducted intensive surveys along the southern NSW coast for the express purpose of clarifying this boundary and noted a contraction of this species south to the Tathra region, but still also in NSW.

The results of Waters *et al.* (2010), however, indicated that the Flindersian differed from the Maugean and the Peronian by more than those provinces differed from one another. This observation contradicts most earlier analyses and it is unclear why this disparity might exist, but as the

Waters *et al.* study was based on herbarium records in AVH, these at the time not including any records from the NSW Herbarium, the Peronian might have been under represented in the analysis. It is perhaps telling that if one presently excludes NSW (Herbarium) records of *D. potatorum* (Bennett and Pope' s Maugean exemplar) in AVH (2014), the species is not recorded for NSW (state). Thus, if the same was true in 2008 when Waters *et al.* conducted their analysis, it is the species composition exclusive to *Durvillaea* driving their suggestion of the northern Maugean boundary being in southern NSW. Given the clear floristic differences between the three provinces shown by Waters *et al.* (2010), it appears best to follow the Bennett and Pope (1953) scheme and recognise them as distinct, but to reassess the reportedly closer link between the Maugean and Peronian by inclusion of NSW records.

The east–west differentiation in southern Australia's marine flora and fauna has been the subject of much interest. Detailed analyses of specific seaweeds are limited, but several fauna studies are particularly illuminating and, if not necessarily directly applicable to the seaweeds, provide a clear framework for future studies. Waters *et al.* (2005) examined the eastern versus western populations of the abundant and dispersive intertidal mollusc *Nerita*, and revealed a clear phylogeographic break in the vicinity of Wilson's Promontory on the Victorian coast. A disjunction at this location was proposed by Hedley (1904) in his original establishment of Australia's marine biogeographic provinces, which he attributed to the Bassian Isthmus that connected northern Tasmania to the Australian mainland during Pleistocene glacial maxima, when sea levels fell more than 80 m below present. This land bridge reduced to a narrow isthmus as sea levels rose approximately 14,000 years BP (Lambeck and Chappell 2001). Later, Waters (2008) expanded his original study and found a similar species turnover along the north coast of Tasmania, essentially mirroring that of the Victorian coast and providing additional evidence for disjunction due to isolation by the low sea levels. He further examined the maintenance of this disjunction, which would be expected to be blurred given the dispersive life history of *Nerita*, and proposed that the species turnover was maintained by surface current systems, the East Australian Current (EAC) travelling south along the coast of NSW and then the east coast of Tasmania, and the Zehan Current along the west coast. These current systems remain essentially independent. Waters *et al.* (2005) and Waters (2008) did not consider the distribution of seaweeds, but a similar phylogeography might explain the maintenance of the east–west seaweed communities. Waters also strongly recommended that biogeographical analyses incorporate abundance measures, noting that historical difficulties in interpreting Australia's marine biogeography may be the result of studies relying largely or exclusively on species range data.

Phillips (2001) also examined the Flindersian seaweed flora, based on published records of species. Her analyses were primarily aimed at examining the high diversity and endemism of the southern Australian flora and to speculate on the reasons behind its establishment and maintenance. Within the extant flora, Phillips (2001) recognised four distinct elements: a southern Australian endemic element, a widely distributed temperate element, a tropical element and a polar element. The relative proportions of each of these elements varied among the Chlorophyta, Phaeophyceae and Rhodophyta, with the southern Australian endemic element and the widely distributed temperate element comprising the majority of species (Phillips 2001). The tropical element included only a small proportion of the flora, and the even smaller polar element was composed largely of species restricted geographically to the cooler coasts of southeast and southern Tasmania. Phillips's conclusions echoed those of other workers (e.g. Hommersand 1986), that the western flora has Tethyan origins and, via the southerly and easterly flowing LC and its extension the Flinders Current, was the source of much of the southern Australian seaweed flora. In contrast, the eastern flora evolved from a different ancestral stock during the Early Cenozoic, when it was separated by a land barrier from the south coast flora (Knox 1980; Hommersand 1986). This flora was also carried south by the EAC, but as this does not impinge on the south coast (Jeffrey *et al.* 1990), intermixing and homogenisation of the flora has not occurred to a great extent. Phillips further suggested that this is also consistent with prevailing theories on biotic interchange, that after a barrier between two biotas with separate histories breaks down (Vermeij 1991), the interchange of species comprises only a small proportion of the resulting biotas. Invasions are thought to be more successful in areas with recent high levels of extinction, something that has seemingly not occurred in the species-rich southern Australian flora.

The Peronian Province

The Peronian Province encompasses the eastern Australian coast from approximately the Queensland/NSW border south to just north of the Victorian border. Its boundaries are seemingly the most precisely delineated of the Australian provinces (Knox 1963; Millar 2007). Regarding the northern boundary, Knox (1963) noted a sharp transition with only a small degree of species overlap in southern Queensland, between Double Island Point (25.93°S) and Point Vernon (25.25°S), these essentially straddling Fraser Island. This boundary was based primarily on the intertidal fauna, but it was noted that the kelp *Ecklonia radiata* (C.Agardh) J.Agardh was a Peronian species absent from the Solanderian. Millar (2007) suggested that the Byron Bay area in northern NSW was possibly the northern limit of *Ecklonia*, as no authenticated vouchers of attached plants were known from Queensland, all records being based on drift specimens. Subsequently, however, an unequivocal visual record (Murray 2011) of the species in the vicinity of Moreton Island (27.134°S) has confirmed the presence of luxuriant attached plants in southern Queensland, close to Knox's northern Peronian boundary.

Knox (1963) regarded the southern boundary as less sharply defined; however, Millar (2007), based on intensive algal surveys in southern NSW, suggested the boundary between the Peronian and Flindersian Provinces (in the broad sense, incorporating the Maugean as a subprovince) lies in the area of Tathra (36.726°S), approximately 90 km north of the Victoria–NSW border. His concept was based primarily on the presence of a dominating species, the bull kelp *D. potatorum*.

While having many species in common (e.g. the kelp *E. radiata*), the flora of the east coast Peronian Province is clearly distinct from that of similar latitudes in the western Dampierian–Flindersian overlap (sometimes called the Baudinian Province) (Millar 2007; Wernberg *et al.* 2013). Congruent ecological differences are also present; as noted by Connell and Irving (2008), the urchin-mediated barrens prevalent in the Peronian Province do not occur in the Flindersian.

Millar (2007) suggested that the offshore Lord Howe (LHI) and Norfolk (NI) Islands formed a distinct and substantial link between the Peronian and Solanderian Provinces. He noted that the affinities of the LHI flora were diverse and differed with taxonomic group; the green algae showed affinities with the tropical Indo-Pacific, the brown algae with the NSW coast and the red algae with the Great Barrier Reef. Nevertheless, both islands had sufficient species in common with the Peronian to be considered a subprovince (Millar 2007: 555). Earlier, Knox (1963) had named the Phillipian and Norfolkian Provinces for, respectively, the two islands. His assessment of the LHI seaweeds was based on Lucas (1935), who noted the presence of tropical species also found on the Queensland coast, but also a distinct temperate element in the presence of species such as *Hormosira banksii* (Turner) Decne., *E. radiata* and *M. pyrifera*. Kraft (2009: 306) commented that the records of the first two (at least) appear to be genuine, but that none of these species were found during his recent extensive surveys of the island. The disappearance of *Hormosira* from LHI has been attributed to overcropping by an unidentified species of grazing fish (Lucas 1935). Both *Hormosira* and *Ecklonia* occur farther north at Norfolk, the latter only in deep water habitats (Millar 1999).

Integrated Marine and Coastal Regionalisation

The most comprehensive attempt to establish marine biogeographic zones was undertaken by the Australian Department of Environment and Heritage and resulted in IMCRA (Commonwealth of Australia 2006), this updated from Thackway and Cresswell (1997). Some 41 zones were recognised, including 24 *core provinces* and 17 *transitions*. Core provinces were based primarily on demersal fish endemicity, with the transitions the areas of overlap between the core provinces. It remains to be seen whether these provinces are of any relevance to the marine algae. The study by Waters *et al.* (2010) questioned the value of the finer-scale zonation of the IMCRA bioregions (2006), but more detailed surveys and a taxonomy incorporating molecular analysis might provide support, particularly if the greater level of endemicity suggested by recent studies (see the following section) proves to be more widespread. As indicated (Commonwealth of Australia 2006), 'Effort should be directed to collating distributional data on additional biological groups that are taxonomically stable, biogeographically informative (i.e. have large numbers of species with narrow distributional ranges) and are well represented in museum [presumably also herbarium] collections.'

Recent Methodological Advances

Australia's Virtual Herbarium

Regarding the distribution patterns of marine algae, ultimately the only reliable source of data (or at least the only one that can be revisited in light of taxonomic updates) are the physical specimens held in Australian and overseas herbaria. In 2001, access to the Australian section of this data became immediately available with the establishment of AVH, presently an online component of the *Atlas of Living Australia* (ALA), which provides the user with locality data for over 5 million plant, algal and fungal specimens stored in Australia's physical herbaria. The relative ease of access and manipulation has led to several 'data mining' exercises that have attempted to assess various aspects of seaweed biogeography based on herbarium records. Previously, these exercises were based largely on literature records, as these were the only readily available contemporaneous source of information. Both methods have limitations that must be considered when assessing results, and these will now be discussed.

Several Major Caveats

Echoing the concerns voiced by Bryan Womersley, it is clear that with all analyses based on organismal distribution, the results can only be as good as the underlying data. Excluding newly generated records based on recent surveys, Australian seaweed distributions are known through two sources: literature records and/or herbarium records. The limitations of these sources can apply to both (e.g. misidentifications, outdated taxonomy, etc.) or are unique to one or the other (e.g. incorrect locality data for herbarium records). It must also be acknowledged that most specimens in herbaria were collected for taxonomic studies and not to provide distributional records. While these specimens can be extremely useful in documenting the presence of particular species at precise locations, they can be misleading if the limitations of the underlying records are not taken into account. These limitations can be numerous, a few examples of which are as follows.

1. *Misidentifications*: The accurate identification of seaweeds is often a difficult task. Many algae display considerable morphological plasticity that can obscure their true identities. Red algae, in particular, are classified based on complex postfertilisation stages that are rarely observed and difficult to interpret. Thus, specimen identification of all but the most common species often requires considerable expertise. Unfortunately, the phycological taxonomic community in Australia – that is, those with a reasonable chance of putting the correct name to a particular specimen – has never been numerically strong. Herbaria, the storage places of specimens, have rarely employed dedicated phycologists, with much of the taxonomic research being conducted at the universities. At present, the research workforce is at a low ebb with an uncertain future, as very few students are prepared to gamble their futures on a career in systematics. Generational succession is not assured. Thus, it is often the case that the identities of herbarium specimens are not routinely checked, and many errors persist.
2. *Incorrect locality data*: Despite best efforts, locality data can be entered incorrectly. The localities of historical specimens with no attached GPS data are occasionally entered incorrectly, particularly with localities of similar names – for example, the Peron Peninsula in Shark Bay and Cape (or Point) Peron south of Perth. These mistakes are often only revealed and corrected serendipitously.
3. *Incomplete databasing*: Databasing of specimens in Australian herbaria is at varying levels of completion and progress. Omissions can undoubtedly influence data analyses and interpretation. For example, Waters *et al.* (2010) reported that the algal collection of the NSW Herbarium was not included in AVH at the time of their analysis. While they suggested that its exclusion was 'unlikely to have influenced our interpretations', NSW is the only herbarium with specimens of the bull kelp *D. potatorum* (a species regarded as indicative of the Maugean Province) from NSW. Furthermore, AVH omits the university herbaria, which includes, for

example, the extensive seaweed collections of Gerald Kraft and his associates at the University of Melbourne.

Similarly, background databases that buttress specimen databases (e.g. nomenclatural, locality) require ongoing maintenance and the incorporation of taxonomic changes if the data are to remain relevant.

4. *Misinterpretation of collection effort*: Perhaps obvious, but seemingly often not appreciated, is that collections can only be made from where botanists/phycologists have visited, and they are often selective in the places they visit and the species they choose to collect. Often, common species, or those that are too large or too small, are ignored; they take up space and facilities and require additional effort to curate.

The Molecular Revolution: One Species Becomes Several, or Many, or Hundreds

What is a species? This was never an easy question for phycologists, tasked with understanding a diverse group whose species limits were essentially untestable and relied largely on the taxonomist's experience and perception, and who were often faced with limited collections, converging morphologies and minor variations. As with all groups, the advent of molecular systematics has had a major impact on our understanding of seaweed species and their distributions (see review by Leliaert *et al.* 2014). In many cases, species thought to be cosmopolitan or widespread have been shown to consist of several species with considerably more limited distributions. On occasion these are cryptic, essentially not distinguishable without molecular characterisation. An excellent example is the recent study by Payo *et al.* (2013). These authors used molecular methods to test the then-current morphology-based assumption that the red algal genus *Portieria* included only a single, widely distributed species in the Indo-West Pacific (*P. hornemannii* (Lyngb.) P.C.Silva). They applied a DNA-based species delimitation method that resulted in the recognition of 21 species within the Philippines. Distributions were found to be highly structured, with most species restricted to island groups within the archipelago. These narrow ranges and high levels of endemism were at variance with the general perception of marine organisms being widely distributed. Payo *et al.* (2013) suggested that their results indicated that speciation in the marine environment may occur at spatial scales smaller than 100 km, comparable with some terrestrial systems. As they appropriately noted, these findings have important consequences for marine conservation and management. Equally, since the distribution patterns of these species feed into biogeographical models, such fine-scale endemism has the potential to markedly alter our perceptions. Leliaert *et al.* (2013) subsequently broadened the Payo *et al.* (2013) study to include specimens collected from additional localities in the Indo-Pacific, and reported a staggering 96 species.

The brown alga *Lobophora variegata* (J.V.Lamour.) E.C.Oliveira was also thought to be widespread in tropical and temperate seas, on most coasts in Australia except for those of cold-temperate Tasmania. The recent study by Sun *et al.* (2012), based on comparative morphology and molecular analyses, recognised considerable species-level variation among specimens that would have previously been identified as part of one morphologically variable species. They described four new species and resurrected another from synonymy. Two upright species were attributed to southern Australia (*L. nigrescens* J.Agardh and *L. australis* Z.Sun, Gurgel and H.Kawai) but the identities of the several upright and encrusting taxa known from tropical Australia are yet to be established. If estimates of species-level diversity of *Lobophora* from New Caledonia are indicative of the diversity of the species across its entire range, there might be well over 100 species of *Lobophora*. Vieira *et al.* (2014), using molecular data, recognised some 20 different lineages of *Lobophora* from New Caledonia alone, 10 of which were formally described. The remainder awaits additional sampling and further study. Although the majority of these species are thus far only known from New Caledonia, at least some are characterised by distributions that may span large areas of the Indo-Pacific Ocean. In this respect, *Lobophora* might be similar to *Padina*, another brown seaweed. Molecular studies of *Padina* (Ni-Ni-Win *et al.* 2010, 2011, 2012; Silberfeld *et al.* 2013) have revealed extensive pseudocryptic diversity, but from a biogeographic point of view many of the molecular lineages have wide-ranging distributions within the Indo-Pacific Ocean. Unlike *Portieria*, it is unclear at present whether indications of range-restricted species are artefacts resulting from incomplete sampling.

The expectation, then, is that molecular analyses will greatly increase the number of species and frequency, at least in some cases, of endemism. While this is not a universal truth – some species such as the red alga *Stylonema alsidii* (Zanardini) K.M.Drew (Zuccarello *et al.* 2008) have been shown to be truly ubiquitous – it would appear to be the most common outcome of the 'molecular revolution'.

Establishing a New Paradigm

The foregoing might justifiably strike fear into the hearts of alpha taxonomists, those who rely on morphology to identify specimens and whose work, often spanning decades, must now be revisited and checked using molecular tools. This revision of already known species is only one of several daunting tasks faced by algal taxonomists. De Clerck *et al.* (2013) estimated the magnitude of effort required before the job of describing the world's algal species is completed. They conservatively estimated the algal flora at approximately 70,000 species, of which some 44,000 have been described. At the current description rate of 150 species/year, they suggested a minimum of some 200 years, but most likely much longer, possibly 'many more centuries'. De Clerck *et al.* (2013) further noted that of the sequences lodged in GenBank, (www.ncbi.nlm.nih.gov/genbank) 90% were identified to species in 1993, this number dropping to <20% in 2010. This decline could be the result of several factors, including a disjunction between those generating sequences and those capable of identifying the source organism. Equally, faced with the increased likelihood of an incorrect identification, phycologists might avoid naming specimens from outside of known ranges, anticipating a later lodgement of 'authentic' (i.e. from type or type locality) specimens that will provide a defensible identification. Also complicating this scenario are species that can only be identified by molecular methods – the 'cryptic diversity'. As pointed out by De Clerck *et al.* (2013), when species fall into this category, linking specimens to existing names can be little more than educated guesswork.

Given the aforementioned difficulties in the seemingly simple documentation of an algal flora, what is the way forward for algal biogeography? Regional biodiversity assessments based on historical, non-molecular methods are likely to be compromised, this amplified when interregional comparisons are attempted. However, molecular tools have allowed phycologists to step beyond the simple distribution patterns that have until recently been their sole avenue for biogeographical speculation. By concentrating on individual taxa, rather than entire floras, molecular methods have allowed the recognition of complex patterns of evolution, diversification and subsequent dispersal. Several landmark studies have been published, most notably those by Heroen Verbruggen and associates, who have used molecular methods to examine the phylogeography of several green algae. The study by Verbruggen *et al.* (2009) examined the largely tropical calcified genus *Halimeda*. Their work demonstrated several examples of previously perceived 'widespread' species being considerably more geographically restricted and representing cryptic or pseudocryptic species. These results have, in part, challenged existing perceptions of Australian species. A pertinent example is that of *H. cuneata* Hering, a species originally described from South Africa and, based on convincing morphological evidence, recorded from southern Western Australia (Womersley 1956, 1984). This and other seemingly similarly distributed species have often been proposed as demonstrating a link between the two regions. The analysis by Verbruggen *et al.* (2009), however, recognised several cryptic species currently identified as *H. cuneata* (including two from South Africa). Although clearly a sister species, the Australian taxon warranted independent recognition. Verbruggen *et al.* (2009) also recognised a fourth *H. cuneata* from the Arabian Sea, the four together forming a clade and, together with molecular clock analyses, allowing reasoned speculation as to the biogeographical pathways that led to these independent but closely related species. Several other species currently named *H. cuneata* (including one from the Atlantic) resolved in entirely different branches of the phylogenetic tree, indicating that similar morphologies have convergently evolved. A comparable study was undertaken by Dixon *et al.* (2014), who used molecular evidence to challenge the previously held perception that the brown algal *Sargassum* subgenera *Arthrophycus* and *Bactrophycus* were restricted, respectively, to the Southern and Northern Hemispheres. Their analyses demonstrated that this was not the case, and they subsumed *Arthrophycus*, resulting in an entirely different perspective on the biogeography of the genus *Sargassum*.

The preceding is only a selection of the many studies that suggest adjustments to species concepts are inevitable, with the clear implication of concomitant changes to biogeographical hypotheses.

If finer-scale distribution and endemism is to be the rule rather than the exception, then a similar adjustment to biogeographic analyses might be required.

Or are we ignoring the big picture for the sake of the detail? It is true that the red algae display markedly greater molecular variation than the brown algae (for example). Widespread brown algae such as *E. radiata* are molecularly relatively uniform across their range (S. M. Boo, unpublished results) and seemingly unlike the example of *Portieria* (Payo *et al.* 2013). The fine-scale endemicity of *Portieria* is also presumably in part due to the archipelagic nature of the Philippines study area, where the opportunities for allopatric speciation are perhaps greater than on, for example, the essentially linear coastline of much of Australia. Whatever the outcome, it is clear that further taxonomic revisions will be necessary, and these might dramatically alter our concepts of Australian seaweed biogeography.

Analytical Tools

All historical and most recent attempts at biogeographical analyses of seaweeds have relied on variations of a centuries-old method, that of comparing species diversity (alpha diversity) between different regions. Generally, these data are analysed using multivariate techniques and traditional similarity indices, with the resulting biogeographical patterns therefore based solely on the common taxa within two or more lists (similarities at the taxon level), and do not take into account similarities in distribution patterns (Schils 2006). To permit a more detailed analysis of biogeographical affinities, Schils (2006) developed the tripartite biogeographical index (TBI), which is calculated for a specific taxon and incorporates several fundamental parameters of presence/absence data in grid cell (block) patterns. The TBI accounts for the relative abundance of a taxon, the average grouping of its occurrences, the average of minimal absence intervals between taxon presences and the largest coherent cluster of taxon occurrences, and also incorporates dispersal aspects.

Similarly, Price *et al.* (2006) compared a range of metrics (species richness [S], average taxonomic distinctness [Δ+, the average degree to which species are related to each other], rarity [R]) based on species records drawn from an extensive (but low-resolution) data set, the Indian Ocean catalogue of Silva *et al.* (1996). They concluded that the metrics showed strong concurrence, suggesting that, in general, a biodiversity hotspot (or coldspot) is so irrespective of which of the three measures is used. Of the 66 localities examined, the Indian Ocean coast of Australia ranked 20th in overall values; however, this did not take into account the many recent studies that have considerably increased the known diversity of the region.

A promising approach to address biogeographical questions consists of modelling the probability of dispersal, extinction and cladogenesis events in a phylogenetic context (Ree and Sanmartín 2009). So far these methods have not been applied to study biogeographic patterns of seaweeds, but with the increasing availability of molecular data it may become feasible in the near future to address flora-wide patterns using statistical biogeographical tools.

Boundary Currents

Various studies have implicated the Australian boundary currents, particularly the LC, as important factors in the distribution of seaweeds (Huisman 1997; Phillips and Huisman 2009), the supporting evidence being primarily the presence of tropical species at higher than expected latitudes. Coleman *et al.* (2011) reported detailed molecular-based connectivity studies of the common kelp *E. radiata* from three current systems of varying strengths: the EAC, the west-coast LC and its extension the south-coast Flinders Current, as listed from strongest to weakest. They found that the level of population connectivity was directly related to current strength, thereby providing strong support for the significant role played by boundary currents in the establishment and maintenance (or not) of seaweed distributions. Wernberg *et al.* (2013) analysed seaweed community structure turnover in the EAC and LC, based on herbarium records in AVH. They found that coasts influenced by the LC (with a consistent flow and weak temperature gradient) have less-variable seaweed communities and lower species turnover across regions than the EAC (with a more inconsistent flow and stronger temperature gradient). They attributed this to a combination of seaweed temperature tolerances and current-driven dispersal. These studies appear to suggest that individual species with broad temperature tolerances, such as *E. radiata*, react differently to the majority of species that comprise regional communities. Coleman *et al.* (2011) found a higher connectivity (essentially greater homogenisation) in the EAC, which they attributed to its stronger flow, whereas in the same current Wernberg *et al.* (2013) reported a high species turnover, which they attributed to its inconsistent flow and strong temperature gradient.

Indicator Species

The Australian biogeographic provinces were established based on substantial similarities/differences in species composition, initially of the molluscan fauna, but later were accepted as seemingly congruent with seaweed distributions. While we do not advocate the recognition of provinces based on the presence of single species, there are several common, habitat-forming species whose distributions align with the accepted boundaries. In Australia, the common kelp *E. radiata* (Figure 4.3), for example, occurs only within the southern temperate zones and is absent from the tropics. Similarly, the large brown seaweeds *D. potatorum* (Figure 4.4) and *M. pyrifera* (formerly *M. angustifolia*) (Figure 4.5) are found only in the cool-temperate Maugean. The common intertidal brown *H. banksii* (Figure 4.6) extends into NSW on the east coast but is not found on the Indian Ocean coast of Western Australia. Its range is therefore

FIGURE 4.3 (See colour insert.) *Ecklonia radiata*, Cape Peron, Western Australia.

FIGURE 4.4 (See colour insert.) *Durvillaea potatorum*, Lorne, Victoria.

FIGURE 4.5 (See colour insert.) *Macrocystis pyrifera*, Point Lonsdale, Victoria.

FIGURE 4.6 (See colour insert.) *Hormosira banksii*, Lorne, Victoria.

more restricted than that of *Ecklonia*, and it possibly serves as an indicator of the high latitude margins of the tropical–temperate overlap zones on both coasts. These common, habitat-forming species would appear to be ideal candidates for monitoring programs looking to assess the impacts of changing environmental conditions. In Australia, the green *C. fastigiata* (Figure 4.7) appears to be restricted to the eastern tropical Solanderian (where it is seemingly rare) and Great Barrier Reef (where it is common) Provinces. Despite extensive surveys by Huisman, this species has not been recorded from the northwest coast. Given the presence of suitable habitat and seemingly similar environmental conditions throughout tropical Australia, it would appear that other, as yet unknown, factors have excluded this species from northern and northwestern Australia.

Future Studies

Australia, for the most part, is still in the 'discovery' phase with regard to the seaweed flora. While some areas such as the south coast are reasonably well known, the recent advent of molecular analyses in taxonomy has shifted the goalposts considerably, and much of the alpha taxonomy that has consumed phycologists for the last two centuries will require testing. Understanding the biogeography of Australian seaweeds, reliant as it is on a defensible taxonomy, will in some respects always be one step behind. Thus, the continued need for ongoing floristic and taxonomic studies is self-evident.

FIGURE 4.7 (See colour insert.) *Chlorodesmis fastigiata*, Lizard Island, Queensland.

While most analyses have investigated provincial similarities based on all available species distribution data (essentially presence/absence), it is clear from the studies of Waters *et al.* (2005) and Waters (2008) that abundance measures are also of great importance. Furthermore, detailed studies of individual species or genera based on molecular analyses, such as those by Verbruggen *et al.* (2005, 2009) and Coleman *et al.* (2011), are proving to be particularly illuminating.

Tools to assist in biogeographic analyses are becoming more readily available – in particular, access to data pertaining to herbarium specimens and species distributions through AVH, barcoding to enable the identification of specimens, DNA sequencing to investigate relationships and phylogenies, and more complex statistical analyses to examine distribution patterns in greater detail. As we enter an age of possibly rapid change, these tools, if used prudently, should enable considerable progress in addressing questions of species' distributions, range shifts and extinctions, and support decision-making regarding conservation needs.

It must be stressed, however, that the bulk of the data underpinning these tools rely on the input of skilled taxonomists, and it is vitally important that the present decline in support for taxonomic research be halted. If not, much of the underlying data could eventually become compromised.

Acknowledgements

John M. Huisman and Roberta A. Cowan thank the Australian Biological Resources Study for its support.

References

Australia's Virtual Herbarium. 2014. Council of Heads of Australasian Herbaria. http://avh.chah.org.au (accessed 17 September 2014).

Belton, G. S., W. F. Prud'homme van Reine, J. M. Huisman, S. G. A. Draisma and C. F. D. Gurgel. 2014. Resolving phenotypic plasticity and species designation in the morphologically challenging *Caulerpa racemosa–peltata* complex (Chlorophyta, Caulerpaceae). *J. Phycol.* 50:32–54.

Bennett, I., and E. C. Pope. 1953. Intertidal zonation of the exposed rocky shores of Victoria, together with a rearrangement of the biogeographical provinces of temperate Australian shores. *Aust. J. Mar. Freshw. Res.* 4:105–59.

Bennett, I., and E. C. Pope. 1960. Intertidal zonation of the exposed rocky shores of Tasmania and its relationship to the rest of Australia. *Aust. J. Mar. Freshw. Res.* 11:182–221.

Breeman, A. M. 1988. Relative importance of temperature and other factors in determining geographic boundaries of seaweeds: Experimental and phenological evidence. *Helgoländer Meeresun.* 42:199–241.

Coleman, M. A., M. Roughan, H. S. Macdonald *et al.* 2011. Variation in the strength of continental boundary currents determines continent-wide connectivity in kelp. *J. Ecol.* 99:1026–32.

Commonwealth of Australia. 2006. *A Guide to the Integrated Marine and Coastal Regionalisation of Australia Version 4.0.* Canberra, Australia: Department of the Environment and Heritage.

Connell, S. D., and A. D. Irving. 2008. Integrating ecology with biogeography using landscape characteristics: A case study of subtidal habitat across continental Australia. *J. Biogeogr.* 35:1608–21.

Cotton, B. C. 1930. Fissurellidae from the 'Flindersian' Region, Southern Australia. *Rec. South Aust. Mus.* 4:219–22.

Cowan, R. A. 2006. *AMANI: Australian Marine Algal Name Index.* Canberra, Australia: Australian Biological Resources Study and Murdoch University. http://www.anbg.gov.au/amanisearch/ (Accessed 12 March 2015).

Cowan, R. A., and S. C. Ducker. 2007. A history of systematic phycology in Australia. *In Algae of Australia: Introduction*, ed. P. M. McCarthy and A. E. Orchard, pp. 1–65. Canberra, Australia: Australian Biological Resources Study.

Dampier, W. 1703. *A New Voyage to New Holland, &c in the Year 1699*. London: James Knapton.

De Clerck, O., M. D. Guiry, F. Leliaert, Y. Samyn and H. Verbruggen. 2013. Algal taxonomy: A road to nowhere? *J. Phycol.* 49:215–25.

Dixon, R. R. M., L. Mattio, J. M. Huisman, C. E. Payri, J. J. Bolton and C. F. Gurgel. 2014. North meets south: Taxonomic and biogeographic implications of a phylogenetic assessment of Sargassum subgenera *Arthrophycus* and *Bactrophycus*. *Phycologia* 53:15–22.

Ducker, S. C. 1967. The genus *Chlorodesmis* (Chlorophyta) in the Indo-Pacific region. *Nova Hedwigia* 13:145–82.

Ducker, S. C. 1979. History of Australian phycology: The significance of early French explorations. *Brunonia* 2:19–42.

Ducker, S. C. 1981a. Australian phycology: The German influence. In *People and Plants in Australia*, ed. D. J. Carr and S. G. M. Carr, pp. 116–38. Sydney, Australia: Academic Press.

Ducker, S. C. 1981b. History of Australian phycology: Early German collectors and botanists. In *History in the Service of Systematics*, ed. A. Wheeler and J. H. Price, pp. 43–51. London: Society for the Bibliography of Natural History, Special Publication no. 1.

Ducker, S. C. 1981c. A history of Australian marine phycology. In *Marine Botany: An Australasian perspective*, ed. M. N. Clayton and R. J. King, pp. 1–14. Melbourne, Australia: Longman Cheshire.

Ducker, S. C. 1983. Port Phillip Heads: A phycological saga. *Phycologia* 22:431–43.

Ducker, S. C. 1986. Lucas, Arthur Henry Shakespeare, ed. B. Nairn and G. Serle, (1853–1936). In *Australian Dictionary of Biography*, Vol. 10, pp. 163–4. Parkville, Australia: Melbourne University Press.

Ducker, S. C. 1988. *The Contented Botanist: Letters of W.H. Harvey about Australia and the Pacific*. Melbourne, Australia: Melbourne University Press.

Endean, R. 1957. The biogeography of Queensland's shallow-water echinoderm fauna (excluding Crinoidea), with a rearrangement of the faunistic provinces of tropical Australia. *Aust. J. Mar. Freshw. Res.* 8:233–73.

Endean, R., W. Stephenson and R. Kenny. 1956. The ecology and distribution of intertidal organisms on certain islands off the Queensland coast. *Aust. J. Mar. Freshw. Res.* 7:317–42.

Harvey, W. H. 1855. Some account of the marine botany of the colony of Western Australia. *Trans. Roy. Irish Acad.* 22(Science):525–66.

Harvey, W. H. 1858. *Phycologia Australica*, Vol. 1. London: Lovell Reeve.

Harvey, W. H. 1859. *Phycologia Australica*, Vol. 2. London: Lovell Reeve.

Harvey, W. H. 1860. *Phycologia Australica*, Vol. 3. London: Lovell Reeve.

Harvey, W. H. 1862. *Phycologia Australica*, Vol. 4. London: Lovell Reeve.

Harvey, W. H. 1863. Phycologia Australica, Vol. 5. London: Lovell Reeve.

Hedley, C. 1904. The effect of the Bassian Isthmus upon the existing marine fauna: A study in ancient geography. *Proc. Linn. Soc. New South Wales* 28:876–83.

Hedley, C. 1926. Zoogeography. In *The Australian Encyclopaedia*, Vol. 2, M to Z, ed. A. W. Jose and H. J. Carter, pp. 743–4. Sydney, Australia: Angus and Robertson.

Hommersand, M. H. 1986. The biogeography of the South African marine red algae: A model. *Bot. Mar.* 29:257–70.

Hommersand, M. H. 2007. Global biogeography and relationships of the Australian marine macroalgae. In *Algae of Australia: Introduction*, ed. P. M. McCarthy and A. E. Orchard, pp. 511–42. Canberra, Australia: Australian Biological Resources Study.

Huisman, J. M. 1997. Marine nenthic algae of the Houtman Abrolhos Islands, Western Australia. In *The Marine Flora and Fauna of the Houtman Abrolhos Islands, Western Australia*, ed. F. E. Wells, pp. 177–237. Perth, Australia: Western Australian Museum.

Huisman, J. M. 2001. *Echinophycus minutus* (Rhodomelaceae, Ceramiales), a new red algal genus and species from northwestern Australia. *Phycol. Res.* 49:177–82.

Huisman, J. M. 2002. The type and Australian species of the red algal genera *Liagora* and *Ganonema* (Liagoraceae, Nemaliales). *Aust. Syst. Bot.* 15:773–838.

Huisman, J. M. 2004. Marine benthic flora of the Dampier Archipelago, Western Australia. *Rec. Western Australian Mus.* Suppl. 66:61–8.

Huisman, J. M. 2007. The Dampierian Province. In *Algae of Australia: Introduction*, ed. P. M. McCarthy and A. E. Orchard, pp. 543–49. Canberra and Melbourne: Australian Biological Resources Study and CSIRO Publishing.

Huisman, J. M. 2015. *Algae of Australia: Marine Benthic Algae of North-Western Australia; 1. Green and Brown Algae*. Canberra, Australia: Australian Biological Resources Study.

Huisman, J. M. *(in prep.)*. *Algae of Australia: Marine Benthic Algae of North-Western Australia; 2. Red Algae.* Canberra, Australia: Australian Biological Resources Study.

Huisman, J. M., and M. A. Borowitzka. 2003. Marine benthic flora of the Dampier Archipelago, Western Australia. In *The Marine Flora and Fauna of Dampier, Western Australia*, ed. F. E. Wells, D. I. Walker and D. S. Jones, pp. 291–344. Perth, Australia: Western Australian Museum.

Huisman, J. M., R. A. Cowan and T. J. Entwisle. 1998. Biodiversity of Australian marine macroalgae: A progress report. *Bot. Mar.* 41:89–93.

Huisman, J. M., O. De Clerck, W. F. Prud'homme van Reine and M. A. Borowitzka. 2011. *Spongophloea*, a new genus of red algae based on *Thamnoclonium* sect. *Nematophorae* Weber-van Bosse (Halymeniales). *Eur. J. Phycol.* 46:1–15.

Huisman, J. M., and F. Leliaert. 2015. Cladophorales. In *Algae of Australia: Marine Benthic Algae of North-Western Australia; 1. Green and Brown Algae*, pp. 32–67. Canberra, Australia: Australian Biological Resources Study.

Huisman, J. M., F. Leliaert, H. Verbruggen and R. A. Townsend. 2009. Marine benthic plants of Western Australia's shelf-edge atolls. *Rec. Western Australian Mus.* Suppl. 77:50–87.

Huisman, J. M., and A. J. K. Millar. 2013. Australian seaweed collections: Use and misuse. *Phycologia* 52:2–5.

Huisman, J. M., and A. J. K. Millar. 2015. Australian seaweed collections: Huisman and Millar respond. *Phycologia* 54:32–4.

Huisman, J. M., and D. I. Walker. 1990. A catalogue of the marine plants from Rottnest Island, Western Australia, with notes on their distribution and biogeography. *Kingia* 1:365–481.

Iredale, T., and W. L. May. 1916. Misnamed Tasmanian chitons. *Proc. Malac. Soc. Lond.* 12:94–117.

Jeffrey, S. W., D. J. Rochford and G. R. Cresswell. 1990. Oceanography of the Australasian Region. In *Biology of Marine Plants*, ed. M. N. Clayton and R. J. King, pp. 243–65. Melbourne, Australia: Longman Cheshire.

Kerswell, A. P. 2006. Global biodiversity patterns of benthic marine algae. *Ecology* 87:2479–88.

Knox, G. A. 1963. The biogeography and intertidal ecology of the Australasian coasts, *Oceanogr. Mar. Biol. Ann. Rev.* 1:341–404.

Knox, G. A. 1980. Plate tectonics and evolution of intertidal and shallow water benthic biota distribution patterns of the south west Pacific. *Palaeogeogr. Palaeoclimatol. Palaeoecol.* 31:267–97.

Kott, P. 1952. The ascidians of Australia: 1. Stolidobranchiata Lahille and Phlebobranchiata Lahille. *Aust. J. Mar. Freshw. Res.* 3:205–335.

Kraft, G. T. 2000. Marine and estuarine benthic green algae (Chlorophyta) of Lord Howe Island, southwestern Pacific. *Aust. Syst. Bot.13*: 509–648.

Kraft, G. T. 2007. *Algae of Australia: Marine Benthic Algae of Lord Howe Island and the Southern Great Barrier Reef; 1. Green Algae*. Canberra, Australia: Australian Biological Resources Study.

Kraft, G. T. 2009. *Algae of Australia: Marine benthic algae of Lord Howe Island and the southern Great Barrier Reef; 2. Brown algae*. Canberra, Australia: Australian Biological Resources Study.

Kraft, G. T. 2011. In memoriam: Tribute to Professor Hugh Bryan Spencer Womersley (19 November 1922–16 January 2011). *Phycologia* 50:439–41.

Kraft, G. T., L. M. Liao, A. J. K. Millar, E. G. G. Coppejans, M. H. Hommersand and D. Wilson Freshwater. 1999. Marine benthic red algae (Rhodophyta) from Bulusan, Sorsogon Province, Southern Luzon, Philippines. *Philipp. Sci.* 36:1–50.

Lambeck, K., and J. Chappell. 2001. Sea level change through the last glacial cycle. *Science* 292:679–86.

Leliaert, F., J. M. Huisman and E. Coppejans. 2007. Phylogenetic position of *Boodlea vanbosseae* (Siphonocladales, Chlorophyta). *Cryptog. Algol.* 28:337–51.

Leliaert, F., D. A. Payo and O. De Clerck. 2013. The geographic scale of speciation in the marine red alga *Portieria*: 10th International Phycological Congress, Orlando, Florida; Abstract 159. *Phycologia* 52 (Suppl.):60–1.

Leliaert, F., H. Verbruggen, P. Vanormelingen *et al.* 2014. DNA-based species delimitation in algae. *Eur. J. Phycol.* 49:179–96.

Lucas, A. H. S. 1935. The marine algae of Lord Howe Island. *Proc. Linn. Soc. New South Wales* 60:194–232.

May, V. 1940. A comparison between marine floras. *Contr. New South Wales Natl Herb.* 1:94–8.

McCarthy, P. M., and A. E. Orchard (eds.). 2007. *Algae of Australia: Introduction*. Canberra, Australia: Australian Biological Resources Study.

Millar, A. J. K. 1990. Marine red algae of the Coffs Harbour region, northern New South Wales. *Aust. Syst. Bot.* 3:293–593.

Millar, A. J. K. 1999. Marine benthic algae of Norfolk Island, South Pacific. *Aust. Syst. Bot.* 12:479–547.

Millar, A. J. K. 2007. The Flindersian and Peronian Provinces. In *Algae of Australia*: *Introduction*, ed. P. M. McCarthy and A. E. Orchard, pp. 554–9. Canberra, Australia: Australian Biological Resources Study.

Murray, B. 2011. Diving at Cherubs Cave. https://www.youtube.com/watch?v=1pg13iCGodY (accessed 2 October 2014).

Ni-Ni-Win, T. Hanyuda, S. Arai *et al.* 2010. Four new species of *Padina* (Dictyotales, Phaeophyceae) from the western Pacific Ocean, and reinstatement of *Padina japonica*. *Phycologia* 49:136–53.

Ni-Ni-Win, T. Hanyuda, S. Arai *et al.* 2011. A taxonomic study of the genus *Padina* (Dictyotales, Phaeophyceae) including the descriptions of four new species from Japan, Hawaii, and the Andaman Sea. *J. Phycol.* 47:1193–209.

Ni-Ni-Win, T. Hanyuda, S. G. A. Draisma *et al.* 2012. Morphological and molecular evidence for two new species of *Padina* (Dictyotales, Phaeophyceae), *P. sulcata* and *P. calcarea*, from the central Indo-Pacific region. *Phycologia* 51:576–85.

Payo, D. A., F. Leliaert, H. Verbruggen, S. D'hondt, H. P. Calumpong and O. De Clerck. 2013. Extensive cryptic species diversity and fine-scale endemism in the marine red alga *Portieria* in the Philippines. *Proc. Roy. Soc. London B Biol. Sci.* 280:20122660.

Phillips, J. A. 2001. Marine macroalgal biodiversity hotspots: Why is there high species richness and endemism in southern Australian marine benthic flora? *Biodivers. Conserv.* 10:1555–77.

Phillips, J. A. 2007. The Solanderian Province. In *Algae of Australia*: *Introduction*, ed. P. M. McCarthy and A. E. Orchard, pp. 550–3. Canberra, Australia: Australian Biological Resources Study.

Phillips, J. A., C. Conacher and J. Horrocks. 1999. Marine macroalgae of the Gulf of Carpentaria, northern Australia. *Aust. Syst. Bot.* 12:449–78.

Phillips, J. C., and J. M. Huisman. 2009. Influence of the Leeuwin Current on the Marine Biota of the Houtman Abrolhos. *J. Roy. Soc. Western Australia* 92:139–46.

Price, A. R. G., L. P. A. Vincent, A. J. Venkatachalam, J. J. Bolton and P.W. Basson. 2006. Concordance between different measures of biodiversity in Indian Ocean macroalgae. *J. Exp. Mar. Biol. Ecol.* 19:85–91.

Ree, R. H., and I. Sanmartin. 2009. Prospects and challenges for parametric models in historical biogeographical inference. *J. Biogeogr.* 36:1211–20.

Saunders, G. W. 2010. ALGA-iBOL's Algal Life Global Audit: Muddled morphologies & molecular mayhem in the topsy-turvy world of algal floristics. Australasian Society for Phycology and Aquatic Botany conference, Rottnest Island, Western Australia 15–18 November: 49.

Schils, T. 2006. The tripartite biogeographical index: A new tool for quantifying spatio-temporal differences in distribution patterns. *J. Biogeogr.* 33:560–72.

Silberfeld, T., L. Bittner, C. Fernández-García *et al.* 2013. Species diversity, phylogeny and large scale biogeographic patterns of the genus *Padina* (Phaeophyceae, Dictyotales). *J. Phycol.* 49:130–42.

Silva, P. C., P. W. Basson and R. L. Moe. 1996. Catalogue of the benthic marine algae of the Indian Ocean. *Univ. Calif. Publ. Bot.* 79:1–1259.

Smale, D. A., and T. Wernberg. 2013. Extreme climatic event drives range contraction of a habitat-forming species. *Proc. Roy. Soc. B* 280:20122829.

Sonder, O. G. 1871. Die Algen des tropischen Australiens. *Abh. Geb. Naturw. Hamburg* 5:33–74.

Sun, Z., T. Hanyuda, P.E. Lim, J. Tanaka, C. F. D. Gurgel and H. Kawai. 2012. Taxonomic revision of the genus *Lobophora* (Dictyotales, Phaeophyceae) based on morphological evidence and analyses [of] *rbc*L and *cox3* gene sequences. *Phycologia* 51:500–12.

Telford, N. 1998. Is there a distinct Dampierian algal flora? Thesis, B.Sc. Hons., University of Western Australia, Perth, Australia.

Thackway, R., and G. R. Cresswell. 1997. *Interim Marine and Coastal Regionalisation for Australia: An Ecosystem-Based Classification for Marine and Coastal Environments*, Version 3.3. Canberra, Australia: Environment Australia.

Verbruggen, H., O. De Clerck, T. Schils, W. H. C. F. Kooistra and E. Coppejans. 2005. Evolution and phylogeography of *Halimeda* section *Halimeda* (Bryopsidales, Chlorophyta). *Mol. Phylogenet. Evol.* 37:789–803.

Verbruggen, H., L. Tyberghein, K. Pauly *et al.* 2009. Macroecology meets macroevolution: Evolutionary niche dynamics in the marine green alga Halimeda. *Global Ecol. Biogeogr.* 18:393–405.

Vermeij, G. T. 1991. When biotas meet: Understanding biotic interchange. *Science* 253:1099–104.

Vieira, C., S. D'Hondt, O. De Clerck and C. E. Payri. 2014. Toward an inordinate fondness for stars, beetles and Lobophora? Species diversity of the genus *Lobophora* (Dictyotales, Phaeophyceae) in New Caledonia. *J. Phycol.* 50:1101–19.

Waters, J. M. 2008. Marine biogeographic disjunction in temperate Australia: Historic landbridge, contemporary currents, or both? *Divers. Distrib.* 14:692–700.

Waters, J. M., T. M. King, P. M. O'Loughlin and H. G. Spencer. 2005. Phylogeographic disjunction in abundant high-dispersal littoral gastropods. *Mol. Ecol.* 14:2789–802.

Waters, J. M., T. Wernberg, S. D. Connell *et al.* 2010. Australia's marine bioregions revisited: Back to the future? *Austral. Ecol.* 35:988–92.

Wells, F. E. 1997. Introduction to the marine environment of the Houtman Abrolhos Islands, Western Australia. In *The Marine Flora and Fauna of the Houtman Abrolhos Islands, Western Australia*, ed. F. E. Wells, pp. 1–10. Perth, Australia: Western Australian Museum.

Wernberg, T., B. D. Russell, C. J. A. Bradshaw *et al.* 2014. Misconceptions about analyses of Australian seaweed collections. *Phycologia* 53:215–20.

Wernberg, T., B. D. Russell, M. S. Thomsen *et al.* 2011. Seaweed communities in retreat from ocean warming. *Current Biol.* 21:1828–32.

Wernberg, T., M. S. Thomsen, S. D. Connell *et al.* 2013. The footprint of continental-scale ocean currents on the biogeography of seaweeds. *PLOS One* 8:e80168.

Westera, M. A., J. C. Phillips, G. T. Coupland, A. J. Grochowski, E. S. Harvey and J. M. Huisman. 2009. Sea surface temperatures of the Leeuwin Current in the Capes region of Western Australia: Potential effects on the marine biota of shallow reefs. *J. Roy. Soc. of Western Australia* 92: 97–210.

Whitley, G. P. 1932. Marine zoogeographical regions of Australasia. *Aust. Nat.* 8:166–7.

Womersley, H. B. S. 1956. A critical survey of the marine algae of southern Australia: I. Chlorophyta. *Aust. J. Mar. Freshw. Res.* 7:343–83.

Womersley, H. B. S. 1958. Marine algae from Arnhem Land, North Australia. In *Records of the American–Australian Scientific Expedition to Arnhem Land*, Vol. 3., ed. R. L. Specht and C. P. Mountford, pp. 139–61. Melbourne, Australia: Melbourne University Press.

Womersley, H. B. S. 1959. The marine algae of Australia. *Bot. Rev.* 25:545–614.

Womersley, H. B. S. 1981a. Aspects of the distribution and biology of Australian marine macro-algae. In *The Biology of Australian Plants*, ed. J. S. Pate and A. J. McComb, pp. 294–306. Nedlands, Australia: University of Western Australia Press.

Womersley, H. B. S. 1981b. Biogeography of Australasian marine macroalgae. In *Marine Botany: An Australasian Perspective*, ed. M. N. Clayton and R. J. King, pp. 292–307. Melbourne, Australia: Longman Cheshire.

Womersley, H. B. S. 1984. *The Marine Benthic Flora of Southern Australia*, Part I. Adelaide, Australia: Government Printer, South Australia.

Womersley, H. B. S. 1987. *The Marine Benthic Flora of Southern Australia*, Part II. Adelaide, Australia: South Australian Government Printing Division.

Womersley, H. B. S. 1994. *The Marine Benthic Flora of Southern Australia*, Part IIIA. Canberra, Australia: Australian Biological Resources Study.

Womersley, H. B. S. 1996. *The Marine Benthic Flora of Southern Australia*, Part IIIB. Canberra, Australia: Australian Biological Resources Study.

Womersley, H. B. S. 1998. *The Marine Benthic Flora of Southern Australia*, Part IIIC. Canberra, Australia: Australian Biological Resources Study.

Womersley, H. B. S. 2003. *The Marine Benthic Flora of Southern Australia*, Part IIID. Canberra, Australia: Australian Biological Resources Study.

Womersley, H. B. S., and S. J. Edmonds. 1958. A general account of the intertidal ecology of South Australian coasts. *Aust. J. Mar. Freshw. Res.* 9:217–60.

Wynne, M. J., and J. Luong-Van Thinh. 1997. A report on collections of benthic marine algae from Darwin, Northern Australia. In *Trochus: Status, Hatchery Practice and Nutrition*. Proceedings of a workshop held at Northern Territory University, 6–7 June 1996, ed. C. L. Lee and P. W. Lynch, pp. 81–7. Canberra, Australia: Australian Centre for International Agricultural Research, Proceedings no. 79.

Zuccarello, G. C., J. A. West and N. Kikuchi. 2008. Phylogenetic relationships within the Stylonematales (Stylonematophyceae, Rhodohyta): Biogeographic patterns do not apply to Stylonema alsidii. *J. Phycol.* 44:384–93.

5

Biogeography of Australian Marine Invertebrates

Shane T. Ahyong

CONTENTS

Introduction

'Anyone who has studied Australian shells, crabs, fishes, or other marine creatures, soon realises that these animals are not uniformly distributed throughout our seas, but restricted to certain areas, governed by fairly definite conditions of environment, temperature, and so on' – so observed Gilbert Whitley in 1932. Such observations drive biogeography, the study of how species are distributed in space and time. In Australia, the marine environment is extensive. At 9 million km^2, the Australian Exclusive Economic Zone is one of the largest in the world. Mainland Australia has a coastline of 36,000 km and spans 35° of latitude and 40° of longitude, separating the two largest ocean basins in the world (Indian and Pacific) and reaching from the tropics to almost the subantarctic. Not surprisingly, the Australian marine invertebrate fauna is massively diverse, with more than 25,000 known species (Butler *et al.* 2010) (some representatives are shown in Figure 5.1).

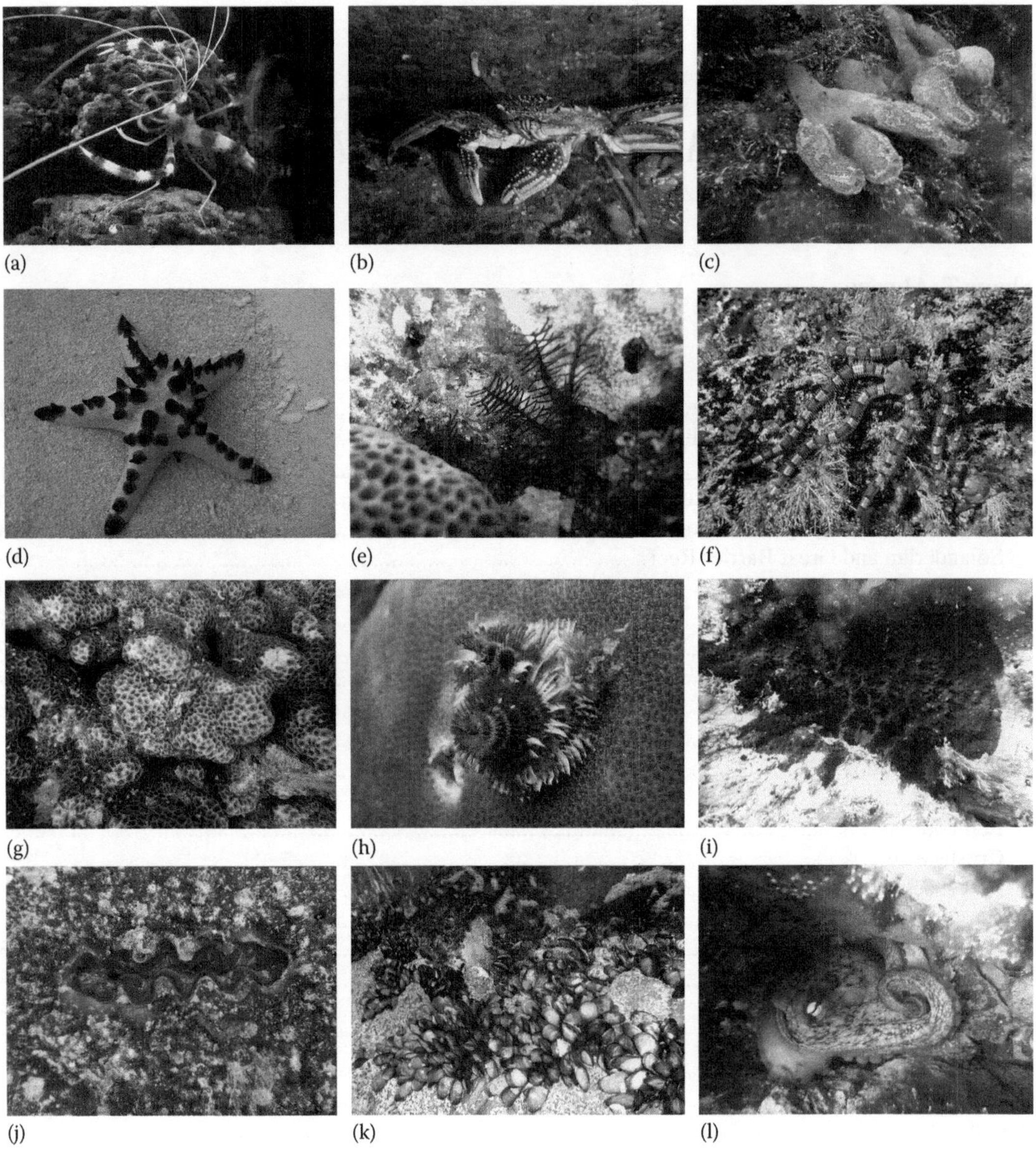

FIGURE 5.1 (See colour insert.) Representative marine invertebrates: (a) Crustacea, shrimp (*Stenopus hispidus*); (b) Crustacea, crab (*Guinusia chabrus*); (c) Tunicata, ascidian (*Botrylloides anceps*); (d) Echinodermata, asteroid (*Protoreaster nodosus*); (e) Echinodermata, crinoid among scleractinian corals and sponge; (f) Echinoderm, ophiuroid (*Ophionereis schayeri*); (g) Cnidaria, scleractinian coral; (h) Polychaeta, Christmas tree worm (*Spirobranchus giganteus*) on *Porites* coral; (i) Porifera, encrusting sponge; (j) Mollusca, burrowing clam (*Tridacna crocea*); (k) Mollusca, mussels (*Mytilus* sp.); (l) Mollusca, octopus (*Octopus* sp.). Courtesy of S. Ahyong.)

Geological Setting and Relationships of Fauna

The current distributions of marine invertebrate species are the result of both contemporary physical-ecological processes and the effects of events in geological history, the former being the domain of ecological biogeography and the latter of historical biogeography. Although the overall latitudinal distributions of modern species are consistent with present-day environmental variables in strong north–south faunal differences, these differences have their roots in the Cretaceous during the breakup of Gondwana. During the Cretaceous, the northern supercontinent Laurasia and southern supercontinent Gondwana were separated by the equatorial Tethys Sea, within which a tropical fauna evolved. The present-day western margin of Australia essentially formed part of the northern Gondwanan coastline bounding the southern Tethys. Conversely, in Early Cretaceous times, present-day eastern Australia was much farther south, contiguous with eastern Antarctica, developing a separate Austral fauna. The sequential breakup of Gondwana saw the separation of Australia from Zealandia about 80 Ma and the beginnings of the break of Australia from Antarctica about 70 Ma. Australia's break from Antarctica started in what is now southwestern Australia, with the progressive opening of a north-facing seaway that increasingly exposed the newly forming 'southern' Australian coastline to a tropical Tethyan biota (Crame 1999). By 40 Ma, Tasmania had also separated from eastern Antarctica, allowing the development of the southern circumpolar current. Australia's northwards movement and rotation towards its present position exposed more of the now northern coastline to warm-water Tethyan faunas, and prompted the transition of the formerly warm southwest towards a cooler-water palaeoaustral fauna that evolved in the southeast. Palaeoaustral remnants, especially in southeast temperate Australia, persist to this day. The collision of Australia with Southeast Asia about 20 Ma during the Miocene finally merged the northern Australian fauna and tropical Indo-West Pacific fauna, whose origin was in the Tethys. Thus, although the distribution of the Australian marine modern biota is governed by present-day physical-ecological conditions, its composition nevertheless carries the distinct signature of the past.

Australia in the Indo-West Pacific

Ekman's (1953) classic *Zoogeography of the Sea* provided a global-scale biogeographic survey of the world's oceans based on the then known distributions of all marine species, designating regions and provinces. Ekman recognised two major warm-water shelf regions: the Atlanto-East Pacific, taking in the Atlantic Ocean and eastern Pacific, and the Indo-West Pacific, spanning the Indian Ocean and the western half of the Pacific Ocean (essentially from Hawaii and French Polynesia to Australia). The Indo-West Pacific is the largest and most speciose warm-water shelf region in the world, and Australia lies almost at its geographic centre. The highest diversity in the Indo-West Pacific is concentrated in the area roughly bounded by the Philippines, Indonesia and New Guinea, the so-called Coral Triangle (Lohman *et al.* 2011). Australia's northern coastline lies along the southern margin of the Coral Triangle, upholding strong faunal affinities. As recognised by Ekman (1953), the northern Australian fauna can thus be understood as an extension of the wider, vast Indo-West Pacific fauna. Ekman (1953) recognised the distinctiveness of the southern Australian fauna as warm temperate in comparison with that of the tropical Indo-West Pacific. He also observed, however, important distinctions between the southern Australian fauna and those of other widely separated southern continents, apparently the result of geological history (noting that plate tectonics were not yet known in Ekman's day). Briggs (1974) reached similar conclusions to those of Ekman, albeit identifying the Tasmanian fauna more closely with the cool-temperate fauna of New Zealand. Although in close proximity to Australia, the marine fauna of New Zealand is largely biogeographically separate, albeit sharing some temperate-water species with southeastern Australia (including Tasmania) (Knox 1960, 1979). In fact, compared with southern Australia, the marine fauna of southern New Zealand and its subantarctic islands appears to generally have more in common with the cold-water Southern Ocean faunas of southern South America and the Indian Ocean subantarctic islands, such as Kerguelen and the Crozets (Crame 1996; Gorny 1999), areas significantly influenced by the West Wind Drift and polar front (De Broyer and Koubi 2014). Northeastern Australia and northern New Zealand also have some species in common, but these elements are part of the wider tropical Indo-West Pacific fauna.

Despite the distinctions between different schemes, the modern Australian marine biota is widely acknowledged as broadly comprising a northern tropical component with strong affinities to the wider tropical Indo-West Pacific fauna and low levels of species endemism, and a southern temperate component with high species endemism.

North and South Biogeographic Regions

A northern (tropical) and southern (temperate) marine fauna has long been recognised across most marine groups (e.g. Gill 1885; Ekman 1953; Wilson and Gillett 1971; Wilson and Allen 1987). As expected, the northern fauna has affinities in the wider tropical Indo-West Pacific and includes a high proportion of widespread genera and species. The southern temperate fauna includes a higher proportion of endemic species and genera, up to 95% for groups such as molluscs (Wilson and Allen 1987). The separation of tropical from temperate faunas, however, is not sharply demarcated on either side of the continent. Instead, there are broad zones of tropical–temperate overlap on both coasts, characterised by mixed tropical and temperate species. Wilson and Gillett (1971) and Wilson and Allen (1987) referred to these faunal regions as the *northern Australian region* and *southern Australian region*, with eastern and western overlap zones (Figure 5.2f). The eastern temperate Peronian region has been recognised as a region of tropical–temperate overlap (Millar 2007), and several earlier works observed a similar situation in southwestern Australia (e.g. Kott 1952; Knox 1963) (Figure 5.2). Recent large-scale analyses of distributions of deep-water species on the west coast have also corroborated the southern Western Australian coast as a transitional zone of mixed tropical and temperate species (McEnnulty *et al.* 2011). A degree of east–west symmetry is achieved in terms of the penetration of tropical species down both eastern and western coastlines owing to southern-flowing boundary currents on each side of the continent: the East Australian Current in the east and the Leeuwin Current in the west.

'Traditional' Biogeographic Regions

The general validity of the broad northern and southern faunal distinction is undisputed, but more finely constrained biogeographic regions have a long history of use (Figures 5.2a–d and 5.3a). These 'traditional' biogeographic regions developed by Hedley (1904), Iredale and May (1916), Whitley (1932), Bennett and Pope (1953), Knox (1963) and others are derived from qualitative consideration of the distributions of primarily shallow-water molluscs, fishes, echinoderms and algae. Interestingly, the common distribution patterns of shallow-water benthic species summarised by Wilson and Allen (1987) (Figure 5.2f) approximate the traditional regions in many respects. The traditional regions have been described under numerous names and variations on theme, variously under the titles *province*, *region* or *subregion*, but five have prominence: Solanderian, Peronian, Maugean, Flindersian and Dampierian (Ebach *et al.* 2013) (Figure 5.3a).

Solanderian and Great Barrier Reef

As proposed by Hedley (1904), who was working with molluscs, and as recently diagnosed by Ebach *et al.* (2013), the Solanderian region is the coastal region of tropical eastern Australia extending from the Torres Strait to southern Queensland near the border of New South Wales, equivalent to the Banksian Province of Whitley (1932) and the NE Coast zone of Hoese *et al.* (2006) based on fish distributions. The offshore region, the Great Barrier Reef, was also recognised (although confusingly termed Solanderian by Whitley [1932]), distinguished from the coastal onshore fauna. Endean (1957) highlighted the strong distinction between the coastal and offshore faunas, noting 44% of shallow-water Queensland echinoderm species were confined to mainland coastal waters and approximately 35% to Great Barrier Reef waters. Endean (1957), however, did not accept a separate eastern and western tropical coastal fauna, merging the coastal Queensland fauna with that of the northwestern coastal fauna (Dampierian) into the Tropical Australian Province.

Peronian

First proposed by Hedley (1904), the Peronian region takes in temperate eastern Australia and extends from Gippsland (Victoria) to Moreton Bay (Queensland). Within the bounds of the Peronian region,

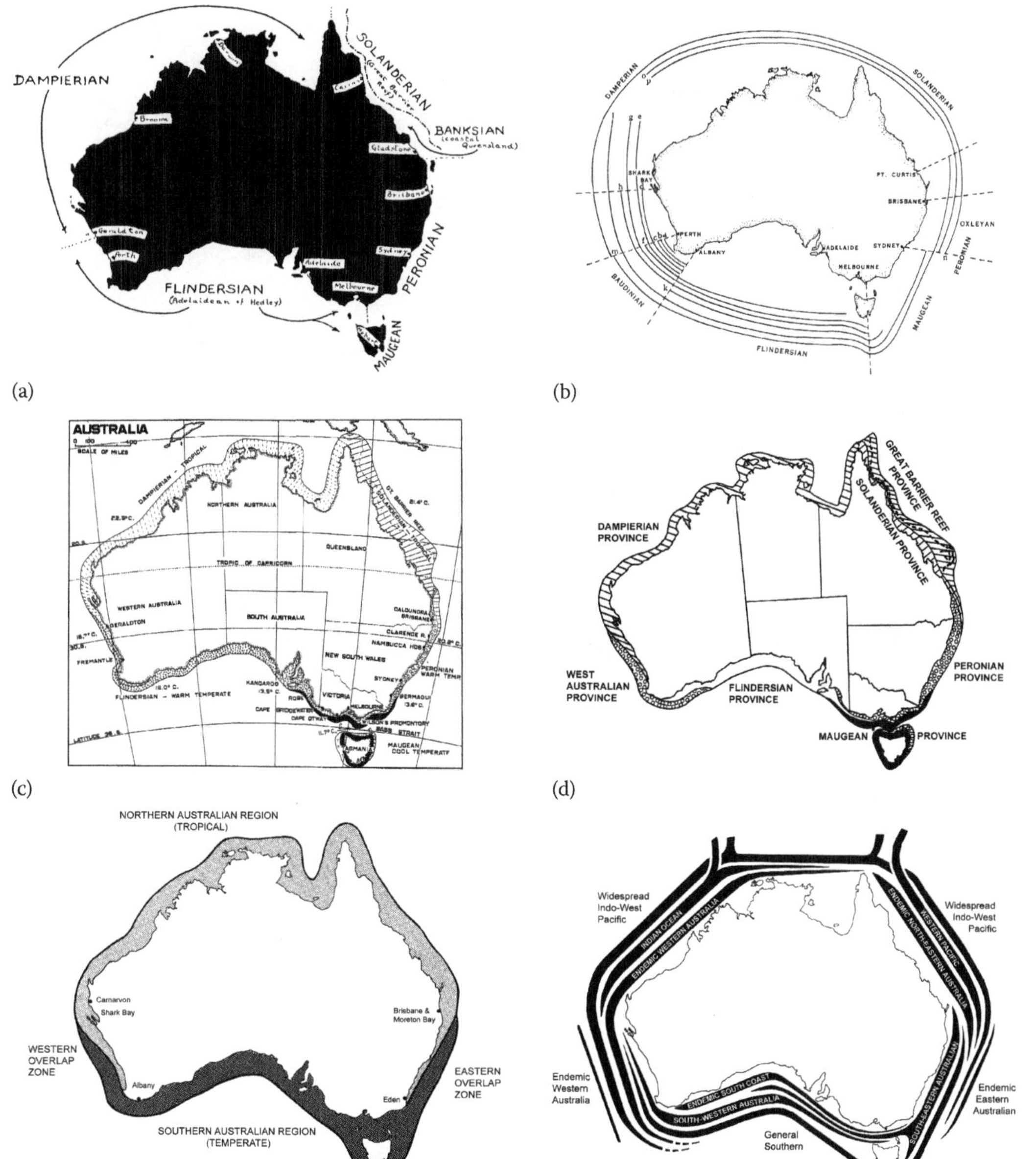

FIGURE 5.2 'Evolution' of marine biogeographical regions in Australia: (a) Whitley (1932); (b) Kott (1952); (c) Bennett and Pope (1953); (d) Knox (1963); (e–f) Wilson and Allen (1987). (Modified from Whitley, G.P., *Australian Naturalist* 8(8), 166–167, 1932: unnumbered figure; Kott, P., *Australian Journal of Marine and Freshwater Research* 3, 205–334, 1952: figure 182; Bennett, I., and E.C. Pope, *Australian Journal of Marine and Freshwater Research* 4(1), 105–159, 1953: figure 5; Knox, G.A., *Oceanography and Marine Biology: An Annual Review* 1, 341–404, 1963: figure 5; Wilson, B.R., and G.R. Allen, *Fauna of Australia General Articles*, Australian Government Publishing Service, Canberra, Australia, 1987: figures 3.8 and 3.13, respectively.)

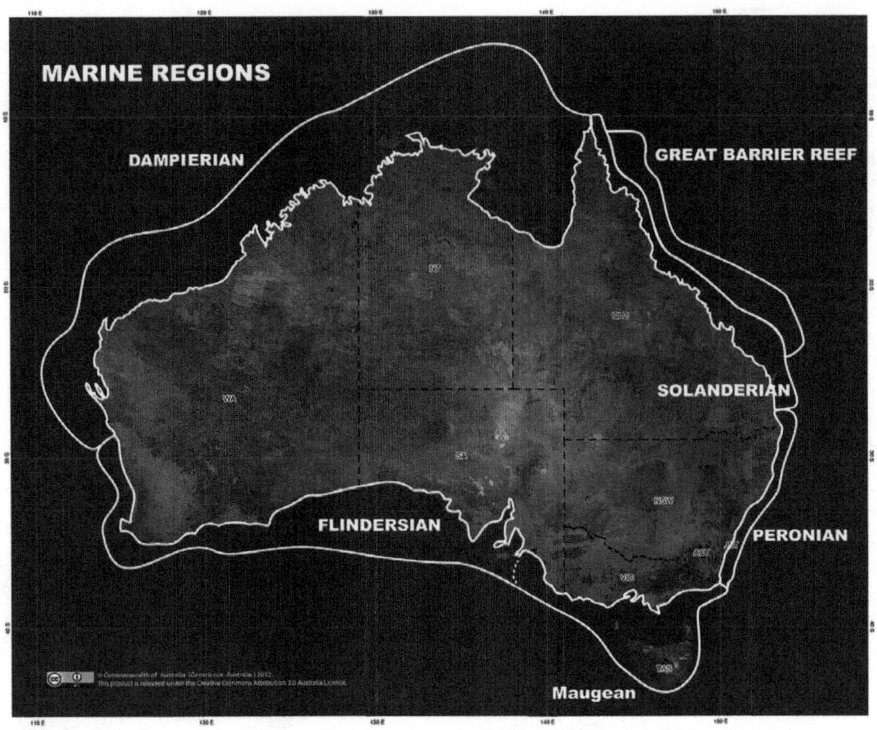

(a)

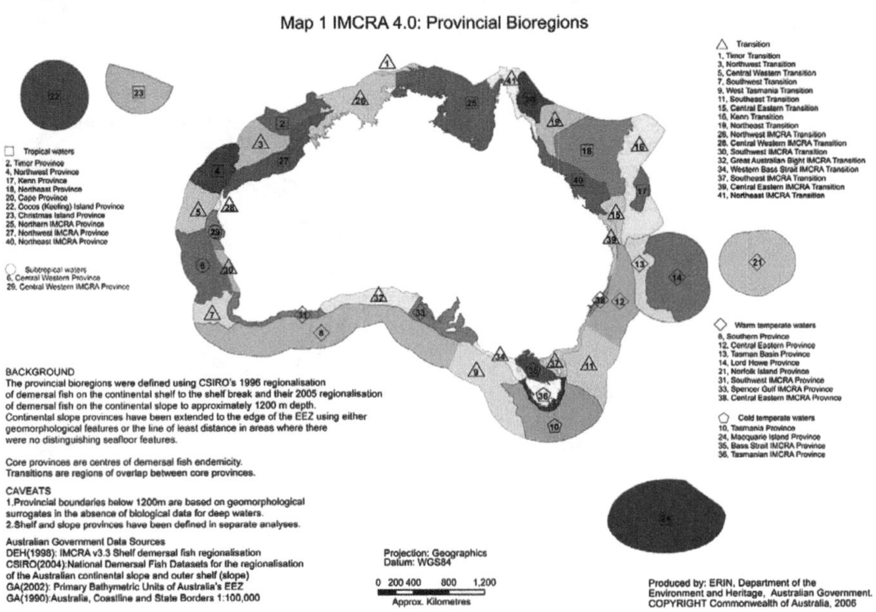

(b)

FIGURE 5.3 (See colour insert.) Recent biogeographic schema: (a) marine biogeographical regions as recognised by Ebach *et al.* (2013) (modified from Ebach, M.C., *et al.*, *Zootaxa* 3619[3], 315–342, 2013: figure 6); (b) IMCRA version 4 (Department of Environment, Australian Government).

Iredale (1937) introduced the Oxleyan subregion, followed by Kott (1952). However, as indicated by Millar (2007) and Ebach *et al.* (2013), the Peronian region is essentially a transitional zone between the tropical northeast and temperate southeast, so an Oxleyan subdivision does not seem meaningful.

Maugean

Iredale and May (1916) recognised the Maugean region primarily for the cold-water habitats of Tasmania and the Bass Strait environs, based largely on molluscan distributions. Kott (1952) extended the Maugean into southern New South Wales (Peronian) to account for ascidian distributions, and Bennett and Pope (1953) extended westwards into South Australia based primarily on molluscan and barnacle distributions. For Ebach *et al.* (2013), the Maugean region (as a subregion) extends from Mallacoota (Victoria) to Victor Harbour (South Australia), including the entire coast of Tasmania.

Flindersian

First proposed by Cotton (1932) in preference over Hedley's (1904) *Adelaidean*, the Flindersian region spans the southern coastline between Geraldton, Western Australia, to the western limit of the Maugean region at Victor Harbour, South Australia (or the southern limit of the Peronian if the Maugean is considered a subregion of the Flindersian [Ebach *et al.* 2013]).

Dampierian

Complementing the Solanderian region in the northeast, the Dampierian region extends westwards of Cape York, Queensland, to Western Australia, south to the Houtman Abrolhos Islands (28°50′S, off the coast from Geraldton), where it meets the Flindersian region (Hedley 1904; Ebach *et al.* 2013). The transitional nature of southwestern Australia has been acknowledged in the Baudinian Province (Kott 1952) (Figure 5.2b) and the West Australian Province (Knox 1963) (Figure 5.2d), both focused on the southwest corner of Western Australia. The Leeuwin Province was recently proposed by Hatcher (1991; modified by Hutchins 1994) for Western Australian biotas under the influence of the Leeuwin Current, a region stretching from Coral Bay (~22°S) south to the Recherche Archipelago (~34°S), not dissimilar to Knox's West Australian Province. This inevitably draws focus on the southwestern fauna, where the signature of the Leeuwin Current is evident in the transitional fauna at the offshore Houtman Abrolhos Islands, in comparison with more distinctly temperate assemblages at the same latitude, where the nearshore influence of the Leeuwin Current is weak. Buchanan and Beckley (2015) detected an epipelagic zoogeographic disjunction between tropical and temperate chaetognath assemblages near 28°S caused by the activity of the Leeuwin Current.

In large part, these 'traditionally' defined regions have not been quantitatively tested, and because of difficulties in defining their boundaries, many recent workers have favoured the more generalised northern and southern biogeographic division (e.g. Wilson and Allen 1987; Morgan and Wells 1991; Poore 1994). A number of recent studies, however, have examined the validity of the traditional biogeographical provinces. The quantitative analysis of marine macroalgal distributions in southern Australia found strong support for the Peronian, Maugean and Flindersian regions (Waters *et al.* 2010). Similarly, numerous phylogeographic studies of intertidal and shallow-water species have identified genetic discontinuities at or near the estimated boundaries between the Peronian and Maugean Provinces in southern New South Wales and between the Maugean and Flindersian Provinces in South Australia (for a review, see Colgan 2015). For these studies, the reality of biogeographic breaks for shallow-water species is not so much the focus as estimating the geographical points of transition and explaining the causes of the observed patterns (e.g. Waters *et al.* 2014). In particular, the role of the Bassian Isthmus and the closure of the Bass Strait during glacial maxima in speciation of southern marine faunas has figured significantly in explanations of biogeographic patterns (e.g. Dartnall 1974; Dawson 2005; Waters 2008; Colgan 2015; not least for Charles Hedley himself [Hedley 1904]).

Bioregionalisation

The aforementioned 'traditional' biogeographic regions essentially followed the observed distributions of intertidal and shallow subtidal species. Originally, biogeographic regions such as the traditional provinces/regions were invoked to account for the distributions of taxa, centres of diversification, modes of dispersal and so forth. That is, they were a means of explaining phenomena: they were explanatory tools, necessary for understanding the processes responsible for what we observe today. More recently, however, bioregionalisations have been developed using combined physical and biological data in order to serve more pragmatic ends, such as resource inventory, resource management and conservation (Ebach 2012). As process models, these bioregionalisations are intended as predictive rather than explanatory tools. This is especially important for nations with large marine exclusive economic zones and by implication extensive marine resources. This in part has motivated the development of the Interim Marine and Coastal Regionalisation for Australia (IMCRA) (Figure 5.3b), with areas defined by a combination of oceanographic, geomorphic and biological data on demersal fishes (Last *et al.* 2005; Commonwealth of Australia 2006), and bioregionalisations continue to be developed to assist with the management of Australian marine resources (Last *et al.* 2010). Spalding *et al.* (2007) developed a similar but larger-scale bioregionalisation of the world coastal and shelf areas known as Marine Ecoregions of the World (MEOW) to aid conservation efforts under the 1992 Convention on Biological Diversity. Recent Australian invertebrate bioregionalisation studies include Schlacher *et al.* (2007; sponges), O'Hara (2008; ophiuroids), McCallum *et al.* (2013; decapods), and Woolley *et al.* (2013; decapods, ophiuroids, polychaetes). A consistent theme among these studies is the high rate of new species discovery, particularly at shelf and slope depths (Williams *et al.* 2010; Poore *et al.* 2015). In this connection, diversity at mid-latitudes is known to be asymmetrical, with notably higher diversity in the Southern Hemisphere (Platnick 1991; Poore and Wilson 1993; Poore and Bruce 2012). Thus, the actual diversity of Australian marine invertebrates is likely to be much higher than currently estimated; by implication, biogeographic patterns can be expected to be more complex than current data show.

Conflicting Classifications

Even a cursory comparison of the regions defined under the traditional regionalisation with IMCRA and MEOW reveals different, even contradictory schemes. For instance, IMCRA and MEOW contain significantly more units than the traditional schema, but these smaller units are not simply finer subdivisions of the same region. Moreover, differences in regions are not merely the result of the superior quantitative over qualitative approaches. The qualitatively derived southern provinces (Peronian, Flindersian and Maugean), for instance, can be quantitatively corroborated (Waters *et al.* 2010). Regional delimitations differ according to data source and analytical method, which in turn depends on the purpose of the analysis. Areas of endemism defined for cladistic biogeographic analysis will not necessarily be congruent with areas based on collective modern species distributions. As such, the results of different classifications may not conflict as much as relate to different questions. The traditional bioregions are not directly comparable to the results of recent quantitative bioregionalisations of the Australian coast (e.g. O'Hara 2008; Woolley *et al.* 2013). We might expect as much given that the former are based primarily on intertidal and shallow subtidal distributions, and the latter on deep subtidal distributions, which are subject to different environmental processes. However, the different classifications also serve different purposes. Whether the traditional regions are biologically and ecologically meaningful can only be determined by testing them in their own right.

As with any type of classification, the approach, results and interpretations will differ depending on the questions at hand. This is true too in the case of biogeographical classifications. The astute observer will recognise, however, that explanatory biogeography and predictive bioregionalisations are not necessarily antithetical; they can be reciprocally illuminating. Predictive tools will be more powerful when the effects of underlying historical processes are considered, and historical explanations have more traction when brought to bear on the present. Common to all approaches, however, is the search for some measure of distinctiveness that differentiates one area from another based on the biota living within those areas – the species that are present or assumed present. Numerous studies, however, have demonstrated a lack of detailed taxonomic knowledge of vast areas of the Australian marine estate, especially in deep water. Recent bioregionalisation studies of extensive stretches of the Western Australian continental margin, for instance, revealed a bathyal fauna in which a remarkable 94.6% of crustaceans were new to science (Poore *et al.* 2015). Biodiversity science in Australian waters is still in the discovery phase for many marine groups. Hooper and Lévi (1994) commented that the 'major task in sponge biology is still to document the biodiversity of living species'. Evidently, this is no less true for most Australian marine invertebrate groups.

Taxonomic Groups

The following sections provide a general overview of the Australian distribution patterns of selected invertebrate groups.

Molluscs

The distribution of molluscs has been the primary basis of the traditional Australian biogeographical regions (Hedley 1904; Iredale and May 1916; Ashby 1926; Cotton 1930). The Australian Mollusca is massively diverse (estimated at around 19,000 species), and most extant marine families are represented in Australian waters. Wilson and Allen (1987) provided an extensive summary of selected endemic and widespread Australian families and recognised two main faunas – northern and southern – with broad overlap zones and the latitudinal attrition of tropical species beyond about 25°S. The tropical northern fauna is considerably more speciose than in the south and has a low level of generic and species endemicity (~10%). More than 70% of the northern molluscan fauna is more widely distributed in the tropical Indo-West Pacific (Wells 1980), although significant east–west faunistic differences may be present for some taxa, such as Volutidae (Wilson and Allen 1987).

In contrast to the tropical northern mollusc fauna, the temperate southern fauna is characterised by a high level of generic and especially species endemicity, estimated to exceed 95% (Wilson and Allen 1987). According to Darragh (1985), the temperate southern fauna is largely derived from the palaeoaustral Tertiary fauna of southeastern Australia, which was itself established chiefly from Tethyan and Zealandian elements. Thus, most of these southern endemics have evolved in situ as relicts of once widespread Tethyan groups or palaeoaustral groups. Consistent with the pattern of separation and formation of the southern Australian coastline, remnants of the Tethyan groups are most evident in the southwest and of Palaeoaustral groups in the southeast.

Many species range across the southern Australian coastline, but significant east–west differentiation is also evident, consistent with the Peronian and Flindersian Provinces. Some exceptions to the north–south faunal distinctions exist. Although the east- and west-coast overlap zones typically have mixed tropical and temperate faunas, a number of endemics occur only in these zones (Ponder and Wells 1998). The endemic mud whelk (*Pyrazus ebeninus*) spans the full north–south latitudinal range, and the relict bivalve and gastropod genera *Neotrigonia* and *Botelloides*, respectively, occur around the entire continent (Ponder 1985).

Echinoderms

The distributions of echinoderms are possibly the best studied of Australian marine invertebrates, having formed much of the basis of traditional marine regions (e.g. H.L. Clark 1946; Endean 1957) as well as recent quantitative studies of large-scale patterns and bioregionalisation (e.g. O'Hara and Poore 2000; O'Hara 2008). Early studies, beginning with a consideration of crinoid distributions (A.H. Clark 1911), supported a division between tropical northern and temperate southern faunas, but later studies (H.L. Clark 1946) considering all Australian echinoderms favoured Hedley's (1904, 1926) multiregion arrangement (Solanderian, Peronian, Flindersian, Dampierian) or included some modification to the northern divisions. For instance, based on overall similarities between the northeastern and northwestern mainland coastal faunas, Endean (1957) favoured a single northern coastal region (Solanderian–Dampierian, with a separate offshore Great Barrier Reef province), as did Marsh and Marshall (1983) and Rowe (1985). Moreover, this northern mainland fauna had the strongest affinities to the fauna of the 'East Indies' to the immediate north, different from the more widespread Indo-West Pacific fauna of the oceanic reefs (Rowe 1985). In contrast, the southern temperate echinoderm fauna is marked by a high level of species endemism (~90%; H.L. Clark 1946), although distributions across the south are not uniform. Southern genera too, such as the asteroids *Parvulastra*, *Patiriella* and *Meridiastra*, may be endemic or have closer affinities to other Austral faunas than to northern Australian taxa (Knox 1979; Waters *et al.* 2004; Waters and Roy 2004). Rowe and Vail (1982) found three common distribution patterns among southern Australian echinoderm genera and species: a general southern group spanning the entire temperate coast, a southeastern group (eastern Tasmania to New South Wales), and a southwestern group (Tasmania to southwestern Australia). The southeastern and southwestern groups approximate the Peronian and Flindersian Provinces, but in overlapping in the Tasmania–Bass Strait area do not accommodate a separate Maugean region. Wilson and Allen (1987), on the basis of multiple phyla, essentially reprised the binary division recognised by A.H. Clark (1911), but also recognised the eastern and western overlap zones, where tropical and temperate faunas mix. Parallel to the demonstrated latitudinal attenuation of asteroid faunas on the west coast (Marsh 1976), analysis of southern offshore reef echinoderm faunas (Lord Howe Island, Elizabeth and Middleton Reefs) demonstrated a clear faunal overlap of temperate and tropical species in the east (Hoggett and Rowe 1988). The emergent general consensus for echinoderm distributions supported the broad tropical–temperate, north–south faunal division with overlap zones on the east and west coasts (Rowe and Gates 1995).

Quantitative bioregionalisation analyses of echinoderms in southern Australia typically found distinct latitudinal and longitudinal gradients in species composition, with notable east–west differentiation (O'Hara and Poore 2000; O'Hara *et al.* 2002). At a global scale, ophiuroid species richness is highest at low latitudes, peaking in the tropical Indo-West Pacific (Stöhr *et al.* 2012). Although Australian ophiuroids also correspond to this pattern, Australia-wide distributions show continual species turnover around the continent, with few definitive biogeographic breaks and only regions of greater or lesser turnover. These patterns are not substantially affected by depth (within the distinctly subtidal 50–1500 m range) and, on the upper slope, species turnover is almost complete between tropical areas and southern Tasmania (O'Hara 2008).

Deep-water patterns recovered by bioregionalisation studies do not correspond with the traditional shallow-water biogeographic regions. Phylogeography of the intertidal seastar *Coscinasterias muricata*, however, is consistent with the Maugean Province of Bennett and Pope (1953, 1960; Waters and Roy 2003).

Crustaceans

Crustaceans are the dominant aquatic arthropods, occurring in every marine habitat. Some groups have been considered from the perspective of the traditional biogeographic regions. The distributions of shore crabs in southeastern Australia reflected the east–west differentiation either side of the Bassian Isthmus (Griffin 1969), as do the Australian species of the deep-sea king crab genus *Lithodes* (Ahyong 2010). Australian stomatopod distributions do not show strong regionalisation apart from 'an indigenous cold water fauna' centred on southeastern Australia and a single expansive northern tropical fauna, with overlap on the east and west coasts (Stephenson and McNeill 1955; Ahyong 2001). Distributions within some genera, such as *Austrosquilla*, however, align closely with traditional regions (Ahyong 2001). Among stomatopods, however, species endemicity is low among both tropical and temperate components, but is notably higher (20%) in the north than the south (11%). This is consistent with the Stomatopoda being a principally Tethyan tropical group that has only moderately expanded into temperate habitats (Ahyong 2001). Most of the endemic southern stomatopod species, however, also belong to endemic genera, whereas all northern genera are widespread in the Indo-West Pacific.

Decapod distributions are similar to those of stomatopods in comprising a large tropical Indo-Pacific element and smaller southern temperate element, albeit with higher southern endemicity, with broad east- and west-coast transition zones (Griffin and Yaldwyn 1968). More than 2400 decapod species are known from Australia, of which more than 800 occur in southern waters (Poore 2004). Overall southern endemicity (approximately 40%), however, is low for decapods in comparison with molluscs and echinoderms. Conversely, the northern fauna typically contains a significant widespread Indo-West Pacific component – for example, 78% for 'Thalassinidea', Anomura and Brachyura (Morgan 1990). Although yet to be specifically analysed for Australia, squat lobsters (Galatheoidea and Chirostyloidea) probably show similar trends to other decapods. Cluster analysis of Pacific squat lobster distributions recognised a single entire eastern Australian province (Macpherson *et al.* 2010), but more extensive analysis of worldwide distributions detected a break between the tropical western Pacific fauna and southeastern Australian/New Zealand fauna, and also recognised a close affinity between the northern Australian fauna and that of the tropical Indo-West Pacific (Schnabel 2009; Schnabel *et al.* 2011). Stephenson (1962) observed the highest diversity of portunid crabs in northern Australia, with a general polewards decline in species richness.

Australian barnacle distributions parallel those of decapods, with high diversity and low endemism in the northern tropical region, and lower diversity but higher endemism in the southern temperate region. Although Bennett and Pope (1953) favoured the multiregional biogeographic classification, Jones (2003, 2012) favoured the more generalised model of Wilson and Allen (1987). Of 279 Australian species recorded, 221 occur in the north and have affinities elsewhere in the tropical Indo-West Pacific, and only eight are endemics. The southern temperate fauna comprised 129 species with 23 endemic (Jones 2012).

Quantitative analysis of southern Australian decapods found distinct latitudinal gradients in species richness, but also longitudinal gradients with significant species turnover, especially near the South Australian gulfs, with only 20% of species shared by the endpoints at Sydney and Perth (O'Hara and Poore 2000). Similar analyses of 340 shelf species along the Western Australian continental margin identified a high percentage of species (68%) with wider Indo-West Pacific affinities and only 4% that could be characterised as Southern Ocean forms (McEnnulty *et al.* 2011). For the Western Australian shelf fauna, 21°S latitude appears to mark the position of high species turnover for decapods as well as ophiuroids and polychaetes (McCallum *et al.* 2013; Woolley *et al.* 2013).

Unlike decapods and stomatopods, whose diversity peaks in the tropics, peracarid crustaceans, such as amphipods and isopods, are abundant even in polar waters. Australian Amphipoda apparently do not show the broad north–south pattern of biogeographic differentiation displayed by other crustaceans such as decapods and stomatopods. Instead, the Australian amphipods show a high level of endemicity, in excess of 80% on the Australian Plate. Moreover, on the basis of parsimony analysis of endemism (PAE), Myers and Lowry (2009) interpret the close relationship between the tropical and temperate fauna of Australia as indicating substantial *in situ* evolution since the breakup of Gondwana (Myers and Lowry 2009).

Poore *et al.* (1994) recorded a very rich asellote isopod fauna from the southeast Australian continental slope, greater than that of comparable habitats in the Northern Hemisphere. Few of the continental slope asellotes, however, occur on the shelf; their affinities are with deep-water and Antarctic fauna, and therefore have little relevance to shallow-water regionalisations.

Polychaetes

The biogeographic distributions of Australian polychaete worms have not been analysed in detail, and to date have not figured largely in studies of Australian, let alone world, biogeography. This owes in part to perceived wide distributions (even if grossly discontinuous) of polychaete taxa at all taxonomic levels and inadequate taxonomic and phylogenetic knowledge in general (Glasby *et al.* 2000). Increasingly, however, revisionary studies have shown these wide and disjunct distributions to be artefacts of insufficient taxonomic resolution; many so-called cosmopolitan or widespread species are revealed as species complexes (e.g. Sun *et al.* 2015). Current knowledge of polychaete faunas around Australia is patchy and uneven. The southern temperate shallow-water fauna is relatively well known, but not the northern tropical fauna. For instance, a single taxonomic workshop recently held at Lizard Island on the northern Great Barrier Reef resulted in the discovery of 91 new species and 67 new distribution records (Hutchings and Kupriyanova 2015). Moreover, the polychaetes from depths beyond about 50 m are very poorly known, with the Bass Strait having the only well-characterised deep-water polychaete fauna, although increasingly, the Western Australian continental shelf fauna is becoming illuminated (Woolley *et al.* 2013; Poore *et al.* 2015). Nevertheless, 72% of the bathyal polychaetes reported by Poore *et al.* (2015) were first records for Australia. Against this backdrop, what can be said of the large-scale patterns of Australian polychaete biogeography? For the calcareous tube worms (Serpulidae), Straughan (1967) recorded low diversity in the south and west of Australia, and the highest diversity particularly between 15° and 30°S. The eastern Australian fauna was said to comprise '(1) a large Indo-West Pacific fauna, and (2) a large percentage of endemic species' (Straughan 1967: 257). Glasby *et al.* (2000) recognised the north–south pattern of two distinct marine faunas recognised by Wilson and Allen (1987). A comparison of the distributions of five major families in the Australian fauna essentially showed a northern tropical fauna comprising primarily widespread Indo-West Pacific species, with endemics in the northeast and northwest, and a southern fauna comprising a higher proportion of endemics than the north, along with a small proportion of species widespread around Australia. The higher level of temperate endemism compared with tropical Australia is corroborated by PAE of world Terebellidae, which retrieved close area relationships among the southern Australian regions, but the northeastern and northwestern Australian regions showed closer relationships to other Indo-West Pacific areas of endemism (Garraffoni *et al.* 2006). PAE of selected taxonomically well-resolved polychaete families showed a proportionally higher level of species endemism in temperate southern Australia (67%) in comparison with other southern faunas (53.2% for southeastern South America, 40% for South Africa) (Glasby and Alvarez 1999), although the level of endemism is lower than for other groups such as echinoderms, with up to 90% in the south.

Few cladistic biogeographic studies of polychaetes have been undertaken to date. Glasby's (1999) analysis of *Namanereis* (Namanereidinae) showed that the genus probably originated in the south, despite there now being only a single relictual species in southern Australia and the majority of known species being tropical.

Corals

Coral diversity is highest in tropical northern Australia, so the latitudinal attenuation of diversity is evident on both the east and west coasts, related to the influence of the south-flowing boundary currents. In Western Australia, the latitudinal attenuation in scleractinian coral diversity follows temperature gradients, being most evident in the southwest, and in the east becoming evident only from the southern end of the Great Barrier Reef (Veron 1995). Off eastern Australia, Lord Howe Island (~31°S) is the southern limit of coral reefs. The presence of coral reefs on the west coast ends at the Houtman Abrolhos Islands (~29°S) owing to the offshore influence of the Leeuwin Current (Hatcher 1991). In addition to favourable water temperatures for coral survival at the Houtman Abrolhos Islands, the offshore flow of the Leeuwin Current is thought to provide an ecologically competitive advantage to corals over macroalgae by suppressing the nutrient upwelling that would otherwise favour macroalgal growth (Hatcher 1991).

Significant coral endemism is not evident across northern Australia. Distributions of species of the staghorn coral genus *Acropora* suggest the eastern Indian Ocean and Arafura Sea faunas are more similar to parts of the Indo-Malay Archipelago than to the Great Barrier Reef and southwest Pacific (Wallace 1999). However, overall numbers of coral species of the Great Barrier Reef and reefs of the North West Shelf overlap by 92%, lending no support to separate eastern and western faunal regions (Veron 1995). Few tropical corals also occur in southern Australia; the majority of southern Australian scleractinians are temperate endemics.

Knowledge of deep-water octocorals (Alcyonacea and Pennatulacea) is rapidly expanding, with species from >80 m depth tripling to 457 between 1997 and 2008 (Alderslade *et al.* 2014); of these, almost 20% were undescribed. This growth in knowledge has come primarily through deep-water bioregionalisation surveys off southeastern, western and northwestern Australia. Among known species, most had Indo-West Pacific affinities and few were cosmopolitan; low levels of endemism were observed (20 species). Five common genera were widely distributed and two were restricted to southeastern Australia. Species richness was not strongly associated with latitude; rather, bathymetry was more significant, with little species overlap between outer-shelf through lower-slope bathyomes. An acknowledged caveat to these results, and an important impediment to biogeographic interpretation of these data, is the poor state of octocoral taxonomy, precluding accurate identification of many species (Alderslade *et al.* 2014).

Sponges

Approximately 1500 named species have been recorded from Australia, but at least 5000 species are estimated at present (Hooper *et al.* 1999; Hooper 2008). In southern Australia, about 800 species are estimated to occur, of which only about 360 species are described. Thus, it is difficult at present to reliably generalise about Australian sponge distributions. Hooper and Lévi (1994), however, provide a detailed assessment of the general distribution patterns of Australian sponges within the framework of the traditional biogeographic provinces. Hooper (2008) estimates more than 2500 sponge species occur in tropical Queensland–Coral Sea waters (i.e. Solanderian and Great Barrier Reef). According to Hooper and Lévi (1994), more than one-third are endemic, with most from shallow water; deep-water diversity (and biomass) is apparently low, with only about 100 species from bathyal and hadal depths. More detailed analysis of northeastern Australian distributions rejected Lendenfeld's (1888, 1889) concept of a homogeneous eastern Australian fauna and instead emphasised regional heterogeneity in sponge distributions, perhaps justifying further subdivision (Hooper *et al.* 1999; Van Soest *et al.* 2011). Moreover, rather than a transitional zone, abrupt boundaries between east-coast tropical and temperate sponge faunas have been identified near 28°S (Tweed River) (Hooper and Lévi 1994).

Like the northeast, northwestern Australian (Dampierian) sponge diversity is very high, with 1500 estimated species, almost 50% endemic. Fromont (2003) recorded 275 species of sponges across all groups from the Dampier Archipelago alone and McEnnulty *et al.* (2011) recorded 372 species of Demospongiae (of which 106 were undescribed) along the western continental margin. Faunal affinities between Indo-Malay and Dampierian sponge faunas are fewer than for other major invertebrate groups. Only about 30% of nonendemic northwestern Australian sponges are shared with Indonesia, compared with 66% for echinoderms (Marsh and Marshall 1983; Hooper and Lévi 1994). Overall, tropical northern Australian sponges of the Solanderian and Dampierian regions do not conform to the homogeneous distribution pattern or have the low level of endemism usually claimed for other pelagically dispersing groups. Sponge endemism is markedly higher than the median 13% of Wilson and Allen (1987) for other marine groups.

Sponges of the southern temperate marine regions have about 40%–60% endemism. The Peronian (southeastern Australia) sponge fauna includes an estimated 600 species, with stronger affinities to the Maugean than Solanderian faunas, and hence may be more Palaeoaustral than Tethyan. An estimated 600 species occur in the Maugean region (39%–50% endemic), with 58% exclusively southern temperate inhabitants. However, a not-insignificant proportion has affinities to tropical Indo-West Pacific faunas (33%) and none to the Antarctic or subantarctic faunas (Wiedenmayer 1989; Hooper and Lévi 1994; Downey *et al.* 2012). The southwestern Flindersian region contains an estimated 800 sponge species, but characteristics of the fauna are as yet insufficiently known. Comparison of the fossil sponge record from sites within the Flindersian region with the present-day fauna, however, suggests Tethyan affinities, given the presence of otherwise geographically distant tropical elements (Łukowiak 2016).

Hooper and Lévi's (1994) cladistic biogeographic analysis of selected Indo-West Pacific sponge families retrieved strong affinities between the two tropical Australia regions and western Indian Ocean areas, and between the temperate Australian regions and southeastern Africa.

Tunicates

Initial studies of Australian ascidian distributions, primarily based on intertidal and shallow subtidal species, applied faunal regions corresponding to those of Hedley (1904) and Iredale and May (1916), but on the basis of nonconforming distributions, Kott (1952) effectively divided the Peronian in half by extending the Maugean from Tasmania to Sydney (New South Wales) instead of stopping near the Victoria–New South Wales border, and adding the Oxleyan (Iredale 1937), ranging from Sydney to Brisbane. She also recognised the Baudinian region in southwestern Australia, ranging from Perth to Albany. Kott, however, found no distinction between the coastal Queensland and offshore Great Barrier Reef ascidian faunas, which she collectively referred to as *Solanderian*. Notably, Kott's (1952) recognition of an Oxleyan subregion, together with transitional zones on the east (Brisbane to Port Curtis) and west coasts (Shark Bay to Fremantle) conspicuously approximates the overlap zones of Wilson and Allen (1987), with broad northern tropical and southern temperate faunas. More recently, Kott (2008) applied the more generalised north–south distribution model recognising distinctive northern and southern faunas. About half (368) of the 726 recorded Australian species occur in southern temperate waters (Shepherd and Kott 2013), with approximately equal proportions of endemic and more widespread Indo-West Pacific species in the north. In the south, however, numbers of endemics are more than twice that of nonendemics (Kott 2008), and the solitary species are in higher proportion than colonials (Shenkar and Swalla 2011). Thus, for ascidians, a latitudinal gradient in species turnover seems more evident than in species richness. Among pelagic tunicates off eastern Australia, little evidence for distinct regionalisation could be found, but instead a southwards attenuation in species richness (Thompson 1948).

As with most other major invertebrate groups, the northern Australian ascidian and pelagic tunicate fauna has the strongest affinities to the more widespread tropical Indo-West Pacific fauna. Interestingly, however, quantitative analysis of temperate Southern Hemisphere ascidian distributions also found the Tasmanian and southern Australian fauna to be more similar to the Indo-West Pacific fauna than to that of other austral regions (Primo and Vázquez 2008). Similarly, Thompson (1948) found no evidence of cold-water influence in the southern faunal composition among pelagic tunicates.

Lophophorates

Bryozoa

The Australian bryozoan fauna, at more than 900 known species, is one of the largest in the world, and similar in size to that of New Zealand. Bryozoans occur at intertidal down to abyssal depths, but most are shelf species with maximum diversity in the first 200 m (Gordon 1999). Notwithstanding that the Australian deep-water bryofauna is poorly known, especially in Western Australia, current studies do not detect latitudinal gradients in species diversity, as is common in many other marine taxa. Although poor dispersers as larvae owing to their short larval period, many bryozoan species are widespread in the Southern Hemisphere, possibly as a result of secondary modes of dispersal such as rafting or transport among fouling. Among southern Australian species, only around 38% are endemic. A combination of the action of the West Wind Drift and the slow evolutionary turnover rate is believed to account for this low level of endemism (Barnes and Griffiths 2008).

Brachiopoda

Although 21 of 33 brachiopod genera (30 Articulata, 3 Inarticulata) known from Australia are endemic, most are widely distributed around the continent. Latitudinal, longitudinal and bathymetric gradients in distribution are not evident; rather, substrate is the principal factor governing distribution (Richardson 1997; Richardson and Shepherd 2013).

Phoronida

Five species of Phoronida are recorded from Australia, all with apparently cosmopolitan distributions in the Pacific and Atlantic oceans (Emig *et al.* 1977). Records from Australia are patchy, often from a single area – for example, Port Phillip Bay. Those species for which records are available from multiple sites, however, occur across a wide latitudinal range spanning temperate and tropical zones.

References

Ahyong, S.T. 2001. Revision of the Australian stomatopod Crustacea. *Records of the Australian Museum Supplement* 26: 1–326.

Ahyong, S.T. 2010. The marine fauna of New Zealand: King crabs of New Zealand, Australia and the Ross Sea (Crustacea: Decapoda: Lithodidae). *NIWA Biodiversity Memoir* 123: 1–196.

Alderslade, P., F. Althaus, F. McEnnulty, K. Gowlett-Holmes and A. Williams. 2014. Australia's deep-water octocoral fauna: Historical account and checklist, distributions and regional affinities of recent collections. *Zootaxa* 3796 (3): 435–452.

Ashby, E. 1926. The regional distribution of Australian chitons (Polyplacophora). *Reports of the Australian and New Zealand Association for the Advancement of Science* 17: 366–393.

Barnes, D.K.A., and H.J. Griffiths. 2008. Biodiversity and biogeography of southern temperate and polar bryozoans. *Global Ecology and Biogeography* 17: 84–99.

Bennett, I., and E.C. Pope. 1953. Intertidal zonation of the exposed rocky shores of Victoria, together with a rearrangement of the biogeographical provinces of temperate Australian shores. *Australian Journal of Marine and Freshwater Research* 4(1): 105–159, pl. 1–4.

Bennett, I., and E.C. Pope. 1960. Intertidal zonation of the exposed rocky shores of Tasmania and its relationship with the rest of Australia. *Australian Journal of Marine and Freshwater Research* 11: 182–221.

Briggs, J.C. 1974. *Marine Zoogeography*. New York, McGraw-Hill. 475 p.

Buchanan, P.J., and L.E. Beckley. 2015. Chaetognaths of the Leeuwin Current system: Oceanographic conditions drive epi-pelagic zoogeography in the south-east Indian Ocean. *Hydrobiologia* 763: 81–96.

Butler, A.J., T. Rees, P. Beesley and N.J. Bax. 2010. Marine biodiversity in the Australian region. *PLoS ONE* 5(8): e11831.

Clark, A.H. 1911. The recent crinoids of Australia. *Memoirs of the Australian Museum* 4: 706–804.

Clark, H.L. 1946. *The Echinoderm Fauna of Australia*. Publication 566. Washington, DC, Carnegie Institution of Washington. 567 p.

Colgan, D.J. 2015. Marine and estuarine phylogeography of the coasts of south-eastern Australia. *Marine and Freshwater Research*. (doi.org/10.1071/MF15106)

Commonwealth of Australia. 2006. *A Guide to the Integrated Marine and Coastal Regionalisation of Australia Version 4.0*. Department of the Environment and Heritage, Canberra, Australia.

Cotton, B.C. 1930. Fissurellidae from the 'Flindersian' region, Southern Australia. No. 1. *Records of the South Australian Museum* 4: 212–222.

Crame, J.A. 1996. Evolution of high-latitude molluscan faunas. In *Origin and Evolutionary Radiation of the Mollusca*, ed. J. Taylor, 119–131. Oxford, Oxford University Press.

Crame, J.A. 1999. An evolutionary perspective on marine faunal connections between southernmost South America and Antarctica. *Scientia Marina* 63 (Suppl 1): 1–14.

Darragh, T.A. 1985. Molluscan biogeography and biostratigraphy of the Tertiary of southeastern Australia. *Alcheringa* 9: 83–116.

Dartnall, A.J. 1974. Littoral biogeography. In *Biogeography and Ecology in Tasmania*, ed. W.D. Williams, Monographiae Biologicae, 171–194. The Hague, the Netherlands, Dr. W. Junk.

Dawson, M.N. 2005. Incipient speciation of *Catostylus mosaicus* (Scyphozoa, Rhizostomeae, Catostylidae), comparative phylogeography and biogeography in south-east Australia. *Journal of Biogeography* 32: 515–533.

De Broyer, C., and P. Koubbi. 2014. The biogeography of the Southern Ocean. In *Biogeographic Atlas of the Southern Ocean*, ed. C. De Broyer, P. Koubbi, H.J. Griffiths, B. Raymond, C.D. Udekem D'acoz, A.P. Van De Putte *et al.*, 1–9. Cambridge, UK, Scientific Committee on Antarctic Research.

Downey, R.V., H.J. Griffiths, K. Linse and D. Janussen. 2012. Diversity and distribution patterns in high southern latitude sponges. *PLoS ONE* 7(7): e41672.

Ebach, M.C. 2012. A history of biogeographical regionalisation in Australia. *Zootaxa* 3392: 1–34.

Ebach, M.C., A.C. Gill, A. Kwan, S.T. Ahyong, D.J. Murphy and G. Cassis. 2013. Towards and Australian Bioregionalisation Atlas: A provisional area taxonomy of Australia's biogeographical regions. *Zootaxa* 3619(3): 315–342.

Ekman, S. 1953. *Zoogeography of the Sea*. London, Sidgwick & Jackson. 417 p.

Emig, C.C., D.F. Boesch and S. Rainer. 1977. Phoronida from Australia. *Records of the Australian Museum* 30(16): 455–474.

Endean, R. 1957. The biogeography of Queensland's shallow-water echinoderm fauna (excluding Crinoidea), with a rearrangement of the faunistic provinces of tropical Australia. *Australian Journal of Marine and Freshwater Research* 8(3): 233–273.

Fromont, J. 2003. Porifera (sponges) of the Dampier Archipelago: Taxonomic affinities and biogeography. In *The Marine Fauna and Flora of Dampier, Western Australia*, eds. F.E. Well, D.I. Walker and D.S. Jones, 405–417. Perth, Australia, Western Australian Museum.

Garraffoni, A.R.S., S.S. Nihei and P.C. Lana. 2006. Distribution patterns of Terebellidae (Annelida: Polychaeta): An application of parsimony analysis of endemicity (PAE). *Scientia Marina* 70(S3): 269–276.

Gill, T.N. 1885. The principles of zoogeography. *Proceedings of the Biological Society of Washington* 2: 1–39.

Glasby, C.J. 1999. The Namanereidinae (Polychaeta: Nereididae): Part 2, Cladistic biogeography. *Records of the Australian Museum Supplement* 25: 131–144.

Glasby, C.J., and B. Alvarez. 1999. Distribution patterns and biogeographic analysis of Austral Polychaeta (Annelida). *Journal of Biogeography* 26(3): 507–533.

Glasby, C.J., P.A. Hutchings and R.S. Wilson. 2000. Biogeography. In *Polychaetes and Allies: The Southen Synthesis Fauna of Australia, Vol. 4A; Polychaeta, Myzostomida, Pogonophora, Echiura, Sipuncula*, eds. P.L. Beesley, G.J.B. Ross and C.J. Glasby, 39–43. Melbourne, Australia. CSIRO Publishing.

Gordon, D.P. 1999. Bryozoan diversity in New Zealand and Australia. In *The Other 99%: The Conservation and Biodiversity of Invertebrates*, eds. W. Ponder and D. Lunney, 199–204. Mosman, Australia, Royal Zoological Society of New South Wales.

Gorny, M. 1999. On the biogeography and ecology of the Southern Ocean decapod fauna. *Scientia Marina* 63: 367–382.

Griffin, D.J.G. 1969. Notes on the taxonomy and zoogeography of the Tasmanian grapsid and ocypodid crabs (Crustacea, Brachyura). *Records of the Australian Museum* 27: 323–347.

Griffin, D.J.G., and J.C. Yaldwyn. 1968. The constitution, distribution and relationships of the Australian decapod Crustacea. *Proceedings of the Linnean Society of New South Wales* 93: 164–183.

Hatcher, B.G. 1991. Coral reefs in the Leeuwin Current: An ecological perspective. *Journal of the Royal Society of Western Australia* 74: 115–127.

Hedley, C. 1904. The effect of the Bassian Isthmus upon the existing marine fauna: A study in ancient geography. *Proceedings of the Linnean Society of New South Wales* 28: 876–883.

Hedley, C. 1926. Zoogeography. In *Australia Encyclopaedia*, eds. A.W. Jose and H.J. Carter. Vol. 2, 743–744. Sydney, Australia, Angus and Roberston.

Hoese, D.F., D.J. Bray, J.R. Paxton, G.R. Allen. 2006. Fishes. In *Zoological Catalogue of Australia*, Vol. 35, eds. P.L. Beesely and A. Wells. Melbourne, Australia, CSIRO Publishing. 2178 p.

Hoggett, A.K., and F.W.E. Rowe. 1988. Zoogeography of the echinoderms on the world's most southern coral reefs. In *Echinoderm Biology: Proceedings of the Sixth International Conference, Victoria, 23–28 August, 1987*, eds. R.D. Burke, P.V. Mladenov, P. Lambert and P.L. Parsley, 370–387. Rotterdam, the Netherlands, Balkema.

Hooper, J.N.A. 2008. Sponges. In *Biology, Environment and Management of the Great Barrier Reef*, eds. P. Hutchings, M. Kingsford and O. Hoegh-Guldberg, 171–187. Melbourne, Australia, CSIRO Publishing.

Hooper, J.N.A., J.A. Kennedy, S.E. List-Armitage, S.D. Cook and R. Quinn. 1999. Biodiversity, species composition and distribution of marine sponges in northeast Australia. *Memoirs of the Queensland Museum* 44: 263–274.

Hooper, J.N.A., and C. Lévi. 1994. Biogeography of Indo-West Pacific sponges: Microcionidae, Raspailiidae, Axinellidae. In *Sponges in Time and Space: Proceedings of the 4th International Porifera Congress*, eds. R.W.M van Soest, T.M.G. van Kempen and J.C. Braekman, 191–212. Rotterdam, the Netherlands, Balkem.

Hutchings, P.A., and E. Kupriyanova, eds. 2015. Coral-reef associated fauna of Lizard Island, Great Barrier Reef: Polychaetes and allies. *Zootaxa* 4019: 1–801.

Hutchins, J.B. 1994. A survey of the nearshore reef fish fauna of Western Australia's west and south coasts: The Leeuwin Province. *Records of the Western Australian Museum*, Suppl 46: 1–66.

Iredale, T. 1937. A basic list of the land Mollusca of Australia. *Australian Zoologist* 8: 287–333.

Iredale, T., and W.L. May. 1916. Misnamed Tasmanian chitons. *Proceedings of the Malacological Society of London* 12: 94–117.

Jones, D.S. 2003. The biogeography of Western Australian shallow-water barnacles. In *The Marine Flora and Fauna of Dampier, Western Australia*, eds. F.E. Wells and D.S. Jones, 479–496. Perth, Australia, Western Australian Museum.

Jones, D.S. 2012. Australian barnacles (Cirripedia: Thoracica), distributions and biogeographical affinities. *Integrative and Comparative Biology* 52(3): 366–387.

Knox, G.A. 1960. Littoral ecology and biogeography of the southern oceans. *Proceedings of the Royal Society of London B* 152: 577–624.

Knox, G.A. 1963. The biogeography and intertidal ecology of the Australasian coasts. *Oceanography and Marine Biology: An Annual Review* 1: 341–404.

Knox, G.A. 1979. Distribution patterns of southern hemisphere marine biotas: Some comments on their origins and evolution. *New Zealand Department of Scientific and Industrial Research Information Series* 137: 43–81.

Kott, P. 1952. The Ascidians of Australia: I. Stolidobranchiata Lahille and Phlebobranchiata Lahille. *Australian Journal of Marine and Freshwater Research* 3: 205–334.

Kott, P. 2008. The Tunicata. In *Biology, Environment and Management of the Great Barrier Reef*, eds. P. Hutchings, M. Kingsford and O. Hoegh-Guldberg, 262–275. Melbourne, Australia, CSIRO Publishing.

Last, P., V. Lyne, A. Williams, R.D. Campbell, A.J. Butler and G.K. Yearsley. 2010. A hierarchical framework for classifying seabed biodiversity with application to planning and managing Australia's marine biological resources. *Biological Conservation* 143: 1675–1686.

Last, P., V. Lyne, G. Yearsley, D. Gledhill, M. Gomon, T. Rees and W. White. 2005. *Validation of National Demersal Fish Datasets for the Regionalisation of the Australian Continental Slope and Outer Shelf (>40 m Depth)*. Hobart, Australia, National Oceans Office. 99 p.

Lendenfeld, R. Von. 1888. *Descriptive Catalogue of the Sponges in the* Australian Museum, *Sydney*. London, Taylor & Francis. 260 p, 12 pls.

Lendenfeld, R. Von. 1889. *A Monograph of the Horny Sponges*. London, Trübner. iii–iv, 1–936, pls 1–50.

Lohman, D.J., M. de Bruyn, T.J. Page, K. von Rintelen, R. Hall, P.K.L. Ng, H.T. Shih, G.R. Carvalho and T. von Rintelen. 2011. Biogeography of the Indo-Australian Archipelago. *Annual Review of Ecology and Systematics* 42: 205–206.

Łukowiak, M. 2016. Fossil and modern sponge fauna of southern Australia and adjacent regions compared: Interpretation, evolutionary and biogeographic significance of the Late Eocene 'soft' sponges. *Contributions to Zoology* 85: 13–35.

Macpherson, E., B. Richer de Forges, K. Schnabel, S. Samadi, M.C. Boisselier and A. Garcia-Rubies. 2010. Biogeography of the deep-sea galatheid squat lobsters of the Pacific Ocean. *Deep-Sea Research I* 57: 228–238.

Marsh, L.M. 1976. Western Australian Asteroidea since H.L. Clark. *Thalassia Jugoslavica* 12: 213–225.

Marsh, L.M., and J.I. Marshall. 1983. Some aspects of the zoogeography of northwestern Australian echinoderms (other than holothurians). *Bulletin of Marine Science* 33: 671–687.

McCallum, A.W., G.C.B. Poore, A. Williams, F. Althaus and T. O'Hara. 2013. Environmental predictors of decapod species richness and turnover along an extensive Australian continental margin (13–35°S). *Marine Ecology* 34(3): 298–312.

McEnnulty, F.R., K.L. Gowlett-Holmes, A. Williams, F. Althaus, J. Fromont, G.C.B. Poore, T.D. O'Hara *et al.* 2011. The deepwater megabenthic invertebrates on the western continental margin of Australia (100–1100 m depths): Composition, distribution and novelty. *Records of the Western Australian Museum*, Suppl 80: 1–191.

Millar, A.J.K. 2007. The Flindersian and Peronian Provinces. In *Algae of Australia: Introduction*, eds. P.M. McCarthy and A.E. Orchard, 554–559. Collingwood, Australia, Australian Biological Resources Study, CSIRO Publishing.

Morgan, G.J. 1990. A collection of Thalassinidea, Anomura and Brachyura (Crustacea: Decapoda) from the Kimberley region of north-western Australia. *Zoologische Verhandelingen* 265: 1–90.

Morgan, G.J., and F.E. Wells. 1991. Zoogeographic provinces of the Humboldt, Benguela and Leeuwin Current systems. *Journal of the Royal Society of Western Australia* 74: 59–69.

Myers, A.A., and J.K. Lowry. 2009. The biogeography of Indo-West Pacific tropical amphipods with particular reference to Australia. *Zootaxa* 2260: 109–127.

O'Hara, T.D. 2008. *Bioregionalisation of Australian Waters Using Brittle Stars (Echinodermata: Ophiuroidea), A Major Group of Marine Benthic Invertebrates*. Canberra, Australia, Department of Environment, Water, Heritage and Arts. 69 p.

O'Hara, T.D., and G.C.B. Poore. 2000. Patterns of distribution for southern Australian marine echinoderms and decapods. *Journal of Biogeography* 27: 1321–1335.

O'Hara, T., G.C.B. Poore, S.T. Ahyong and D.A. Staples. 2002. Rapid assembly of invertebrate data for the SE Regional Marine Plan. Museum Victoria Technical Report, 1–37.

Platnick, N.I. 1991. Patterns of biodiversity: Tropical vs. temperate. *Journal of Natural History* 25: 1083–1088.

Ponder, W.F. 1985. A revision of the genus Botelloides (Mollusca: Gastropoda: Trochacea). *Department of Mines and Energy*, South Australia, Special Publication 5: 301–327.

Ponder, W.F., and F.E. Wells. 1998. Distribution and relationships of marine and estuarine fauna. In *Mollusca: The Southern Synthesis Fauna of Australia*, Vol. 5, eds. P.L. Beesley, G.J.B. Ross and A. Wells, 77–80. Melbourne, Australia, CSIRO Publishing.

Poore, G.C.B. 1994. Marine biogeography of Australia. In *Marine Biology*, eds. L.S. Hammond and R.N. Synnot, 189–212. Melbourne, Australia, Longman Cheshire.

Poore, G.C.B. 2004. *Marine Decapod Crustacea of Southern Australia: A Guide to Identification with Chapter on Stomatopoda by Shane Ahyong*. Melbourne, Australia, CSIRO Publishing.

Poore, G.C.B., L. Avery, M. Blaźewicz-Paszkowycz, J. Browne, N.L. Bruce, S. Gerken, C.J. Glasby *et al.* 2015. Invertebrate diversity of the unexplored marine western margin of Australia: Taxonomy and implications for global biodiversity. *Marine Biodiversity* 45(2): 271–286.

Poore, G.C.B., and N.L. Bruce. 2012. Global diversity of marine isopods (except Asellota and crustacean symbionts). *PLoS ONE* 7 (8):e43529.

Poore, G.C.B., J. Just and B.F. Cohen. 1994. Composition and diversity of Crustacea Isopoda of the southeastern Australian continental slope. *Deep Sea Research I* 41(4): 677–693.

Poore, G.C.B., and G.D.F. Wilson. 1993. Scientific correspondence: Marine species richness (with reply from R.M. May). *Nature* 361: 597–598.

Primo, C., and E. Vázquez, 2015. Zoogeography of the southern New Zealand, Tasmanian and southern African ascidian fauna. *New Zealand Journal of Marine and Freshwater Research* 42(2): 233–256.

Richardson, J.R. 1997. Brachiopods (Phylum Brachiopoda). In *Marine Invertebrates of Southern Australia Part III*, eds. S.A. Shepherd and M. Davies, 999–1027. Adelaide, Australia, SARDI Aquatic Sciences.

Richardson, J.R., and S.A. Shepherd. 2013. Brachiopods. In *The Ecology of Australian Temperate Reefs*, eds. S. Shepherd and G. Edgar, 47–47. Melbourne, Australia, CSIRO Publishing.

Rowe, F.W.E. 1985. Preliminary analysis of distribution patterns of Australia's non-endemic, tropical echinoderms. In *Proceedings of the 5th Echinoderm Conference, Galway, Ireland 24–29 September, 1984,* eds. B.F. Keegan and B.D.S. O'Connor, 91–98. Rotterdam, the Netherlands, Balkema.

Rowe, F.W.E., and J. Gates. 1995. Echinodermata. In *Zoological Catalogue of Australia* Vol. 33, ed. A. Wells, 510 p. Melbourne, Australia, CSIRO Publishing.

Rowe, F.W.E., and L.L. Vail. 1982. The distributions of Tasmanian echinoderms in relation to southern Australian biogeographic provinces. In *Echinoderm: Proceedings of the International Conference, Tampa Bay*, ed. J.M. Lawrence, 219–225. Rotterdam, the Netherlands, Balkema.

Schlacher, T.A., M. Schlacher-Hoenlinger, A. Williams, F. Althaus, J.N.A. Hooper and R.J. Kloser. 2007. Richness and distribution of sponge megabenthos in continental margin canyons off southeastern Australia. *Marine Ecology Progress Series* 340: 73–88.

Schnabel, K. 2009. A review of the New Zealand Chirostylidae (Anomura: Galatheoidea) with description of six new species from the Kermadec Islands. *Zoological Journal of the Linnean Society* 155: 542–582.

Schnabel, K.E., P. Cabezas, A. McCallum, E. Macpherson, S.T. Ahyong and K. Baba. 2011. World-wide distribution patterns of squat lobsters. In *The Biology of Squat Lobsters: Crustacean Issues*, Vol. 19, eds. G.C.B. Poore, S.T. Ahyong and J. Taylor, 149–182. Melbourne, Australia, CSIRO Publishing.

Shenkar, N., and B.J. Swalla. 2011. Global diversity of Ascidiacea. *PLoS ONE* 6(6): e20657.

Shepherd, S., and P. Kott. 2013. Ascidians or sea squirts. In *The Ecology of Australian Temperate Reefs*, eds. S. Shepherd and G. Edgar, 56–57. Melbourne, Australia, CSIRO Publishing.

Spalding, M.D., H.E. Fox, G.R. Allen, N. Davidson, Z.A. Ferdaña, M. Finlayson, B.S. Halpern *et al.* 2007. Marine ecoregions of the world: A bioregionalization of coastal and shelf areas. *Bioscience* 57(7): 573–583.

Stephenson, W. 1962. Evolution and ecology of portunid crabs, with special reference to Australian species. In *The Evolution of Living Animals*, ed. C.W. Leeper, 311–327. Melbourne, Australia, Melbourne University Press.

Stephenson, W., and F.A. McNeill. 1955. The Australian Stomatopoda (Crustacea) in the collections of the Australian Museum, with a check list and key to the known Australian species. *Records of the Australian Museum* 23(5): 239–265.

Stöhr, S., T.D. O'Hara and B. Thuy B. 2012. Global diversity of brittle stars (Echinodermata: Ophiuroidea). *PLoS ONE* 7 (3):e31940.

Straughan, D. 1967. Marine Serpulidae (Annelida: Polychaeta) of eastern Queensland and New South Wales. *Australian Journal of Zoology* 15: 201–261.

Sun, Y., E. Wong, H.A. Ten Hove, P.A. Hutchings, J.E. Williamson and E.K. Kupriyanova. 2015. Revision of the genus *Hydroides* (Annelida: Serpulidae) from Australia. *Zootaxa* 4009(1): 1–99.

Thompson, H. 1948. *Pelagic Tunicates of Australia*. Melbourne, Australia, CSIRO. 196 p., 75 pls.

Van Soest, R.W.M., N. Boury-Esnault, J. Vacelet, M. Dohrmann, D. Erpenbeck, N.J. De Voogd *et al.* 2012 Global diversity of sponges (Porifera). *PLoS ONE* 7(4): e35105.

Veron, J.E.N. 1995. *Corals in Space and Time:* The Biogeography and *Evolution of the Scleractinia*. Sydney, Australia, University of New South Wales Press. 321 p.

Wallace, C.C. 1999. *Staghorn Corals of the World: A Revision of the Genus Acropora*. Collingwood, Australia, CSIRO Publishing. 422 p.

Waters, J.M. 2008. Marine biogeographical disjunction in temperate Australia: Historical landbridge, contemporary currents, or both? *Diversity and Distributions* 14: 692–700.

Waters, J.M., S.A. Condie and L.B. Beheregaray. 2014. Does coastal topography constrain marine biogeography at an oceanographic interface? *Marine and Freshwater Research* 65: 969–977.

Waters, J.M., P.M. O'Loughlin and M.S. Roy. 2004. Molecular systematics of some Indo-Pacific asterinids (Echinodermata, Asteroidea): Does taxonomy reflect phylogeny? *Molecular Phylogenetics and Evolution* 30: 872–878.

Waters, J.M., and M.S. Roy. 2003. Marine biogeography of southern Australia: Phylogeographical structure in a temperate sea-star. *Journal of Biogeography* 30: 1787–1796.

Waters, J.M., and M.S. Roy. 2004. Out of Africa: The slow train to Australasia. *Systematic Biology* 53: 18–24.

Waters, J.M., T. Wernberg, S.D. Connell, M.S. Thomsen, G.C. Zuccarello, G.T. Kraft, J.C. Sanderson, J.A. West and C.F.D. Gurgel. 2010. Australia's marine biogeography revisited: Back to the future? *Austral Ecology* 35: 988–992.

Wells, F.E. 1980. The distribution of shallow-water marine prosobranch gastropods along the coastline of Western Australia. *Veliger* 22: 232–247.

Whitley, G.P. 1932. Marine zoogeographical regions of Australasia. *The Australian Naturalist* 8(8): 166–167.

Wiedenmayer, F. 1989. Demospongiae (Porifera) from northern Bass Strait, southern Australia. *Memoirs of the Museum of Victoria* 50: 1–242.

Williams, A., F. Althaus, P.K. Dunstan, G.C.B. Poore, N.J. Bax, R.J. Kloser and F.J. McEnnulty. 2010. Scales of habitat heterogeneity and megabenthos biodiversity on an extensive Australian continental margin (100–1100 m depths). *Marine Ecology* 31(1): 222–236.

Wilson, B.R., and G.R. Allen. 1987. Major components and distribution of marine fauna. In *Fauna of Australia General Articles*, Vol. 1A, eds. G.R. Dyne and D.W. Walton, 43–68. Canberra, Australia, Australian Government Publishing Service.

Wilson, B.R., and K. Gillett. 1971. *Australian Shells*. Sydney, Reed. 168 p.

Woolley, S.N.C., A.W. McCallum, R. Wilson, T.D. O'Hara and P.K. Dunstan. 2013. Fathom out: Biogeographical subdivision across the Western Australian continental margin: A multispecies modelling approach. *Diversity and Distributions* 19(12): 1506–1517.

6

Biogeography of Australian Marine Fishes

Anthony C. Gill and Randall D. Mooi

CONTENTS

Introduction

The biogeography of Australian fishes has been the subject of several recent reviews, including treatments of marine fishes by Wilson and Allen (1987), Paxton *et al.* (2006), Hoese *et al.* (2006) and Last *et al.* (2011; based largely on earlier Australian marine bioregionalisation efforts: Interim Marine and Coastal Regionalisation Technical Group 1998; Commonwealth of Australia 2006; Last *et al.* 2005, 2010; and Lyne *et al.* 2009), and of freshwater fishes by Williams and Allen (1987), Unmack (2001, 2013), Merrick (2006) and Hoese *et al.* (2006). We confine our current treatment to Australian marine fishes, with a focus on demersal (bottom-living) shorefishes. We present an overview of the history of biogeographic research on Australian marine fishes, summarise current ideas and highlight topics for future research.

In addition to the studies previously noted, there have been several reviews of broader geographic areas that address various aspects of the biogeography of Australian shorefishes. These include Mead (1970), Briggs (1974), Springer (1982), Helfman *et al.* (1997), Randall (1998), Mooi and Gill (2002), Spalding *et al.* (2007), Allen (2008), Briggs and Bowen (2012) and Kulbicki *et al.* (2013). Springer (1982), for example, though concentrating on evidence for Pacific Plate endemism, provided an important summary of the distribution of all warm-water Indo-Pacific shorefish families and brought together information on Indo-Pacific geology.

Species and Their Distributions

Several factors are fundamental to developing an understanding of the biogeography of Australian fishes. First and foremost is the knowledge of Australian species and their distribution within and outside Australia. Information on Australian fishes and their distributions has in particular accumulated through the production of checklists and taxonomic monographs.

The history of Australian fish checklists was reviewed by Hoese *et al.* (2006; see also Paxton *et al.* 2006). The first list of Australian fishes was compiled by John White (1790), who reported on 12 species from New South Wales. Styled largely on Albert Günther's *Catalogue of Fishes in the British Museum* (Günther 1859–1870), William John Macleay's *Descriptive Catalogue of the Fishes of Australia* (Macleay 1881a–d) and its subsequent supplement (Macleay 1884) dramatically increased the number of species to 1291, although he included species from 'all seas from the South Coast of New Guinea to the South of Tasmania, and from the West Coast of New Holland eastward as far as Norfolk Island' (Macleay 1881a: 302). In addition to a list of species, Macleay included brief descriptions of all species (or at least references to descriptions) and – more importantly for biogeography – information on the distribution of each species in Australian waters. The next Australian checklist was by Allan R. McCulloch (1930), completed by Gilbert P. Whitley after McCulloch's death in 1925; it listed 2023 species. Although not providing descriptions of each species, it included species synonymies and both Australian (by state) and foreign distributions. Later, Whitley (1964a) provided an important summary and indexed bibliography of Australian ichthyology, including a list of 2447 species. The most recent checklist was by Hoese *et al.* (2006), with 4449 native species from the Australian Exclusive Economic Zone (of which 4107 are marine), including detailed taxonomic synonymies and collection-based distributions for each species. Since the publication of that work, additional species have been added to a list of the Australian fauna at the rate of about one per week (D.F. Hoese, pers. comm.).

Much of our increasing knowledge of Australian fish biodiversity has come from efforts to document state faunas, primarily by fish curators in state natural history museums, based on active sampling programs and associated collections development: New South Wales – Tenison-Woods (1882), Ogilby (1886), McCulloch (1922) and Whitley (1934); Victoria – Castelnau (1872); Tasmania – Lord (1923, 1927), Lord and Scott (1924) and Last *et al.* (1983); Northern Territory – Larson *et al.* (2013); Queensland – McCulloch and Whitley (1925), Marshall (1964) and Grant (1965, 1975, 1982, 2002); South Australia – Waite (1921, 1923), Scott (1962) and Scott *et al.* (1974); Western Australia – Whitley (1948), Allen (1985, 1996) and Hutchins (1994, 2001). Accounts of Australian territorial waters include Lord Howe Island (Allen *et al.* 1976; Francis 1993; Francis and Randall 1993; Speare *et al.* 2004), Middleton and Elizabeth Reefs (Gill and Reader 1992; Oxley *et al.* 2004), Ashmore Reef and Cartier Island (Allen 1993), Cocos-Keeling Island (Allen and Smith-Vaniz 1994), Christmas Island (Allen and Steene 1979, 1988), Macquarie Island (Williams 1988b) and Norfolk Island (Francis 1993).

Important regional guides, though varying in comprehensiveness, have contributed also to our understanding of fish distribution in Australia. Freshwater fishes, including some estuarine or diadromous species, have been treated by Whitley (1964b), Lake (1971, 1978), McDowall (1980, 1996), Cadwallader and Backhouse (1983), Merrick and Schmida (1984), Leggett and Merrick (1987), Allen (1989), Larson and Martin (1990), Allen *et al.* (2002), Pusey *et al.* (2004) and Kuiter (2013). Major reviews of marine and estuarine fishes include Gloerfelt-Tarp and Kailola (1984), Sainsbury *et al.* (1985), Hutchins and Swainston (1986), Allen and Swainston (1988), Randall *et al.* (1990, 1997), Kuiter (1993, 1996), Last and Stephens (1994, 2009) and Gomon *et al.* (1994, 2008).

Perhaps of even greater importance have been the numerous revisionary and monographic studies which have attempted to revise and document all known species within genera or higher taxa. These are the source of state-of-the-art information on fish species, their synonymy and their distributions, both within and outside Australia, often accompanied by phylogenetic analyses. Monographic studies have not only allowed more complete understanding of diversity and distribution than localised studies, but by virtue of their more global approach have also increasingly brought in a larger pool of foreign researchers. Such studies are difficult and laborious, however, and often require dedicated commitment over long periods of time. For example, much of the Smithsonian Institution researcher V.G. Springer's efforts concentrated on a single large fish suborder (Blennioidei), with targeted collecting and research output spanning over 60 years (e.g. Springer 1955, 1967, 1988; Smith-Vaniz and Springer 1971; Springer and Gomon 1975; George and Springer 1980; Springer and Williams 1994; Hastings and Springer 1994; Springer and Allen 2004; Attaran-Farimani *et al.* 2016). It is perhaps not surprising, then, that the production of monographic works has declined considerably under the 'publish or perish' attitudes that currently pervade scientific research.

Delimitation of Areas within Australia

Gill (1885: 8) noted, 'It has been shown that the Australian Realm is divisible into temperate and tropical portions, and also that the land surface is separable into zones of even still narrower limits, corresponding in a general way with those recognised by Dana for marine life.'

The general makeup of Australia's fish fauna is captured in several statistics by Hoese *et al.* (2006), who noted that 3029 species (68.1% of total Australian native fish species) are found in tropical seas, 1222 (27.5%) are found in temperate seas and 390 (8.8%) occur in either subtropical or both temperate and tropical areas. They noted that overall (i.e. including freshwater fishes, which make up only 5.8% of the total fauna), 24.1% of Australia's native fish species are endemic.

Although we here place emphasis on the distribution of endemic species and the biogeographic areas they imply, various methods have been used to delimit biogeographic areas for fishes in Australia, such as the overall composition of faunas (regardless of whether areas circumscribe known ranges of endemic species), or on entirely physical parameters rather than biotic evidence (such as drainage basins). In recent years these efforts have been extended to related – though distinct – concepts, such as conservation and resource management areas.

A general history of biogeographic regionalisation in Australia is given by Ebach (2012). Ebach *et al.* (2013a,b) provide diagnoses, nomenclatural synonymies and type localities for Australian biogeographic regions following criteria outlined in the International Code of Area Nomenclature (ICAN; Ebach *et al.* 2008). Despite increasingly more sophisticated analytical methods, the essence of our understanding of biogeographic areas for Australian fishes was captured early in several papers by Gilbert P. Whitley (Whitley 1932, 1959; Iredale and Whitley 1938; see also Waters *et al.* 2010). Although Whitley's areas and their names (Figure 6.1a) have been used only intermittently by Australian biogeographers, they offer a convenient standpoint from which to discuss the history of and evidence for bioregionalisation in Australian fishes. We therefore present our discussion in the context of these areas (as refined and diagnosed by Ebach *et al.* 2013a; Figure 6.1d). We provide examples of endemic taxa for each area, based mainly on distribution data provided by Hoese *et al.* (2006). Hoese *et al.* also provided a number of tables documenting the levels of endemism of fish species within Australia, including by state, biogeographic zone (adapted from Commonwealth of Australia 2005), drainage system and climatic zone. Unfortunately, their estimates of percentage endemism and other summaries can't be readily applied to the areas listed in the following, because their areas are not directly comparable. The biogeographic zones, for example, are generally much smaller, and because data are not presented for combined zones (which would often equate to Whitley's areas), adding counts from each area would tend to underestimate the numbers of endemic species in Whitley's areas (though in some cases overestimate the percentages of endemism).

We note that levels of endemism do not necessarily identify historically relevant biogeographic areas; any geographic region can be delimited and have the numbers of endemic taxa enumerated. Australian marine regions are defined currently by a combination of ecological and taxonomic characteristics, but the historical reality of these regions and their possible interrelationships have yet to be thoroughly tested through comparative analyses of areagrams based on repeated phylogenetic patterns of taxa endemic to these regions.

Whitley (1932) provided a map of marine regions within Australia, basing it partly on earlier marine invertebrate studies by Hedley (1904), Hull (1911), Iredale and May (1916) and Cotton (1930), as well as his own knowledge of fish distributions. Whitley recognised six marine regions ('provinces') for Australia (Figure 6.1a), as well as additional ones for adjacent areas in the southwest Pacific (which in turn called on studies by Hull 1911; Finlay 1925, 1926; and Iredale and Hull 1929). The biogeographic reality of the Australian regions can be gauged to some extent by reference to the distribution of endemic marine fish species within selected shorefish families (Serranidae, Pseudochromidae, Opistognathidae, Plesiopidae, Pempheridae, Chaetodontidae, Pomacanthidae, Labridae; Table 6.1). Additional examples of endemic taxa are mentioned under each region.

The Great Barrier Reef (GBR) region (Knox 1963) was diagnosed by Ebach *et al.* (2013a: 337) as the 'marine region comprising the Great Barrier Reef and associated islands'. They listed Lizard Island as

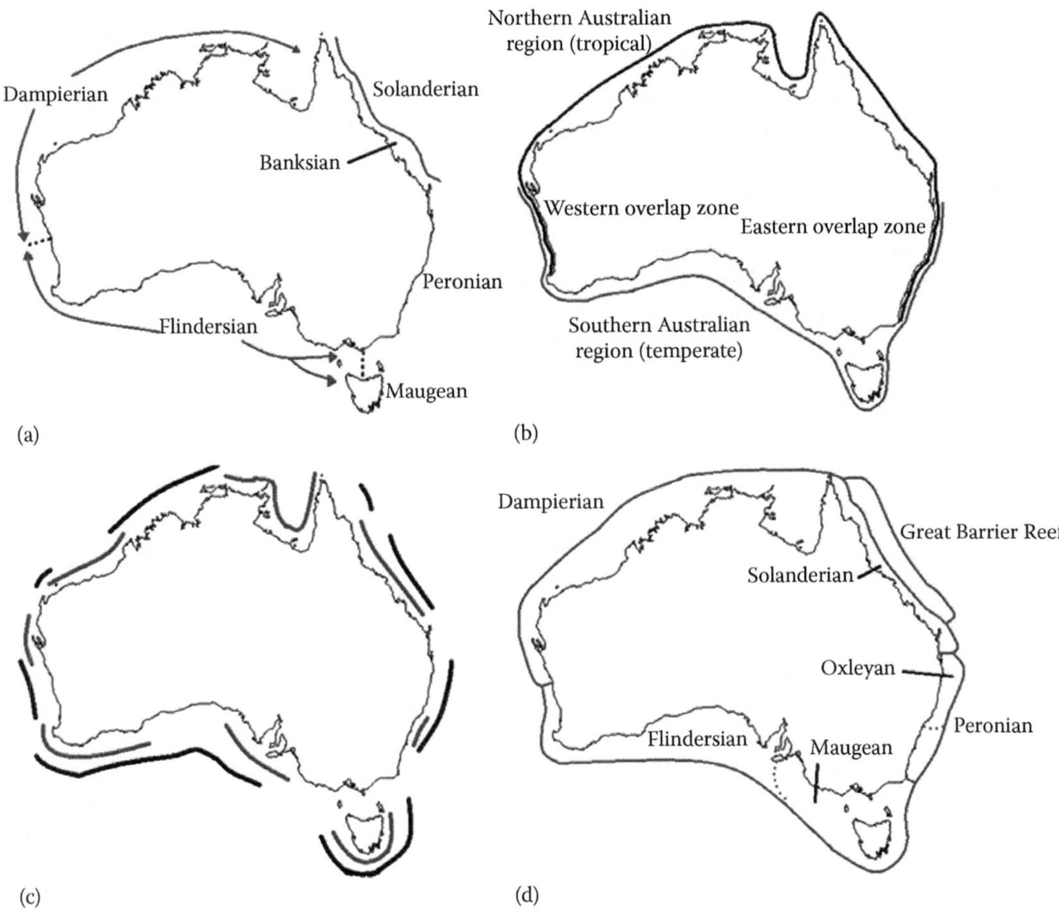

FIGURE 6.1 Summary of biogeographical maps for Australian marine fishes: (a) Whitley, G.P., Marine zoogeographical regions of Australia, *The Australian Naturalist* 8:166–167, 1932, based on his map, excluding non-Australian areas; (b) Wilson, B.R., and G.R. Allen, Origins and adaptations of the fauna of inland waters, In *Fauna of Australia, vol. 1A: General Articles*, ed. G.R. Dyne and D.W. Walton, 184–201, Canberra, Australia, Australian Government Publishing, 1987, based on their figure 3.8; (c) Last, P.R., *et al.*, Biogeographic structure and affinities of the marine demersal ichthyofauna of Australia, *Journal of Biogeography* 38:1484–1496, 2011, showing continental slope (in black) and shelf (grey) zones, based on their Figure 1; (d) Ebach *et al.* (2013a), based on their figures 6.6 and 6.7.

the type locality. Whitley (1932) referred to it as the *Solanderian Province*, and provided a new name (*Banksian Province*) for the adjacent Queensland coast. However, Hedley's (1904) original usage of *Solanderian* refers to coastal Queensland, and is thus equivalent to Whitley's *Banksian*. The GBR is home to the greatest diversity of fishes in Australia. Hoese *et al.* (2006) recorded 1625 species from the region, but noted that endemism in the region is relatively low, with only 20 species.

The Solanderian Region (Hedley 1904) was diagnosed by Ebach *et al.* (2013a: 337) as the 'coastal marine region extending from the Torres Strait (10°S latitude) to southern Queensland (possibly Moreton Bay or 25°S) or near the border of New South Wales'. They selected Magnetic Island as the type locality. Originally defined as 'the Queensland coast from Moreton Bay to Torres Strait' (Hedley 1904: 880), it was later defined as *Queensland* by malacologists Schilder and Schilder (1939). Whitley (1932: 166) noted that the region (his Banksian Province) possesses 'a marine fauna quite distinct from that of the Great Barrier Reef'. The region is equivalent to Hoese *et al.*'s (2006) *NE Coast* zone. Hoese *et al.* noted that it included four endemic fish species. No endemic species are found among the families listed in Table 6.1, though some species have distributions circumscribed by the Solanderian plus GBR,

TABLE 6.1

Distribution of Endemic Australian Marine Species within Selected Shorefish Families

Family	Species	D	G	S	P	Fe	Fw
Serranidae	*Acanthistius ocellatus*	–	–	–	+	+	–
	Acanthistius paxtoni	–	–	–	+	–	–
	Acanthistius serratus	–	–	–	–	–	+
	Caesioperca razor	–	–	–	–	+	+
	Chelidoperca margaritifera	–	+	–	–	–	–
	Epinephelides armatus	–	–	–	–	–	+
	Epinephelus darwiniensis	+	–	–	–	–	–
	Epinephelus ergastularius	–	+	–	+	–	–
	Epinephelus trophis	+	–	–	–	–	–
	Hypoplectrodes annulatus	–	–	–	+	+	–
	Hypoplectrodes cardinalis	–	–	–	–	–	+
	Hypoplectrodes jamesoni	–	–	+	+	–	–
	Hypoplectrodes maccullochi	–	–	–	+	+	–
	Hypoplectrodes nigroruber	–	–	–	+	+	+
	Hypoplectrodes wilsoni	–	–	–	–	–	+
	Lepidoperca brochata	–	–	–	+	–	–
	Lepidoperca caesiopercula	–	–	–	+	–	–
	Lepidoperca filamenta	–	–	–	–	–	+
	Lepidoperca occidentalis	–	–	–	–	–	+
	Lepidoperca pulchella	–	–	–	+	+	–
	Lepidoperca tasmanica	–	–	–	–	+	–
	Othos dentex	–	–	–	–	+	+
	Plectranthias robertsi	–	+	–	–	–	–
	Pseudanthias sheni	+	–	–	–	–	–
Pseudochromidae	*Assiculoides desmonotus*	+	–	–	–	–	–
	Assiculus punctatus	+	–	–	–	–	–
	Congrogadus winterbottomi	+	–	–	–	–	–
	Labracinus lineatus	+	–	–	–	–	–
	Ogilbyina novaehollandiae	–	+	+	–	–	–
	Ogilbyina queenslandiae	–	+	–	–	–	–
	Oxycercichthys veliferus	–	+	–	–	–	–
	Pseudochromis flammicauda	–	+	–	–	–	–
	Pseudochromis howsoni	+	–	–	–	–	–
	Pseudochromis quinquedentatus	+	+	+	–	–	–
	Pseudochromis reticulatus	+	–	–	–	–	–
	Pseudochromis wilsoni	+	+	+	–	–	–
Opistognathidae	*Opistognathus alleni*	+	–	–	–	–	–
	Opistognathus darwiniensis	+	–	–	–	–	–
	Opistognathus eximus	–	+	–	–	–	–
	Opistognathus inornatus	+	–	–	–	–	–
	Opistognathus jacksoniensis	–	+	+	+	–	–
	Opistognathus reticeps	+	–	–	–	–	–
	Opistognathus reticulatus	+	–	–	–	–	–
	Opistognathus seminudus	–	+	–	–	–	–

(Continued)

TABLE 6.1 (CONTINUED)

Distribution of Endemic Australian Marine Species within Selected Shorefish Families

Family	Species	D	G	S	P	Fe	Fw
	Opistognathus stigmosus	–	+	–	–	–	–
	Opistognathus verecundus	+	–	–	–	–	–
	Stalix flavida	+	–	–	–	–	–
Plesiopidae	*Assessor flavissimus*	–	+	–	–	–	–
	Beliops xanthocrossos	–	–	–	–	–	+
	Fraudella carassiops	–	+	–	–	–	–
	Notograptus gregoryi	+	–	–	–	–	–
	Paraplesiops alisonae	–	–	–	–	+	–
	Paraplesiops bleekeri	–	–	–	+	–	–
	Paraplesiops meleagris	–	–	–	–	+	+
	Paraplesiops poweri	–	+	+	–	–	–
	Paraplesiops sinclairi	–	–	–	–	–	+
	Plesiops genaricus	–	+	–	–	–	–
	Trachinops brauni	–	–	–	–	–	+
	Trachinops caudimaculatus	–	–	–	–	+	–
	Trachinops noarlungae	–	–	–	–	–	+
	Trachinops taeniatus	–	–	–	+	–	–
Pempheridae	*Parapriacanthus elongatus*	–	–	–	–	+	+
	Pempheris affinis	–	–	+	+	–	–
	Pempheris compressa	–	–	–	+	–	–
	Pempheris klunzingeri	+	–	–	–	–	+
	Pempheris multiradiata	–	–	–	+	+	+
	Pempheris ornata	–	–	–	–	–	+
	Pempheris ypsilychnus	+	+	+	–	–	–
Chaetodontidae	*Chaetodon assarius*	+	–	–	–	–	+
	Chelmon marginalis	+	–	–	–	–	–
	Chelmon muelleri	+	+	+	–	–	–
	Chelmonops curiosus	–	–	–	–	–	+
	Chelmonops truncatus	–	–	–	+	–	–
	Roa australis	+	–	–	–	–	–
Pomacanthidae	*Chaetodontoplus meredithi*	–	+	+	–	–	–
	Chaetodontoplus personifer	+	–	–	–	–	–
Labridae	*Achoerodus gouldii*	–	–	–	–	–	+
	Achoerodus viridis	–	–	–	+	+	–
	Austrolabrus maculatus	–	–	–	+	–	+
	Anampses lennardi	+	–	–	–	–	–
	Bodianus frenchii	–	–	–	+	–	+
	Bodianus unimaculatus	–	–	–	+	+	–
	Bodianus vulpinus	–	–	–	–	–	+

(Continued)

TABLE 6.1 (CONTINUED)

Distribution of Endemic Australian Marine Species within Selected Shorefish Families

Family	Species	D	G	S	P	Fe	Fw
	Choerodon cauteroma	+	–	–	–	–	–
	Choerodon cephalotes	+	+	+	+	–	–
	Choerodon cyanodus	+	+	+	+	–	–
	Choerodon frenatus	+	+	+	+	–	–
	Choerodon rubescens	+	–	–	–	–	–
	Choerodon sugillatum	+	+	–	–	–	–
	Choerodon venustus	+	+	+	–	–	–
	Cirrhilabrus morrisoni	+	–	–	–	–	–
	Cirrhilabrus randalli	+	–	–	–	–	–
	Conniella apterygia	+	–	–	–	–	–
	Coris auricularis	–	–	–	–	–	+
	Coris aurilineata	–	–	+	+	–	–
	Dotalabrus alleni	–	–	–	–	–	+
	Dotalabrus aurantiacus	–	–	–	–	+	+
	Eupetrichthys angustipes	–	–	–	+	+	+
	Halichoeres brownfieldi	+	–	–	–	–	+
	Haletta semifasciata	–	–	–	–	+	+
	Heteroscarus acroptilus	–	–	–	+	+	+
	Neodax balteatus	–	–	–	+	+	+
	Notolabrus parilus	–	–	–	–	–	+
	Notolabrus tetricus	–	–	–	+	+	–
	Olisthops cyanomelas	–	–	–	+	+	+
	Ophthalmolepis lineolata	–	–	–	+	+	+
	Pseudocoris aequalis	–	+	+	+	–	–
	Pictilabrus brauni	–	–	–	–	–	+
	Pictilabrus laticlavius	–	–	–	+	+	+
	Pictilabrus viridis	–	–	–	–	–	+
	Pseudolabrus biserialis	–	–	–	–	–	+
	Pseudolabrus guentheri	–	+	+	+	–	–
	Pseudolabrus rubricundus	–	–	–	+	+	+
	Siphognathus argyrophanes	–	–	–	–	+	+
	Siphognathus attenuatus	–	–	–	–	+	+
	Siphognathus beddomei	–	–	–	–	+	+
	Siphognathus caninus	–	–	–	–	+	+
	Siphognathus radiatus	–	–	–	–	+	+
	Siphognathus tanyourus	–	–	–	–	+	+
	Suezichthys bifurcatus	–	–	–	–	–	+
	Suezichthys cyanolaemus	+	–	–	–	–	+
	Suezichthys devisi	–	+	+	+	–	–
	Thalassoma septemfasciatum	+	–	–	–	–	+

Note: D: Dampierian; G: Great Barrier Reef (including western Coral Sea); S: Solanderian; P: Peronian; Fe: Flindersian east (= Maugean); Fw: Flindersian west.

Dampierian or Peronian regions. For example, Paxton *et al.* (2006: Table 4) listed 29 coral reef fishes that are endemic to the Solanderian plus GBR regions (as 'Cape York south to the Capricorn Group'). Wilson and Allen (1987) suggested that the distinction between the two eastern Queensland areas was largely ecological and noted that the entire area is commonly grouped under a single GBR area. Last *et al.* (2011) did not recognise a separate region that corresponded with the Solanderian, but instead recognised three provinces that span the Solanderian and GBR areas plus the western Coral Sea: a single continental shelf province and two continental slope provinces, one in the north and one in the south (Figure 6.1c). The need for additional subdivision of the western Coral Sea was suggested by Last *et al.* (2014) based on subsequent analysis of continental slope collections.

The Dampierian region (Hedley 1904) was defined by Ebach *et al.* (2013a: 339) as the 'marine region extending from the Torres Strait west to the Houtman Abrolhos Islands (28°50′S, off the coast from Geraldton), Western Australia'. Aside from misspelling Dampierian 'Damperian' (also misspelt by Kott 1952 and Paxton *et al.* 2006), Ebach *et al.*'s definition is identical to Hedley's original definition. Ebach *et al.* designated Dampier, Western Australia, as the type locality. As diagnosed, the region spans Hoese *et al.*'s (2006) Gulf of Carpentaria (two endemic fish species), *N Coast* (26 endemic fish species), *NW Coast* (27 endemic fish species) and *Central W Coast* zones (seven endemic fish species). Paxton *et al.* (2006: Table 3) listed 49 endemic coral reef fish species from the region. Last *et al.* (2011) recognised three continental shelf provinces (one spanning the Gulf of Carpentaria and the northern coast of the Northern Territory, one equivalent to Hoese *et al.*'s NW Coast zone, and a subtropical one from Shark Bay to Geraldton) and two continental slope provinces (one off the Northern Territory and the Kimberley Coast of Western Australia, and the other off North West Cape and Barrow Island) (Figure 6.1c). Moore *et al.* (2014) documented shorefish distributions for the Kimberley Coast, noting differences between inshore and offshore localities; the latter referred to 'the shelf edge atolls, which arise from deeper waters (200–400 m) along the continental margin' (2014: 162). Similar inshore and offshore differences had been noted earlier (e.g. Allen and Russell 1986; Hutchins 1999).

Based on their analysis of the distribution patterns of various fish and marine invertebrates (molluscs, corals and echinoderms), Wilson and Allen (1987) recognised a single, broad *Northern Tropical Zone* that spanned the areas covered by the Dampierian, Solanderian and GBR regions (Figure 6.1b). However, they noted (1987: 51), 'There is a significant endemic element in the northern fish fauna, with about 13% of the species and 9% of the genera in this category. Of these, about 40% are restricted to the eastern coast and Great Barrier Reef, 30% are from the seas west of Cape York and 30% have widespread distributions encompassing both of these regions.' Wilson and Allen (1987: Table 3.4) noted that three fish families are endemic to northern Australia (though with two extending into New Guinea), all of which are monotypic: Tetrabranchiidae, Leptobramidae and Rhinoprenidae. The Rhinoprenidae was placed in the synonymy of the widely distributed marine family Ephippidae by Cavalluzzi (2000). The Notograptidae has also been regarded as endemic to northern Australian and southern New Guinea (e.g. Springer 1982; Nelson 2006), but is now placed in the synonymy of the Plesiopidae (Mooi and Gill 2004).

The Flindersian region (Cotton 1930) was defined by Ebach *et al.* (2013a: 337) as the marine region 'from Geraldton, Western Australia, to the New South Wales–Victorian border including Tasmania'. They designated the Gulf of St. Vincent off Adelaide, South Australia, as type locality. The area was originally termed *Adelaidean* by Hedley (1904: 880) to include the 'marine fauna which extends from Melbourne along the south coast of Australia', but was later modified by Cotton (1930) and renamed *Flindersian*. However, Hedley's (1904: 879) map includes Tasmania west of its southern tip (west of South Cape Bay). Ebach *et al.* included the Maugean Province (Iredale and May 1916) as a subregion within the Flindersian, defining it as 'the coast of Victoria (from Mallacoota) to Tasmania and eastern South Australia to the vicinity of Victor Harbour' (Ebach *et al.* 2013a: 338). Whitley (1932) recognised the Maugean as a separate province, but with vague northern boundaries (Figure 6.1a). The Maugean was one of the few Australian marine areas of endemism acknowledged by Briggs and Bowen (2012), largely as a consequence of their emphasis on percentage endemism rather than the distribution of endemic species. Ebach *et al.* (2013a,b) overlooked Hatcher's (1991) recognition of a Leeuwin Province for the Western Australian areas under the influence of the warm Leeuwin Current. The province was later restricted by Hutchins (1994) to include only southwestern Australia, extending from Coral Bay to the Recherche Archipelago, thus incorporating the western part of the Flindersian and the southern part of the Dampierian. Species endemic to this area include the apogonid *Vincentia punctata*, the pempherid *Pempheris klunzingeri* and the plesiopids *Trachinops brauni* and *Paraplesiops sinclairi* (Figure 6.2a,b,d; Table 6.1).

As defined by Ebach *et al.* (2013a), the Flindersian spans the following coastal zones of Hoese *et al.* (2006): *Lower W Coast*, *SW Coast*, Great Australian Bight, *S Gulfs Coast*, Bass Strait and Tasmania. The

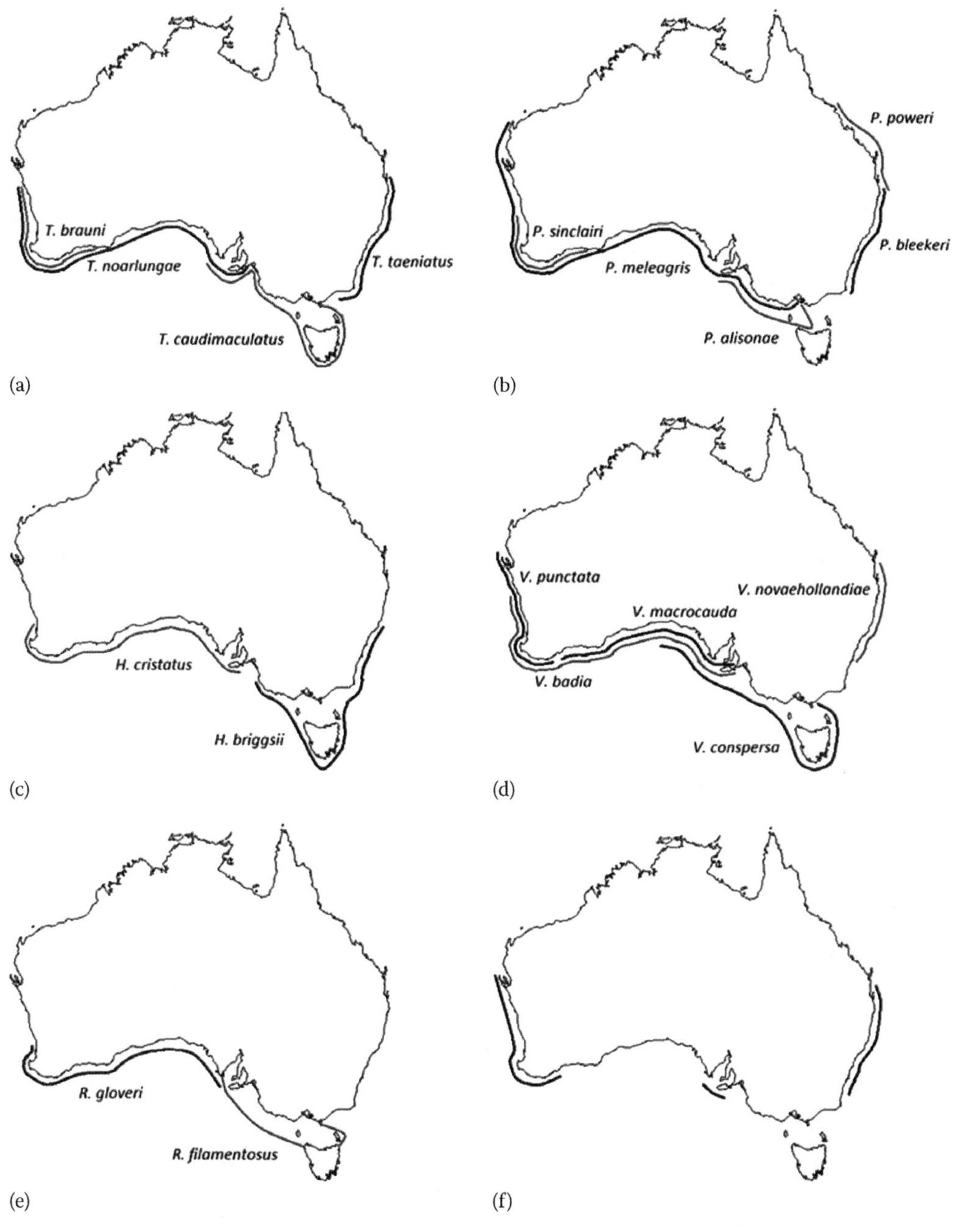

FIGURE 6.2 Distributions of select southern Australian endemic shorefish taxa: (a) species in the endemic plesiopid genus *Trachinops*; (b) species in the endemic plesiopid genus *Paraplesiops*; (c) species in the endemic syngnathid genus *Histiogamphelus*; (d) species in the endemic apogonid genus *Vincentia*; (e) species in the endemic antennariid genus *Rhycherus*; (f) the endemic pataecid *Pataecus fronto*.

last two essentially circumscribe the Maugean subregion. Last *et al.* (2011) recognised three continental shelf provinces, one in the southwest, one centred on the gulfs area and the other essentially equivalent to Ebach *et al.*'s Maugean subregion. They also recognised two large continental slope provinces, one extending from off the southwest tip of Australia to off Kangaroo Island and other off the Maugean area. In addition, they recognised a subtropical continental slope province from off Perth to Shark Bay, thus spanning the southern part of the Dampierian and the northern part of the Flindersian (Figure 6.1c).

As can be gleaned from Table 6.1, there is significant endemism within the Flindersian region. Although there is often overlap in the vicinity of the gulf area of South Australia, many species can be classified as either having eastern (Maugean) or western (Leeuwin) Flindersian distributions. Genera with probable sister species in the eastern and western Flindersian include the antennariid *Rhycherus*, the apogonid *Vincentia* and the plesiopid *Trachinops* (e.g. Figure 6.2a,d,e). Taxa endemic to the overlap (gulf) area include the syngnathids *Kaupus costatus* and *Vanacampus versoi* and the tripterygiid *Trinorfolkia cristata*. Several species have isolated populations in the gulf area. For example, the pataecid *Pataecus fronto* is known from three isolated warm-temperate to subtropical populations, one in the east, one in the west and one in the gulf area (Figure 6.2f).

The Peronian region (Hedley 1904) was defined by Ebach *et al.* (2013a: 339) as the 'marine region extending from Gippsland, Victoria, to Moreton Bay, Queensland'. They designated Salamander Bay, Port Stephens, New South Wales, as the type locality. Hedley's (1904) original description of the region extended farther south to include the east coast of Tasmania. Ebach *et al.* (2013a: 339) further recognised the Oxleyan (Iredale 1937) as a subregion within the Peronian, defining it as the 'subregion extending from Port Jackson to Moreton Bay'. The Peronian spans Hoese *et al.*'s (2006) *Central E Coast* and *Lower E Coast* zones, for which they reported 4 and 10 endemic fish species, respectively. Last *et al.* (2011) recognised two provinces, a continental shelf one that coincides with Ebach *et al.*'s Oxleyan subregion, and a continental slope one that spans most of the Peronian. Aside from the various serranid, plesiopid and chaetodontid species listed in Table 6.1, fish species endemic to the Peronian include the pomacentrids *Mecaenichthys immaculatus*, *Parma microlepis* and *P. unifasciata*, the clinids *Heteroclinus nasutus* and *H. whiteleggii* and the gobiesocids *Alabes parvulus*, *Cochleoceps orientalis* and *Kopua kuiteri*.

Wilson and Allen (1987) did not recognise separate Peronian and Flindersian regions, and instead combined them into a large *Southern Temperate Zone*, which they contrasted with their Northern Tropical Zone (Figure 6.1c). They noted that 'of an estimated 600 inshore species, about 85% are endemics' (Wilson and Allen 1987: 51). They further noted the presence of subtropical transition zones on both coasts. Their eastern zone (from Seal Rocks, New South Wales, to Yeppoon, Queensland, and the Capricorn-Bunker Group), mostly agrees with the Oxleyan subregion. They also noted that there are 'two main centres of endemism which correspond to the southwestern and southeastern corners of the continent' (Wilson and Allen 1987: 53), and listed apparent east–west sister species in southern Australia (e.g. Figures 6.2 and 6.3).

High endemism has long been noted for southern Australian fishes. Endemic families include the Gnathanacanthidae, Dinolestidae, Brachionichthyidae (though also known from Eocene deposits in Monte Bolca, Italy; Carnevale and Pietsch 2010), Enoplosidae (also regarded as being represented in the Monte Bolca deposits, but the basis for this – the fossil species *Enoplosus pygopterus* – bears only a superficial resemblance to *Enoplosus* and is undoubtedly incorrectly classified) and Pataecidae, and there as well are several enigmatic genera that are inconclusively assigned to families (e.g. *Caesioscorpis*, *Schuettea* and *Percalates*). Several other families, though more widespread, have high levels of diversity and endemism in southern Australia (e.g. Gobiesocidae, Gobiidae, Platycephalidae, Clinidae, Monacanthidae, Antennariidae, Syngnathidae, Ostraciidae, Plesiopidae and Labridae).

Wilson and Allen's (1987) justification for recognising only two broad zones, despite acknowledging endemism within those zones, was based largely on difficulties in defining discrete borders for the areas. Their eastern and western subtropical transition zones, resulting largely from the seasonal influences of the warm southward-flowing East Australian and Leeuwin Currents (e.g. Hutchins 1993, 1994; Kuiter 1993), in particular make identification of area borders problematic. This seasonal presence of tropical species in warm-temperate or temperate waters has long been noted. For example:

> The fish market of Sydney represents two different aspects; in winter, it contains only a few sorts, of dark colour, and almost all the same as those found in the Melbourne sea, and exclusively Australian. In the warm months of the year appear denizens of the Indian and Pacific oceans, adorned with all the splendid hues that nature seems so apt to lavish on the tropical sorts. (Castelnau 1879: 349)

Last *et al.* (2011) largely circumvented the problem of overlap zones by restricting their areas of endemism (provinces) to small, isolated areas (Figure 6.1c) separated from each other by broad transition zones. Although such transition zones are issues for delimiting ecological regions, historical analyses

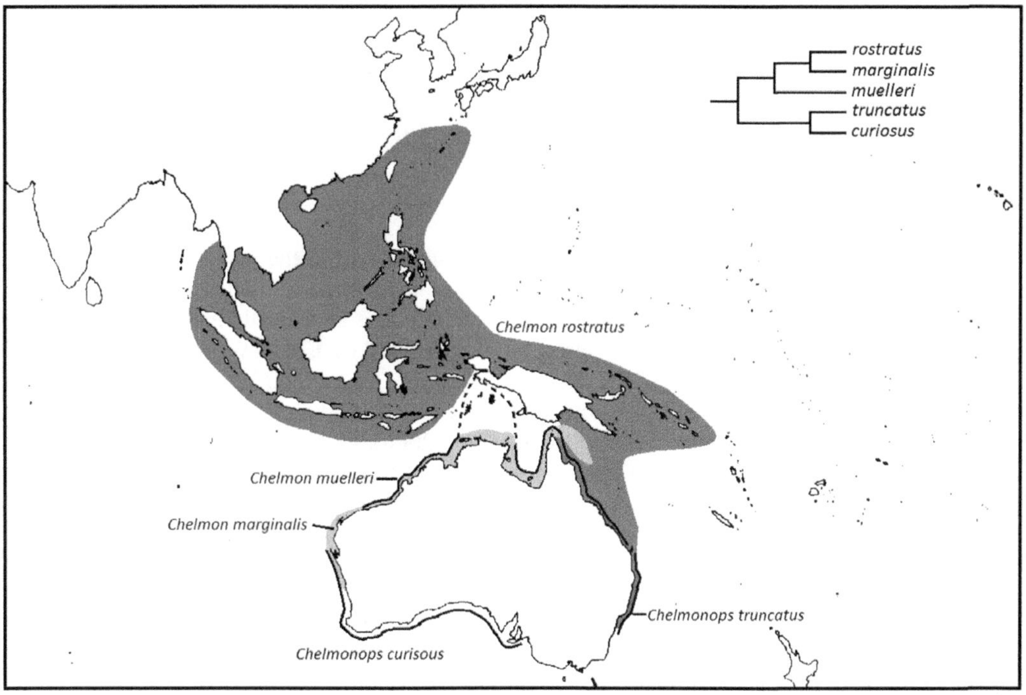

FIGURE 6.3 Phylogenetic relationships (based on Bellwood *et al.* 2010) and distribution of species of the chaetodontid genera *Chelmon* and *Chelmonops*. Dotted line indicates possible distribution of *Chelmon muelleri* based on isolated record from Fak Fak, West Papua. (From Allen, G.R., and M.V. Erdmann, *Reef Fishes of the East Indies*, Perth, Australia, Tropical Reef Research, 2012.)

based on shared patterns of taxon and area relationships are not likely to be hampered by apparent overlaps, or at least they can be accommodated (Harold and Mooi 1994).

Wilson and Allen (1987) also provided an important summary of oceanographic and geological events that have contributed to the distributions of Australian marine species. These include (but are not limited to) the fragmentation of Gondwana and the northward drift of Australia, leading to the closure of the Tethys Ocean and the opening of the Southern Ocean seaway, with associated oceanographic and climatic changes; the periodic closure of the Torres and Bass Straits during glaciation events, leading to the isolation of eastern and western populations; and the influence of tropical (South Equatorial, East Australia and Leeuwin) and circumpolar (West Wind Drift) ocean currents. They also noted various environmental factors that may serve as barriers to certain species, such as low or high sea temperatures and the broad absence of reef habitat in some areas.

Australian Fishes in a Global Context

Efforts to understand Australian fishes from a global perspective can be generally grouped as either studies based on individual species distributions or studies based on the phylogenetic relationships of taxa.

Analyses Based on Species Distributions

With few exceptions, almost all attempts at putting Australian fishes in a global biogeographic context have been based on the distribution of individual species. On the basis of a relatively low percentage of species endemism, Briggs (1974), for example, considered the eastern tropical portion of Queensland (i.e. the GBR and Solanderian regions of Ebach *et al.* 2013a) as part of his large Indo-Polynesian

Province. Similarly, Wilson and Allen (1987) noted that many species from their Northern Tropical Zone are widespread throughout the Indo-West Pacific Province (which spans East Africa to central Oceania). They further noted that some species in the western portion of their Northern Tropical Zone are otherwise known only from the Indian Ocean, whereas in the eastern portion, some are known only from the West Pacific. For their Southern Temperate Zone, they noted, in particular, that many species are shared with New Zealand (11% of Australia's inshore fishes or about 30% of New Zealand's inshore fish fauna). They also noted that some species have antitropical distributions (discussed further on pp. 117–118).

Hoese *et al.* (2006: Table 3) provided a more detailed breakdown of the non-Australian distribution patterns of Australian fish species, listing counts of species under some 25 mutually exclusive distribution patterns (clarified in their Figure 6). Their 10 highest categories accounted for nearly 90% (88.1%) of Australian species: endemic (24.1%), Indo-West Central Pacific (12.6%), Indo-West Pacific (12.3%), West Pacific (12.3%), Southwest Pacific (6.1%), circumglobal (5.7%), East Indo-West Pacific (5.7%), West-Central Pacific (3.4%), New Guinea (3.4%) and Southern circumglobal (2.5%).

Using similar non-Australian categories to Hoese *et al.* (2006), Last *et al.* (2011: Table 2) provided a considerably more detailed evaluation of the non-Australian distributions of 3734 Australian demersal species grouped under their 16 continental slope and shelf 'provinces'. They also analysed species distributions across depth (*bathome*) and province categories within Australia in order to evaluate biogeographic areas based on endemic species distributions. The primary aim of the analyses, however, was to develop an understanding of the faunal composition of areas for resource and conservation management.

Kulbicki *et al.* (2013) attempted a hierarchical classification for the biogeography of the world's tropical reef fishes based on similarity analyses of the distribution of individual species. Depending on the selection of different treatments of data, they produced four different classifications. In each, however, they recognised three global realms, Indo-Pacific, Tropical Eastern Pacific and Atlantic, with Australian fishes classified in the first. Classification of tropical Australian reef fishes differed in the analyses. In one, for example, all were classified in a single region extending from Western Australia across the southwest Pacific to the Kermadec Islands, whereas in the other northern and west Australian fishes were grouped under a broad Central Indo-Pacific region that extends westward to the Indian subcontinent, north to Japan and east to the Solomon Islands, while central eastern Australian fishes were grouped in a broad Central Pacific region that covers much of the southern and central Pacific.

Although area classifications based on species ranges are often compared with those based on taxon relationships, they are not directly comparable. At most, species range approaches produce area phenograms, which lack information about the phylogenetic (thus historical) relationships about either areas or taxa (Parenti and Ebach 2009). Another obvious point of contrast is in how the two approaches treat widespread species (i.e. species that occur across two or more areas under consideration). In species-based analyses, such species are the sole source of information for grouping areas, whereas in approaches based on taxon relationships they are viewed as a source of potential error (Nelson and Ladiges 1991; Gill and Kemp 2002). Gill and Kemp further note that apparent widespread species may be an artefact of poor taxonomic understanding or practice.

Analyses Based on Taxon Relationships

The basis for this approach is that biogeographic history will be reflected in the phylogenetic relationships and distributions of taxa. Ideally, taxon relationship methods involve the production of a cladogram of taxa, then replacing the taxa on the phylogeny with their distributions to produce an areagram (Parenti and Ebach 2009). Areagrams from various taxa are then compared to explore general area relationships. Such studies are relatively recent. However, because higher taxa such as genera and families are often (and ideally) monophyletic, studies of their distributions may also give insight into area relationships. Analyses based on higher taxa distributions have a much longer history.

One of the earliest and most important attempts at a global overview was Theodore Nicholas Gill's (1885) analysis of the distributions of genera and families of (mostly) freshwater fishes. Gill reviewed earlier global biogeographic classifications by Wallace (1876) and Allen (1878) and expanded on his earlier contribution (Gill 1875). He placed Australia in his *Austrogaean* realm, 'limited northward by

Wallace's line or strait, which separates Lombok from Bali and Celebes from Borneo, including Papua or New Guinea and the Solomon Islands to the eastward, and southward embraces Tasmania or Van Diemen's Land' (Gill 1885: 23). He classified it with South America (consisting of his *Dendrogaean* or *Neotropical* and *Amphigaean* or *Temperate South American* realms), Africa, Madagascar and the Mascarene Islands (together forming his *Afrogaean* realm) and New Zealand (his *Ornithogaean* realm) in a relationship he termed *Eogaea*, which he contrasted with one consisting of Eurasia (his *Eurygaean* realm), North America (his *Anglogaean* realm) and India and Southeast Asia (together forming his *Indogaean* realm), termed *Cenogaea*. He had proposed both Eogaea and Cenogaea in his earlier study of fish family distributions (Gill 1875). With the exception of the placement of India (which he noted was problematic), his classification agrees with our current concepts of the supercontinents Gondwana and Laurasia.

Certain area relationships based on shared families or genera have long been noted, mostly in temperate and subtropical fishes. For example, New Zealand shares several families and numerous genera with temperate Australia. Additional relationships based on shared families and genera are discussed later in the chapter.

Since the 1970s and the refinement of methods for estimating phylogenetic relationships (e.g. Hennig 1966), there has been a notable increase in the production of relationship-based biogeographic studies. One of the earliest was Vari's (1978) cladistic study of the Indo-Pacific family Terapontidae, a group with high endemism in Australia and New Guinea (particularly in freshwater). He employed vicariance biogeography methods to analyse area relationships and attributed distribution patterns to the fragmentation of Gondwana. Subsequent taxon relationship-based studies have employed various biogeographic methods, some of which might be better viewed as dispersalist (e.g. centre of origin) rather than area relationship methods. Nevertheless, there has been a growing body of phylogenetic information, particularly with the advent of molecular systematics. For example, fish families and higher taxa with Australian marine endemic species that have featured in relationship-based biogeographic studies include Atherinidae (Crowley 1990; Unmack and Dowling 2010), Blenniidae (Springer 1988, 1999; Springer and Williams 1994; Williams 1988a), Chaetodontidae (Blum 1989; Bellwood *et al.* 2010; Cowman and Bellwood 2013; Bellwood and Pratchett 2014), Cirrhitoidei (Burridge 1999, 2000a,b; Burridge and White 2000; Burridge and Smolenksi 2004; Burridge *et al.* 2006), Gobioidei (Murdy 1989; Larson 2001), Labridae (Gomon and Paxton 1986; Bellwood 1994; Clements *et al.* 2004; Bernardi *et al.* 2004; Barber and Bellwood 2005; Gomon 2006; Read *et al.* 2006; Cowman *et al.* 2009; Hodge *et al.* 2012; Cowman and Bellwood 2013; Puckridge *et al.* 2015), Terapontidae (Vari 1978; Davis *et al.* 2012), Pentacerotidae (Kim 2012), Pholidichthyidae (Springer and Larson 1996), Plesiopidae (Hardy 1985; Hutchins 1987; Smith-Vaniz and Johnson 1990; Mooi 1995; Mooi and Gill 2004), Pomacanthidae (Bellwood *et al.* 2004), Pomacentridae (Santini and Polacco 2006; Timm *et al.* 2008; Cowman and Bellwood 2013) and Pseudochromidae (Winterbottom 1986; Mooi and Gill 2004).

Despite the ever-increasing volume of phylogenetic and associated biogeographic studies, there have been few attempts to explore general area relationships (e.g. Santini and Winterbottom 2002; Mooi and Gill 2004; Halas and Winterbottom 2009; Parenti and Ebach 2009). Instead, most studies have taken the form of narratives that discuss mechanisms for the distribution of individual study taxa. Moreover, rather than placing emphasis on area relationships, for most studies the emphasis has been placed on issues such as the origins of high biodiversity, habitat and trophic shifts, speciation modes, lineage dating and centres of origin.

General Area Relationship Patterns and Associated Narratives

Even without comparative area relationship analyses, several general patterns are suggested by cursory comparison. As noted in the section on areas within Australia, there are some examples involving the east–west separation of closely related tropical marine fishes (e.g. the pomacanthids *Chaetodontoplus meredithi* and *C. personifer*). Separation is usually around the Torres Straits area (separating Dampierian from Solanderian and GBR distributions), and most explanations for east–west vicariance involve the periodic formation of land bridges between Australia and New Guinea during the Pliocene and Pleistocene (e.g. Wilson and Allen 1987; Bellwood and Pratchett 2014).

However, the situation may be more complex and require a more global perspective. Many tropical Australian east–west relationships also involve distributions that extend to the north (to Indonesia or farther) or east (to the Coral Sea or farther). In the chaetodontid genus *Chelmon*, for example, there is a break near the Torres Straits between the sister species *C. marginalis* (Dampierian) and *C. rostratus* (mostly GBR and Solanderian). The distribution of the latter species extends eastward to the Solomon Islands, northward to the Ryukyu Islands, and westward through the Indo-Malaysian Archipelago to the Andaman Sea (Figure 6.3). A different situation is found in the sister genera *Ogilbyina* and *Labracinus* (Pseudochromidae). Though the two genera occur either side of the Torres Straits, *Ogilbyina* has two endemic species in the Solanderian and GBR, which together form the sister of an additional species in New Caledonia, and *Labracinus* has a southern Dampierian endemic species, which is sister to two species that occur farther north (collectively, the northern Kimberley Coast of the Dampierian region to the Ryukyu Islands) (Figure 6.4). Similar northward relationships of Dampierian taxa may or may not involve east–west relationships. For example, the sister group of a clade of Dampierian, Indonesian and Philippine pseudochromids is widely distributed in northern Australia (Figure 6.5). Species of the labrid genus *Choerodon* also do not fall into east–west endemic pairs across northern Australia; instead, the sister taxa of the seven endemic Australian species are more widely distributed to the north, east or west of Australia (Puckridge *et al.* 2015). Similarly, the Dampierian–Leeuwin endemic labrid *Thalassoma septemfasciatum* is sister species to the Hawaiian endemic *T. ballieui* (an antitropical distribution) and these together form the sister group to a circumtropical clade made up of the remaining members in the genus (Bernardi *et al.* 2004). The Dampierian pseudochromid subfamilies Assiculinae and Assiculoidinae are successive sister groups to a clade consisting of Anisochrominae, Congrogadinae and Pseudoplesiopinae (Mooi and Gill 2004; Gill 2013); genera in these three subfamilies have either western Indian Ocean or eastern Indian Ocean to Pacific distributions (Winterbottom 1986; Gill and Edwards 1999; Gill and Fricke 2001). The northern Australian–southern New Guinea plesiopid genus *Notograptus* is sister to the genus *Acanthoplesiops*, which ranges from Tonga to the Red Sea but is absent from Australia and southern New Guinea (Mooi and Gill 2004; Gill *et al.* 2013).

Mooi and Gill (2004) noted that area relationships that place Australia as sister to wider tropical distributions may be general in nature and that they may also involve temperate Australian components (e.g. subfamilies of Plesiopidae, Mooi 1993; *Parma* and *Mecaenichthys* in the pomacentrid subfamily Stegastinae, Cooper *et al.* 2009; *Chelmon* and *Chelmonops*, Figure 6.3; *Choerodon* and Odacini, Figure 6.6), some of which imply early diversification within Australia. Heads (2014a) noted numerous similar examples involving Tasman and Coral Sea marine and terrestrial organisms, and additional examples are discussed below. Analyses that employ ancestral distribution reconstruction methods have identified the Indo-Australian area as a *centre of origin* (e.g. Cowman and Bellwood 2013; Bellwood and Pratchett 2014). Bellwood *et al.* (2010, p. 344) suggest that, within the Chaetodontidae, 'the bannerfish clade (*Chelmon*, *Chelmonops*, *Amphichaetodon*, *Coradion*, *Forcipiger*, *Johnrandallia*, *Heniochus* and *Hemitaurichthys*) has close links with Australia and temperate or subtropical waters, and a subtropical Australian origin for this clade remains a distinct possibility'. Rather than an Australian origin, we instead suggest that such distributions indicate early divergence and isolation of Australian taxa within a more broadly distributed older taxon.

Widespread distribution of such groups outside of Australia has been attributed variously to periods of range expansion in the Oligocene and Palaeocene/Eocene prior to the disruption of the Tethys Ocean or to more recent Miocene and Pliocene dispersion (e.g. Cowman and Bellwood 2013). In contrast, Winterbottom (1986) invoked Mesozoic Gondwanan fragmentation to explain the distribution of congrogadine pseudochromid genera on African and Indo-Australian plates. Springer (1988) also invoked much older (Cretaceous) rifting of the Lord Howe Rise from the east Australian coast to explain the distribution of species within the *Ecsenius opsifrontalis* complex (Blenniidae). Mooi (1995) offered a similar mechanism to explain the distribution of the plesiopid species pair *Plesiops genaricus* (GBR) and *P. insularis* (Lord Howe Rise and New Caledonia).

As previously noted in the section on areas within Australia, there are numerous examples of east–west divergence of temperate shorefishes (Figure 6.2). Mechanisms for divergence usually involve fluctuating sea temperatures and periodic closure of the Bass Strait (the formation of the Bassian Isthmus) during the Late Quaternary (Wilson and Allen 1987; Hutchins 1987).

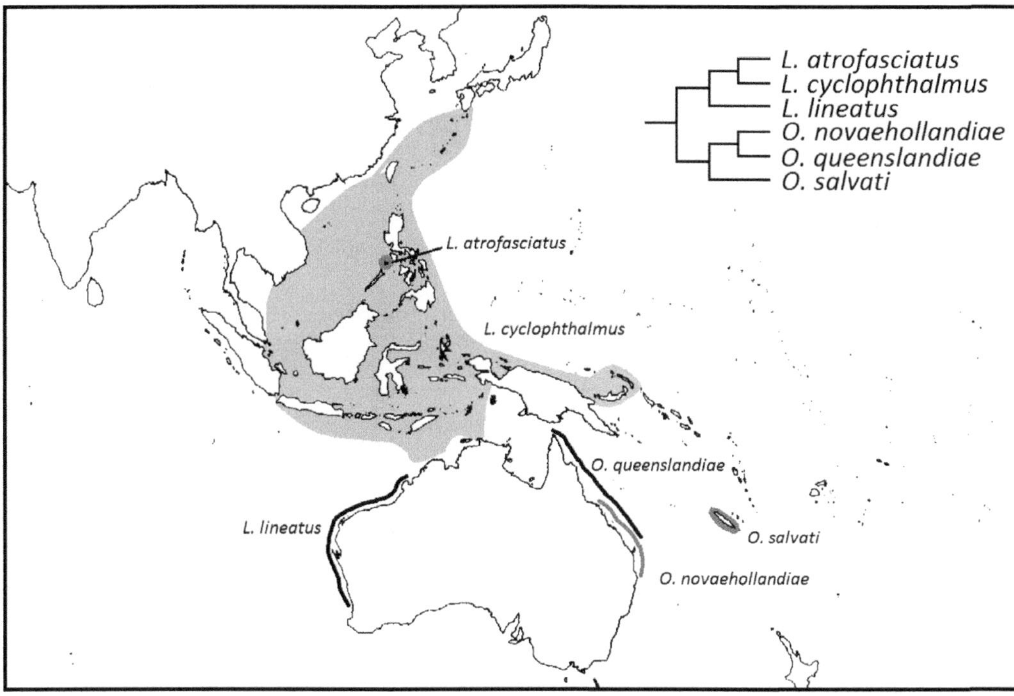

FIGURE 6.4 Phylogenetic relationships and distribution of species of the pseudochromid genera *Labracinus* and *Ogilbyina*. (Based on Gill, A.C., *Smithiana Monograph* 1, 1–214, 2004.)

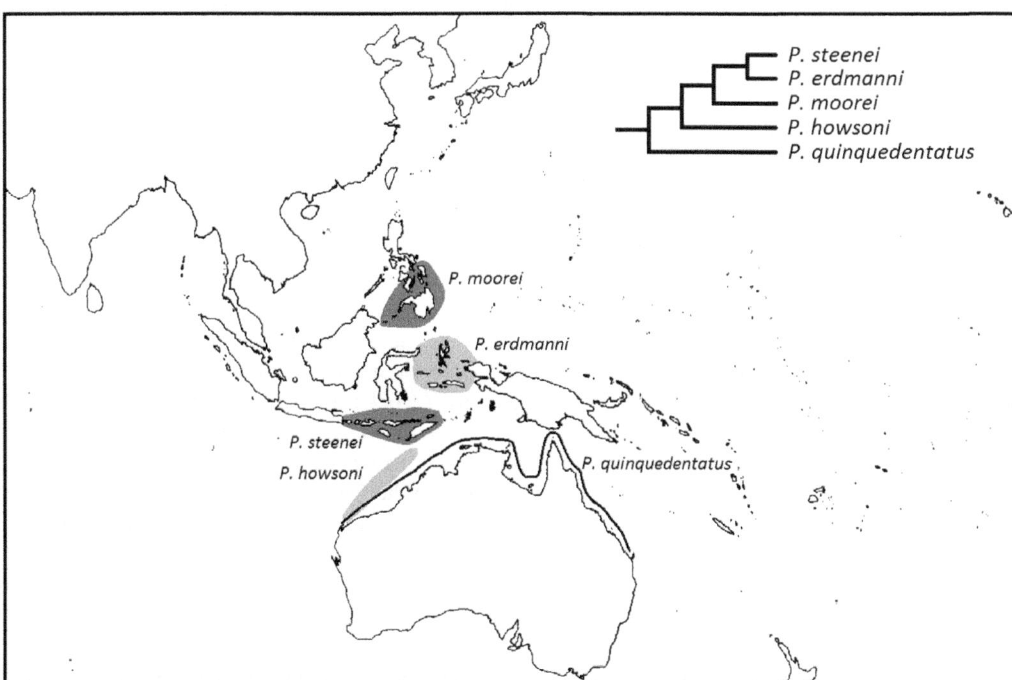

FIGURE 6.5 Phylogenetic relationships and distribution of species of the pseudochromid *Pseudochromis quinquedentatus* complex. (Based on Gill, A.C., and G.R. Allen, *Zootaxa* 2924, 57–62, 2011.)

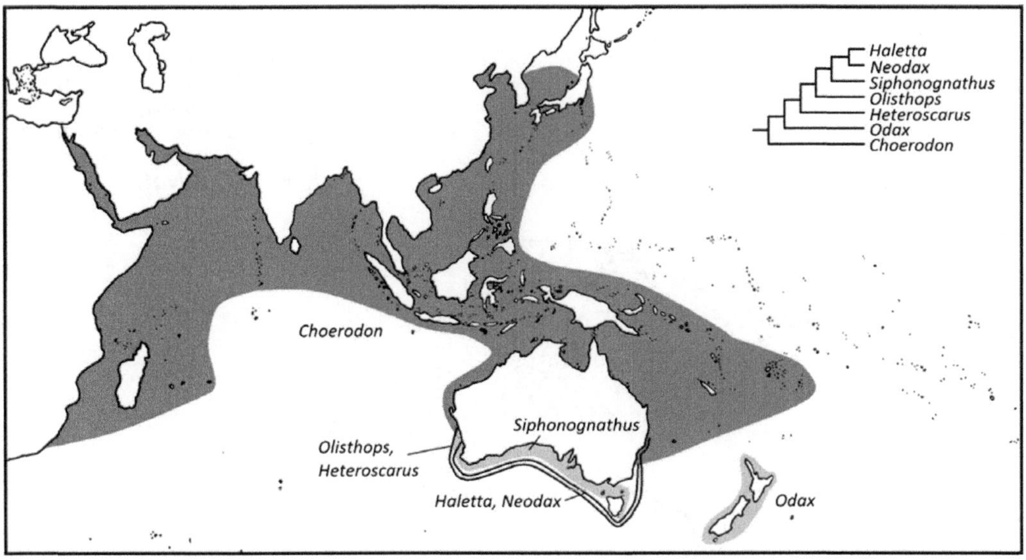

FIGURE 6.6 Phylogenetic relationships. (Based on Clements, K.D., *et al.*, *Molecular Phylogenetics and Evolution* 32, 575–587, 2004) and the distribution of *Choerodon* and odacin genera (family Labridae).

A close relationship between New Zealand and temperate Australia has also long been noted. New Zealand shares several fish families with temperate Australia, including Leptoscopidae and Thalasseleotrididae. With just two genera and three species, the latter family was recently described as the sister group of the large (ca. 2000 species) worldwide family Gobiidae (Gill and Mooi 2012), which perhaps suggests the divergence is relatively old. With the exception of a southwest Atlantic species, the family Rhombosoleidae is also endemic to temperate Australia and New Zealand. Other taxa shared exclusively by temperate or subtropical Australia and New Zealand include *Notolabrus* (Labridae), Odacini (Labridae, Figure 6.6), *Caesioperca* (Serranidae), *Optivus* (Trachichthyidae), *Stigmatopora* (Syngnathidae), *Pseudophycis* (Moridae), *Upeneichthys* (Mullidae), *Contusus* (Tetraodontidae) and *Kopua* (Gobiesocidae). Other families and genera range a little beyond the two areas – for example, to Lord Howe Island and the nearby Elizabeth and Middleton Reefs, Norfolk Island or the Kermadec Islands (e.g. the family Arripidae, the clinid genus *Cristiceps*, the pomacentrid genus *Parma* and the microcanthid genus *Atypichthys*). In almost all cases, such families and genera involve different species in the different areas, thereby implying vicariance between the areas.

The association of Australian and New Zealand biotas has been attributed either to Gondwanan fragmentation or to more recent dispersal across the Tasman Sea, and the alternative mechanisms have been the subject of ongoing debate among biogeographers (e.g. Heads 2014a). Gomon and Paxton (1986) discussed, then dismissed, Gondwanan fragmentation to explain the distribution of odacin labrids in Australia and New Zealand, and instead favoured more recent trans-Tasman dispersal. In contrast, Clements *et al.* (2004) invoked mid-Eocene diversification of the Odacini in warm-temperate to subtropical conditions, when Australia and New Zealand were closer to Antarctica, prior to the development of the modern Southern Ocean circulation. Relationships within the Odacini suggest diversification within Australia subsequent to vicariance with New Zealand, though much of the distribution pattern associated with the Australian diversification is now obscured by range expansion and sympatry (Figure 6.6).

Temperate Australian relationships may extend around much of the Southern Ocean. The family Chironemidae has endemic species in Australia, New Zealand and in the Juan Fernández and Desventuradas (San Felix) Islands off Chile. A similar distribution pattern is found in the chaetodontid genus *Amphichaetodon*, with one species (*A. howensis*) in Peronian Australia and northern New Zealand to the Kermadec Islands and the other (*A. meldae*) in the Desventuradas Islands, as well as in various marine and terrestrial organisms (Heads 2014a). In their study of chironemid biogeography, Burridge *et al.* (2006) invoked Miocene to Eocene eastward range expansion from Australia and New Zealand to older, now submerged or subducted islands in the eastern Pacific, followed by oceanographic changes that led to the vicariance of the eastern Pacific species. In contrast, Burridge (1999) invoked periodic changes in the West Wind Drift during the Pleistocene–Pliocene to explain distribution patterns throughout the Southern Ocean in the cheilodactylid genus *Nemadactylus*, a genus in which member species have a lengthy (7–12 month) pelagic larval stage. We note, however, that long pelagic larval duration is not necessarily correlated with widespread distribution (Wellington and Victor 1989; Victor and Wellington 2000; Gill and Kemp 2002; Zapata and Herrón 2002; Heads 2005).

Some Australian marine fishes are involved in disjunct, antitropical distributions (Hubbs 1952). Randall (1982) noted that the disjunction may only occur in equatorial waters and added the related concept of antiequatorial distribution. He reviewed Indo-Pacific examples and listed 56 species with isolated distributions in the north (e.g. Hawaii, Midway Atoll, southern Japan) or south (e.g. southern Africa, Madagascar, Mascarene Islands, southern Australia, New Zealand, South Pacific). For example, the microcanthid species *Microcanthus strigatus* has disjunct populations in the Leeuwin and Peronian regions of Australia, the southwest Pacific, Hawaii, Midway Atoll and southern Japan to Taiwan (Figure 6.7a). Many of Randall's species are now recognised as sister species; for example, the northern population of the labrid *Coris picta* is now considered a separate species, *C. musume* (Figure 6.7b). Other sister species with antitropical or antiequatorial distributions include the southwest Pacific *Pseudochromis jamesi* and northwest Pacific *P. luteus* (Figure 6.7c). The two species in the pseudochromid genus *Cypho*

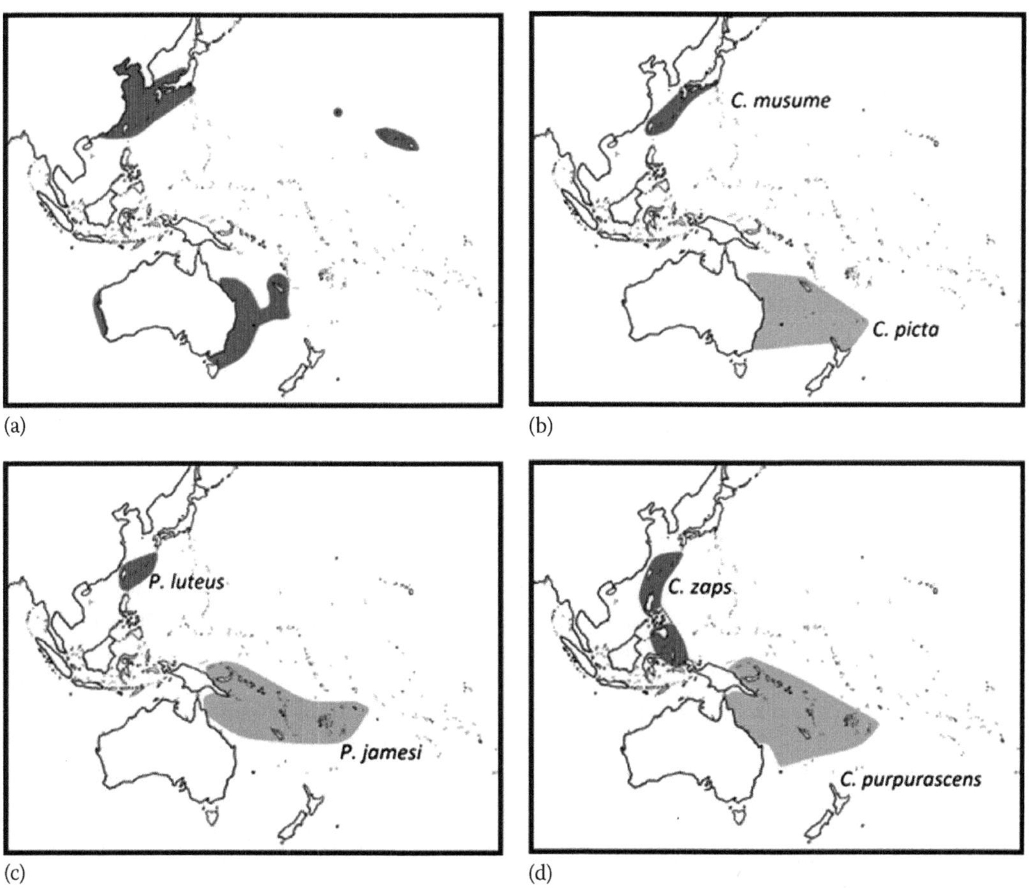

FIGURE 6.7 Antitropical and antiequatorial distributions of shorefishes: (a) the microcanthid *Microcanthus strigatus*; (b) the labrids *Coris musume* and *C. picta*; (c) the pseudochromids *Pseudochromis jamesi* and *P. luteus*; (d) species in the pseudochromid genus *Cypho* (the dotted line represents the probable presence of *C. zaps* in the poorly collected eastern Philippines).

have an almost identical distribution, except for the presence of the 'northern' species in equatorial Indonesian waters (Figure 6.7d). As such, the distribution approaches patterns shown in tropical species and doesn't qualify as antiequatorial. However, it suggests a continuum between antitropical and tropical patterns, which in turn suggests that similar explanatory mechanisms should be investigated.

Antitropical and antiequatorial distributions are apparent in higher taxa as well, including the families Cheilodactylidae, Girellidae, Oplegnathidae, and Neosebastidae, the subfamily Aracaninae (Ostraciidae), various genera such as *Gonorynchus* (Gonorynchidae), *Pentaceros* (Pentacerotidae), *Caprodon* (Serranidae), *Callanthias* (Callanthiidae), *Beliops* (Plesiopidae), *Pseudocaranx* (Carangidae), *Pseudolabrus* (Labridae), *Maroubra* (Syngnathidae), *Lissocampus* (Syngnathidae) and *Urocampus* (Syngnathidae) and the subgenus *Verreo* (Labridae, *Bodianus*).

Various mechanisms have been put forward to explain antitropical and antiequatorial distributions, including the Mesozoic fragmentation of Pangaea or Pacifica (e.g. Nelson 1985, 1986; Humphries and Parenti 1986), competitive exclusion from low-latitude regions by more recently evolved taxa (e.g. Briggs 1987a,b), global Eocene/Oligocene cooling followed by mid-Miocene equatorial warming (e.g. White 1986, 1989) and dispersal during cool Pleistocene glaciation events (e.g. Berg 1933; Randall 1982). Based on molecular clock estimates, Burridge and White (2000) proposed three transequatorial divergences for antitropical cheilodactylids during the mid-Miocene and Late Miocene to Pleistocene.

Consideration too should be extended to various freshwater taxa with antitropical distributions. Based on analyses of sequence data, Chen *et al.* (2014) identified a large freshwater fish clade, Percichthyoidea, consisting of a northern North American (Centrarchidae and Elassomatidae) and Asian (Sinopercidae) clade, and a southern Australian and South American clade (Percichthyidae). Although Chen *et al.* were able to hypothesise reasonable Late Mesozoic to Early Cenozoic scenarios for the diversification of the northern and southern clades, they were unable to explain the disjunction between the two clades, and appealed to the possible existence of a widespread marine ancestor. The frequent appeal to extinct, unknown marine ancestors to explain disjunct freshwater distributions (e.g. for certain atheriniform fishes by Aarn and Ivanstoff 1997) is both problematic and untestable; such explanations need to be put in the context of broader marine biogeographic patterns.

Discussion and Conclusions

Progress in our understanding of the biogeography of Australian marine fishes has been hampered by a set of issues. Perhaps the most significant has been the lack of a unifying goal, particularly for the classification of biogeographic areas. Although we feel an emphasis on evaluating area relationships will lead to a better understanding of the historical factors that underpin distribution patterns (e.g. Parenti and Ebach 2009; Heads 2015), most researchers have instead placed emphasis on ecological classifications. These may, however, contribute to improved understanding of other goals, such as biodiversity assessment and conservation and resource management. A lack of appreciation of the different approaches in biogeography has led to misleading comparisons between (and amalgamations of) different classifications, where, for example, ahistorical schemes based on the distribution of individual species are compared with those based on area and taxon relationships (e.g. Kulbicki *et al.* 2013). Though clearly not yet resolved by the discipline, the separation of ecology from history has been recognised as important for some time: 'The confusion between these two classes of ideas is one of the causes that have most retarded the science, and that have prevented it from acquiring exactitude' (de Candolle 1820: 383; translation from Nelson 1978: 281). Clarifying the goals of particular studies from the outset to avoid such confusion will permit the results of ecological, conservational and historical approaches to biogeography to remain distinct and complementary.

Scant attention has been given to the testing of Australian marine biogeographic areas from a cladistic biogeographic perspective. Do areas reflect shared biogeographic histories of their component biotas, or are they composites, reflecting two or more distinct histories? In other words, are there multiple relationships for a given area, indicating that two or more areas have been incorrectly lumped together? Conversely, are the areas a reflection of biogeographic history, or do they reflect ecological partitioning within historical areas? Is, for example, the distinction made between the Solanderian and GBR regions (and likewise, the various western Coral Sea slope areas proposed by Last *et al.* 2014) a reflection of current ecology or history? These questions imply the need for an iterative approach to assess whether areas have been correctly circumscribed (Harold and Mooi 1994).

Many of these questions stem from the choice of taxon and the identification of areas. As noted by Platnick (1991), biogeographers should have an advantage because they can choose taxa that exhibit maximal endemism and are most likely to provide results for the regions under investigation. Unfortunately, rather than allowing the distributions of the study taxa to delineate areas, regions are frequently predetermined based on geography, ecology or broadly recognised zones of biodiversity. Hence, areas analysed frequently have no endemic study taxa associated with them. Even if these predetermined areas are historically relevant, how could we expect that a taxon that is not endemic to them will have any information to contribute? Platnick (1991: xii) called this approach 'the triumph of hope over evidence'. Just as different character systems will provide evidence for different taxonomic levels, taxa and their distributions will provide area relationships at different levels of geographic space and time. Careful choice of taxonomic groups for comparative analysis that exhibit endemism among Australian areas and those that demonstrate endemic patterns between Australia and other regions will be an important step forward.

There is a particular need for an investigation of comparative area relationship studies to search for general area relationships – that is, area relationships shared by multiple groups of taxa (fish or otherwise). General area relationships are important because they are patterns that require general process explanations. It is important also that area relationship studies include both Australian and non-Australian areas, so that Australian areas can be put into a global context. The paucity of general area relationship studies for Australian marine fishes is noteworthy. Although there has been a rapid increase in phylogenetic studies (particularly as a result of advances in molecular sequencing methods), the emphasis from a biogeographic perspective has been on the generation of individual narratives rather than the search for general patterns. Such studies often appeal to unique events (such as chance dispersal across barriers) to explain distributions, rather than the shared general processes required to explain general area relationships.

Associated with the single taxon narrative has been an emphasis on lineage dating in phylogenies. Disparate dates are frequently used to dismiss otherwise shared distribution patterns, thus justifying ad hoc narratives based on single taxa. However, the various dating methods, which are primarily based on molecular clock hypotheses calibrated with fossil specimens (which may be incorrectly identified and at best provide only minimum ages for taxa), are liable to error and have been seriously questioned (Graur and Martin 2004; Heads 2012; Nelson and Ladiges 2009; Morrison 2014).

Advances in our understanding of the historical biogeography of Australian marine fishes are also impeded by the nature of the questions that biogeographers formulate. In summing up Willi Hennig's (1913–1976) influence on systematics, palaeoichthyologist Colin Patterson noted, 'Our mistake was thinking in terms of origins rather than relationships' (Patterson 2011: 124). This mistake extends to a large part of current historical biogeography, where interest is often focused on origins of species, origins of biodiversity, origins of habitat and trophic preferences, centres of origin and so on, rather than relationships between areas of endemism. Further to this issue has been the application of methods that impose centre-of-origin explanations on area trees, thus excluding the discovery of alternative explanations (e.g. ancestral distribution methods, reviewed in Kodandaramaiah 2010; Herrera *et al.* 2015; critiqued in Ladiges *et al.* 2012; Heads 2014b).

A final impediment is our limited understanding of the distribution of Australian marine fish species. This is in part to an incomplete understanding of the taxonomy of species, particularly widespread species (Gill and Kemp 2002). More important, however, it is due to the lack of adequate sampling of marine habitats. Cryptic reef fishes from deeper than about 50 m, are particularly poorly sampled. Moreover, comprehensive collecting is sparse for many areas. For example, even the relatively well-documented fish fauna of the GBR is far from comprehensively collected, being based primarily on two extensively collected areas (Lizard Island in the northern third of the reef and One Tree Island near the southern end). A global understanding of Australian fish distributions is also dependent on collecting efforts from non-Australian areas.

Despite these impediments, however, the growing body of information on the taxonomy and distribution of Australian marine fishes combined with robust hypotheses of their relationships will ultimately lead to a better understanding of their biogeography. These will provide the foundation for synthetic iterative studies that can identify historically relevant areas of endemism and investigate their relationships. Research programs to this end will finally attend to Sclater's (1858, p. 131) ancient lament that 'little or no attention is given to the fact that two or more of these geographical divisions may have much closer relations to each other than to any third'.

References

Aarn and W. Ivantsoff. 1997. Descriptive anatomy of *Cairnsichthys rhombosomoides* and *Iriatherina werneri* (Teleostei: Atheriniformes) and a phylogenetic analysis of Melanotaeniidae. *Ichthyological Exploration of Freshwaters* 8:107–150.

Allen, J.A. 1878. The geographical distribution of the Mammalia, considered in relation to the principal ontological regions of the earth, and the laws that govern the distribution of animal life. *Bulletin of the US Geological Survey* 4:1–376.

Allen, G.R. 1985. Fishes of Western Australia. In *Pacific Marine Fishes*, book 9, ed. W.E. Burgess and H.R. Axelrod, 2207–2534. Neptune City, NJ, T.F.H. Publications.

Allen, G.R. 1989. *Freshwater Fishes of Australia*. Neptune City, NJ, T.F.H. Publications, 240 p.

Allen, G.R. 1993. Fishes of Ashmore Reef and Cartier Island. *Records of the Western Australian Museum*, Suppl 44:67–86.

Allen, G.R. 1996. New records of reef and shore fishes from northwestern Australia. *Records of the Western Australian Museum* 18:109–112.

Allen, G.R. 2008. Conservation hotspots of biodiversity and endemism for Indo-Pacific coral reef fishes. *Aquatic Conservation: Marine and Freshwater Ecosystems* 18:541–556.

Allen, G.R., and M.V. Erdmann. 2012. *Reef Fishes of the East Indies*. Perth, Australia, Tropical Reef Research, 1292 p.

Allen, G.R., D.F. Hoese, J.R. Paxton, J.E Randall, B.C. Russell, W.A. Starck, F.H. Talbot and G.P. Whitley. 1976. Annotated checklist of the fishes of Lord Howe Island. *Records of the Australian Museum* 30:365–454.

Allen, G.R., S.H. Midgley and M. Allen. 2002. *Field Guide to the Freshwater Fishes of Australia*. Perth, Australia, Western Australian Museum, 394 p.

Allen, G.R., and B.C. Russell. 1986. Faunal survey of the Rowley Shoals, Scott Reef and Seringapatam Reef north-western Australia: Part VII. Fishes. *Records of the Western Australian Museum*, Suppl 25:75–103.

Allen, G.R., and W.F. Smith-Vaniz. 1994. Fishes of the Cocos (Keeling) Islands. *Atoll Research Bulletin* 412:1–21.

Allen, G.R., and R.C. Steene. 1979. The fishes of Christmas Island, Indian Ocean. *Australian National Parks and Wildlife Service*, Special Publication 2:1–81.

Allen, G.R., and R.C. Steene. 1988. *Fishes of Christmas Island, Indian Ocean*. Christmas Island, Australia, Christmas Island Natural History Association, 197 p.

Allen, G.R., and R. Swainston. 1988. *The Marine Fishes of North-Western Australia*. Perth, Australia, Western Australian Museum, 201 p.

Attaran-Farimani, G., S. Estekani, V.G. Springer, O. Crimmen, G.D. Johnson and C.C. Baldwin. 2016. Validation of the synonymy of the teleost blenniid fish species *Salarias phantasticus* Boulenger 1897 and *Salarias anomalus* Regan 1905 with *Ecsenius pulcher* (Murray 1887) based on DNA barcoding and morphology. *Zootaxa* 4072:171–184.

Barber, P.H., and D.R. Bellwood. 2005. Biodiversity hotspots: Evolutionary origins of biodiversity in wrasses (*Halichoeres*: Labridae) in the Indo-Pacific and new world tropics. *Molecular Phylogenetics and Evolution* 35:235–253.

Bellwood, D.R. 1994. A phylogenetic study of the parrotfishes family Scaridae (Pisces: Labroidei), with a revision of genera. *Records of the Australian Museum*, Suppl 20:1–86.

Bellwood, D.R., L. van Herwerden and N. Konow. 2004. Evolution and biogeography of marine angelfishes (Pisces: Pomacanthidae). *Molecular Phylogenetics and Evolution* 33:140–155.

Bellwood, D.R., S. Klanten, P.F. Cowman, M.S. Pratchett, N. Konow and L. van Herwerden. 2010. Evolutionary history of the butterflyfishes (f: Chaetodontidae) and the rise of coral feeding fishes. *Journal of Evolutionary Biology* 23:335–349.

Bellwood, D.R., and M.S. Pratchett. 2014. The origins and diversification of coral reef butterflyfishes. In *Biology of butterflyfishes*, ed. M.S. Pratchett, M.L. Berumen and D.G. Kapoor, 1–18. Boca Raton, FL, CRC Press.

Berg, L.S. 1933. Die bipolar Vebreitung der Organismen under die Eiszeit. *Zoogeografica* 1:444–484.

Bernardi, G., G. Bucciarelli, D. Costagliola, D.R. Robertson and J.B. Heiser. 2004. Evolution of coral reef fish *Thalassoma* spp. (Labridae): 1. Molecular phylogeny and biogeography. *Marine Biology* 144:369–375.

Blum, S.D. 1989. Biogeography of the Chaetodontidae: An analysis of allopatry among closely related species. *Environmental Biology of Fishes* 25:9–31.

Briggs, J.C. 1974. *Marine Zoogeography*. New York, McGraw-Hill, 475 p.

Briggs, J.C. 1987a. Antitropical distribution and evolution in the Indo-West Pacific Ocean. *Systematic Zoology* 36:237–247.

Briggs, J.C. 1987b. Antitropicality and vicariance. *Systematic Zoology* 36:206–207.

Briggs, J.C., and B.W. Bowen. 2012. A realignment of marine biogeographic provinces with particular reference to fish distributions. *Journal of Biogeography* 39:12–30.

Burridge, C.P. 1999. Molecular phylogeny of *Nemadactylus* and *Acantholatris* (Perciformes: Cirrhitoidea: Cheilodactylidae), with implications for taxonomy and biogeography. *Molecular Phylogenetics and Evolution* 13:93–109.

Burridge, C.P. 2000a. Biogeographic history of geminate cirrhitoids (Perciformes: Cirrhitoidea) with east–west allopatric distributions across southern Australia, based on molecular data. *Global Ecology and Biogeography* 9:517–525.

Burridge, C.P. 2000b. Molecular phylogeny of the Aplodactylidae (Perciformess: Cirrhitoidei), a group of Southern Hemisphere marine fishes. *Journal of Natural History* 34:2173–2185.

Burridge, C.P., C.R. Meléndez and B. Dyer. 2006. Multiple origins of the Juan Fernández kelpfish fauna and evidence for frequent and unidirectional dispersal of cirrhitoid fishes from the South Pacific. *Systematic Biology* 55:566–578.

Burridge, C.P., and A.J. Smolenski. 2004. Molecular phylogeny of the Cheilodactylidae and Latridae (Perciformes: Cirrhitoidea) with notes on taxonomy and biogeography. *Molecular Phylogenetics and Evolution* 30:118–127.

Burridge, C.P., and R.W.G. White. 2000. Molecular phylogeny of the antitropical subgenus *Gonistius* (Perciformes: Cheilodactylidae: *Cheilodactylus*): Evidence for multiple transequatorial divergences and non-monophyly. *Biological Journal of the Linnean Society* 70:435–458.

Cadwallader, P.L., and G.N. Backhouse. 1983. *A Guide to the Freshwater Fishes of Victoria*. Melbourne, Australia, F.D. Atkinson Government Printer, 249 p.

Carnevale, G., and T.W. Pietsch. 2010. Eocene handfishes from Monte Bolca, with description of a new genus and species, and a phylogeny of the family Brachionichthyidae (Teleostei: Lophiiformes). *Zoological Journal of the Linnean Society* 160:621–647.

Castelnau, F.L. de. 1872. Contributions to the ichthyology of Australia: 1. The Melbourne fish market. *Proceedings of the Zoological and Acclimatisation Society of Victoria* 1: 29–242,1 pl.

Castelnau, F.L. de. 1879. Essay on the ichthyology of Port Jackson. *Proceedings of the Linnean Society of New South Wales* 3:347–402.

Cavalluzzi, M. 2000. Osteology, phylogeny and biogeography of the marine fish family Ephippidae (Perciformes, Acanthuroidei), with comments on sister group relationships. PhD dissertation, School of Marine Science, College of William and Mary, Williamsburg, VA, 213 p.

Chen, W.J., S. Lavoué, L.B. Beheregaray and R.L. Mayden. 2014. Historical biogeography of a new antitropical clade of temperate freshwater fishes. *Journal of Biogeography* 41:1806–1818.

Clements, K.D., M.E. Alfaro, J.L. Fessler and M.W. Westneat. 2004. Relationships of the temperate Australasian labrid fish tribe Odacini (Perciformes; Teleostei). *Molecular Phylogenetics and Evolution* 32:575–587.

Commonwealth of Australia. 2005. *Interim Bioregionalisation of Australia*. Canberra, Australia, Department of Environment and Heritage, 142 p.

Commonwealth of Australia. 2006. *A Guide to the Integrated Marine and Coastal Regionalisation of Australia Version 4.0*. Canberra, Australia, Department of the Environment and Heritage, 16 p.

Cooper, W.J., L.L. Smith and M.W. Westneat. 2009. Exploring the radiation of a diverse reef fish family: Phylogenetics of the damselfishes (Pomacentridae), with new classifications based on molecular analyses of all genera. *Molecular Phylogenetics and Evolution* 52:1–16.

Cotton, B.C. 1930. Fissurellidae from the 'Flindersian' region, southern Australia. *South Australian Museum Records* 4:219–222.

Cowman, P.F., and D.R. Bellwood. 2013. The historical biogeography of coral reef fishes: Global patterns of origination and dispersal. *Journal of Biogeography* 40:209–224.

Cowman, P.F., D.R. Bellwood and L. van Herwerden. 2009. Dating the evolutionary origins of wrasse lineages (Labridae) and the rise of trophic novelty on coral reefs. *Molecular Phylogenetics and Evolution* 52:621–631.

Crowley, L.E.L.M. 1990. Biogeography of the endemic freshwater fish *Craterocephalus* (family Atherinidae). *Memoirs of the Queensland Museum* 28:89–98.

Davis, A.M., P.J. Unmack, B.J. Pusey and R.G. Pearson. 2012. Marine-freshwater transitions are associated with the evolution of dietary diversification in terapontid grunters (Teleostei: Terapontidae). *Journal of Evolutionary Biology* 25:1163–1179.

de Candolle, A.P. 1820. Géographie botanique. In *Dictionnaire des sciences naturelles XVIII*. Strasbourg and Paris.

Ebach, M.C. 2012. A history of bioregionalisation in Australia. *Zootaxa* 3392:1–34.

Ebach, M.C., A.C. Gill, S.T. Ahyong, A. Kwan, D.J. Murphy and G. Cassis. 2013a. Towards an Australian Bioregionalisation Atlas: A provisional area taxonomic revision of Australia's biogeographical regions. *Zootaxa* 3619:315–342.

Ebach, M.C., A.C. Gill, A. Kwan, S.T. Ahyong, D.J. Murphy and G. Cassis. 2013b. Corrections to a recently published area taxonomy of Australia. *Zootaxa* 3652:299–300.

Ebach, M.C., J.J. Morrone, L.R. Parenti and Á.L. Viloria. 2008. International Code of Area Nomenclature. *Journal of Biogeography* 35:1153–1157.

Finlay, H.J. 1925. Some modern conceptions applied to the study of the caenozoic Mollusca of New Zealand. *Dr R.D.M. Verbeek Memorial Birthday* 3:161–172.

Finlay, H.J. 1926. A further commentary on New Zealand molluscan systematics. *Transactions of the New Zealand Institute* 57:320–485.

Francis, M.P. 1993. Checklist of the coastal fishes of Lord Howe, Norfolk, and Kermadec Islands, southwest Pacific Ocean. *Pacific Science* 47:136–170.

Francis, M.P., and J.E. Randall. 1993. Further additions to the fish fauna of Lord Howe and Norfolk Islands, southwest Pacific Ocean. *Pacific Science* 47:118–135.

George, A., and V.G. Springer. 1980. Revision of the clinid fish tribe Ophiclinini, including five new species, and definition of the family Clinidae. *Smithsonian Contributions to Zoology* 307:i–iii, 1–31.

Gill, A.C. 2004. Revision of the Indo-Pacific dottyback fish subfamily Pseudochrominae (Perciformes: Pseudochromidae). *Smithiana Monograph* 1: i–ii,1–214, pls 1–12.

Gill, A.C. 2013. Classification and relationships of *Assiculus* and *Assiculoides* (Teleostei: Pseudochromidae). *Zootaxa* 3718:128–136.

Gill, A.C., and G.R. Allen. 2011. *Pseudochromis erdmanni*, a new species of dottyback with medially placed palatine teeth from Indonesia (Teleostei: Perciformes: Pseudochromidae). *Zootaxa* 2924:57–62.

Gill, A.C., S. Bogorodsky and A. Mal. 2013. *Acanthoplesiops cappuccino*, a new species of acanthoclinine fish from the Red Sea (Teleostei: Plesiopidae). *Zootaxa* 3750:216–222.

Gill, A.C., and A.J. Edwards. 1999. Monophyly, interrelationships and description of three new genera in the dottyback fish subfamily Pseudoplesiopinae (Teleostei: Perciformes: Pseudochromidae). *Records of the Australian Museum* 52:141–160.

Gill, A.C., and R. Fricke. 2001. Revision of the western Indian Ocean fish subfamily Anisochrominae (Perciformes, Pseudochromidae). *Bulletin of the Natural History Museum, London, Zoology Series* 67:191–207.

Gill, A.C., and J.M. Kemp. 2002. Widespread Indo-Pacific shore-fish species: A challenge for taxonomists, biogeographers, ecologists, and fishery and conservation managers. *Environmental Biology of Fishes* 65:165–174.

Gill, A.C., and R.D. Mooi. 2012. Thalasseleotrididae, new family of marine gobioid fishes from New Zealand and temperate Australia, with a revised definition of its sister taxon, the Gobiidae (Teleostei: Acanthomorpha). *Zootaxa* 3266:41–52.

Gill, A.C., and S.E. Reader. 1992. Fishes. In *Kowari* 3, *Reef Biology: A Survey of Elizabeth and Middleton Reefs, South Pacific*, ed. R. Longmore, 90–93, 193–228. Canberra, Australia, Australian National Parks and Wildlife Service.

Gill, T.N. 1875. On the geographical distribution of fishes. *Annals and Magazine of Natural History* (series 4) 15:251–255.

Gill, T.N. 1885. The principles of zoogeography. *Proceedings of the Biological Society of Washington* 2:1–39.

Gloerfelt-Tarp, T., and P.J. Kailola. 1984. *Trawled Fishes of Southern Indonesia and Northwest Australia*. Jakarta, the Australian Development Assistance Bureau, Australia, the Directorate General of Fisheries, Indonesia, and the German Agency for Technical Cooperation, Germany, 406 p.

Gomon, M.F. 2006. A revision of the labrid fish genus *Bodianus* with descriptions of eight new species. *Records of the Australian Museum*, Suppl 30:1–133.

Gomon, M.F., D.J. Bray and R.H. Kuiter (ed.). 2008. *Fishes of Australia's Southern Coast*. Melbourne, Australia, New Holland Publishers, 928 p.

Gomon, M.F., C.J.M. Glover and R.H. Kuiter (ed.). 1994. *The Fishes of Australia's South Coast*. Adelaide, State Print, 992 p.

Gomon, M.F., and J.R. Paxton. 1986. A revision of the Odacidae, a temperate Australian–New Zealand labroid fish family. *Indo-Pacific Fishes* 8: 1–57,pls 1–6.

Grant, E.M. 1965. *Guide to Fishes*. Brisbane, Australia, Department of Primary Industries, 280 p.

Grant, E.M. 1975. *Guide to Fishes*, 3rd edn. Brisbane, Australia, Queensland Government Co-ordinator General's Department, 640 p.

Grant, E.M. 1982. *Guide to Fishes*, 5th edn. Brisbane, Australia, Department of Harbours and Marine, 896 p.

Grant, E.M. 2002. *Guide to Fishes*, 9th edn. Redcliffe, Australia, E.M. Grant Pty, 880 p.

Graur, D., and W. Martin. 2004. Reading the entrails of chickens: Molecular timescales of evolution and the illusion of precision. *Trends in Genetics* 20:80–86.

Günther, A. 1859–1870. *Catalogue of the Fishes in the British Museum*. 8 vols. London, British Museum.

Halas, D., and R. Winterbottom. 2009. A phylogenetic test of multiple proposals for the origins of the East Indies coral reef biota. *Journal of Biogeography* 36:1847–1860.

Hardy, G.S. 1985. Revision of the Acanthoclinidae (Pisces: Perciformes), with descriptions of a new genus and five new species. *New Zealand Journal of Zoology* 11:357–393.

Harold, A.S., and R.D. Mooi. 1994. Areas of endemism: Definition and recognition criteria. *Systematic Biology* 43:261–266.

Hastings, P.A., and V.G. Springer. 1994. Review of *Stathmonotus*, with redefinition and phylogenetic analysis of the Chaenopsidae (Teleostei: Blennioidei). *Smithsonian Contributions to Zoology* 558:i–iii, 1–48.

Hatcher, B.G. 1991. Coral reefs in the Leeuwin Current: An ecological perspective. *Journal of the Royal Society of Western Australia* 74:115–127.

Heads, M. 2005. Towards a panbiogeography of the seas. *Biological Journal of the Linnean Society* 84:675–723.

Heads, M. 2012. Bayesian transmogrification of clade divergence times: A critique. *Journal of Biogeography* 39:1749–1756.

Heads, M. 2014a. *Biogeography of Australasia: A Molecular Analysis*. Cambridge, UK, Cambridge University Press, 493 p.

Heads, M. 2014b. Biogeography by revelation: Investigating a world shaped by miracles. *Australian Systematic Botany* 27:282–304.

Heads, M. 2015. The relationship between biogeography and ecology: Envelopes, models, predictions. *Biological Journal of the Linnean Society* 115:456–486.

Hedley, C. 1904. The effect of the Bassian Isthmus upon the existing marine fauna: A study in ancient geography. *Proceedings of the Linnean Society of New South Wales* 28:876–883.

Helfman, G.S., B.B. Collette and D.E. Facey. 1997. *The Diversity of Fishes*. Oxford, UK, Blackwell Science, 528 p.

Hennig, W. 1966. *Phylogenetic Systematics*. Urbana, University of Illinois Press, 263 p.

Herrera, N.D., J.J. ter Poorten, R. Bieler, P.M. Mikkelsen, E.E. Strong, D. Jablonski and S.J. Steppan. 2015. Molecular phylogenetics and historical biogeography amid shifting continents in the cockles and giant clams (Bivalvia: Cardiidae). *Molecular Phylogenetics and Evolution* 93:94–106.

Hodge, J.R., C.I. Read, L. van Herwerden and D.R. Bellwood. 2012. The role of peripheral endemism in species diversification: Evidence from the coral reef fish genus *Anampses* (family: Labridae). *Molecular Phylogenetics and Evolution* 62:653–663.

Hoese, D.F., D.J. Bray, J.R. Paxton and G.R. Allen. 2006. Fishes. In *Zoological Catalogue of Australia*, vol. 35, ed. P.L. Beesley and A. Wells, 1–2178. Canberra, Australia, ABRS and CSIRO Publishing.

Hubbs, C.L. 1952. Antitropical distribution of fishes and other organisms. *Proceedings of the 7th Pacific Science Congress* 3:324–329.

Hull, A.F.B. 1911. The birds of Lord Howe and Norfolk Islands. *The Emu* 11:58–61.

Humphries, C.J., and L.R. Parenti. 1986. *Cladistic Biogeography*. Oxford, UK, Clarendon Press, 98 p.

Hutchins, J. B. 1987. Description of a new plesiopid fish from south-western Australia, with a discussion of the zoogeography of *Paraplesiops*. *Records of the Western Australian Museum* 13:231–240.

Hutchins, J.B. 1993. Dispersal of tropical fishes to temperate seas in the southern hemisphere. *Journal of the Royal Society of Western Australia* 74:79–84.

Hutchins, J.B. 1994. A survey of the nearshore reef fish fauna of Western Australia's west and south coasts: The Leeuwin Province. *Records of the Western Australian Museum*, Suppl 46:1–66.

Hutchins, J.B. 1999. Biogeography of the nearshore marine fish fauna of the Kimberley, Western Australia. In *Proceedings of the 5th Indo-Pacific Fish Conference (Noumea, 3–8 November 1997)*, ed. B. Séret and J.Y. Sire, 99–108. Paris, France, Société Française d'Ichtyologie.

Hutchins, J.B. 2001. Checklist of the fishes of Western Australia. *Records of the Western Australian Museum*, Suppl 63:9–50.

Hutchins, J.B., and R. Swainston. 1986. *Sea Fishes of Southern Australia*. Perth, Australia, Swainston Publishing, 180 p.

Interim Marine and Coastal Regionalisation Technical Group. 1998. *Interim Marine and Coastal Regionalisation of Australia: An Ecosystem Classification for Marine and Coastal Environments, Version 3.3*. Canberra, Australia, Environment Australia, 104 p.

Iredale, T. 1937. A basic list of the land Mollusca of Australia. *The Australian Zoologist* 8:287–333.

Iredale, T., and A.F.B. Hull. 1929. The loricates of the Neozelanic Region. *The Australian Zoologist* 5: 305–323,pl. 34.

Iredale, T., and W.L. May. 1916. Misnamed Tasmanian chitons. *Proceedings of the Malacological Society of London* 12:94–117.

Iredale, T., and G.P. Whitley. 1938. The fluvifaunulae of Australia. *South Australian Naturalist* 18:64–68.

Kim, S.Y. 2012. Phylogenetic systematics of the family Pentacerotidae (Actinopterygii: Order Perciformes). *Zootaxa* 3366:1–111.

Kodandaramaiah, U. 2010. Use of dispersal-vicariance analysis in biogeography: A critique. *Journal of Biogeography* 37:3–11.

Knox, G.A. 1963. The biogeography and intertidal ecology of the Australasian coasts. *Oceanography and Marine Biology, Annual Review* 1:341–404.

Kott, P. 1952. The ascidians of Australia: I. Stolibranchiata Lahille and Phlebobranchiata Lahille. *Australian Journal of Marine and Freshwater Research* 3:205–334.

Kuiter, R.H. 1993. *Coastal Fishes of South-Eastern Australia*. Bathurst, Australia, Crawford House Press, 437 p.

Kuiter, R.H. 1996. *Guide to Sea Fishes of Australia*. Sydney, Australia, New Holland, 433 p.

Kuiter, R.H. 2013. *Victoria's Freshwater Fishes*. Seaford, Australia, Aquatic Photographics, 178 p.

Kulbicki M., V. Parravicini, D.R. Bellwood, E. Arias-Gonzàlez, P. Chabanet, S.R. Floeter, A. Friedlander, *et al.* 2013. Global biogeography of reef fishes: A hierarchical quantitative delineation of regions. *PLoS ONE* 8(12): e81847.

Ladiges, P., M.J. Bayly, and G. Nelson. 2012. Searching for ancestral areas and artifactual centers of origin in biogeography: With comment on east–west patterns across southern Australia. *Systematic Zoology* 61:703–708.

Lake, J.S. 1971. *Freshwater Fishes and Rivers of Australia*. Melbourne, Australia, Nelson, 61 p.

Lake, J.S. 1978. *Australian Freshwater Fishes*. Melbourne, Australia, Thomas Nelson, 160 p.

Larson, H.K. 2001. A revision of the gobiid fish genus *Mugilogobius* (Teleostei: Gobioidei), and its systematic placement. *Records of the Western Australian Museum*, Supplement 62: i–iv, 1–233.

Larson, H.K., and K.C. Martin. 1990. *Freshwater Fishes of the Northern Territory*. Darwin, Australia, Northern Territory Museum of Arts and Sciences, 102 p.

Larson, H.K., R.S. Williams and M.P. Hammer. 2013. An annotated checklist of the fishes of the Northern Territory, Australia. *Zootaxa* 3696:1–293.

Last, P.R., and D.C. Gledhill. 2009. A revision of the Australian handfishes (Lophiiformes: Brachionichthyidae), with descriptions of three new genera and nine new species. *Zootaxa* 2252:1–77.

Last, P., V. Lyne, G. Yearsley, D. Gledhill, M. Gomon, T. Rees and W. White. 2005. *Validation of National Demersal Fish Datasets for the Regionalisation of the Australian Continental Slope and Out Shelf (> 40m depth)*. Hobart, Australia, National Oceans Office, 99 p.

Last, P.R., V.D. Lyne, A. Williams, C.R. Davies, A.J. Butler and G.K. Yearsley. 2010. A hierarchical framework for classifying seabed biodiversity with application to planning and managing Australia's marine biological resources. *Biological Conservation* 143:1675–1686.

Last, P.R., J.J. Pogonoski, D.C. Gledhill, W.T. White and C.J. Walker. 2014. The deepwater demersal ichthyofauna of the western Coral Sea. *Zootaxa* 3887:191–224.

Last, P.R., E.O.G. Scott and F.H. Talbot. 1983. *Fishes of Tasmania*. Hobart, Australia, Tasmanian Fisheries Development Authority, 563 p.

Last, P.R., and J.D. Stephens. 1994. *Sharks and Rays of Australia*. Australia, CSIRO, 513 p, 84 pls.

Last, P.R., and J.D. Stephens. 2009. *Sharks and Rays of Australia*, 2nd edn. Australia, CSIRO, 656 p.

Last, P.R., W.T. White, D.C. Gledhill, J.J. Pogonoski, V. Lyne and N.J. Bax. 2011. Biogeographic structure and affinities of the marine demersal ichthyofauna of Australia. *Journal of Biogeography* 38:1484–1496.

Leggett, R., and J.R. Merrick. 1987. *Australian Native Fishes for Aquariums*. Artarmon, Australia, J.R. Merrick Publications, 241 p.

Lord, C.E. 1923. A list of the fishes of Tasmania. *Papers and Proceedings of the Royal Society of Tasmania* 1922:60–73.

Lord, C.E. 1927. A list of the fishes of Tasmania. *Journal of the Pan-Pacific Research Institute* 2(4):11–16.

Lord, C.E., and H.H. Scott. 1924. *A Synopsis of the Vertebrate Animals of Tasmania*. Hobart, Australia, Oldham, Beddome and Meredith, 96 p.

Lundberg, J.G., M. Kottelat, G.R. Smith, M.L.J. Stiassny and A.C. Gill. 2000. So many fishes, so little time: An overview of recent ichthyological discovery in continental waters. *Annals of the Missouri Botanical Garden* 87:26–62.

Lyne, V.D., W.T. White, D.C. Gledhill, P.R. Last, T. Rees and R. Porter-Smith. 2009. *Analysis of Australian Continental Shelf Provinces and Biomes Based on Fish Data*. Hobart, Australia, CSIRO Marine and Atmospheric Research, 44 p.

McCulloch, A.R. 1922. Checklist of the fish and fish-like vertebrates of New South Wales. *Australian Zoological Handbook* 1: 1–104,pls. 1–43.

McCulloch, A.R., and G.P. Whitley. 1925. A list of the fishes known from Queensland waters. *Memoirs of Queensland Museum* 8:125–182.

McCulloch, A.R. 1930. A checklist of the fishes recorded from Australia. *Memoirs of the Australian Museum* 5:1–534.

McDowall, R.M. (ed.). 1980. *Freshwater Fishes of South-Eastern Australia*. Sydney, Australia, A.H. and A.W. Reed, 208 p.

McDowall, R.M. (ed.). 1996. *Freshwater Fishes of South-Eastern Australia*, 2nd edn. Sydney, Australia, Reed Books, 247 p.

Macleay, W. 1881a. Descriptive catalogue of the fishes of Australia: Part I. *Proceedings of the Linnean Society of New South Wales* 5:302–444.

Macleay, W. 1881b. Descriptive catalogue of the fishes of Australia: Part II. *Proceedings of the Linnean Society of New South Wales* 5: 510–629,pls. 13–14.

Macleay, W. 1881c. Descriptive catalogue of the fishes of Australia: Part III. *Proceedings of the Linnean Society of New South Wales* 6: 1–138, 202–387.

Macleay, W. 1881d. A descriptive catalogue of Australian fishes: Part IV. *Proceedings of the Linnean Society of New South Wales* 6:202–387.

Macleay, W. 1884. Supplement to the descriptive catalogue of the fishes of Australia. *Proceedings of the Linnean Society of New South Wales* 9:2–64.

Marshall, T.C. 1964. *Fishes of the Great Barrier Reef and Coastal Waters of Queensland*. Sydney, Australia, Angus and Robertson, 566 p, 136 pls.

Mead, G.W. 1970. A history of South Pacific fishes. In *Scientific Explorations of the South Pacific*, ed. W.S. Wooster, 236–251. Washington, DC, National Academy of Sciences.

Merrick, J.R. 2006. Australasian freshwater fish faunas: Diversity, interrelationships, radiations and conservation. In *Evolution and Biogeography of Australian Vertebrates*, ed. J.R. Merrick, M. Archer, G.M. Hickey and M.S.Y. Lee, 195–224. Oatlands, Australia, Auscipub.

Merrick, J.R., and G.E. Schmida. 1984. *Australian Freshwater Fishes Biology and Management*. Sydney, Australia, J.R. Merrick, 409 p.

Mooi, R.D. 1993. Phylogeny of the Plesiopidae (Pisces: Perciformes) with evidence for the inclusion of the Acanthoclinidae. *Bulletin of Marine Science* 52:284–326.

Mooi, R.D. 1995. Revision, phylogeny, and discussion of biology and biogeography of the fish genus *Plesiops* (Perciformes: Plesiopidae). *Royal Ontario Museum Life Sciences Contributions* 159:1–108.

Mooi, R.D., and A.C. Gill. 2002. Historical biogeography of fishes. In *Handbook of Fish Biology and Fisheries, vol. 1: Fish Biology*, ed. P.J.B. Hart and J.D. Reynolds, 43–68. Oxford, UK, Blackwell Science.

Mooi, R.D., and A.C. Gill. 2004. Notograptidae, sister to *Acanthoplesiops* Regan (Teleostei: Plesiopidae: Acanthoclininae), with comments on biogeography, diet and morphological convergence with Congrogadinae (Teleostei: Pseudochromidae). *Zoological Journal of the Linnean Society* 141:179–205.

Moore, G.I., S.M. Morrison, J.B. Hutchins, G.R. Allen and A. Sampey. 2014. Kimberley marine biota: Historical data; Fishes. *Records of the Western Australian Museum*, Suppl 84:161–206.

Morrison, D. 2014. Review of 'The monkey's voyage: How improbable journeys shaped the history of life', by Alan de Queiroz. *Systematic Biology* 63:847–849.

Murdy, E.O. 1989. A taxonomic revision and cladistic analysis of the oxudercine gobies (Gobiidae: Oxudercinae). *Records of the Australian Museum*, Suppl 11:1–93.

Nelson, G. 1978. From Candolle to Croizat: Comments on the history of biogeography. *Journal of the History of Biology* 11:269–305.

Nelson, G.J. 1985. A decade of challenge: The future of biogeography. *Earth Science History* 4:187–196.

Nelson, G. 1986. Models and prospects of historical biogeography. *UNESCO Technical Papers in Marine Science* 49:214–218.

Nelson, G., and P. Ladiges. 1991. Standard assumptions for biogeographic analysis. *Australian Systematic Botany* 4:41–58.

Nelson, G., and P.Y. Ladiges. 2009. Biogeography and the molecular dating game: A futile revival of phenetics? *Bulletin de la Société Géologique de France* 180:39–43.

Nelson, J.S. 2006. *Fishes of the World*, 4th edn. Hoboken, NJ, Wiley, 601 p.

Ogilby, J.D. 1886. *Catalogue of the Fishes of New South Wales, with Their Principal Synonyms*. Sydney, Australia, Government Printer, 67 p.

Oxley, W.G., A.M. Ayling, A.J. Cheal and K. Osborne. 2004. *Marine Surveys Undertaken in the Elizabeth and Middleton Reefs* Marine Nature Reserve, December 2003. Townsville, Australia, Australian Institute of Marine Science, 64 p.

Parenti, L.R., and M.C. Ebach. 2009. *Comparative Biogeography*. Berkeley, University of California Press, 295 p.

Patterson, C. (ed. and intro. D.M. Williams and A.C. Gill). 2011. *Adventures in the fish trade. Zootaxa* 2946:118–136.

Paxton, J.R., G.R. Allen and D.F. Hoese. 2006. Australian marine fishes: Zoogeography, endemics and conservation. In *Evolution and Biogeography of Australian Vertebrates*, ed. J.R. Merrick, M. Archer, G.M. Hickey and M.S.Y. Lee, 185–194. Oatlands, Australia, Auscipub.

Platnick, N.I. 1991. Commentary: On areas of endemism. *Australian Systematic Botany* 4: xi–xii.

Puckridge, M., P.R. Last and N. Andreakis. 2015. The role of peripheral endemism and habitat associations in the evolution of the Indo-West Pacific tuskfishes (Labridae: Choerodon). *Molecular Phylogenetics and Evolution* 84:64–72.

Pusey, B.J., M.J. Kennard and A.H. Arthington. 2004. *Freshwater Fishes of North-Eastern Australia*. Collingwood, Australia, CSIRO Publishing, 684 p.

Randall, J.E. 1982. Examples of antitropical and antiequatorial distribution of Indo-West-Pacific fishes. *Pacific Science* 35:197–209.

Randall, J.E. 1998. Zoogeography of shore fishes of the Indo-Pacific region. *Zoological Studies* 37:227–268.

Randall, J.E., G.R. Allen and R.C. Steene. 1990. *Fishes of the Great Barrier Reef and Coral Sea*. Bathurst, Australia, Crawford House Press, 507 p.

Randall, J.E., G.R. Allen and R.C. Steene. 1997. *Fishes of the Great Barrier Reef and Coral Sea*, 2nd edn. Bathurst, Australia, Crawford House Press, 557 p.

Read, C.I., D.R. Bellwood and L. van Herwerden. 2006. Ancient origins of the Indo-Pacific coral reef fish biodiversity: A case study of the leopard wrasse (Labridae: *Macropharyngodon*). *Molecular Phylogenetics and Evolution* 38:808–819.

Sainsbury, K.J., P.J. Kailola and G.G. Leyland. 1985. *Continental Shelf Fishes of Northern and North-Western Australia*. Canberra, Australia, Fisheries Information Service, 375 p.

Santini, F., and R. Winterbottom. 2002. Historical biogeography of Indo-Western Pacific coral reef biota: Is the Indonesian region a centre of origin? *Journal of Biogeography* 29:189–205.

Santini, S., and G. Polacco. 2006. Finding Nemo: Molecular phylogeny and evolution of the unusual life style of anemonefish. *Gene* 385:19–27.

Schilder, F.A., and M. Schilder, M. 1939. Prodome of a monograph on living Cypraeidae. *Proceedings of the Malacological Society of London* 23:181–252.

Sclater, P.L. 1858. On the general geographical distribution of the members of the class Aves. *Journal of the Proceedings of the Linnean Society London (Zoology)* 2:131–136.

Scott, T.D. 1962. *The Marine and Fresh Water Fishes of South Australia*. Adelaide, Australia, Government Printer, 338 p.

Scott, T.D., C.J.M. Glover and R.V. Southcott. 1974. *The Marine and Fresh Water Fishes of South Australia*, 2nd edn. Adelaide, Australia, Government Printer, 392 p.

Smith-Vaniz, W.F., and G.D. Johnson. 1990. Two new species of Acanthoclininae (Pisces: Plesiopidae) with a synopsis and phylogeny of the subfamily. *Proceedings of the Academy of Natural Sciences of Philadelphia* 142:211–260.

Smith-Vaniz, W.F., and V.G. Springer. 1971. Synopsis of the tribe Salariini, with description of five new genera and three new species (Pisces: Blenniidae). *Smithsonian Contributions to Zoology* 73:1–72.

Spalding, M.D., H.E. Fox, G.R. Allen, N. Davidson, Z.A. Ferdaña, M. Finlayson, B.S. Halpern, *et al.* 2007. Marine ecoregions of the world: A bioregionalization of coastal and shelf areas. *BioScience* 57:573–583.

Speare, P., M. Cappo, M. Rees, J. Brownlie and W. Oxley. 2004. *Deep Water Fish and Benthic Surveys in the Lord Howe Island Marine* Park *(Commonwealth Waters): February 2004*. Townsville, Australia, Australian Institute of Marine Science, 30 p.

Springer, V.G. 1955. Western Atlantic fishes of the genus Paraclinus. *Texas Journal of Science* 6:422–441.

Springer, V.G. 1967. Revision of the circumtropical shorefish genus *Entomacrodus* (Blenniidae: Salariinae). *Proceedings of the United States National Museum* 122(3582):1–150, pls 1–30.

Springer, V.G. 1982. Pacific Plate biogeography, with special reference to shorefishes. *Smithsonian Contributions to Zoology* 367:1–182.

Springer, V.G. 1988. The Indo-Pacific blenniid fish genus Ecsenius. *Smithsonian Contributions to Zoology* 465:1–134, pls 1–14.

Springer, V.G. 1999. *Ecsenius polystictus*, new species of blenniid fish from Mentawai Islands, Indonesia, with notes on other species of *Ecsenius. Revue française d'aquariologie herpētologie* 26:39–48.

Springer, V.G., and G.R. Allen 2004. *Ecsenius caeruliventris* and *E. shirleyae*, two new species of blenniid fishes from Indonesia, and new distribution records for other species of *Ecsenius Zootaxa* 791:1–12.

Springer, V.G., and M.F. Gomon. 1975. Revision of the blenniid fish genus *Omobranchus* with descriptions of three new species and notes on other species of the tribe Omobranchini. *Smithsonian Contributions to Zoology* 177:i–iii, 1–135.

Springer, V.G., and H.K. Larson. 1996. *Pholidichthys anguis*, a new species of pholidichthyid fish from Northern Territory and Western Australia. *Proceedings of the Biological Society of Washington* 109:353–365.

Springer, V.G., and J.T. Williams. 1994. The Indo-Pacific blenniid fish genus *Istiblennius* reappraised: A revision of *Istiblennius, Blenniella*, and *Paralticus*, new genus. *Smithsonian Contributions to Zoology* 565:i–iv, 1–193.

Tenison-Woods, J.E. 1882. *Fish and Fisheries of New South Wales*. Sydney, Australia, Thomas Richards, Government Printer, 213 p.

Timm, J., M. Figiel and M. Kochzius. 2008. Contrasting patterns in species boundaries and evolution of anemonefishes (Amphiprioninae, Pomacentridae) in the centre of marine biodiversity. *Molecular Phylogenetics and Evolution* 49:268–276.

Unmack, P.J. 2001. Biogeography of Australian freshwater fishes. *Journal of Biogeography* 28:1053–1089.

Unmack, P.J. 2013. Biogeography. In *Ecology of Australian Freshwater Fishes*, ed. P. Humphries and K. Walker, 25–48. Collingwood, Australia, CSIRO Publishing.

Unmack, P.J., and T.E. Dowling. 2010. Biogeography of the genus *Craterocephalus* (Teleostei: Atherinidae) in Australia. *Molecular Phylogenetics and Evolution* 55:968–984.

Vari, R.P. 1978. The terapon perches (Percoidei, Teraponidae). A cladistic analysis and taxonomic revision. *Bulletin of the American Museum of Natural History* 159:175–340.

Victor, B.C., and G.M. Wellington. 2000. Endemism and pelagic larval duration of reef fishes in the eastern Pacific Ocean. *Marine Ecology Progress Series* 205:241–248.

Waite, E.R. 1921. Illustrated catalogue of the fishes of South Australia. *Records of the South Australian Museum* 2: 1–208,pl. 1.

Waite, E.R. 1923. *Fishes of South Australia*. Adelaide, Government Printer, 243 p.

Wallace, A. 1876. *The Geographical Distribution of Animals*, 2 vols. New York, Harper and Brothers.

Waters, J.M., T. Wernberg, S.D. Connell, M.S. Thomsen, G.C. Zuccarello and G.T. Kraft. 2010. Australia's marine biogeography revisited: Back to the future? *Austral Ecology* 35:988–992.

Wellington, G.M., and B.C. Victor. 1989. Planktonic larval duration of one hundred species of Pacific and Atlantic damselfishes (Pomacentridae). *Marine Biology* 101:557–567.

White, B.N. 1986. The isthmian link, antitropicality and American biogeography: Distributional history of the Atherinopsinae (Pisces: Atherinidae). *Systematic Zoology* 35:176–194.

White, B.N. 1989. Antitropicality and vicariance: A reply to Briggs. *Systematic Zoology* 38:77–79.

White, J. 1790. *Journal of a Voyage to New South Wales*. London, Debrett, 299 p., 65 pls.

Whitley, G.P. 1932. Marine zoogeographical regions of Australia. *The Australian Naturalist* 8:166–167.

Whitley, G.P. 1934. Supplement to the check-list of the fishes of New South Wales. In *The Fishes and Fish-like Vertebrates of New South Wales*, 3rd edn, ed. A.R. McCulloch, 1–12. Sydney, Australia, Royal Zoological Society of New South Wales.

Whitley, G.P. 1948. A list of the fishes of Western Australia. *Western Australia Fisheries Department, Fisheries Bulletin* 2: 1–35, foldout map.

Whitley, G.P. 1959. The freshwater fishes of Australia. *Monographiae Biologicae* 8:136–149.

Whitley, G.P. 1964a. A survey of Australian ichthyology. *Proceedings of the Linnean Society of New South Wales* 89:11–127.

Whitley, G.P. 1964b. *Freshwater Fishes of Australia*, revised edn. Brisbane, Australia, Jacaranda Press, 127 p.

Williams, J.T. 1988a. Revision and phylogenetic relationships of the blenniid fish genus *Cirripectes*. *Indo-Pacific Fishes* 17:1–78, pls 1–7.

Williams, R. 1988b. The nearshore fishes of Macquarie Island. *Papers and Proceedings of the Royal Society of Tasmania* 122:233–245.

Williams, W.D., and G.R. Allen. 1987. Origins and adaptations of the fauna of inland waters. In *Fauna of Australia, vol. 1A: General Articles*, ed. G.R. Dyne and D.W. Walton, 184–201. Canberra, Australia, Australian Government Publishing.

Wilson, B.R., and G.R. Allen. 1987. Major components and distribution of marine fauna. In *Fauna of Australia, vol. 1A: General articles*, ed. G.R. Dyne and D.W. Walton, 43–68. Canberra, Australia, Australian Government Publishing.

Winterbottom, R. 1986. Revision and vicariance biogeography of the subfamily Congrogadinae (Pisces: Perciformes: Pseudochromidae). *Indo-Pacific Fishes* 9:1–34.

Zapata, F.A., and P.A. Herrón. 2002. Pelagic larval duration and geographic distribution of tropical eastern Pacific snappers (Pisces: Lutjanidae). *Marine Ecology Progress Series* 230:295–300.

7

Australian Comparative Phytogeography: A Review

Daniel J. Murphy and Darren M. Crayn

CONTENTS

Introduction

This review covers the historical biogeography of flowering plants within Australia. Comprehensive reviews of this topic appeared in the first and second editions of the introductory volume of the *Flora of Australia* (Barlow 1981; Crisp *et al.* 1999; the latter hereafter referred to as CWL). These reviews dealt comprehensively with the plant biogeography literature at the time. The present review is intended as an update and a critical review of the development of terrestrial plant biogeography in Australia over the past two decades, with a particular focus on developments after CWL. It contrasts with CWL in focusing on methods that deal predominantly with comparative approaches to plants and their distributional data rather than providing an analysis of single-taxon studies. While an updated analysis of single-taxon studies is necessary, it is beyond the scope of the current chapter as it involves novel reinterpretation of the original data sets and the results of very different methodological approaches. While globally the number of studies focused on intracontinental plant biogeography is somewhat smaller than those with an emphasis on intercontinental biogeography, there remains a considerable literature to cover, even when looking only at the past 20 years.

Australian Plant Biogeography: A Selected History

... the need to find a way to better understanding.

Nelson and Ladiges (2001, p. 399)

Australia's flora strongly reflects its Gondwanan heritage, and many Australian lineages, such as Atherospermataceae, Proteaceae, Monimiaceae, *Nothofagus* and southern conifer groups, are shared with other Gondwanan landmasses (Raven and Axelrod 1974; Weston and Hill 2013; Kooyman *et al.* 2014). But separation from Antarctica and the northward migration of the Australian craton from around the end of the Eocene (ca. 35.5 mya; McLoughlin 2001), together with changes in the position and size of neighbouring landmasses, provided opportunities for the expansion or contraction of existing biomes and the emergence of new ones, as well as floristic exchanges with other regions such as New Caledonia, New Zealand and the Sunda Shelf. For most of its history, Australian plant biogeography has progressed through empirical study of the distribution of plants within the continent, and the relationships of the flora to that of other continents.

Perhaps the most influential figure in the development of terrestrial plant biogeography in Australia was Nancy Burbidge. In her seminal 1960 paper, Burbidge analysed the distributions of Australian plant genera and from these data defined three phytogeographical zones: tropical, temperate and Eremaean. She noted areas of 'special phytogeographic interest', including the South-west province of Western Australia, Tasmania, northeast Queensland and a region of overlap for the tropical and temperate zones called the *MacPherson–Macleay Overlap* (note original spelling of this overlap region), and recognised three additional *interzones*. Her concepts of the tropical, temperate and Eremaean zones form the basis of the biomes used in many recent analyses of terrestrial plant biogeography (e.g. Crisp *et al.* 2004, and see Figure 7.1) and have largely withstood empirical testing of their boundaries (e.g. González-Orozco *et al.* 2014a).

Further to defining biogeographical areas, Burbidge (1960) developed explanatory hypotheses for the origins of the Australian flora. In relation to the rainforest flora she was strongly influenced by Hooker's (1860) and van Steenis's (1950) earlier analyses in formalising the invasion hypothesis, which postulated that Australia's rainforest was assembled mainly from lineages that immigrated via land bridges (for which there was little evidence; see Ladiges 1998 and Crisp *et al.* 1999 for discussion). Importantly, like many of her time, Burbidge was not satisfied by continental drift as an explanatory process for Southern Hemisphere plant distributions. In contrast, Leon Croizat (1962), who admired Burbidge and agreed with much of her 'objectively elaborating the phytogeography of Australia', presented patterns in terms of generalised tracks, invoking the process of vicariance and accepting continental drift (Ladiges 1998, p. 233, Figure 7.2).

While Burbidge's invasion hypothesis was somewhat derivative, her hypothesis regarding the origins of the arid zone was a major original contribution (Crisp *et al.* 1999). She argued that the arid zone flora originated from ancestors with cosmopolitan coastal distributions that were preadapted to arid environments. Examples include the families Aizoaceae, Amaranthaceae, Asteraceae, Brassicaceae, Chenopodiaceae, Convolvulaceae, Frankeniaceae and Portulacaceae. Schodde (1989) adopted this idea and subsequent studies have provided some empirical support (Crisp *et al.* 2004; reviewed in Byrne *et al.* 2008).

CWL (1999) comprehensively summarised the history of biogeographical study of the Australian flora up to that time (and as such that history is not repeated here, but is briefly summarised). Since 1999, plant biogeography in Australia has become increasingly concerned with the development and application of methodological innovations. In the time since CWL there has been a stronger recognition of the long-standing notion (e.g. Schodde 1989) that the taxon and the research question dictate the evolutionary timescale that must be addressed in biogeographical analysis. As a consequence there has been a proliferation of approaches that fall under the modern remit of biogeography, and inevitably a sorting into various subdisciplines that attempt to answer questions at different levels of temporal patterning. Notably, some authors have moved away from comparative biogeography, perhaps disillusioned by the prospect that each individual taxon has its own biogeographic history such that the search for general

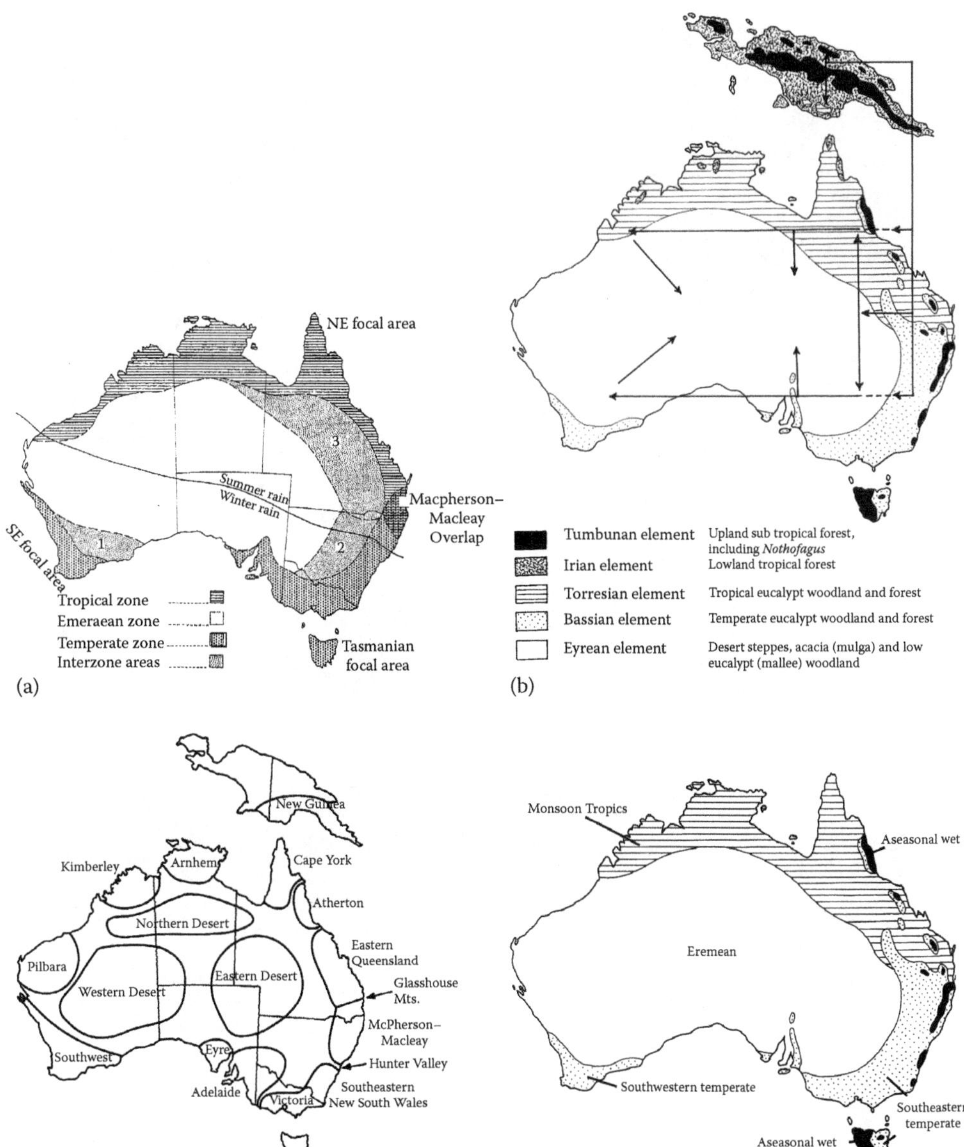

FIGURE 7.1 Comparison of maps of Australian biotic areas: (a) Burbidge (1960) (reproduced from Burbidge, N.T., *Australian Journal of Botany*, 8, 75–209, 1960, with permission from CSIRO Publishing); (b) Schodde (1989) (reproduced with permission from the author and publisher); (c) Crisp *et al.* (1999) CWL areas (as first used in Crisp, M.D., *et al.*, *Systematic Biology*, 44, 457–473, 1995, reproduced with permission from the author and publisher); (d) Crisp *et al.* (2004) biomes (the authors note that these are modified from Burbidge, N.T., *Australian Journal of Botany*, 8, 75–209, 1960; Schodde, R., *Australian Systematic Botanical Society Newsletter*, 60, 2–11, 1989; hence the similarities, reproduced with permission from the publisher.)

patterns of plant distributions may be futile (Crisp *et al.* 2011). However, it is the opinion of the current authors that this has not been adequately demonstrated. A worthy question to examine in modern historical biogeography is whether biogeographic patterns can be discerned for the distributional ranges of different taxa within continental Australia. This also includes evaluating the methods by which such comparative studies can be undertaken.

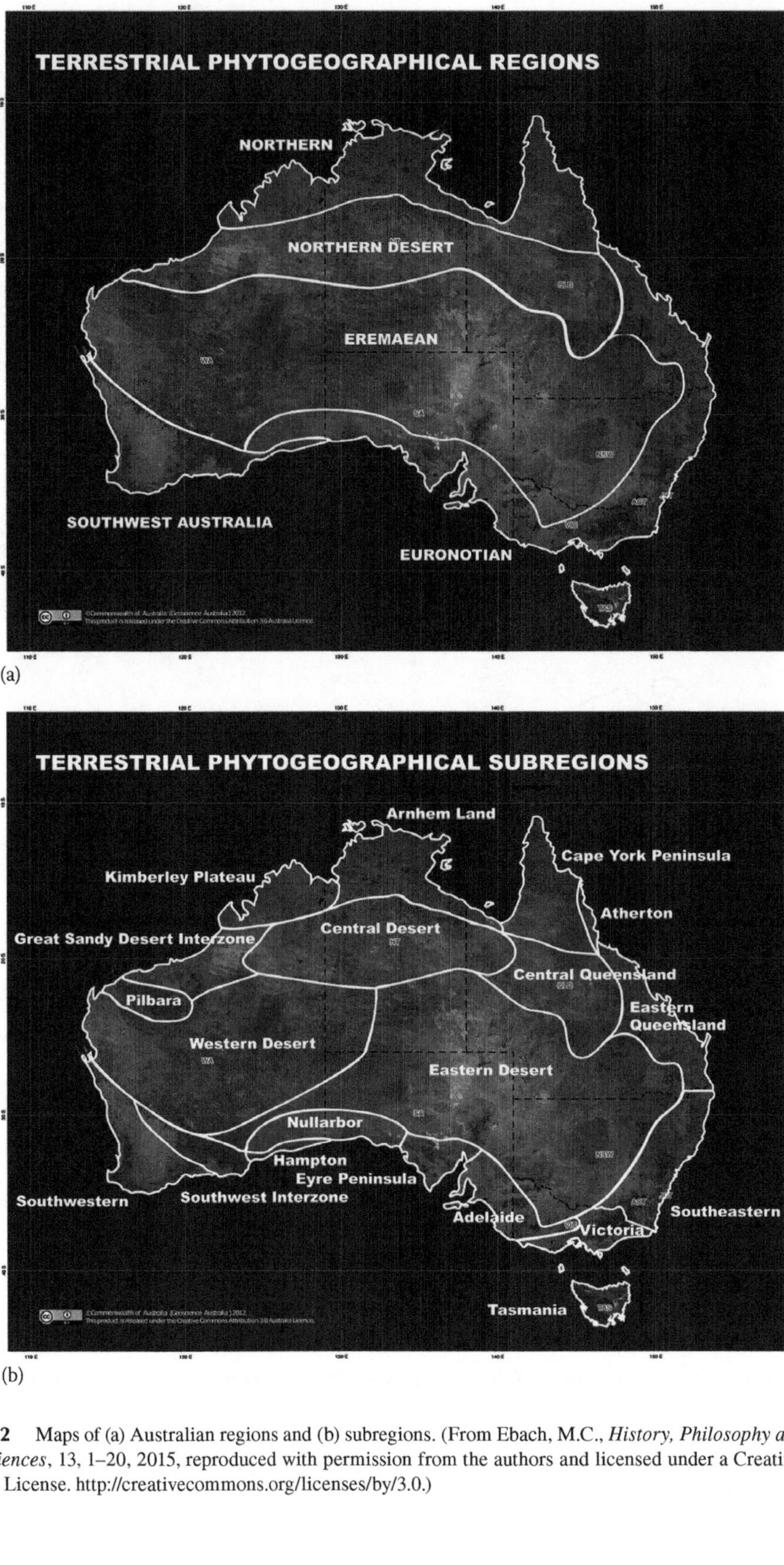

(a)

(b)

FIGURE 7.2 Maps of (a) Australian regions and (b) subregions. (From Ebach, M.C., *History, Philosophy and Theory of the Life Sciences*, 13, 1–20, 2015, reproduced with permission from the authors and licensed under a Creative Commons Attribution License. http://creativecommons.org/licenses/by/3.0.)

After 1999, biogeographers began to seize the opportunity provided by the molecular biological revolution to estimate phylogenies and attempt molecular dating. A simultaneous increase in the availability of DNA sequence data (due to the availability of universal primers and improved laboratory techniques in molecular biology) and advances in analytical and computational methods (e.g. *relaxed clocks* and software such as *r8s*) led to an increased focus on the reconstruction of the biogeographical histories of single lineages and less focus on comparative historical approaches. In conjunction, the emphasis of some biogeographical studies shifted towards using ecologically defined biomes, rather than smaller biotic areas (*areas of endemism*) in Australia. The recent history of intracontinental biogeographic analyses of plants in Australia reflects the diversity of approaches that have become available, and this chapter reviews this literature. There have been several recent reviews that deal with the general history of Australian plant biogeography (e.g. Weston and Hill 2013) or take a regional or biome focus (e.g. Byrne *et al.* 2008; Bowman *et al.* 2010; Wannan 2014; and from a panbiogeographic perspective, Heads 2014). The aim of this review is to summarise what is known so far about the comparative phytogeography of Australia, and also to look to the future prospects of this research area.

Discovering Areas and Area Relationships

Historical biogeography seeks general patterns of relationships between biotic areas, and interprets these in the context of evolution and earth history. These patterns can be discovered using many methods (Crisci *et al.* 2003). Two of the most influential approaches in the Australian context over the last 15 years have been the *comparative approach* and the *quantitative geospatial* approach.

Comparative Approach

Area relationships are sometimes used to infer or hypothesise origins or ancestral areas, particularly of individual taxa, and recently this has generally been undertaken using molecular divergence dating approaches. General area relationships, or commonalities in the area relationships among different groups, have been sought using comparative methods or geographic congruence (in the sense of Ladiges 1998). This *comparative* approach to biogeography may allow the inference of events that have had a broad evolutionary consequence for a biota, such as the emergence of a biotic break or barrier that affects many lineages (e.g. Nelson and Platnick 1981; Humphries and Parenti 1999; Parenti and Ebach 2009). If natural biotic areas are identified this can lead to the erection of hypotheses of area histories too – the combined pattern of life and earth. Such an approach has been less commonly applied to areas within a continental landmass such as Australia, due to the difficulties in defining areas to compare.

A criticism of the comparative approach has been that it favours vicariance explanations and minimises the possibility of dispersal, particularly when molecular dating analysis suggests clades post-date geological events. It is less commonly acknowledged that there is no reason why a comparative approach to historical biogeography using multiple taxa could not identify general patterns that are a consequence of dispersal (e.g. Sanmartín *et al.* 2007). It is the pattern that is the pivotal first step in understanding biogeography. This realisation has led to calls for a more integrative approach to biogeography that incorporates distributional data, ecological processes, molecular divergence dating of phylogenies and palaeontological information in a hypothesis testing framework (Crisp *et al.* 2011). There have been some developments in a comparative phylogeographic framework to incorporate multiple intraspecies lineages (e.g. Moritz *et al.* 2009; Hodel *et al.* 2016). Other model-based methods (e.g. the software *BioGeoBears* or Markov Chain Monte Carlo approaches for ancestral area Bayesian reconstructions) that allow the analysis of multiple clades of co-distributed taxa have recently been developed and applied – for example, in island biogeography (Wen *et al.* 2013). These newer methods are often called *event-based* methods, to distinguish them from other historical biogeographic methods (Ronquist and Sanmartín 2011).

Despite such progress, there are potential pitfalls in using fossil-calibrated molecular divergence dates in an uncritical manner (Heads 2005; Sauquet *et al.* 2012; Wilf and Escapa 2015), as minimum ages for some lineages are likely to be much older than the earliest known fossils.

Quantitative Geospatial Approach

In the last 15 years, a new methodological approach to biogeography has emerged and begun to influence the study of plant distributions in Australia; we term this approach the *quantitative geospatial biogeography* (QGB) approach (sensu Crisp *et al.* 2001). The QGB approach has its roots in species-level distributional data analysis (e.g. Hopper 1979; Hopper and Gioia 2004) but incorporates phylogenetic information, often by way of metrics, such as phylogenetic diversity (Faith 1992), phylogenetic endemism (Rosauer *et al.* 2009) and their derivatives (e.g. relative phylogenetic diversity and endemism; Mishler *et al.* 2014). This approach has not generally been recognised (Ebach 2015), but its use and importance is increasing worldwide, particularly in Australia (Ebach 2012).

The development and application of QGB methods occurred early in Australia, enabled by the improved availability of digitally verifiable spatial records of species occurrences since the 2000s, through the continent-wide Australian herbarium record digitisation project *Australia's Virtual Herbarium* (http://avh.ala.org.au). Using these data, many biogeographical studies primarily focus on geospatial methods and often do not define areas under study a priori, but rather divide the continent into arbitrary grids of varying scales. Notably, these types of methods are not new; Burbidge (1960) used a similar methodology to quantify species occurring in different areas (e.g. *Acacia* and *Eucalyptus*), and prior to her work, the nineteenth-century botanists Hooker (1860) and Engler (1882) used a similar approach on the Australian flora. However, what is new is the recent availability of huge species occurrence data sets and explicit algorithms to analyse these data. Generally, these studies have focused on quantifying biodiversity in areas. This forms a nexus with historical biogeographic studies, but may aim to answer fundamentally different questions from historical biogeography, such as what areas should be conserved to maximise phylogenetic diversity or phylogenetic endemism (and related metrics), or to examine current ecological units (e.g. Costion *et al.* 2015; Laity *et al.* 2015).

The maturity of tools related to QGB, such as more sophisticated biodiversity metrics (Laffan *et al.* 2016), the software *Biodiverse* (Laffan *et al.* 2010) and methods (Mishler *et al.* 2014) that incorporate the older concepts of paleo- and neoendemism (Wulff 1943), have enabled a series of studies on the spatial distribution of the diversity of different Australian plant taxa; these include studies of *Acacia* (Mishler *et al.* 2015), ferns (Nagalingum *et al.* 2015), Asteraceae (Schmidt-Lebuhn *et al.* 2015) and angiosperm genera overall (Thornhill *et al.* 2016). The QGB methods have also been applied to explorations of plant palaeodistributions (e.g. Jordan *et al.* 2016), setting conservation priorities (Laity *et al.* 2015; Pollock *et al.* 2015) and more objective determinations of biogeographic area boundaries (e.g. González-Orozco *et al.* 2015; results summarised in Ebach *et al.* 2013, 2015, and see herein Figure 7.2).

Defining Biogeographical Areas

A necessary first step in historical biogeographical analysis is the definition of areas to be considered. While the literature on the definition of areas for biogeographical studies is extensive, the approaches applied in the Australian context can be grouped into four main categories based on the use of different distributional or geographical data – namely,

1. Endemic areas
   a. Predefined (e.g. Cracraft 1991; Crisp *et al.* 2004)
   b. Defined by species with 'restricted distributional boundaries' (e.g. Weston and Crisp 1994)
2. Interim Biogeographic Regionalisation for Australia (IBRA) regions (Thackway and Cresswell 1995) or a combination of biotic areas and IBRA (e.g. Ladiges *et al.* 2006)
3. Point distributions (e.g. González-Orozco *et al.* 2014a)

Each of these approaches is briefly discussed in the following sections.

Endemic Areas

Most historical biogeographic approaches use areas of endemism, defined a priori, as the units of comparison. CWL review how these areas have been defined within the Australian continent but note that defining areas of endemism may be problematic because there are few sharp boundaries and barriers. This has led Bowman *et al.* (2010), for example, to call for more research 'to determine biogeographical zonation within the AMT [Australian Monsoon Tropics]'. Ebach (2012) explores the history of biogeographical regionalisation more broadly in the Australian context, and in Table 7.1 we provide a comparison of the major area classification schemes since Burbidge (1960), focusing on intra-Australian biotic areas.

Some historical biogeographic studies of Australian plants have not used predefined areas. An example is the landmark study of Crisp and Weston (1995), a comparative study of *Lechenaultia* (Goodeniaceae) and *Leptosema* (Fabaceae), with areas based on the distributions of endemic species. However, soon after, Crisp *et al.* (1995) recognised the potential drawbacks of such an approach when comparing many taxa and complex areas, and settled on a modified version of Cracraft's (1991) biotic areas, which were subsequently modified in CWL. These areas have served as the standard for plant historical biogeography within Australia for almost 20 years.

IBRA Regions

Within Australia, as in all continents, the difficulties in identifying biogeographic boundaries for areas can render area definition problematic. However, some authors have attempted the description of fine-scale regional differentiation and the naming of areas. For example, Ladiges *et al.* (2006) regarded the Western Desert sensu Crisp *et al.* (1995) as an 'oversimplification' and defined subareas (e.g. Gascoyne, Little Sandy Desert) using IBRA regions (IMCRA Technical Group 1998). However, while the IBRA system seems a useful way of uniformly applying a bioregionalisation to Australia for all organisms, Ebach (2012) critiques the opaque nature of this bioregionalisation and advocates for an *area taxonomy* that can be modified and is comparative. Ebach *et al.* (2013, with subsequent modifications in Ebach *et al.* 2015) outline an area taxonomy for Australia. The use and implication of differing or uncertain area boundaries is evident in most plant studies of biogeography within Australia, even though many authors have used modified versions of Cracraft (1991), Crisp *et al.* (1995) or IBRA. Even the slightly differing naming schemes of areas can have a confusing or confounding effect, such as Cape York (Crisp *et al.* 1995) versus the Cape York Peninsula (Cracraft 1991). Ebach *et al.* (2013, 2015) break down the areas into regions and subregions (Figures 7.2a,b, respectively), which in effect seems to combine the use of larger biomes (similar to regions) and the previously more commonly used biotic areas (similar to subregions) of Crisp *et al.* (1995). Table 7.1 is an attempt to compare equivalent areas used in different schemes, noting that some areas nest satisfactorily within and between different schemes, whereas others conflict or do not overlap clearly in a linearly arranged table.

Point Distributional Data

Crisp *et al.* (2001), in seeking to identify biogeographic areas more objectively by reducing logical circularity or potential biases in area definition, published an early example of the QGB approach. Their analysis divided the Australian continent into one-degree grid squares and for each calculated species richness and endemism using data for over 8000 plant species. As a result, 12 distinct centres of diversity were identified: (1) South-west Western Australia, (2) north Kimberley, (3) Kakadu–Alligator Rivers, (4) the Central Australian Ranges, (5) Adelaide–Kangaroo Island, (6) the Iron Range–McIlwraith Range (Cape York Peninsula), (7) the Wet Tropics, (8) the Border Ranges (McPherson–Macleay), (9) Sydney Sandstone, (10) the Australian Alps, (11) Tasmania and (12) New England–Dorrigo. Crisp *et al.* (2001) noted that most of the centres of endemism they identified are coastal, and interpreted this as a signature of periods of extreme aridity in the Pleistocene in central areas of Australia, limiting the range of narrow endemics. They also highlighted the Adelaide–Kangaroo Island region as a centre that had been overlooked altogether by previous authors; although an Adelaide biotic region (including Kangaroo

TABLE 7.1

Comparison of the Major Area Classification Schemes, Focusing on Intra-Australian Biotic Areas

Burbidge (1960)	Schodde (1989)	Cracraft (1991)	Crisp *et al.* (1995)	Crisp *et al.* (1999)*	(Crisp *et al.* 2004): Biomes	Ebach *et al.* (2013): Regions	Ebach *et al.* (2013): Subregions
N/A	Irian	N/A	New Guinea	New Guinea	N/A	Northern Australia	Southern New Guinea
Tropical zone	Torresian element	Kimberley Plateau	Kimberley		Monsoon Tropics		Kimberley Plateau
		Arnhem Land					Arnhem Land
		Cape York Peninsula	Arnhem				Cape York Peninsula
			Cape York				
		Atherton Plateau	Atherton	Wet Tropics		Euronotian	Atherton
		Eastern Queensland (includes McPherson–Macleay of Crisp *et al.* 1995)	Eastern Queensland				Eastern Queensland
Temperate zone	Bassian	Southeastern Forest	Southeastern NSW	Southeast	Southeastern temperate		Southeast
			Victoria				Victoria
		Adelaide	Adelaide				Adelaide
		Eyre Peninsula	Eyre				Eyre Peninsula
Macpherson–Macleay Overlap			McPherson–Macleay				Nullabor
Tasmanian focal area		Tasmania	Tasmania	Tasmania			Tasmania
							Hampton
Southwestern focal area		Southwestern Forest	South-west	South-west	Southwestern temperate	Southwest Australia	South-west
Interzone 1							Southwest interzone
Eremaean zone	Eyrean			Central Australia	Eremean	Eremean	
		Western Desert	Western Desert				Western Desert
Interzone 2		Eastern Desert	Eastern Desert				Eastern Desert
		Pilbara	Pilbara			Northern Desert	Pilbara
		Northern Desert	Northern Desert				Great Sandy Desert interzone
Interzone 3							
							Central Desert
							Central Queensland
N/A	Tumbuna				Aseasonal wet		

Source: Burbidge, N.T., *Australian Journal of Botany*, 8, 75–209, 1960; Schodde, R., *Australian Systematic Botanical Society Newsletter*, 60, 2–11, 1989; Cracraft (1991); Crisp, M.D., *et al.*, *Systematic Biology* 44, 457–473, 1995; Crisp, M.D., *et al.*, *Flora of Australia* 1, 321–367, 1999; Crisp, M.D., *et al.*, *Philosophical Transactions of the Royal Society B: Biological Sciences* 359, 1551–1571, 2004; Ebach, M.C., *et al.*, *Phytotaxa* 208, 261–277, 2015.

Note: Equivalent or similar areas are aligned from left to right across the table where possible. Some areas overlap, some are not directly comparable and others are not applicable across schemes. Note the complexity of integrating climatically based biomes, because Schodde (1989) and Crisp *et al.* (2004 and elsewhere) have smaller areas nested within other areas. Note spelling: Burbidge used '*Mac*Pherson–Macleay', whereas most subsequent authors use '*Mc*Pherson–Macleay', with some exceptions (e.g. Ladiges 1998).

* Areas Used in 'Phytogeography within Australia–New Guinea' Section.

Island) had previously been identified (Cracraft 1991; used by Crisp *et al.* 1995), that older definition was somewhat more extensive than the grid cells identified in Crisp *et al.* (2001). Comparisons between the areas conceived by others are always somewhat difficult to convey unless clearly mapped and defined.

Recently, access to point distribution data for Australian plants has been greatly facilitated by the delivery of herbarium specimen record data through *Australia's Virtual Herbarium* and the *Atlas of Living Australia* (http://ala.org.au). These developments have enabled a new generation of biogeographical studies that analyse large data sets of point records for many species. González-Orozco *et al.* (2014a) used this approach to systematically determine the boundaries of biogeographical areas by searching for the landscape positions of significant changes in species composition turnover in nine major plant groups (*Acacia*, Asteraceae, eucalypts, ferns, hornworts, liverworts, *Melaleuca*, mosses and orchids). This analysis recovered six areas: northern, Northern Desert, Eremaean, eastern Queensland, Euronotian and South-western. These areas were substantially spatially congruent with earlier analyses based on *Acacia* (González-Orozco *et al.* 2013) and eucalypts (González-Orozco *et al.* 2014b). The areas of González-Orozco *et al.* (2014a) differ substantially from those defined in CWL and are similar in some ways to the biomes identified by Schodde (1989) and Crisp *et al.* (2004).

The QGB approach exemplified by González-Orozco *et al.* (2014a) has the advantage of being algorithmic and therefore repeatable. It allows for rigorous quantitative analysis of the relationship between species distributions and physical variables such as climate and soils. For example, González-Orozco *et al.* (2014a) found strong congruence between areas resolved by their analysis and the Köppen climate zones. Furthermore, this approach can provide a sound basis for the systematic classification and nomenclature of areas, such as has been developed by Ebach *et al.* (2013, 2015). However, it remains unclear how the differing methods for defining areas in historical biogeography relate to each other, or even whether they are applicable to the questions posed by researchers in different fields of biogeography.

Understanding Relationships between Areas

In Australia, workers have employed the comparative approach in an attempt to understand the evolutionary history of the Australian flora through discovering the area relationships of biotic areas (should they exist). While this aim has remained constant, the methodology used to achieve it has fragmented in the past 20 years.

In 1994, Peter Weston and Michael Crisp initiated a series of comparative plant biogeographic studies for Australia. Their paper (Weston and Crisp 1994) treated two clades in the Proteaceae family, the *Lomatia* clade and its sister taxon Embothriinae (waratahs), and was an early example of morphological cladistic biogeographic analysis using the computer program *COMPONENT* (Page 1993). Using species with restricted ranges, the boundaries of six small areas of endemism along the east coast of Australia were defined and southern New Guinea was included (Figure 11 of Weston and Crisp 1994). The major result was a summary area cladogram of Australian areas (their Figure 11) comprising (Tasmania, (New Guinea + northern Queensland), ((Sydney + McPherson–Macleay), (Gippsland + Budawangs))).

In their next study, Crisp and Weston (1995) used *Lechenaultia* (Goodeniaceae) and *Leptosema* (Fabaceae) to define areas of endemism within Australia based directly on the species distribution of endemic taxa included in the study. These areas are notable for being quite different from those of Cracraft (1991) that were used in most other studies up to that time. The resultant biogeographic analysis, again using *COMPONENT*, comprised a summary area cladogram (their Figure 11), with two main clades and a polytomy. In clade 1, Arnhem Land was sister to the Great Sandy Desert, and clade 2 resolved the area of Irwin in Western Australia as sister to a clade comprising the Gibson Desert and Coolgardie areas. A polytomy of three areas – Charters Towers, the Simpson Desert, and Carnarvon – was also found. Taken together, the results suggested that the arid areas of Australia are not monophyletic but rather are sister to geographically close nonarid areas, a finding which supports the *arid-peripheral* hypothesis (sensu Burbidge).

Crisp *et al.* (1995) then undertook a comparative study of 11 angiosperms, again by combining components (nodes) of individual area cladograms into a summary tree. The areas defined in their map (Figure 7.1c; see Table 7.1 herein) represented a modified version of Cracraft's (1991) areas.

This study is a significant comparative biogeographical study of Australian plants and Australian areas, at least in a way that combines and summarises different phylogenetic trees (i.e. using cladistic biogeographical methods). A general area cladogram, with widespread taxa removed, was the result (their Figure 2). Crisp *et al.* (1995) found Tasmania was sister to three major clades: (1) a polytomy comprising Kimberley + Arnhem + Cape York + New Guinea + the Northern Desert, (2) a clade of the South-west + the Western Desert which is sister to (3) a clade comprising, in phyletic sequence, Atherton + east Queensland + Adelaide + Victoria + (southeastern New South Wales [NSW] + Macpherson–Macleay).

The general area cladogram was interpreted by Crisp *et al.* (1995) as showing two different histories. The first comprised the northern monsoonal areas, including southern New Guinea, and the second linked the northern Atherton area to areas of successive differentiation along the east coast and continuing to the south coast of Australia, which collectively are sister to areas of the South-west (and the Western Desert). Tasmania was interpreted as the area of earliest vicariance and differentiation, in a sister relationship with all other Australian areas; the timing and mechanism for this separation was speculated on but noted as obscured by repeated cycles of separation and reconnection of Tasmania and mainland Australia. As found in another study (Crisp and Weston 1995), different arid areas of Australia related more closely to their proximal coastal areas than to each other.

In an insightful and reflective contribution to the field of cladistic biogeography, Ladiges (1998) explored the theory and methods for combining areagrams via *subtree* analysis into general area cladograms, and the importance of removing geographic paralogy in summary area cladograms. As Ladiges explained, geographic paralogy is the overlap of geographic distributions of taxa that are related at a node in a cladogram (analogous to gene duplications tracking different evolutionary histories). This research was comparative in approach, as Ladiges noted that comparing subclades within a single lineage really is no different from comparing different plant lineages. Ladiges analysed *Banksia* using subtree analysis of three clades, ser. *Salicinae*, sect. *Banksia* and ser. *Spicigerae* (with the results summarised as Figure 10 therein). The results provided evidence of geographic paralogy in the overlapping distributions of the three main *Banksia* clades, and subtree analysis produced six subtrees that were summarised into an area cladogram relating Atherton and eastern Queensland and also southeastern NSW to Victoria. The McPherson–Macleay area was found to either relate to southeastern NSW–Victoria or to Atherton–Eastern Queensland, both equally parsimonious solutions. Adelaide was related to the areas Atherton, eastern Queensland, southeastern NSW, McPherson–Macleay and Victoria, although in an unresolved polytomy. The South-west, Eyre and Tasmania were in a basal polytomous position in the area cladogram, interpreted as evidence for the early vicariance of the South-west (due to a preponderance of endemics).

Following on from the demonstration of the method of cladistic biogeographic subtree analysis was the application of the approach in studies by Ladiges and collaborators. For example, Brown *et al.* (2001) in a study of the biogeography of *Melaleuca*, *Callistemon* and related genera, and using the biotic areas defined by Crisp *et al.* (1995) plus New Caledonia in the outgroup, found that subtree analysis resulted (their Figure 7) in an unresolved basal node of New Caledonia, New Guinea, eastern Queensland and the Northern Desert. A node also related the northern areas of Australia (Kimberley, Arnhem, Cape York, Atherton and Pilbara) and included the Western and Eastern Deserts, although the relationships of these areas were not resolved with respect to each other or external areas. However, a further subclade did relate the southern regions into a clade within which the South-west is related to Eyre/Adelaide (designated the *South*) and Tasmania is related to the southeast and McPherson–Macleay. Brown *et al.* (2001) interpreted this result as indicative of an early vicariance event possibly in response to major climatic change (from the Early Tertiary) between northern and southern Australia.

A Comparative Phytogeographic Analysis of Australia: CWL

By the end of the twentieth century, few plant biogeographical studies in Australia had progressed beyond a narrative approach or had invoked general explanations involving multiple taxa. CWL (p. 332) opined that 'general patterns and general explanations will never be found if they are not sought', and it was at that time clear that understanding the biogeographic history of areas in Australia would require drawing together the accumulating phylogenetic and distributional data into comparative studies to allow general patterns of taxon–area relationships to be identified (Ladiges *et al.* 2012). While acknowledging that the data accumulated by botanists specialising in different individual taxa were invaluable, CWL saw a major opportunity to undertake comparative biogeographic analyses of multiple taxa in the Australian flora.

At the time, cladistic biogeography (often referred to as *vicariance biogeography*) was at the height of its influence (Humphries and Parenti 1999). Indeed, CWL focus heavily on results using cladistic methods, while acknowledging other methods that were then emerging, including molecular divergence dating approaches for reconstructing biogeographic scenarios. However, much of CWL is devoted to an original analysis of area relationships within Australia (although somewhat limited by a lack of appropriate phylogenies available at that time) using a combination of track analysis and cladistic biogeographic methods. They noted the potential difficulty of using tracks within a continent, recognising that there may be no sharp boundaries or barriers between biotic areas. Further, they were cognisant that over time a succession of taxa may have differentiated in the same areas within the continental land mass. This methodological difficulty has been addressed by the development of paralogy-free subtree methods (Nelson and Ladiges 1996) and, more recently, quantitative geospatial methods (Crisp *et al.* 2001; Laffan and Crisp 2003; González-Orozco *et al.* 2014a).

Up to that time, there was only one published cladistic analysis of the whole of Australia involving multiple plant taxa: Crisp *et al.* (1995). CWL analysed a modified set of those data, which included 11 taxa from five angiosperm families, plus the legume genus *Cullen* and minus the stringy-bark eucalypts and *Banksia*, the latter two having been found to have no relevant paralogy-free area relationships (Ladiges 1998). The biotic areas used were revised from Crisp *et al.* (1995) and included some new areas (in bold) which were taken up by many subsequent workers (Crisp *et al.* 1995 areas in parentheses): New Guinea (New Guinea), Tasmania (Tasmania), the **Wet Tropics** (Atherton), the **Monsoon Tropics** (Kimberley, Arnhem, Cape York), the southeast (Eyre, Adelaide, Victoria, southeastern NSW, McPherson–Macleay, eastern Queensland), the **South-west** (South-west) and **Central Australia** (Pilbara, Western Desert, Eastern Desert, Northern Desert).

CWL noted that Crisp *et al.* (1995) found conflicting area relationships between taxa and no single nonconflicting area relationship, a result interpreted as a failure to reduce the complexity of nonparalogous subtree results, rather than an absence of general area relationships per se. To overcome this CWL used the methods of Nelson and Ladiges (1996) to generate paralogy-free subtrees (Figure 80 of CWL, reproduced here as Figure 7.3), thereby discovering four area cladograms (cliques). These are as follows, described from the basal node towards the tips in phyletic sequence:

1. Tasmania sister to all; then the southeast, then the Monsoon Tropics, then the South-west and Central Australia sister (see Figure 7.3a). This 'largest clique' includes five subtrees supported by the following taxa: *Daviesia latifolia* group, *Leptosema*, *Persoonia* and *Cullen*.
2. Tasmania sister to all; then the southeast, then Central Australia, then the Monsoon Tropics and South-west sister (see Figure 7.3b), supported by three of the same taxa as in 1. (Figure 7.3a) – namely, the *Daviesia latifolia* group, *Leptosema* and *Cullen*, with *Persoonia* replaced by *Lechenaultia*. Note there is a conflict with the first area cladogram.
3. Monsoon Tropics sister to all; then Central Australia, then the southeast and South-west sister (Figure 7.3c). This clique is supported by relationships in *Gossypium* and *Leptosema*.
4. Tasmania sister to all; then the South-west, then the Wet Tropics and New Guinea sister (Figure 7.3d). This clique is supported by the proteaceous taxa *Lomatia* and subtribe Embothriinae.

Due to the conflicting area relationships among the four area cladograms (cliques), these results were interpreted as demonstrating multiple overlapping tracks within Australia. However, CWL did discover some combinable subtrees which may represent more general patterns that could be tested using other taxa. In addition, the following issues were highlighted (pp. 352–353), followed by some comments in light of selected recent studies.

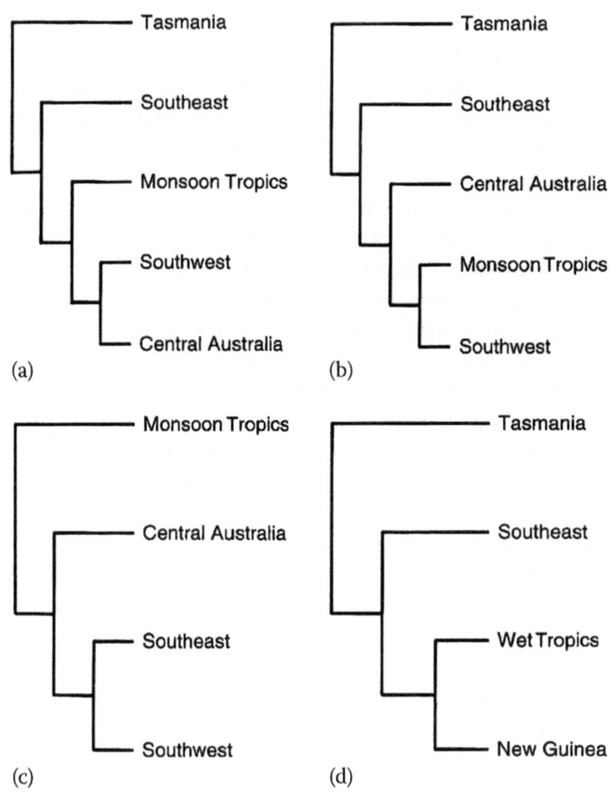

FIGURE 7.3 Summary Area cladograms of Australian plants from CWL (1999). These are described by CWL as follows: 'Four partial area cladograms for Australia. Each is a hand-built clique of combinable non-paralogous subtrees extracted from area-cladograms of Australian angiosperm taxa. None of the partial area cladograms except (c) and (d) can be combined, i.e. they conflict. (a), *Daviesia latifolia* group, *Leptosema* (first subtree), *Cullen* (Fabaceae) and *Persoonia* (Proteaceae). (b), as for (a) except that *Lechenaultia* (Goodeniaceae) is substituted for *Persoonia*. (c), *Gossypium* (Malvaceae) and the second subtree from *Leptosema*. (d), Lomatia and Embothriinae (Proteaceae)'. (Reproduced from Crisp, M.D., *et al.*, *Flora of Australia*, 1, 351, 1999, figure 80, with permission from the authors and publisher.)

1. *Wet Tropics as sister to New Guinea*: The discovery of this relationship begs the question, how does this *external track* relate to *internal tracks* if the age of the relationship is post-Gondwanan? Appropriate phylogenies are required (several legumes were suggested) to examine area relationships both within and without Australia, in order to distinguish whether the tracks within Australia are pre- or post- Gondwanan. Numerous more recent studies show close relationships between taxa in New Guinea and northern Australia (e.g. Crayn *et al.* 2006) or even conspecific taxa (e.g. *Acacia*; Brown *et al.* 2012). It is clear that at least the southern areas of New Guinea should be treated as part of Australia for intracontinental biogeographic analyses, as southern New Guinea is part of the Australian plate geologically and has been repeatedly reconnected through the Quaternary. Crisp *et al.* (1995) recognised this by including New Guinea as a biotic area, which was one of the main developments over Cracraft's (1991) biotic areas.
2. *Tasmania as sister to the rest in area cladograms, representing its early differentiation*: This general result was contradicted by the patterns shown by some eucalypts and legumes, suggesting the possibility that it may be an artefact of the choice of taxa and data for analysis. As CWL note, these contradictory patterns could reflect later, iterative contact between Tasmania and mainland Australia.
3. *Conflicting hypotheses of Central Desert relationships*: Different subtrees show the alternative relationships of the Central Australia biotic area. The result was interpreted as supporting the

desert peripheral hypothesis of Burbidge and Schodde, rather than the Central Australian arid areas being monophyletic. Alternatively, the inclusion of the desert areas may render the summary area cladogram polytomous. Some taxa may have radiated in the arid zone (which would make the area subtrees for those taxa monophyletic for the desert areas). Arid zone specialists such as Chenopodiaceae, *Triodia* and subclades of *Acacia* were highlighted as important taxa to examine. While relatively few studies with a focus on the arid zone have been published, Ladiges *et al.* (2011a) identified three biogeographical tracks across the interzone and Eremaean zone in *Nicotiana*, and inferred that desert regions differentiated first in the north; and Byrne *et al.* (2008), reviewing animal and plant studies, suggested that the assembly of the arid zone has occurred in two major phases: an old origin and a recent diversification.

4. *Southwestern and southeastern Australia are related and show consistent generic level disjunctions*: Several notable Australian genera (namely, *Eucalyptus*, *Acacia*, Mirbelieae (Leguminosae), *Banksia*, *Grevillea*, *Hakea* and *Boronia*) are common and diverse in the temperate areas of South-western and southeastern Australia (separated by the Nullarbor Plain), but there are few species shared between west and east (see Burbidge 1960). This genus-level disjunction is perhaps the most significant finding in CWL and has generated interest and subsequent research (e.g. Crisp and Cook 2007). Most of these disjunct groups tend to have higher numbers of species in the west and multiple basal nodes in cladograms for which the area is most parsimoniously reconstructed as Western Australia: for example, (WA,(WA,(WA,EA))). Examples summarised in Ladiges *et al.* (2012) include the *Phebalium* group (Mole *et al.* 2004), the *Tremandra* group (Butcher *et al.* 2007) and *Eucalyptus* (Ladiges *et al.* 2010). This has led to speculation that taxa with this biogeographic pattern originated in the west and subsequently dispersed to the east (e.g. Crisp *et al.* 1995). However, in their subtrees CWL did not find evidence for this. As Ladiges *et al.* (2012) point out, the pattern of (WA,(WA,(WA,EA))) is an example of geographic paralogy. One explanation for this pattern is that the South-west, with very high levels of endemicity, is an area of old/ancient differentiation due to long periods of isolation, rather than a centre of origin. Heads (2014) developed this argument too – that widespread southern taxa repeatedly budding off of subpopulations on the western periphery would give rise to the general pattern (WA,(WA,(WA,EA))). However, this model contains a bold assumption: that while active diversification was generating lineage after lineage in the west, there was no diversification in the east; widespread southern and eastern populations were in cladogenetic stasis across perhaps thousands of kilometres and millions of years, and only later in the evolution of the group did eastern cladogenesis occur. Contrary to Heads' model, some recent studies of widespread Australian groups conclude, based on nodes in phylogenies resolving WA in a 'basal' position, a western origin for those groups (e.g. *Lechenaultia/Anthotium/Dampiera*, Jabaily *et al.* 2014; *Tetratheca/Tremandra/Platytheca*, Butcher *et al.* 2007; *Hakea*, Lamont *et al.* 2016). However, this interpretation remains contentious (Downing *et al.* 2008; Ladiges *et al.* 2012).

CWL remains the most comprehensive review of the biogeography of Australia's terrestrial flora so far undertaken. However, the number of citations of this review is surprisingly low, in comparison to the equivalent earlier review (Barlow 1981) in the first edition of the *Flora of Australia*. We do not believe this in any way reflects the relative merits of the two works, but rather the much greater rate of change in the research field since CWL. This change has mostly been manifest in two ways: (1) a shift from a focus on biotic areas to biomes in analyses of the Australian flora; and (2) a fragmentation of *historical biogeography* caused by the rapid development and adoption of new methods, such as molecular divergence dating, geospatial analysis techniques (the QGB approach) and ancestral area reconstruction. This *diversification of approaches* was flagged at the time by CWL.

Since 1999, single taxon analyses have proliferated, due largely to advances in molecular divergence dating approaches, yet there have been surprisingly few comparative studies exploiting this wealth of new data. Nonetheless, the contribution comparative analyses have made to our understanding of Australian biogeography is profound.

Historical Biogeography of Biomes: Understanding the Evolutionary Assembly of Australian Plant Communities

Crisp *et al.* (2004) took a novel approach to Australian plant biogeography. They divided Australia into *biomes*, which, although reflected in the ideas of Burbidge and Schodde, was a significant departure from contemporary methods applied to Australian biogeography, as (Crisp *et al.* 2004) explicitly recognised that biomes can move in space and time. To undertake this comparative study, (Crisp *et al.* 2004) compared molecular phylogenies across multiple taxa in order to answer questions about the evolution of biomes. They recognised five biomes in Australia (Figure 7.1d), modified from Burbidge (1960; see Figure 7.1a) and using the boundaries mapped by Schodde (1989; see Figure 7.1b): southeastern temperate, South-western temperate, Monsoonal Tropics, Eremaean and aseasonal wet (the latter in disjunct patches along the east coast and Tasmania).

In detail, Crisp *et al.* (2004) focused on questions of plant evolution during the Cenozoic, with associated fossil and geological evidence for concurrent climatic shifts, to address major questions of Australian plant biogeography and evolution. The questions of particular relevance to biogeography included (1) Do similar evolutionary patterns of divergence occur among taxa within biomes? and (2) Do major radiations occur within biomes as indicated by the monophyly of lineages? Crisp *et al.* (2004) found several noteworthy results using the results of molecular divergence dating on four main groups: *Eucalyptus*, Mirbelieae + Bossiaeeae, Casuarinaceae and *Banksia*. They also compiled the results of a further 89 molecular phylogenetic studies of Australian plant taxa. (Crisp *et al.* 2004) interpreted some currently species-poor lineages in aseasonal rainforests, such as *Nothofagus* and conifers, had abruptly declined in fossil abundance and diversity after 25 mya. By comparing these rainforest lineages with those found in sclerophyll habitats such as *Eucalyptus*, *Banksia*, *Allocasuarina* and pea-flowered legumes, they found that between 10 and 25 mya, as the climate became drier and more seasonal, the diversification rate of these sclerophyll lineages increased. In the arid biome Chenopodiaceae and Brassicaceae taxa radiated rapidly and recently, from 3 mya, which Crisp and Cook hypothesised as indicating the preadaptation of some taxa to the conditions that characterise this biome.

It is worth noting that despite the very different approaches to the analysis of the *Banksia* phylogeny, Ladiges (1998) and (Crisp *et al.* 2004) arrived at similar interpretations. Crisp and Cook inferred multiple radiations within the South-west and suggested that this indicates radiations have taken place within each biome, with virtually no exchange between the South-west and southeast after 25–30 mya, rather than geographic paralogy.

Crisp and Cook (2007) further explored the historical biogeography of southern Australian biomes with a specific focus on the Nullarbor Plain in southern Australia as a putative biogeographic barrier. To do this they compared the timing of congruence of disjunct distributions in divergence dating patterns in 23 lineages, using multiple phylogenies. They found that 16 of the divergences of sister taxa on either side of the Nullarbor Plain were broadly temporally congruent with the origin of the Nullarbor Plain, dated to 13–14 mya. However, some taxa such as *Scaevola* and the peas *Aotus* and *Chorizema* postdated this time. The authors suggested that a greater tolerance to either aridity or calcareous soils allowed these taxa to maintain South-west–southeast connectivity later than others, until extreme aridification 2–4 mya finally separated them. On the other hand, *Banksia*, some Restionaceae and *Allocasuarina* diverged earlier than the origin of the Nullarbor, which was interpreted as a consequence of earlier climatic shifts and marine transgressions. Importantly, the authors recognised that the subsequent re-expansion of some of these lineages had the potential to obscure signatures of earlier vicariance.

In a more expansive study, Crisp and Cook (2013) compiled data from 85 dated plant clades to address broad questions in Australian plant evolution in a biome context. These questions included the origins of major Australian lineages in space and time in light of evolutionary drivers such as fire, climate, landform geophysiology and ecological opportunity. To do this, Crisp and Cook mapped biomes to their dominant vegetation types – for example, *Acacia*, Chenopodiaceae, eucalypts, *Casuarinaceae* and *Triodia*. One question examined was how many clades date back to Gondwana, with the following underlying assumptions: (1) if the split from a sister taxon outside Australia predates the timing of connectivity to a Gondwanan continent the origin is *Gondwanan*; or (2) if the split is dated to after separation, it is likely to be a result of taxon dispersal. They posited 38 (45%) clades likely originated during the time of Australian landmass connectivity – that is, prior to 33 mya. These lineages included Proteaceae, pea-flowered legumes, *Nothofagus* and eucalypts. Another 41 living clades were regarded as immigrant, including *Livistona*, *Cycas* and *Brachychiton* (summarised in Figure 7 of Crisp and Cook 2013); the geographic origins of *Acacia* were unresolved.

Synthetic Approaches to Australian Biogeography

Starting in 2008, a series of papers emerged that examined the origins and assembly of each of the major Australian biomes. These papers reported comparative analyses of both plant and animal data for the arid, mesic (which combined the temperate eastern and aseasonal wet) and northern monsoonal biomes.

The first paper (Byrne *et al.* 2008) used dated phylogenies of arid and mesic lineages to evaluate scenarios of arid biome origin and assembly. There was evidence for a single origin of some arid lineages – that is, monophyletic arid only clades – in *Tetratheca* (Crayn *et al.* 2006), *Lepidium* (Mummenhoff *et al.* 2014) and *Holosarcia* (Shepherd *et al.* 2004). However, for chenopods the initial diversification of the Australian clade probably occurred in coastal areas (Kadereit *et al.* 2006). Further insights were limited by the lack of suitable studies to synthesise, prompting a call for further research on lineages with arid zone representatives.

The second paper (Bowman *et al.* 2010) focused on the AMT, defined as a region of seasonal moist climate, evidence of which was found to date back to the Eocene. While acknowledging that biogeographical zonation in the AMT requires further work, the authors inferred the origins of a number of taxa. Those that originated in situ, the signature of which was sister taxa in other biomes that diverge prior to the estimated age of the AMT (at or about 40 mya), including eucalypts, peas in the endemic Australian tribe Mirbelieae, *Melaleuca*, *Syzygium*, *Casuarina* and *Allocasuarina*. Conversely, plant taxa inferred to have immigrated into the AMT biome included legumes in the genera *Crotolaria*, *Tephrosia*, *Indigofera*, *Cullen* and *Kennedia/Vandasina*; *Cycas*; *Adansonia* and *Bombax ceiba*; *Bambusa*; *Bauhinia* and *Erythrophleum*.

Of these groups the eucalypts and pea-flowered legumes were examined in more detail. The biogeographic patterns were inferred to show older patterns of differentiation of the Kimberley and Arnhem land from Cape York/northern Queensland (and presumably PNG), subsequently overlain with younger patterns reflecting dispersal from Southeast Asia and range expansions and contractions. In *Jacksonia*, narrow-range endemics were noted to occur across the AMT but with lower species richness than in South-west Western Australia, indicating deeper division in the east coast Queensland Monsoon Tropics and northwestern Monsoon Tropics. As for the arid biome, the number of relevant lineages for which suitable data were available was very limited and the authors appealed for comparative biogeographical analysis of additional data sets.

The final paper (Byrne *et al.* 2011) examined the mesic biome, for which the most data were available of the three major biomes analysed. The study was framed as a series of testable hypotheses for the origins of the biome and the lineages it contains. For plants, three major historical biogeographic scenarios were outlined and the taxonomic groups that supported these patterns listed. These were (1) mesic ancestral, supported by Casuarinaceae, *Callitris*, *Eucalyptus* and Proteaceae; (2) rainforest ancestral, supported by Elaeocarpaceae, Casuarinaceae, *Alectryon*, Styphelieae and the Proteaceae subfamily Grevilleoideae; and (3) extra-Australian areas ancestral, supported by *Livistona*, *Citrus*, *Aglaia*, *Rhododendron* and *Ficus*. The overarching scenario illustrated by this analysis was of a widespread mesic biota comprising rainforest and sclerophyll taxa contracting to southeastern, South-western and eastern coastal refugia, with rainforest elements as ancestral. These patterns were later influenced by recent climatic oscillations, which left signatures of fragmentation at different phylogenetic depths. In effect, Byrne *et al.* (2011) supported a composite of the various ideas of the evolution of the Australian flora since Burbidge (1960).

Heads (2014) devoted a book to Australasian biogeography using data resulting from molecular studies of relevant taxa. Heads's book is probably the most comprehensive and significant contribution to the biogeography of Australia in the past two decades. In it, Chapter 4 comprises a review of biogeographic patterns within Australia. By summarising both fauna and flora results, with a panbiogeographic perspective and method, Heads presents an interpretation of the biogeography of the Australian continent that is controversial – namely, that phylogenetic groups are generally older than molecular divergence dating allows (due to artefacts of how nodes are constrained by molecular divergence dating analyses) and that most distribution patterns are interpreted as demonstrative of older vicariance events, rather than younger long-distance dispersal. Overall, Heads notes that Australia is not a monophyletic unit, with parts more closely related to areas outside Australia. Major patterns that Heads highlights (when plant examples in particular are important evidence) are as follows:

1. The link between coastal taxa and central desert (arid) lineages, including those associated with salt lakes – for example, *Salicornia* (and other Chenopodiaceae) and certain clades of *Acacia* (e.g. Ladiges *et al.* 2006)
2. The high diversity of the South-west, as evidenced by endemic clades, such as *Nuytsia* as sister to cosmopolitan Loranthaceae
3. The difference between taxa in South-western Australia to those in southeastern Australia
4. Northwestern lineages allied to Asian groups (rather than distinct, isolated endemics)
5. An inference that Tasmania and the mainland of Australia were separated prior to New Zealand and Tasmania
6. The McPherson–Macleay Overlap as an area of endemism and where a number of northern and southern groups have overlapping distributions
7. Cairns region as an important area with distinct endemics
8. A distributional break within southern New Guinea (this pattern is found in some *Acacia* clades; Brown *et al.* 2012)

Heads's (2014) book is a valuable contribution to the literature, not least in that it raises a number of explicit scenarios that can be examined with further phylogenetic studies of taxa with distributions that are appropriate to investigate the biogeography of the highlighted areas.

In a notable example of integrating different approaches to biogeography, Cook *et al.* (2015) undertook a comparative study of two legume clades, *Bossiaea* and *Daviesia*. Their study revisited the relationship between the southeastern and South-western Australian flora examined in earlier work (Crisp and Cook 2007), but combined a historical divergence dating approach with geospatial techniques. The premise is clearly different from the historical biogeography undertaken under the framework of cladistic biogeography, and combines divergence dating with ancestral area reconstruction and the geospatial analysis of range sizes, using the *Biodiverse* software. Two areas – the South-west Australian Floristic Region (SWAFR) and southeast Australia (SEA) – were selected with the aim of comparing a biodiversity hotspot (SWAFR) with an appropriate 'non-hotspot' (SEA) in order to explore the ecological and climate correlates and the timing of diversification. The result is interpreted as showing, for these pea-flowered legumes, that the time of occupancy in a region contributes to the level of diversity and to the geographic overlap of clades. Factors such as steep climatic gradients were considered to contribute to high species turnover and high endemism, with the proviso that other environmental factors may play a part. Rossetto *et al.* (2015) developed another alternative approach to understanding biome assembly using multiple taxa mapped at a fine geographical scale; in their case they examined Australian rainforest assembly. Such integrative approaches are at an early stage, and the scale of the studies is such that it is still too soon to expand to the broadest questions of the evolution of the Australian flora; however, they are an interesting development as they bring many process-driven hypotheses into biogeography.

Role of Northern Immigration

Three principal processes have influenced the assembly and maintenance of Australian biodiversity at the continental scale: Gondwanan vicariance, in situ diversification (speciation and extinction) and immigration. However, the contribution of immigration, particularly from the north, to the Australian flora has long been debated.

Similarities between the floras and faunas of Australia and neighbouring regions to the north were documented through the comparative analysis of plant distributions as early as the 1860s by Joseph Hooker (Hooker 1860) and Alfred Wallace (Wallace 1869). Hooker identified an 'Indomalayan element' in the Australian flora but did not attempt an explanation of its origins. Van Steenis (1950) published a comprehensive genus-by-genus biogeographical assessment of the Indo-Malayan flora and showed complex patterns of relationships, which was a major advance over Hooker's much more limited assessment. This was a landmark work, which outlined in great detail the distribution and presumed geographical affinity of the Indo-Malayan–Australasian flora generally, and strongly influenced the development of Burbidge's (1960) invasion hypothesis for Australian rainforests.

Webb and Tracey (1981), however, turned on its head the notion of Australia having received its rainforest flora from elsewhere. From an analysis that incorporated ecological surveys, fossils and plate tectonics, they argued that Australian rainforests (including monsoon rainforests) are remnants of ancient widespread Gondwanan flora, and that they survive today as fragmented relics, resulting from changing environmental conditions. This view, while echoing that of Hooker in acknowledging an ancient widespread flora, differs by incorporating a geological mechanism – plate tectonics – for the fragmentation of this ancient flora. Thus Webb and Tracey were able to ascribe the origins of Australia's rainforest flora to a tangible entity: Gondwana. This was effectively the death of the invasion hypothesis (sensu Burbidge).

Subsequently, Thorne (1986) and Barlow and Hyland (1988) proposed that there has been a recent remixing of the Australian flora with the Indo-Malesian flora after a long period of separation, a view which reconciled previous opposing views by incorporating Gondwanan vicariance and Indo-Malayan intrusion in a model of biome assembly. This synthetic theory, which was informed by the author's unrivalled knowledge of the world flora in Thorne's case and the Australian flora in Barlow and Hyland's, did not impress Werren and Sluiter (1991), however, who argued from palynological evidence that the Indo-Malayan intrusion has been insignificant at best.

Recently, the role of the immigration of extra-Australian lineages in the assembly of the flora of Australian biomes has been addressed using phylogenetic methods. The biome-specific studies previously discussed (Byrne *et al.* 2008, 2011; Bowman *et al.* 2007) considered immigration as one of a number of processes influencing the evolution of biomes. Starting in 2011, however, a group of studies sought, using the phylogenies of relevant lineages, to isolate and quantify the contribution of the immigration of Northern Hemisphere lineages alone to the assembly of the Australian flora (Sniderman and Jordan 2011; Richardson *et al.* 2012; Crayn *et al.* 2015).

These synthetic studies are consistent in affirming the views of Thorne (1986) and Barlow and Hyland (1988) and cast considerable light on the dynamics of this exchange and the identity of the lineages involved. They draw on numerous single-lineage analyses that have inferred an Asian origin for many Australian rainforest taxa including members of Annonaceae, Araceae, Arecaceae, Caprifoliaceae, Cucurbitaceae, Euphorbiaceae, Meliaceae and Piperaceae. Synthetic analysis of those studies that incorporated the molecular dating of clades has proposed a general model in which the influx of Asian lineages began about 33 mya but increased in rate after about 12 mya when the Sunda and Sahul Shelves had neared their current positions (Crayn *et al.* 2015). Work published subsequent to this (e.g. van Welzen *et al.* 2015; de Boer *et al.* 2015; Wei *et al.* 2015) agrees in broad terms with that model. A number of lineages including *Elaeocarpus* and *Sloanea* (Elaeocarpaceae: Crayn *et al.* 2006), *Gymnostoma* (Casuarinaceae; Steane *et al.* 2003) and members of Monimiaceae (Renner *et al.* 2010) migrated in the opposite direction. However, these northwestward dispersals were much less frequent (Sniderman and Jordan 2011; Richardson *et al.* 2012; Crayn *et al.* 2015). A bias in dispersal direction or *asymmetry of invasion* has also been demonstrated for the Sahulian fauna, and for many other cases of biotic interchange globally (Vermeij 1991).

Historically, most of the debate about northern floristic exchange has concerned the rainforest flora. Evidence for exchange in other biomes such as the monsoonal, arid, wetland, montane, mangrove and sclerophyll biomes is weaker, because detailed biogeographical analyses of lineages that characterise these biomes are fewer (as previously mentioned). However, it can be surmised that of these biomes the monsoonal biome has probably been most strongly influenced by exchanges with northern floras.

Summary of the Contribution of Comparative Approaches to Australian Plant Biogeography

Discovery of General Patterns

Comparative biogeographical analyses have resolved some broad plant biogeographical patterns within continental Australia. These include the following:

1. *General temperate southeast–South-west disjunct pattern*: SWAFR is notably phylogenetically distinct, and is a geologically ancient island-like region (Hopper and Gioia 2004). Many of the taxa in the South-west have evidence of sister taxa in groups from mesic or rainforest environments (reviewed in Byrne *et al.* 2014). For example, Tremandraceae/Elaeocarpaceae (Crayn *et al.* 2006; Butcher *et al.* 2007; Downing *et al.* 2008) show that the South-west harbours the earliest diverging lineages, a pattern seen also in *Eucalyptus* subgenus *Eucalyptus*, and *Banksia* (Ladiges *et al.* 2012). It is likely that for many Australian plant groups these patterns can be interpreted as the early overlap of lineages (paralogous nodes, sensu Ladiges 1998) with little exchange of taxa after 25–30 mya (Cook and Crisp 2004). The relative roles of the biogeographical processes that gave rise to this pattern are the subject of debate and a worthy subject for future in-depth analysis. Notably this was a major result in the CWL track analysis (see point 4 of 'A Comparative Phytogeographic Analysis of Australia: CWL' in this chapter).
2. *The arid zone has a complex history*: In some studies there is evidence that the arid zone is not monophyletic, with multiple peripheral-coastal relationships interpreted as relatively recent patterns (e.g. Cook and Crisp 2004; Byrne *et al.* 2008). However, for some clades the arid area is a single monophyletic region of radiation and endemism (such as the *Corymbia* and *Eucalyptus* subgenus *Eudesmia* [Ladiges *et al.* 2011b] and certain clades of *Acacia* [Ladiges *et al.* 2006]). For *Nicotiana*, Ladiges *et al.* (2011) identified three biogeographical tracks across the interzone and Eremaean zone, suggesting that the northern track is older than the central and southern tracks, indicating that desert regions differentiated first in the north. González-Orozco *et al.* (2014a) also recover a similar pattern, and in a result that harks back to CWL, they separate the arid biome into two regions, recognising a Northern Desert and separate Eremaean. How this change of interpretation will impact the understanding and implementation of Australian historical biogeography is as yet unknown. In CWL, there was some support for the desert peripheral pattern, but subsequent studies have not yet resolved the historical biogeography of the arid zone in any clear way. For example, using QGB approaches, the arid parts of Australia are not identified as strong centres of endemism (Crisp *et al.* 2001; González-Orozco *et al.* 2014a).
3. *The McPherson–Macleay zone is contentious*: Burbidge (1960) described this as a zone of overlap between tropical and temperate genera, and it has tended to be recognised as a significant biotic area since then. There is no equivalent zone in the west of Australia. Heads (2014) recognised the area as an area of endemism, rather than an area where there is an overlap of distributions, and invoked a complex geological composite history. Interestingly, the McPherson–Macleay area has not been strongly identified using geospatial methods (e.g. González-Orozco *et al.* 2014a) nor was it resolved in the analysis of CWL.
4. *Tasmania as sister to the rest of Australia*. This was a major result in CWL and mentioned in Heads (2014) but has otherwise been surprisingly little investigated. A number of studies have investigated within-species genetic variation and have often found higher structure in Tasmania compared with populations in mainland Australia (see Byrne *et al.* 2011), which may reflect secondary contact between the island and the mainland and/or the genetic impacts of glaciation. Increased attention at a deeper phylogenetic scale, however, is warranted, and may be addressed by including taxa found in Tasmania with divergences from mainland taxa of an appropriate age.
5. *Northern and southern Australia show an early break in connectivity*, possibly in response to Early Cenozoic climatic change, rather than more recent climatic cycling (e.g. Brown *et al.*

2001). There remains much uncertainty about the biogeography of the AMT, and how best to subdivide that region for analysis (Bowman *et al.* 2010).

6. *Northern immigration is strong and mostly recent.* A strong pattern of relationships with areas to the north and northwest of Australia indicates a common temporal and directional pattern of floristic exchange (Richardson *et al.* 2012; Sniderman and Jordan 2011; Crayn *et al.* 2015).

It is noteworthy that while approaches to biogeography have diversified, these major patterns have to a large extent been consistently recovered in analyses since CWL (1999).

Eucalyptus is one of the few groups for which various authors have analysed biogeographic data in different ways. For example, Ebach and Parenti (2015) compare the *Eucalyptus* areagrams from Ladiges *et al.* (2011) and Crisp *et al.* (2004) to investigate whether or not the Eremaean is a natural (monophyletic) area. It seems that the overlapping of different temporal patterns in the distributional ranges of Australian plants requires deeper understanding and perhaps more explicit methodologies to tease out these patterns in greater detail. There is a lack of clarity in many studies regarding the impact of multiple overlapping distributions at different phylogenetic depths. Molecular chronograms alone have not explicitly dealt with this due to confounding factors such as extinction and widespread taxa. As an example, Ladiges *et al.* (2012) showed that the earliest lineages in *Eucalyptus* subgenus *Eucalyptus* are in the South-west of Australia and that east–west divergence is at a shallower node. They argued against interpreting the South-west as the centre of origin of subgenus *Eucalyptus* because at an even lower node, the sister subgenus *Idiogenes* and other taxa are endemic to eastern Australia, which would imply under this logic a dispersal first from east to west, then back from west to east. Ladiges *et al.* (2012) lays the groundwork for potential comparative area analyses; however, there still remain many questions and areas that have received little or no attention.

Until recently it was apparent that a major goal of plant biogeographers studying intracontinental biogeography was to accumulate studies of a broad range of taxa in order to undertake comparative synthetic analyses. While there were some studies published, such as the landmark paper by Crisp *et al.* (1995) and a series of studies by Ladiges *et al.* (as pointed out previously and in CWL), overall general biogeographic patterns within Australia have remained largely elusive. CWL stated in 1999 that 'efforts to trace general floristic tracks within the continent, and to find general relationships among areas, have not been very successful'; this remains true. While there is a possibility that general biogeographic patterns may not exist and biogeographic structure is taxon specific, a more optimistic view is that insufficient taxa have been studied with consistent comparative approaches, a consequence perhaps of the lack of unifying biogeographic methods. This is an issue that has plagued Australian biogeography since the 1990s (Ladiges *et al.* 2012). For example, there has been a lack of recognition of paralogous nodes, leading to spurious reconstructions of ancestral areas and the difference in the scale of biogeographic barriers for different taxa (Chapple *et al.* 2011). Whatever the cause, it appears that the comparative historical biogeography research program has stalled somewhat in recent years and has been replaced by other approaches or an emphasis on single-taxon studies. As Bowman *et al.* (2010) conclude, there is a need to synthesise biogeographical analyses. How this is to be done, and the methods to best do so, are as yet unclear. However, as noted previously, the emergence of the QGB approaches are providing an alternative way to recognise and define general biogeographic patterns, for which later process-led explanations may arise or on which area definitions can be based.

Future Prospects

In the last decade the attention of Australian plant biogeographers has become increasingly focused on technical (e.g. genomics) and methodological advances (e.g. molecular dating approaches and model-based analyses). The most popular methods of biogeographical analysis are phylogeny-based, rendering progress in the field limited by the availability of suitable cladograms. Advances in supertree and supermatrix methods have enabled data and results from disparate studies to be amalgamated to generate detailed phylogenetic hypotheses de novo. However effective these methods have been, there remains many key taxa for which data suitable for molecular phylogenetic analysis are lacking. In part this is due to the availability of samples, but the advent of shotgun sequencing methods that are more robust to sample degradation has increased the range of herbarium specimens that are amenable to DNA analysis, reducing this impediment. The development of sequencing technologies is accelerating, bringing nearer the possibility of a complete phylogeny of all Australian plant species. Beyond species, it is hoped that advances in genomics, bioinformatics and analytical approaches based on the concept of the coalescent (Kingman 1982; Harding 1996) will enable the routine use of least-inclusive clades, rather than groups corresponding to arbitrary taxonomic ranks such as genera or species, as the units of comparison in biogeographical analyses.

The digitisation and online delivery of herbarium specimen records has vastly increased the availability of geospatial data. In response, biogeographers have developed and deployed new approaches that utilise these data for defining areas (phytogeographical regionalisation) and discovering phylogeographical patterns. Quality issues with geospatial data are well known and can hamper their use, but as the exposure and use of these data increases so does their quality. Vigilant users can be excellent data quality checkers, but more efficient processes to feed back error detection and implement corrections at the source (herbarium database) are required to derive maximum advantage.

Much progress has also been made on conceptual fronts. For example, the issue of paralogous nodes clouding biogeographic inferences (Ladiges 1998) has been accounted for in some recent studies (e.g. Byrne *et al.* 2008; Ladiges *et al.* 2012). This issue has been somewhat controversial and no doubt debate will continue. Some authors have deployed integrative approaches that combine several methods to deal with distributional and phylogenetic data, such as Toon *et al.* (2015) for *Triodia* and Cook *et al.* (2015) for pea-flowered legumes.

While all of these technical and methodological developments undoubtedly enrich the science of biogeography and provide valuable new insights, they have coincided with a decline in the popularity of comparative biogeographical analysis in the past 10–15 years. This seems to be a missed opportunity, particularly as it comes at a time when distributional data and phylogenies have never been more available and methods to analyse the role of processes that have been traditionally difficult to study (such as extinction and dispersal) are proliferating. Will phytogeographical studies reveal general biogeographic patterns within Australia, other than a broad east–west disjunction? There is some evidence for this in single-taxon studies; however, comparative studies remain somewhat inconclusive, as there are few studies within Australia that have used multiple taxa (Ladiges *et al.* 2011). We believe the time is right for a renewed interest in comparative approaches to biogeography and maintain an optimism that much remains to be discovered regarding general biogeographical patterns in the Australian flora.

References

Barlow, B.A. (1981). The Australian flora: Its origin and evolution. In *Flora of Australia* (Ed. A.S. George), Vol. 1, Introduction, pp. 25–75. Australian Government Publishing Service, Canberra, Australia.

Barlow, B.A. (1994). Phytogeography of the Australian region. *Australian Vegetation* 2, 3–36.

Barlow, B.A., and Hyland, B.P.M. (1988). The origins of the flora of Australia's Wet Tropics. *Proceedings of the Ecological Society of Australia* 15, 1–17.

Bowman, D.M.J.S., Brown, G.K., Braby, M.F., Brown, J.R., Cook, L.G., Crisp, M.D., Ford, F., *et al.* (2010). Biogeography of the Australian Monsoon Tropics. *Journal of Biogeography* 37, 201–216.

Brown, G.K., Murphy, D.J., Kidman, J., and Ladiges, P.Y. (2012). Phylogenetic connections of phyllodinous species of *Acacia* outside Australia are explained by geological history and human-mediated dispersal. *Australian Systematic Botany* 25, 390–403.

Brown, G.K., Udovicic, F., and Ladiges, P.Y. (2001). Molecular phylogeny and biogeography of *Melaleuca, Callistemon* and related genera (Myrtaceae). *Australian Systematic Botany* 14, 565–585.

Burbidge, N.T. (1960). The phytogeography of the Australian Region. *Australian Journal of Botany* 8, 75–209.

Butcher, R., Byrne, M., and Crayn, D.M. (2007). Evidence for convergent evolution among phylogenetically distant rare species of *Tetratheca* (Elaeocarpaceae, formerly Tremandraceae) from Western Australia. *Australian Systematic Botany* 20, 126–138.

Byrne, M., Coates, D.J., Forest, F., Hopper, S.D., Krauss, S.L., Sniderman, J.K., and Thiele, K.R. (2014). A diverse flora–species and genetic relationships. In *Plant Life on the Sandplains in Southwest Australia*, pp. 81–99. UWA Publishing, Perth, Australia.

Byrne, M., Steane, D.A., Joseph, L., Yeates, D.K., Jordan, G.J., Crayn, D., Aplin, K., *et al.* (2011). Decline of a biome: Evolution, contraction, fragmentation, extinction and invasion of the Australian mesic zone biota. *Journal of Biogeography* 38, 1635–1656.

Byrne, M., Yeates, D.K., Joseph, L., Kearney, M., Bowler, J., Williams, M.A.J., Cooper, S., *et al.* (2008). Birth of a biome: Insights into the assembly and maintenance of the Australian arid zone biota. *Molecular Ecology* 17, 4398–4417.

Chapple, D.G., Chapple, S.N., and Thompson, M.B. (2011). Biogeographic barriers in south-eastern Australia drive phylogeographic divergence in the garden skink, *Lampropholis guichenoti*. *Journal of Biogeography* 38(9), 1761–1775.

Cook, L.G., Hardy, N.B., and Crisp, M.D. (2015). Three explanations for biodiversity hotspots: Small range size, geographical overlap and time for species accumulation; An Australian case study. *New Phytologist* 207, 390–400.

Costion, C.M., Edwards, W., Ford, A.J., Metcalfe, D.J., Cross, H.B., Harrington, M.G., Richardson, J.E., Hilbert, D.W., Lowe, A.J., and Crayn, D.M. (2015). Using phylogenetic diversity to identify ancient rain forest refugia and diversification zones in a biodiversity hotspot. *Diversity and Distributions* 21, 279–289.

Cracraft, J. (1991). Patterns of diversification within continental biotas: Hierarchical congruence among the areas of endemism of Australian vertebrates. *Australian Systematic Botany* 4(1), 211–227.

Crayn, D.M., Costion, C., and Harrington, M.G. (2015). The Sahul–Sunda floristic exchange: Dated molecular phylogenies document Cenozoic intercontinental dispersal dynamics. *Journal of Biogeography* 42, 11–24.

Crayn, D.M., Rossetto, M., and Maynard, D.J. (2006). Molecular phylogeny and dating reveals an Oligo-Miocene radiation of dry-adapted shrubs (former Tremandraceae) from rainforest tree progenitors (Elaeocarpaceae) in Australia. *American Journal of Botany* 93, 1328–1342.

Crisci, J.V., Katinas, L., Posadas, P. (2003). *Historical Biogeography: An Introduction*. Harvard University Press, Cambridge, MA.

Crisp, M.D., and Cook, L.G. (2007). A congruent molecular signature of vicariance across multiple plant lineages. *Molecular Phylogenetics and Evolution* 43, 1106–1117.

Crisp, M.D., and Cook, L.G. (2013). How was the Australian flora assembled over the last 65 million years? A molecular phylogenetic perspective. *Annual Review of Ecology, Evolution, and Systematics* 44, 303–324.

Crisp, M.D., Cook, L.G., and Steane, D.A. (2004). Radiation of the Australian flora: What can comparisons of molecular phylogenies across multiple taxa tell us about the evolution of diversity in present-day communities? *Philosophical Transactions of the Royal Society B: Biological Sciences* 359, 1551–1571.

Crisp, M.D., Laffan, S., Linder, H.P., and Monro, A. (2001). Endemism in the Australian flora. *Journal of Biogeography* 28, 183–198.

Crisp, M.D., Linder, H.P., and Weston, P.H. (1995). Cladistic biogeography of plants in Australia and New Guinea: Congruent pattern reveals two endemic tropical tracks. *Systematic Biology* 44, 457–473.

Crisp, M.D., Trewick, S.A., and Cook, L.G. (2011). Hypothesis testing in biogeography. *Trends in Ecology and Evolution* 26, 66–72.

Crisp, M.D., West, J.G., and Linder, H.P. (1999). Biogeography of the terrestrial flora. *Flora of Australia* 1, 321–367.

Crisp, M.D., and Weston, P.H. (1995). Mirbelieae: Advances in legume systematics, Part Seven. In *Phylogeny*, pp. 245–282. Royal Botanic Gardens, Kew, UK.

Croizat, L. (1962). *Space, Time and Form: The Biological Synthesis*. Caracas, Venezuela. Published by the author.

de Boer, H.J., Steffen, K., and Cooper, W.E. (2015). Sunda to Sahul dispersals in *Trichosanthes* (Cucurbitaceae): A dated phylogeny reveals five independent dispersal events to Australasia. *Journal of Biogeography* 42, 519–531.

Downing, T.L., Ladiges, P.Y., and Duretto, M.F. (2008). Trichome morphology provides phylogenetically informative characters for *Tremandra, Platytheca* and *Tetratheca* (former Tremandraceae). *Plant Systematics and Evolution* 271, 199–221.

Ebach, M.C. (2012). A history of bioregionalisation in Australia. *Zootaxa* 3392, 1–34.

Ebach, M.C. (2015). Origins of biogeography. *History, Philosophy and Theory of the Life Sciences* 13, 1–20.

Ebach, M.C., Gill, A.C., Kwan, A., Ahyong, S.T., Murphy, D.J., and Cassis, G. (2013). Towards an Australian Bioregionalisation Atlas: A provisional area taxonomy of Australia's biogeographical regions. *Zootaxa* 3619, 315–342.

Ebach, M.C., González-Orozco, C.E., Miller, J.T., and Murphy, D.J. (2015). A revised area taxonomy of phytogeographical regions within the Australian Bioregionalisation Atlas. *Phytotaxa* 208, 261–277.

Ebach, M.C., and Parenti, L.R. (2015). The dichotomy of the modern bioregionalization revival. *Journal of Biogeography* 42, 1801–1808.

Engler, A. (1882). *Versuch einer Entwicklungsgeschichte der Pflanzenwelt, insbesondere der florengegiete seit der tertiärperiode*. Wilhelm Engelmann, Leipzig, Germany.

Faith, D.P. (1992). Conservation evaluation and phylogenetic diversity. *Biological Conservation* 61, 1–10.

González-Orozco, C.E., Laffan, S.W., Knerr, N., and Miller, J.T. (2013). A biogeographical regionalization of Australian *Acacia* species. *Journal of Biogeography* 40, 2156–2166.

González-Orozco, C.E., Mishler, B.D., Miller, J.T., Laffan, S.W., Knerr, N., Unmack, P., Georges, A., Thornhill, A.H., Rosauer, D.F., and Gruber, B. (2015). Assessing biodiversity and endemism using phylogenetic methods across multiple taxonomic groups. *Ecology and Evolution* 5, 5177–5192.

González-Orozco, C.E., Thornhill, A.H., Knerr, N., Laffan, S., and Miller, J.T. (2014b). Biogeographical regions and phytogeography of the eucalypts. *Diversity and Distributions* 20, 46–58.

Harding, R.M. (1996). New phylogenies: An introductory look at the coalescent. In *New Uses for New Phylogenies* (Eds. P.H. Harvey, A.J.L. Brown, J. Maynard Smith, S. Nee), pp. 15–22. Oxford University Press, Oxford, UK.

Heads, M. (2014). *Biogeography of Australasia: A Molecular Analysis*. Cambridge University Press, Cambridge, UK.

Heads, M. (2005). Dating nodes on molecular phylogenies: A critique of molecular biogeography. *Cladistics* 21(1), 62–78.

Hodel, R.G., de Souza Cortez, M.B., Soltis, P.S., and Soltis, D.E. (2016). Comparative phylogeography of black mangroves (*Avicennia germinans*) and red mangroves (*Rhizophora mangle*) in Florida: Testing the maritime discontinuity in coastal plants. *American Journal of Botany* 103, 730–739.

Hooker, J.D. (1860). On the origination and distribution of species: Introductory essay to the flora of Tasmania. *American Journal of Science* 85, 1–25.

Hopper, S.D. (1979). Biogeographical aspects of speciation in the South-west Australian flora. *Annual Review of Ecology and Systematics* 10, 399–422.

Hopper, S.D., and Gioia, P. (2004). The South-west Australian floristic region: Evolution and conservation of a global hot spot of biodiversity. *Annual Review of Ecology, Evolution, and Systematics* 623–650.

Humphries, C.J., and Parenti, L.R. (1999). *Cladistic Biogeography: Interpreting Patterns of Plants and Animal Distributions*, 2nd edn. Oxford University Press, Oxford, UK.

IMCRA (Interim Marine and Coastal Regionalisation for Australia) Technical Group (1998). *Interim Marine and Coastal Regionalisation for Australia: An Ecosystem-Based Classification for Marine and Coastal Environments Version 3.3*. Environment Australia, Commonwealth Department of the Environment, Canberra, Australia, 102 p.

Jabaily, R.S., Shepherd, K.A., Gardner, A.G., Gustafsson, M.H.G., Howarth, D.G., and Motley, T.J. (2014). Historical biogeography of the predominantly Australian plant family Goodeniaceae. *Journal of Biogeography* 41, 2057–2067.

Jordan, G.J., Harrison, P.A., Worth, J.R., Williamson, G.J., and Kirkpatrick, J.B. (2016). Palaeoendemic plants provide evidence for persistence of open, well-watered vegetation since the Cretaceous. *Global Ecology and Biogeography* 25, 127–140.

Kadereit, G., Mucina, L., and Freitag, H. (2006). Phylogeny of Salicornioideae (Chenopodiaceae): Diversification, biogeography, and evolutionary trends in leaf and flower morphology. *Taxon* 55, 617–642.

Kingman, J.F.C. (1982). On the genealogy of large populations. *Journal of Applied Probability* 19, 27–43.

Kooyman, R.M., Wilf, P., Barreda, V.D., Carpenter, R.J., Jordan, G.J., Sniderman, J.M.K., Allen, A., *et al.* (2014). Paleo-Antarctic rainforest into the modern Old World tropics: The rich past and threatened future of the 'southern wet forest survivors'. *American Journal of Botany* 101, 2121–2135.

Ladiges, P.Y. (1998). Biogeography after Burbidge. *Australian Systematic Botany* 11, 231–242.

Ladiges, P.Y., Ariati, S.R., and Murphy, D.J. (2006). Biogeography of the *Acacia victoriae, pyrifolia* and *murrayana* species groups in arid Australia. *Journal of Arid Environments* 66, 462–476.

Ladiges, P.Y., Bayly, M.J., and Nelson, G.J. (2010). East–west continental vicariance in *Eucalyptus* subgenus *Eucalyptus*. In *Beyond Cladistics: The Branching of a Paradigm*, pp. 267–302. University of California Press, Berkeley.

Ladiges, P.Y., Bayly, M.J., and Nelson, G. (2012). Searching for ancestral areas and artifactual centers of origin in biogeography: With comment on east–west patterns across southern Australia. *Systematic Biology* 61, 703–708.

Ladiges, P.Y., Marks, C.E., and Nelson, G. (2011a). Biogeography of *Nicotiana* section Suaveolentes (Solanaceae) reveals geographical tracks in arid Australia. *Journal of Biogeography* 38, 2066–2077.

Ladiges, P.Y., Parra-O.C., Gibbs, A., Udovicic, F., Nelson, G., and Bayly, M.J. (2011b). Historical biogeographic patterns in continental Australia: Congruence among areas of endemism of two major clades of eucalypts. *Cladistics* 27, 29–41.

Laffan, S.W., and Crisp, M.D. (2003). Assessing endemism at multiple spatial scales, with an example from the Australian vascular flora. *Journal of Biogeography* 30, 511–520.

Laffan, S.W., Lubarsky, E., and Rosauer, D.F. (2010) Biodiverse, a tool for the spatial analysis of biological and related diversity. *Ecography* 33, 643–647.

Laffan, S.W., Rosauer, D.F., Di Virgilio, G., Miller, J.T., González-Orozco, C.E., Knerr, N., Thornhill, A.H., and Mishler, B.D. (2016). Range-weighted metrics of species and phylogenetic turnover can better resolve biogeographic transition zones. *Methods in Ecology and Evolution* 7(5), 580–588.

Laity, T., Laffan, S.W., González-Orozco, C.E., Faith, D.P., Rosauer, D.F., Byrne, M., Miller, J.T., *et al.* (2015). Phylodiversity to inform conservation policy: An Australian example. *Science of the Total Environment* 534, 131–143.

Lamont, B.B., He, T., and Lim, S.L. (2016). *Hakea*, the world's most sclerophyllous genus, arose in South-western Australian heathland and diversified throughout Australia over the past 12 million years. *Australian Journal of Botany* 64, 77–88.

McLoughlin, S. (2001). The breakup history of Gondwana and its impact on pre-Cenozoic floristic provincialism. *Australian Journal of Botany* 49, 271–300.

Mishler, B.D., Knerr, N., González-Orozco, C.E., Thornhill, A.H., Laffan, S.W., and Miller, J.T. (2014). Phylogenetic measures of biodiversity and neo- and paleo-endemism in Australian Acacia. *Nature Communications* 5, 4473.

Mole, B.J., Udovicic, F., Ladiges, P.Y., and Duretto, M.F. (2004). Molecular phylogeny of *Phebalium* (Rutaceae: Boronieae) and related genera based on the nrDNA regions ITS 1 + 2. *Plant Systematics and Evolution* 249, 197–212.

Moritz, C., Hoskin, C.J., MacKenzie, J.B., Phillips, B.L., Tonione, M., Silva, N., VanDerWal, J., Williams, S.E., and Graham, C.H. (2009). Identification and dynamics of a cryptic suture zone in tropical rainforest. *Proceedings of the Royal Society B: Biological Sciences* 276, 1235–1244.

Mummenhoff, K., Linder, P., Friesen, N., Bowman, J.L., Lee, J.Y., and Franzke, A. (2004). Molecular evidence for bicontinental hybridogenous genomic constitution in *Lepidium* sensu stricto (Brassicaceae) species from Australia and New Zealand. *American Journal of Botany* 91, 254–261.

Nagalingum, N.S., Knerr, N., Laffan, S.W., González-Orozco, C.E., Thornhill, A.H., Miller, J.T., and Mishler, B.D. (2015). Continental scale patterns and predictors of fern richness and phylogenetic diversity. *Frontiers in Genetics* 6, 132.

Nelson, G., and Platnick, N. (1981). Systematics and Biogeography: Cladisticsand Vicariance. Columbia University Press, New York, US.

Nelson, G., and Ladiges, P.Y. (1996). Paralogy in cladistic biogeography and analysis of paralogy-free subtrees. *American Museum Novitates* 3167, 1–58.

Nelson, G., and Ladiges, P.Y. (2001). Gondwana, vicariance biogeography and the New York School revisited. *Australian Journal of Botany* 49, 389–409.

Page, R.D. (1993). COMPONENT: Tree comparison software for Microsoft Windows, version 2.0. *The Natural History Museum, London.*

Parenti, L.R., and Ebach, M.C. (2009). *Comparative Biogeography: Discovering and Classifying Biogeographical Patterns of a Dynamic Earth*, Vol. 2. University of California Press, Berkeley.

Pollock, L.J., Rosauer, D., Thornhill, A.H., Kujala, H., Crisp, M.D., Miller, J.T., and McCarthy, M.A. (2015). Phylogenetic diversity meets conservation policy: Small areas are key to preserving eucalypt lineages. *Philosophical Transactions of the Linnean Society, Series B* 370, 20140007.

Raven, P.H., and Axelrod, D.I. (1974). Angiosperm biogeography and past continental movements. *Annals of the Missouri Botanical Garden* 61(3), 539–673.

Renner, S.S., Strijk, J.S., Strasberg, D., and Thebaud, C. (2010). Biogeography of the Monimiaceae (Laurales): A role for East Gondwana and long-distance dispersal, but not West Gondwana. *Journal of Biogeography* 37, 1227–1238.

Richardson, J.E., Costion, C., and Muellner, A.N. (2012). The Malesian floristic interchange: Plant migration patterns across Wallace's Line. In *Biotic Evolution and Environmental Change in Southeast Asia* (Eds. D. Gower, K. Johnson, J.E. Richardson, B. Rosen, L. Rüber, S. Williams), pp. 138–163. Cambridge University Press, Cambridge, UK.

Ronquist, F., and Sanmartín, I. (2011). Phylogenetic methods in biogeography. *Annual Review of Ecology, Evolution, and Systematics* 42(1), 441.

Rosauer, D.F., Laffan, S.W., Crisp, M.D., Donnellan, S.C., and Cook, L.G. (2009). Phylogenetic endemism: A new approach for identifying geographical concentrations of evolutionary history. *Molecular Ecology* 18, 4061–4072.

Rossetto, M., McPherson, H., Siow, J., Kooyman, R., Merwe, M., and Wilson, P.D. (2015). Where did all the trees come from? A novel multispecies approach reveals the impacts of biogeographical history and functional diversity on rainforest assembly. *Journal of Biogeography* 42, 2172–2186.

Sanmartin, I., Wanntorp, L., and Winkworth, R.C. (2007). West Wind Drift revisited: Testing for directional dispersal in the Southern Hemisphere using event-based tree fitting. *Journal of Biogeography* 34, 398–416.

Sauquet, H., Ho, S.Y., Gandolfo, M.A., Jordan, G.J., Wilf, P., Cantrill, D.J., Bayly, M.J., *et al.* (2012). Testing the impact of calibration on molecular divergence times using a fossil-rich group: The case of *Nothofagus* (Fagales). *Systematic Biology* 61, 289–313.

Schmidt-Lebuhn, A.N., Knerr, N.J., Miller, J.T., and Mishler, B.D. (2015). Phylogenetic diversity and endemism of Australian daisies (Asteraceae). *Journal of Biogeography* 42, 1114–1122.

Schodde, R. (1989). Origins, radiations and sifting in the Australasian biota: Changing concepts from new data and old. *Australian Systematic Botanical Society Newsletter* 60, 2–11.

Shepherd, K.A., Waycott, M., and Calladine, A. (2004). Radiation of the Australian Salicornioideae (Chenopodiaceae): Based on evidence from nuclear and chloroplast DNA sequences. *American Journal of Botany* 91, 1387–1397.

Sniderman, J.M.K., and Jordan, G.J. (2011). Extent and timing of floristic exchange between Australian and Asian rain forests. *Journal of Biogeography* 38, 1445–1455.

Steane, D.A., Wilson, K.L., and Hill, R.S. (2003). Using *matK* sequence data to unravel the phylogeny of Casuarinaceae. *Molecular Phylogenetics and Evolution* 28, 47–59.

Thackway, R., and Cresswell, I.D. (Eds.) (1995). *An Interim Biogeographic Regionalisation for Australia: A Framework for Setting Priorities in the National Reserves System Cooperative Program*. Australian Nature Conservation Agency, Canberra, Australia.

Thorne, R.F. (1986). Summary statement on relationships of Australasian rainforest floras. *Telopea* 2, 697–704.

Thornhill, A., Mishler, B., Knerr, N., González-Orozco, C., Costion, C., Crayn, D., Laffan, S., and Miller, J. (2016). Continental scale phylogenetic diversity and endemism in Australian angiosperms. *Journal of Biogeography*.

Toon, A., Crisp, M.D., Gamage, H., Mant, J., Morris, D.C., Schmidt, S., and Cook, L.G. (2015). Key innovation or adaptive change? A test of leaf traits using Triodiinae in Australia. *Scientific Reports* 5, 12398.

van Steenis, C.G.G.F. (1950). The delimitation of Malaysia and its main plant geographical divisions. In *Flora Malesiana* (Ed. C.G.G.F. van Steenis.), Vol. 1, pp. 70–75. Noordhoff-Kolff, Jakarta, Indonesia.

van Welzen, P.C., Pruesapan, K., Telford, I.R.H., and Bruhl, J.J. (2015). Historical biogeography of Breynia (Phyllanthaceae): What caused speciation? *Journal of Biogeography* 42, 1493–1502.

Vermeij, G.J. (1991). When biotas meet: Understanding biotic interchange. *Science* 253, 1099–1104.

Wallace, A.R. (1869). *The Malay Archipelago*. Periplus, Hong Kong.

Wannan, B. (2014). Review of the phytogeography of Cape York Peninsula: A flora that illustrates the development of the Australian sclerophyll biota. *Australian Journal of Botany* 62(2), 85–113.

Webb, L.J., and Tracey, J.G. (1981). Australian rainforests: Patterns and change. In *Ecological Biogeography of Australia* (Ed. A. Keast.), Vol. 1, pp. 605–694. Dr. W. Junk, The Hague, the Netherlands.

Wei, R., Xiang, Q., Schneider, H., Sundue, M.A., Kessler, M., Kamau, P.W., Hidayat, A., and Zhang, X. (2015). Eurasian origin, boreotropical migration and transoceanic dispersal in the pantropical fern genus *Diplazium* (Athyriaceae). *Journal of Biogeography* 42, 1809–1819.

Wen, J., Ree, R.H., Ickert-Bond, S.M., Nie, Z., and Funk, V. (2013). Biogeography: Where do we go from here? *Taxon* 62, 912–927.

Werren, G.L., and Sluiter, I.R. (1991). Australian rainforests: A time for reappraisal. In *The Rainforest Legacy: Rainforest History, Dynamics and Management* (Eds. G.L. Werren, A.P. Kershaw), pp. 31–43. Australian Government Publishing Service, Canberra, Australia.

Weston, P.H., and Crisp, M.D. (1994). Cladistic biogeography of waratahs (Proteaceae, Embothrieae) and their allies across the pacific. *Australian Systematic Botany* 7, 225–249.

Weston, P.H., and Hill, R.S. (2013). Southern (Austral) Ecosystems. In *Encyclopedia of Biodiversity* (Ed. S.A. Levin), Vol. 6, 2nd edn., pp. 612–619. Academic Press, Waltham, MA.

Wilf, P., and Escapa, I.H. (2015). Green Web or megabiased clock? Plant fossils from Gondwanan Patagonia speak on evolutionary radiations. *New Phytologist* 207, 283–290.

Wulff, E.V. (1943). *An Introduction to Historical Plant Geography.* The Chronica Botanica Company, Waltham.

8

Biogeography of Australasian Fungi: From Mycogeography to the Mycobiome

Tom W. May

CONTENTS

Introduction

Fungi are megadiverse, ubiquitous and carry out significant ecosystem roles, yet have been largely overlooked in accounts of biogeography in the Australasian region. The purpose of this chapter is to summarise previous syntheses and recent literature in order to demonstrate that fungi are tractable organisms for exploring hypotheses about distribution in space and time.

In this introduction the scope is established and some background is provided on fungal biogeography (mycogeography) in general and fungal biology as relevant to biogeography, especially spore dispersal. The second section, deals with 'Previous Syntheses of Austral Mycogeography', summarising three key syntheses (from 1983, 1996 and 2001) along with available information on distribution within land masses. The third section, 'Molecular Mycology Reviewed' deals with the literature on mycogeography from the mid-1990s onwards is reviewed, with a focus on studies including material from Australasia and an emphasis on the use of molecular data. This third section commences with a description of the impact of molecular data on fungal classification and species delimitation and identification; presents case studies that show that most phylogenetic species of fungi are geographically restricted (with an examination of exceptions); provides a precis of rebuttals of the *Everything is everywhere* (EIE) concept; discusses the role of host specificity and hosts shifts in geographic patterning; summarises the dating of fungal phylogenies and the fossil evidence; explores analyses of fungal distribution in relation to area and/or time, including *Cyttaria* as a case study for themes of time, space and host; examines infraspecific patterning of genetic variation; and wraps up with an assessment of the potential contribution of metagenomics to mycogeography. The final section 'Synthesis and Prospects highlights the frequency of inferred dispersal events and hosts shifts, draws attention to issues around geographic and taxon sampling, and advocates the benefits of better integration of mycogeography into biogeography in general, particularly for fungi that are symbiotic with other organisms.

Previous Reviews on Fungal Biogeography

The topic of fungal biogeography is in need of up-to-date review globally. The latest edition of the *Dictionary of the Fungi* (Kirk *et al.* 2008) has an entry on 'Geographical distribution', but lacks treatments of biogeography or phylogeography'. Walker (1996) included 'A brief history of fungal biogeography'. Key works include contributions by Bisby (1943) and Pirozynski (1968) on the 'Geographical Distribution of Fungi' and Pirozynski and Weresub (1979) on 'A Biogeographic View of the History of Ascomycetes', focused on pleomorphism. Arnolds (2007) reviewed 'Biogeography and Conservation' for the series *The Mycota* focusing on the mapping of fungi and distribution patterns. Key themes across these reviews were that fungi are generally more widely distributed that plants, saprotrophs are more widely distributed than parasites (due to dependence on host distribution) and climate is an important influence on fungal distribution.

At the continental and regional level, useful treatments include Redhead (1989) for North America, notable for the use of maps; Knudsen and Ryman (2000) for the Scandinavian countries covered by 'Nordic Macromycetes', who listed fungi associated with different phytogeographic zones. Petersen and Hughes (2003), who dealt with the phylogeography of Asian mushrooms, including Eurasian and trans-Beringian patterns; Yang (2005), who described mycogeographic elements and their distribution among regions in China and elsewhere; and Petersen and Hughes (2007), who reviewed distributions around the Pacific Ocean. The relationship between populations and species of fungi in North America, Europe and Asia and the origin and timing of supposed migration has attracted interest, especially in relation to the Bering and North Atlantic land bridges and to climatic fluctuations (Wu *et al.* 2000; Mueller *et al.* 2001; Methven *et al.* 2000; Petersen and Hughes 2003, 2007; Geml 2011; Geml *et al.* 2012a,b; Sánchez-Ramírez *et al.* 2014). For the tropics, several chapters in *Aspects of Tropical Mycology* (Isaac *et al.*, 1993) deal with distribution patterns.

Scope and Conventions

Scope and Out-of-Scope Groups and Themes

The scope of this review is terrestrial true Fungi, concentrating on species of native ecosystems, rather than pathogens of agriculture or forestry. The focus is on the distribution of individual species, especially the interplay between species delimitation and biogeography. Geographically, Australasia is as covered by May (2001), including Australia, New Zealand, New Caledonia and New Guinea, although most examples come from the first two land masses.

Other Ecological Groups

For ecological groups not covered in detail, the biogeography of marine fungi (mostly Ascomycota, growing on dead wood or living organisms such as algae) was reviewed by Kohlmeyer (1983) and Hughes (1986). Kohlmeyer (1983) concluded that, as for marine plants and animals, 'the geographic distribution of ... marine fungi is controlled mainly by temperature', although he noted that some parasites have more limited temperature requirements than their hosts, such as for *Spathulospora* occurring on the red alga *Ballia*. A recent global metabarcoding study of ocean samples found environmental factors such as depth and oxygen rather than geography were the significant influences on community structure of marine fungi (Tisthammer *et al.* 2016).

The biogeography of freshwater fungi, specifically aquatic hyphomycetes, was reviewed by Duarte *et al.* (2015), who found that sampling efforts and species richness in this group are correlated, but detected a species richness peak at mid-latitudes and a distance–decay relationship in community similarity (albeit with considerable noise). However, they also found high similarity between communities that were geographically separated but in similar climate zones, with samples from Australia and New Zealand more similar to those from temperate areas in the northern hemisphere than to samples from the tropics. For wood-inhabiting freshwater fungi, Hyde *et al.* (2016) found higher diversity in the tropics along a north–south latitudinal gradient from Asia to Australia.

Other Organismal Groups

Most fungi belong to the kingdom Fungi. Slime moulds are fungus-like organisms in the Protista, typically with an amoeboid stage, the plasmodium, that produces a sporotheca, with powdery spores. The biogeography of slime moulds is not considered further, although they are the subject of informative biogeographical studies such as those by Aguilar *et al.* (2014), who used environmental niche models to test the EIE hypothesis for *Badhamia*, and Estrada-Torres *et al.* (2013), who applied parsimony analysis of endemicity to biogeographic patterns among Myxomycetes from the Americas. There are also fungus-like organisms in the Chromista, such as *Phytophthora* and other water moulds, that are only mentioned incidentally herein.

Another group not covered is lichens (lichenised fungi), which are a stable association between a mycobiont (usually an ascomycete fungus) and a small range of photobionts (algae or cyanobacteria). In general, each lichen species is a unique fungus and lichen classification is based on the fungal component. Due to their distinct morphological and physiological characteristics, lichens are often dealt with separately to other, non-lichenised fungi, even though lichenised fungi do not form a monophyletic taxonomic group. Pirozynski (1983) pointed out that lichens are not as strongly associated with hosts as are biotrophic fungi, and that a major issue in understanding lichen distribution was 'insufficient knowledge of the circumstances surrounding resynthesis of a lichen from independently dispersed components'. Informative syntheses of the biogeography of Australasian lichens, with detailed discussion of geographic patterns and elements, include Rogers and Stevens (1981), Rogers (1992), Jørgensen (1983) and Galloway (1991, 2008a,b).

Other Themes

Island and ecological biogeography are not covered. For examples of research on 'habitat island' biogeography for fungi, see Peay *et al.* (2007, 2010), who found a strong species–area relationship and an effect of habitat patch isolation for diversity of ectomycorrhizal (ECM) fungi tree islands in grassland. From an ecological biogeography perspective across suites of species, analyses have been carried out, for example, on macrofungi (Mueller *et al.* 2007), detecting endemism at continental scales, and on lichens (Arcadia 2013), grouping geographic areas (with the main split between Laurasia and Gondwana). Indications from such analyses, which intergrade with community ecology (synecology) at local and regional scales, need to be repeated once there is more certainty about the phylogenetic basis of species concepts as applied used across different regions.

Conventions

The term *species complex* or the appending of *complex* after a species name is used to designate species, originally defined on phenotypic characters such as morphology, which contain multiple phylogenetic species. Taxon names are given as in the publications cited and generally have not been updated. Where part of a classification, elements or distribution patterns are italicised, as in *arid–semiarid* or *Gondwanan*. Range is given as the greatest distance between any two locations. Age is indicated in millions of years ago (mya). Ages for geological events are as provided by the cited publications.

Fungal Biology and Diversity

The aspects of fungal biology relevant to biogeography are that fungi reproduce by spores that are minute (mostly in the range 3–50 μm) and have considerable variation across species in shape, surface characters, wall thickness and degree of melanisation (Wyatt *et al.* 2013; Halbwachs and Bässler 2015). Vast numbers of spores are produced by many fungi. For sexually reproducing fungi, asexual stages are common, and some fungi only reproduce asexually. There is no sexual differentiation macroscopically, and sex is controlled by mating types, with compatible matings between different mating types, of which there may be two to many (Burnett 2003; James 2015). Depending on the mating system, some siblings have compatible mating types. There are also homothallic (self-fertile) species. Vegetative growth is typically by hyphae in the substratum, forming a mycelium, on which spores may be produced directly, or in or on sporophores (fruit bodies), such as mushrooms and truffles.

Fungi are heterotrophs, with a variety of nutritional strategies including parasitism and saprotrophism and the frequent formation of mutualisms such as lichens and mycorrhizas, the latter including ECM associations, which are prevalent on forest trees in the region such as *Allocasuarina*, *Eucalyptus* and *Nothofagus*. Fungi occupy numerous niches across all continents. They are especially common in soil and in or on all parts of living and dead plants, and also frequently associate with invertebrates. Fungi may be important components of the diet of invertebrates and mammals.

Fungal Diversity in Australasia

There are numerous species of fungi, most of which are yet to be collected and formally described. A commonly accepted estimate of global fungal diversity is 1.5 million species, based on projected ratios of fungi to plants, but not taking into account invertebrate associates (Hawksworth 1991). For Australia, around 15,000 species are known, of which around 4,000 are lichens, with between 50,000 and 250,000 species expected (Chapman 2009). For New Zealand, approximately 7,400 species are known, with around 1,700 lichenised species and 22,000 species expected (McKenzie 2004). For one host genus, *Nothofagus*, with five New Zealand taxa, McKenzie *et al.* (2000) recorded around 900 associated species of fungi, and for *Eucalyptus* Sankaran *et al.* (1995) listed more than 1,000 associated species. May (2001) summarised resources on fungal biodiversity for Australasia, including a tabulation of checklists for countries in the region.

Spore Dispersal

Fungus spores are released passively or propelled actively over mostly microscopic distances, rarely to several metres. Mushrooms induce the alteration of their microclimate through evaporative cooling, creating convective airflow that transports released spores into the air above (Dressaire *et al.* 2016). Natural modes of further dispersal include air or water currents and animals, both invertebrates and vertebrates (such as by the ingestion of truffles).

Spore travel can be traced through spore-trapping devices or the analysis of dust or rainwater, or by the use of trap plants (for ECM fungi), with the identification of the fungi involved through morphology, or increasingly through use of molecular tools (Viljanen-Rollinson *et al.* 2007; Peay *et al.* 2012). For individual species, trap cultures can be used, where falling spores dikaryotise monokaryotic cultures (James and Vilgalys 2001).

For dispersal by animals, mycophagy is common among Australian mammals, and spores in scats are viable (Claridge and May 1994). Various invertebrates are implicated in fungal spore dispersal (Halbwachs and Bässler 2015). As examples of transport by birds, Bailey and James (1979) found royal albatross to carry thallus fragments and spores of lichenised fungi, and O'Donnell *et al.* (2000a) suggested graminivorous migratory birds carrying infected seed as plausible dispersal vectors for *Fusarium*.

For dispersal in the air, Walker (1996) pointed out that fungal spores around 10 μm in size have low gravitational sedimentation rates and thus 'their pattern of movement will be determined largely by the mass flow of the body of turbulent air', with the distance travelled depending on air speed and turbulence. Walker (1996) estimated that small spores (around 5 μm) could potentially travel more than 10,000 km under certain conditions, but spores four times as large would only travel 16 km under the same conditions. Certainly, when large numbers of spores are produced, such as by rust fungi on agricultural monocultures, spores can travel thousands of kilometres (Schmale and Ross 2015).

The rapid spread of introduced fungi, whether from deliberate (biocontrol) or accidental initiation, demonstrates significant dispersal capabilities (Walker 1996). Examples of rapid expansion within Australia include *Puccinia chondrillina* on skeleton weed (*Chondrilla juncea*) (Cullen *et al.* 1973), *Melampsora* on poplars (*Populus*) (Walker *et al.* 1974) and Myrtle rust *Puccinia psidii* on various native and cultivated Myrtaceae (Kriticos *et al.* 2013). There is also evidence for the intercontinental movement of various rust species from Africa to Australia (Watson and de Sousa 1983) and Australia to New Zealand (Viljanen-Rollinson and Cromey 2002), although long-distance dispersal over oceanic barriers to Australia has been rare, estimated at three times over 60 years for wheat stem rust (Watson and de Sousa 1983).

For macrofungi, many sporophores are sporadic in appearance and short-lived, and most spores are considered to land within short distances in the order of tens of metres (Galante *et al.* 2011; see also Dam 2013 and Horton *et al.* 2013). Diverse factors affect the distance and effectiveness of spore dispersal, including fecundity, initial overcoming of the laminar boundary layer, the speed and direction of prevailing winds, local air drainage, the role of rain in facilitating spore deposition and the effects of ultraviolet radiation on spore viability (Walker 1996; Schmale and Ross 2015).

James and Vilgalys (2001) pointed out the dichotomy between the potential for very long-distance dispersal and the likelihood of most spores being deposited locally. They trapped spores of the mushroom *Schizophyllum commune* around the Caribbean and, after genetic characterisation using population markers, concluded that spores were 'primarily reflective of the local, established population'. Molecular analysis to investigate the biogeography of fungi in the atmosphere has been applied by Fröhlich-Nowoisky *et al.* (2012), who detected differences in the ratio of Ascomycota to Basidiomycota across samples taken over land versus sea. Metagenomic sampling (see 'Metagenomic Biogeography' in this chapter) offers promise for further characterisation of the air spora.

Previous Syntheses of Austral Mycogeography

Fungi are notable for their omission from textbooks and synthetic works on biogeography. Pirozynski (1983) lamented 'few biogeographers have ventured from the well-trodden tracks of zoo- and phytogeography into the microbial underworld'. Walker (1996) highlighted that fungi had been largely overlooked in all key general works on biogeography in the region, including landmark works such as Kuschel (1975) on *Biogeography and Ecology in New Zealand*. The only treatments of fungi in any such synthetic works were the chapters by Shaw (1982) in *Biogeography and Ecology of New Guinea*, which was ecological in focus and dealt with mainly plant pathogens of exotic hosts, and the brief treatment of lichens by Rogers and Stevens (1981) in *Ecological Biogeography of Australia* (Keast 1981). In his classic work *Phylogeography*, Avise (2000) mentions a single fungal study on *Coccidioides* (Koufopanou *et al.* 1997). Heads (2014) included but one fungal case in his 'molecular analysis' of the biogeography of Australasia, and Wallis and Trewick (2009) in their review of 'New Zealand Phylogeography' omitted fungi entirely. Beheregaray (2008) reviewed '20 Years of Phylogeography', focusing on the Southern Hemisphere. From a sample of 3049 articles on phylogeography, he identified only 1.8% about fungi, and nearly all were on species of the Northern Hemisphere. Beheregaray (2008) pointed out that comparative phylogeography of fungi (as opposed to studies on one or closely related species) has been very rare. It is demonstrated in this section and the next that despite the omission of fungi from most synthetic works on biogeography and phylogeography, there are pertinent examples across various groups of fungi that utilise a variety of analytical methods.

Three key syntheses of fungal biogeography in the Australasian region to the end of the twentieth century are (1) the collection of symposium papers published as 'Pacific Mycogeography' (Pirozynski and Walker 1983), (2) the chapter on biogeography in *Fungi of Australia* (Walker 1996), and (3) the volume of conference papers published as 'Biodiversity and Biogeography of Australasian Fungi' (May and Farrer 2001; May and Simpson 2001). In this section, the content of these key syntheses is summarised and the theme of distribution within land masses is also explored. Other sources, mostly contemporaneous, that deal with similar issues are referred to as appropriate.

A more recent collation of papers, 'Fungal Phylogeography and Biogeography' (Lumbsch *et al.* 2008), arising from a symposium at the International Mycological Congress in Cairns in 2006, mostly deals with cases from the Southern Hemisphere, but with a molecular focus, and therefore relevant papers are referred to in the section 'Molecular Mycogeography Reviewed'.

Pacific Mycogeography 1983

The 13th International Botanical Congress, held in Sydney in 1981, included a symposium on 'Pacific Mycogeography'. Symposium papers were published as 'Pacific Mycogeography: A Preliminary Approach', edited by Pirozynski and Walker (1983). Five articles treated different taxonomic or ecological groups of fungi, with a concluding 'appraisal' by Pirozynski (1983). The contributions by Horak (1983) and Walker (1983) are considered in detail below, while the paper by Korf (1983) on the coevolution of *Cyttaria* and *Nothofagus* is dealt with in the section 'Cyttaria: Evolution of Ideas about Coevolution of Fungi in Relation to Time and Area'.

Horak on Basidiomycete Macrofungi

Horak (1983) considered the distribution of macrofungi in the Basidiomycota, mainly agarics (mushrooms). His focus was indicated by the opening quote from Darlington (1969): 'The key to the history of terrestrial life in the far south may be *Nothofagus*.' Noting the close connection between fungi and living and dead plants, Horak (1983) posited a number of questions, including 'Are the agarics … associated with *Nothofagus* … also "living fossils"?' and 'Where does the mycoflora of the myrtaceous genera *Eucalyptus* and *Leptospermum* originate from?'. Rather than discuss answers, the article elaborates the distribution of 75 fungi, with distributions strongly associated with that of *Nothofagus*, across New Guinea, New Caledonia, Australia, New Zealand and southwest South America (Figure 8.1). Such fungi occurred in various combinations of these areas, such as *Mycena interrupta* in Argentina, Chile, New Zealand and Australia. Horak (1983) also included some species that occurred in the Southern Hemisphere, but not strictly with *Nothofagus*, such as *Tubaria rufofulva* in both western and eastern Australia as well as in New Zealand. In addition, he identified a circum-Pacific distribution, around the Pacific Rim, for species such as *Cystoagaricus strobilomyces* from southern United States and Argentina on the east and several locations in the west from Japan to New Zealand. Very few distribution patterns included localities in Africa. A low overlap between the mycota of Africa and Australia was also detected by Berndt (2008), who found only 1% of African rust fungi to also occur naturally in Australia.

Horak (1983) explained Gondwanan macrofungal distributions as resulting from coevolution with *Nothofagus*, with distinct taxa in different areas being the result of vicariance. However, gradual dispersal was advanced as an explanation for some taxa. For *Phialocybe improvisa*, Horak (1983) hypothesised a Gondwanan origin for the present-day occurrence in Australia, New Zealand, New Guinea and Chile, with spread to the Northern Hemisphere (Europe) 'after collision between the Asian and Australian plates' about 10–12 mya. This spread was suggested as facilitated by the co-occurrence in New Guinea of *Nothofagus* with other Fagales such as *Lithocarpus*, leading to subsequent association with *Fagus* in

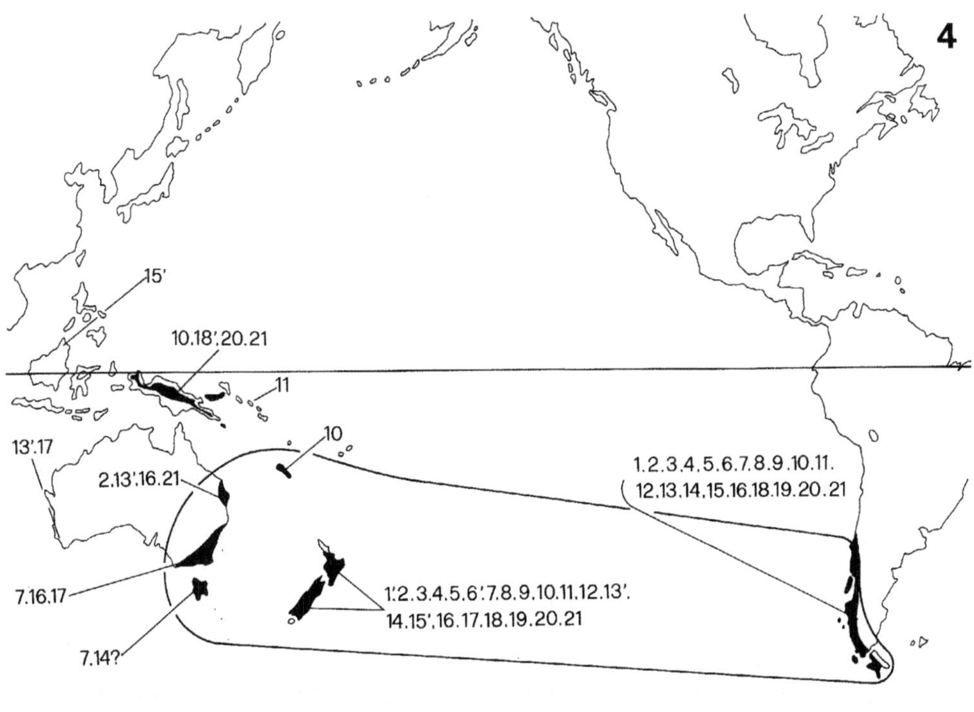

FIGURE 8.1 Distribution of 21 species of fungi (numbered, including 7, *Mycena interrupta* and 17, *Tubaria rufofulva*) in relation to the present distribution of *Nothofagus* (black shading). Originally published as Figure 4 in Horak, E., *Australian Journal of Botany* 10, 1–41, 1983; see original caption for further details. Reproduced with permission.

the Northern Hemisphere. Dispersal and host-switching was also suggested to explain presence of the otherwise Gondwanan genus *Descolea* in Asia.

The inclusion of suggested hosts for ECM fungi by Horak (1983) was groundbreaking. Knowledge of host had been recognised as important for the identification of saprotrophic fungi and was included in monographs such as Cunningham (1965) on New Zealand polypores, and of course also for plant parasitic fungi (Walker 1983), but hosts had not been elaborated for ECM fungi in works such as Cleland (1934). Indeed, Pirozynski (1983) pointed out that ECM fungi had often merely been noted as occurring 'on ground in woods'. The focus by Horak (1983) on *Nothofagus* stimulated the collecting of fungi from cool-temperate rainforests across the region, as can be seen from the collections utilised, for example, in monographs of *Mycena* (Grgurinovic 2003) and Hygrophoraceae (Young 2005).

Walker on Plant Parasitic Fungi

Walker (1983) focused on 'deficiencies and irregularities in the distribution of plant parasitic fungi'. In particular, he documented the relative lack in the Southern Hemisphere, especially on native hosts, of (1) the plant-associated chytrids *Physoderma* and *Synchytrium*, (2) downy mildews such as *Peronospora*, and within the Ascomycota, (3) *Taphrina* (causing leaf blister or leaf curl) and (4) powdery mildews (Erysiphales), including a relative lack of species of the latter group forming sexual spores. These deficiencies were not able to be explained, apart from the possibility of undersampling and the fact that the host families of some of the deficient taxa were poorly represented in Australia, as were hosts of Taphrinales, such as Betulaceae, Salicaceae and Ulmaceae. Takamatsu (2013) has recently suggested, from molecular phylogeny, an origin of the Erysiphales in the Late Cretaceous in the Northern Hemisphere in association with broad-leaved deciduous trees.

Influenced by the approach of Savile (1979) in using fungi as aids in plant classification, Walker (1983) pointed out that 'the close association of some [fungi] with particular host plant families' has the potential to 'throw light on the origin and distribution of the plants'. For rust fungi (Uredinales) on Myrtaceae, Proteaceae and the genus *Acacia*, Walker (1983) pointed out the predominance of rust species on certain taxonomic groups of their host plants and related this to the distribution of the hosts. For example, in the Myrtaceae, he documented the prevalence of rust fungi on the subfamily Myrtoideae in Central and South America but an apparent scarcity in Australia on Myrtoideae and Leptospermoideae – there being but one species in Australia in the latter subfamily, the newly described *Uromyces xanthostemonis*. In the Proteaceae, only three rust fungi were recorded, from Australia, New Guinea and Costa Rica, all on Grevilleoideae, one of the five subfamilies. For rusts on *Acacia*, *Uromycladium* was more or less confined to hosts occurring in Australia, while *Ravenelia* was mostly on hosts in the Americas, Asia and Africa, with the exception of one species on *Acacia farnesiana* (this host and fungus were considered exotic in Australia, and were suggested to have travelled together from their native distribution in Central America). For the *Uromyces* recorded on *Acacia*, Walker (1983) noted that this fungal genus was prevalent on Faboideae and the few species on Mimosoideae may well be better placed in other genera. Recently, McTaggart *et al.* (2015) included in *Endoraecium* a number of the *Acacia* rusts originally described in *Uromyces*, along with nine novel species from Australia.

Walker (1983) discussed several examples of host jumps. For the downy mildew *Plasmopara viticola*, a well-known pathogen of grape that also occurs on native Vitaceae, such as *Cissus*, in Australia, the possibility of a host jump from the exotic crop to native hosts was suggested. Walker (1983) noted that *Puccinia psidii* on the myrtaceous host *Psidium* (Guava) in South America had the ability to infect several Australian Leptospermoideae, such as *Eucalyptus* and *Melaleuca*, when these hosts were grown in South America. In reverse, *Aulographina eucalypti*, an Australian fungus on *Eucalyptus*, could infect the South American plant *Fejoa* (Myrtoideae) when it was grown in Australia. Walker (1983) also noted that in some cases native hosts that had evolved in the absence of some groups of fungi were not susceptible, as with native grasses, a very rare host for *Erysiphe graminis*, despite the occurrence of this fungus on numerous grass hosts around the world, including grass genera with native Australian species. The lack of predictability of the infectiveness of exotic fungi led Walker (1983) to emphasise the importance of strict quarantine, rather presciently, given the introduction and subsequent rapid spread into Australia of *Puccinia psidii* (Kriticos *et al.* 2013).

Biogeographic links to southern continents were provided by the Cryptobasidiaceae, with *Drepanoconis nesodaphnes* on Lauraceae newly recorded by Walker (1983) from Australia, adding to the eight other species of the family (in *Drepanoconis* and other genera) on South American Lauraceae and the one species of *Coniodictyon* on Rhamnaceae in Africa. Links between Australia and cooler areas of the Northern Hemisphere were also elaborated, such as for *Puccinia crucifearum*, where Walker (1983) introduced a new subspecies from Australia to add to the four subspecies known from Europe and North America. Ancient co-migration of host and fungus to the Southern Hemisphere was suggested, followed by the evolution of local taxa.

It is notable that a significant portion of the contribution by Walker (1983) is taken up with taxonomic descriptions, including the documentation of a number of fungi newly recorded from Australia. Walker (1983) suggested that, as well as general collecting to fill gaps in distribution information, the intensive collecting of fungi from plant families well represented in the region and with biogeographic patterns, such as Epacridaceae, Goodeniaceae and Restionaceae, could be of value in interpreting the current distribution of both fungi and their hosts.

Synthesis of Pacific Mycogeography

Pirozynski and Walker (1983) considered that the examples dealt with in the volume demonstrated 'the importance of fungi to the plant geographer as well as to the mycologist'. Pirozynski (1983) concluded the volume with an overview of Pacific mycogeography. His emphasis was on relationships between fungi and other biota. Rather than avoid fungi in biogeography due to a lack of knowledge about distribution and relationships, Pirozynski (1983) encouraged a focus on 'ecologically and edaphically specialized mutualistic and parasitic biotrophs'. He pointed out the complexity of relationships, such as in the Pacific Northwest of America, a hot spot of truffle diversity, where these hypogeal fungi are mostly ECM and rely on dispersal by mammals (Trappe 1977). Therefore, the dispersal of the tripartite set of organisms was thought likely to require overland routes for migration. Another complex system where concomitant dispersal was considered necessary was the ECM fungi of forests trees that are also hosts for heteroecious rust fungi that have an alternate host within the same ecosystem. Pirozynski (1983) emphasised that such biotic relationships impose constraints on the 'freedom of movement of components either individually or in partnership'.

Walker in *Fungi of Australia* 1996

In 1996, the first two volumes of the *Fungi of Australia* series were published by the Australian Biological Resources Study. They contained essays on a variety of topics, including classification, biology, fossil records and Aboriginal usage, as well as a history of the taxonomic study of Australian fungi and an extensive treatment by Walker, 'Biogeography of Fungi with Special Reference to Australia' (1996), which built on and significantly expanded his 1983 essay, 'Pacific Mycogeography' (1983). Some 60 volumes of *Fungi of Australia* were planned, admittedly a flexible projection, perhaps requiring several parts in some groups of fungi, due to imprecision as to the eventual number of species. In the nearly two decades since the series was commenced, two catalogue volumes have appeared (May and Wood 1997, May *et al.* 2003) but only three taxonomic treatments, for Hygrophoraceae (Young 2005), *Septoria* (Priest 2006) and smut fungi (Vánky and Shivas 2008; maps are on accompanying CD-ROM). Distribution maps were provided in the three *Fungi of Australia* volumes and in the monograph on Australian Phyllachoraceae (Pearce and Hyde 2006). Priest (2006) noted the absence or paucity of species of *Septoria* on *Eucalyptus*, *Acacia* and Proteaceae, all highly diverse in Australian ecosystems, but otherwise these monographic works included little discussion about biogeography.

Distribution Patterns of Australian Fungi

In his comprehensive review on 'Biogeography of Fungi with special reference to Australia', Walker (1996) summarised the distribution of fungi occurring in the region, considered distributional differences within Australia (see 'Distribution within Land Masses' in this chapter) and discussed six general patterns (summarised in the following sections). Walker (1996), also organised the material presented by particular groups of fungi, often on particular host plant families, so that some patterns included examples overlapping with other patterns.

Northern Hemisphere: Southern Hemisphere Differences

Under this heading, Walker (1996) extended his earlier demonstration (Walker 1983) that some groups of plant parasites are depauperate in the Southern Hemisphere. Putatively primitive rust genera such as *Hyalospora*, *Milesina* and *Uredinopsis*, which have uredinia and telia on fern hosts but pycnia and aecia on Pinaceae, were noted as much less diverse in Australia and New Zealand and elsewhere in the Southern Hemisphere, occurring only on ferns, suggesting an origin of these rust fungi in the Northern Hemisphere. Rust fungi of conifers in general were also noted as depauperate in the Southern Hemisphere.

Circum- and Trans-Pacific Distributions

These were exemplified by rust fungi with a particular type of pycnia (Type 12) that all have pycnia and aecia on conifers in the Cupressaceae, Araucariaceae and Podocarpaceae (the latter two hosts exclusively or mainly in the Southern Hemisphere) and, where present in the life cycle, uredinia and telia on Fagales, including *Nothofagus*; *Melanodothis caricis*, an ascomycete on *Carex*, known from around the western Pacific Rim, including Australia and New Guinea, and in Alaska and western Canada; the sclerotium-forming agaric *Hypholoma tuberosum*, known from Canada and New South Wales (NSW); and the Rhytismatales (see 'Austral Fungi and Broader Distributions' in this chapter.) For further analysis of distributions in relation to the Pacific Ocean and the Pacific Rim, see Petersen and Hughes 2007 and Halling *et al.* 2008 on boletes.

Australasian and South American Distributions

These were exemplified by rusts of Myrtaceae, expanding the treatment in Walker (1983), and also by mycorrhizal associates of *Nothofagus* and Myrtaceae, such as those elaborated by Horak (1983) and extended for the genus *Rozites* by Bougher *et al.* (1994; see also 'Initial Phylogenetic Studies Transform Fungal Classification' in this chapter). Some polypores analysed by Rajchenberg (1989) also fall within this distribution pattern, with a number of *Palaeoaustral* species occurring across South America and New Zealand and/or Australia, with some characterised as *Subantarctic*, only in these regions, while others designated *Gondwanic*, also extended to India and/or Africa. Ryvarden (1991) provided a mycogeographical classification for distribution patterns shown by the approximately 100 genera of noncosmopolitan polypores that were considered to have climate-dependent distributions, including *Phaeotrametes*, with a *Gondwanaland* distribution across Australia, Africa and South America. In relation to the genera of polypores, Ryvarden (1991) commented that 'most … have a wide distribution either because they were evolved before Gondwanaland broke up … or because they have effective spore dispersal mechanisms which we do not understand'. Watling (1993) compared the macromycete biota of Africa and Australia, highlighting differences in assemblages (rather than individual species) such as the lack of annulate species of *Lactarius* in Australia and the rarity of Cortinariaceae in Africa.

The North American–South American Interhemispheric Corridor

For this pattern, Walker (1996) extended his 1983 analysis of the rust *Puccinia crucifearum* on *Cardamine*, recognising a further subspecies from Chile, and noting that there was a progression in the chromosome number of the host species of *Cardamine*, from the Northern to the Southern Hemisphere. Walker (1996) suggested dispersal from boreal areas of cold-adapted fungi such as *Puccinia crucifearum* along with their hosts may have occurred along the mountain chains of North and South America and then across the Pacific to Australasia.

West Gondwana: Africa and America and Relationships with Asia and Australia

In addition to the examples among Cryptobasidiaceae and rusts on *Acacia* detailed by Walker (1983), this pattern was exemplified by the large bolete *Phlebopus marginatus* and by Parmulariaceae, as elaborated by Pirozynski and Weresub (1979), who showed the greatest diversification in South America and Africa as well as occurrence in Australia and New Zealand, but very low diversity in the corresponding climatic belt of the Northern Hemisphere. For *Acacia* rusts, Walker (1996) discussed the relationship between the host range of the fungi and issues around the circumscription and evolution of higher-level taxonomic groupings within *Acacia* in the broad sense (these issues eventually led to the restriction of *Acacia* in the strict sense to mainly Australian species).

Other World Distribution Patterns

For this pattern, Walker (1996) noted that Coryneliales, as discussed by Pirozynski and Weresub (1979), included conifer-inhabiting species with a predominantly Gondwanan distribution and extensions to the Philippines and Japan, along with two genera on the fungus *Cyttaria* and the plant *Drimys* in South America, but also genera with a variety of other distribution patterns in tropical and boreal areas. For some fungi, such as canker pathogens of *Eucalyptus* and other Myrtaceae (e.g. *Endothia gyrosa* and *Cryphonectria cubensis*), the status as native in Australia and natural distribution elsewhere remained enigmatic.

Other Patterns

Subsequently, May and Simpson (1997), focusing on eucalypt ecosystems, distinguished broad distribution patterns similar to those of Walker (1996), with the addition of *Pantropical* and *Australia–Malesia* patterns, the latter overlapping with the *West Gondwanan* pattern. It is notable that there are examples of species found on either side of Wallace's line in boletes (Halling *et al.* 2008) and polypores (Núñez *et al.* 2002), and Arcadia (2013) concluded that Wallace's line was not a significant boundary for lichens. Rajchenberg (1989) provided examples of *Bipolar* and *Pantropical* polypore species. May and Simpson (1997) also included a *Cosmopolitan* element for species such as the mushroom *Schizophyllum commune* and the mould *Cladosporium herbarum*. Bougher (1983) considered that species of dung-inhabiting *Coprinus* in Western Australia were cosmopolitan for both native (kangaroo) and introduced (horse) animals. Döbbeler and Hertel (2013), for bryophilous ascomycetes, mapped examples of *Austral-Subantarctic*, *Australasian* (some extending to Borneo and Mindanao), *Bipolar*, *Cosmopolitan* and *Pantropical* patterns, as well as one species, *Epibryon pogonati-urnigeri*, that was widespread in the Northern Hemisphere and occurred in Australasia.

Factors Affecting Distributions

Having described the various patterns, Walker (1996) examined the factors affecting distributions. He considered that the 'presence of suitable substrata is of utmost significance in the distribution of most fungi' and concluded that 'many physical, chemical and biological properties of soil, water and air … play … a major ecological role in determining … distribution … on local and regional scales'. These factors were only briefly discussed, although some examples were provided of climatic limitations of plant pathogenic fungi. Walker (1996) acknowledged that 'slow coevolution and migration with plant hosts over geological time' in association with vicariance due to plate tectonics had been advanced as an explanation for Gondwanan and other distributions. However, drawing on his considerable experience in plant pathology, he extensively discussed examples of dispersal (see 'Spore Dispersal' in this chapter) and host jumps, primarily observed on plant pathogens of crops. In relation to host jumps, he quoted Savile (1990), who considered that in rust and smut fungi evolution had generally been 'a complex mixture of radiation with the evolving host and jumps to ecologically associated [but taxonomically unrelated] plants'. Overall, Walker (1996) emphasised the potential for long-distance spore dispersal, even in groups such as the ECM fungi of *Nothofagus*, where vicariance had been widely assumed. While recognising the rarity of effective long-distance dispersal, he pointed out that even a rate of one successful dispersal over 10,000 years, as calculated for Australasian alpine plants by Smith (1986), could be biogeographically significant.

Biogeography of Exotic Organisms

Walker (1996) also discussed the biogeography of exotic organisms, pointing out that assessing the origin status of fungi may be difficult. While determining origin status is critical in investigations of fungal biogeography and for practical considerations of biosecurity, trade and quarantine (Coetzee *et al.* 2011), the subject of exotic fungi is not dealt with further herein. However, see Gladieux *et al.* (2015) for a review of the population biology of fungal invasions and Pringle *et al.* (2009) on the death cap (*Amanita phalloides*) in North America as an example of the utility of molecular data in determining origin status as exotic or native. An example of uncertain origin status is *Heterobasidion araucariae*. This species occurs in Australia and New Zealand, but the rest of the genus is almost entirely from the Northern Hemisphere. The origin of *H. araucariae* in Australasia has been suggested as either by dispersal in geological time or through the alternative and rather different possibility of 'recent, human-mediated introductions' (Chen *et al.* 2015).

Biodiversity and Biogeography of Australian Fungi 2001

May and Farrer (2001) edited a series of papers called 'Biodiversity and Biogeography of Australian Fungi' arising from presentations at the 9th International Congress of Mycology, held in Sydney in 1999 elaborated or added examples to the distribution patterns distinguished by Horak (1983), Walker (1996) and May and Simpson (1997). Contributions included those on Agaricales (Grgurinovic 2001), Aphyllophorales (Buchanan 2001), Boletales (Watling 2001a), Rhytismatales (Johnston 2001), sequestrate (truffle-like) fungi (Bougher and Lebel 2001) and yeasts (Fleet 2001). Further examples of Gondwanan fungi included the ascomycete *Bivallum* from New Zealand, Australia and southern South America (Johnston 2001), and the sequestrate basidiomycetes *Torrendia* (Australia and Africa), *Thaxterogaster* and *Weraroa* (predominantly Southern Hemisphere) (Bougher and Lebel 2001). Some further (nonmolecular) literature about distribution patterns appearing subsequently to May and Farrer (2001) is integrated into the following paragraphs, and see also Johnston (2006) for distribution patterns and origin of New Zealand fungi.

For several groups, it was noted that there were very few examples of species naturally found in the temperate areas of both the Southern and Northern Hemispheres (Bougher and Lebel 2001; Grgurinovic 2001; Johnston 2001). The same point was made later by Soop and Gasparini (2011) for the ECM mushroom *Cortinarius*, a genus of more than 2000 species worldwide, with no species in common between the Northern and Southern Hemispheres in natural vegetation. Bougher and Lebel (2001) noted that there were no sequestrate fungi in common between Australasia and Europe or North America, but *Mycoamaranthus auriorbis* from Australia, Malaysia and Thailand was highly unusual in extending across the equator. At the generic level, Johnston (2001) pointed out another unusual transhemisphere connection: the similarity between the ascomycete *Gelineostroma* on Taxodiaceae endemic to Tasmania and the morphologically similar *Ceratophacidium* also on Taxodiaceae but in America. The mushroom *Cruentomycena*, with one species in Australia and one in Far Eastern Russia, turns out to have another unusual transhemisphere distribution (Petersen *et al.* 2008).

For Boletales, Watling (2001a,b) noted the connections between the bolete mycotas of Australia and Asia and hypothesised southward migration, eventually reaching Australia via Pleistocene land bridges. Wolfe and Bougher (1993), in relation to the mycogeography of the *Tylopilus* subgenus *Roseoscabra*, which occurs in North America, China and Japan, had also postulated a Laurasian origin with subsequent southward migration to Australia in the Pleistocene via land bridges. They suggested the absence of this group in Europe resulted from the extinction of hosts during glaciation events. Wolfe and Bougher (1993) and Watling (2001b) emphasised that for migration from the Northern Hemisphere across land bridges to be effective for ECM fungi, host switching must have occurred across geological time. Bougher *et al.* (1994) pointed out that in *Descolea*, another ECM genus associated mainly with *Nothofagus*, the Western Australian species *D. maculata* (not naturally found with *Nothofagus*) will form ectomycorrhizas with both Myrtaceae and *Nothofagus* in the laboratory. As an example of northward migration, Núñez *et al.* (2002) suggested that some fungi of temperate Australasia that also occur in Japan, such as *Laccocephalum hartmannii*, may have originated in Australasia with subsequent migration to Asia.

Augmenting the analysis of May and Simpson (1987), Grgurinovic (2001) demonstrated low overlap between the Agaricales of Australia and New Zealand, not only for ECM genera such as *Amanita*, but also for saprotrophs in Hygrophoraceae and *Mycena*. In the latter genus, only 4 of the 65 Australian species were also found in New Zealand. For sequestrate fungi, endemism was high in both Australia (95%) and New Zealand (80%), explained by the predominance of ECM species and the mostly hypogeal habit that precludes the aerial discharge of spores (Bougher and Lebel 2001).

For Rhytismatales, Johnston (2001) noted that until recent taxonomic activity in austral regions, the order had been considered as predominantly occurring in the Northern Hemisphere, but on current knowledge the order is 'probably as diverse in the tropics and the Southern Hemisphere as it is in the Northern Hemisphere'. Species of Rhytismatales were highly host specific, with about two-thirds of species occurring on one host family, and those species with broad host ranges usually had correspondingly broad geographic ranges. In *Coccomyces*, 4 of the 13 species from the region are found in both Australia and New Zealand on the same host families, except for one species, *C. globosus*, which occurs on several hosts in different families. For Rhytismatales on Epacridaceae, the pattern of occurrence of the fungi, with 19 species in Australia and 21 in New Zealand, but only 5 in common between the two land masses, was suggested to be due to recent independent bursts of speciation in the two countries (Johnston 2001).

Also appearing in 2001, Halling (2001) considered the distribution and host specificity of ECM fungi worldwide and discussed three processes: ancestral generalist fungi distributed over large geographic areas being split during geological time, with possible allopatric speciation after isolation; the switching of plant hosts by fungi as a consequence of vegetation history; and the co-migration of mycorrhizal communities over land or via 'island hopping'. Halling *et al.* (2008) discussed the same processes, adding long-distance dispersal.

Taxonomic and Geographic Uncertainty

A constant theme across all reviews and collations about fungal biogeography, as in those by Pirozynski and Walker (1983), Walker (1996) and May and Farrer (2001), has been that uncertain taxonomy coupled with incomplete knowledge of distribution makes the analysis of fungal biogeography difficult. Misidentifications may place species in areas where they do not actually occur, and there is always a caveat that further collecting may change distributional boundaries, especially for supposed endemics. A striking extension of geography and associated vegetation is demonstrated by *Fusarium babinda*, which was originally isolated from the Wet Tropics at Mt. Lewis in Queensland (Summerell *et al.* 1993, as *Fusarium* sp. Mt. Lewis) and subsequently found in warm-temperate rainforest in New South Wales, as well as in cool-temperate rainforest in the Barrington Tops (NSW), Otways (Vic) and Tasmania (Summerell *et al.* 1995). The potential for new lineages and new diversity within lineages remains high, as exemplified by *Chlorociboria*, a highly distinctive disc-shaped ascomycete associated with blue-green wood stains, where a thorough revision of New Zealand material by Johnston and Park (2005), integrating morphological and molecular data, added 13 species to the 5 known previously worldwide. Some regions of Australasia remain comparatively underexplored mycologically. In particular, Hosaka (2009) pointed out that despite the interest of New Caledonia to the biogeographer, very few collections of fungi from that country had been included in phylogeographic studies. Collecting is also particularly poor for tropical Australia, especially for fleshy fungi such as agarics (Grgurinovic 2001).

Distribution within Land Masses

Patterns within Australia and New Zealand

In relation to distributional differences within Australia, patterns within land masses have been noted by various authors, but low-intensity and regionally biased collecting meant that it has been difficult to be sure of precise distributions. May and Avram (1997) noted that among 724 species of macrofungi with collections in the National Herbarium of Victoria, there were only around four collections per species, and around 40% of species were represented by a single collection.

May and Simpson (1997) distinguished the following patterns within Australia: (1) *widespread*, (2) *arid–semiarid*, (3) *eastern*, (4) *eastern plus southwestern*, (5) *southwestern*, and (6) *alpine–subalpine*. They noted, however, that there were few fungi in the latter two categories, and could find little evidence for a distinct alpine element, apart from a few fungi restricted to alpine host plants. Walker (1996) provided examples of several leaf-spot fungi of eucalypts, including *Kirramyces epicoccoides* and *Aulographina eucalypti* that were common in eastern Australia on particular subgenera of *Eucalyptus* (and for the latter, also *Angophora*) but absent from Western Australia, despite the presence there of species of the relevant host subgenus. Walker (1996) also noted some differences between the assemblages of leaf parasites on *Kennedia* in eastern and western Australia. Bougher and Lebel (2001) and Grgurinovic (2001) gave examples of species found in either or both regions for (1) eastern and western Australia and (2) the North and South Islands of New Zealand. Bougher and Lebel (2001) also noted a suite of 'desert truffles', restricted to the arid and semiarid interior of Australia (Claridge *et al.* 2014).

Given the highly diverse endemic plant flora in southwest Australia, making it a global biodiversity hotspot, May (2002) found little evidence for species of fungi restricted to this region. Among 491 macrofungi from southwest Australia, around 90% were also found in eastern Australia and/or in other countries. Species with small ranges (<100 km) were mostly those that had been described recently and/or known from single collections. Only the sequestrate *Torrendia grandis* and *T. inculta* were known from several collections and very localised in distribution (see also Bougher and Lebel 2001). For Australian smut fungi, Shivas and Vánky (2003) tabulated the distribution of 115 endemic species across five phytogeographic regions, modified from those of Doing (1981), finding that the highest diversity was in the *Eastern forest region*, probably due to the intensity of collecting there. There were few species in the *Southwestern forest and heath region*, despite the high diversity of plant hosts, although Shivas and Vánky (2003) pointed out that this region is not rich in grasses, a major host of smut fungi elsewhere in the continent. For *Sebacina* associating with Australian orchids, all four well-sampled phylogenetic species occurred on both sides of the continent (Davis *et al.* 2015). In addition to the mycogeographic significance of this finding, the widespread nature of these orchid symbionts means that the restricted distribution of some of the orchids is not due to parallel restricted distributions of their fungal partners.

For *Fusarium*, sampling across environmental gradients in Australia showed that species were characteristically found either in dry, semiarid soils or wetter, tropical soils (Burgess and Summerell 1992; Summerell *et al.* 1993; Sangalang *et al.* 1995b). In vitro cultures of species from hot, arid regions grew and survived better at higher temperatures (Sangalang *et al.* 1995a). For natural occurrences, Backhouse *et al.* (2001) distinguished host-associated and climate-mediated distributions, while pointing out that plant-colonising species ultimately depend on climates suitable for their hosts. Summerell *et al.* (2011) considered that rainfall, temperature, soil type and the associated vegetation were key factors influencing the distribution of *Fusarium* within Australia. They also suggested that the ability to survive drought was critical, and that species occurring in drier areas often have features such as chlamydospores (resting spores) or the ability to grow as endophytes (within plant tissues).

Bioclimatic Modelling

Bioclimatic modelling has been applied to explore climatic limits, particularly for exotic fungi in terms of identifying the potential area of spread, as for *Puccinia psidii* (myrtle rust) (Elith *et al.* 2013; Kriticos *et al.* 2013), *Batrachochytrium dendrobatidis*, the cause of chytridiomycosis of amphibians (Drew *et al.* 2006), and various species of *Fusarium* (Backhouse and Burgess 1995, 2002; Backhouse *et al.*, 2001). Otherwise, modelling has been rarely carried out for native fungi, as by Young (1996), who generated predicted distributions for several species of *Hygrocybe* utilising temperature and soil moisture indices. At the landscape scale, Claridge *et al.* (2000) found that climatic factors, as well as microhabitat characteristics, were important in determining the occurrence of sequestrate fungi in far southeast Australia.

Austral Fungi and Broader Distributions

There has been very little investigation of the relationship between distributions within the land masses of the region and broader distribution patterns. For *Fusarium*, Summerell *et al.* (2010) noted that several species in northern Australia were also found in Southeast Asia. For Rhytismatales, Johnston (1993 and pers. comm.) distinguished New Zealand species that extended (1) to Asia and Central America, but not Australia (*Pacific Ocean*) versus (2) to Australia (sometimes as a vicariant species pair) and in some cases also to South America (*Southern Ocean*). For the Pacific Ocean pattern, the distribution within New Zealand was restricted to the north and west of the North Island and the west of the South Island, whereas among the 13 species with the Southern Ocean pattern, only one was restricted to the north and west. It would be of considerable interest to expand this type of analysis to other groups, with the inclusion of phylogeny.

Mobilisation of Fungal Distribution Data

Several advances over the last decade have led to unprecedented access to and increase in distribution records for fungi in the region. Firstly, there has been significant digitisation (including georeferencing) of reference collections of fungi (for both specimens and cultures) and these data have been aggregated via *Australia's Virtual Herbarium* (avh.chah.org.au) and the *New Zealand Virtual Herbarium* (www.virtualherbarium.org.nz). Secondly, citizen science initiatives, particularly the Australian Fungimap scheme, are rapidly increasing observations of readily recognisable macrofungi. Results can be seen from maps of the 100 target species of Fungimap in *Fungi Down Under* (Grey and Grey 2005), the first fungal field guide worldwide to include maps of various species. Thirdly, the *Atlas of Living Australia* (www.ala.org.au) aggregates various sources of distribution data (both specimens and observations) in one readily accessible portal. As a result, common and widespread fungi, such as *Mycena interrupta*, are represented by several thousand records in the *Atlas of Living Australia*, giving precise shape to distributions (Figure 8.2). There is much scope for analysis of aggregated distribution data for fungi, to establish distribution patterns and their boundaries, and the factors controlling distributions.

Most Fungi Have Wide Distributions within Larger Land Masses

Where there are multiple records resulting from targeted sampling over wide areas, such as for *Fusarium* (Backhouse and Burgess 1995, 2002) and for the Fungimap target species (Grey and Grey 2005), most fungi mapped for Australia have wide distribution patterns that can often be explained by combinations of climate variables (Backhouse and Burgess 1995, 2002; May and Harris 2014). The impression from the inspection of such maps is that many species of fungi have distributions that occupy large areas within bioregions at the highest level defined in biological regionalisation (Ebach 2012). Examples of such 'top level' bioregions for Australia are the regions of Tate (1889) and Doing (1970) and the zones of Burbidge (1960) for plants, and the subregions of Spencer (1896) for animals. Exactly how mycogeographic regions within Australia and New Zealand correspond to regions as defined on other biota remains to be determined. There is little evidence for fungi restricted to lower-level bioregions, such as the areas delimited in the *Interim Biogeographic Regionalisation for Australia* (Thackway and Cresswell 1995).

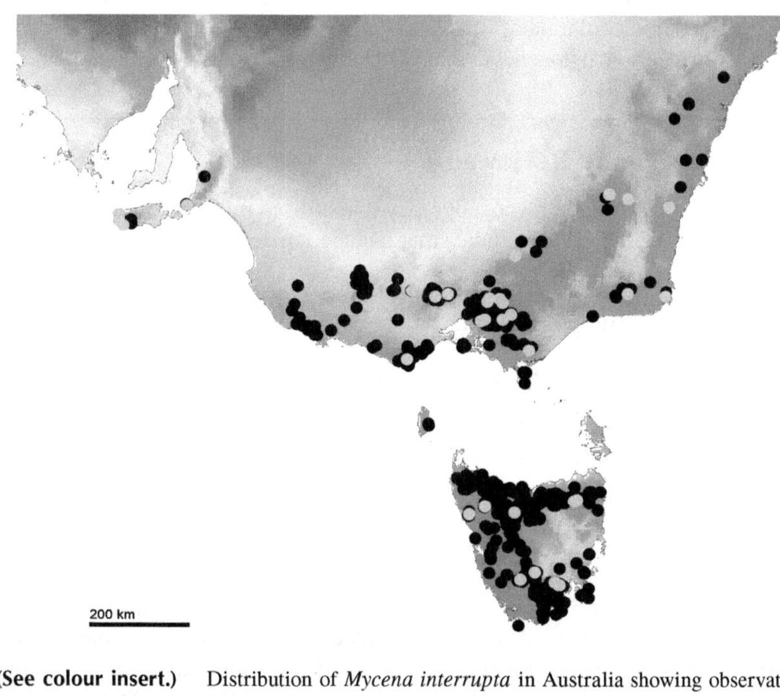

FIGURE 8.2 (See colour insert.) Distribution of *Mycena interrupta* in Australia showing observational (black) and herbarium (green) records, background layer is mean annual precipitation. (Produced using the *Atlas of Living Australia* http://www.ala.org.au. Accessed 22 March, 2016). CC-BY.

Predicted ranges from modelling also usually span whole top-level bioregions (Young 1996; May and Harris 2014). Habitat breadth also appears to be wide for many fungi, in terms of occurrence in different ecological communities as defined by dominant plant species. The wide distribution of most Australasian fungi is one reason why so few fungi are formally listed on conservation schedules, whereas small geographic ranges automatically make many plant species eligible for listing (May 2002).

However, there is an important caveat to the generalisation that fungi have wide distribution patterns at the regional level. Some fungi may well be geographically limited when they are highly host specific and the host has a limited range or when the habitat of the host is limited in some way. For example, the rust fungus *Puccinia embergeriae* is only known from the rare host *Embergeria grandifolia*, which occurs only in the Chatham Islands (Buchanan and May 2003). Among the Fungimap target species, one of the very few species with a narrow distribution is the rare *Hypocreopsis amplectens*, which is usually found in a presumably mycoparasitic association with another fungus (*Hymenochaete*) in long unburnt stands of *Leptospermum* woodland in southern Victoria within a 150 km range, although intriguingly, also at one site in New Zealand in cool-temperate rainforest (Johnston *et al.* 2007). There has been little targeted collecting on rare hosts (plant or animal). Range-limited fungi are less likely to have been discovered as yet, but metagenomic approaches (see 'Metagenomic Biogeography' in this chapter) may assist in detecting such species.

Summary of Twentieth-Century Mycogeography

The three syntheses previously summarised and other literature from the twentieth century documented a variety of distribution patterns among Australasian fungi. Various explanations were advanced, often involving the breakup of ancient land masses. However, the perspectives from plant pathology integrated by Walker (1983; 1996) also highlighted the potential for dispersal and for human-mediated transport. With rare exceptions (see '*Cyttaria*: Evolution of Ideas about Coevolution of Fungi in Relation to Time and Area' in this chapter), considerations of mycogeography in this period did not utilise the knowledge of evolutionary relationships gained from analytical methods such as parsimony. Nor was there analysis of distribution data, such as by clustering of areas. Pirozynski (1983) recognised that the 'pursuit of biogeography requires a more nearly phylogenetic systematics and a more complete record of geographic distribution'. Indeed, a caveat to earlier analyses of the distribution of austral fungi, once there was an awareness of the utility of molecular data, was that all taxonomic relationships supposed from morphology needed to be confirmed by phylogenetic analysis (Johnston 2001).

Molecular Mycogeography Reviewed

This section reviews the literature on mycogeography that utilised molecular data, which appeared from the mid-1990s onwards. The focus is on studies that included material from Australasia.

The Impact and Relevance of Molecular Phylogeny and Species Delimitation for Biogeography

Initial Phylogenetic Studies Transform Fungal Classification

From the 1990s onwards, the analysis of molecular data led to major reconfigurations of fungal classification. This had significant ramifications for other disciplines such as biogeography in two main areas: firstly, higher taxa such as genera were often found to be unfounded phylogenetically (such as by being polyphyletic), and secondly, the application of the *phylogenetic species concept* (PSC) to molecular data led to the discovery of otherwise cryptic species in numerous cases (see 'Most Phylogenetic Species of Fungi Are Geographically Restricted' in this chapter).

Phylogenetic analyses of fungi were greatly stimulated by the development of primers that amplified specific portions of the nuclear-encoded ribosomal RNA cistron. This region is flanked at either end by an intergenic spacer (IGS) and codes for the nuclear small (18S) and large (28S) subunits (nLSU and nSSU) interspersed with internal transcribed spacer (ITS) regions ITS1 and ITS2, separated by the 5.8S gene (White *et al.*, 1990; see also 'Molecular Species Identification: ITS as the Fungal Barcode' in this chapter). The newly available molecular data were analysed by various methods of phylogenetic analysis. Previously, there had been very limited application of cladistic analysis to morphological data on fungi.

Early molecular studies immediately revealed surprising relationships. For example, in the phylogenetic analysis of homobasidiomycete Basidiomycota by Hibbett *et al.* (1997), various genera with *gasteroid* sporophores, both puffballs (e.g. *Lycoperdon*) and birds nest fungi (e.g. *Cyathus*), were placed well within a clade otherwise composed of genera with a mushroom sporophore (e.g. *Agaricus*). Moncalvo *et al.* (2002) used nLSU sequences from an extensive sample of more than 800 species to define 117 monophyletic clades of mushrooms and their relatives. Some long-established taxa were supported, but many genera and families recognised on morphology were polyphyletic.

Essentially, these and numerous studies that followed found that while some taxa (genera, families, orders, etc.) traditionally defined on morphology were supported in molecular phylogenies, many others were not. There was a high degree of parallel evolution across a wide range of morphological characters, from fruit body form to spore features. Morphology may be mapped on to molecular phylogeny after the fact, but is now rarely used as a primary character in reconstructing the evolutionary history of fungi. A major synthesis of fungal classification was published in 2007, resulting from the Assembling the Fungal Tree of Life (AFTOL) project, which used multigene data sets to reconstruct phylogenies across most major lineages of fungi, with many new relationships revealed at all levels of classification (Blackwell *et al.* 2007 and two-dozen papers in the same volume). Ongoing initiatives using genome-wide information for fungal phylogeny aim to fill gaps in taxon sampling and clarify the deeper branches of the fungal tree of life (McLaughlin *et al.* 2009).

There are several examples where new classifications based on molecular phylogenies challenged biogeographic hypotheses that had been put forward from classifications based on morphology. The genus *Rozites*, distinguished from *Cortinarius* by the membranous rather than arachnoid partial veil, had been an iconic genus in Southern Hemisphere biogeography (Horak 1983; Bougher *et al.* 1994). The 15 species of *Rozites* were found in Australia, New Zealand, New Caledonia and southern South America, mostly associated with *Nothofagus*, but there was one species in each of Europe, Asia and northern South America (Figure 8.3). Horak (1983) concluded that *Rozites* was ‘of Gondwanaland origin’ and

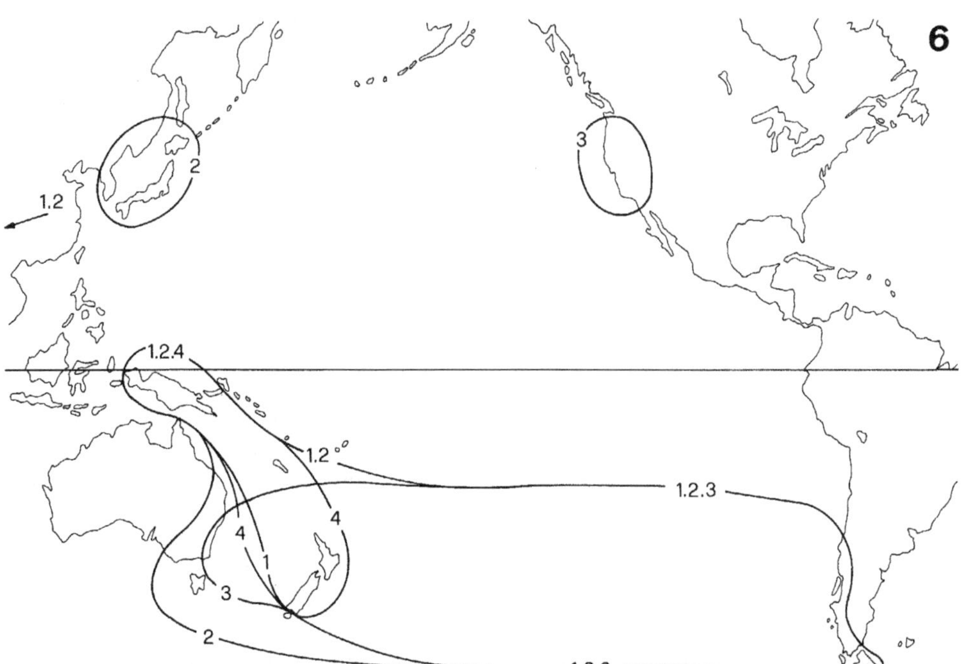

FIGURE 8.3 Distribution of the genera: (a) *Rozites*; (b) *Descolea*; (c) *Thaxterogaster*; and (d) *Cuphocybe*. *Descolea* remains as delimited by Horak (1983), the other three genera are now synonyms of *Cortinarius*, a cosmopolitan genus. Originally published as Figure 6 in Horak, E., *Australian Journal of Botany* 10, 1–41, 1983. Reproduced with permission.

'coevolved with *Nothofagus*', and Bougher *et al.* (1994) postulated an origin in the Cretaceous with fagalean plants in the Asian–Australian region, followed by speciation in the Southern Hemisphere as *Nothofagus* diversified. With the use of molecular data, *Rozites* was found to be polyphyletic, with all nine sampled species falling within *Cortinarius*, in five different clades (Peintner *et al.* 2002). Similarly, rearrangements of species and generic boundaries mean that explanations advanced by Wolfe and Bougher (1993) for the distribution of members of *Tylopilus* subgenus *Roseoscabra* also need revisiting, now that molecular data have led to members of the group being dispersed among several genera, including *Australopilus* and *Harrya* (Halling *et al.* 2012).

Species Delimitation: Morphology to Molecules

Reviews of fungal biogeography often highlight the critical importance of accurate species taxonomy for meaningful biogeographic analysis. Walker (1996) gave the plant pathogenic mushroom *Armillaria* as an example. In this genus, Australasian material was originally assigned to *Armillaria mellea* (described from Europe) using a morphological concept. Interfertility tests and closer examination of morphology led to the recognition of at least seven species from Australia and New Zealand, and *A. mellea* was in fact confirmed as absent from native forests in the region (Kile and Watling 1983, 1988; Coetzee *et al.* 2001, 2003; Maphosa *et al.* 2006). Walker (1996) concluded 'biogeographic conclusions drawn up ... on the basis of an assumed ubiquitous *A. mellea* would have been meaningless'.

Historical Approaches to Species Delimitation of Austral Fungi

Since the first fungi were formally described from the Southern Hemisphere, a tension has been apparent between two assumptions: (1) austral fungi that look similar to Northern Hemisphere species are the same, leading to the incorrect recording of boreal species from Australasia, as in *Armillaria* and many other mushrooms; and (2) minor variations in morphology between collections from different locations are sufficient to erect new species, as happened for widespread species of polypores and stinkhorns such as *Hexagonia tenuis* and *Aseroe rubra* (May *et al.* 2003). Such assumptions about splitting or lumping were rarely articulated and the degree to which geography was consciously a character utilised by nineteenth- and twentieth-century taxonomic mycologists is a subject that merits further investigation by historians of science, as indeed does the basis for their species delimitation per se. Essentially though, until late in the twentieth century, the practice of species delimitation in fungi relied on morphological characters. At first, macroscopic features of the sporophore were used. Then, from the late nineteenth century, microscopic characters such as spores were also important. For those fungi readily grown in pure culture, characters of the cultures were utilised, and in addition, interfertility could be utilised to delimit biological species. For biotrophic fungi, such as rust fungi, the host is also an important component of species delimitation, but some fungi can have wide host ranges, and care must therefore be taken in using the host as a primary character in delimiting fungi.

Phylogenetic Species and Genealogical Concordance Phylogenetic Species Recognition

Taylor *et al.* (2000) usefully distinguished between theoretical *species concepts* and the practical operation of *species recognition*. The same distinction was made by Giraud *et al.* (2008) between, respectively, *species definition* and *species criteria*. For fungi, species recognition has primarily been morphological, biological (utilising interfertility) or phylogenetic. Taylor *et al.* (2000) published an influential discussion of the theory of the PSC as applied to fungi, but also outlined a practical approach, which they called *genealogical concordance phylogenetic species recognition* (GCPSR). This practice involved the use of multiple independent genes, for which separate gene trees were constructed and then compared, searching for the point at which there was a transition from concordance among gene trees (matching the divergent phylogenetic relationships of the species tree) to incongruity between gene trees (representative of the reticulate *tokogenetic* associations between individuals within a species) (Hennig 1966; Avise and Wollenberg 1997; Taylor *et al.* 2000). In other words, 'Within species, the mixing effects of recombination cause unlinked loci to have distinct genealogical histories, but [phylogenetic] divergence leads to concordant genealogical histories [across multiple] loci' (Leavitt *et al.* 2015).

From the start of the application of GCPSR to fungi, it was recognised that incomplete lineage sorting (due to deep coalescence) meant that reciprocal monophyly would not be expected across all regions (Taylor *et al.* 2000; Baum and Smith 2013). Methodologically, the recognition of concordance was initially by visual comparison of trees often accompanied by the presentation of a concatenated tree; although strictly, well-supported clades in such a tree are not of themselves evidence of concordance. There is now a plethora of methods for species delimitation, including those that integrate coalescent theory (Brito and Edwards 2008; Fujita *et al.* 2012; Carstens *et al.* 2013; Baum and Smith 2013; Leavitt *et al.* 2015), and such methods have been applied to fungi (Carbone and Kohn 2001; Geml *et al.* 2006).

GCPSR has become widely used among mycologists as the method by which a phylogenetic species concept is applied practically; but its application is by no means universal, and many new species of fungi are still being described on morphology alone or with use of single gene regions. Leavitt *et al.* (2015), in reviewing species delimitation in lichenised fungi, argued that while single-gene data can be useful for initial sorting of collections, it is only the analysis of independent genomic regions that 'can provide robust hypotheses of species boundaries'. Species recognition in fungi will increasingly use genome-wide data, as has been introduced for higher-level classification (Dentinger *et al.* 2015). The use of a barcode region is another approach, but only works once the region has been properly calibrated (see the following section).

Hybridisation leading to introgression is a potentially complicating factor in species delimitation (Baum and Smith 2013) and has been demonstrated between phylogenetic species in some fungal genera such as *Coccidioides* (Neafsey *et al.* 2010) and *Coniophora* (Kauserud *et al.* 2006), the latter after human-mediated movement of allopatric species, which do not necessarily develop prezygotic reproductive barriers. However, in *Neurospora*, where hybrids had been supposed, molecular phylogenetic analysis did not detect any evidence of hybridisation, and putative hybrids were in fact representatives of new phylogenetic species segregated within biological species (Dettman *et al.* 2003a).

The application of GCPSR has been useful in clarifying the host range, as with *Diaporthe*, in which numerous species were initially described on different hosts. However, the genus contains a mix of host-specific and host-generalist phylogenetic species (Udayanga *et al.* 2014). The host should not automatically be used as a criterion for species delimitation unless host specificity in the particular group is well understood.

Sometimes, the recognition of phylogenetic species led to the lumping of morphologically distinct species, as with the *Tylopilus chromapes* group, where six of the distinct species recognised on morphology by Wolfe and Bougher (1993; using a phenetic analysis), albeit with rather subtle differences, collapsed into two species, in two novel genera *Harrya* and *Australopilus*, once molecular data became available (Halling *et al.* 2012). More often, phylogenetic species were morphologically cryptic (splitting existing species into two or more) or could only be distinguished by very subtle differences in morphology that made sense after the fact, but of themselves would not be clearly sufficient for species delimitation. Even the most iconic and readily recognised mushroom, the red- and white-spotted fly agaric (*Amanita muscaria*), whose native distribution is in the Northern Hemisphere (although it also occurs in Australian and New Zealand as an exotic), is a complex of eight phylogenetic species, based on a genealogical concordance approach (Geml *et al.* 2008).

Molecular Species Identification: ITS as the Fungal Barcode

One particular gene region is proving to have general utility for species identification in fungi. The ITS of the nuclear ribosomal DNA is present in multiple copies, and there are primers that amplify this region readily in most fungi. A comparison between genetic regions commonly utilised for fungi confirmed the ITS as the official barcode for fungi, showing that it provided correct identification 70% of the time for species otherwise delimited on multigene data (Schoch *et al.* 2012). Secondary barcodes are also being investigated for use in groups where ITS does not adequately distinguish species (Stielow *et al.* 2015).

It is important to make a distinction between the use of the ITS as a barcode region to (1) routinely identify fungi, as opposed to (2) provide the only evidence for species delimitation. Although many studies apply a 97% similarity level for ITS as a cutoff for the identification of species-level taxa, it is ideal to establish the precise cutoff level at which there is a gap in the frequency distribution of the infra- and interspecific distances, because this level varies across different groups. The barcode cutoff should be calibrated from multigene studies, as in *Cortinarius*, where a 97.8% similarity level correctly assigned species in 99.8% of pairwise comparisons among 21 phylogenetic species defined by concordance across seven loci (Stefani *et al.* 2014). Once a cutoff has been established as effective across a particular lineage, only then is it appropriate to use the barcode region for delimiting new species in the absence of multigene data. The relevance for biogeography of the appropriate use of barcode regions is both in the day-to-day delimitation of phylogenetic species, but also in the identification of metagenomic samples (see 'Metagenomic Biogeography' in this chapter).

Most Phylogenetic Species of Fungi Are Geographically Restricted

Studies of fungi utilising phylogenetic reconstructions with a geographic component may include intensive sampling within one or a few closely related species or otherwise tend to use single or a few samples across a range of species. The former approach falls under phylogeography (Avise 2000), while the latter can be characterised as 'molecular biogeography'. There is a burgeoning literature on fungi in relation to both phylogeography and molecular biogeography, and many studies have included samples from Australasia.

Early applications of molecular techniques to fungal taxonomy at the species level, often on economically important fungi such as crop and human pathogens or cultivated mushrooms, frequently recovered cryptic species. Indeed, Pringle *et al.* (2005) considered that 'in all cases in which the PSC has been used to describe species of global isolates of a morphological fungal species, cryptic species have been recognized'. Furthermore, phylogenetic diversity was often geographically structured within species long thought to be widely distributed. Seven case studies, utilising at least some isolates from the Australasian region, where cryptic phylogenetic species were found to be geographically restricted, are presented for *Pleurotus*, *Lentinula*, *Schizophyllum*, *Panellus*, *Pisolithus*, *Neurospora* and *Fusarium*. Further examples are discussed in the section 'Further Examples of Geographic Structure in Phylogenetic Species of Fungi'. Cases where phylogenetic species of fungi were found to be widespread, such as in some species of *Aspergillus*, are dealt with in 'Widespread Phylogenetic Species'.

Case Studies on Geographic Structure of Phylogenetic Species

Pleurotus

Vilgalys and Sun (1994) generated sequences for the LSU and ITS regions of samples from around the world of eight intersterility groups (biological species) in the mushroom *Pleurotus*, some species of which are cultivated for food. They found either a one-to-one correspondence between biological and phylogenetic species or, in three cases, that intersterility groups comprised geographically separated phylogenetic species, such as Asian, European and North American lineages within *Pleurotus ostreatus*. They postulated that lineages found in the Northern and Southern Hemispheres had wider distributions due to their older origin, in contrast to more recently evolved lineages restricted to the Northern Hemisphere. Vilgalys and Sun (1994) concluded that speciation had been associated with geographic isolation and emphasised the 'importance of understanding biogeography for understanding speciation in *Pleurotus*'.

Further studies on *Pleurotus* added geographically distinct phylogenetic species. Isikhuemhen *et al.* (2000) found two distinct *evolutionary lineages* based on ITS sequences within otherwise intercompatible isolates of *Pleurotus tuber-regium*, consisting of those from (1) Africa and (2) Australasia (including Australia, New Caledonia and New Guinea) and Indonesia. Variation within the second lineage did not correspond to geography (suggestive of gene flow across the region), and there was low variation among African isolates, suggesting 'an origin of diversification for *P. tuber-region* … in the Australasia–Pacific region' (Isikhuemhen *et al.* 2000). For the *Pleurotus cystidiosus* complex, Zervakis *et al.* (2004), using ITS sequences, found four lineages that 'correspond largely with geographical groupings', with a deep split between two New World and two Old World lineages. Although there were high levels of interfertility among isolates of the complex (except for one species), they invoked a PSC in recognising the four lineages at species rank. They also recognised *P. australis* (restricted to Australia and New Zealand), which was almost completely intersterile with members of the *Pleurotus cystidiosus* complex.

Lentinula

Lentinula edodes (shiitake) is a mushroom cultivated for food, originally domesticated in Asia. Hibbett *et al.* (1995) examined collections from Asia, New Guinea, Australia and New Zealand that were assigned to three morphological species, noting that all were intercompatible and thus could be considered as belonging to a single biological species. Phylogenetic analysis of the ITS region, using a variety of methods and coding, recovered four clades: I (East and Southeast Asia), II (New Guinea and Australia), III (New Zealand) and IV (Papua New Guinea). The two clades with representatives from New Guinea (II and IV) did not show up as sister taxa in any analyses, and clade IV was usually basal to the others. There was a mismatch between morphological species and phylogenetic species, with some phylogenetic species comprised of isolates that would be assigned to more than one morphological species, or conversely, the same morphological species was present in more than one phylogenetic species (Hibbett and Donoghue 1996). The mapping of geography onto phylogenies suggested Borneo–New Guinea as the likely ancestral area for the *Lentinula edodes* complex (Hibbett *et al.* 1995). However, when further isolates from Asia and Australasia were added to the original set, a further lineage (V) was recovered, present in China and Nepal, sister to the New Zealand lineage III (Hibbett *et al.* 1998). In an analysis of the biogeography of the genus *Lentinula*, including Old World isolates of the *Lentinula edodes* complex and New World isolates of the *L. boryana* complex, Hibbett (2001) suggested an ancestral range for the genus in Eurasia and North America and an Asian ancestral range for the *L. edodes* complex. Phylogenies with Australasian lineages (II, III and IV), constrained to monophyly or not, all required two or three dispersal events to explain current distributions (Hibbett 2001).

Schizophyllum

Splitgill (*Schizophyllum commune*) is a model organism for studies on sex in larger fungi. It forms bracket-shaped sporophores on a variety of woody substrates. Raper *et al.* (1958) showed that a series of isolates from around the world (including Australia) were all sexually compatible and there was no geographic structure among the alleles of the mating-type loci. However, James *et al.* (1999) examined 11 allozyme loci among a global sample of *S. commune* isolates and found significant geographic differentiation, particularly between the Eastern and Western Hemispheres. James *et al.* (2001) confirmed from a sample of nearly 200 collections that there was significant geographic structure within *S. commune*. Using the IGS1 region as well as a more restricted sampling of the ITS and IGS2 regions, they found three 'evolutionarily distinct lineages': (1) North and Central America, (2) South America and the Caribbean (SAM) and (3) the Eastern Hemisphere (EAS) (including samples from Australia, New Guinea and New Zealand, most in one subclade, but others scattered throughout the clade). Nested clade analysis of the IGS1 phylogeny suggested a long-distance colonisation of America from Europe in the EAS clade, and recent contiguous range expansion in the SAM clade (James *et al.* 2001).

Fuller *et al.* (2013) expanded the sampling of *Schizophyllum commune* from New Zealand and the Cook Islands to determine the source of collections, due to the observation of the species on imported timber products. IGS1 sequences of all 18 isolates fell within the EAS clade that had been delimited by James *et al.* (2001) and there was no evidence of recent introductions of other phylogenetic species. Among the New Zealand collections there were four haplotypes, in two main groups, that fell into different parts of the EAS clade. A haplotype network suggested two natural dispersal events in prehuman times, one from Australia, and one from the hypothetical ancestral population of the EAS clade (Fuller *et al.* 2013).

Panellus

In the cosmopolitan mushroom *Panellus stypticus*, isolates from various parts of the Northern Hemisphere and New Zealand were all intercompatible (Petersen and Bermudes 1992). An ITS phylogeny presented by Jin *et al.* (2001) had two main clades, one for collections from mainland Australia, Tasmania and New Zealand, the other including all collections from the Northern Hemisphere. Within the latter clade there was a clear separation between collections from eastern North America and those from Eurasia and western North America. There was also geographic structure among haplotypes of restriction fragment length polymorphisms (RFLPs). Southern Hemisphere occurrences were considered to be of later origin due to deletions that were not present either in the Northern Hemisphere collections or in the sister taxon *Panellus pusillus* (as *Dictyopanus pusillus*). The presence of an nLSU intron in all Northern Hemisphere collections and in one of the Australian collections, along with shared RFLP haplotypes between Australia and Eastern Russia, suggested that 'dispersal between the Northern and Southern Hemispheres may have been through Southeast Asia to Australia' (Jin *et al.* 2001). Within Australasia, the establishment of the New Zealand population was suggested as being due to long-distance dispersal from Australia because of the relatively small genetic divergence in relation to the separation of the land masses 60–70 mya (Jin *et al.* 2001).

Pisolithus

Members of the ECM genus *Pisolithus* (known as *dyeballs*) form a puffball sporophore that erodes to release powdery spores. *Pisolithus* is cosmopolitan in warm-temperate regions and widespread throughout drier areas of Australia, where it is especially common in the arid interior. For more than 50 years, the view of Cunningham (1944) that there were few species in Australia was generally accepted. One of the Australian species was identified as *Pisolithus arhizus* (sometimes called *P. tinctorius*), originally described from the Northern Hemisphere. Indeed, this name was applied to most collections of the genus from around the world. Martin *et al.* (2002) analysed the ITS sequences of *Pisolithus* in order to apply phylogeny to 'understanding ... host specificity and biogeographical patterns'. However, as a result of the comprehensive set of 148 collections assembled for the phylogeny, the study also provided significant insights into species delimitation. Phylogenetic analysis revealed two main lineages and 11 species-level clades, only 3 of which had been formally named. The phylogenetic species were restricted to 'specific geographic regions', often with hosts confined to these regions, as with 'species 4' from pine and oak in Spain, 'species 3' from *Cistus* in Spain and 'species 1' from *Afzelia* in Kenya. One of the clades was identified as *P. tinctorius*, and this occurred only in the Northern Hemisphere with pines and oaks. Five clades were exclusively comprised of collections from Australia (along with a few collections from planted *Eucalyptus* outside of Australia) and several of these clades contained samples originating from both *Acacia* and *Eucalyptus*. The presence of Australian species in both major lineages led Martin *et al.* (2002) to suggest Australia as 'the centre of diversification' of the genus *Pisolithus*, with the origin of the genus dating back to the Triassic.

Moyersoen *et al.* (2003) used the taxonomic framework for *Pisolithus* established by Martin *et al.* (2002) to investigate the origin of that genus in New Zealand, where it occurs as an ECM associate of *Kunzea*, restricted to localised vegetation islands in the Taupo volcanic zone on the North Island, which is characterised by extremely acidic and low-nutrient soils. Three New Zealand taxa of *Pisolithus* were recognised on the basis of ITS sequences, each matching a widespread phylogenetic species previously identified from Australia. The presence of these three species in New Zealand was suggested to be the result of several recent long-distance, transoceanic dispersal events, made possible by a combination of (1) the 'development of west-wind drift related to separation of Australia from Antarctica in the Paleocene', (2) the 'presence of suitable plant hosts belonging to Leptospermoideae ... since the Tertiary and their capacity to cope with particularly harsh edaphic conditions such as are found in geothermal areas', and (3) the 'development of volcanoes subsequent to the Pliocene in the Central North Island ... with associated geothermal areas' (Moyersoen *et al.* 2003; see therein for further references). While the origin was considered 'geologically recent', no particular date was calculated. Hosaka (2009) added new samples of *Pisolithus*, particularly from New Caledonia, and from a data set of more than 300 ITS sequences detected several New Caledonian 'taxa' that were sister to Australian taxa. However, the exact boundaries of phylogenetic species were not specified, and in some cases New Zealand collections were also present in the clades. Given the more recent geological connection between New Zealand and New Caledonia (in comparison with the earlier split with Australia) further investigation of these samples will be of interest.

Neurospora

Neurospora is a model eukaryote with a sexual stage producing asci in perithecia and with some species also producing asexual spores. It is commonly found after fire or in association with substrates that have been heated, such as steamed timber (Turner *et al.* 2001). Species delimitation was originally on morphology, such as the number of spores per ascus and later by the use of interfertility tests. Using a *biological species concept*, Turner *et al.* (2001) showed that the five outbreeding species of *Neurospora* were widespread, but not everywhere. Within these five biological species, Dettman *et al.* (2003a) used four loci to delimit eight phylogenetic species, three of which contained significant infraspecific phylogenetic structure. Some of the phylogenetic species were widespread and/or confined to certain substrates, while others were geographically restricted, such as PS3 from Africa, PS2 from Madagascar and Mexico and PS1 from Haiti. For phylogenetic species such as *Neurospora crassa* in the strict sense, there was evidence of some intersterility between infraspecies phylogenetic groups (not otherwise supported as species by genealogical concordance), which was suggestive of 'incipient speciation' (Dettman *et al.* 2003a). *Neurospora discreta* had the most genetic diversity and a subsequent study using 73 strains delimited at least eight phylogenetic species, including PS9 from New Zealand, PS10 from New Zealand and South America and PS7 from North America (Dettman *et al.* 2006). The most widespread phylogenetic species within *Neurospora discreta* (PS4) contained two subgroups, one from New Guinea and the other from Asia, Europe, Africa and North America. Further isolates were integrated into the phylogenetic species of *Neurospora* by Villalta *et al.* (2009) without much alteration to the range of those species that were geographically restricted.

Fusarium

For plant and animal pathogens, biogeography is complicated by the human-mediated movement of hosts (along with their fungi) around the globe. Nevertheless, even in microfungi such as *Fusarium*, a genus with many species that are pathogens of agricultural crops, multigene studies applying a *genealogical concordance* approach detected phylogenetic species or groups of such species with restricted geographic distributions. In the *Gibberella fujikuroi* complex (*Gibberella* was a name used for the sexual stage of *Fusarium*), O'Donnell *et al.* (1998) recovered three clades, consistent with species radiations in South America, Africa and Asia. When species with widespread distributions (in at least part of their distribution, often on hosts growing outside of their native areas) were ignored, the African and Asian clades were the most closely related. For the 26 phylogenetic species of the *Fusarium solani* complex, O'Donnell (2000) also found clades containing species from New Zealand, South America, Africa and India–Sri Lanka, and Nalim *et al.* (2011) added further phylogenetic species to the South American clade, from Sri Lanka. In both the *Gibberella fujikuroi* complex and the *Fusarium solani* complex, apart from recent distributional changes associated with the movement of economically important plants, current distributions were postulated as arising from events associated with the fragmentation of Gondwana from the Late Cretaceous, with only one putative long-distance trans-Pacific dispersal (O'Donnell *et al.* 1998; O'Donnell 2000; cf. 'Analyses Including Time' in this chapter). *Fusarium graminearum*, the cause of head blight in wheat and barley, was composed of seven geographically structured species, mostly found on noncereal hosts in the Southern Hemisphere. At least two independent long-distance dispersals were invoked to explain the phylogeographic pattern in the Southern Hemisphere (O'Donnell *et al.* 2000a). Two years after their initial study on the *Gibberella fujikuroi* complex, O'Donnell *et al.* (2000b) added 10 new species to the group, all consistent with their original geographically distinct clades, once the movement of cultivated hosts was taken into account. Kvas *et al.* (2009) and Summerell *et al.* (2011) critique some of the aforementioned studies, particularly O'Donnell *et al.* (1998) on the *Gibberella fujikuroi* complex, arguing that distributions as originally circumscribed are not so clear cut once further isolates are taken into account. Summerell *et al.* (2011) document several range extensions to Australia, and consequently suggest that the *African* clade be renamed *Gondwanan*. Nevertheless, phylogenetic species of *Fusarium* from natural habitats usually have noncosmopolitan distributions. Those species previously categorised as cosmopolitan (Backhouse *et al.* 2001) are in need of critical examination of species boundaries (Summerell *et al.* 2010).

Further Examples of Geographic Structure in Phylogenetic Species of Fungi

It was a remarkable finding of these early studies on fungal molecular biogeography that when species were delimited phylogenetically, there was little evidence for globally widespread species, except when human-mediated dispersal was likely. In addition to the studies described in 'Case Studies on

Geographic Structure of Phylogenetic Species' in this chapter, most other studies applying a PSC to widespread fungi, when sampled across their range, found geographically restricted phylogenetic species – for example, the human pathogen *Histoplasma*, with some of the seven phylogenetic species of the *H. capsulatum* complex occurring only in North America, South and Central America, Africa or Australia (Kasuga *et al.* 1999, 2003); *Coccidioides* and related genera, also human pathogens, where three different species complexes yielded a total of eight phylogenetic species (Koufopanou *et al.* 2001); and the *Artomyces pyxidatus* group, with six species occurring in Australasia, most confined to Australia, New Guinea and/or New Zealand, but with one species extending to Malaysia, one to Indonesia and one also found in southern South America, and with Australasian species mostly having sister relationships with species of South or Central America, rather than Asia or North America (Lickey *et al.* 2003; Petersen and Hughes 2007). For macrofungi occurring across Asia and North America, Wu *et al.* (2000) and Mueller *et al.* (2001) found molecular data to suggest both splitting and lumping, depending on the group, although in the case of lumping the resulting distributions were still restricted to those regions.

Multigene Taxonomic Studies Yield Biogeographic Structure

The use of molecular data for fungal biogeography is now the norm. While molecular tools facilitate studies with explicit biogeographical hypotheses, discussions of biogeographic patterns are also often a spin-off from taxonomic revisions on the basis of multigene data that yield geographically distinct phylogenetic species.

Research into taxonomy is frequently confirming that fungi are 'not as ubiquitous as we thought' (Salgado-Salazar *et al.* 2013). Recent studies on fungi found in Australasia that have recovered geographically restricted phylogenetic species (many often previously unnamed) in multigene molecular phylogenies include those on the following:

1. Mushrooms of *Lactarius* subgenus *Gerardii*, with 30 phylogenetic species, most occurring within a continent, and no clustering of species according to continent, but rather with frequent sister relationships among pairs of species from America, Asia or Australasia, such as North America/Thailand and North America/Indonesia, with Australasian species most closely related to species from Asia, and with long-distance dispersal indicated for the two species occurring in Australia and New Zealand, due to comparatively low infraspecific variation (Stubbe *et al.* 2010).
2. *Pycnoporus*, a polypore with Australian, Neotropical, Palaeotropical (including New Caledonia) and East Asian clades (Lesage-Meessen *et al.* 2011).
3. The lichenised agaric *Lichenomphalia*, in which Geml *et al.* (2012a) identified two new phylogenetic species from subantarctic islands of New Zealand, each sister to taxa or clades of taxa with predominantly Northern Hemisphere distributions, suggesting rare transequatorial dispersal in this genus.
4. The *Geastrum triplex* group, puffballs with six phylogenetic species in either the Northern or Southern Hemisphere, and in the latter with two species occurring in both Australasia and southern South America (Kasuya *et al.* 2012).
5. Mushrooms in *Entoloma* subgenus *Entoloma*, in which there were 'strong phylogeographical partitions'. No collection from Australasia was conspecific with Northern Hemisphere species, and species from Australasia were found in many subclades of a major clades, in 'basal positions to their closest relatives in the Holarctic', suggesting that 'Australasia might be an ancestral area' for the major clade (Morgado *et al.* 2013).
6. The ascomycete *Thelonectria discophora* (with sexual spores in perithecia and/or asexual spores), previously considered cosmopolitan, but discovered to contain 16 lineages treated as 'putative species' (Figure 8.4), many with a restricted distribution at the continent or regional level, such as clades III and IV from New Zealand and clade VIII from Australia and Indonesia, but with the sizes of distributions apparently not correlated with whether species had asexual stages or were saprobes versus plant pathogens (Salgado-Salazar *et al.* 2013).

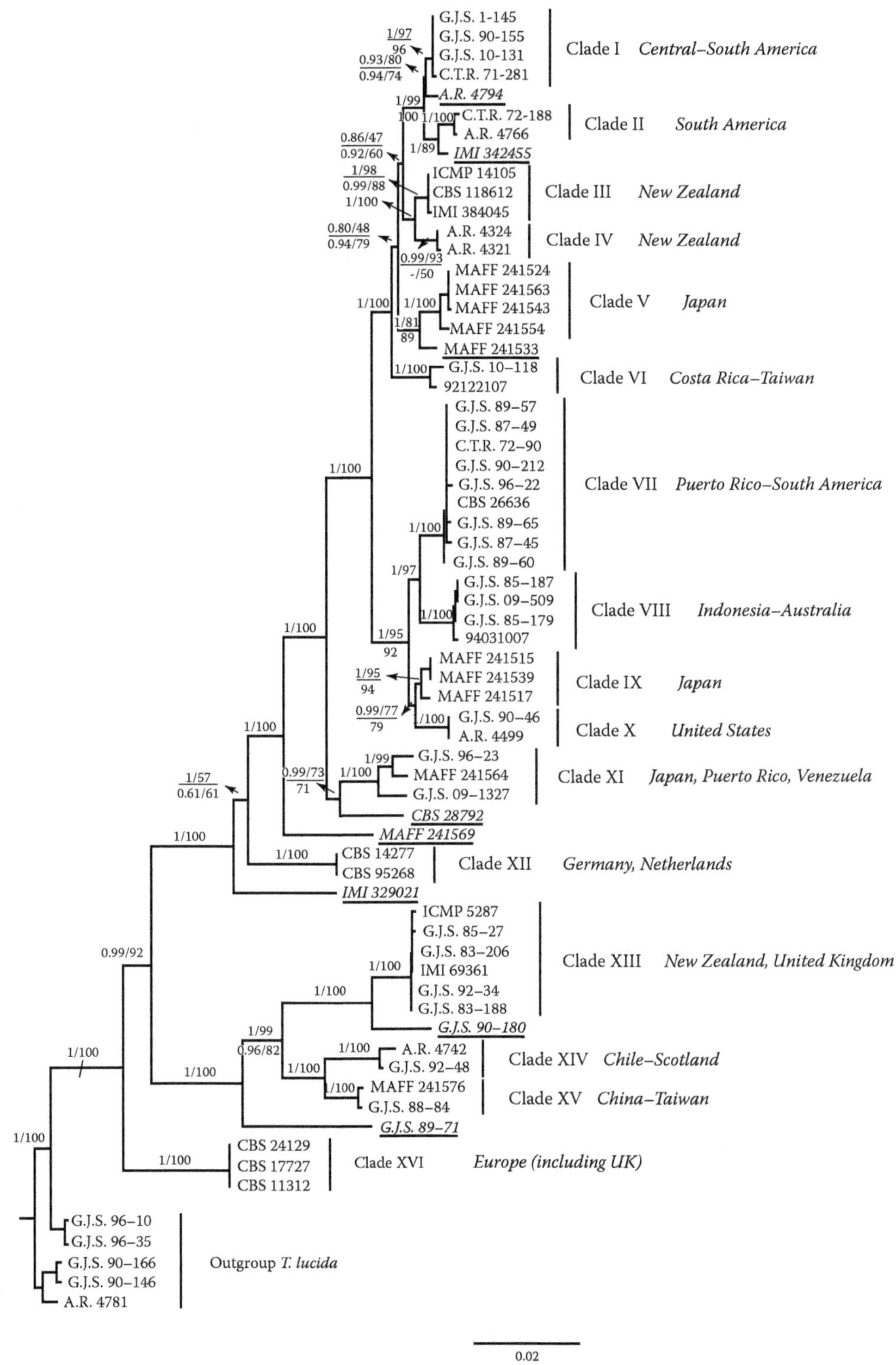

FIGURE 8.4 Bayesian phylogram showing relationships among isolates in the *Thelonectria discophora* complex based on concatenated analysis of six loci. Originally published as Figure 2 (http://dx.doi.org/10.1371/journal.pone.0076737.g002) in Salgado-Salazar, C., *et al.*, *PLoS ONE* 8(10), 1–12, 2013; see original caption for further details. Accessed 22 March, 2016. CC-BY.

7. The mushroom *Singerocybe*, in which Qin *et al.* (2014) delimited seven species, each geographically restricted at about the continent level, with *Singerocybe clitocyboides* occurring in Australia, and all other species occurring in the Northern Hemisphere, in either Europe, North America, East Asia or Sri Lanka.
8. *Stephanospora*, a false truffle, with 16 species, mostly segregates of *Stephanospora flava*, occurring in either the Northern or Southern Hemisphere, and in Australasia, where species were mostly restricted to either Australia, New Zealand, New Guinea or New Caledonia, with only one species extending across Australia and New Zealand (Lebel *et al.* 2014).
9. Vertebrate pathogen species of *Sporothrix*, in which five phylogenetic species were mapped from 14,000 human and animal case studies by Zhang *et al.* (2015), who found *S. schenckii* was the only species in Australia, and was otherwise common in warm-humid climates of the Southern Hemisphere, while *S. globosa* was the predominant species in North and East Asia, mostly in areas with a cool climate.

Geographic Patterns within Phylogenetic Species

In relation to geographic patterns for phylogenetic species within the land masses of Australasia there has usually been insufficient sampling to confirm distributions. However, there are indications that at least some phylogenetic species have wide distributions within continents. For the phylogenetic species of *Pisolithus* identified by Martin *et al.* (2002), where there were multiple samples, the species were usually widely distributed within Australia, as with species 2 (WA and NSW), species 7 (WA, Qld, NSW and Vic) and species 9 (WA, SA, Qld and NSW). Lebel *et al.* (2014) noted that three of the four phylogenetic species of *Stephanospora* occurring in New Zealand were found in the northwest of the North Island, with the fourth species found on the northern tip of the South Island – this pattern revealing 'an apparent gap in the presence of species in the central and southern parts of the North Island'. Within Australia, some species were localised, such as *Stephanospora flava* from Tasmania, but *S. sheoak* occurred in Queensland, Victoria and Tasmania. (See also 'Infraspecific Geographic Structure and Population Genetics' in this chapter.)

Sampling Issues

It is probable that where sample numbers are low, ranges for some phylogenetic species will be extended with additional sampling. There is certainly a high probability of extensions at the regional level, where often there are samples from only one or few countries within a continent. However, some of the aforementioned studies analysed numerous samples and strong geographic patterns held up, as with *Neurospora*, where PS4 was the only one of eight phylogenetic species within the *N. discreta* complex to be found in western North America, a focus of sampling for the genus (Dettman *et al.* 2006; Taylor *et al.* 2006).

Widespread Phylogenetic Species

In contrast to the foregoing studies, the lack of geographic structure among fungal species has been detected infrequently. Among the few examples of widespread phylogenetic species, most are associated with habitats modified by humans or their distribution is human mediated.

Aspergillus are moulds that produce colonies covered by minute, asexual spores, and several cosmopolitan species have been confirmed as lacking phylogeographic structure. *Aspergillus fumigatus* is found globally, in soil, compost and other decomposing organic material, and also as a human pathogen, facilitated by the ability to grow at higher temperatures. Using five loci, Pringle *et al.* (2005) found two phylogenetic species within a global sample of *A. fumigatus*, but there was no correlation between genotype and geography in either species. The lack of geographic structure within the more common phylogenetic species was corroborated, using three different loci, by Rydholm *et al.* (2006). Among fungi, Pringle *et al.* (2005) considered that *A. fumigatus* was the 'first to be shown by population genetic evidence to have a global population structure with no endemism'. Although *A. fumigatus* has spores at the small end of the range for fungi, at around 2 μm in diameter, the lack of local adaptation was suggested as not so much due to dispersal ability (which appears to exist in many other fungi) but rather was connected to the ability to grow in composting vegetation, a habitat greatly expanded by human activity (Pringle *et al.* 2005; Taylor *et al.* 2006); and indeed, the species may have only recently spread throughout the world (Rydholm *et al.* 2006). Another member of the genus, *Aspergillus flavus*, has been isolated from a wide variety of terrestrial, marine and freshwater habitats, including from corals, soil, seeds and dust, and is also a human pathogen. Using amplified fragment length polymorphisms (AFLPs) across a worldwide sample, Ramírez-Camejo *et al.* (2012) found no phylogeographic structure in *Aspergillus flavus* and no correlation between genetic and geographic distances, and nor was there any differentiation between marine and terrestrial isolates. Nevertheless, they concluded that 'everything is everywhere, but not all the time' because at the local level they found examples of predominance of particular subclades. However, the sampling was unbalanced across continents and local sites and further analysis of a wider data set is warranted. Although many *Aspergillus*, including *A. fumigatus* and *A. flavus*, have no known sexual stage and most reproduction is expected to be clonal, some evidence of recombination was recovered in both species (Pringle *et al.* 2005; Ramírez-Camejo *et al.* 2012). Selective sweeps are a potential cause of reduced genetic variation, but were considered unlikely in *A. fumigatus*, and the lack of population genetic differentiation on a global scale was considered more consistent with a bottleneck event or recent spread and/or continual gene flow (Rydholm *et al.* 2006).

Another example of a widespread mould is *Wallemia sebi*, a xerotolerant basidiomycete commonly found on food products and in house dust, within which Jančič *et al.* (2015) and Nguyen *et al.* (2015) delimited four phylogenetic species, including three that were novel. Although one species was found in temperate regions and another in the tropics, the other two were distributed worldwide, and within species there was no 'strong link between geography and haplotype' (Nguyen *et al.* 2015). In the moulds of the *Trichoderma harzianum* complex, Chaverri *et al.* (2015) delimited 14 phylogenetic species, many of which were segregated according to continent, although two species were cosmopolitan, both of which are utilised in commercially available biocontrol applications. Cosmopolitan moulds in genera such as *Aspergillus* and *Wallemia* are often strongly associated with human-modified or created habitats, and in the case of *Trichoderma*, deliberately introduced into ecosystems. They offer model systems for exploring spore dispersal dynamics at a global scale as well as the timing and pattern of evolution in relation to human activity.

An interesting exception to geographic structure in widespread fungi is the *Phialocephala fortinii* complex (also including *Acephala applanata*), endophytic ascomycetes that are commonly found in the roots of conifers and Ericaceae in a wide range of ecosystems. Some 21 phylogenetic species have been delimited on the basis of multigene analyses. Across the Northern Hemisphere there is no biogeographic pattern and the community of species in this complex does not show a distance decay, even though some sampling sites were separated by as much as 12,000 km (Queloz *et al.* 2011). Human-mediated movement arising from the planting of colonised nursery plants was suggested as a possible dispersal factor, but the lack of a distance–decay relationship was present even when plantation sites were excluded from analysis (Queloz *et al.* 2011).

Phylogenetic Species and Speciation

The fact that many phylogenetic species of fungi turned out to have restricted distributions, at least at continental scales, assists in resolving issues about the stability of fungal species over time raised by earlier authors. Redhead (1989) observed that species listed as occurring across southern hemisphere landmasses by Horak (1983) would be at least 100 million years old if they resulted from Gondwanan vicariance. Rajchenberg (1989) considered it 'astonishing' that there was low or absent morphological variation between what appeared to be the same species across Australasia and South America, separated since the Cretaceous. Johnston (2010) suggested that species with such wide distributions probably represent examples of long-distance dispersal. Re-examination of all cases of Gondwanan species as elaborated by Horak (1983) and others is clearly required to delimit phylogenetic species, before advancing biogeographic explanations. Data from one or two DNA regions can be sufficient to show that supposed trans-Pacific species cannot be conspecific, as with material from New Zealand identified as the South American *C. magellanicus* (Peintner *et al.* 2004). However, unless there is sufficient sampling of taxa and loci, analyses can be inconclusive, as with the phylogeography of the *Tylopilus balloui* group around the Pacific Rim (Halling *et al.* 2008).

Some initial studies on fungal phylogeny that utilised molecular data relied on data from a single gene region. In such cases it must be emphasised that even though lineages were considered 'phylogenetic species' by the authors of studies (e.g. Martin *et al.* 2002; Zervakis *et al.* 2004) or by others such as Taylor *et al.* (2006), molecular data comprising only one or two gene regions do not allow for the detection of genealogical concordance. The ITS region is often among those sampled, and because this region does satisfactorily act as a barcode in many fungal genera, it may be that true phylogenetic species were being detected, but it will be instructive to revisit these early studies using multigene data to more rigorously delimit phylogenetic species, and/or to check the divergence of the ITS against barcode calibrations.

The recognition of phylogenetic species sometimes led to taxonomic changes through the formal application of existing or novel names (where required) to each lineage, as with *Fusarium*, *Neurospora* and *Pleurotus*. However, in other cases, the phylogenetic species were not formally named, apart from by the use of whatever names already existed, as with the *Lentinula edodes* group, *Schizophyllum commune* and *Panellus stypticus*. When formal names were applied to phylogenetic species, the strongly correlated geographic patterning was often explicitly mentioned as providing support for such an action (Zervakis *et al.* 2004).

In relation to speciation modes (Kohn 2005; Giraud *et al.* 2008), many of the examples of geographic structure among sibling phylogenetic species of fungi have nonoverlapping distributions, consistent with allopatric speciation, either as a result of vicariance or dispersal followed by diversification and range expansion. However, there are some cases where sister phylogenetic species are sympatric, such as that of *Amanita muscaria*, in which Geml *et al.* (2006) initially delimited three phylogenetic species that co-occurred in Alaska and subsequently (Geml *et al.* 2008) added five more, three of which were found only on Santa Cruz Island off the coast of California. Geml *et al.* (2008) suggested that the evolution of phylogenetic species in *Amanita muscaria* in sympatry could be explained, for at least some of the species, by adaptation to different ecological niches (such as altitudinal range) within the same broad distributions, and possibly also by microallopatric separation, as with different hosts, soil horizons or microclimates. Taylor *et al.* (2006) discussed intrinsic barriers to mating between sympatric members of different species, and pointed out that matings between different phylogenetic species of *Neurospora* were less successful in sympatric matings in comparison with allopatric matings, and that the barriers occurred postfertilisation, as elaborated by Dettman *et al.* (2003b).

Fungi Are Not Everywhere

Taylor *et al.* (2006) reviewed the geographic limits of fungi as a reaction to the contention of some authors such as Finlay (2002) that both prokaryotic and eukaryotic microbes had global ranges. The concept of wide ranges was first articulated for bacteria by Baas Becking (1934) as '*Everything is everywhere*, but, *the environment selects*', building on the ideas of Beijerinck (1913). The concept is conveniently shortened as *Everything is everywhere*. However, de Wit and Bouvier (2006) pointed out that the 'but' (sometimes omitted) and the following phrase 'the environment selects' are critical in interpreting Baas Becking's aphorism, and while microbes may well have massive populations with the potential to disperse widely, it is selection by the environment that contributes to spatial variation in microbial distribution (Martiny *et al.* 2006; Bass and Boenigk 2011).

Despite the nuances of the original EIE concept, Finlay (2002) proposed a transition between microscopic *ubiquitous species* and macroscopic *species with biogeographies*, with the transition between microbes and macrobes at a body size of around 1–10 mm. While individual fungal organisms vary from unicellular to very large (with mycelia of *Armillaria* extending over hectares), propagules are generally minute, and therefore would fall within the ubiquitous species in the model of Finlay (2002). However, Foissner (2006) pointed out that eukaryotes such as macrofungi occupy distinct areas, despite minute propagules. Moreover, the frequent intimate relationships between fungi and macrobes should mitigate against such fungi being globally distributed (Taylor *et al.* 2006).

Taylor *et al.* (2006) used case studies on fungi such as *Lentinula*, *Neurospora* and *Schizophyllum* to demonstrate that species recognition criteria are key to understanding biogeography, and that phylogenetic species of fungi often have noncosmopolitan distributions (see also 'Case Studies on Geographic Structure of Phylogenetic Species' and 'Further Examples of Geographic Structure in Phylogenetic

Species of Fungi' in this chapter). They concluded that apparent ubiquity in eukaryotic microbes such as fungi could be a consequence of recognising morphological or biological species, rather than phylogenetic species, which were often more narrowly circumscribed. Indeed, morphological species in fungi have been suggested to be closer to higher-level taxa in macrobes (Martiny *et al.* 2006; Taylor *et al.* 2006; Bass and Boenigk 2011). Smaller and less complex organisms (with fewer cells and cell types) were considered by Taylor *et al.* (2006) to have fewer morphologically differentiated characters that evolved more slowly than in more complex organisms, which explained mismatches between morphological and phylogenetic species in fungi. Taylor *et al.* (2006) also concluded that genetic isolation (leading to phylogenetic species) precedes reproductive isolation (which defines biological species). Among the case studies discussed in the section above on 'Most Phylogenetic Species are Geographically Restricted' there are several where biological species were found to contain more than one, and as many as eight, phylogenetic species.

Further rebuttals of the EIE concept can be found in the volumes *Protist Diversity and Geographical Distribution* (Foissner and Hawksworth 2009), including contributions on protists, in which a 'moderate endemicity model' was proposed (Foissner 2009), and on slime moulds (a reprint of Stephenson *et al.* 2007); and *Biogeography of Microscopic Organisms: Is Everything Small Everywhere?* (Fontaneto 2011), including contributions on lichen fungi and their photobionts (Werth 2011), cactophilic yeasts (Gantner 2011) and mushrooms (Geml 2011). The latter study used coalescent analyses to demonstrate a strikingly greater level of intercontinental migration in arctic–alpine species compared with boreal–temperate species (Geml 2011). In addition, Sato *et al.* (2012) provided an interesting perspective by studying sequences from the mushrooms at one locality in Japan and inferring their global distribution using sequences in GenBank of known geographic origin (taking into account the undersampling of taxa and regions). When locations with similar climate and host vegetation to the Japanese reference site were compared, there was a marked distance decay for the suite of 56 fungi. Overall, the distribution of the ECM species was less extensive than saprotrophs, the geographic constraint of the former suggested as being due to host specificity (Sato *et al.* 2012).

Martiny *et al.* (2006) offered a more complex interpretation of microbial biogeography than a simple EIE concept. They encouraged the investigation of hypotheses about the relative contributions of environmental and historical effects and how such contributions might be reflected in relationships between biotic similarity and, respectively, environmental similarity and geographic distance. Bass and Boenigk (2011) also emphasised the 'crucial but potentially elusive distinction between historical and ecological biogeography'. In addition, Martiny *et al.* (2006) suggested that, rather than a strict allometric (size-dependent) relationship, there may be wider variation for passively dispersed microbes in the rates of underlying processes such as colonisation, diversification and extinction, leading to wider variation in dispersal rate, population density and range size, compared with macrobes. There is considerable scope for the testing of such hypotheses, especially with metagenomic data (see 'Metagenomic Biogeography' in this chapter).

The blanket application of EIE to both prokaryotes and eukaryotes in the 2000s by authors such as Finlay (2002) did stimulate a flurry of reactions by Taylor *et al.* (2006) and the other authors cited previously, but apart from these rebuttals of EIE as applied to fungi, the EIE concept actually had little influence on fungal biogeography during the twentieth century. The most important paradigm shift for fungal biogeography was the rigour in species delimitation introduced by the application of a PSC in combination with the availability of molecular data as material for GCPSR. Coincidentally, the fact that the same molecular data sets could be used for both species delimitation and for investigations of biogeography was a boost for fungal biogeography.

Frequent Host Shifts in Ectomycorrhizal Fungi

In mutualistic associations, hosts complicate the interpretation of fungal distributions, because fungal symbionts may be restricted to a certain area due to host specificity with geographically restricted hosts or due to historical factors, such as when the fungi evolved within an already isolated geographic area. In addition, host shifting has been advanced as part of explanations for how particular fungi came to be in particular areas. Malloch *et al.* (1980) postulated that fungi associated with *Eucalyptus* and other Myrtaceae in the Southern Hemisphere spread there with *Nothofagus* or its ancestors, and Horak (1983) suggested that switching between *Nothofagus* and other Fagales in New Guinea facilitated the northward spread of agarics of southern origin. A mix of host tracking (cospeciation or coevolution) and host shifting has been detected in both mycorrhizal (see Rochet *et al.* 2011 for ECM fungi and Betulaceae) and parasitic fungi (see Roy 2001 for rust fungi on crucifers; McTaggart *et al.* 2015 for *Endoraecium* on *Acacia*; and '*Cyttaria*: Evolution of Ideas about Coevolution of Fungi in Relation to Time and Area' in this chapter). Several studies have explored the roles of host specificity and historical events by using the dense sampling of ECM lineages containing austral representatives, in particular the Hysterangiales, Inocybaceae and Sclerodermatineae.

Hysterangiales

Hysterangiales is an order of truffle-like fungi that are dispersed primarily by mycophagous mammals. Hysterangiales are found worldwide and are mostly ECM with trees, with each fungus species usually associated with a single plant family, genus or species. Using a five-gene phylogeny, Hosaka *et al.* (2008) recovered six main clades of Hysterangiales with strong biogeographic patterns (Figure 8.5). The most basal clade comprised non-mycorrhizal species from both hemispheres; all other species were mycorrhizal, indicating a single origin for this trophic mode. Within the ECM clade, three basal clades (Gallaceaceae, Mesophelliaceae and *Hysterangium* I) were exclusively of Southern Hemisphere species, while Northern Hemisphere taxa were restricted to two terminal clades: *Hysterangium* I (Northern Hemisphere only) and *Aroramyces* (Northern and Southern Hemisphere). With regard to hosts, many closely related species did not share host families, and only one clade strictly associated with one host (*Mesophelliaceae* with Myrtaceae). Frequent host shifts were inferred (not necessarily between closely related plant families) and shifts between distantly related plant families, such as Pinaceae and Fagaceae, were observed frequently (Figure 8.6). Given earlier hypotheses about host jumps from *Nothofagus* (Horak 1983), it was interesting that there were no shifts between Nothofagaceae and Fagaceae. Hosaka *et al.* (2008) conclude that the 'biogeography of Hysterangiales has been shaped by host availability, that is co-occurrence with host trees in a common geographic area, and not by a pattern of speciation based on strict host tracking'.

Inocybaceae

Inocybaceae is a family of ECM mushrooms (and a few truffles) with more than 700 species globally and associates with at least 19 plant families. Individual species are nearly always regionally endemic in the tropics and Southern Hemisphere, but more widespread in the Northern Hemisphere (Matheny *et al.* 2009). For 186 species of Inocybaceae, Matheny *et al.* (2009) concluded that early members of the family were Palaeotropical in origin and were associated with angiosperms, most likely during the Cretaceous. Some 21 unambiguous transitions were inferred between host associations coded as angiosperm, conifer or generalist. However, switches to conifer or generalist associations did not occur at the earliest until the Palaeogene, even though conifers were present at the time of origin of the family. Switches to Nothofagaceae and Dipterocarpacae were also inferred as relatively recent.

Sclerodermatineae

Sclerodermatineae is a suborder of the Boletales, containing mostly ECM genera that produce either puffball or bolete (pored mushroom) sporophores. Wilson *et al.* (2012) used a supermatrix approach to construct a phylogram and dispersal–extinction–cladogenesis (DEC) analysis for ancestral range reconstruction under restricted or relaxed (allowing for dispersal between nonadjacent areas) models and with different constraints on the ancestral range (less than two or three areas). The more than 150 species of the suborder worldwide are almost always found in no more than two areas (at the continental scale) and are associated with 16 host families, including Myrtaceae and Nothofagaceae in Australasia. The reconstruction of an ancestral host yielded Pinaceae or one of the two groups within eurosids, with an angiosperm host considered most likely, and with frequent host switches. Wilson *et al.* (2012) concluded, from a synthesis of divergence dating and ancestral range and host reconstruction, that the Sclerodermatineae originated in Asia and North America during the Late Cretaceous and diversified in the Early Cenozoic.

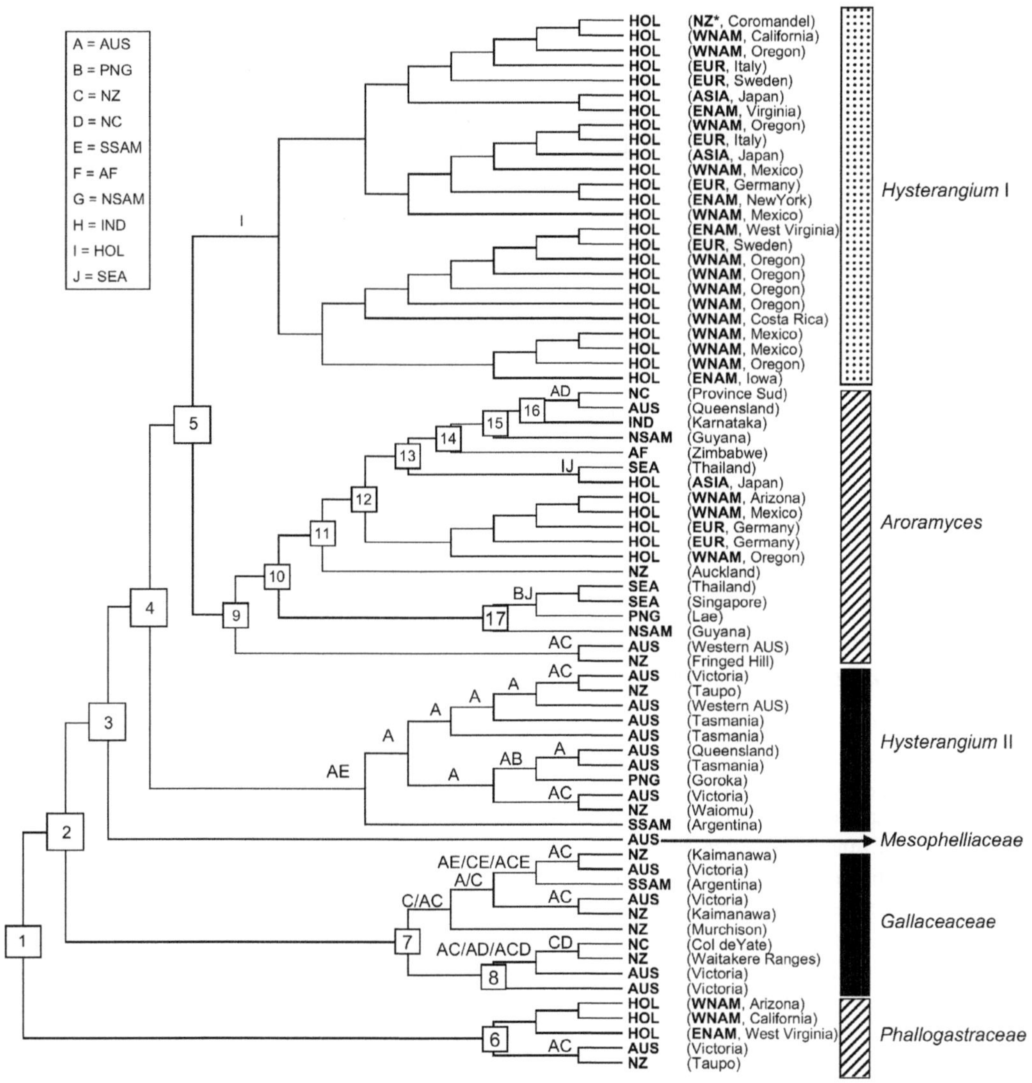

FIGURE 8.5 Ancestral area reconstruction in Hysterangiales using a taxon–area cladogram. Terminal branches are individual taxa of *Hysterangium*, *Aroramyces*, Gallaceaceae and Phallogastraceae, except that 23 taxa of Mesophelliaceae (all AUS) are collapsed into one branch. Terminal branches are labelled with areas of endemism, followed by more specific locality information. Holarctic is subdivided into four unit areas, but these were not used for the analyses. Characters above branches indicate results of ancestral area reconstructions using DIVA (see original caption for further information). Numbers in squares (1–17) indicate nodes that had different reconstructions with different settings. Area abbreviations are as follows: AF: Africa; AUS: Australia; HOL: Holarctic; IND: India; NC: New Caledonia; NSAM: northern South America; NZ: New Zealand; PNG: New Guinea; SEA: Southeast Asia; SSAM: southern South America. Divisions of HOL are as follows: WNAM: western North America; ENAME: eastern North America; Asia and EUR: Europe. Originally published as Figure 3 in Hosaka, K., *et al.*, *Mycological Research* 112, 448–462, 2008. Reproduced with permission.

Distributions outside of the ancestral area, such as *Calostoma* with Myrtaceae and Nothofagaceae in Australasia, occurred no more than 33 mya (Wilson *et al.* 2012, from their Figure 3), postdating the breakup of Gondwana, and were explained by long-distance dispersal, which required host switching. A 'generalist' ECM habit, as earlier suggested for ancestral *Pisolithus* by Martin *et al.* (2002), was considered to have facilitated the wide distribution of many groups within the suborder. However, for

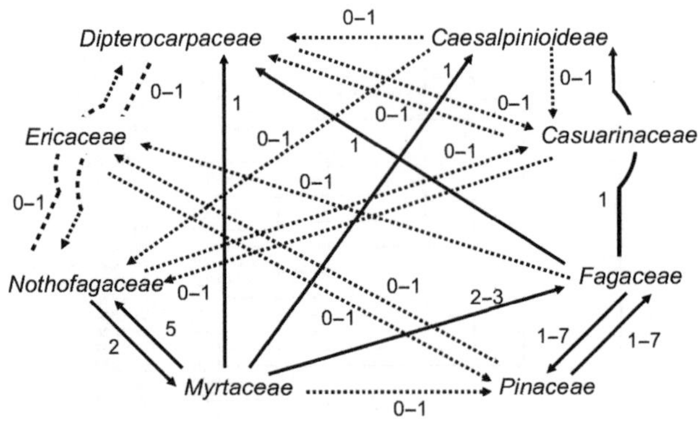

FIGURE 8.6 Frequency of ECM host shifts in Hysterangiales inferred from unweighted parsimony reconstructions. Arrows indicate the direction of host shift. Numbers on arrows indicate the minimum and maximum possible steps. Host shifts with possible zero occurrences are indicated by dotted arrows. ECM hosts are coded in eight states, each state corresponding to host family. Originally published as Figure 5 in Hosaka, K., *et al.*, *Mycological Research* 112, 448–462, 2008. Reproduced with permission.

Calostoma, the limited host range (only three families) was suggested as a factor in the more limited distribution of the genus in comparison with others in the suborder (Wilson *et al.* 2012).

Host Range and Host Switching

While the term *host generalist* has been used, such as for lineages in Sclerodermatineae (Martin *et al.* 2002; Wilson *et al.* 2012), individual species often have narrow host ranges at the plant family level, as far as is known, and so it is probably the ability of species to switch hosts that would have contributed to the ability to colonise new areas, rather than a preexisting wide host range. However, it should be noted that many ECM fungi are known from single collections and knowledge of hosts is by no means complete. Confirmation of hosts in the field is especially difficult for ECM fungi whose sporophores are only connected to host roots by microscopic hyphae. For Sclerodermatineae, Wilson *et al.* (2012) used stringent coding with hosts, such as when an association was confirmed by in vitro synthesis or the molecular identification of fungi directly from host ECM roots, but in some analyses also allowed for liberal coding when an association was observed in the field but not proven. The distinction between realised and fundamental niches in host associations needs to be explored further, given reports of successful in vitro syntheses of some ECM fungi with hosts not encountered in the wild. Genomic approaches are uncovering the underlying mechanisms of mycorrhizal symbiosis (Martin *et al.* 2008), and research on the genetic basis of host specificity and its evolution through time will be instructive.

These global studies on Hysterangiales, Sclerodermatineae and Inocybaceae show that host relationships have not constrained the distribution of lineages over time as much as initially conceived in Southern Hemisphere fungal biogeography. Another global study, on *Russula*, concluded that host switching at the level of Pinaceae versus angiosperms has been 'an important driver for diversification' (Looney *et al.* 2016). At a local scale, Roy (2001) pointed out that the millions of spores produced by an individual rust fungus are most likely to travel to the nearest nonhost plant, rather than more geographically distant, but phylogenetically related, plant species, thus constantly providing opportunities for host switching at a local scale. Host switching in ECM fungi also appears to have been occurring over large distances, so that when long-distance dispersal did occur, in at least some cases, establishment with a new host in the new area was possible.

Dating Fungal Phylogenies

Before discussing the dating of biogeographic events (in the following sections), it is necessary to consider the methods of dating fungal phylogenies, which has been carried out by placing dates on nodes derived from (1) fossils, (2) geological events and (3) symbiont lineages.

Fossil Fungi

There are fossilised fungi, albeit less common and often rather difficult to place taxonomically in comparison with other biota such as plants. In *Fossil Fungi*, Taylor *et al.* (2015) extensively catalogued and illustrated fossils from all major lineages of fungi and fungoid organisms, including a discussion of taxonomic placement and likely trophic modes. Truswell (1996) reviewed the fossil record of fungi in Australia, noting that fossil fungi from Australia are mostly from the Cretaceous or younger, and are often associated with plant leaves. Walker (1996) also discussed fossil fungi, highlighting several groups of biogeographic interest, including the leaf-inhabiting ascomycete *Vizella* and seemingly close relatives in genera such as *Manginula*. These two genera are almost exclusively found in the tropics and the Southern Hemisphere, on the basis of living and fossils examples, including several of both from Australia. Authors writing about fossil fungi often emphasise the interconnectedness of fungi with other organisms (Walker 1996; Taylor *et al.* 2015), and indeed, the movement of plants to land has been suggested as being assisted by the formation of symbiosis with fungi (Pirozynski and Malloch 1975).

Examples of fossil fungi include Chytridiomycota from the Devonian, hyphae with clamp connections attributable to the Basidiomycota as early as the Carboniferous and mushrooms such as *Archaeomarasmius* from amber as old as the Early Cretaceous (Taylor *et al.* 2015). A key fossil belonging to the Ascomycota is *Palaeopyrenomycites* from Devonian Rhynie chert dated at approximately 410 mya. It was found within the tissue of an early lycopod-like plant *Asteroxylon* and appears to have an associated asexual stage (Taylor *et al.* 2015).

Calibrations

There are significant issues around the use of the limited fossil record for fungi in calibrating phylogenies, including the dating of associated rocks, attaching fossils to the correct node in phylogenies and substitution rate variation across phylogenies (Berbee and Taylor 1993, 2001, 2010; Taylor and Berbee 2006; Taylor *et al.* 2015). Relaxed molecular clocks can address rate heterogeneity, although O'Donnell *et al.* (2013) point out the lack of relatively recent fossils for calibration.

Molecular clocks constructed using a variety of techniques and calibration points have yielded widely different ages for the origin of fungi, between 660 million and 2.14 billion years ago, and estimations for more recent nodes such as the age of the Ascomycota also vary widely (Lücking *et al.* 2009 and see reference therein). While *Palaeopyrenomycites* is clearly attributable to the Ascomycota, it has been placed in various lower-level taxa including the Pezizomycotina, and with variant options within this subphylum as to whether it is in the Pezizomycetes or Sordariomycetes. After a detailed analysis of its position, Lücking *et al.* (2009) place it at either the stem base or crown of Pezizomycotina, yielding an origin of Fungi between 750 and 1090 mya. Their most conservative placement at the stem base suggests an origin of early chytrids and Zygomycota around 600 mya (in parallel to the evolution of freshwater algae), Glomeromycota at 590 mya, and Ascomycota and Basidiomycota at around 500 mya (more or less in parallel to primitive land plant fossils) (Lücking *et al.* 2009).

Calibrations in the studies discussed in the sections 'Analyses Including Time' through '*Cyttaria*: Evolution of Ideas about Coevolution of Fungi in Relation to Time and Area' used a small range of fungi fossils, and often applied secondary calibrations from a set representative of higher-level taxa, transferring an age and associated error for a node in this first tree (in their own or a previous study) to a node on the phylogram for the set of taxa of interest. Dating from plant associates was utilised by Takamatsu and Matsuda (2004), who calibrated a phylogeny for Erysiphales (powdery mildews) by the time of appearance of key host groups in the Asteraceae (ultimately derived plant fossils). For the ECM Tuberaceae, Bonito *et al.* (2013) set the maximum age constraint in the prior distribution to that of Pinaceae, the oldest known lineage of obligate ECM hosts. Comparisons of fungal and plant ages have also been used to cross-check calibration methods, as with Matheny *et al.* (2009), who rejected one potential calibration because it produced dates for some nodes that were significantly older than the angiosperm families associated with the particular lineages (from ancestral state reconstruction). Similarly, Cai *et al.* (2014) chose to use *Palaeopyrenomycites* for calibration, rather than the more recent Basidiomycete fossil *Quatsinoporites*, because using the latter not only gave an age for the Ascomycota/

Basidiomycota divergence considered too young, but also an age for the origin of *Amanita* around 80 mya, which was much younger than the appearance of hosts such as Pinaceae. While the ages of geological events are often compared with inferred biogeographic events, the dates of such events are less often used as calibrations in the first place, although Matheny and Bougher (2006) applied the date of formation of the Nullarbor Plain to the node subtending two species of *Auritella*, restricted to western and eastern Australia, respectively.

Some studies have detected area-specific differences in the rate of evolution of specific loci and in the rate of speciation: the ITS evolution rate in the /sebacina lineage was higher in the tropics (Tedersoo *et al.* 2014b), while the speciation rate was higher in temperate regions for *Amanita* section *Caesareae* (Sánchez-Ramírez *et al.* 2015) and *Clavulina* (Kennedy *et al.* 2012), and there was a 'higher net diversification rate in extratropical lineages' in *Russula* (Looney *et al.* 2016).

Large confidence limits that often occur for inferred dates on nodes in fungal chronograms are a reminder that care should always be taken when interpreting absolute ages in such chronograms, a sentiment reinforced by Berbee and Taylor (2010), who conclude that 'fossil evidence remains scanty and substitution rates change chaotically from lineage to lineage, and together these two factors conspire to produce artefacts that skew divergence time estimates'.

Time and Area Analyses in Fungal Molecular Biogeography

Some of the initial studies utilising molecular data that showed geographic patterns, as previously described (see 'Most Phylogenetic Species of Fungi Are Geographically Restricted' in this chapter), did discuss historical factors that may have led to current distributions of taxa and genetic variation within taxa, such as vicariance, long-distance dispersal, dispersal via land bridges or expansion from glacial refugia (e.g. Vilgalys and Sun 1994; O'Donnell *et al.* 1998; Jin *et al.* 2001; Zervakis *et al.* 2004). However, hypotheses about the origins of distributions remained speculative in the absence of specific analyses in relation to area and time.

Analyses of Areas

In terms of areas, many of the molecular phylogenetic studies elaborated previously merely indicated distribution for the tips of molecular phylogenies, either for individual collections or for clades, which certainly demonstrated striking patterns, but did not interrogate process. However, the mapping of distribution onto the branches of cladograms and the determination of the likely geographic origin of clades using parsimony was carried out for *Pleurotus* (Vilgalys and Sun 1994), *Fusarium* (O'Donnell *et al.* 1998; O'Donnell 2000) and *Lentinula* (Hibbett *et al.* 1995), with Hibbett (2001) also applying geographic constraints to clades in the latter genus. Morgado *et al.* (2013) calculated phylogenetic beta diversity using the UniFrac metric to investigate geographic structure across clades in *Entoloma*. Nested clade analysis of haplotype networks was applied to investigate the relationship between phylogenetic diversity and geography in *Schizophyllum* (James *et al.* 2001), for *Histoplasma* (Kasuga *et al.* 2003), in Southern Hemisphere *Ganoderma*, where isolation-by-distance was suggested (Moncalvo and Buchanan 2008), and within and between species of the Northern Hemisphere *Amanita muscaria* group, where contiguous range expansion and allopatric fragmentation were inferred (Geml *et al.* 2006). For further studies that integrate area and time, see 'Analyses Integrating Area and Time' and '*Cyttaria*: Evolution of Ideas about Coevolution of Fungi in Relation to Time and Area' in this chapter.

In the Hysterangiales, Hosaka *et al.* (2008) utilised dispersal–vicariance analysis, with 10 biogeographic areas, including Australia (with the highest diversity of 44 species), New Zealand, New Caledonia and New Guinea. Even with varying costs for events (vicariance, duplication, dispersal and extinction), the observed distribution patterns could not be fully explained by Pangaean breakup. Hosaka *et al.* (2008) concluded that the ECM lineages of Hysterangiales originated in Australia or Eastern Gondwana with subsequent extension to the Northern Hemisphere. Most scenarios resulted in the inference of a greater number of dispersal events in comparison with vicariance events, which was surprising given that most members of the lineage form hypogeal, truffle-like sporophores that are currently dispersed by mammals (and perhaps originally by insects) and thus might be expected to have limited spore dispersal compared with epigeous mushrooms. Within one of the two main clades of *Hysterangium*, several New Zealand taxa were deeply nested within clades otherwise comprising Australian taxa, suggestive of long-distance dispersal from Australia to New Zealand. Hosaka *et al.* (2008) did not directly add the time element to their analyses, but noted that Myrtaceae was reconstructed as the most ancient host of ECM Hysterangiales, and suggested that the fungi are no older than the origin of their hosts, around 90 mya, which is inconsistent with Gondwanan breakup contributing to current geographical patterns in this group of fungi.

Areas used in ancestral area reconstruction for fungi tended to be broad areas particular to the group under study, at the scale of continents (O'Donnell *et al.* 1998; Vilgalys and Sun 1994) or countries (Hibbett *et al.* 1995) or a mix of the two (O'Donnell 2000; Hosaka *et al.* 2008). In their study of *Amanita* section *Caesareae*, Sánchez-Ramírez *et al.* (2014) utilised more precise areas 'marked by the presence of natural barriers deemed to present limits to dispersal' and with areas being 'continuous forested regions' within which 'dispersal is possible', so that, for example, four areas were recognised in North America, including the Pacific Northwest and the Cordillera de Talamanca in Costa Rica. There have been no comparisons of areas used to describe fungal distribution as applied to different groups of fungi.

Analyses Including Time

Initial analyses calculated divergence time based on the assumption of a constant molecular clock, calibrated by supposed vicariance events. For *Flammulina*, where the New Zealand *F. stratosa* was reconstructed as sister to other taxa in the genus, all from the Northern Hemisphere, Hughes *et al.* (1999) calibrated a molecular clock for the ITS region based on the divergence of *F. stratosa*, assumed to correspond to the opening of the Tasman Sea around 70 mya. Once calibrated, while noting that the use of ITS as a molecular clock was 'fraught with suppositions and uncertainties', Hughes *et al.* (1999) used the clock to date the disjunction between North American and Eurasian lineages at 9 mya. For sister lineages of *Pleurotus tuber-regium* in Australasia/Indonesia and Africa, Isikhuemhen *et al.* (2000) calculated the rate of divergence of the ITS, calibrated by the split of Africa from Gondwana 150 mya, but found that the rate was surprisingly low, suggesting that the divergence was of more recent origin. Further studies using phylogenies calibrated by various means have found that divergence dates for fungal lineages in the Southern Hemisphere are almost always too late to invoke vicariance due to Gondwanan breakup.

In studies on divergence ages in *Lentinula* and *Ganoderma*, a clock-like model was explicitly tested, and could not be rejected for at least some of the analysed markers. For *Lentinula*, Hibbett (2001) examined alternate hypotheses for the origin of an Old World–New World disjunction in the genus, resulting from the fragmentation of either a Gondwanan or Laurasian ancient range. When nodes were constrained to ages of hypothesised Gondwanan vicariance (for the South America–Australia split at 100 mya and Australia–New Zealand split at 80 mya) the ages of basal nodes were inconsistent with dates from fossils (*Archaeomarasmius* for the euagarics) or the molecular clock estimate of the age of the homobasidiomycetes from Berbee and Taylor (1993). Consequently, a Laurasian origin was favoured, with dispersal by diffusion suggested as occurring from North to South America and from Asia to Australasia, along with at least one long-distance dispersal from Australia to New Zealand. For the white-rot polypore *Ganoderma*, Moncalvo and Buchanan (2008) identified eight clades in the *Ganoderma applanatum-australe* group, all strongly associated with geography, including one from the Southern Hemisphere (including isolates from Australia, New Zealand, South Africa and South America) and others from the Northern Hemisphere. When the basal node was constrained to the time of the Gondwana–Laurasia separation (120 mya), estimates for the diversification time of the entire polypore clade were twice those suggested by Berbee and Taylor (2001). However, the use of the nucleotide substitution rate from Berbee and Taylor (2001) suggested an origin of the group around 30 mya. This date, along with the finding of identical ITS haplotypes in South America and in South Africa, Australia and/or New Zealand, led Moncalvo and Buchanan (2008) to conclude that 'only dispersal can explain the phylogenetic patterns'.

For the *Histoplasma capsulatum* group, Kasuga *et al.* (2003) estimated the radiation of the group commencing at 3.2–13.0 mya, based on calibration using the *Blastomyces–Histoplasma* split as estimated by Kasuga *et al.* (2002) at 31.8 or 127.8 mya, depending on whether the algorithm used for divergence estimation assumed rate constancy. The origin of the group was suggested in South/Central America at a time of warmer climate, with dispersal to other areas such as Australia and Africa. Species in the group are currently found in tropical rainforests or temperate forests, and low genetic variation in the Australian phylogenetic species was suggested to have been caused by a population bottleneck during the Last Glacial Maximum, when the modern Australian sites would have been semidesert or arid scrub (Kasuga *et al.* 2003).

For *Fusarium*, O'Donnell *et al.* (1998) and O'Donnell (2000), without detailing calculations, postulated a 110 mya date of origin for the genus, consistent with their Gondwanan vicariance hypothesis in the *Gibberella fujikuroi* and *Fusarium solani* complexes. O'Donnell *et al.* (2013) utilised a dated phylogeny under a relaxed molecular clock, calibrated by seven fungal taxa as exemplars of groups dated by Sung *et al.* (2008) and Gueidan *et al.* (2011). The origin of *Fusarium* was inferred at 91 mya, and diversification of the fungal lineage was considered to have been associated with angiosperm radiations in the Cretaceous and Miocene. The *Gibberella fujikuroi* species complex originated at 8.8 mya, much too late to be associated with Gondwanan vicariance, and instead the current distribution was suggested to result from long-distance dispersal (O'Donnell *et al.* 2013). For the discomycete *Torrendiella*, which has multiple species on both *Metrosideros* (within New Zealand) and *Nothofagus* (across the Southern Hemisphere), Johnston (2006, 2010) suggested an initial diversification 80 mya, consistent with Gondwanan vicariance, based on the timing of the radiation of *Metrosideros*-inhabiting species in New Zealand, assumed to be concurrent with divergence of the host genus around 30 mya.

Matheny and Bougher (2006) introduced the genus *Auritella* (Inocybaceae) for several species from Australia and Africa. A dated phylogeny was calibrated by placing the time of the formation of the Nullarbor Plain at 15 mya, at the split between the older of two pairs of disjunct east–west sister taxa: *A. geoaustralis* (WA) and *A. serpentinocystis* (NSW). The single African species (*A. aureoplumosa*) was sister to a clade containing the Australian species, and a nucleotide substitution rate derived from the calibration point was extrapolated to yield a divergence between the African species and the Australian clade of 86 mya. Matheny and Bougher (2006) concluded that this Late Cretaceous divergence was 'consistent with a vicariance event between the two southern hemisphere land masses'.

Analyses Integrating Area and Time

Recent analyses of fungal biogeography have combined area and time, commonly mapping area and ancestral area reconstructions onto chronograms (dated phylogenies). Matheny *et al.* (2009) applied this approach to the Inocybaceae, suggesting initial diversification in the Palaeotropics during the Cretaceous. Ancestral area reconstruction yielded several transitions from North Temperate and Palaeotropic areas to the Southern Hemisphere but none from the Southern Hemisphere to other areas. Current distributions were considered to have arisen through 'a mixture of several ancient vicariance events involving Africa, India and Australia, followed by Late Cretaceous to Palaeogene movement into the north temperate zone, and then later into southern temperate areas'. Several Southern Hemisphere lineages, including five in Australia, arose recently (in the Palaeogene); consequently, dispersal from the Northern Hemisphere was invoked (Matheny *et al.* 2009). New Zealand taxa all diversified at less than 35 mya, and were considered to have arisen after immigration from Australia or, in one instance, the Northern Hemisphere. The Inocybaceae chronogram was calibrated using a secondary calibration for the split between African and Australian species of *Auritella*, derived from a separate analysis of several subphyla of Basidiomycota, itself calibrated with two options for the split between Ustilaginomycota and the Agaricomycotina of 430 mya (Berbee and Taylor 2001) or 966 mya (Heckman *et al.* 2001). The temporally earlier calibration was rejected because inferred host associations of eight lineages were older than the age of the plant families concerned. Using the later calibration gave the origin of *Auritella* at 68 mya, reasonably close to the dating at 86 mya by Matheny and Bougher (2006). More recently, a new species of *Auritella* sister to all the other species of the genus was discovered in India, which 'pushes back in time the age … of the crown group node' of the genus (Matheny *et al.* 2012), exemplifying the effect that the discovery of new taxa from new areas can have on existing interpretations.

The more than 50 species of *Armillaria* occur either in the Holarctic or in the Southern Hemisphere. Within Australasia, seven species were distinguished by Coetzee *et al.* (2001, 2003) that mostly occur across wide areas – for example, *Armillaria novae-zelandiae* in Australia, New Zealand and South America, and *A. luteobubalina* in Western Australia, South Australia and Victoria as well as South

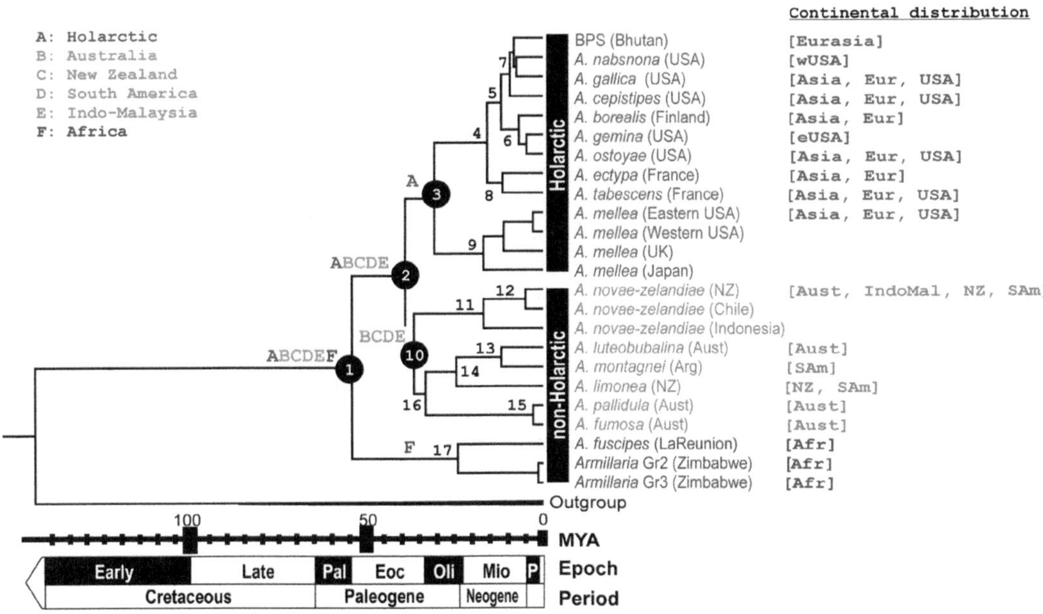

FIGURE 8.7 (See colour insert.) Bayesian chronogram for species of *Armillaria*, generated from analysis of combined ITS, LSU and EF-1α sequences, with optimum reconstruction of area distributions indicated (by numerals) at the main ancestral nodes on the tree. Key nodes are 1: Most recent common ancestor of extant *Armillaria* species; 3: Holarctic lineage; and 10: South American–Australasian lineage. The origin of the isolates is denoted in brackets next to the species name. The biogeographical distribution for each species is indicated in square brackets. Abbreviations are as follows: wUSA: western United States; eUSA: eastern United States; Eur: Europe; Aust: Australia; IndoMal: Indo-Malaysia; NZ: New Zealand; SAm: South America; Afr: Africa; P: Pliocene; Mio: Miocene; Oli: Oligocene; Eoc: Eocene; Pal: Palaeocene. Originally published as Figure 2 (doi:10.1371/journal.pone.0028545.g002) in Coetzee, M.P., *et al.*, *PLoS ONE* 6, 1–9, 2011; see original caption for further details. CC-BY.

America. Coetzee *et al.* (2003) initially proposed on the basis of phylogenetic relatedness that species occurring in both Australasia and South America originated 'before the breakup of Gondwana'. Subsequently, Coetzee *et al.* (2011) created a chronogram utilising a relaxed molecular clock with a secondary calibration procedure similar to that of Matheny *et al.* (2009) but utilising additional taxa (Figure 8.7). Ancestral area reconstruction gave all areas as possible at the root of the *Armillaria* phylogeny, but there was strong geographic structure, including lineages from Africa, Australasia–South America and the Holarctic, with the African clade diverging first and the latter two lineages being sister clades. The radiation of the genus was estimated to commence at 54 mya, 'within the Early Paleogene, postdating the tectonic breakup of Gondwana' and divergence of the Holarctic and Australasia–South America clade was estimated at 39 mya. The authors interpret the chronogram as supporting 'the radiation of *Armillaria* from the Southern Hemisphere, well after the fragmentation of Gondwana' and invoked stepping-stone and transoceanic long-distance dispersal, rather than vicariance, to explain current distributions in light of the phylogeny (Coetzee *et al.* 2011).

In the Serpulaceae, which contains saprotrophic fungi with resupinate (flat) sporophores such as dry rot (*Serpula lacrymans*) but also some ECM mushrooms and false truffles, Skrede *et al.* (2011) utilised dated phylogenies based on five nuclear DNA regions to suggest a Late Cretaceous origin, consistent with the Cretaceous origin of Pinaceae, a major host of the fungus. In the ECM *Austropaxillus*, they found a predominantly South American clade and an Australasian clade, as well as a species pair in Australia and New Zealand, all of which were suggested as originating via long-distance dispersal, due to recent divergence dates. While some divergences were consistent with Beringian vicariance scenarios, several other long-distance dispersal events were invoked as an explanation for the spread of two separate lineages of the family to southern South America from the hypothesised ancestral area in western North America (Skrede *et al.* 2011).

For the ascomycete truffle *Tuber* and allies, Bonito *et al.* (2013) found high levels of regional and continental endemism and identified a novel Southern Hemisphere lineage /gymnohydnotrya as sister to the Tuberaceae (almost all Northern Hemisphere). A chronogram constructed with a relaxed molecular clock was calibrated with an absolute rate of molecular evolution for the LSU locus taken from Takamatsu and Matsuda (2004), who derived this rate by calibrating the phylogeny of the Erysiphales (powdery mildews) based on the time of appearance of key host groups in the Asteraceae (the timing of which, in turn, came from the application of a molecular clock ultimately derived from an analysis with a minimum age calibrated by plant fossils). The divergence of /gymnohydnotrya and the Tuberaceae commenced 160 mya, early enough to be consistent with the splitting of Gondwana and Laurasia, but the divergence between the Australasian /labyrinthomyces and Northern Hemisphere / hoiromyces lineages was much later, around 94 mya, and thus dispersal was invoked to explain the latter distribution (Bonito *et al.* 2013).

For mushrooms in the *Cortinarius violaceus* group, Harrower *et al.* (2015) delimited eight phylogenetic species, each restricted to areas within Australasia (one species in Australia, one in New Zealand and one in both) or to various combinations of areas within Central and northern South America, or otherwise with *C. violaceus* in the strict sense occurring across Europe and North America. The diversification age of the group at 12.1 mya was taken from the analysis of Ryberg and Matheny (2012), which had used a set of Basidiomycota analysed by Ryberg and Matheny (2011) that was calibrated by the origin of the phylum at 430 mya from Berbee and Taylor (2001). Ancestral area analysis suggested the initial diversification of the group in Australasia, with long-distance dispersal to the Neotropics followed by migration to the Holarctic. DEC modelling suggested that founder event speciation was an important factor in the evolution of the group (Harrower *et al.* 2015).

For *Amanita* section *Caesareae*, Sánchez-Ramírez *et al.* (2014) used DEC modelling (under various permissive and restrictive models) and a chronogram calibrated with fossil Agaricomycetes (*Archaeomarasmius* and *Quatsinoporites*). These analyses suggested a Palaeotropical origin between the Palaeocene and the Eocene, followed by separate dispersal events to temperate areas of Europe, North America and Australia, mostly during the Miocene and Pliocene. The two separate dispersals to Australia were dated at 12 and 7 mya.

Further studies utilising dated phylogenies, often combined with ancestral area reconstruction, to investigate fungal biogeography include those on *Morchella* (O'Donnell *et al.* 2010), Porcini mushrooms in the *Boletus edulis* group (Dentinger *et al.* 2010; Feng *et al.* 2012), lethal *Amanita* mushrooms of section *Phalloideae* (Cai et *al.* 2014), *Cudonia* and *Spathularia* (Ge *et al.* 2014) and *Heterobasidion* (Chen *et al.* 2015; see also Dalman *et al.* 2010). Phylogenetic species included in these studies were often novel, and always structured by geography. Where Australasian taxa were included, they were often basal to clades that were the focus of the studies, and Gondwanan origins were raised as a possibility, but such interpretations were qualified by the likelihood of incomplete taxon sampling, particularly for Southern Hemisphere areas (Dentinger *et al.* 2010; Feng *et al.* 2012; Cai *et al.* 2014).

Cyttaria: Evolution of Ideas about Coevolution of Fungi in Relation to Time and Area

Cyttaria, an obligate, gall-forming fungal parasite of *Nothofagus* (Figure 8.8), presents a compelling subject for biogeography because it exhibits a variety of patterns of geographic distribution and host specificity, in combination with occurrence on a host that has been the focus of evolutionary and biogeographic studies. Therefore, a case study is presented across the history of biogeographic investigations on this genus.

Berkeley (1847) observed when introducing the genus that it was 'peculiar to the Southern Hemisphere', known at that time from South America and Australia and predicted to occur in New Zealand. According to the latest account, there are 12 species of *Cyttaria*, occurring on 11 host species, in Australia, New Zealand and southern South America, but not in New Guinea or New Caledonia (Petersen *et al.* 2010). For the *Nothofagus–Cyttaria* association, host–parasite relations are complex and there is not a one-to-one correspondence between the two. There are up to four *Cyttaria* species on the one host, and while the five *Cyttaria* species in Australia and New Zealand each infect one host, those in South America all infect multiple (up to five) hosts. *Cyttaria* infects all species of *Nothofagus* subgenus *Nothofagus* (all of

FIGURE 8.8 (See colour insert.) *Cyttaria gunnii* growing on gall on a branch of *Nothofagus cunninghamii*, Victoria, Australia. Photo by Paul George. Reproduced with permission.

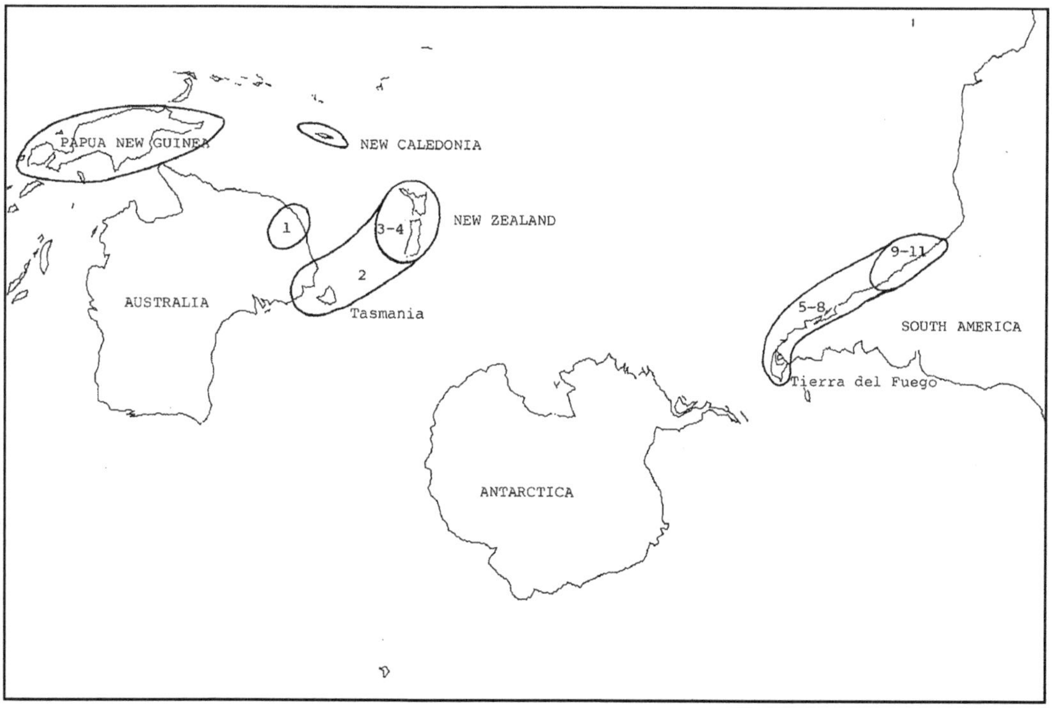

FIGURE 8.9 Distribution of *Nothofagus* (bounded by lines) and *Cyttaria* (indicated by numbers). Originally published as Figure 1 in Korf, R.P., *Australian Journal of Botany* 10, 77–87, 1983; see original caption for further details. Reproduced with permission.

which are South American) and subgenus *Lophozonia* (found in Australasia and South America), but does not occur on any species of the other two subgenera (*Brassospora* and *Fuscospora*). *Cyttaria* is not known from the fossil record, while *Nothofagus* is richly represented.

Korf (1983) proposed an evolutionary tree for *Nothofagus*, on which the species of *Cyttaria* were indicated. The tree was not based on an explicit analytical method, but rather was created to 'account for the distribution [both host and geographic] of *Cyttaria* species and their *Nothofagus* hosts' (Figure 8.9).

Korf (1983) proposed two monophyletic groups within *Cyttaria*, corresponding to the species of Australasia and those of South America, and put forward a number of postulates arising from the tree, including that *Cyttaria* arose later than *Nothofagus*, because no *Cyttaria* species infect the presumed 'ancestral stock' of the host with *brassii*-type pollen.

Humphries *et al.* (1986) compared phylogenies of *Nothofagus* and three of its parasites, including *Cyttaria*, to test for coevolution (association by descent) versus colonisation by parasites. They analysed host and parasite cladograms (all on the basis of morphology), seeking to locate components in common, while minimising colonisation events. For *Cyttaria*, at best only half the speciation events were ascribable to coevolution. They also concluded that at least some components of this coevolution were significantly correlated with biogeographic patterns, because disjunct distributions analysed as area cladograms and visualised as *tracks* showed common patterns between host and parasite. These patterns were considered most likely the result of vicariance.

Crisci *et al.* (1988) carried out another cladistic analysis of *Cyttaria*, largely based on morphology (and otherwise on phenology and the nature of the gall). They found 'little congruence between the current classification of *Nothofagus* and the monophyletic groups of *Cyttaria*'. They provided information on geography and host, without further discussion, but did note that that among the host–parasite connections tabulated by Humphries *et al.* (1986), some were incorrect or missing. The cladistic analyses of Humphries *et al.* (1986) and Crisci *et al.* (1988), utilising morphology, were pioneering applications of phylogenetic analyses to fungi, but subsequently there was little take-up of such analyses onto morphological data, especially in comparison with the frequent use in other biota such as vascular plants. This was most likely due to early awareness of the high level of parallel evolution of morphological character states in fungi. It was only once molecular data were available for fungi that explicit phylogenetic analysis of fungi became widespread (see 'Initial Phylogenetic Studies Transform Fungal Classification' in this chapter).

Peterson and Pfister (2010) utilised molecular data to produce a phylogeny for *Cyttaria* using four loci. This molecular phylogeny differed in several respects from earlier evolutionary schemes (Korf 1983) and phylogenies (Humphries *et al.* 1986; Crisci *et al.* 1988) based on morphology. Peterson and Pfister (2010) recovered three major clades, one on each of subgenera *Lophozonia* and *Nothofagus*, while the other was mostly on *Nothofagus*, but with one species (*C. berteroi*) on *Lophozonia*. In relation to geography, there was a nonmonophyletic grade of South American species and a monophyletic group of Australasian species. Species critical to interpretations were the South American *C. berteroi* and *C. espinosae*, which occur on several species of subgenus *Lophozonia*. The former clustered with other South American species that occur on subgenus *Nothofagus*, while the latter clustered phylogenetically with all the other species of *Cyttaria* on subgenus *Lophozonia* that are from Australasia. Peterson and Pfister (2010) concluded that 'the phylogenetic history of *Cyttaria* cannot be explained solely by geography or host association'.

The newly developed *Cyttaria* phylogeny was utilised by Peterson *et al.* (2010), along with a de novo phylogeny for the host, based on molecular and morphological data (Figure 8.10). They found highly significant cophylogeny and advanced two scenarios, which needed to also explain the absence of *Cyttaria* on subgenera *Brassospora* and *Fuscospora*. In the first scenario, there was a concurrent origin of *Cyttaria* and the genus *Nothofagus*, and extensive extinction explained its absence on subgenera *Brassospora* and *Fuscospora*; duplication of the *Cyttaria* lineage on *Lophozonia* was postulated, and one host jump was required from subgenus *Lophozonia* to subgenus *Nothofagus*. In the second scenario, *Cyttaria* evolved later on subgenus *Nothofagus*, and a host jump was required to subgenus *Lophozonia*, but also back to subgenus *Nothofagus*.

Peterson *et al.* (2010) utilised divergence time dating, incorporating the fossil *Palaeopyrenomycites* from the Rhynie chert as calibration, either for Ascomycota or Pezizomycotina. Dates were consistent with a vicariant origin of Australasian and South American clades of *Cyttaria*, but the divergence of Australian and New Zealand *Cyttaria* clades was suggested as having occurred no more than 44.6 mya, and therefore long-distance dispersal between Australia and New Zealand was suggested. Dates of origin of the genus *Cyttaria* predated the presumed origin of *Nothofagus*, which the authors suggested was a result of 'the problem of using a fossil [for calibration] far removed from our group of interest'. Humphries *et al.* (1986) noted that extant parasites that are older than extant hosts may have colonised

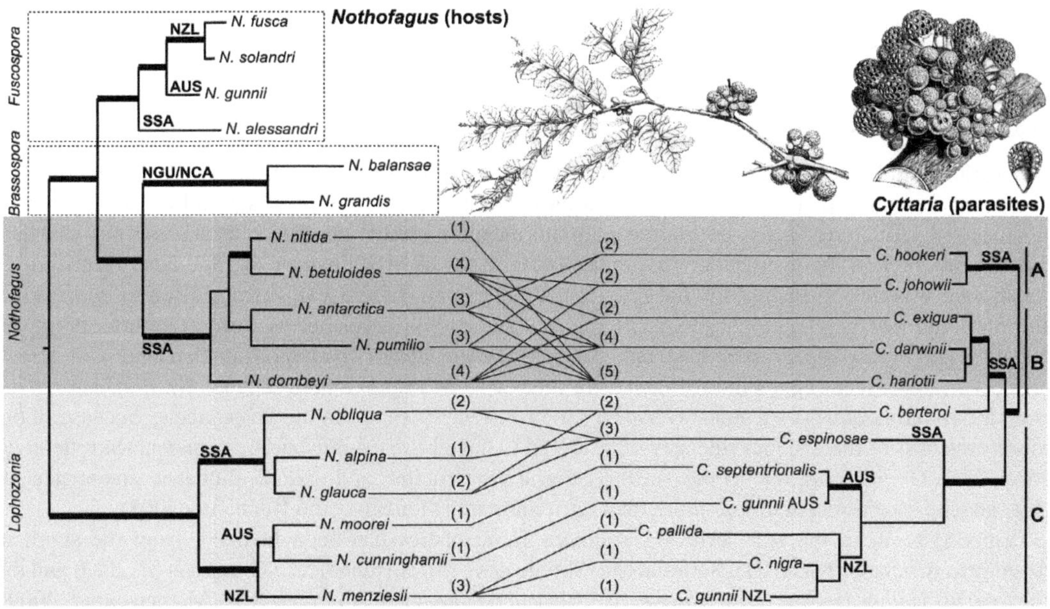

FIGURE 8.10 Relationships between the phylogeny of *Nothofagus* and *Cyttaria* and original illustration of sporophores of *Cyttaria gunnii* on *Nothofagus cunninghamii*. From Berkeley, M.J., *The Botany of the Antarctic Voyage of H.M. Discovery Ships Erebus and Terror, in the Years 1839–1843 … I. Flora Antarctica: Part II. Botany of Fuegia, the Falklands, Kerguelen's Land, etc.*, Reeve Brothers, London, 1847. Abbreviations are as follows: AUS: Australia; NCA: New Caledonia; NGU: New Guinea; NZL: New Zealand; SSA: southern South America. Reproduced from the composite Figure 1 of Peterson, K.R., *et al.*, *Mycologia* 102, 1417–1425, 2010; see original caption for further details. Reprinted with permission from *Mycologia*. © The Mycological Society of America.

from now-extinct hosts. Sauquet *et al.* (2012) discussed the considerable impact of different calibration strategies on divergence dates for *Nothofagus*, and also present a slightly different phylogeny for *Nothofagus* to that of Peterson *et al.* (2010). No doubt, comparisons of the phylogeny and biogeography of *Cyttaria* and *Nothofagus* will continue to be of interest, especially with refinements to the calibration of fungal phylogenies.

In addition to these intensive studies, *Cyttaria* remains one of the very few examples of Australasian fungi that have been included in broad-scale analyses of biogeographic patterns across biota, such as those by Humphries *et al.* (1986), Seberg (1991) and Sanmartín and Ronquist (2004); the only other examples have been among lichens, such as the inclusion by Heads (1998) of species of *Anzia*, *Pseucocyphellaria* and *Xanthoparmelia* in his investigation of the biogeographical disjunction along the Alpine fault in New Zealand.

Metastudies

Lately, metastudies have combined environmental sequences, increasingly derived from metagenomic sampling (see 'Metagenomic Biogeography' in this chapter) with sequences in reference databases derived from specimens or cultures. Dated global phylogenies constructed in this way reveal geographic structure for *Clavulina* (Kennedy *et al.* 2012) and the /sebacina lineage (Tedersoo *et al.* 2014b), for which multiple recent dispersal events were postulated, including a relatively recent origin for Australian species associated with both *Nothofagus* and Myrtaceae, consistent with their low phylogenetic diversity. For *Russula*, Looney *et al.* (2016) combined vouchered sequences with from environmental samples, to create a global phylogeny, inferring an origin around 40 mya in temperate regions of the Northern Hemisphere, and detecting frequent switches to the tropics and subsequent reversals to temperate areas. Their study also confirmed high continental endemicity for ECM fungi, with around 93% of molecular operational taxonomic units (mOTUs) endemic to a single continent;

overlap was mainly between Europe and North America, while among 74 Australasian mOTUs, only two extended to other regions (excluding obvious exotics) (Looney *et al.* 2016).

Dispersal and Its Direction

The various studies using dated phylogenies discussed in the sections 'Analyses of Areas' through '*Cyttaria*: Evolution of Ideas about Coevolution of Fungi in Relation to Time and Area', as well as Wilson *et al.* (2012) (see 'Sclerodermatineae' in this chapter), usually invoked at least some long-distance dispersal to explain the occurrence of fungal lineages across areas within the Southern Hemisphere. Vicariance was only plausible for the /gymnohydnotrya and Tuberaceae lineage (Bonito *et al.* 2013) and for some lineages in Inocybaceae (Matheny *et al.* 2009). Nevertheless, several authors point out that dispersal events appear to be rare, either because of the lack of cosmopolitan distribution of genera or species across seemingly suitable habitat for otherwise widespread taxa such as *Armillaria* (Coetzee *et al.* 2011) and *Lactarius* subgenus *Gerardii* (Stubbe *et al.* 2010); or, in the Serpulaceae, because of the 'distinctiveness of the disjunct lineages' considered to have been established by dispersal (Skrede *et al.* 2011); or in *Ganoderma*, due to the 'strong geographic structure and overall allopatric divergence' of phylogenetic species occurring in more than one continent (Moncalvo and Buchanan 2008).

Dispersal between the Northern and Southern Hemispheres has been invoked, from the south in *Armillaria* (Coetzee *et al.* 2011), Serpulaceae (Skrede *et al.* 2011), Tuberaceae (Bonito *et al.* 2013) and the *Cortinarius violaceus* group (Harrower *et al.* 2015), and from the north in *Inocybe* (Matheny *et al.* 2009), and in both directions for *Ganoderma* (Moncalvo and Buchanan 2008). Johnston (2010) also suggests rare cross-hemisphere long-distance dispersal, from the south in *Chlorociboria* (most species in New Zealand, but with two Northern Hemisphere species in terminal clades) and from the north in *Armillaria* (one austral species, *Armillaria hinnulea*, is nested within an otherwise Northern Hemisphere clade).

Infraspecific Geographic Structure and Population Genetics

So far, there is little evidence for infraspecific geographic structure among the DNA sequences commonly used for phylogeny, such as the ITS region. However, these regions may not evolve fast enough to manifest infraspecific geographic variation, and in addition there are usually too few samples. There is a variety of approaches available for analysing infraspecific genetic variation, exemplified by applications to fungi of coalescent analyses by Geml *et al.* (2012a) on arctic *Lichenomphalia*, where geographic structure is lacking; haplotype networks and isolation-with-migration modelling by Oono *et al.* (2014) on *Lophodermium* endophytes in North America, detecting infraspecific genetic structure in one of two species; and population genomics by Ellison *et al.* (2011) to *Neurospora crassa* from the Americas and Africa, detecting geographic variation in genes that could be related to the ability to cope with different temperature regimes.

Walker (1996) used examples about the pathogenicity of exotic plant pathogenic fungi in Australia to demonstrate the potential for geographically structured genetic variation within species, pointing out the influence of the degree of isolation and the extent to which sexual and parasexual processes operate to create a recombination of genetic variation, as well as the significance of the presence or absence of alternate hosts for the sexual stages of pleomorphic fungi. Population genetic studies of crop pathogens are vital in tracking and understanding the spread of pathogens and their virulence genes. Such studies often have very wide and deep sampling, as in the global study of the origin of the barley scald pathogen *Rhynchosporium secalis* (Zaffarano *et al.* 2006).

In contrast, there are very few studies on native fungi in native ecosystems looking at the partitioning of genetic variation across space. Douhan *et al.* (2011) reviewed the population genetics of ECM fungi, showing that infraspecific genetic variation existed at scales from local to continental. There are too few studies on Australasian native fungi to detect patterns in relation to either biology or potential geographic barriers. For *Schizophyllum commune* in New Zealand, three of four IGS1 haplotypes were found in both the North and South Islands; the other, rarest haplotype occurred in the South Island and the Cook Islands; and overall there was no 'haplotype distribution pattern related to indigenous or modified habitats' (Fuller *et al.* 2013). Similarly, for the puffball *Pisolithus microcarpus*, Hitchcock *et al.*

(2011) found a high level of genetic homogeneity across geographical populations from Victoria and New South Wales, using microsatellite markers. In contrast, for the Australian ECM mushroom *Laccaria* sp. *A*, which is exclusively associated with *Nothofagus cunninghamii* in Tasmania and Victoria, Sheedy *et al.* (2015) found geographically patterned genetic variation in microsatellite markers, across the entire range of the fungus, that was highly correlated with genetic structure of the host.

There is much that remains of interest about infraspecific genetic variation in Australasian fungi, and there is scope for a variety of investigations in relation to geography, especially where there are barriers, such as oceans (Australia and New Zealand) or currently inhospitable regions (the Nullarbor plain between western and eastern Australia), or for species that occur on large land masses plus isolated oceanic islands such as Lord Howe Island, Norfolk Island or the Chatham Islands. For the many species of fungi that appear to be in common between southwest Western Australia and eastern Australia, the directionality of gene flow will be of interest, given the dominant eastward movement of high- and low-pressure systems in southern continental Australia around 30° S, although there is potential for reversals in large-scale meteorological features such as the Walker circulation over the Pacific, which can occur with El Niño events (Whetton 1997). It is essential that investigations of infraspecific genetic variation are carried out in concert with the rigorous delimitation of phylogenetic species, due to the frequent occurrence of morphologically cryptic phylogenetic species (Oono *et al.* 2014; Sheedy *et al.* 2013, 2015).

Metagenomic Biogeography

Metagenomics (also referred to as *environmental genomics*, *ecogenomics* or *community genomics*) is the study of genomic material directly from the environment. Metagenomics was first applied to unculturable microorganisms (Handelsman 2004) but has been increasingly utilised in studies across all organisms in a wide variety of environments. A key advance in metagenomics has been the availability of next-generation sequencing (NGS) technologies, such as 454 and Illumina, that rapidly and economically produce high numbers of sequences. Metagenomics covers approaches from whole genome shotgun sequencing to *metabarcoding*, which targets specific, short 'barcode' regions from environmental samples in order to characterise mOTUs (Orgiazzi *et al.* 2015).

The term *metagenomic biogeography* describes studies that utilise mass sequence data recovered directly from environmental samples to establish and evaluate the distribution of organisms. Due to the utility of the ITS region as an effective barcode for many groups of fungi (Schoch *et al.* 2012), metabarcoding studies of fungi that characterise the entire mycobiome of a sample are providing new insights into the way that fungal communities vary geographically, at unprecedented scales – both in terms of taxonomic and geographic coverage and sampling density. In a global metabarcoding study of soil fungi across 365 sites, Tedersoo *et al.* (2014a) recovered around 44,000 nonsingleton fungal mOTUs (about half the total number of fungal species described so far) and demonstrated that, consistent with Rapoport's rule, the geographic range of fungi increased towards the poles. Metagenomic studies often detect significant distance decay relationships in community similarity. The metastudy of Meiser *et al.* (2013) found very few mOTUs in common between 10 published studies, mostly from the Northern Hemisphere. In North America, soil fungi associated with one host (*Pinus ponderosa*) were well separated at the regional scale (ca. 1000 km) with 85% of mOTUs unique to one of three bioregions (Talbot *et al.* 2014), and fungi in dust settled on external surfaces of houses exhibited a regional-scale geographic pattern, related to variation in soil properties and climate (Barberán *et al.* 2015). In New Zealand, significant geographic structure was detected among the suite of fungal mOTUs associated with grapevines (Taylor *et al.* 2014). Talbot *et al.* (2014) conclude that, at least for soil fungi, 'large-scale geographic processes like dispersal limitation are first-order determinants of … regional species pools'.

There are methodological challenges for metagenomic analyses of DNA, including a lack of distinction between life history stages (such as resting spores, sclerotia or active hyphae) or living and dead material, although NGS does provide the option of assaying the transcriptome. In metabarcoding studies, mOTUs are generated by a binning approach at a given level of similarity, usually 97% when ITS is utilised. It is generally conceded that mOTUs are species level rather than mapping exactly to species, as would be determined by genealogical concordance. The majority of mOTUs do not match sequences from named species in reference databases, partly an effect of the incomplete sequencing of reference collections, but otherwise mostly due to detecting as yet uncollected fungi. Further challenges include relatively short read lengths (often only ITS1 or ITS2, but likely to increase in length with technological developments), chimeric sequences, the effect of primer choice and annealing temperatures on the recovery of taxa, and distance versus phylogenetic methods of mOTU-calling – all issues under active investigation (Schmidt *et al.* 2013; Nilsson *et al.* 2015; Ryberg 2015). The most controversial aspect at present is the choice of genomic region, exemplified by studies on the Glomeromycota (microfungi that form arbuscular mycorrhizas), where, when a part of the 18S region was utilised, one-third of 246 virtual taxa were globally distributed (Davison *et al.* 2015) and there was no effect of geography on the distribution of 563 mOTUs within continents (Kivlin *et al.* 2011). However, a critique of the use of the 18S region by Bruns and Taylor (2016) contended that this region distinguished taxa at above the species level in Glomeromycota.

For individual species, metabarcoding data have great potential to confirm and extend known distributions combining unprecedented sampling intensity with molecular confirmation of the identifications. A vast bycatch from metabarcoding studies targeting the fungal community rather than individual species is rapidly accumulating and awaits analysis and mapping. Matching mOTUs from one study to another is complicated by the difficulty of access to the entire set of sequences (often only exemplar sequences are available for mOTUs) and nomenclatural issues (the current International Code of Nomenclature does not allow the description of new species on ITS sequences alone in the absence of a type specimen). Nevertheless, innovative taxonomic approaches such as the automated UNITE reference database of ITS 'species hypotheses' (Nilsson *et al.* 2015) provide options for taxon matching across studies. The Biome of Australian Soil Environments (BASE), managed by Bioplatforms Australia, is a metabarcoding project for soil organisms, including fungi. Some 600 samples across Australia have yielded around 44,000 mOTUs of Fungi, defined at a 97% cutoff in the ITS1 region (Bioplatforms Australia 2016), with less than 5,000 mOTUs matching known species in sequence reference databases. BASE data already suggest significant extensions to the distribution of species known only from the type or from a few collections, such as the truffle *Ulurua nonparaphysata* (Figure 8.11).

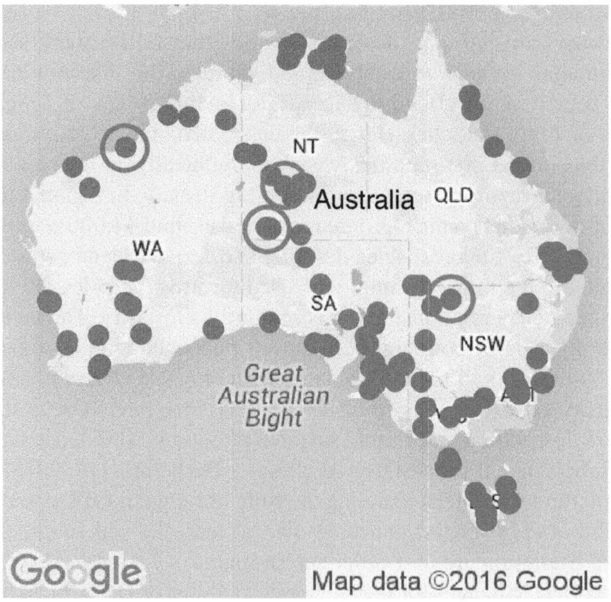

FIGURE 8.11 (See colour insert.) Extension of the distribution of *Ulurua nonparaphysata* by metagenomic samples. Sampling sites for the Biome of Australian Soils Project (blue markers), with red circles indicating sample sites yielding mOTU 3683, which has a 99.3% match in the ITS1 region to a sequence from the type and only known specimen of the ascomycete truffle *Ulurua nonparaphysata*, whose type locality is Uluru, corresponding to the lower red circle in the Northern Territory. (Produced using the *Atlas of Living Australia* http://www.ala.org.au. Accessed 22 March 2016.). CC-BY.

Synthesis and Prospects

Mycogeography in Australasia began with observations on *Cyttaria*, an iconic Southern Hemisphere fungus on an iconic plant, *Nothofagus* (Berkeley 1847). It took more than a century before analyses of the distribution of austral fungi began in earnest (Pirozynski and Walker 1983; Walker 1996). The Gondwanan nature of the distribution of *Cyttaria* contributed to a focus on distributions spanning southern continents and an emphasis on the seemingly tight associations with *Nothofagus* as a host. Although observations on plant pathogens showed the potential of long-distance dispersal, early explanations often invoked ancient vicariance. The advent of molecular methods allowed the meaningful analysis of phylogeny, resulting in numerous rearrangements of morphological classifications. In parallel, fungi of economic importance as cultivated food or as plant or human pathogens showed unexpected geographic structure as a byproduct of species delimitation.

Recent investigations of explicit biogeographic hypotheses with multigene phylogenies have yielded a variety of patterns, many of which have divergences that are too young to involve vicariance. In addition, host shifts appear to be frequent. The dispersal of fungi has been inferred from a number of sources both to and from Australasia, at a variety of times since connections between land masses were severed. Sister relationships between Australasian and African fungi are rare, but connections with Asia are common. With more studies, it will be interesting to compare the timing and direction of dispersal routes and how this might relate to biology, with respect to life cycle (asexual stages) and dispersal mode (e.g. wind-dispersed mushrooms versus animal-vectored truffles). In particular, comparative studies across groups with different spore characteristics (size, shape, melanisation, etc.) in terms of the size of distribution areas and historical dispersal patterns will be instructive.

Geographic sampling remains an issue. Phylogenetic analyses often use a single sample per species to represent the distribution, especially for austral taxa, even when they may occur across several land masses. Denser sampling is required to establish the shape of ranges and to underpin the investigation of infraspecies phylogeographic patterning. Some attention should be paid to locate the most geographically restricted fungi, which are likely to be highly host-specific fungi limited by associations with geographically restricted symbionts. For establishing distribution, metagenomic environmental sampling is being used to rapidly increase the knowledge of distribution within and between continents, but will benefit from the expansion of specimen- and culture-based sequence reference libraries to aid the interpretation of environmental sequences. Where phylogenetic species of larger fungi can be confirmed as distinctive in the field, citizen science recording in concert with the mass digitisation of georeferenced specimen data also has potential to provide a much finer-grained picture of fungal distribution for a subset of target species.

Taxon sampling is also often patchy across the Southern Hemisphere, and the addition of further taxa to analyses may have significant implications for existing explanations of the timing and direction of origins. Just as taxonomy is regularly revised as new collections, characters and analyses become available, biogeographical investigations should not be regarded as static. It would be worthwhile building on existing studies that have extensively sampled lineages such as *Cyttaria*, Hysterangiales and Inocybaceae. Greater sampling of microfungi is also needed, given most existing studies are on macrofungi such as mushrooms. Sampling across a wider range of hosts across their entire distribution will be of interest, especially for ECM and pathogenic associates of symbionts that are highly diverse in the region, such as eucalypts.

There are benefits to both mycogeographers and biogeographers from the better integration of studies on distribution focused on fungi with studies across biota. For example, the study of biomes has generally lacked mycological input. It will be of interest to investigate mycogeographic breaks, hotspots and provinces and to compare patterns with those of other biota. The apparent breadth of fungal distributions within Australia suggests that mycogeographic areas will be more similar to zoogeographic rather than phytogeographic areas, but critical comparison is required.

For symbiotic fungi, there is much scope for integrated investigation of their biogeography in tandem with that of their symbionts. Indeed, Hacquard (2015) calls for the study of the holobiont (plant and fungus) and provides an example where host biogeography explains a significant proportion of the variance in the associated fungal assemblage. The significance of symbiont constraints on host distribution (such as through different bioclimatic niches) should also be considered, especially when assessing the potential effects of climate change.

The studies enumerated in this review demonstrate that fungi are not everywhere but exhibit various patterns of distribution across space and time. A sophisticated approach to exploring the distribution of fungi is now based on rigorous phylogenetic species delimitation, feasible through a multigene concordance approach. Further studies will add more genes and more thorough geographic and taxon sampling. Origin status (indigenous or exotic) remains an issue with the increased unintentional movement of fungal propagules around the world. Metagenomics offers much promise in creating dense samples for biogeographic analysis across species and geography.

Acknowledgements

Thanks to Sara Maroske for her support and insightful editing, Elizabeth Sheedy for her extensive comments on drafts, Julia Simon Cardoso and Travis Heafield for their databasing of literature, Sapphire McMullan-Fisher and Alison Pouliot for their useful feedback, Peter Johnston for sharing his unpublished analyses of New Zealand fungi distributions, Dan Murphy for his helpful discussions and Malte Ebach for his patient encouragement.

References

Aguilar, M.; Fiore-Donno, A.M.; Lado, C.; Cavalier-Smith, T. 2014. Using environmental niche models to test the 'Everything is everywhere' hypothesis for *Badhamia*. *ISME Journal* 8:737–745.

Arcadia, L. 2013. Lichen biogeography at the largest scales. *Lichenologist* 45:565–578.

Arnolds, E.J.M. 2007. Biogeography and conservation. In *The Mycota: IV. Environmental and Microbial Relationships*, 2nd edn., eds. Kubicek, C.P.; Druzhinina, I.S., 105–124. Springer: Berlin, Germany.

Avise, J.C. 2000. *Phylogeography: The History and Formation of Species*. Harvard University Press: Cambridge, MA.

Avise, J.C.; Wollenberg, K. 1997. Phylogenetics and the origin of species. *Proceedings of the National Academy of Sciences of the United States of America* 94:7748–7755.

Baas-Becking, L.G.M. 1934. *Geobiologie of Inleiding tot de Milieukunde*. Van Stockkum & Zoon: The Hague, the Netherlands.

Backhouse, D.; Burgess, L.W. 1995. Mycogeography of *Fusarium*: Climatic analysis of the distribution with in Australia of *Fusarium* species in the section *Gibbosum*. *Mycological Research* 99:1218–1224.

Backhouse, D.; Burgess, L.W. 2002. Climatic analysis of the distribution of *Fusarium graminearum, F. pseudograminearum* and *F. cumorum* on cereals in Australia. *Australasian Plant Pathology* 31:321–327.

Backhouse, D.; Burgess, L.W.; Summerell, B.A. 2001. Biogeography of *Fusarium*. In *Fusarium: Paul E. Nelson Memorial Symposium*, eds. Summerell, B.A.; Leslie, J.F.; Backhouse, D.; Bryden, W.L.; Burgess, L.W., 122–137. APS Press: St. Paul, MN.

Bailey, R.H.; James, P.W. 1979. Birds and the dispersal of lichen propagules. *Lichenologist* 11:105–106.

Barberán, A.; Ladau, J.; Leff, J.W.; Pollard, K.S.; Menninger, H.L.; Dunn, R.R.; Fierer, N. 2015. Continental-scale distributions of dust-associated bacteria and fungi. *Proceedings of the National Academy of Sciences of the United States of America* 112:5756–5761.

Bass, D.; Boenigk, J. 2011. Everything is everywhere: A twenty-first century de-/reconstruction with respect to protists. In *Biogeography of Microscopic Organisms: Is Everything Small Everywhere?*, ed. Fontaneto, D., 88–110. Cambridge University Press: Cambridge, UK.

Baum, D.A.; Smith, S.D. 2013. *Tree Thinking: An Introduction to Phylogenetic Biology*. Roberts: Greenwood Village, CO.

Beheregaray, L.B. 2008. Twenty years of phylogeography: The state of the field and the challenges for the Southern Hemisphere. *Molecular Ecology* 17:3754–3774.

Beijerinck, M.W. 1913. *De Infusies en de Ontdekking der Backteriën*. Jaarboek van de Koninklijke Akademie van Wetenschappen. Müller: Amsterdam, the Netherlands.

Berbee, M.L.; Taylor, J.W. 1993. Dating the evolutionary radiations of the true fungi. *Canadian Journal of Botany* 71:1114–1127.

Berbee, M.L.; Taylor, J.W. 2001. Fungal molecular evolution: Gene trees and geologic time. In *The Mycota: VII, part B. Systematics and Evolution*, eds. McLaughlin, D.J.; McLaughlin, E.G.; Lemke, P.A., 229–245. Springer: Berlin, Germany.

Berbee, M.L.; Taylor, J.W. 2010. Dating the molecular clock in fungi: How close are we? *Fungal Biology Reviews* 24:1–16.

Berkeley, M.J. 1847. Fungi. In *The Botany of the Antarctic Voyage of H.M. Discovery Ships Erebus and Terror, in the Years 1839–1843 … I. Flora Antarctica: Part II. Botany of Fuegia, the Falklands, Kerguelen's Land, etc.*, ed. Hooker, J.D., 447–545. Reeve Brothers: London, UK.

Berndt, R. 2008. The rust mycobiota of southern Africa: Species richness, composition, and affinities. *Mycological Progress* 112:463–471.

Bioplatforms Australia. Biome of Australian soil environments. https://ccgapps.com.au/bpa-metadata/base (visited 22 March 2016).

Bisby, G.R. 1943. Geographical distribution of fungi. *Botanical Review* 9:466–482.

Blackwell, M.; Hibbett, D.S.; Taylor, J.W.; Spatafora, J.W. 2007. Research coordination networks: A phylogeny for kingdom fungi (Deep Hypha). *Mycologia* 98:829–937.

Bonito, G.; Smith, M.E.; Nowak, M.; *et al.* 2013. Historical biogeography and diversification of truffles in the Tuberaceae and their newly identified southern hemisphere sister lineage. *PLoS ONE* 8(1):e52765:1–15.

Bougher, N.L. 1983. Western Australian *Coprinus* as part of a cosmopolitan flora. *Transactions of the British Mycological Society* 81:147–149.

Bougher, N.L.; Fuhrer, B.A.; Horak, E. 1994. Taxonomy and biogeography of Australian *Rozites* species mycorrhizal with *Nothofagus* and Myrtaceae. *Australian Systematic Botany* 7:353–375.

Bougher, N.L.; Lebel, T. 2001. Sequestrate (truffle-like) fungi of Australia and New Zealand. *Australian Systematic Botany* 14:439–484.

Brito, P.H.; Edwards, S.V. 2008. Multilocus phylogeography and phylogenetics using sequence-based markers. *Genetica* 135:439–455.

Bruns, T.D.; Taylor, J.W. 2016. Comment on global assessment of arbuscular mycorrhizal fungus diversity reveals very low endemism. *Science* 351:826-b.

Buchanan, P.K. 2001. Aphyllophorales in Australasia. *Australian Systematic Botany* 14:417–437.

Buchanan, P.K.; May, T.W. 2003. Conservation of New Zealand and Australian fungi. *New Zealand Journal of Botany* 41:407–421.

Burbidge, N. 1960. The phytogeography of the Australian region. *Australian Journal of Botany* 8:75–211.

Burgess, L.W.; Summerell, B.A. 1992. Mycogeography of *Fusarium*: Survey of *Fusarium* species in subtropical and semi-arid grassland soils from Queensland, Australia. *Mycological Research* 96:780–784.

Burnett, J. 2003. *Fungal Populations and Species*. Oxford University Press: Oxford, UK.

Cai, Q.; Tulloss, R.E.; Tang, L.P.; *et al.* 2014. Multi-locus phylogeny of lethal amanitas: Implications for species diversity and historical biogeography. *BMC Evolutionary Biology* 14:143:1–16.

Carbone, I.; Kohn, L.M. 2001. A microbial population-species interface: Nested cladistic and coalescent inference with multilocus data. *Molecular Ecology* 10:947–964.

Carstens, B.C.; Pelletier, T.A.; Reid, N.M.; Satler, J.D. 2013. How to fail at species delimitation. *Molecular Ecology* 22:4369–4383.

Chapman, A.D. 2009. *Numbers of Living Species in Australia and the World*, 2nd edn. Report for the Australian Biological Resources Study: Canberra, Australia.

Chaverri, P.; Branco-Rocha, F.; Jaklitsch, W.; Gazis, R.; Degenkolb, T.; Samuels, G.J. 2015. Systematics of the *Trichoderma harzianum* species complex and the re-identification of commercial biocontrol strains. *Mycologia* 107:558–590.

Chen, J.-J.; Cui, B.-K.; Zhou, L.-W.; Korhonen, K.; Dai, Y.-C. 2015. Phylogeny, divergence time estimation, and biogeography of the genus *Heterobasidion* (Basidiomycota, Russulales). *Fungal Diversity* 71:185–200.

Claridge, A.W.; Barry, S.C.; Cork, S.J.; Trappe, J.M. 2000. Diversity and habitat relationships of hypogeous fungi: II. Factors influencing the occurrence and number of taxa. *Biodiversity and Conservation* 9:175–199.

Claridge, A.W.; May, T.W. 1994. Mycophagy among Australian mammals. *Australian Journal of Ecology* 19:251–275.

Claridge, A.W.; Trappe, J.M.; Paull, D.J. 2014. Ecology and distribution of desert truffles in the Australian outback. In *Desert Truffles: Phylogeny, Physiology, Distribution and Domestication*, eds. Kagan-Zur, V.; Roth-Bejerano, N.; Sitrit, Y.; Morte, A., 203–214. Springer: Berlin, Germany.

Cleland, J.B. 1934. *Toadstools and Mushrooms and Other Larger Fungi of South Australia, Part 1*. Harrison Weir, Government Printer: Adelaide, Australia.

Coetzee, M.P.; Bloomer, P.; Wingfield, M.J.; Wingfield, B.D. 2011. Paleogene radiation of a plant pathogenic mushroom. *PLoS ONE* 6:1–9.

Coetzee, M.P.; Wingfield, B.D.; Bloomer, P.; Ridley, G.S.; Wingfield, M.J. 2003. Molecular identification and phylogeny of *Armillaria* isolates from South America and Indo-Malaysia. *Mycologia* 95:285–293.

Coetzee, M.P.A.; Wingfield, B.D.; Ridley, G.S.; Wingfield, M.J.; Bloomer, P.; Kile, G.A. 2001. Phylogenetic relationships of Australian and New Zealand *Armillaria* species. *Mycologia* 93:887–896.

Crisci, J.V.; Gamundi, I.J.; Caballo, M.N. 1988. A cladistic analysis of the genus *Cyttaria* (Fungi – Ascomycotina). *Cladistics* 4:279–290.

Cullen, J.M.; Kable, P.F.; Catt, M. 1973. Epidemic spread of a rust imported for biological control. *Nature* 244:462–464.

Cunningham, G.H. 1944. *The Gasteromycetes of Australia and New Zealand*. Published by the author. Auckland, New Zealand.

Cunningham, G.H. 1965. *Polyporaceae of New Zealand*. New Zealand Department of Scientific and Industrial Research Bulletin 164. R.E. Owen, Government Printer: Wellington, New Zealand.

Dalman, K.; Olson, A.; Stenlid, J. 2010. Evolutionary history of the conifer root rot fungus *Heterobasidion annosum* sensu lato. *Molecular Ecology* 19:4979–4993.

Dam, N. 2013. Spores do travel. *Mycologia* 105:1618–1622.

Darlington, P.J. 1969. *Biogeography of the Southern End of the World*. Harvard University Press: Cambridge, MA.

Davis, B.J.; Phillips, R.D.; Wright, M.; Linde, C.C.; Dixon, K.W. 2015. Continent-wide distribution in mycorrhizal fungi: Implications for the biogeography of specialized orchids. *Annals of Botany* 116:413–421.

Davison, J.; Moora, M.; Öpik, M.; *et al.* 2015. Global assessment of arbuscular mycorrhizal fungus diversity reveals very low endemism. *Science* 349:970–973.

de Wit, R.; Bouvier, T. 2006. 'Everything is everywhere, but, the environment selects': What did Baas Becking and Beijerinck really say? *Environmental Microbiology* 8:755–758.

Dentinger, B.T.M.; Ammirati, J.F.; Both, E.E.; *et al.* 2010. Molecular phylogenetics of porcini mushrooms (*Boletus* section *Boletus*). *Molecular Phylogenetics and Evolution* 57:1276–1292.

Dentinger, B.T.M.; Gaya, E.; O'Brien, H.; *et al.* 2015. Tales from the crypt: Genome mining from fungarium specimens improves resolution of the mushroom tree of life. *Biological Journal of the Linnean Society* 117:11–32.

Dettman, J.R.; Jacobson, D.J.; Taylor, J.W. 2003a. A multilocus genealogical approach to phylogenetic species recognition in the model eukaryote *Neurospora*. *Evolution* 57:2703–2720.

Dettman, J.R.; Jacobson, D.J.; Taylor, J.W. 2006. Multilocus sequence data reveal extensive phylogenetic species diversity within the *Neurospora discreta* complex. *Mycologia* 98:436–446.

Dettman, J.R.; Jacobson, D.J.; Turner, E.; Pringle, A.; Taylor, J.W. 2003b. Reproductive isolation and phylogenetic divergence in *Neurospora*: Comparing methods of species recognition in a model eukaryote. *Evolution* 57:2721–2741.

Döbbeler, P.; Hertel, H. 2013. Bryophilous ascomycetes everywhere: Distribution maps of selected species on liverworts, mosses and Polytrichaceae. *Herzogia* 26:361–404.

Doing, H. 1970. Botanical geography and chorology in Australia. *Mededelingen van de Botanische Tuinen en het Belmonte Arboretum der Landbouwhogeschool te Wageningen* 13:81–89.

Doing, H. 1981. Phytogeography of the Australian floristic kingdom. In *Australian Vegetation*, ed. Groves, R.H., 3–25. Cambridge University Press: Melbourne, Australia.

Douhan, G.W.; Vincenot, L.; Gryta, H.; Selosse, M.-A. 2011. Population genetics of ectomycorrhizal fungi: From current knowledge to emerging directions. *Fungal Biology* 115:569–597.

Dressaire, E.; Yamada, L.; Song, B.; Roper, M. 2016. Mushrooms use convectively created airflows to disperse their spores. *Proceedings of the National Academy of Sciences of the United States of America* 113:2833–2838.

Drew, A.; Allen, E.J.; Allen, L.J.S. 2006. Analysis of climatic and geographic factors affecting the presence of chytridiomycosis in Australia. *Diseases of Aquatic Organisms* 68:245–250.

Duarte, S.; Bärlocher, F.; Pascoala, C.; Cássioa, F. 2015. Biogeography of aquatic hyphomycetes: Current knowledge and future perspectives. *Fungal Ecology* 19:169–181.

Ebach, M.C. 2012. A history of biogeographical regionalisation in Australia. *Zootaxa* 3392:1–34.

Elith, J.; Simpson, J.; Hirsch, M.; Burgman, M.A. 2013. Taxonomic uncertainty and decision making for biosecurity: Spatial models for myrtle/guava rust. *Australasian Plant Pathology* 42:43–51.

Ellison, C.E.; Hall, C.; Kowbel, D.; *et al.* 2011. Population genomics and local adaptation in wild isolates of a model microbial eukaryote. *Proceedings of the National Academy of Sciences of the United States of America* 108:2831–2836.

Estrada-Torres, A.; Basanta, D.W.D.; Lado, C. 2013. Biogeographic patterns of the myxomycete biota of the Americas using a parsimony analysis of endemicity. *Fungal Diversity* 59:159–177.

Feng, B.; Xu, J.; Wu, G.; *et al.* 2012. DNA sequence analyses reveal abundant diversity, endemism and evidence for Asian origin of the porcini mushrooms. *PLoS ONE* 7(5)e37567:1–12.

Finlay, B.J. 2002. Global dispersal of free-living microbial eukaryote species. *Science* 296:1061–1063.

Fleet, G.H. 2001. Biodiversity and ecology of Australasian yeasts (Fungi). *Australian Systematic Botany* 14:501–511.

Foissner, W. 2006. Biogeography and dispersal of micro-organisms: A review emphasizing protists. *Acta Protozoologica* 45:111–136.

Foissner, W. 2009. Protist diversity and distribution: Some basic considerations. In *Protist Diversity and Geographical Distribution*, eds. Foissner, W.; Hawksworth, D.L., 1–8. Springer: Berlin, Germany.

Foissner, W.; Hawksworth, D.L. 2009. *Protist Diversity and Geographical Distribution*. Springer: Berlin, Germany.

Fontaneto, D. 2011. *Biogeography of Microscopic Organisms: Is Everything Small Everywhere?* Cambridge University Press: Cambridge, UK.

Fröhlich-Nowoisky, J.; Burrows, S.M.; Xie, Z.; *et al.* 2012. Biogeography in the air: Fungal diversity over land and oceans. *Biogeosciences* 9:1125–1136.

Fujita, M.K.; Leaché, A.D.; Burbrink, F.T.; McGuire, J.A.; Moritz, C. 2012. Coalescent-based species delimitation in an integrative taxonomy. *Trends in Ecology and Evolution* 27:480–488.

Fuller, R.J.M.; Johnston, P.R.; Pearson, M.N. 2013. *Schizophyllum commune*: A case study for testing the potential introduction of non-native strains into New Zealand. *New Zealand Journal of Botany* 51:286–296.

Galante, T.E.; Horton, T.R.; Swaney, D.P. 2011. 95% of basidiospores fall within 1 m of the cap: A field-and modeling-based study. *Mycologia* 103:1175–1183.

Galloway, D.J. 1991. Phytogeography of Southern Hemisphere lichens. In *Quantitative Approaches to Phyteogeography*, eds. Nimis, P.L.; Crovello, T.J., 233–262. Kluwer Academic Publishers: Dordrecht, the Netherlands.

Galloway, D.J. 2008a. Austral lichenology: 1690–2008. *New Zealand Journal of Botany* 46:433–521.

Galloway, D.J. 2008b. Lichen biogeography. In *Lichen Biology*, 2nd edn., ed. Nash, T.I, 315–335. Cambridge University Press: Cambridge, UK.

Gantner, P.F. 2011. Everything is not everywhere: The distribution of cactophilic yeasts. In *Biogeography of Microscopic Organisms: Is Everything Small Everywhere?*, ed. Fontaneto, D., 130–174. Cambridge University Press: Cambridge, UK.

Ge, Z.-W.; Yang, Z.L.; Pfister, D.H.; Carbone, M.; Bau, T.; Smith, M.E. 2014. Multigene molecular phylogeny and biogeographic diversification of the earth tongue fungi in the genera *Cudonia* and *Spathularia* (Rhytismatales, Ascomycota). *PLoS ONE* 9(8):e103457:1–13.

Geml, J. 2011. Coalescent analyses reveal contrasting patterns of intercontinental gene flow in arctic-alpine and boreal-temperate fungi. In *Biogeography of Microscopic Organisms: Is Everything Small Everywhere?*, ed. Fontaneto, D., 178–190. Cambridge University Press: Cambridge, UK.

Geml, J.; Kauff, F.; Brochmann, C.; *et al.* 2012a. Frequent circumarctic and rare transequatorial dispersals in the lichenised agaric genus *Lichenomphalia* (Hygrophoraceae, Basidiomycota). *Fungal Biology* 116:388–400.

Geml, J.; Laursen, G.A.; O'Neill, K.; Nusbaum, H.C.; Taylor, D.L. 2006. Beringian origins and cryptic speciation events in the fly agaric (*Amanita muscaria*). *Molecular Ecology* 15:225–239.

Geml, J.; Timling, I.; Robinson, C.H.; Lennon, N.; Nusbaum, H.C.; Brochmann, C.; Noordeloos, M.E.; Taylor, D.L. 2012b. An arctic community of symbiotic fungi assembled by long-distance dispersers: Phylogenetic diversity of ectomycorrhizal basidiomycetes in Svalbard based on soil and sporocarp DNA. *Journal of Biogeography* 39:74–88.

Geml, J.; Tulloss, R.E.; Laursen, G.A.; Sazanova, N.A.; Taylor, D.L. 2008. Evidence for strong inter- and intracontinental phylogeographic structure in *Amanita muscaria*, a wind-dispersed ectomycorrhizal basidiomycete. *Molecular Phylogenetics and Evolution* 48:694–701.

Giraud, T.; Refrégier, G.; Le Gac, M.; de Vienne, D.M.; Hood, M.E. 2008. Speciation in fungi. *Fungal Genetics and Biology* 45:791–802.

Gladieux, P.; Feurtey, A.; Hood, M.E.; *et al.* 2015. The population biology of fungal invasions. *Molecular Ecology* 24:1969–1986.

Grey, P.; Grey, E. 2005. *Fungi Down Under: The Fungimap Guide to Australian Fungi*. Fungimap: South Yarra, Australia.

Grgurinovic, C.A. 2001. Agaricales in Australasia. *Australian Systematic Botany* 14:395–406.

Grgurinovic, C.A. 2003. *The Genus* Mycena *in South-eastern Australia*. Fungal Diversity Press and Australian Biological Resources Study: Canberra, Australia.

Gueidan, C.; Ruibal, C.; de Hoog, G.S.; Schneider, H. 2011. Rock-inhabiting fungi originated during periods of dry climate in the Late Devonian and Middle Triassic. *Fungal Biology* 115:987–996.

Hacquard, S. 2015. Disentangling the factors shaping microbiota composition across the plant holobiont. *New Phytologist* 209:454–457.

Halbwachs, H.; Bässler, C. 2015. Gone with the wind: A review on basidiospores of lamellate agarics. *Mycosphere* 6:78–112.

Halling, R. 2001. Ectomycorrhizae: Co-evolution, significance and biogeography. *Annals of the Missouri Botanical Garden* 88:5–13.

Halling, R.E.; Nuhn, M.; Osmundson, T.; *et al.* 2012. Affinities of the *Boletus chromapes* group to *Royoungia* and the description of two new genera, *Harrya and Australopilus. Australian Systematic Botany* 25:418–431.

Halling, R.E.; Osmundson, T.W.; Neves, M.A. 2008. Pacific boletes: Implications for biogeographic relationships. *Mycological Research* 112:437–447.

Handelsman, J. 2004. Metagenomics: Application of genomics to uncultured microorganisms. *Microbiology and Molecular Biology Reviews* 68:669–685.

Harrower, E.; Bougher, N.L.; Henkel, T.W.; Horak, E.; Matheny, P.B. 2015. Long-distance dispersal and speciation of Australasian and American species of *Cortinarius* sect. *Cortinarius. Mycologia* 107:697–709.

Heads, M. 1998. Biogeographic disjunction along the Alpine fault, New Zealand. *Biological Journal of the Linnean Society* 63:161–176.

Heads, M. 2014. *Biogeography of Australasia: A Molecular Analysis*. Cambridge University Press: Cambridge, UK.

Hennig, W. 1966. *Phylogenetic Systematics*. University of Illinois Press: Urbana, IL.

Hibbett, D.S. 2001. Shiitake mushrooms and molecular clocks: Historical biogeography of *Lentinula. Journal of Biogeography* 28:231–241.

Hibbett, D.S.; Donoghue, M.J. 1996. Implications of phylogenetic studies for conservation of genetic diversity in shiitake mushrooms. *Conservation Biology* 10:1321–1327.

Hibbett, D.S.; Fukumasa-Nakai, Y.; Tsuneda, A.; Donoghue, M.J. 1995. Phylogenetic diversity in shiitake inferred from nuclear ribosomal DNA sequences. *Mycologia* 87:618–638.

Hibbett, D.S.; Hansen, K.; Donoghue, M.J. 1998. Phylogeny and biogeography of *Lentinula* inferred from an expanded rDNA dataset. *Mycological Research* 102:1041–1049.

Hibbett, D.S.; Pine, E.M.; Langer, E.; Langer, G.; Donoghue, M.J. 1997. Evolution of gilled mushrooms and puffballs inferred from ribosomal DNA sequences. *Proceedings of the National Academy of Sciences, USA* 94:12002–12006.

Hitchcock, C.J.; Chambers, S.M.; Cairney, J.W.G. 2011. Genetic population structure of the ectomycorrhizal fungus *Pisolithus microcarpus* suggests high gene flow in south-eastern Australia. *Mycorrhiza* 21:131–137.

Horak, E. 1983. Mycogeography in the South Pacific region: Agaricales, Boletales. *Australian Journal of Botany*, Supplementary Series, 10:1–41.

Horton, T.R.; Swaney, D.P.; Galante, T.E. 2013. Dispersal of ectomycorrhizal basidiospores: The long and short of it. *Mycologia* 105:1623–1626.

Hosaka, K. 2009. Phylogeography of the genus *Pisolithus* revisited with some additional taxa from New Caledonia and Japan. *Bulletin of the National Science Museum, Tokyo* 35:151–167.

Hosaka, K.; Castellano, M.A.; Spatafora, J.W. 2008. Biogeography of Hysterangiales (Phallomycetidae, Basidiomycota). *Mycological Research* 112:448–462.

Hughes, G.C. 1986. Biogeography and the marine fungi. In *The Biology of Marine Fungi*, ed. Moss, S.T., 275–295. Cambridge University Press: Cambridge, UK.

Hughes, K.W.; McGhee, L.L.; Johnson, J.E.; Methven, A.S.; Petersen, R.H. 1999. Patterns of geographic speciation in the genus *Flammulina* based on sequences of the ribsomal ITS1-5.8S-ITS2 area. *Mycologia* 91:978–986.

Humphries, C.J.; Cox, J.M.; Nielsen, E.S. 1986. *Nothofagus* and its parasites: A cladistic approach to coevolution. In *Coevolution and Systematics*, eds. Stone, A.R.; Hawksworth, D.L., 55–76. Claredon Press: Oxford, UK.

Hyde, K.D.; Fryar, S.; Tian, Q.; Bahkali, A.H.; Xu, J.C. 2016. Lignicolous freshwater fungi along a north–south latitudinal gradient in the Asian/Australian region: Can we predict the impact of global warming on biodiversity and function? *Fungal Ecology* 19:190–200.

Isaac, S.; Frankland, J.C.; Watling, R.; Whalley, A.J.S.; eds. 1993. *Aspects of Tropical Mycology*. Cambridge University Press: Cambridge, UK.

Isikhuemhen, O.S.; Moncalvo, J.-M.; Nerud, F.; Vilgalys, R. 2000. Mating compatibility and phylogeography in *Pleurotus tuberregium*. *Mycological Research* 104:732–737.

James, T.Y. 2015. Why mushrooms have evolved to be so promiscuous: Insights from evolutionary and ecological patterns. *Fungal Biology Reviews* 29:167–178.

James, T.Y.; Moncalvo, J.-M.; Li, S.; Vilgalys, R. 2001. Polymorphism at the ribosomal DNA spacers and its relation to breeding structure of the widespread mushroom *Schizophyllum commune*. *Genetics* 157:149–161.

James, T.Y.; Porter, D.; Hamrick, J.L.; Vilgalys, R. 1999. Evidence for limited intercontinental gene flow in the cosmopolitan mushroom, *Schizophyllum commune*. *Evolution* 53:1665–1677.

James, T.Y.; Vilgalys, R. 2001. Abundance and diversity of *Schizophyllum commune* spore clouds in the Caribbean detected by selective sampling. *Molecular Ecology* 10:471–479.

Jančič, S.; Nguyen, H.D.T.; Frisvad, J.C.; *et al.* 2015. A taxonomic revision of the *Wallemia sebi* species somplex. *PLoS ONE* 10 (5):1–25.

Jin, J.; Hughes, K.W.; Petersen, R.H. 2001. Biogeographical patterns in *Panellus stypticus*. *Mycologia* 93:309–316.

Johnston, P.R. 1993. Biogeography of New Zealand Rhytismataceae. In *Abstracts, 'Southern Temperate Ecosystems: Origin and Diversification'*, ed. Hill, R.S., 53. A combined conference of Southern Connection, the Australian Systematic Botany Society and the Ecological Society of Australia: Hobart, Tasmania, January 1993.

Johnston, P.R. 2001. Rhytismatales of Australasia. *Australian Systematic Botany* 14:377–384.

Johnston, P.R. 2006. New Zealand's nonlichenised fungi — where they came from, who collected them, where they are now. In *Proceedings of the Seventh and Eighth Symposia on Collection Building and Natural History Studies in Asia and the Pacific Rim*, eds. Tomida, Y.; Kubodera, T.; Akiyama, S.; Kitayama, T., 37–49. National Science Museum Monographs 34, National Science Museum: Tokyo, Japan.

Johnston, P.R. 2010. Causes and consequences of changes to New Zealand's fungal biota. *New Zealand Journal of Ecology* 34:175–184.

Johnston, P.R.; May, T.W.; Park, D.; Horak, E. 2007. *Hypocreopsis amplectens* sp. nov., a rare fungus from New Zealand and Australia. *New Zealand Journal of Botany* 45:715–719.

Johnston, P.R.; Park, D. 2005. *Chlorociboria* (Fungi, Helotiales) in New Zealand. *New Zealand Journal of Botany* 43:679–719.

Jørgensen, P.M. 1983. Distribution patterns of lichens in the Pacific region. *Australian Journal of Botany*, Supplementary Series, 10:43–66.

Kasuga, T.; Taylor, J.W.; White, T.J. 1999. Phylogenetic relationships of varieties and geographical groups of the human pathogenic fungus *Histoplasma capsulatum* Darling. *Journal of Clinical Microbiology* 37:653–663.

Kasuga, T.; White, T.J.; Koenig, G.; *et al.* 2003. Phylogeography of the fungal pathogen *Histoplasma capsulatum*. *Molecular Ecology* 12:3383–3401.

Kasuga, T.; White, T.J.; Taylor, J.W. 2002. Estimation of nucleotide substitution rates in Eurotiomycete fungi. *Molecular Biology and Evolution* 19:2318–2324.

Kasuya, T.; Hosaka, K.; Uno, K.; Kakishima, M. 2012. Phylogenetic placement of *Geastrum melanocephalum* and polyphyly of *Geastrum triplex*. *Mycoscience* 53:411–426.

Kauserud, H.; Svegården, I.B.; Decock, C.; Hallenberg, N. 2006. Hybridization among cryptic species of the cellar fungus *Coniophora puteana* (Basidiomycota). *Molecular Ecology* 16:389–399.

Keast, A. 1981. *Ecological Biogeography of Australia*. Dr. W. Junk: The Hague, the Netherlands.

Kennedy, P.G.; Matheny, P.B.; Ryberg, K.M.; Henkel, T.W.; Uehling, J.K.; Smith, M.E. 2012. Scaling up: Examining the macroecology of ectomycorrhizal fungi. *Molecular Ecology* 21:4151–4154.

Kile, G.A.; Watling, R. 1983. *Armillaria* species from south-eastern Australia. *Transactions of the British Mycological Society* 81:129–140.

Kile, G.A.; Watling, R. 1988. Identification and occurrence of Australian *Armillaria* species, including *A. pallidula* sp. nov. and comparative studies between them and non-Australian tropical and Indian *Armillaria*. *Transactions of the British Mycological Society* 91:305–315.

Kirk, P.M.; Cannon, P.F.; Minter, D.W.; Stalpers, J.A. 2008. *Dictionary of the Fungi*, 10th edn. CABI Publishing: Wallingford, UK.

Kivlin, S.N.; Hawkes, C.V.; Treseder, K.K. 2011. Global diversity and distribution of arbuscular mycorrhizal fungi. *Soil Biology and Biochemistry* 43:2294–2303.

Knudsen, H.; Ryman, S. 2000. Vegetation zones and mycogeography of the area. In *Nordic Macromycetes*, vol. 1, eds. Hansen, L.; Knudsen, H., 21–33. Nordsvamp: Copenhagen, Denmark.

Kohlmeyer, J. 1983. Geography of marine fungi. *Australian Journal of Botany*, Supplementary Series, 10:67–76.

Kohn, L.M. 2005. Mechanisms of fungal speciation. *Annual Review of Phytopathology* 43:279–308.

Korf, R.P. 1983. *Cyttaria* (Cyttariales): Coevolution with *Nothofagus*, and evolutionary relationship to the Boedijnopezizeae (Pezizales, Sarcoscyphaceae). *Australian Journal of Botany*, Supplementary Series, 10:77–87.

Koufopanou, V.; Burt, A.; Szaro, T.; Taylor, J.W. 2001. Gene genealogies, cryptic species and molecular evolution in the human pathogen *Coccidioides immitis* and relatives (Ascomycota, Onygenales). *Molecular Biology and Evolution* 18:1246–1258.

Koufopanou, V.; Burt, A.; Taylor, J.W. 1997. Concordance of gene genealogies reveals reproductive isolation in the pathogenic fungus *Coccidioides immitis*. *Proceedings of the National Academy of Sciences of the United States of America* 94:5478–5482.

Kriticos, D.J.; Morin, L.; Leriche, A.; *et al.* 2013. Combining a climatic niche model of an invasive fungus with its host species distributions to identify risks to natural assets: *Puccinia psidii* sensu lato in Australia. *PLoS ONE* 8(5):e64479:1–13.

Kuschel, G. 1975. *Biogeography and Ecology in New Zealand*. Dr. W. Junk: The Hague, the Netherlands.

Kvas, M.; Marasas, W.F.O.; Wingfield, B.D.; Wingfield, M.J.; Steenkamp, E.T. 2009. Diversity and evolution of *Fusarium* species in the *Gibberella fujikuroi* complex. *Fungal Diversity* 34:1–21.

Leavitt, S.D.; Moreau, C.S.; Lumbsch, H.T. 2015. The dynamic discipline of species delimitation: Progress toward effectively recognizing species boundaries in natural populations. In *Recent Advances in Lichenology: Modern Methods and Approaches in Lichen Systematics and Culture Techniques*, vol. 2, eds. Upreti, D.K.; Divakar, P.K.; Shukla, V.; Bajpai, R., 11–44. Springer India: New Delhi.

Lebel, T.; Castellano, M.A.; Beever, R.E. 2014. Cryptic diversity in the sequestrate genus *Stephanospora* (Stephanosporaceae: Agaricales) in Australasia. *Fungal Biology* 119:201–228.

Lesage-Meessen, L.; Haon, M.; Uzan, E.; *et al.* 2011. Phylogeographic relationships in the polypore fungus *Pycnoporus* inferred from molecular data. *FEMS Microbiology Letters* 325:37–48.

Lickey, E.B.; Hughes, K.W.; Petersen, R.H. 2003. Phylogenetic and taxonomic studies in *Artomyces* and *Clavicorona* (Homobasidiomycetes: Auriscalpiaceae). *Sydowia* 55:181–254.

Looney, B.P.; Ryberg, M.; Hampe, F.; Sánchez-García, M.; Matheny, P.B. 2016. Into and out of the tropics: Global diversification patterns in a hyperdiverse clade of ectomycorrhizal fungi. *Molecular Ecology* 25:630–647.

Lücking, R.; Huhndorf, S.; Pfister, D.H.; Plata, E.R.; Lumbsch, H.T. 2009. Fungi evolved right on track. *Mycologia* 101:810–822.

Lumbsch, H.T.; Buchanan, P.K.; May, T.W.; Mueller, G.M. 2008. Phylogeography and biogeography of fungi. *Mycological Research* 112:423–424.

Malloch, D.W.; Pirozynski, K.A.; Raven, P.H. 1980. Ecological and evolutionary significance of mycorrhizal symbiosis in vascular plants (a review). *Proceedings of the National Academy of Sciences of the United States of America* 77:2113–2118.

Maphosa, L.; Wingfield, B.D.; Coetzee, M.P.A.; Mwenje, E.; Wingfield, M.J. 2006. Phylogenetic relationships among *Armillaria* species inferred from partial elongation factor 1-alpha DNA sequence data. *Australasian Plant Pathology* 35:513–520.

Martin, F.; Aerts, A.; Ahren, D.; *et al.* 2008. The genome sequence of the basidiomycete fungus *Laccaria bicolor* provides insights into the mycorrhizal symbiosis. *Nature* 452:88–92.

Martin, F.; Diez, J.; Dell, B.; Delaruelle, C. 2002. Phylogeography of the ectomycorrhizal *Pisolithus* species as inferred from nuclear ribosomal DNA ITS sequences. *New Phytologist* 153:345–357.

Martiny, J.B.; Bohannan, B.J.; Brown, J.H.; *et al.* 2006. Microbial biogeography: Putting microorganisms on the map. *Nature Reviews: Microbiology* 4:102–112.

Matheny, P.B.; Aime, M.C.; Bougher, N.L.; *et al.* 2009. Out of the Palaeotropics? Historical biogeography and diversification of the cosmopolitan mushroom family Inocybaceae. *Journal of Biogeography* 36:577–592.

Matheny, P.B.; Bougher, N.L. 2006. The new genus *Auritella* from Africa and Australia (Inocybaceae, Agaricales): Molecular systematics, taxonomy and historical biogeography. *Mycological Progress* 5:2–17.

Matheny, P.B.; Pradeep, C.K.; Vrinda, K.B.; Varghese, S.P. 2012. *Auritella foveata*, a new species of Inocybaceae (Agaricales) from tropical India. *Kew Bulletin* 67:119–125.

May, T.W. 2001. Documenting the fungal biodiversity of Australasia: From 1800 to 2000 and beyond. *Australian Systematic Botany* 14:329–356.

May, T.W. 2002. Where are the short-range endemics among Western Australian macrofungi? *Australian Systematic Botany* 15:501–511.

May, T.W.; Avram, J. 1997. *The Conservation Status and Distribution of Macrofungi in Victoria*. National Herbarium of Victoria, Royal Botanic Gardens Melbourne: South Yarra, Australia.

May, T.W.; Farrer, S.L.; eds. 2001. Biodiversity and biogeography of Australasian fungi. *Australian Systematic Botany* 14(3).

May, T.W.; Harris, G. 2014. Australian macrofungi have wide distributions that are explained by climate, *Atlas of Living Australia Science Symposium*, Canberra, Australia. http://www.ala.org.au/about-the-atlas/communications-centre/presentations/2014-symposium-presentations. Accessed 22 March, 2016.

May, T.W.; Milne, J.; Shingles, S.; Jones, R.H. 2003. *Catalogue and Bibliography of Australian Fungi: 2. Basidiomycota p.p. & Myxomycota p.p.* Fungi of Australia, vol. 2B. CSIRO Publishing: Melbourne, Australia.

May, T.W.; Simpson, J. 2001. Preface [to 'Biodiversity and Biogeography of Australasian Fungi']. *Australian Systematic Botany* 14:i–iii:329–511.

May, T.W.; Simpson, J.A. 1997. Fungal diversity and ecology in eucalypt ecosystems. In *Eucalypt Ecology: Individuals to Ecosystems*, eds. Williams, J.E.; Woinarski, J.C.Z., 246–277. Cambridge University Press: Cambridge, UK.

May, T.W.; Wood, A.E. 1997. *Catalogue and Bibliography of Australian Macrofungi: 1. Basidiomycota p.p.* Fungi of Australia, vol. 2A. Australian Biological Resources Study: Canberra, Australia.

McKenzie, E.H.C. 2004. *Fungi of New Zealand (Ngā Harore o Aotearoa), Vol. 1: Introduction to fungi of New Zealand*. Fungal Diversity Press: Hong Kong.

McLaughlin, D.J.; Hibbett, D.S.; Lutzoni, F.; Spatafora, J.W.; Vilgalys, R. 2009. The search for the fungal tree of life. *Trends in Microbiology* 17:488–497.

McTaggart, A.R.; Doungsa-Ard, C.; Geering, A.D.; Aime, M.C.; Shivas, R.G. 2015. A co-evolutionary relationship exists between *Endoraecium* (Pucciniales) and its *Acacia* hosts in Australia. *Persoonia* 35:50–62.

Meiser, A.; Bálint, M.; Schmitt, I. 2013. Meta-analysis of deep-sequenced fungal communities indicates limited taxon sharing between studies and the presence of biogeographic patterns. *New Phytologist* 201:623–635.

Methven, A.; Hughes, K.W.; Petersen, R.H. 2000. *Flammulina* RFLP patterns identify species and show biogeographical patterns within species. *Mycologia* 92:1064–1070.

Moncalvo, J.-M.; Buchanan, P.K. 2008. Molecular evidence for long distance dispersal across the Southern Hemisphere in the *Ganoderma applanatum-australe* species complex (Basidiomycota). *Mycological Research* 112:425–436.

Moncalvo, J.-M.; Vilgalys, R.; Redhead, S.A.; *et al.* 2002. One hundred and seventeen clades of euagarics. *Molecular Phylogenetics and Evolution* 23:357–400.

Morgado, L.N.; Noordeloos, M.E.; Lamoureux, Y.; Geml, J. 2013. Multi-gene phylogenetic analyses reveal species limits, phylogeographic patterns, and evolutionary histories of key morphological traits in *Entoloma* (Agaricales, Basidiomycota). *Persoonia* 31:159–178.

Moyersoen, B.; Beever, R.E.; Martin, F. 2003. Genetic diversity of *Pisolithus* in New Zealand indicates multiple long-distance dispersal from Australia. *New Phytologist* 160:569–579.

Mueller, G.M.; Schmit, J.P.; Leacock, P.R.; *et al.* 2007. Global diversity and distribution of macrofungi. *Biodiversity and Conservation* 16:37–48.

Mueller, G.M.; Wu, Q.-W.; Huang, Y.-Q.; Guo, S.-Y.; Aldana-Gomez, R.; Vilgalys, R. 2001. Assessing biogeographical relationships between North American and Chinese macrofungi. *Journal of Biogeography* 28:271–281.

Nalim, F.A.; Samuels, G.J.; Wijesundera, R.L.; Geiser, D.M. 2011. New species from the *Fusarium solani* species complex derived from perithecia and soil in the Old World tropics. *Mycologia* 103:1302–1330.

Neafsey, D.E.; Barker, B.M.; Sharpton, T.J.; *et al.* 2010. Population genomic sequencing of *Coccidioides* fungi reveals recent hybridization and transposon control. *Genome Research* 20:938–946.

Nguyen, H.D.T.; Jančič, S.; Meijer, M.; *et al.* 2015. Application of the phylogenetic species concept to *Wallemia sebi* from house dust and indoor air revealed by multi-locus genealogical concordance. *PLoS ONE* 10(5):1–17.

Nilsson, R.H.; Tedersoo, L.; Ryberg, M.; *et al.* 2015. A comprehensive, automatically updated fungal ITS sequence dataset for reference-based chimera control in environmental sequencing efforts. *Microbes and Environments* 30:145–150.

Núñez, M.; Suhirman; Stokland, J. 2002. Patterns of polypores distribution in the Lesser Sunda Islands, Indonesia. Is Wallace's line significant? In *Tropical Mycology, Vol. 1: Macromycetes*, eds. Watling, R.; Frankland, J.C.; Ainsworth, A.M.; Isaac, S.; Robinson, C.H., 73–86. CABI Publishing: Wallingford, UK.

O'Donnell, K. 2000. Molecular phylogeny of the *Nectria haematococca–Fusarium solani* species complex. *Mycologia* 92:919–938.

O'Donnell, K.; Cigelnik, E.; Nitenberg, H.I. 1998. Molecular systematics and phylogeography of the *Gibberella fujikuroi* species complex. *Mycologia* 90:465–493.

O'Donnell, K.; Kistler, H.C.; Tacke, B.K.; Casper, H.H. 2000a. Gene genealogies reveal global phylogeographic structure and reproductive isolation among lineages of *Fusarium graminearum*, the fungus causing wheat scab. *Proceedings of the National Academy of Sciences of the United States of America* 97:7905–7910.

O'Donnell, K.; Nirenberg, H.I.; Aoki, T.; Cigelnik, E. 2000b. A multigene phylogeny of the *Gibberella fujikuroi* species complex: Detection of additional phylogenetically distinct species. *Mycoscience* 41:61–78.

O'Donnell, K.; Rooney, A.P.; Mills, G.L.; Kuo, M.; Weber, N.S.; Rehner, S.A. 2010. Phylogeny and historical biogeography of true morels (*Morchella*) reveals an Early Cretaceous origin and high continental endemism and provincialism in the Holarctic. *Fungal Genetics and Biology* 48:252–265.

O'Donnell, K.; Rooney, A.P.; Proctor, R.H.; *et al.* 2013. Phylogenetic analyses of RPB1 and RPB2 support a Middle Cretaceous origin for a clade comprising all agriculturally and medically important fusaria. *Fungal Genetics and Biology* 52:20–31.

Oono, R.; Lutzoni, F.; Arnold, A.E.; *et al.* 2014. Genetic variation in horizontally transmitted fungal endophytes of pine needles reveals population structure in cryptic species. *American Journal of Botany* 101:1362–1374.

Orgiazzi, A.; Dunbar, M.B.; Panagos, P.; de Groot, G.; Lemanceau, P. 2015. Soil biodiversity and DNA barcodes: Opportunities and challenges. *Soil Biology and Biochemistry* 80:244–250.

Pearce, C.A.; Hyde, K.D. 2006. *Phyllachoraceae of Australia*. Fungal Diversity Press: Hong Kong.

Peay, K.G.; Bruns, T.D.; Kennedy, P.G.; Bergemann, S.E.; Garbelotto, M. 2007. A strong species-area relationship for eukaryotic soil microbes: Island size matters for ectomycorrhizal fungi. *Ecology Letters* 10:470–480.

Peay, K.G.; Garbelotto, M.; Bruns, T.D. 2010. Evidence of dispersal limitation in soil microorganisms: Isolation reduces species richness on mycorrhizal tree islands. *Ecology* 91:3631–3640.

Peay, K.G.; Schubert, M.G.; Nguyen, N.H.; Bruns, T.D. 2012. Measuring ectomycorrhizal fungal dispersal: Macroecological patterns driven by microscopic propagules. *Molecular Ecology* 21:4122–4136.

Peintner, U.; Horak, E.; Moser, M.M.; Vilgalys, R. 2002. Phylogeny of *Rozites, Cuphocybe* and *Rapacea* inferred from ITS and LSU rDNA sequences. *Mycologia* 94:620–629.

Peintner, U.; Moncalvo, J.-M.; Vilgalys, R. 2004. Toward a better understanding of the infrageneric relationships in *Cortinarius* (Agaricales, Basidiomycota). *Mycologia* 96:1042–1058.

Petersen, R.H.; Bermudes, D. 1992. *Panellus stypticus*: Geographically separated interbreeding populations. *Mycologia* 84:209–213.

Petersen, R.H.; Hughes, K.W. 2003. Phylogeographic examples of Asian biodiversity in mushrooms and their relatives. *Fungal Diversity* 13:95–109.

Petersen, R.H.; Hughes, K.W. 2007. Some agaric distribution patterns involving Pacific landmasses and Pacific Rim. *Mycoscience* 48:1–14.

Petersen, R.H.; Hughes, K.W.; Lickey, E.B.; Kovalenko, A.E.; Morozova, O.V.; Psurtseva, E.B. 2008. A new genus, *Cruentomycena*, with *Mycena viscidocruenta* as type species. *Mycotaxon* 105:119–136.

Peterson, K.R.; Pfister, D.H. 2010. Phylogeny of *Cyttaria* inferred from nuclear and mitochondrial sequence and morphological data. *Mycologia* 102:1398–1416.

Peterson, K.R.; Pfister, D.H.; Bell, C.D. 2010. Cophylogeny and biogeography of the fungal parasite *Cyttaria* and its host, *Nothofagus*, southern beech. *Mycologia* 102:1417–1425.

Pirozynski, K.A. 1968. Geographical distribution of fungi. In *The Fungi: An Advanced Treatise*, eds. Ainsworth, G.C.; Sussman, A.S., 487–504. Academic Press: New York.

Pirozynski, K.A. 1983. Pacific mycogeography: An appraisal. *Australian Journal of Botany*, Supplementary Series, 10:137–159.

Pirozynski, K.A.; Malloch, D.W. 1975. The origin of land plants: A matter of mycotrophism. *Biosystems* 6:153–164.

Pirozynski, K.A.; Walker, J. 1983. Preface [to 'Pacific Mycogeography: A Preliminary Approach']. *Australian Journal of Botany*, Supplementary Series, 10:v–vi.

Pirozynski, K.A.; Weresub, L.K. 1979. A biogeographic view of the history of Ascomycetes and the development of their pleomorphism. In *The Whole Fungus: The Sexual-Asexual Synthesis*, ed. Kendrick, B., 93–123. National Museum of Natural Sciences, National Museums of Canada and the Kananaskis Foundation: Ottawa, Canada.

Priest, M.J. 2006. *Fungi of Australia: Septoria*. CSIRO Publishing: Melbourne, Australia.

Pringle, A.; Adams, R.I.; Cross, H.B.; Bruns, T.D. 2009. The ectomycorrhizal fungus *Amanita phalloides* was introduced and is expanding its range on the west coast of North America. *Molecular Ecology* 18:817–833.

Pringle, A.; Baker, D.M.; Platt, J.L.; Wares, J.P.; Latgé, J.P.; Taylor, J.W. 2005. Cryptic speciation in the cosmopolitan and clonal human pathogenic fungus *Aspergillus fumigatus*. *Evolution* 59:1886–1899.

Qin, J.; Feng, B.; Yang, Z.L.; *et al.* 2014. The taxonomic foundation, species circumscription and continental endemisms of *Singerocybe*: Evidence from morphological and molecular data. *Mycologia* 106:1015–1026.

Queloz, V.; Sieber, T.N.; Holdenrieder, O.; McDonald, B.A.; Grünig, C.R. 2011. No biogeographical pattern for a root-associated fungal species complex. *Global Ecology and Biogeography* 20:160–169.

Rajchenberg, M. 1989. Polyporaceae (Aphyllophorales, Basidiomycetes) from southern South America: A mycogeographical view. *Sydowia* 41:277–291.

Ramírez-Camejo, L.A.; Zuluaga-Montero, A.; Lázaro-Escudero, M.; Hernández-Kendall, V.; Bayman, P. 2012. Phylogeography of the cosmopolitan fungus *Aspergillus flavus*: Is everything everywhere? *Fungal Biology* 116:452–463.

Raper, J.R.; Krongelb, G.S.; Baxter, M.G. 1958. The number and distribution of incompatibility factors in *Schizophyllum*. *American Naturalist* 92:221–232.

Redhead, S.A. 1989. A biogeographical overview of the Canadian mushroom flora. *Canadian Journal of Botany* 67:3003–3062.

Rochet, J.; Moreau, P.-A.; Manzi, S.; Garde, M. 2011. Comparative phylogenies and host specialization in the alder ectomycorrhizal fungi *Alnicola, Alpova* and *Lactarius* (Basidiomycota) in Europe. *BMC Evolutionary Biology* 11(40):1–14.

Rogers, R.W. 1992. Lichen ecology and biogeography. In *Flora of Australia, Vol. 54: Lichens; Introduction; Lecanorales 1*, ed. George, A.S., 30–42. Australian Government Publishing Service: Canberra, Australia.

Rogers, R.W.; Stevens, G.N. 1981. Lichens. In *Ecological Biogeography of Australia*, ed. Keast, A., 593–603. Dr. W. Junk: The Hague, the Netherlands.

Roy, B.A. 2001. Patterns of association between crucifers and their flower-mimic pathogens: Host jumps are more common than coevolution or cospeciation. *Evolution* 55:41–53.

Ryberg, M. 2015. Molecular operational taxonomic units as approximations of species in the light of evolutionary models and empirical data from Fungi. *Molecular Ecology* 24:5770–5777.

Ryberg, M.; Matheny, P.B. 2011. Incomplete taxon sampling and diversification in a large clade of mushroom-forming fungi. *Evolution* 65:1862–1878.

Ryberg, M.; Matheny, P.B. 2012. Asynchronous origins of ectomycorrhizal clades of Agaricales. *Proceedings of the Royal Society B* 279:2003–2011.

Rydholm, C.; Szakacs, G.; Lutzoni, F. 2006. Low genetic variation and no detectable population structure in *Aspergillus fumigatus* compared to closely related *Neosartorya* species. *Eukaryotic Cell* 5:650–657.

Ryvarden, L. 1991. *Genera of Polypores: Nomenclature and Taxonomy*. Fungiflora: Oslo, Norway.

Salgado-Salazar, C.; Rossman, A.Y.; Chaverri, P. 2013. Not as ubiquitous as we thought: Taxonomic crypsis, hidden diversity and cryptic speciation in the cosmopolitan fungus *Thelonectria discophora* (Nectriaceae, Hypocreales, Ascomycota). *PLoS ONE* 8(10):1–12.

Sánchez-Ramírez, S.; Etienne, R.S.; Moncalvo, J.-M. 2015. High speciation rate at temperate latitudes explains unusual diversity gradients in a clade of ectomycorrhizal fungi. *Evolution* 69:2196–2209.

Sánchez-Ramírez, S.; Tulloss, R.E.; Amalfi, M.; Moncalvo, J.-M. 2014. Paleotropical origin, boreotropical distribution, and increased rates of diversification in a clade of edible ectomycorrhizal mushrooms (*Amanita* section *Caesareae*). *Journal of Biogeography* 42:351–363.

Sangalang, A.E.; Backhouse, D.; Burgess, L.W. 1995a. Survival and growth in culture of four *Fusarium* species in relation to occurrence in soils from hot climatic regions. *Mycological Research* 99:529–533.

Sangalang, A.E.; Burgess, L.W.; Duff, J.; Backhouse, D.; Wurst, M. 1995b. Mycogeography of *Fusarium* species in soils from tropical, arid and Mediterranean regions of Australia. *Mycological Research* 99:523–528.

Sankaran, K.V.; Sutton, B.C.; Minter, D.W. 1995. A checklist of fungi recorded on *Eucalyptus*, Mycological Papers No. 170. CABI Publishing: Wallingford, UK.

Sanmartín, I.; Ronquist, F. 2004. Southern hemisphere biogeography inferred by event-based models: Plant versus animal patterns. *Systematic Biology* 53:216–243.

Sato, H.; Tsujino, R.; Kurita, K.; Yokoyama, K.; Agata, K. 2012. Modelling the global distribution of fungal species: New insights into microbial cosmopolitanism. *Molecular Ecology* 21:5599–5612.

Sauquet, H.; Ho, S.Y.W.; Gandolfo, M.A.; *et al.* 2012. Testing the impact of calibration on molecular divergence times using a fossil-rich group: The case of *Nothofagus* (Fagales). *Systematic Biology* 61:289–313.

Savile, D.B.O. 1979. Fungi as aids in higher plant classification. *Botanical Review* 45:377–503.

Savile, D.B.O. 1990. Coevolution of Uredinales and Ustilaginales with vascular plants. *Reports of the Tottori Mycological Institute* 28:15–24.

Schmale, D.G., 3rd; Ross, S.D. 2015. Highways in the sky: Scales of atmospheric transport of plant pathogens. *Annual Review of Phytopathology* 53:591–611.

Schmidt, P.-A.; Bálinta, M.; Greshakea, B.; Bandowa, C.; Römbkea, J.; Schmitt, I. 2013. Illumina metabarcoding of a soil fungal community. *Soil Biology and Biochemistry* 65:128–132.

Schoch, C.L.; Seifert, K.A.; Huhndorf, S.; *et al.* 2012. Nuclear ribosomal internal transcribed spacer (ITS) region as a universal DNA barcode marker for Fungi. *Proceedings of the National Academy of Sciences of the United States of America* 109:6241–6246.

Seberg, O. 1991. Biogeographic congruence in the South Pacific. *Australian Systematic Botany* 4:127–136.

Shaw, D.E. 1982. Ecology of fungi in New Guinea. In *Biogeography and Ecology of New Guinea*, Vol. 1, ed. Gressitt, J.L., 475–496. Dr. W. Junk: The Hague, the Netherlands.

Sheedy, E.M.; Van de Wouw, A.P.; Howlett, B.J.; May, T.W. 2013. Multigene sequence data reveal morphologically cryptic phylogenetic species within the genus *Laccaria* in southern Australia. *Mycologia* 105:547–563.

Sheedy, E.M.; Van de Wouw, A.P.; Howlett, B.J.; May, T.W. 2015. Population genetic structure of the ectomycorrhizal fungus *Laccaria* sp. A resembles that of its host tree *Nothofagus cunninghamii. Fungal Ecology* 13:23–32.

Shivas, R.G.; Vánky, K. 2003. Biodiversity of Australian smut fungi. *Fungal Diversity* 13:137–152.

Skrede, I.; Engh, I.B.; Binder, M.; Carlsen, T.; Kauserud, H.; Bendiksby, M. 2011. Evolutionary history of Serpulaceae (Basidiomycota): Molecular phylogeny, historical biogeography and evidence for a single transition of nutritional mode. *BMC Evolutionary Biology* 11(230):1–13.

Smith, J.M.B. 1986. Origins of Australasian tropicalpine and alpine floras. In *Flora and Fauna of Alpine Australasia, Ages and Origins*, ed. Barlow, B. 109–128. CSIRO: Melbourne, Australia.

Soop, K.; Gasparini, B. 2011. Europe et Pacifique Sud: Une comparaison de deux flores de *Cortinarius*. *Journal des JEC* 13:34–51.

Spencer, W.B. 1896. Report on the work of the Horn Scientific Expedition to Central Australia: Part I. Introduction, narrative, summary of results, supplement to zoological report, map. Melville, Mullen & Slade: Melbourne, Australia.

Stefani, F.O.P.; Jones, R.H.; May, T.W. 2014. Concordance of seven gene genealogies compared to phenotypic data reveals multiple cryptic species in Australian dermocyboid *Cortinarius* (Agaricales). *Molecular Phylogenetics and Evolution* 71:249–260.

Stephenson, S.L.; Schnittler, M.; Novozhilov, Y.K. 2007. Myxomycete diversity and distribution from the fossil record to the present. *Biodiversity and Conservation* 17:285–301.

Stielow, J.B.; Lévesque, C.A.; Seifert, K.A.; *et al.* 2015. One fungus, which genes? Development and assessment of universal primers for potential secondary fungal DNA barcodes. *Persoonia* 35:242–263.

Stubbe, D.; Nuytinck, J.; Verbeken, A. 2010. Critical assessment of the *Lactarius gerardii* species complex (Russulales). *Fungal Biology* 114:271–283.

Summerell, B.A.; Laurence, M.H.; Liew, E.C.Y.; Leslie, J.F. 2010. Biogeography and phylogeography of *Fusarium*: A review. *Fungal Diversity* 44:3–13.

Summerell, B.A.; Leslie, J.F.; Liew, E.C.Y.; *et al.* 2011. *Fusarium* species associated with plants in Australia. *Fungal Diversity* 46:1–27.

Summerell, B.A.; Rugg, C.A.; Burgess, L.W. 1993. Mycogeography of *Fusarium*: Survey of *Fusarium* species associated with forest and woodland communities in north Queensland, Australia. *Mycological Research* 97:1015–1019.

Summerell, B.A.; Rugg, C.A.; Burgess, L.W. 1995. Characterization of *Fusarium babinda* sp. nov. *Mycological Research* 99:1345–1348.

Sung, G.-H.; Poinar, G.O.; Spatafora, J.W. 2008. The oldest fossil evidence of animal parasitism by fungi supports a Cretaceous diversification of fungal–arthropod symbioses. *Molecular Phylogenetics and Evolution* 49:495–502.

Takamatsu, S. 2013. Origin and evolution of the powdery mildews (Ascomycota, Erysiphales). *Mycoscience* 54:75–86.

Takamatsu, S.; Matsuda, S. 2004. Estimation of molecular clocks for ITS and 28S rDNA in Erysiphales. *Mycoscience* 45:340–344.

Talbot, J.M.; Bruns, T.D.; Taylor, J.W.; *et al.* 2014. Endemism and functional convergence across the North American soil mycobiome. *Proceedings of the National Academy of Sciences of the United States of America* 111:6341–6346.

Tate, R. 1889. On the influence of physiological changes in the distribution of life in Australia, Report of the First Meeting of the Australian Association for the Advancement of Science, pp. 312–326.

Taylor, J.W.; Berbee, M.L. 2006. Dating divergences in the Fungal Tree of Life: Review and new analyses. *Mycologia* 98:838–849.

Taylor, J.W.; Jacobson, D.J.; Kroken, S.; *et al.* 2000. Phylogenetic species recognition and species concepts in fungi. *Fungal Genetics and Biology* 31:21–32.

Taylor, J.W.; Turner, E.; Townsend, J.P.; Dettman, J.R.; Jacobson, D. 2006. Eukaryotic microbes, species recognition and the geographic limits of species: Examples from the kingdom Fungi. *Philosophical Transactions of the Royal Society of London, Series B* 361:1947–1963.

Taylor, M.W.; Tsai, P.; Anfang, N.; Ross, H.A.; Goddard, M.R. 2014. Pyrosequencing reveals regional differences in fruit-associated fungal communities. *Environmental Microbiology* 16:2848–2858.

Taylor, T.N.; Krings, M.; Taylor, E.L. 2015. *Fossil Fungi*. Elsevier: Amsterdam, the Netherlands.

Tedersoo, L.; Bahram, M.; Põlme, S.; *et al.* 2014a. Global diversity and geography of soil fungi. *Science* 346:1078.

Tedersoo, L.; Bahram, M.; Ryberg, M. 2014b. Global biogeography of the ectomycorrhizal/sebacina lineage (Fungi, Sebacinales) as revealed from comparative phylogenetic analyses. *Molecular Ecology* 23:4168–4183.

Thackway, R.; Cresswell, I.D. 1995. *An Interim Biogeographic Regionalisation for Australia: A Framework for Establishing the National System of Reserves*, Version 4.0. Australian Nature Conservation Agency: Canberra, Australia

Tisthammer, K.H.; Cobian, G.M.; Amend, A.S. 2016. Global biogeography of marine fungi is shaped by the environment. *Fungal Ecology* 19:39–46.

Trappe, J.M. 1977. Biogeography of hypogeous fungi: Trees, mammals, and continental drift. In *Abstracts 2nd International Mycological Congress*, eds. Bigelow, H.E.; Simmons, E.E., 675. University of South Florida: Tampa, FL.

Truswell, E.M. 1996. The fossil record of the fungi in Australia and the Australasian region. In *Fungi of Australia Volume 1A*, eds. Grgurinovic, C.; Mallett, K., 321–340. CSIRO: Canberra, Australia.

Turner, B.C.; Perkins, D.D.; Fairfield, A. 2001. *Neurospora* from natural populations: A global study. *Fungal Genetics and Biology* 32:67–92.

Udayanga, D.; Castlebury, L.A.; Rossman, A.Y; Chukeatirote, E.; Hyde, K.D. 2014. Insights into the genus *Diaporthe*: Phylogenetic species delimitation in the *D. eres* species complex. *Fungal Diversity* 67:203–229.

Vánky, K.; Shivas, R.G. 2008. *Fungi of Australia: The Smut Fungi*. CSIRO: Melbourne, Australia.

Vilgalys, R.; Sun, B.L. 1994. Ancient and recent patterns of geographic speciation in the oyster mushroom *Pleurotus* revealed by phylogenetic analysis of ribosomal DNA sequences. *Proceedings of the National Academy of Sciences of the United States of America* 91:4599–4603.

Viljanen-Rollinson, S.L.H.; Cromey, M.G. 2002. Pathways of entry and spread of rust pathogens: Implications for New Zealand's biosecurity. *New Zealand Plant Protection* 55:42–48.

Viljanen-Rollinson, S.L.H.; Parr, E.L.; Marroni, M.V. 2007. Monitoring long-distance spore dispersal by wind: A review. *New Zealand Plant Protection* 60:291–296.

Villalta, C.F.; Jacobson, D.J.; Taylor, J.W. 2009. Three new phylogenetic and biological *Neurospora* species: *N. hispaniola, N. metzenbergii* and *N. perkinsii*. *Mycologia* 101:777–789.

Walker, J. 1983. Pacific mycogeography: Deficiencies and irregularities in the distribution of plant parasitic fungi. *Australian Journal of Botany*, Supplementary Series, 10:89–136.

Walker, J. 1996. Biogeography of fungi with special reference to Australia. In *Fungi of Australia Volume 1A*, eds. Grgurinovic, C.; Mallett, K., 263–320. CSIRO: Canberra, Australia.

Walker, J.; Hartigan, D.; Bertus, A.L. 1974. Poplar rusts in Australia with comments on potential conifer rusts. *European Journal of Forest Pathology* 4:100–118.

Wallis, G.P.; Trewick, S.A. 2009. New Zealand phylogeography: Evolution on a small continent. *Molecular Ecology* 18:3548–3580.

Watling, R. 1993. Comparison of the macromycete biotas in selected tropical areas of Africa and Australia. In *Aspects of Tropical Mycology*, eds. Isaac, S.; Frankland, J.C.; Watling, R.; Whalley, A.J.S., 171–191. Cambridge University Press: Cambridge, UK.

Watling, R. 2001a. Australian boletes: Their diversity and possible origins. *Australian Systematic Botany* 14:407–416.

Watling, R. 2001b. The relationships and possible distributional patterns of boletes in South-East Asia. *Mycological Research* 105:1440–1448.

Watson, I.A.; De Sousa, C.N.A. 1983. Long distance transport of spores of *Puccinia graminis tritici* in the southern hemisphere. *Proceedings of the Linnean Society of New South Wales* 106:311–321.

Werth, S. 2011. Biogeography and phylogeography of lichen fungi and their symbionts. In *Biogeography of Microscopic Organisms: Is Everything Small Everywhere?*, ed. Fontaneto, D., 191–208. Cambridge University Press: Cambridge, UK.

Whetton, P. 1997. Floods, droughts and the Southern Oscillation connection. In *Windows on Meteorology: Australian Perspective*, ed. Webb, E.K. 180–199. CSIRO: Collingwood, Australia.

White, T.J.; Bruns, T.; Lee, S.; Taylor, J.W. 1990. Amplification and direct sequencing of fungal ribosomal RNA genes for phylogenetics. In *PCR Protocols: A Guide to Methods and Applications*, eds. Innis, M.A.; Gelfand, D.H.; Sninsky, J.J.; White, T.J., 315–322. Academic Press: New York.

Wilson, A.W.; Binder, M.; Hibbett, D.S. 2012. Diversity and evolution of ectomycorrhizal host associations in the Sclerodermatineae (Boletales, Basidiomycota). *New Phytologist* 194:1079–1095.

Wolfe, C.B., Jr.; Bougher, N.L. 1993. Systematics, mycogeography, and evolutionary history of *Tylopilus* subg. *Roseoscabra* in Australia elucidated by comparison with Asian and American species. *Australian Systematic Botany* 6:187–213.

Wu, Q.-X.; Mueller, G.M.; Lutzoni, F.M.; Huang, Y.Q.; Guo, S.Y. 2000. Phylogenetic and biogeographic relationships of eastern Asian and eastern North American disjunct *Suillus* species (fungi) as inferred from nuclear ribosomal RNA ITS sequences. *Molecular Phylogenetics and Evolution* 17:37–47.

Wyatt, T.T.; Wösten, H.A.B.; Dijksterhuis, J. 2013. Fungal spores for dispersion in space and time. In *Advances in Applied Microbiology*, Vol. 85, eds. Sariaslani, S.; Gadd, G.M., 43–91. Academic Press: Burlington, MA.

Yang, Z.L. 2005. Diversity and biogeography of higher fungi in China. In *Evolutionary Genetics of Fungi*, ed. Xu, J., 35–62. Horizon Bioscience: Norfolk, UK.

Young, A.M. 1996. Macrofungi and CLIMEX: A computer program with applications for modelling species' distributions. *Australasian Mycological Newsletter* 15:30–34.

Young, A.M. 2005. *Fungi of Australia: Hygrophoraceae*. CSIRO Publishing: Melbourne, Australia.

Zaffarano, P.L.; McDonald, B.A.; Zala, M.; Linde, C.C. 2006. Global hierarchical gene diversity analysis suggests the Fertile Crescent is not the centre of origin of the barley scald pathogen *Rhynchosporium secalis*. *Phytopathology* 96:941–950.

Zervakis, G.I.; Moncalvo, J.-M.; Vilgalys, R. 2004. Molecular phylogeny, biogeography and speciation of the mushroom species *Pleurotus cystidiosus* and allied taxa. *Microbiology* 150:715–726.

Zhang, Y.; Hagen, F.; Stielow. B.; *et al.* 2015. Phylogeography and evolutionary patterns in *Sporothrix* spanning more than 14,000 human and animal case reports. *Persoonia* 35:1–20.

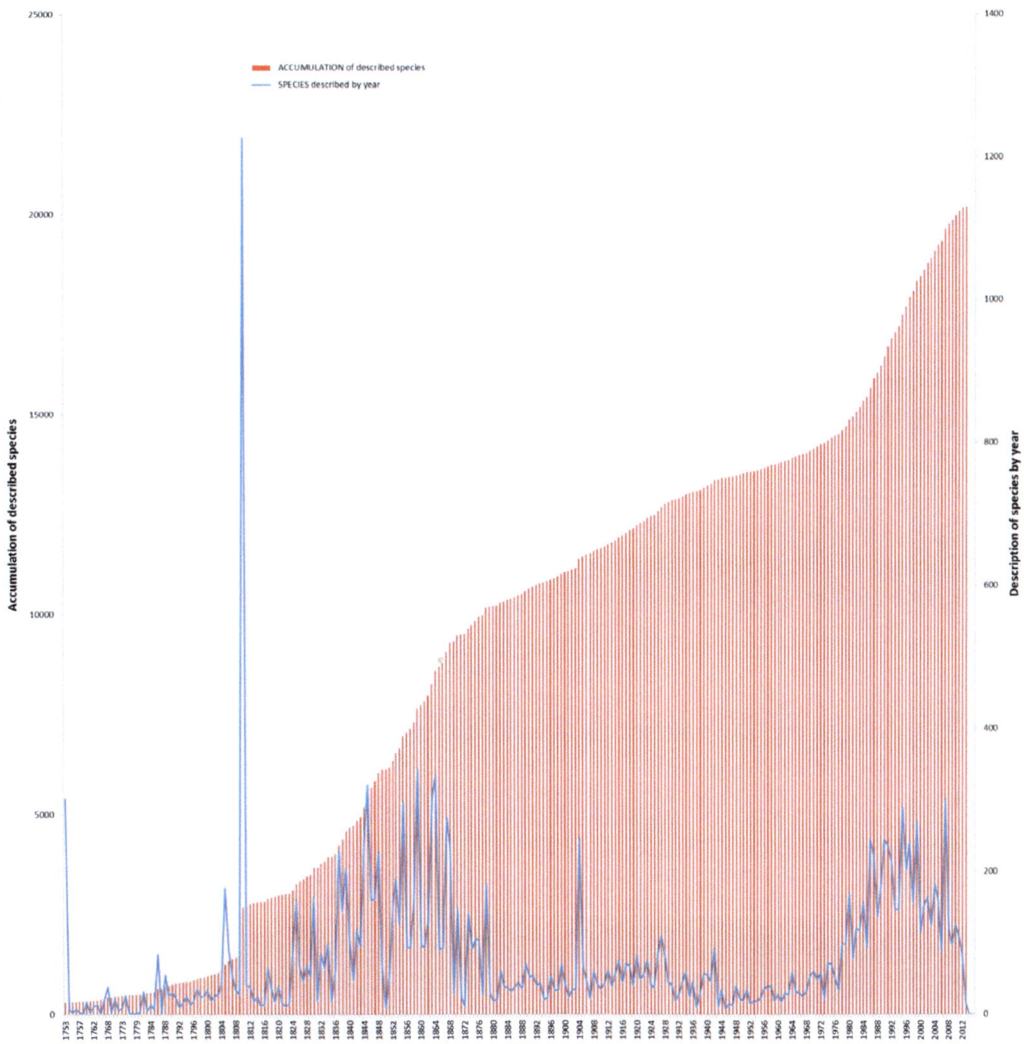

ACCUMULATION of described species
SPECIES described by year
Accumulation of described species
Description of species by year
Year

FIGURE 1.1

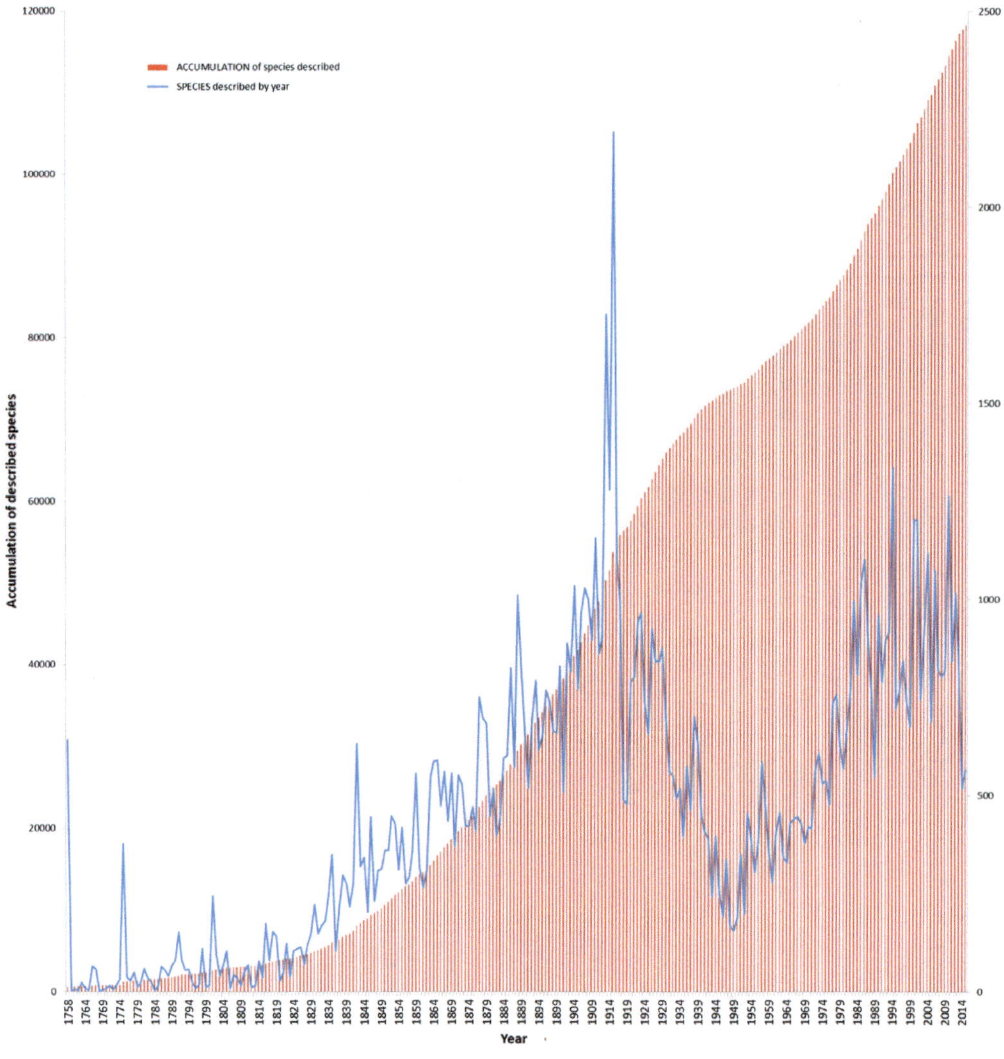

ACCUMULATION of species described
SPECIES described by year
Accumulation of described species
Description of species by year
Year

FIGURE 1.2

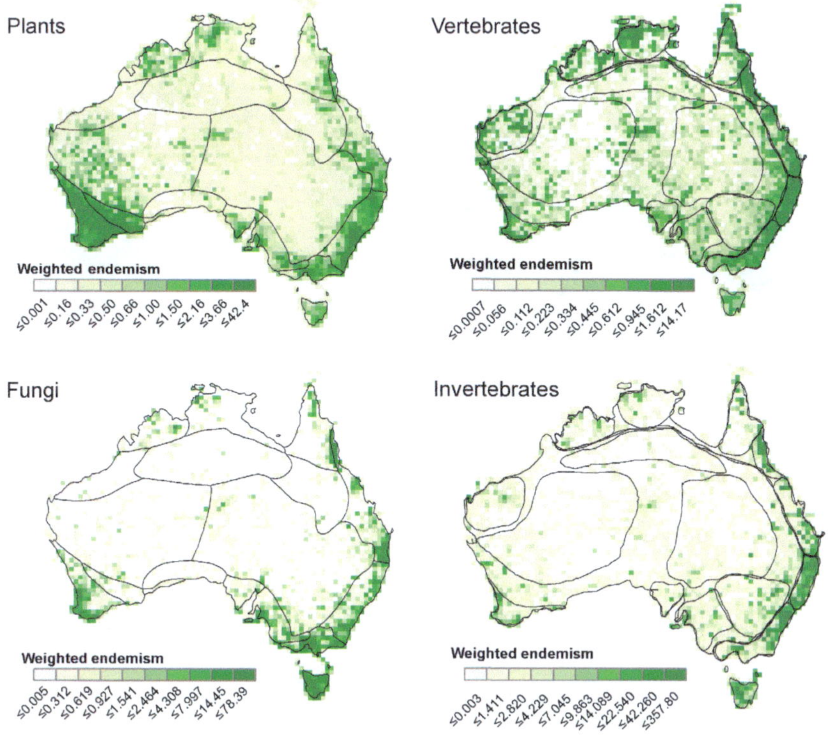

FIGURE 1.3

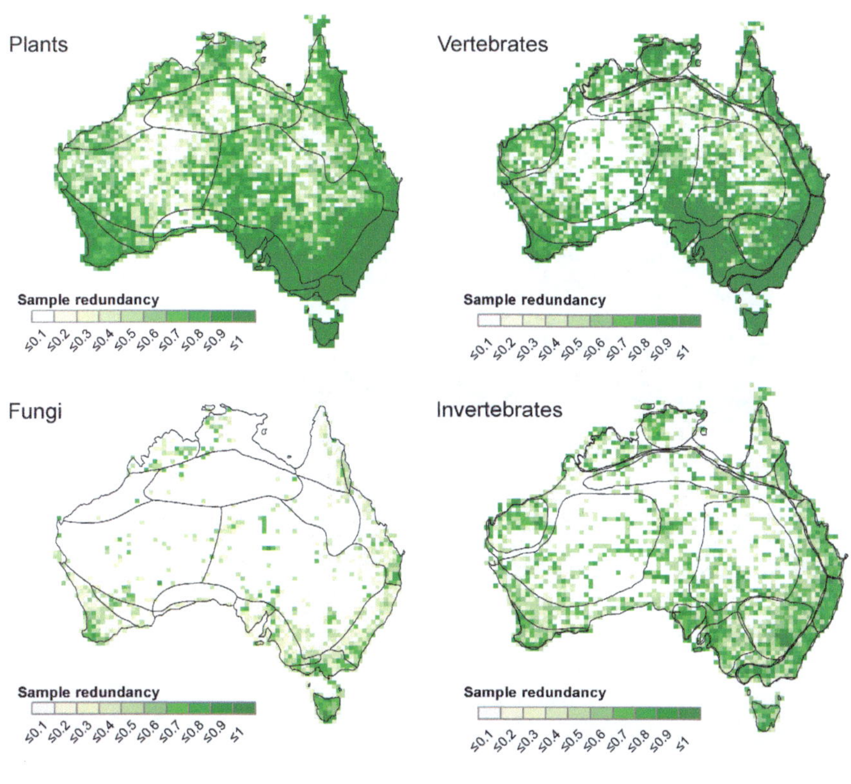

FIGURE 1.4

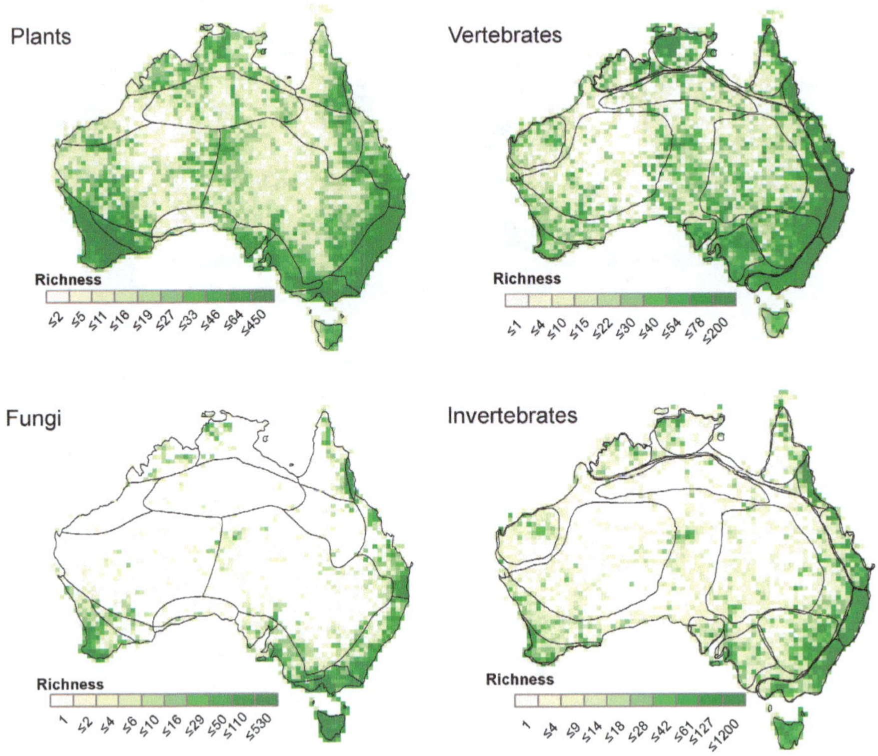

FIGURE 1.5

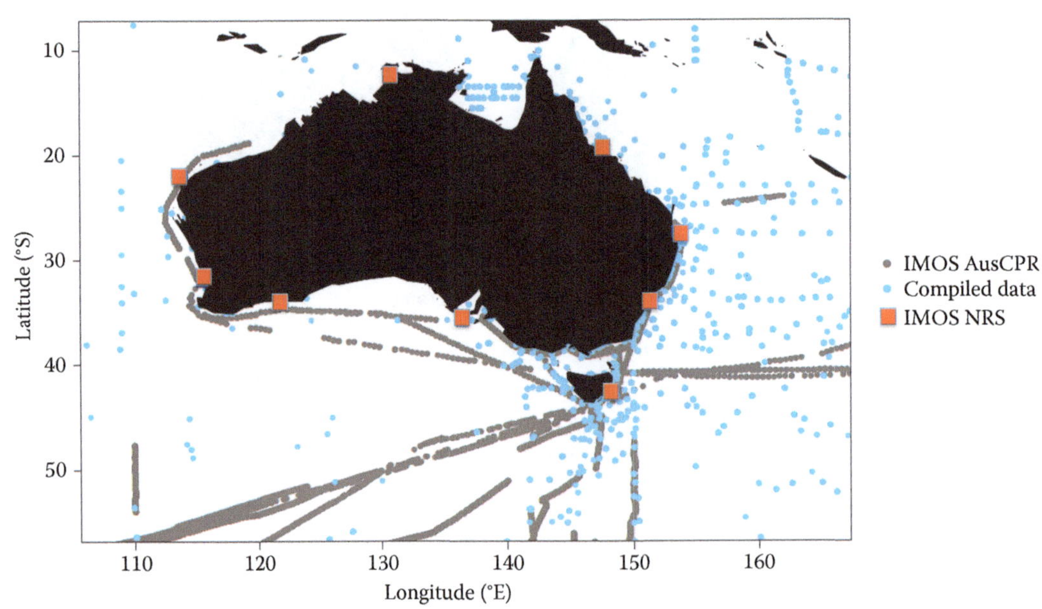

FIGURE 3.2

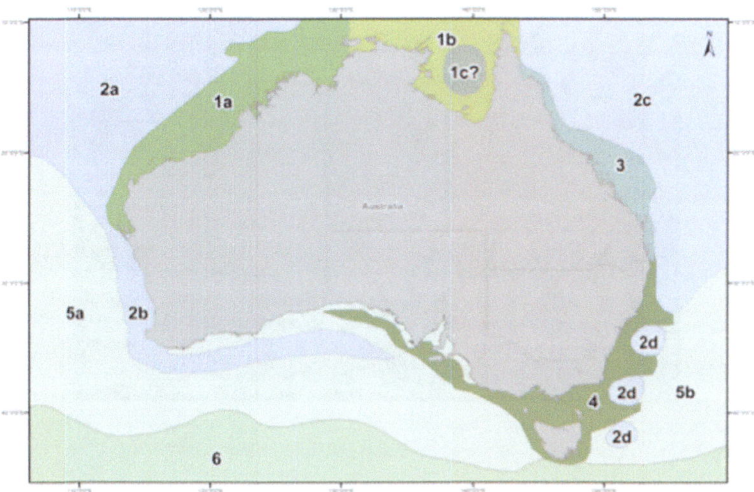

FIGURE 3.4

FIGURE 4.3

FIGURE 4.4

FIGURE 4.5

FIGURE 4.6

FIGURE 4.7

FIGURE 5.1

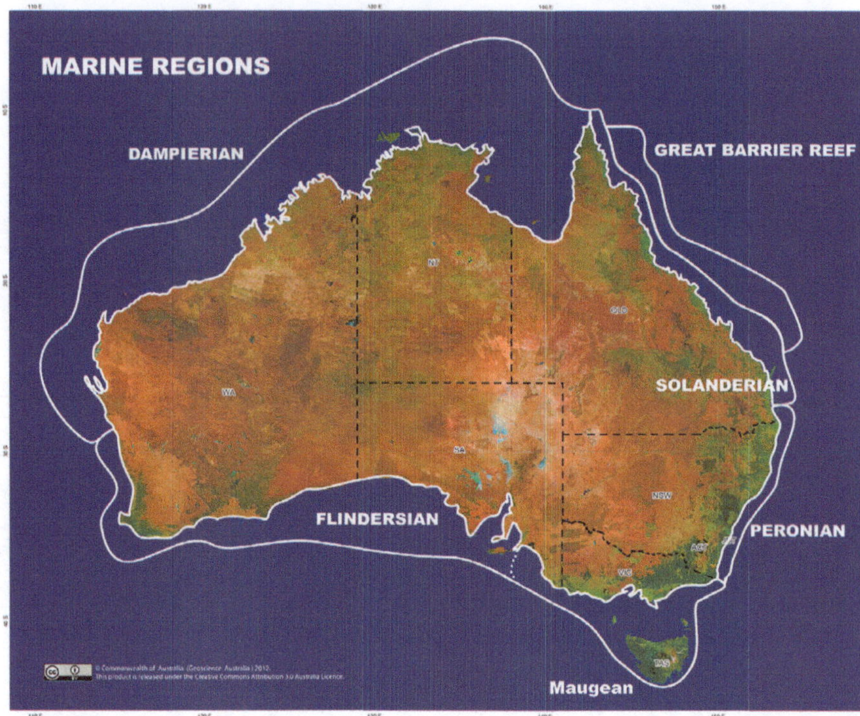
MARINE REGIONS
DAMPIERIAN
GREAT BARRIER REEF
SOLANDERIAN
FLINDERSIAN
PERONIAN
Maugean
© Commonwealth of Australia (Geoscience Australia) 2012.
This product is released under the Creative Commons Attribution 3.0 Australia Licence.

(a)

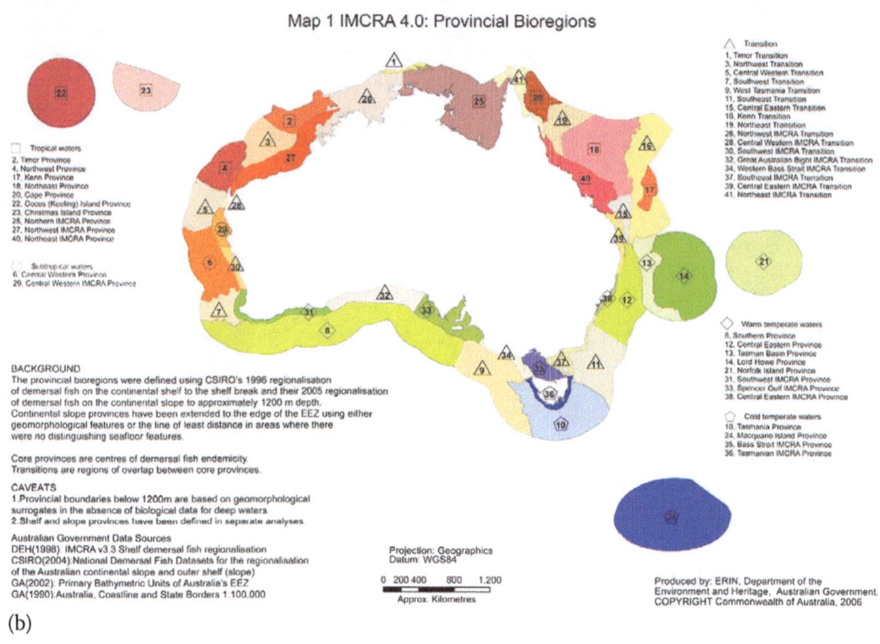
Map 1 IMCRA 4.0: Provincial Bioregions
Tropical waters
2, Timor Province
4, Northwest Province
17, Kenn Province
18, Northeast Province
20, Cape Province
22, Cocos (Keeling) Island Province
23, Christmas Island Province
25, Northern IMCRA Province
27, Northwest IMCRA Province
40, Northeast IMCRA Province
Subtropical waters
6, Central Western Province
29, Central Western IMCRA Province
Transition
1, Timor Transition
3, Northwest Transition
5, Central Western Transition
7, Southwest Transition
9, West Tasmania Transition
11, Southeast Transition
15, Central Eastern Transition
16, Kenn Transition
19, Northeast Transition
26, Northwest IMCRA Transition
28, Central Western IMCRA Transition
30, Southwest IMCRA Transition
32, Great Australian Bight IMCRA Transition
34, Western Bass Strait IMCRA Transition
37, Southeast IMCRA Transition
39, Central Eastern IMCRA Transition
41, Northeast IMCRA Transition
Warm temperate waters
8, Southern Province
12, Central Eastern Province
13, Tasman Basin Province
14, Lord Howe Province
21, Norfolk Island Province
31, Southwest IMCRA Province
33, Spencer Gulf IMCRA Province
38, Central Eastern IMCRA Province
Cold temperate waters
10, Tasmania Province
24, Macquarie Island Province
35, Bass Strait IMCRA Province
36, Tasmanian IMCRA Province
BACKGROUND
The provincial bioregions were defined using CSIRO's 1996 regionalisation of demersal fish on the continental shelf to the shelf break and their 2005 regionalisation of demersal fish on the continental slope to approximately 1200 m depth.
Continental slope provinces have been extended to the edge of the EEZ using either geomorphological features or the line of least distance in areas where there were no distinguishing seafloor features.
Core provinces are centres of demersal fish endemicity.
Transitions are regions of overlap between core provinces.
CAVEATS
1.Provincial boundaries below 1200m are based on geomorphological surrogates in the absence of biological data for deep waters.
2.Shelf and slope provinces have been defined in separate analyses.
Australian Government Data Sources
DEH(1998): IMCRA v3.3 Shelf demersal fish regionalisation
CSIRO(2004):National Demersal Fish Datasets for the regionalisation of the Australian continental slope and outer shelf (slope)
GA(2002): Primary Bathymetric Units of Australia's EEZ
GA(1990):Australia, Coastline and State Borders 1:100,000
Projection: Geographics
Datum: WGS84
0 200 400 800 1,200
Approx. Kilometres
Produced by: ERIN, Department of the Environment and Heritage, Australian Government.
COPYRIGHT Commonwealth of Australia, 2006

(b)

FIGURE 5.3

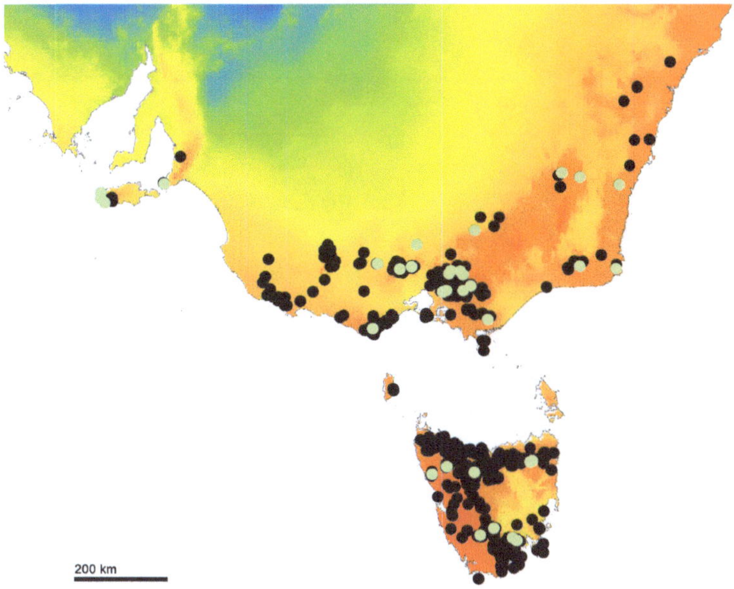

FIGURE 8.2

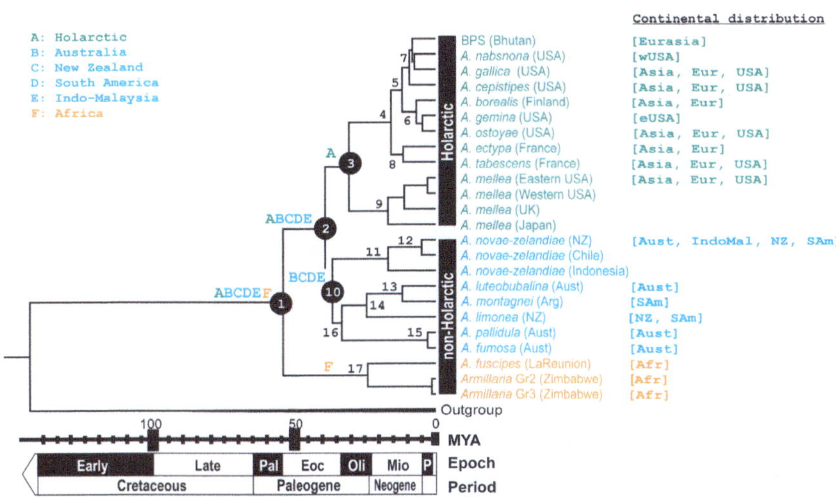

FIGURE 8.7

FIGURE 8.8

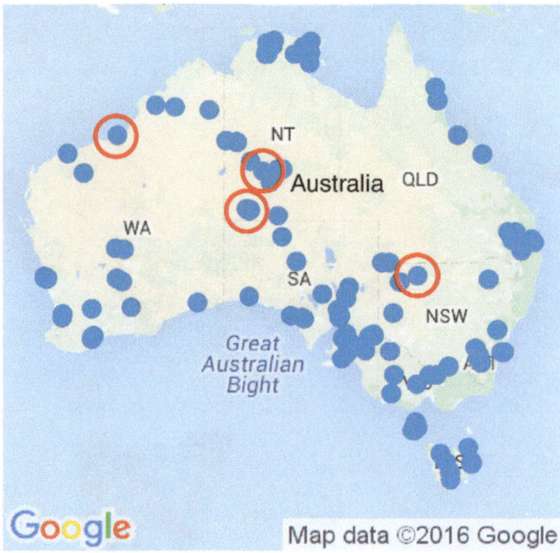

FIGURE 8.11

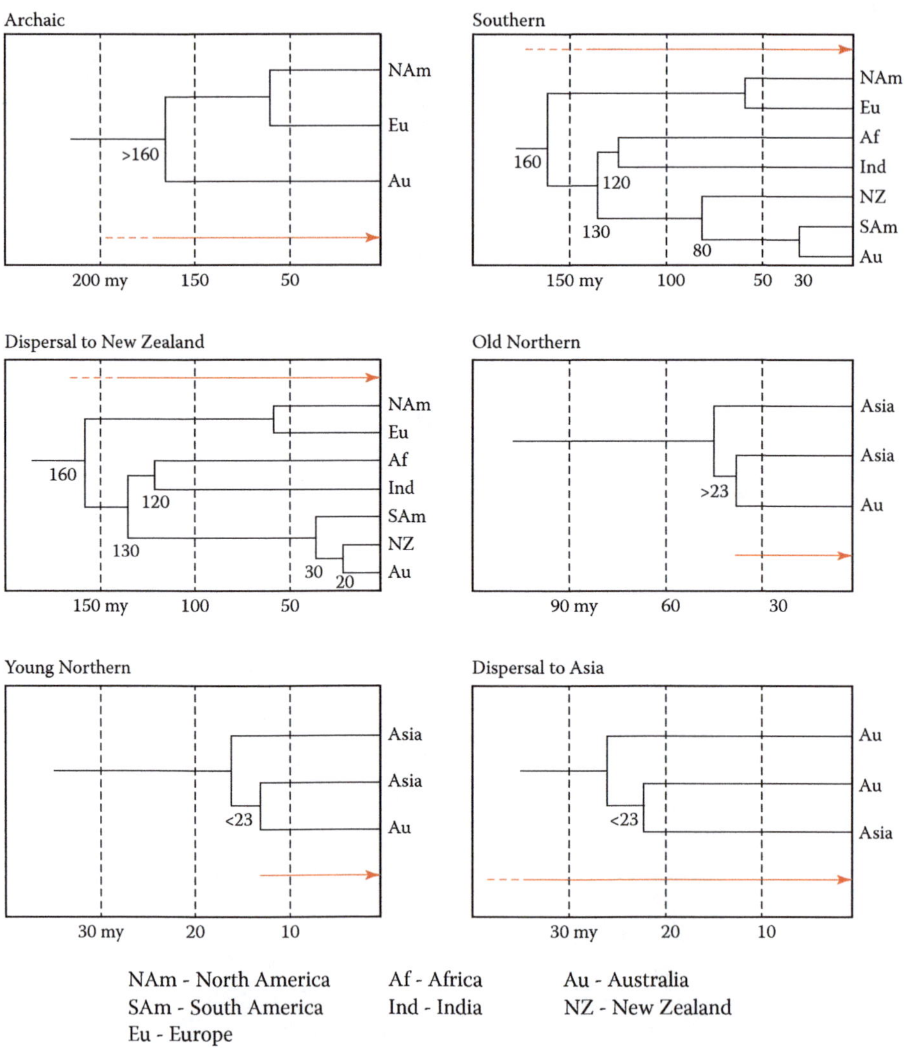

FIGURE 9.3

FIGURE 10.1

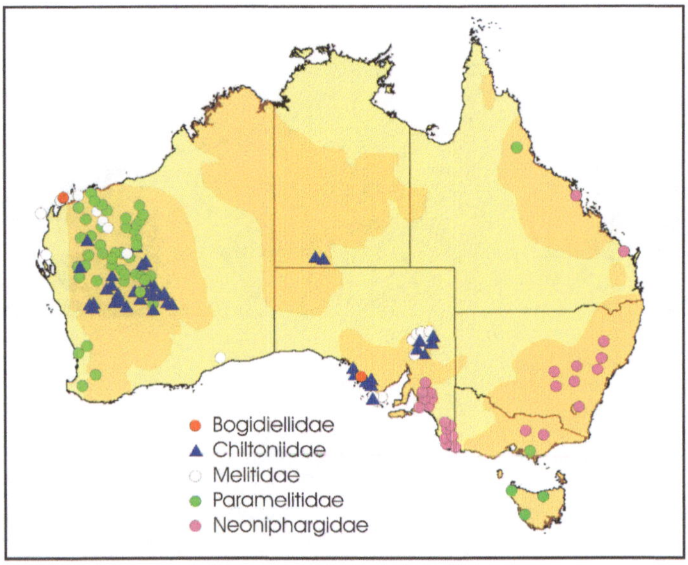

FIGURE 11.3

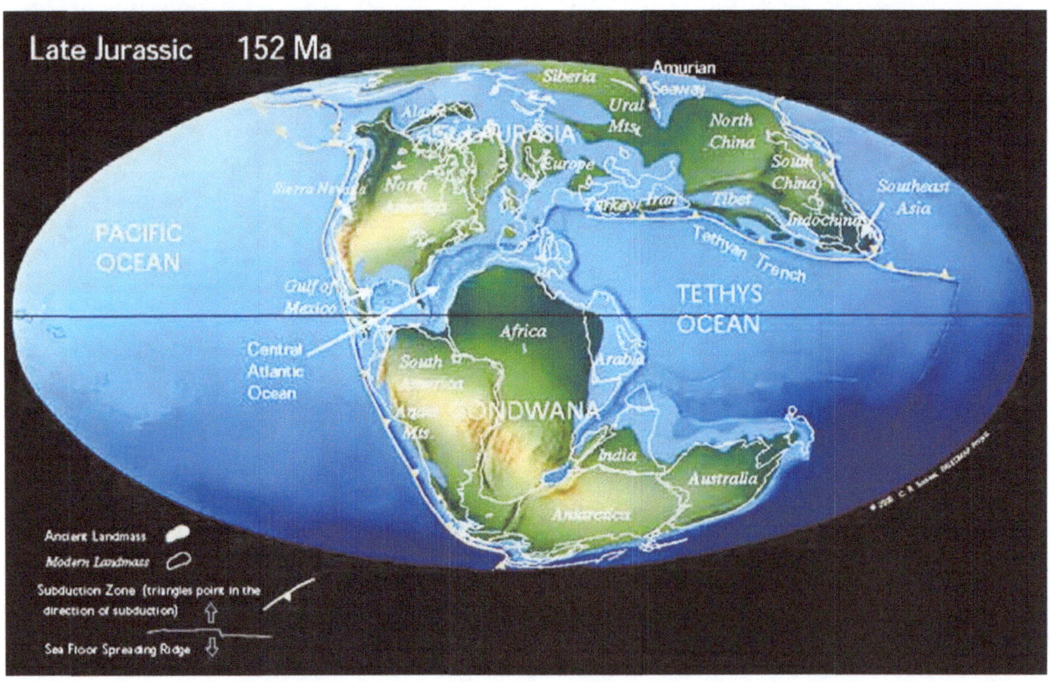

FIGURE 11.5

(a) (b) (c) (d) (e) (f) (g) (h)

FIGURE 12.1

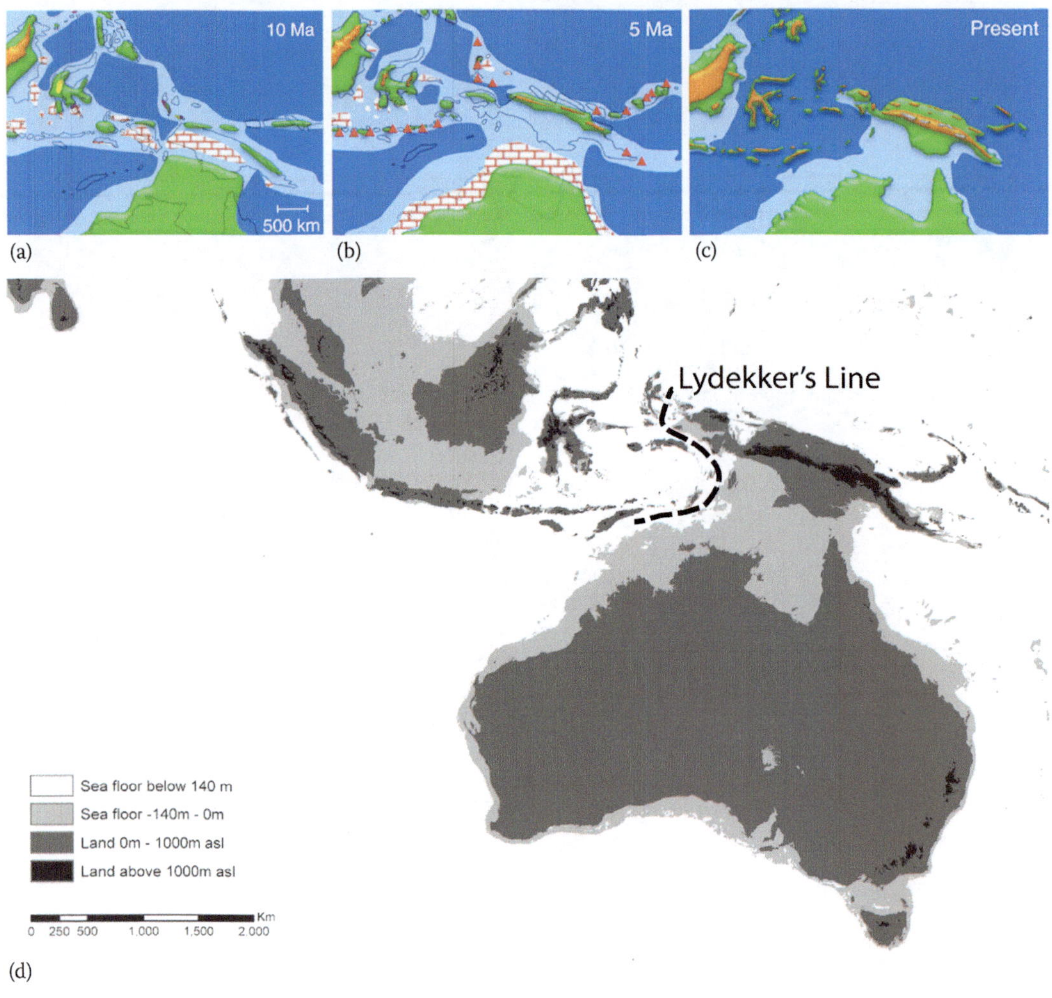

10 Ma
500 km
5 Ma
Present
(a)
(b)
(c)
Lydekker's Line
Sea floor below 140 m
Sea floor -140m - 0m
Land 0m - 1000m asl
Land above 1000m asl
Km
0 250 500 1,000 1,500 2,000
(d)

FIGURE 13.1

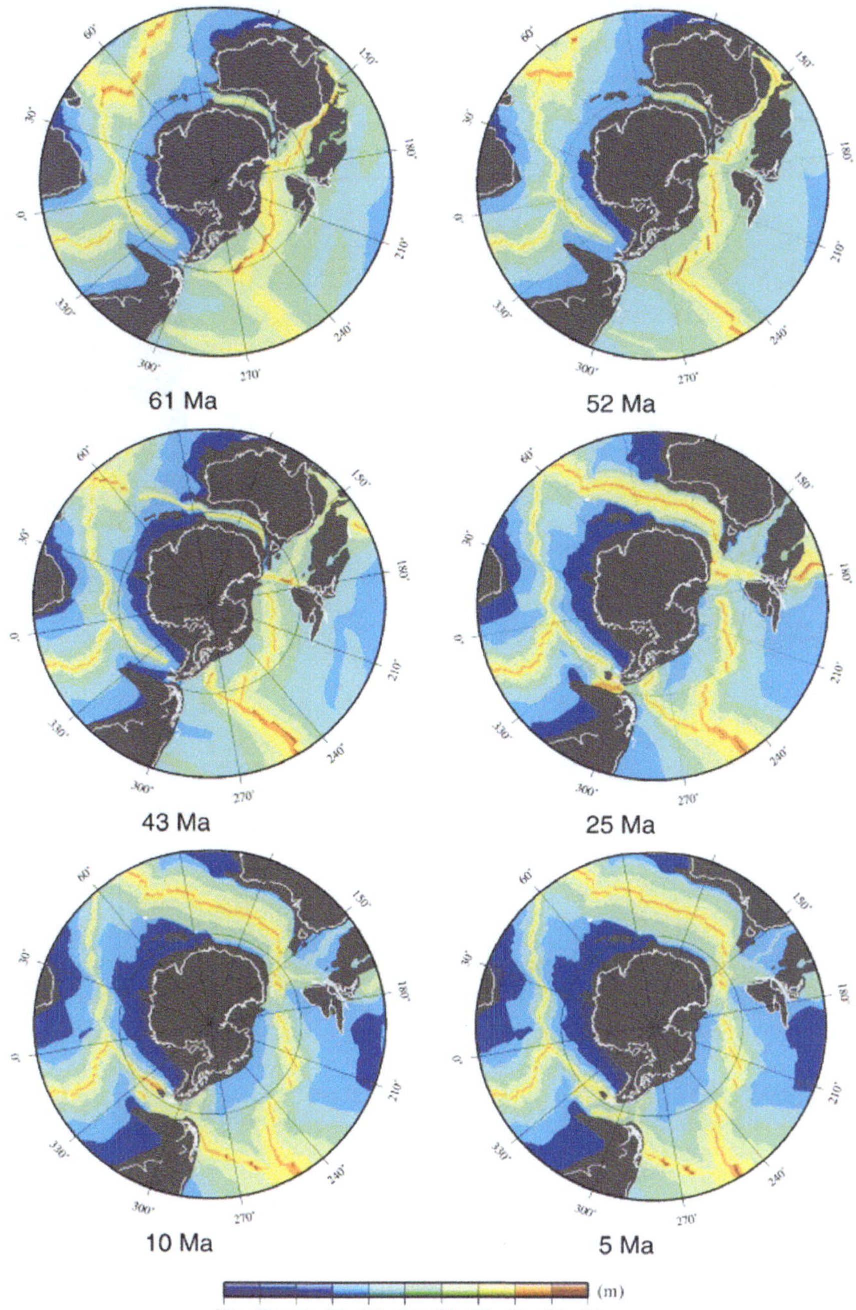

FIGURE 13.2

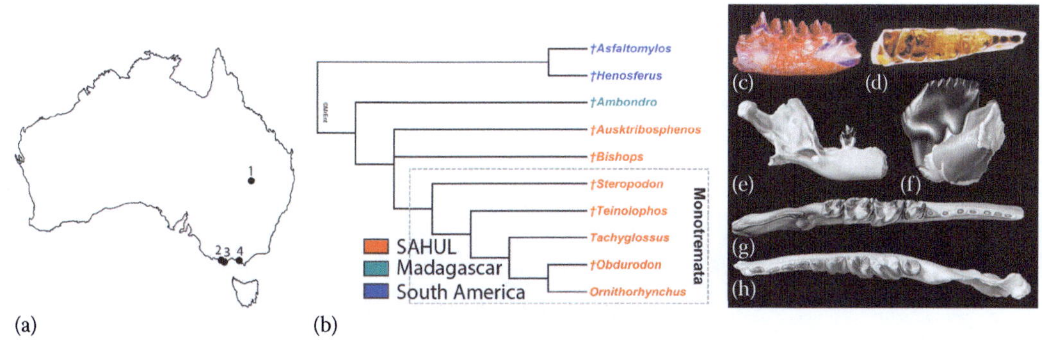

FIGURE 13.3

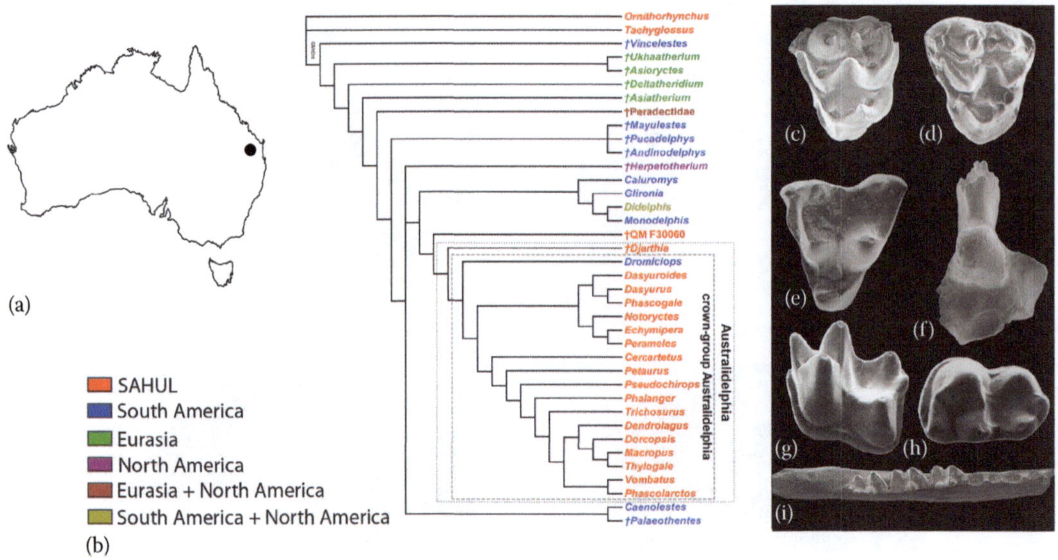

FIGURE 13.4

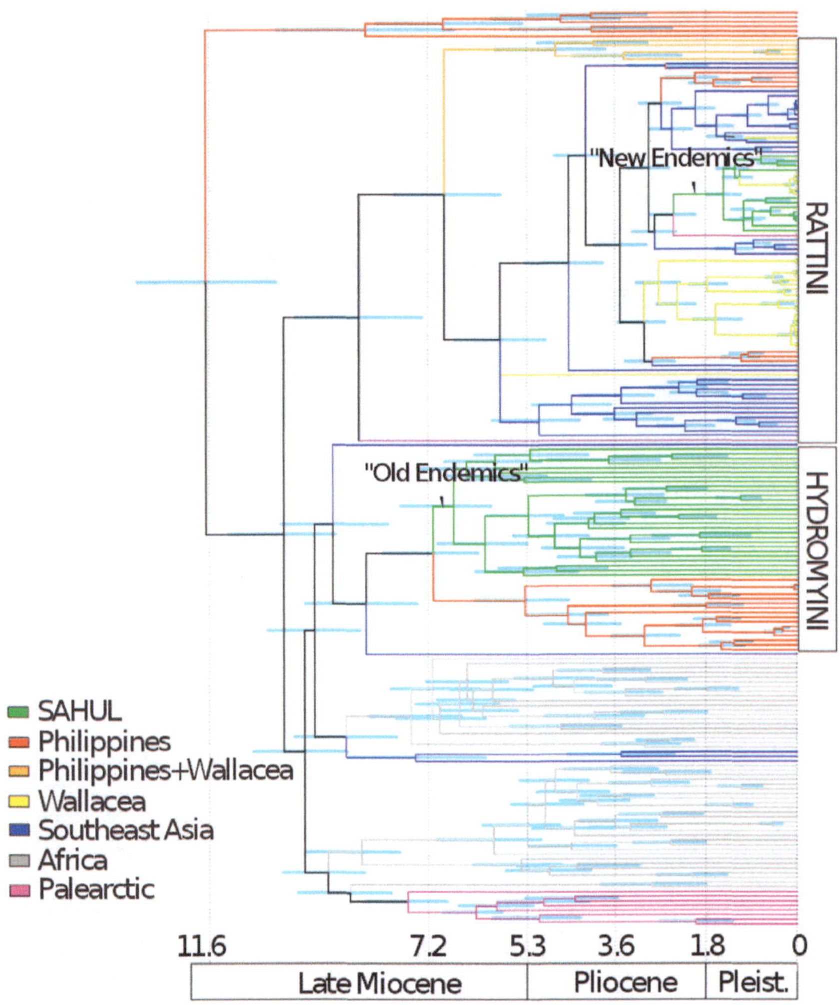

FIGURE 13.5

9

Australian Insect Biogeography: Beyond Faunal Provinces and Elements towards Processes

David K. Yeates and Gerasimos Cassis

CONTENTS

> The origin and composition of the Australian fauna is a time-worn subject. It is an unsatisfactory one, too, because it is so difficult to test the hypotheses that may be built on distributional and morphological evidence. Nevertheless – perhaps indeed for this reason – those who touch on it become addicts, and cannot leave it alone.
>
> **– Mackerras (1950, p. 157)**

Introduction

The insect fauna of the island continent of Australia is rich and diverse (Austin *et al.* 2004; Cranston 2010), with over 200,000 species (Yeates *et al.* 2003; Raven and Yeates 2007), and entomologists have attempted to explain the biogeography of the fauna for the past two centuries. Insect origins have now been traced back almost half a billion years (Misof *et al.* 2014), and it should be no surprise that almost all insect orders, except the small and unusual polyneopteran orders Zoraptera, Grylloblattodea, Mantophasmatodea and the holometabolous order Rhaphidioptera, are present in Australia. The crown groups of almost all insect orders were in place from the beginning of the Carboniferous to the beginning of the Triassic, 350–250 mya, and during the vast temporal range of insect evolution Australia has been broadly connected to other continents through Antarctica until 35 mya, and interaction with the Asian plate through Wallacea began around 10 million years later. The recent Tertiary isolation of Australia represents a tiny fraction of the branch length joining modern Australian insect lineages to the root of the insect phylogenetic tree, essentially a footnote in the annals of insect diversification. Thus, the emphasis on the unique insect fauna of Australia needs to be balanced with the strong links that the fauna has with other parts of the world, derived from this long history of connectedness. Modern biogeographic inquiry is possible only in the few groups that have well understood taxonomy, distribution and phylogenetic relationships. With only about 25% of the insect fauna described at species level (Yeates *et al.* 2003; Raven and Yeates 2007), much more basic descriptive research, phylogenetic reconstruction and surveys of distribution will underpin our future biogeographic understanding.

Significant insect family-level diversification had occurred in most orders during the period of Gondwanan breakup beginning 160 mya (Grimaldi and Engel 2005). Consistent with this is the observation that many family-level lineages are absent or depauperate in Australia, whereas others have radiated to an extent not seen on other continents (Austin *et al.* 2004; Cranston 2010). The Australian insect fauna is a global hotspot for *relictual* and *disjunct* groups; relicts are lineages whose distributions are very restricted from a wider distribution in the past, and disjuncts are close relatives that are now widely separated, either because of long-distance colonisation or the separation of ancestral ranges by continental drift (Grimaldi and Engel 2005). Relictual groups have drawn significant attention, such as the damselfly *Hemiphlebia*, dragonfly Petaluridae, *Apteropanorpa* and *Austromerope* in the Mecoptera, the hairy cicada *Tettigarcta*, the termite *Mastotermes*, beetle family Hygrobiidae, and the ant genera *Nothomyrmecia* and *Myrmecia*. In addition, groups with disjunct distributions that included South America and often New Zealand also gained early attention, such as the mecopteran Nannochoristidae, the bug families Peloridiidae, Aradidae (*Isodermus*) and Idiostolidae, the eustheniid and austroperlid stone flies, the fly families Xylophagidae (*Exeretonevra*), Apioceridae and Pelecorhynchidae, and the beetle families Protocucujidae and Hobartiidae.

We will concentrate this review on the Australian continent, excluding the recently separated island of New Guinea for practical reasons, even though we acknowledge it as part of the Australian continental plate. The biogeographical understanding of Australia from an entomological viewpoint began at the end of the *colonial* period (1820–1910) of biogeographic regionalisation and continued through the *postfederation* period (1920s–1940s) and into the *ecogeographical* (1950s–1980s) and *systematic* (1980s–2000) periods (Ebach 2012).

Entomologists have attempted to provide biogeographic explanations of the fauna over the past century based on the detailed understanding of one or a few groups (e.g. Sloane 1915; Mackerras 1950; Evans 1959) or to provide more synthetic overviews (e.g. Mackerras 1970; Main 1987; Cranston and Naumann 1991). Most of the synthetic overviews, and especially those produced in the ecogeographical and systematic periods, consist of three basic elements: (1) an account of geological history and broad climatic patterns found in Australia (the general acceptance of continental drift has profoundly affected this section from the 1970s onwards), (2) a summary of faunal provinces, or subregions, usually borrowing heavily from provinces defined by authors studying other zoological or botanical groups, (3) an assessment of insect faunal elements, defined by different origins or routes of arrival in Australia.

We do not provide a summary of the paleogeography of Australia here, and we take that much of the basic history of Pangaean and Gondwanan fragmentation as understood. Useful technical summaries of this material can be found in Cranston and Naumann (1991), Sanmartín and Ronquist (2004), Grimaldi and Engel (2005, pp. 624–634) and in the other chapters in this book. More recent updates can be found in Hall (2013) and Metcalfe (2013) for the interaction of the Australian and Asian plate since the Jurassic, and the early history of Gondwana in Torsvik and Cocks (2013). Free animated resources for Pangaean and Gondwanan breakup are now widely available on the Internet. Recent summaries of our understanding of palaeoclimates during the Australian Tertiary are available in Byrne *et al.* (2008), Byrne *et al.* (2011) and Bowman *et al.* (2010).

Methods Used in Australian Insect Biogeography

The methods used in Australian insect biogeographic analyses have evolved with time, from descriptive narratives based on an understanding of geological history and insect species distribution patterns, to modern analyses relying on molecular phylogenetic reconstructions and divergence time estimation using fossil calibrations. At the broadest level, methods can be divided into two major types: those that seek to derive their understandings based on the geological history of the fauna and those that seek to understand patterns based primarily on modern-day ecology. Clearly, both processes operate together and impact on the relationships and distributions of the insect fauna in Australia as they do elsewhere. For example, many Gondwanan areas share a geological *and* ecological history, such as Patagonia, New Zealand and western Tasmania ; it is not a simple, or even useful, process to separate these two factors out.

Some of the most advanced methods of historical biogeography in Australia are those that use cladistic biogeography to derive general area relationships from a consensus of area cladograms derived from individual studies (Nelson and Platnick 1981). An area cladogram is derived from a taxon cladogram by replacing the taxonomic group in the tree with the area or areas that it occupies (Crisci *et al.* 2003). Multiple taxon cladograms can then be reconciled to produce consensus areagrams (Parenti and Ebach 2009) with the purpose of finding general explanations about the distribution and diversification of biotas. In order to derive a geographically coherent consensus in this process, the units of area used in different taxonomic groups must be comparable. Missing and duplicated areas provide analytical complexity, and there are also challenges in deriving a consensus area cladogram from a number of individual (and different) plant or animal area cladograms. The concept of *areas of endemism* has developed at least in part to provide a common currency of areas for historical biogeography (Harold and Mooi 1994; Morrone and Crisci 1995; Parenti and Ebach 2009). Areas of endemism have been defined as the 'congruent distributional limits of two or more species' (Platnick 1991), and traditional biogeographic subregions as defined in Australia are composed of many areas of endemism defined this way. This is especially so for insects, with most groups having many allopatric or partially overlapping species occupying each of the subregions discussed in this chapter. These in turn have overlap with noninsect taxa, including vertebrates (Cracraft 1991) and vascular plant surrogates (Crisp *et al.* 1999). Ebach *et al.* (2013) recently reviewed area classifications for plants and animals, including freshwater and terrestrial biotas, in an attempt to establish a testable framework for analysing area relationships.

Some of the most sophisticated ecological biogeographic methods include those that model the distributions of individual species in relation to abiotic physical aspects of the environment such as temperature and rainfall (for an early Australian example, see Nix 1991). These methods have been developed in recent years with impetus from climate change research and have become increasingly sophisticated. One recent approach, generalised dissimilarity modelling (GDM), uses abiotic and biotic factors in its modelling of regional species assemblages (beta diversity), an emergent property of biodiversity (Ferrier *et al.* 2007). Increasingly, invertebrate and especially insect distributional data are used in these models (Williams *et al.* 2010). These models can be used to retrodict the past and estimate palaeodistributions, or predict the future, and can now incorporate the ability of organisms to evolve as the environment changes. Challenges to these approaches include the use of presence-only data from collections to estimate distributions. These data cannot tell us where species don't occur, just where they have been found at some time in the past.

Can the historical and ecological biogeographic approaches be reconciled? Perhaps the residuals in a GDM model represent the signal of historical biogeography. Hugall and Stanisic's (2011) study of land snails offers a model for future research because they use data from both organismal history and ecology to infer evolutionary processes. In their study they use densely sampled molecular-population genetic and phylogenetic data sets to infer the process of population dynamics, the mechanisms of speciation and past demographics. These methods can be used to discover *arks* or refugial areas that contain a large proportion of relatively old, deeply diverging lineages, as well as cradles of recent diversification. A common theme in such studies is to address the evolutionary processes that give rise to these patterns: is habitat stability critical for the establishment of refugia? Do strong ecological gradients provide the basis for concerted speciation in a particular area?

Faunal Provinces

> The biological regions and subregions are essentially unquantified descriptions of relative biotic similarities and differences and are nonhierarchical, lacking explanation of their relationships or origins.
>
> **– (Cranston and Naumann 1991: p. 196)**

Australian insect biogeography is always conducted in the context of prevailing views on the biotic elements and provinces derived from other groups of plants and animals. One of the first themes of Australian insect biogeography was to divide the continent up into biogeographic regions or provinces. The method and basis of regionalisation in an Australian faunal context has been discussed and critiqued by Moore (1961), Littlejohn (1981), Ebach (2012) and Ebach *et al.* (2013).

Thomas Sloane (1905, 1915), relying heavily on the earlier works of Tate (1889) on the flora and Spencer (1896) on the fauna, addressed the faunal subregionalisation of Australia from the perspective of his studies on ground beetles of the family Carabidae. He accepted Spencer's four faunal elements in the Australian fauna: the New Holland (Tate's endemic element), Pacific, Antarctic and Austro-Malayan. Sloane's major innovation was to divide Spencer's three regions, the Torresian, Eyrean and Bassian, into a total of 11 subregions (which he termed *districts*) (Figure 9.1). Curiously, his map figure only showed 10 subregions because he neglected to include Tasmania, its own subregion in his scheme.

The Torresian region was divided into three western, middle and southern subregions. The western subregion was divided from the remainder by a line running directly south from the Gulf of Carpentaria, and was typified by the fauna around the Daly River (Northern Territory). The middle subregion continued east and south to near the Tropic of Capricorn in Queensland and was typified by the fauna of Mt. Bellenden Ker just south of Cairns (Queensland). The southern subregion continued from the tropic to the Clarence River in northern New South Wales, and typified by the faunas found between the Burnett River in southeastern Queensland and the Richmond River in northern New South Wales.

The Bassian region was also divided into three by Sloane: a northern subregion, a middle subregion and Tasmania. There was no true line of demarcation between the northern and middle Bassian subregions, but the northern subregion was typified by the fauna of the Sydney Basin and the middle subregion by the fauna of the Australian Alps.

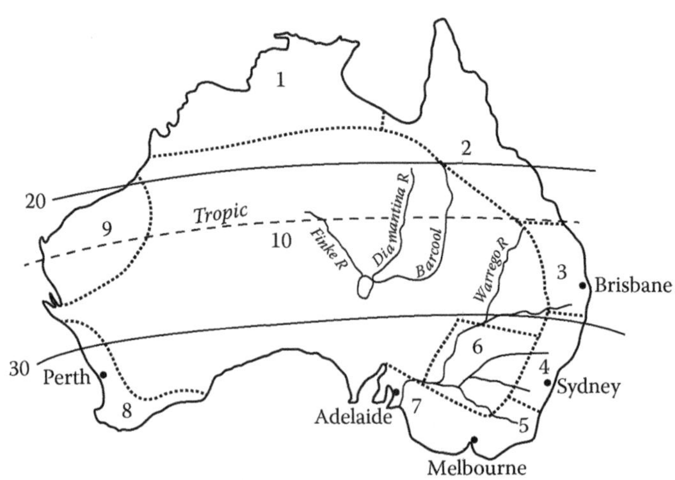

FIGURE 9.1 Thomas Sloane's (1915) entomological districts of Australia, based largely on the distribution of Carabidae (Coleoptera) shows remarkable similarity to the scheme of Matthews and Bouchard (2008) published almost 100 years later for Tenebrionidae (Coleoptera) (Figure 9.2).

The Eyrean region was divided into five subregions, four of these being peripheral sections excised from the central Australian subregion, centred on the MacDonnell Ranges. The Riverina subregion consisted of the basin of the Darling River, separated from the central Australian subregion by the watershed between the Darling and Barcoo Rivers. The south Australian subregion was centred on the Mt. Lofty and Flinders Ranges, and includes the Victorian Mallee and extends around the head of Spencer's Gulf to take in the Eyre Peninsula. The southwest Australian subregion was separated from the central Australian subregion by the 20 in (500 mm) rainfall line. The northwest subregion, poorly defined but centred on the Pilbara and including the watersheds of the De Grey, Ashburton, Gascoyne and Murchison Rivers, is the last in Sloane's system.

Sloane included many carabid genera that typified each subregion, but listed the Riverina district as characterised by immigrants from surrounding subregions. The carabid fauna of the northwest Australian subregion was too poorly understood to identify typical carab faunal elements. Interestingly, Sloane was only confident that the southwest Australian district, elevated to the Autochthonian region in Tate's botanical system, was the only area in his system that could be surely defined. However, Spencer eliminated this region from his fauna scheme.

Sloane's regions and districts were differentiated to variable degrees, and the demarcation between them was weak or strong. He commented that it was impossible to draw a line of demarcation between the Torresian and Eyrean subregions; however, subregions were often separated by river basins or the Bass Strait, in the case of Tasmania. The Eyrean subregion had rainfall less than 25 in (635 mm); however, the southwest district was demarcated from the Eyrean by the 20 in (500 mm) rainfall line.

Allan Keast's monumental *Biogeography and Ecology in Australia* (1959) included four chapters on insects, auchenorrhynchan Hemiptera (bugs) by Evans, Diptera (flies) by Paramonov, Orthoptera (grasshoppers and locusts) by Key and Isoptera (termites) by Calaby and Gay. These chapters mark the dawn of the ecogeographical period of Australian insect biogeography, and deal in a general narrative way with faunal elements and origins, but were not consistent.

Evans (1959) made a strong argument for the presence of a distinct southern or Gondwanan faunal element, as evidenced by a number of groups with aquatic immature stages, leafhoppers and the moss bug family Peloridiidae, that had Gondwanan origins. In the leafhopper fauna, he distinguished an older Jurassic element with connections to India, an element derived from Asia at the close of the Tertiary, and an endemic Australian element.

Focusing on the Diptera (fly) fauna, Paramonov (1959) accepted Spencer and Sloane's biogeographic regions, but with a novel nomenclature, and he elevated the southwestern Australian subregion to full region status, termed *Australia westralica*. Paramonov found a number of archaic, endemic elements in the Diptera fauna, but seemed reticent to divide the fauna into elements. He recognised Papuan and Malayan elements across the north from Darwin to Cairns and down the east coast of Queensland. He also found evidence that there may have been direct contact between Australia and South America in the past, and introduced a novel nomenclature for the Australian faunal provinces, which has not been adopted subsequently.

Key (1959) took a distinctly ecological and descriptive view, relating Spencer's zoogeographical regions with climatic areas and assigning elements of the grasshopper fauna to different climate zones. He found that a large bulk of the fauna was endemic, but there were connections with New Guinea. Key could find no evidence of a connection to other Southern Hemisphere continents, except a single species closely related to one in New Zealand. Calaby and Gay (1959) discussed the general ecology and distinctive features of the termite fauna, but did not divide the continent into provinces, or the termite fauna into elements, as other authors in the volume had done.

By the late 1960s Wegener's theory of continental drift was becoming mainstream, and the different lines of evidence for the theory occupied a large section of Mackerras's chapter on the biogeography of the insect fauna in *Insects of Australia* (1970). Mackerras accepted the regions (called *provinces*) and nomenclature developed by Tate and Spencer, and discarded Tate's Autochthonian element for the fauna, it being based on the distinctive old endemic flora found in southwestern Australia. He noted that the three major regions can be defined by ecological as well as faunal characteristics, as a number of the chapters in Keast *et al.* (1959) had stressed, particularly Key's.

TABLE 9.1

Comparison of Nomenclature for Australian Faunal Provinces/Biomes Used by Authors Mentioned in the Text

Tate 1889	Spencer 1896	Sloane 1915	Mackerras 1970	Crisp *et al.* 2004[a]	Biome Equivalent[b]
Euronotian	Bassian	Bassian	Bassian	Southeastern temperate	Mesic
Eremian	Eyrean	Eyrean	Eyrean	Eremaean	Arid
Part of Euronotian	Torresian	Torresian	Torresian	Monsoon Tropic	Monsoon Tropic
Autochthonian	Part of Eyrean	Part of Eyrean	Part of Bassian	Southwestern temperate	Part of Mesic

Note: (a) Crisp *et al.* (2004) also included an aseasonal wet biome (rainforest) inside the southeastern temperate and Monsoonal Tropics biomes. (b) This biome nomenclature is used in Byrne *et al.* (2008), Byrne *et al.* (2011) and Bowman *et al.* (2010).

Analysing the environmental determinants of biogeography, Nix (1982) found that thermal regimes, rainfall seasonality and light regimes could be integrated into a plant growth index that classified Australia into three main bioclimatic regions. These three regions, the megatherm seasonal, mesotherm/microtherm seasonal and megatherm/mesotherm arid regions, correspond to the Torresian, Bassian and Eremaean Provinces of the biogeographic literature. Thus, the biogeographic provinces of Australia identified by earlier generations of biogeographers could also be identified by coupling basic plant eco-physiological responses to climatic data, underlining the important role that modern ecology plays in biogeographic patterns.

Over the past few decades it has become more commonplace to refer to the traditional three main faunal provinces or faunal regions of the older literature as *biomes*. This evolution can be seen in the works of Crisp *et al.* (1999), Crisp *et al.* (2004), Byrne *et al.* (2008), Bowman *et al.* (2010) and Byrne *et al.* (2011). We will use the Monsoon Tropics biome for the Torresian Province of the older literature, the arid zone biome for the Eremaean province of the older literature and the mesic zone biome for the Bassian Province of the older literature. Table 9.1 provides a comparative nomenclature for Australia's faunal regions.

Faunal Elements

> Although most of these elements can be easily identified in the fauna, the origins and makeup of specific higher-level taxa usually comprise a mosaic of several elements that have interacted in complex ways with long- and short-term climate change, mediated through changes in vegetation.
>
> **– Austin *et al.* (2004, p. 217)**

Harrison (1928) divided the fauna into four elements, largely following Tate (1889), Hedley (1894) and Spencer (1896): the ancient Autochthonian element in southern Western Australia, the Euronotian element, which entered Australia from the south before the Miocene, the Pantropical element introduced during the Pliocene, and the Papuan element, which has entered Australia since the Pliocene.

Tillyard (1924) brought with him a deep knowledge of palaeoentomology and geological history to divide the Australian insect fauna into a total of 10 sources, with 3 kinds of Antarctic elements, 2 Austro-Gondwanan, 3 Austro-Malayan, a single Western Australian and an Autochthonous element. Each of the elements was characteristic of a particular geological time of appearance, from the Upper Permian to the Late Tertiary. It is difficult to relate Tillyard's scheme to more recent authors' attempts to divide the fauna into a smaller number of elements. The older Austro-Gondwanan and Antarctic elements probably map to the *Archaic* and *Southern* elements of Mackerras (1970), respectively, and the Austro-Malayan elements to the *Older* and *Younger Northern* elements of Mackerras (1970). Tillyard (1924) had a very different concept of the Autochthonous element than other authors such as Harrison (1928), who used this term in relation to an archaic, endemic group centred in Western Australia with very distant relationships to groups in other regions of the world. Tillyard used this term to refer to a more recent, Tertiary-derived and xeric-adapted faunal group that was dominant in the Quaternary fauna.

Mackerras (1950) could find no evidence for an ancient Autochthonous fauna restricted to southwestern Australia in the Diptera (fly) fauna. He found that the distinctive fly fauna of the southwest could be easily derived from the Bassian fauna. He termed the Euronotian element *Bassian*, and listed five properties of these groups: they were relatively primitive, had relatives in South America and sometimes New Zealand, in Australia they occurred predominantly in the southeast from Tasmania with a tongue extending north along the Great Dividing Range, and are generally on the wing in the spring in the north and along the coast, but in summer in the south and at altitude. He believed the weight of evidence supported their entry of Bassian groups into Australia from the south, and nominated *Pelecorhynchus* (Pelecorhynchidae) as an exemplary member of this group. Mackerras complicated the nomenclature of the faunal elements considerably, also referring to the Bassian element as *Antarctic* or *Southern* elsewhere (e.g. Mackerras 1970), used *Lemurian* for Harrison's Pantropical, and *Indo-Malayan* for the Papuan (Mackerras 1950).

Mackerras's influential chapter on the composition and distribution of the fauna in *The Insects of Australia* (1970) contained Harrison's faunal elements, in a slightly modified form, with a different nomenclature. The ancient Autochthonian element in southern Western Australia was eliminated, but replaced with an *Archaic* element consisting of primitive relictual animals that had survived since the Palaeozoic or Early Mesozoic. In the insect fauna the mecopteran Meropeidae and dipteran *Exeretonevra* (Xylophagidae) were good examples. This group is very similar to Harrison's Autochthonian, except they were not necessarily restricted to southern Western Australia. The *Southern* element was equivalent to Harrison's Euronotian, and had two characteristics: they were generally primitive and were often also found in southern South America, New Zealand and South Africa. Mackerras renamed Harrison's Pantropical element the *Older Northern* element, a minor, older group with relatives in Africa, India or Madagascar. He renamed Harrison's Papuan element the *Younger Northern* element, a conspicuous, clearly defined element with recent connections to Asia. In the interests of stability we use Mackerras's (1970) nomenclature for faunal elements.

Testing the Provinces and Elements

A series of chapters in Allan Keast's monumental three-volume *Ecological Biogeography of Australia* (1981) represent the most significant contributions at the end of the ecogeographical period of Australian insect biogeography. Within 10 years the first cladistic biogeographical assessments of the Australian vertebrate fauna would be published (Cracraft 1991). The contributions in Keast's monograph examined the reality of elements and regions for particular taxonomic groups, and Main (1981) provided a summary for the terrestrial groups, classifying them into Harrison's faunal elements with Mackerras's (1970) nomenclature, using extralimital distribution, and distribution within Australia, as evidence.

Zwick (1981) found that the four stonefly families (Plecoptera) had southern affinities, with two genera, *Stenoperla* and *Notonemoura*, shared with New Zealand. Taken together they provide strong evidence of continental drift, and virtually all stoneflies are found in streams of southeastern Australia. The net-winged midge *Edwardsina*, with species in Australia and South America, was also considered a good example of a Gondwanan group (Zwick 1981).

Tindale's chapter drew attention to the concentration of early diverging lineages of Lepidoptera (moths and butterflies), including Micropteygidae, and new fossil evidence for the antiquity of Lepidoptera, as well as the diversification of one of the early diverging lineages, the swift moths (Hepialidae), in Australia. This represented evidence of the long history of moths and butterflies in Australia, and there was also evidence of Gondwanan links with other Southern Hemisphere continents in groups such as the Castniidae and also Hepialidae again. Tindale also drew attention to the dozen or so families that have recently arrived from Asia.

Kitching (1981; updated in Kitching and Dunn 1999) took an entirely different and ecological approach in his chapter on the geography of the butterflies. He calculated isoclines of species densities across the continent, noting a general decrease in species densities from north to south, and a more dramatic decline from east to west. The highest species densities occur on Cape York Peninsula (Iron Range–McIlwraith Range area) and the Wet Tropics in Queensland. Similar patterns of distribution can be found in *Drosophila* (Parsons and Bock 1981). These patterns can largely be explained by the modern-day ecology of these almost exclusively phytophagous and fungivorous groups.

Howden's chapter (1981) focused on the scarabaeoid beetles, noted in the fauna a commonly repeated southern temperate pattern in the Lucanidae, Buprestidae and Belidae in particular. He noted that Australian arid zone scarabaeoids were most often of ancient Mesozoic origin, and discussed the fauna in terms of older Mesozoic groups and more modern Cenozoic elements, and could find little evidence in the scarabaeoid beetles of speciation being driven by Pleistocene climate change. Matthews (2000) also found considerable evidence for older Mesozoic groups in the Tenebrionidae of the arid zone.

Geological, Evolutionary and Ecological Processes

Contemporary biogeographic analyses of the fauna in the systematic era (Ebach 2012) beginning in the 1980s used quantitative hypotheses of relationships in a vicariance biogeographic paradigm. Perhaps one of the first works to address the subject in the systematic era was Cranston and Naumann's (1991) chapter in *Insects of Australia*. These authors contrasted the ecological, dispersal and vicariance approaches to biogeographical explanation, and explained the process of deriving an area cladogram from a phylogeny, noting that these were only as good as the underlying evidence for relationships. These authors reviewed what we know of Gondwanan vicariance and discussed its importance in structuring the biogeography of the Southern Hemisphere, as have more recent authors (e.g. Sanmartín and Ronquist 2004; Grimaldi and Engel 2005).

Even more recent studies use densely sampled molecular phylogenetic analyses that provide testable hypotheses of relationships, in the context of temporal diversification and our increasingly detailed knowledge of the geological and palaeoclimatic influences on the Australian biota. Various causal processes can be invoked in these studies, integrating and combining the ecological, dispersal and vicariance explanations discussed by Cranston and Naumann (1991) into the historical biogeography of a single lineage.

A useful modern entry into this literature is provided by three recent reviews of the biogeography of the arid and mesic zones and the Monsoon Tropics (Bowman *et al.* 2010; Byrne *et al.* 2008, 2011), corresponding to the Torresian, Eyrean and Bassian Provinces of the ecogeographic period, respectively. These reviews focus on the Tertiary, and synthesise modern geological and palaeoclimatic evidence, palaeontological data and molecular phylogenetic results. While these studies reviewed the few insect examples available, almost all of the biological studies included were from the broader botanical and zoological literature. As such, they provide a broad sweep of hypotheses to frame future insect biogeographical studies, and we have drawn on the approach used in these studies and the work of Rix *et al.* (2014) to provide testable hypotheses for Australian insect biogeography.

Byrne *et al.* (2011) find evidence that the mesic biome is ancestral when mapped onto plant and animal phylogenies that include mesic and nonmesic zone organisms that have had a long Mesozoic history in Australia. They also found that within this zone, rainforest habitat is ancestral, and that Asian immigrant lineages only appear in the mesic biota after 20 mya. Levels of endemism and phylogeographic structuring were high in the mesic zone, consistent with the idea that many elements of the mesic biota have shown limited mobility and localised persistence through climatic fluctuations beginning in the Miocene (25 mya) and amplified in the Quaternary. The evidence available is consistent with the continued fragmentation and increased isolation of both rainforest and other mesic habitat types in Australia from the Late Eocene.

In a review of the biogeography of the southwestern portion of the mesic zone, Rix *et al.* (2014) derive four testable models of historical biogeography for animal taxa in this region, based on the strongest patterns seen in many animal groups. They find ancient relicts restricted to the high rainfall zone in the far southeast of Western Australia. These may have relatives in Australia or other parts of the world but will have diverged from them more than 35 mya. Lineages that are the result of vicariance between groups that were once widespread across southern Australia would have divergence dates in the mid-Miocene (14–16 mya), when the Nullarbor Plain was inundated from the Southern Ocean. Those groups that have responded to the continued xerification of southwest Western Australia by speciation into more mesic patches, or have adapted in situ, would produce different phylogenetic signatures over the Late Miocene and Pliocene. Their effort to provide hypotheses crafted as spatially and temporally explicit patterns linked to known biogeographic processes represents a major step forwards in the science of Australian biogeography.

In contrast to the ancient mesic environments of Australia, the aridification of Australia probably began in the mid-Miocene (15 mya), and the current arid zone and fully arid landforms appeared much more recently in Australia, 2–4 mya (Byrne *et al.* 2008). The closest relatives of arid zone taxa occur in Australia's mesic zone, and the relatively recent and rapid expansion of the arid zone has led to high rates of diversification. Regions of high topographic relief embedded in the current arid zone have become refugia for some groups, including insects.

Evidence of monsoon tropical climates in northern Australia date back to the Late Eocene (40 mya), implying that at least some endemic elements of the biota have a long history (Bowman *et al.* 2010). This region is characterised by the vicariant separation of eastern Cape York elements from those farther west in the 'Top End' of the Northern Territory via the Gulf of Carpentaria. Significant areas of endemism occur in the sandstone plateaus of Arnhem Land and the Kimberley. The biota of the region shows both links to New Guinea and farther west in Asia as well as relationships to the Australian mesic zone and arid zone groups.

Insect Biogeography in the Systematic Period: Processes Revealed Using Syntheses of Distribution, Phylogenetic Relationships and Divergence Times

In contrast to the reviews of different biomes in Australia, most biogeographic studies are focused on particular taxonomic groups, and many of these use molecular phylogenies coupled with divergence time estimation to understand the origin of the Australian biota and the contribution of Australian lineages to the faunas of other regions. Some of the very oldest of these lineages include groups that are now relictual in Australia, and comprise members of the Archaic element in the older literature from the ecogeographical period.

Probably the best recent example is the study of petaltail dragonflies of the family Petaluridae (Odonata; Ware *et al.* 2014). This family has only 11 extant species worldwide, distributed in Australia, New Zealand, Japan, Chile and North America. The family originated in the mid-Jurassic (ca. 150 mya) and diverged into a Laurasian and Gondwanan clade shortly thereafter. The Australian species may have been derived from original stock in New Zealand, or possibly Antarctica, and the Australian and Chilean species are more closely related to each other than they are to the New Zealand species. The majority of species exist for 70–75 million years, unusual for such ecologically specialised groups. The biogeography of this ancient group spans almost the entire history of Gondwana, and their phylogenetic relationships are consistent with the breakup of Gondwana, but the ages of the lineages indicate that they diverged well before this breakup, with, for example, the Australian and Chilean groups diverging at around 120 mya, almost 90 mya before the vicariant separation of Australia from Chile through Antarctica.

Peloridiidae (moss bugs) are mostly wingless, moss-inhabiting Hemiptera belonging to their own suborder (Coleorhyncha) and are the sister group of the suborder Heteroptera. Moss bugs have a rich fossil history extending back into the Permian and were once distributed worldwide, and they are often cited as an Archaic element in Australia. In a morphological cladistic analysis, Burckhardt (2009) found relationships of the extant species to be consistent with Gondwanan breakup, with Chilean and Australian species more closely related to each other than they were to the New Zealand species. New Caledonian representatives were most closely associated with New Zealand, and Lord Howe Island species with those from adjacent areas of Australia. The group is so old that the Lord Howe Island lineages could reflect relationships established before Zealandia separated from the east coast of Australia 60–80 mya.

The scorpionfly family Meropeidae is often considered a member of the archaic element because of their widely disjunct distribution. They are found in eastern North America, Western Australia and the Atlantic Forest of southeastern Brazil, and have a fossil record going back to the mid-Jurassic in Russia. There are no modern molecular phylogenetic analyses, but the recent discovery of a mid-Cretaceous fossil in Burmese (Myanmar) amber lends credence to the idea that their current disjunct distribution is the result of extinction (Grimaldi and Engel 2013).

Austral Groups, Extinction and Dispersal: Emerging Complexity of Australian Insect Biogeography over the Past 160 Million Years

> Austral distributions occur only in those groups that occurred in the Cretaceous and earlier, which indicates some involvement of continental drift, but widespread fossils indicate that drift had a minor impact on distributions compared with climatic and biotic change.
>
> **– Grimaldi and Engel (2005, p. 634)**

Australian biogeographers attribute the source of older, southern elements in the fauna to Gondwana, and interpret their phylogenetic relationships according to the vicariant breakup of the supercontinent. As our knowledge of the insect fossil record becomes more complete, it is increasingly obvious that many groups that are now only found in the Southern Hemisphere had more widespread distributions that extended to Laurasia during the Cretaceous and Tertiary. In these cases their current distributions are probably based on a combination of ecological and environmental stability in the southern temperate regions over very long periods of time, combined with extinction in Laurasian-derived landmasses. Evidence is also beginning to emerge that many of these groups are old enough to have diversified on Pangaea prior to its breakup in the mid-Jurassic. These worldwide fossil distributions mean a Gondwanan origin for these groups cannot be assumed.

A few examples will serve to illustrate this point. The giant termite *Mastotermes* (Mastotermitidae) shares many anatomical features with related cockroaches (Blattodea) and now only occurs naturally in Australia; however, it has been introduced to New Guinea. *Mastotermes* is considered the earliest branching lineage of termites, with estimates of divergence from other termites in the Late Jurassic around 150 mya (Misof *et al.* 2014; Bourguignon *et al.* 2014). The fossil record of the family Mastotermitidae is almost global in the Tertiary, and a few fossil representatives are known through the Cretaceous (Grimaldi and Engel 2005). Clearly extinction has caused their distribution to shrink through the Tertiary, and the biogeography of this group may be best attributed to the Pangaean union of all continents that exited 300 mya until they began to separate into Gondwanan and Laurasian supercontinents around 175 mya.

The highly autapomorphic bug family Thaumastocoridae is another case in point. The family is divided into three subfamilies: the palm-inhabiting American subfamily Xylastodorinae, the largely Australian subfamily Thaumastocorinae, and the Oriental and monotypic Thaicorinae (Heiss and Popov 2002; Noack *et al.* 2011). The Xylastodorinae are known from Baltic amber (*Proxylastodoris*), and recently, a modern species of this genus was found on palms on New Caledonia, establishing an ancestral Pangaean distribution for the family and the first record of the subfamily in the Eastern Hemisphere (van Doesburg *et al.* 2010).

The nonbiting midge subfamily Aphroteniinae has members in both Australia and South Africa, and this divergence was considered to have occurred at least 120 mya based on the vicariance dates for Australia and Africa (Cranston and Edward 1992). Subsequent molecular analyses estimated the divergence of this subfamily around 220 mya, with even the confidence interval for the extant sampled crown group divergence extending back 110 mya (Cranston *et al.* 2012).

The Bombyliidae genus *Comptosia* has long been considered a Gondwanan group, with close relatives in the subfamily Lomatiinae in southern South America. However, fossil wing impressions of *Comptosia* have now been found in Germany and the United States from the Tertiary, indicating that the group's current distribution is due to Cenozoic extinction (Wedmann and Yeates 2008). The family Bombyliidae of the lower Brachycera is likely to be just old enough to have predated the breakup of Pangaea into Laurasia and Gondwana (Wiegmann *et al.* 2011).

The aquatic beetle family Hygrobiidae has a relictual distribution, with one species in Europe, one in China and the remaining four species in the mesic zone of Australia. One species extends from Cape York to adjacent areas of the monsoon tropical biome in the Northern Territory. Molecular phylogeny and divergence time analysis suggest an age for the family in the Late Triassic, 220 mya (Hawlitschek *et al.* 2012a). This corresponds to the period just before the breakup of Pangaea, and is consistent with the current distribution of the family on both Laurasian and Gondwanan continental fragments. The Australian species diverged from the European species in the Early Cretaceous about 125 mya. The four Australian species diverged from the Late Eocene, 45 mya onwards. The blind, flightless, terrestrial water beetle *Terradessus* from the Wet Tropics of north Queensland may also be explained using similar reasoning; its closest relatives are in the Himalayas (Watts 1982).

Myrmeciine (*Myrmecia* and *Nothomyrmecia*, Formicidae) ants are distinctive elements of the Australian ant fauna (there is also a single species in New Caledonia), considered to be relicts of the early diversification of ants in the mid-Cretaceous. Modern phylogenetic and fossil evidence support this age, but not their Gondwanan origin. Myrmeciine ants probably diverged from other ants 100–120 mya (Moreau *et al.* 2006; Brady *et al.* 2006), and myrmeciine fossils are known from Argentina in the Eocene and are widespread in Baltic amber during the Eocene (Ward and Brady 2003).

Gondwanan Breakup, Vicariance and Dispersal

Entomologists were early adopters of the idea that drifting continents may have shaped the distribution of insects in the Southern Hemisphere. Mackerras (1950) was an early adopter, as was Brundin (1966), but not so Hennig (1960). As geological evidence has accumulated we now have quite a detailed understanding of the breakup sequence of the former continents of Gondwana, and this sequence provides the framework for a well-structured Southern Hemisphere biogeography with vicariance as an overarching causal process (Sanmartín and Ronquist 2004).

Some recent biogeographic examples are consistent with the traditional view of Gondwanan breakup, with the exceptions usually finding that the New Zealand clades are too young to have been in place since Zealandia separated from eastern Australia around 80 mya. Many of these examples suggest recent long-distance over-water dispersal from Australia to New Zealand.

The lower brachyceran Diptera families Apioceridae and Mydidae probably date from the mid-Jurassic, and both include clades that have distributions that span both Gondwana and Laurasia. A morphological cladistic analysis found that the Laurasian lineages were sister to the Gondwanan clades in both *Apiocera* and the megasceline Mydidae, and in both cases the relationships of the Gondwana groups were as expected by vicariance. Both clades included an African representative, but both also lack a New Zealand lineage (Yeates and Irwin 1996).

In a biogeographic analysis of a phyline mirid clade (Heteroptera: Miridae) using Brooks parsimony analysis, area relationships were found between New Zealand and Australia, and South American taxa being the sister to that area (Weirauch and Schuh 2010); this conforms to the *plant southern pattern* found in several plant groups (Sanmartín and Ronquist 2004). In this case, low host plant specificity may facilitate long-distance over-ocean dispersal to New Zealand. In a study with similar biogeographic conclusions, molecular data were used to show that *Nothofagus* feeding scale insects from New Zealand were more closely related to Australian species than South American, incongruent with the standard vicariance model (Hardy *et al.* 2008). Thus it appears that long-distance over-ocean dispersal from Australian may have been important for both new Zealand *Nothofagus* (Cook and Crisp 2005; McGlone 2005) and their associated scale insects.

Some examples that provide more equivocal support for Gondwanan vicariance as a primary mechanism to explain Southern Hemisphere insect relationships come from the true flies (Diptera). Nonbiting midges (family Chironomidae) have obligate immature stages in freshwater, and the family is likely to have originated in the Late Permian or Early Triassic, around 250 mya (Cranston *et al.* 2012). An extensively sampled molecular study of the genus *Stictocladius* found that the New Zealand species were most closely related to Australian ones, not South American groups (Krosch and Cranston 2013). In addition, the divergence times of the New Zealand species, around 50 mya, with confidence intervals for the relevant nodes spanning 70–31 mya, postdated the separation of Zealandia from eastern Australia at around 80 mya. One explanation is there was an archipelagic connection between Australia and Zealandia after initial rifting. The monophyletic South American clade diverged from its Australian sister group around 25 mya, consistent with a vicariance explanation. The young divergences ages of New Zealand lineages may be an artefact caused by high extinction rates in small, old clades. This could have happened to lineages in New Zealand that survived the Oligocene (Sharma and Wheeler 2013).

A more compelling case for a vicariant explanation comes from pangoniine horse flies (Tabanidae) of the tribe Scionini (Lessard *et al.* 2013). In this case the South American and Australian clades were not reciprocally monophyletic, and were more closely related to each other than the New Zealand species. Divergence times between the Australian and South American genera (35–25 mya) were consistent with a vicariant explanation; however, the mean divergence time for the New Zealand genus is 53 mya, and the confidence interval for this node (74–32 mya) just overlaps the separation of Zealandia from Australia in the Cretaceous. Again, an archipelagic connection between Australia and Zealandia in the Late Cretaceous may have facilitated this dispersal.

Spitfires, the gregarious larvae of the sawfly family Pergidae (Hymenoptera) are iconic insects of Australian eucalypt forests, and the family is also found in South America and New Guinea, with a few species in North America. Two subfamilies (Perginae and Pterygophorinae) have developed mechanisms to detoxify eucalypt oils and include large numbers of species. A calibrated molecular phylogeny shows that the Pergidae diverged from their sister family Argidae in the Late Jurassic, about 153 mya. The Australian and South American lineages of the family diverged within a narrow time frame, 90–105 mya, and the faunas of each continent are not reciprocally monophyletic. This is consistent with regional diversification on Gondwana before the full separation of South America and Australia much later in the Oligocene, 32 mya. Accelerated radiation within the eucalypt feeding subfamilies occurred at the beginning of the Tertiary, around 60 mya, when their host species began radiating (Crisp *et al.* 2004). In contrast, the three pergid subfamilies in Australia that do not feed on eucalypts each have only a handful of species and are restricted to the rainforests of eastern Australia. This strongly suggests a pattern of adaptive radiation. Once pergid sawflies had gained detoxification mechanisms in the Early Tertiary, they then were able to exploit eucalypts and expand their range beyond the ancestral rainforest habitat to other areas of the mesic biome, and other biomes, as the eucalypts diversified and spread in Australia (Schmidt and Walter 2014).

There are a number of examples of insect groups that apparently originated on or near Australia in the Early Tertiary, dispersed, often more than once, and occasionally returned. Butterflies of the subtribe Aporiina (*Delias* and its relatives Pieridae) are widespread globally, most likely originated in southern Gondwana in the Early Tertiary, 50–60 mya (Braby *et al.* 2007); however, to achieve their current distribution would require a combination of vicariance, extinction, three long-distance dispersal events and nine range expansions. These dispersal events may have been facilitated as fragments of the northern margin of the Australian plate accreted onto the Southeast Asian margin, notably in Sulawesi (Hall 2009). Focusing exclusively on *Delias*, Müller *et al.* (2013) confirmed the finding of Braby and Pierce (2007) that the genus originated in Australia with Lagrange and Bayes-DIVA ancestral area analyses, and five clades dispersed to Wallacea and Southeast Asia between 15 and 10 mya, diversifying in both areas. Troidine butterflies (Papilionidae) known as *pipevine swallowtails* originated in the Australian region (including New Guinea) up to 60 mya and *Troides* dispersed across Wallacea into the Oriental region around 20 mya (Braby *et al.* 2005; Simonsen *et al.* 2010).

Gall-inducing sycophagine Agaonidae (Hymenoptera) known as *fig wasps* have a worldwide distribution, but members of the earliest evolving genus *Eukoebelea* are endemic to Australia. Molecular phylogenetic analysis combined with divergence time estimation shows that the group originated in Australia in the Eocene and dispersed twice out of Australia, once via the Ninety East Ridge (40–30 mya) to India and subsequently Africa, and then back to Australia in the Pliocene, and a different lineage dispersed later via the microplates that were beginning to accrete onto Southeast Asia 20–25 mya (Cruaud *et al.* 2011).

The Australian bee fauna is highly endemic, with two families, the Colletidae and Halictidae, comprising approximately 80% of the estimated 2000 bee species in Australia. Two families that are common elsewhere in the world, the Andrenidae and Melittidae, are absent from Australia (Michener 2007). Bees most likely arose in western Gondwana with the rise of flowering plants some 120 mya (Danforth *et al.* 2006), just as the southern continents were fragmenting. The bee fauna of Australia has attracted significant biogeographical attention using modern molecular phylogenies, worldwide lineage sampling, divergence time analyses and ancestral area reconstructions using Lagrange and Bayes-DIVA. Colletidae comprise about 50% of the Australian fauna, with many deeply diverging lineages, and arose in Australia (Almeida *et al.* 2012) about 70 mya (Almeida *et al.* 2012). Much of their early diversification was in Australia and South America mediated though Antarctica, between the Late Cretaceous and the Late Oligocene. The movement of these bees among the Southern Hemisphere landmasses ended with the development of the circumpolar current during the Oligocene. New Zealand obtained its single colletid genus *Nesocolletes* via long-distance over-ocean dispersal at the end of the Pliocene, about 14 mya, and the exclusively African Scrapterinae probably arrived by long-distance over-ocean dispersal in the Eocene, around 54 mya.

An example of dispersal from Africa to Australia, probably mediated through Antarctica, is provided by the allodapine bees (Apidae). This lineage originated in Africa and arrived in Australia between 42 and 34 mya (Schwarz *et al.* 2006; Tierney *et al.* 2008; Chenoweth and Schwarz 2011). This is most likely to have been through long-distance over-ocean dispersal because Africa had long since departed the other Gondwanan continents on its passage north. The allodapine genus *Braunsapis* probably arrived in Australia from Asia much more recently, around 10 mya, as the Australian plate interacted significantly with that from the Asian plate (Fuller *et al.* 2005; Chenoweth and Schwarz 2011). Thus, the allodapines reached Australia from Africa at two different times and likely using two quite different routes. The apid genus *Lestis* and the Halictinae are hypothesised to have reached Australia from Asia around 30 mya (Leys *et al.* 2002; Danforth *et al.* 2004); however, palaeogeographical models suggest this would require significant long-distance over-ocean dispersal.

Bees of the genus *Hylaeus* (Colletidae) contain more than 500 described species. The genus diversified rapidly in Australia after their origin about 30 mya and dispersed out of Australia twice, both shortly after the initial diversification; one dispersal event to New Zealand around 23 mya resulted in only a small radiation of eight species, but the second, presumably through Wallacea to the Oriental region at around 22 mya, resulted in a worldwide radiation (Kayaalp *et al.* 2013). The authors ponder what ecological forces have simultaneously promoted the diversification of the genus outside Australia but prevented further dispersals.

Coptotermes (Rhinotermitidae) contains some of the world's most damaging and invasive termites, and the Australian species are the only ones to build mounds. This genus appears to have followed a similar route of arrival into Australia as the bee *Braunsapis*, originating in Africa, expanding its range into Asia, and then crossing into Australia in the Late Miocene, 13 mya, as faunal interactions between the Australian and Asian plates intensified (Lee *et al.* 2015).

Biogeographic Relationships in the Mesic Biome (Bassian Province)

Insects have been the subject of a number of biogeographic studies within the mesic biome, principally examining the relationships and divergences of groups that occupy the rainforests and adjacent wet sclerophyll forests of eastern Australia. The mesic zone in Australia is very large, spanning tropical to cool-temperate latitudes from 11°S to 44°S in Tasmania, and including the southwestern region of Western Australia, isolated 2000 km from the nearest mesic components on the east coast.

Fourteen species of the wingless, predatory ground beetle genus *Pamborus* (Carabidae) are found in rainforest patches from the Wet Tropics in north Queensland to central New South Wales. Dated molecular phylogeny suggests that *Pamborus* began differentiating at the end of the Oligocene, 30 mya, and the beginning of the Miocene (Sota *et al.* 2005). This coincides with the onset of cooler, dryer conditions in

Australia and the fragmentation of rainforests in Australia from continuous habitat to isolated fragments in favourable situations (Byrne *et al.* 2011). The pattern of differentiation of *Pamborus* is not coherent geographically, and it appears that the group has differentiated and expanded its range repeatedly in eastern Australia during the Late Tertiary. In a study with a similar geographic scope, Baker *et al.* (2008) used molecular data to reconstruct the phylogeny of the nine species of glow worms (*Arachnocampa*, Mycetophilidae: Diptera) in Australia and New Zealand. These insects are well-known cave dwellers, but many populations can also be found in rainforest gullies. They are restricted to the mesic zone from Tasmania to north Queensland. The deepest divergences in the *Campara* clade were between north Queensland species and the remainder, and these were dated mid-Miocene (6 mya). Most of the divergence times between the species from New South Wales and Victoria were dated to the very end of the Pliocene and beginning of the Pleistocene (3 mya).

A detailed analysis of biogeography and diversification dynamics in the *Necterosoma* group of 11 predaceous diving beetle (Dytiscidae) genera in Australia demonstrates the impact of Cenozoic climate change on Australian insect diversity and distributions. The group originated in mesic east-coast forests in the Oligocene, and in the Early-to-Middle Miocene two clades diversified in the eastern and northern parts of the continent. As conditions became cooler and dryer in the mid-Miocene 14 mya, these beetles expanded towards central Australia and began colonising groundwater systems in the western half of the continent (Toussaint *et al.* 2015). The Pilbara and southwestern Western Australia were colonised twice in the Late Miocene, 6–10 mya, and during this period New Caledonia, Fiji and New Zealand were colonised, probably by long-distance dispersal. The tempo of diversification differed between the subterranean and terrestrial clades, the groundwater systems fostering an adaptive radiation with a diversity-dependent pattern, whereas the terrestrial ecosystems shaped a damped and slower diversification rate. This is consistent with the mesic-adapted diversity declining as the mesic zone continued to contract and fragment from the mid-Miocene onwards (Byrne *et al.* 2011; Toussaint *et al.* 2015).

Sternopriscus diving beetles (Dytiscidae) of the *tarsalis* group are confined to the southeastern mesic zone of Australia, and many species are sympatric. They are morphologically distinct but cannot be reliably distinguished using nuclear or mitochondrial DNA. They represent one of the few examples of a very rapid Pleistocene species radiation of up to 2.4 species per million years, perhaps caused by the repeated isolation of populations in glacial refugia (Hawlitschek *et al.* 2012b).

Patterns of Endemism and Levels of Divergence in the Mesic Biome

Biogeographically, one of the best-studied regions in the mesic zone is the Wet Tropics of north Queensland (Stork and Turton 2008), and the invertebrate fauna has been the subject of some detailed taxonomic and biogeographic research (Yeates and Monteith 2008). The region has large areas of intact rainforest from sea level to the highest mountains, high topographic relief, with mountains rising above 1600 m, combined with strong endemism and geographic structuring of species distributions. These distributions tend to be focused around 14 rainforest blocks known to have been refugia in the Late Pliocene and Pleistocene. These blocks are distributed from just south of Townsville to Mt. Finnigan just south of Cooktown. Climate change is a major threat to many beetles in the Wet Tropics, with many species predicted to have their ranges reduced by 80% or more by 2080 (Staunton *et al.* 2014).

Yeates *et al.* (2002) used a large data set of the distributions of 274 flightless rainforest insects to examine the patterns and levels of endemism in the Wet Tropics. Half of the species endemic to the Wet Tropics were also only found in a single rainforest block, a much higher percentage than that found in vertebrates, and this may be related to the smaller size and home range of insects. The largest numbers of endemic species were found on Mt. Finnigan, the Carbine Tableland, the adjacent Mt. Bellenden Ker/Bartle Frere mountains and the Atherton Tableland. Rainforest area and shape best explained the variance in numbers of endemic species between rainforest blocks. Those rainforest blocks with a large number of endemic species were found to be Plio-/Pleistocene refugia, but those with few endemic species were replaced with open forest during the same period. These rainforest regions with low endemism were most likely recolonised from adjacent refugia once more mesic conditions returned. Similar results were found for a large darkling beetle (Tenebrionidae) data set analysed using Brooks parsimony analysis (Bouchard *et al.* 2005).

Dung beetles of the genus *Temnoplectron* (Scarabaeidae) have been studied because they have 19 species in the Wet Tropics, are obligate in rainforest, show patterns of narrow species endemism in many cases, and include a mix of winged and wingless species. These factors suggest that substantial speciation has occurred in situ. Many sister species pairs have allopatric distributions, and the biogeographic breaks are consistent with our knowledge of the breaks in rainforest during the cooler, drier climates of the last few million years, and are consistent with the patterns found in vertebrate studies – for example, the Black Mountain barrier (Bell *et al.* 2004). Further studies indicated that two of the winged sister species pairs had diverged in the Late Pliocene (Bell *et al.* 2007), consistent with the separation of the rainforest blocks these beetles occupy in cooler, drier periods at and after the end of the Pliocene.

While no other region has been studied as intensively as the Wet Tropics of north Queensland, some studies are emerging from the Australian Alps, in New South Wales and the adjacent upland regions of Victoria, with mountains and plateau regions rising above 2000 m. This area lacks rainforest and hosts the small region of alpine vegetation communities in Australia. The habitat is fragmented, with high mountain regions separated by valleys and rivers, providing barriers to gene flow and facilitating diversification. Pleistocene climate cycles would have repeatedly joined and separated the alpine populations of many species, with the snow line predicted to be 800 m lower during the Last Glacial Maximum (Slayter *et al.* 2014). Recent studies have revealed deep Pleistocene genetic differentiation at very fine spatial scales in a suite of insects (two beetles, a grasshopper, a springtail and a millipede) and other invertebrates, regardless of species vagility, and show that genetic differentiation has been maintained in these species through the Pleistocene climatic cycling (Endo *et al.* 2015).

A small number of unusual Australian butterflies have evolved to live inside ants' nests and prey on the ant larvae. Species of the *idmo* group in the lycaenid genus *Ogyris* have a patchy distribution across southern Australia, with disjunct populations in Victoria/South Australia in the southeast and in southwestern Western Australia. In a detailed molecular study of these rare and threatened insects, Schmidt *et al.* (2011) found the first diversification in the group dated to 16 mya, around the time of marine incursions in the Nullarbor. This divided the group of species into two pairs of clades with species on either side of the Nullarbor. Divergences across the Nullarbor within these groups were dated to 3–6 mya, later than the Miocene marine transgressions in the Nullarbor. More coastal, mesic-adapted species pairs diverged earlier (5–6 mya) than the more arid-adapted species (2–3 mya).

Biogeographic Relationships and Endemism in the Arid Biome (Eremaean Province)

The aridification of Australia probably began in the mid-Miocene, and the current arid zone and fully arid landforms appeared much more recently in Australia, 1–4 mya (Byrne *et al.* 2008). There has been remarkably little modern biogeography of the entomofauna of Australia's arid zone, and this is surprising given the extent of the biome. A compelling exception is Eric Matthews's work (2000) on the origins of the darkling beetle (Tenebrionidae) fauna of arid Australia. He found three distinct faunal elements, an Archaic element, a Bassian element and a Younger Northern element. Each were characterised by different phylogenetic relationships, relative depths in phylogenetic trees and signature geographic distributions. The Younger Northern element is exemplified primarily by the tribe Opatrini, who have congeneric relatives in Asia, and all are xerophylic. They have not reached very high levels of endemism in Australia, and Matthews (2000) suggests that they arrived in Australia relatively recently, after the arid zone was formed here, since the mid-Miocene.

The Bassian element consists of seven tribes of Tenebrionidae, including the two tribes (Heleini and Adeliini) that are the most diverse at genus and species levels, with a combined 750 named species. The Heleini has close relatives in South America and likely derived from ancestors on Gondwana, and morphological phylogenetic analysis indicates that there is a progression from rainforest-inhabiting genera through sclerophyll genera, with the most recent groups being in the arid zone. The flattened form of these beetles has evolved so that they can live under eucalypt bark, but in the arid zone they are used to protect the head and legs from attack by spiders, scorpions, carab beetles and ants during nocturnal foraging. In the subtribe Heleina there are three different adaptations for living in deserts: daytime shelter in mammal burrows, living in caves, or living underground, facilitated by fossorial legs.

Matthews (2000) placed six tribes with 55 described species in the Archaic element. This group has relatives elsewhere in the world that live in deserts or coastal dunes, do not have relatives in the mesic biome and have no relatives in South America or Indonesia. The tribe Belopini is a good example, with the Australian genera being related to groups in either central or western North America, or the Mediterranean. These Northern Hemisphere localities were all on the shores of the Tethys Sea in the Jurassic, and Matthews suggests that the Australian fauna may have been adapted to dune environments in the Jurassic, preadapting them to desert environments that developed in Australia in the Tertiary. It would be exciting to test this idea with an appropriately sampled and dated molecular phylogeny, which requires that the Australian groups arose at least 170 mya. Many of the archaic element species have adapted to live in ants nests (myrmecophily), a very common habitat in arid Australia.

The ranges of high topographic relief embedded in the current arid zone have become refugia for some groups (Byrne *et al.* 2008) and we should expect so in insects. Matthews and Bouchard (2008) compare their areas of endemism with the 14 proposed by Cracraft (1991) for vertebrates and adopted by Crisp *et al.* (1999) for plants. They find only 10 regions (Figure 9.2), 5 in the mesic zone, 2 in the Monsoon Tropics and 3 in the arid zone. All three of the areas of endemism in the arid zone are significant mountain ranges with high topographic relief: the Flinders Ranges, MacDonnell Ranges and the Pilbara. Cassis and Symonds (2008, 2011) also used an amalgam of the areas of endemism of Cracraft (1991) and Crisp *et al.* (1999) for the lacebug genera *Inoma* and *Lasiacantha*, with area relationships that were incongruent to the former two studies.

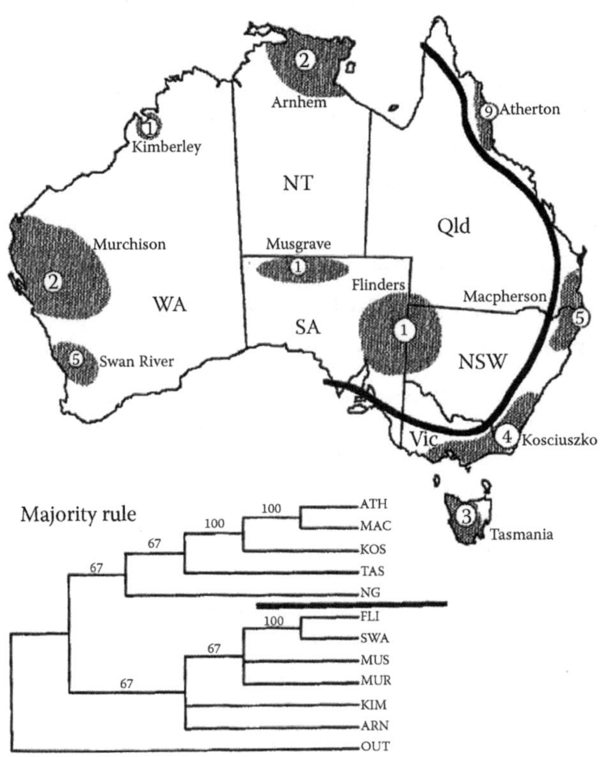

FIGURE 9.2 Areas of generic endemism for Tenebrionidae, (from Matthews and Bouchard, 2008). This shows remarkable similarity to the areas of endemism proposed by Cracraft (1991) for Australian vertebrates. The main difference between the schemes is the arid zone (Eremaean Province).

Biogeographic Relationships and Endemism in the Monsoon Tropics Biome (Torresian Province)

Evidence of monsoon tropical climates in northern Australia date back to the Late Eocene, 45 mya, implying that at least some endemic elements of the biota have a long history (Bowman *et al.* 2010). This region is characterised by the vicariant separation of eastern Cape York elements from those farther west in the Northern Territory via the Gulf of Carpentaria. Significant areas of endemism occur in the sandstone plateaus of Arnhem Land and the Kimberley. The biota of the region shows links both to New Guinea and farther west to Asia as well as relationships to Australian mesic zone and arid zone groups.

A detailed analysis of the biogeography of the butterflies in the Australian Monsoon Tropics (Braby 2008) revealed a rich fauna of 265 species (over half the Australian total) but very low endemism (6% of species). Western Cape York and the Top End show closer biogeographic relationships than the Kimberley region, and the Gulf of Carpentaria is a significant biogeographic gap; however, the Bonaparte Gap, separating the Top End from the Kimberley, has not been a significant vicariant gap.

Testable Models of Historical Biogeography in Australia

A primary aim of this chapter is to assess quantitative biogeographic analyses of the Australian insect fauna and to generate testable models of historical biogeography for Australia. These models are generated from common themes in empirical biogeographic studies and aim to provide temporally and spatially explicit hypotheses to test in future studies. We present six models, four of which correspond to a faunal element of the literature prior to the systematic period (1980–present), and we add two variants. These models present what we would expect of a taxonomic group that belonged to either the Archaic, Southern, Older Northern and Younger Northern faunal elements, in terms of phylogeny, divergence time, fossil evidence and distribution, both within Australia and for relatives elsewhere. The two variant models we have included describe long-distance over-ocean dispersal to New Zealand since the Oligocene and another showing dispersal out of Australia to Asia since the Late Miocene. Summaries of the models are provided in Figure 9.3 and explained in the following paragraphs. While the distinctions between these elements and models are effectively attempts to arbitrarily cut the continuous historical process of faunal assembly and evolution in Australia, we believe they offer significant heuristic value.

Archaic elements in the Australian insect fauna are Pangaean, so they should diverge from their relatives prior to the separation of Laurasia and Gondwana in the Jurassic, 160–180 mya. Because of this antiquity they are likely to be an early diverging lineage in their respective order, relative to the extant crown group. They may have relatives on other Gondwanan continents, but they should primarily have relatives in parts of Laurasia, either North America, Europe or Asia. They are most likely to be relictual in distribution across their range and have few species, most likely to occur in the mesic biome of Australia, but not always. They should have a long fossil record across their range extending back to the Jurassic.

Southern groups are Gondwanan, so they should diverge from their relatives between the time of separation of Gondwana and Laurasia (but no older) and the breakup of Gondwana beginning in the Late Jurassic (160 mya). Because of their relative age they should occupy middle-ranking positions in their orders, relative to their crown group. They should have relatives on some or all of the other Gondwanan fragments: Africa, India, New Zealand and South America. Apparently, because of dramatic ecological change, Gondwanan elements are often absent in India and Africa. They are most likely to occur in the mesic biome of Australia, but not always. Their fossil record should extend back to the Late Jurassic, but they should not have fossils on Laurasian continents, except if the group dispersed there in post-Gondwanan times. The distinction between Archaic and Southern elements can be difficult to determine if their divergence age is uncertain or their fossil record is poor. For example, the hypothesis that a group is southern is falsified by a single Laurasian fossil older than 160 million years. This group could be classified as archaic if divergence time estimates indicate it dates from Pangaean times.

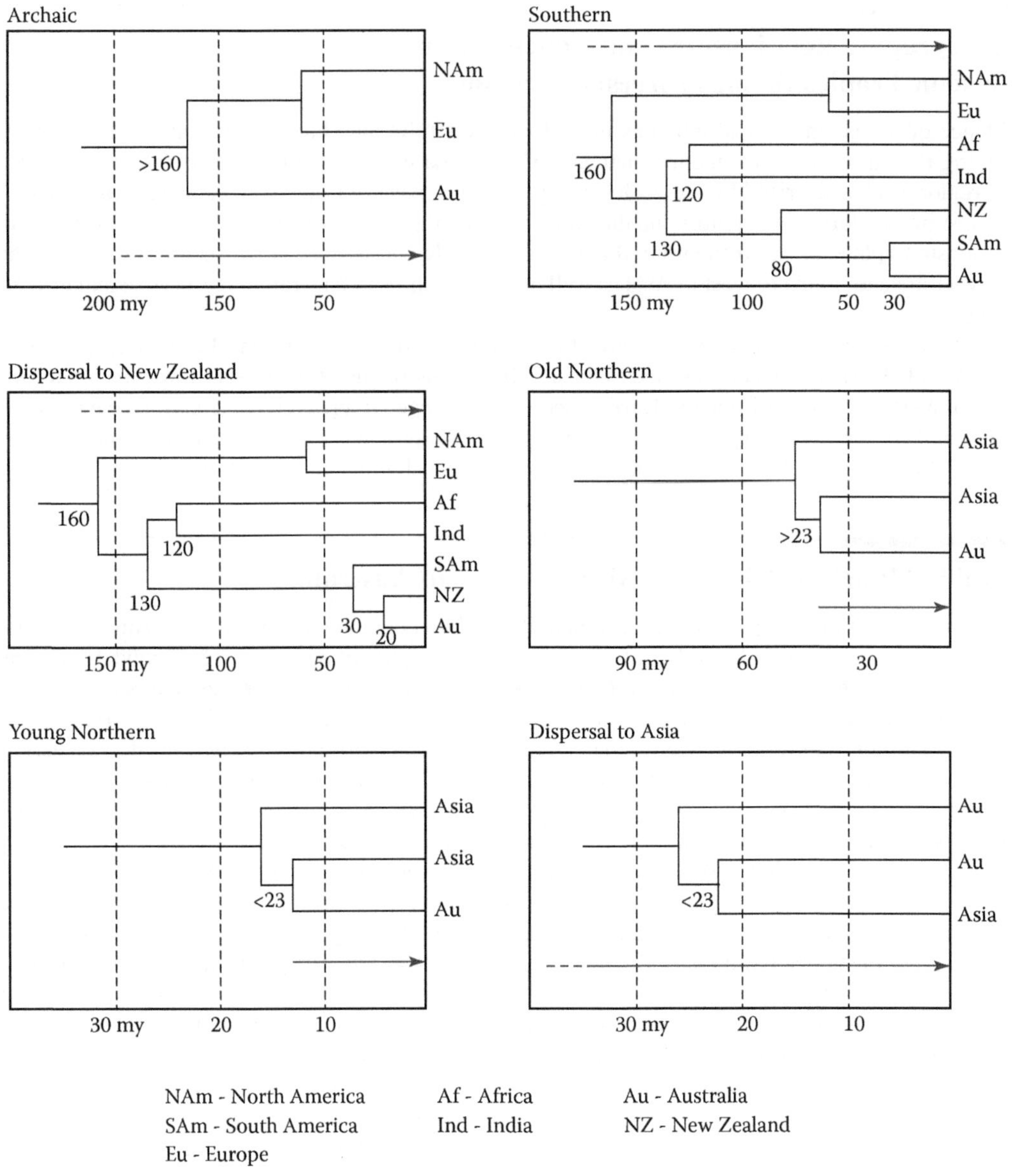

FIGURE 9.3 (See colour insert.) Six different testable models of historical biogeography for Australian insects. Red arrows indicate the duration of fossils in Australia predicted by the model. See text for further details.

Recent evidence shows that New Zealand offers an interesting variant on the traditional southern model offered here. Contemporary quantitative biogeographical evidence for the southern element in the New Zealand insect fauna is poor. For a New Zealand insect group to be truly southern, it must show divergences with Australian relatives dating at least to the rifting of Zealandia from Australia 80 mya. Most evidence to date suggests that divergences are younger than this, either dating to before the Oligocene inundation of New Zealand or after. Those dates of, say, 50–80 mya may represent archipelagic dispersal from Australia to New Zealand via the Lord Howe rise or over-ocean dispersal. Those groups with divergences dating to the mid-Oligocene or younger must have arrived in New Zealand via long-distance over-ocean dispersal. This pattern is very likely in phytophagous insect groups whose hosts also colonised New Zealand from Australia using a similar route and process.

The *Older Northern* element of the Australian insect fauna has always posed a geological puzzle, and it all depends on how old is *old*. These groups should be older than the Younger Northern element, which have arrived in Australia since the Australian and Asian plates began to interact, with land providing the possibility of faunal interchange, dated to the mid-Tertiary at the earliest, 23 mya (Hall 2013). Older Northern groups should have diverged from their Eurasian sister groups prior to 23 mya, and they may have an older fossil record on Laurasia, but not in Gondwana unless they are African or Indian. They will usually occupy reasonably distal positions in their orders relative to the crown group. They should have no fossils in Australia older than 23 million years and may occupy any biome. There are a number of possible process explanations for the Older Northern element. It is now clear that fragments of Sundaland (east Java, west Sulawesi and southwest Borneo) have been derived from the Australian plate in the Jurassic, and these pose the most compelling vicariance-based explanations for the Older Northern element (Metcalfe 2013). Long-distance over-ocean dispersal is also a possibility.

The *Younger Northern* element should have arrived in Australia in very recent times, since the mid-Tertiary 23 mya, and most of them have arrived since the end of the Miocene 15 mya, as the Australian and Asian plates began to interact and build significant terrestrial areas in Wallacea for stepping-stone range expansion. These groups should have diverged from their Asian sister groups 23 mya, and often much more recently. They should occupy quite distal positions in comparison with the crown group of their orders, and will often be lineages of the major families that dominate the four megadiverse insect orders Diptera, Hymenoptera, Coleoptera and Lepidoptera, which have diversified rapidly in the Tertiary. They most often occupy northern Australia or form components of the rainforest fauna of the northern mesic zone or of the Monsoon Tropics. They should be so recent to have not left much of a fossil record in Australia, and, if so, definitely younger than 23 mya.

It is important to note that traffic via this Wallacean route was not one way: Australia has generated lineages that have also dispersed out of Australia through Wallacea over the past 23 million years via this northern route, as illustrated in the sixth model (Figure 9.3).

Conclusions

The methods used in Australian insect biogeography have evolved considerably over the past century. Early efforts were largely qualitative and descriptive, assigning broad areas of Australia to three different faunal provinces (now termed *biomes*), and attributing different taxonomic groups to four different faunal elements. Gradually, over the past 30 years the methods have become more spatially, phylogenetically and temporally explicit, synthetic and quantified. The most sophisticated recent biogeographic studies of the Australian insect fauna have combined robust, multilocus molecular phylogenies with divergence time estimation calibrated using fossil evidence, and quantitative ancestral area reconstruction. They have been conducted within the framework of our growing understanding of the geological and palaeoclimatic history of Australia. These new methods allow for the integration of causal processes (vicariance, dispersal and ecological phenomena such as niche conservatism and adaptive diversification) into the biogeographic hypothesis for a single lineage. There are no examples in insect biogeography of a single causal process being a sufficient explanation. We are now at a point where we can pose biogeographical questions as hypotheses with strict phylogenetic and temporal expectations.

The insect orders had undergone significant diversification prior to the separation of Pangaea into Laurasia and Gondwana around 170–180 mya. There is growing evidence that many old Australian relict groups and disjuncts with living representatives also in Europe and/or North America are remnants of the Pangaean fauna. These groups tend to now reside in the mesic biome, but not always (e.g. *Mastotermes*). More of these can be expected to be identified, especially among the Paraneoptera, as the systematics and phylogeny of these groups become better understood.

Significant diversification of the insect orders also occurred while Australia was part of Gondwana. There are very few really good examples in the insect fauna of groups that show the phylogenetic relationships and divergence times expected under a Gondwanan vicariance scenario. Many of the inconsistencies occur with New Zealand clades either diverging early, but too late (ca. 50 mya) to relate to the rifting of Zealandia (e.g. *Stictocladius* and the Scionini), or very recently, up to 20 mya. This last category of divergences is likely to have occurred in plant-feeding groups if the host plants have colonised New Zealand recently via long-distance over-ocean dispersal. Again, these groups are most likely to occur in the mesic zone, but not always. Much more research needs to be conducted on low-vagility Southern Hemisphere groups to test the Gondwanan vicariance hypothesis.

Due to its proximity to the Australian plate for all but the last 30 million years, the insect fauna of Antarctica was the most closely related to the Australian fauna. The ice cap that obliterated the Antarctic fauna over the past 30 million years significantly hampers our ability to understand the relationships of the Australian fauna to the other Southern Hemisphere continents.

Diversification during Gondwanan times was much more complex that the orthodox vicariance hypothesis would suggest. Recent evidence suggests that groups that diversified in Australia during Gondwanan times dispersed out of Australia during Gondwanan fragmentation, either through Africa or through Asia. Some of this dispersal occurred by what must have been considerable long-distance over-ocean dispersal, and much of it was likely mediated through Antarctica. There is also evidence of dispersal both into and out of Australia beginning at around 23 mya, just as the northern edge of the Australian plate began interacting with the Asian plate. In a number of cases (pergid sawflies and scionine horse flies), the Australian and South American lineages are not reciprocally monophyletic, indicating regional diversification on Gondwana prior to breakup.

Within the Australian mesic zone there is significant evidence of diversification beginning in the mid-Miocene. This occurs in obligate rainforest groups, and often the deepest divergences here include Tasmania and the Wet Tropics of north Queensland. Due to its significant altitude, topography and conservation, the Wet Tropics remains a significant hot spot for Mesozoic insect lineages on a continental scale, not just in a regional context. Mesic zone insects provide evidence of both deep phylogenetic structuring within species that predates the Pleistocene, as well as species diversification in the Pleistocene. Some evidence is emerging from insect examples that rainforest is the plesiomorphic biome in Australia, and sclerophyll and arid biomes are derived from this community embedded in the mesic biome.

Much more phylogenetic research is required on Australian insects, especially those occupying the arid zone and Monsoon Tropics, if we are to build a more complete understanding of Australian insect biogeography. We would expect that the most pronounced areas of endemism in the arid zone will be in the areas of high topographic relief, such as the Flinders and MacDonnell Ranges and the Pilbara (Byrne *et al.* 2008; Matthews 2000; Matthews and Bouchard 2008), and that most of the arid zone fauna has arisen relatively recently from ancestors in the mesic zone. Conversely, Matthews (2000) has provided tantalising evidence that at least some of the arid zone fauna may in fact be archaic, and have been pre-adapted to the aridification of Australia through their association with Tethyan coastlines in the Jurassic. The Monsoon Tropics is perhaps the least understood of our faunal provinces, and further focus on this region will no doubt help us understand Australia's interactions with the Asian plate and their mediation through New Guinea.

We have provided six testable models to explain the historical biogeography of Australian insects in terms of the assembly and dispersal of the biota since Pangaean times. As far as possible we have provided the information required to operationalise the testing of these models in terms of phylogeny, divergence time, distribution and fossils. We hope that by testing these models with more empirical data we can evaluate the extent to which these models are useful in explaining what we see in nature in terms of known processes. We expect that we will learn far more about Australian insect biogeography from empirical data sets that don't conform to the models than those that do.

References

Almeida, E.A.B., Pie, M.R., Brady, S.G., and Danforth, B.N. (2012). Biogeography and diversification of colletid bees (Hymenoptera: Colletidae): Emerging patterns from the southern end of the world. *Journal of Biogeography*, 39, 526–544.

Austin, A.D., Yeates, D.K., Cassis, G., Fletcher, M.J., La Salle, J., Lawrence, J.F., McQuillan, P.B., *et al.* (2004). Insects 'Down Under': Diversity, endemism and evolution of the Australian insect fauna; Examples from select orders. *Australian Journal of Entomology*, 43, 216–234.

Baker, C.H., Graham, G., Scott, K., Yeates, D.K., and Merritt, D.J. (2008). Distribution and phylogenetic relationships of Australian glow-worms *Arachnocampa* (Diptera, Keroplatidae). *Molecular Phylogenetics and Evolution*, 48, 506–514.

Bell, K.L., Moritz, C., Moussalli, A., and Yeates, D.K. (2007). Comparative phylogeography and speciation of dung beetles from the Australian Wet Tropics rainforest. *Molecular Ecology*, 16, 4984–4998.

Bell, K.L., Yeates, D.K., Moritz, C., and Monteith, G.B. (2004). Molecular phylogeny and biogeography of the dung beetle genus *Temnoplectron* Westwood (Scarabaeidae: Scarabaeinae) from Australia's Wet Tropics. *Molecular Phylogenetics and Evolution*, 31, 741–753.

Bouchard, P., Yeates, D.K., and Brooks, D. (2005). Biogeography of the terrestrial arthropods of the Wet Tropics, pp. 425–462 in E. Bermingham, C.W. Dick and C. Moritz (Eds.), *Tropical Rainforests: Past, Present and Future*. Chicago University Press, Chicago, IL.

Bourguignon, T., Lo, N., Cameron, S.L., Sobotník, J., Hayashi, Y., Shigenobu, S., Watanabe, D., Roisin, Y., Miura, T., and Evans, T.A. (2014). The evolutionary history of termites as inferred from 66 mitochondrial genomes. *Molecular Biology and Evolution*, 32, 406–421.

Bowman, D.M.J.S., Brown G.K., Braby, M.F., Brown, J.R., Cook, L.G., Crisp, M.D., Ford, F., *et al.* (2010). Biogeography of the Australian Monsoon Tropics. *Journal of Biogeography*, 37, 201–216.

Braby, M.F. (2008). Biogeography of butterflies in the Australian Monsoon Tropics. *Australian Journal of Zoology*, 56, 41–56.

Braby, M.F., and Pierce, N.E. (2007). Systematics, biogeography and diversification of the Indo-Australian genus *Delias* Hübner (Lepidoptera: Pieridae): Phylogenetic evidence supports an 'out-of-Australia' origin. *Systematic Entomology*, 32, 2–25.

Braby, M.F., Pierce, N.E., and Vila, R. (2007). Phylogeny and historical biogeography of the subtribe Aporiina (Lepidoptera: Pieridae): Implications for the origin of Australian butterflies. *Biological Journal of the Linnean Society*, 90, 413–440.

Braby, M.F., Trueman, J.W.H., and Eastwood, R. (2005). When and where did troidine butterflies (Lepidoptera: Papilionidae) evolve? Phylogenetic and biogeographic evidence suggests an origin in remnant Gondwana in the Late Cretaceous. *Invertebrate Systematics*, 19, 113–143.

Brady, S.G., Schultz, T.R., Fisher, B.L., and Ward, P.S. (2006). Evaluating alternative hypotheses for the early evolution and diversification of ants. *Proceedings of the National Academy of Sciences*, 103, 18172–18177.

Brundin, L. 1966. Transantarctic relationships and their significance, as evidenced by chironomid midges with a monograph of the subfamilies Podonominae and Aphroteniinae and the austral Heptagyiae. *Kunglica Svenska vetenskapsakademiens handlingar* 11(1), 472, +30 plates.

Burckhardt, D. (2009). Taxonomy and phylogeny of the Gondwanan moss bugs or Peloridiidae (Hemiptera, Coleorrhyncha). *Deutsche Entomologische Zeitschrift*, 56, 173–235.

Byrne, M., Steane, D.A., Joseph, L., Yeates, D.K., Jordan, G.J., Crayn, D., Aplin, K., *et al.* (2011). Decline of a biome: Evolution, contraction, fragmentation, extinction and invasion of the Australian mesic zone biota. *Journal of Biogeography*, 38, 1635–1656.

Byrne, M., Yeates, D.K., Joseph, L., Kearney, M., Bowler, J., Williams, M.A., Cooper, S., *et al.* (2008) Birth of a biome: Insights into the assembly and maintenance of the Australian arid zone biota. *Molecular Ecology*, 17, 4398–4417.

Calaby, J.H., and Gay, F.J. (1959). Aspects of the distribution and ecology of Australian termites, pp. 211–223 in A. Keast, R.L. Crocker and C.S. Christian (Eds.), *Biogeography and Ecology in Australia*. Dr. W. Junk, The Hague, the Netherlands.

Cassis, G., and Symonds, C. (2008). Systematics, biogeography and host associations of the lace bug genus *Inoma* (Hemiptera: Heteroptera: Tingidae). *Acta Entomologica Musei Nationalis Pragae*, 48, 433–484.

Cassis, G., and Symonds, C. (2011). Systematics, biogeography and host plant associations of the lacebug genus *Lasiacantha* Stal (Insecta: Heteroptera: Tingidae). *Zootaxa*, 2818, 1–63.

Chenoweth, L.B., and Schwarz, M.P. (2011). Biogeographical origins and diversification of the exoneurine allodapine bees of Australia (Hymenoptera, Apidae). *Journal of Biogeography*, 38, 1471–1483.

Cook, L.G., and Crisp, M.D. (2005). Not so ancient: The extant crown group of *Nothofagus* represents a post-Gondwanan radiation. *Proceedings of the Royal Society of London B*, 272, 2535–3544.

Cracraft, J. (1991). Patterns of diversification within continental biotas: Hierarchical congruence among the areas of endemism of Australian vertebrates. *Australian Systematic Botany*, 4, 211–27.

Cranston, P.S. (2010). Insect biodiversity and conservation in Australasia. *Annual Review of Entomology*, 55, 55–75.

Cranston, P.S., and Edward, H.D. (1992). A systematic reappraisal of the Australian Aphroteniinae (Diptera: Chironomidae) with dating from vicariance biogeography. *Systematic Entomology*, 17, 41–54.

Cranston, P.S., Hardy, N.B., and Morse, G.E. (2012). A dated molecular phylogeny for the Chironomidae (Diptera). *Systematic Entomology*, 37, 172–188.

Cranston, P.S., and Naumann, I.D. (1991). Biogeography, pp. 180–197 in I.D. Naumann (Ed.), *The Insects of Australia*, 2nd Edn. Melbourne University Press, Melbourne, Australia.

Crisci, J.V., Katinas, L., and Posadas, P. (2003). *Historical Biogeography, An Introduction*. Harvard University Press, Cambridge, MA.

Crisp, M., Cook, L., and Steane, D. (2004) Radiation of the Australian flora: What can comparisons of molecular phylogenies across multiple taxa tell us about the evolution of diversity in present-day communities? *Philosophical Transactions of the Royal Society of London B*, 359, 1551–1571.

Crisp, M.D., West, J.G., and Linder, H.P. (1999). Biogeography of the terrestrial flora, pp. 321–367 in A.E. Orchard and H.S. Thompson (Eds.), *Flora of Australia*, Vol. 1, 2nd Edn. CSIRO Publishing, Melbourne, Australia.

Cruaud, A., Jabbour-Zahab, R., Genson, G., Couloux, A., Yan-Qiong, P., Da Rong, Y., Ubaidillah, R., *et al.* (2011). Out of Australia and back again: The world-wide historical biogeography of non-pollinating fig wasps (Hymenoptera: Sycophaginae). *Journal of Biogeography*, 38, 209–225.

Danforth, B.N., Brady, S.G., Sipes, S.D., and Pearson, A. (2004). Single copy nuclear genes recover Cretaceous-age divergences in bees. *Systematic Biology*, 53, 309–326.

Danforth, B.N., Sipes, S., Fang, J., and Brady, S.G. (2006). The history of early bee diversification based on five genes plus morphology. *Proceedings of the National Academy of Sciences of the USA*, 103, 15118–15123.

Ebach, M.C. (2012). A history of biogeographical regionalisation in Australia. *Zootaxa*, 3392, 1–34.

Ebach, M.C., Gill, A.C., Kwan, A., Ahyong, S.T., Murphy, D.J., and Cassis, G. (2013). Towards an Australian Bioregionalisation Atlas: A provisional area taxonomy of Australia's biogeographical regions. *Zootaxa*, 3619(3),315–342.

Endo, Y., Nash, M., Hoffmann, A.A., Slatyer, R., and Miller, A.D. (2015). Comparative phylogeography of alpine invertebrates indicates deep lineage diversification and historical refugia in the Australian Alps. *Journal of Biogeography*, 42, 89–102.

Evans, J.W. (1959). The zoogeography of some Australian insects, pp. 150–163 in A. Keast, R.L. Crocker and C.S. Christian (Eds.), *Biogeography and Ecology in Australia*. Dr. W. Junk, The Hague, the Netherlands.

Ferrier, S., Manion, G., Elith, J., and Richardson, K. (2007). Using generalized dissimilarity modelling to analyse and predict patterns of beta diversity in regional biodiversity assessment. *Diversity and Distributions*, 13, 252–264.

Fuller, S., Schwarz, M., and Tierney, S. (2005). Phylogenetics of the allodapine bee genus *Braunsapis*: Historical biogeography and long-range dispersal over water. *Journal of Biogeography*, 32, 2135–2144.

Grimaldi, D.A., and Engel, M.S. (2005). *Evolution of the Insects*. Cambridge University Press, Cambridge, UK.

Grimaldi, D.A., and Engel, M.S. (2013). The relict scorpionfly family Meropeidae (Mecoptera) in Cretaceous amber. *Journal of the Kansas Entomological Society*, 86, 253–263.

Hall, R. (2009). Southeast Asia's changing paleogeography. *Blumea*, 54, 148–161.

Hall, R. (2013). The paleogeography of Sundaland and Wallacea since the Late Jurassic. *Journal of Limnology*, Supplement 2, 72, 1–17.

Hardy, N.B., Gullan, P.J., Henderson, R.C., and Cook, L.G. (2008). Relationships among felt scale insects (Hemiptera: Coccoidea: Eriococcidae) of southern beech, *Nothofagus* (Nothofagaceae), with the first descriptions of Australian species of the *Nothofagus*-feeding genus *Madarococcus* Hoy. *Invertebrate Systematics*, 22, 365–405.

Harold, A.S., and Mooi, R.D. (1994). Areas of endemism: Definition and recognition criteria. *Systematic Biology*, 43, 261–266.

Harrison, L. (1928). The composition and origins of the Australian fauna, with special reference to the Wegener hypothesis. *Report of the Australasian Association for the Advancement of Science*, 18, 332–396.

Hawlitschek, O., Hendrich, L., and Balke, M. (2012a). Molecular phylogeny of the squeak beetles, a family with disjunct Palearctic–Australian range. *Molecular Phylogenetics and Evolution*, 62, 550–554.

Hawlitschek, O., Hendrich, L., Espeland, M., Toussaint, E.F.A., Genner, M.J., and Balke, M. (2012b). Pleistocene climate change promoted rapid diversification of aquatic invertebrates in Southeast Australia. *BMC Evolutionary Biology*, 12, 142.

Hedley, C. (1894). The faunal regions of Australia. *Reports of the Australian Association for the Advancement of Science*, 5, 444–446.

Heiss, E., and Popov, Y.A. (2002). Reconsideration of the systematic position of Thaicorinae with notes on fossil and extant Thaumastocoridae (Hemiptera: Heteroptera). *Polskie Pismo Entomologiczne*, 71, 247–259.

Hennig, W. (1960) Die Dipteren-Fauna von Neuseeland als systematisches und tiergeographisches Problem. *Beitrage Entomologie*, 10, 221–329.

Howden, H.F. (1981). Zoogeography of some Australian Coleoptera as exemplified by the Scarabaeoidea, pp. 1007–1035 in A. Keast (Ed.), *Ecological Biogeography of Australia*. Dr. W. Junk, The Hague, the Netherlands.

Hugall, A.F., and Stanisic, J. (2011). Beyond the prolegomenon: A molecular phylogeny of the Australian camaenid land snail radiation. *Zoological Journal of the Linnean Society*, 161, 531–572.

Kayaalp, P., Schwarz, M.P., and Stevens, M.I. (2013). Rapid diversification in Australia and two dispersals out of Australia in the globally distributed bee genus, *Hylaeus* (Colletidae: Hylaeinae). *Molecular Phylogenetics and Evolution*, 66, 668–678.

Keast, A. (1981). *Ecological Biogeography of Australia*. Dr. W. Junk, The Hague, the Netherlands.

Keast, A., Crocker, R.L., and Christian, C.S. (Eds.). (1959). *Biogeography and Ecology in Australia*. Dr. W. Junk, The Hague, the Netherlands.

Key, K.H.L. (1959). The ecology and biogeography of Australian grasshoppers, pp. 192–210 in A. Keast, R.L. Crocker and C.S. Christian (Eds.), *Biogeography and Ecology in Australia*. Dr. W. Junk, The Hague, the Netherlands.

Kitching, R.L. (1981). The geography of Australian Papilionoidea, pp. 977–1005 in A. Keast (Ed.), *Ecological Biogeography of Australia*. Dr. W. Junk, The Hague, the Netherlands.

Kitching, R.L., and Dunn, K.L. (1999). The biogeography of Australian butterflies, pp. 53–74 in R.L. Kitching, R.E. Jones, E. Scheermeyer and N.E. Pierce (Eds.), *Biology of Australian Butterflies*. Monographs of Australian Lepidoptera, Vol 6. CSIRO Publishing, Melbourne, Australia.

Knowles, L.L. (2009). Statistical phylogeography. *Annual Review of Ecology, Evolution, and Systematics*, 40, 593–612.

Krosch, M., and Cranston, P.S. (2013). Not drowning, (hand)waving? Molecular phylogenetics, biogeography and evolutionary tempo of the 'Gondwanan' midge *Stictocladius* Edwards (Diptera: Chironomidae). *Molecular Phylogenetics and Evolution*, 68, 595–603.

Lee, T.R.C., Cameron, S.L., Evans, T.A., Ho, S.Y.W., and Lo, N. (2015). The origins and radiation of Australian *Coptotermes* termites: From rainforest to desert dwellers. *Molecular Phylogenetics and Evolution*, 82, 234–244.

Lessard, B.D., Cameron, S.L., Bayless, K.M., Wiegmann, B.M., and Yeates, D.K. (2013). The evolution and biogeography of the austral horse fly tribe Scionini (Diptera: Tabanidae: Pangoniinae) inferred from multiple mitochondrial and nuclear genes. *Molecular Phylogenetics and Evolution*, 68, 516–540.

Leys, R., Cooper, S.J.B., and Schwarz, M.P. (2002). Molecular phylogeny and historical biogeography of the large carpenter bee, genus *Xylocopa* (Hymenoptera: Apidae). *Biological Journal of the Linnean Society*, 77, 249–266.

Littlejohn, M.J. (1981). The amphibia of mesic South Australia: A zoogeographic perspective, pp. 1305–1330 in A. Keast (Ed.), *Ecological Biogeography of Australia*. Dr. W. Junk, The Hague, the Netherlands.

Main, A.R. (1987). Evolution and radiation of the terrestrial fauna, pp. 136–155 in *Fauna of Australia*, Vol 1A, *General Articles*. Bureau of Flora and Fauna, Canberra. Australian Government Publishing Service, Canberra, Australia.

Main, B.Y. (1981). A comparative account of the biogeography of terrestrial invertebrates in Australia: Some generalisations, pp. 1055–1077 in A. Keast (Ed.), *Ecological Biogeography of Australia*. Dr. W. Junk, The Hague, the Netherlands.

Mackerras, I.M. (1950). The zoogeography of the Diptera. *The Australian Journal of Science*, 12, 157–161.

Mackerras, I.M. (1970). Composition and distribution of the fauna, pp. 187–203 in CSIRO, *The Insects of Australia*. Melbourne University Press, Melbourne Australia.

Matthews, E.G. (2000). Origins of Australian arid-zone tenebrionid beetles. *Invertebrate Taxonomy*, 14, 941–951.

Matthews, E.G., and Bouchard, P. (2008). *Tenebrionid Beetles of Australia*. Australian Biological Resources Study, Canberra, Australia.

McGlone, M.S. (2005). Goodbye Gondwana. *Journal of Biogeography*, 32, 739–740.

Metcalfe, I. (2013). Gondwana dispersion and Asian accretion: Tectonic and palaeogeographic evolution of eastern Tethys. *Journal of Asian Earth Sciences*, 66, 1–33.

Michener, C.D. (2007). *The Bees of the World*, 2nd Edn. Johns Hopkins University Press, Baltimore, MD.

Misof, B., Liu, S., Meusemann, K., Peters, R.S., Donath, A., Mayer, C., Frandsen, P.B., *et al.* (2014). Phylogenomics resolves the timing and pattern of insect evolution. *Science*, 346, 763–767.

Moreau, C.S., Bell, C.D., Vila, R., Archibald, S.B., and Pierce, N.E. (2006). Phylogeny of the ants: Diversification in the age of angiosperms. *Science*, 312, 101–104.

Moore, J.A. (1961). The frogs of eastern New South Wales. *Bulletin of the American Museum of Natural History*, 121, 149–386.

Morrone, J.J., and Crisci, J.V. (1995). Historical biogeography: Introduction to methods. *Annual Review of Ecology and Systematics*, 26, 373–401.

Müller, C.J., Matos-Maraví, P.F., and Beheregaray, L.B. (2013). Delving into *Delias* Hübner (Lepidoptera: Pieridae): Fine-scale biogeography, phylogenetics and systematics of the world's largest butterfly genus. *Journal of Biogeography*, 40, 881–893.

Nelson, G., and Platnick, N.I. (1981). *Systematics and Biogeography: Cladistics and Vicariance*. Columbia University Press, New York.

Nix, H.A. (1982). Environmental determinants of biogeography and evolution in Terra Australis, pp. 47–66 in W.R. Barker and P.J.M. Greenslade (Eds.), *Evolution of the Flora and Fauna of Arid Australia*. Peacock Publications, Adelaide, Australia.

Nix, H.A. (1991). Biogeography: Pattern and process, pp. 11–112 in H.A. Nix and M.A. Switzer (Eds.), *Rainforest Animals: Altas of Vertebrates Endemic to Australia's Wet Tropics*. Australian National Parks and Wildlife Service, Canberra, Australia.

Noack, A.E., Cassis, G., and Rose, H.A. (2011). Systematic revision of *Thaumastocoris* Kirkaldy (Hemiptera: Heteroptera: Thaumastocoridae). *Zootaxa*, 3121, 1–60.

Paramonov, S.J. (1959). Zoogeographical aspects of the Australian dipterofauna, pp. 164–191 in A. Keast, R.L. Crocker and C.S. Christian (Eds.), *Biogeography and Ecology in Australia*. Dr. W. Junk, The Hague, the Netherlands.

Parenti, L.R., and Ebach, M.C. (2009). *Comparative Biogeography: Discovering and Classifying Biogeographical Patterns of a Dynamic Earth*, Vol. 2. University of California Press, Berkeley.

Parsons, P.A., and Bock, I.R. (1981). Australian Drosophila: Diversity, resource utilization and radiations, pp. 1037–1054 in A. Keast (Ed.) *Ecological Biogeography of Australia*. Dr. W. Junk, The Hague, the Netherlands.

Platnick, N.I. (1991). On areas of endemism, pp. xi–xii in P.Y. Ladiges, C.J. Humphries and L.W. Martinelli (Eds.), *Austral Biogeography*. CSIRO, Melbourne, Australia. Reprinted from *Australian Systematic Botany*, 4(1).

Raven, P.H., and Yeates, D.K. (2007). Australian biodiversity: Threats for the present, opportunities for the future. *Australian Journal of Entomology*, 46, 177–187.

Rix, M.G., Edwards, D.L., Byrne, M., Harvey, M.S., Joseph, L., and Roberts, J.D. (2014). Biogeography and speciation of terrestrial fauna in the south-western Australian biodiversity hotspot. *Biological Reviews*, 90(3),762–793.

Sanmartín, I., and Ronquist, F. (2004). Southern Hemisphere biogeography inferred by event-based models: Plant versus animal patterns. *Systematic Biology*, 53, 216–243.

Schmidt, D.J., Grund, R., Williams, M.R., and Hughes, J.M. (2011). Australian parasitic *Ogyris* butterflies: East–west divergence of highly-specialized relicts. *Biological Journal of the Linnean Society*, 111, 473–484.

Schmidt, S., and Walter, G.H. (2014). Young clades in an old family: Major evolutionary transitions and diversification of the eucalypt-feeding pergid sawflies in Australia (Insecta, Hymenoptera, Pergidae). *Molecular Phylogenetics and Evolution*, 74, 111–121.

Schwarz, M.P., Fuller, S., Tierney, S.M., and Cooper, S.J.B. (2006). Molecular phylogenetics of the exoneurine allodapine bees reveal an ancient and puzzling dispersal from Africa to Australia. *Systematic Biology*, 55, 31–45.

Sharma, P.P., and Wheeler, W.C. (2013). Revenant clades in historical biogeography: The geology of New Zealand predisposes endemic clades to root age shifts. *Journal of Biogeography*, 40, 1609–1618.

Simonsen, T.J., Zakharov, E.V., Djernaes, M., Cotton, A.M., Vane-Wright, R.I., and Sperling, F.A.H. (2010) Phylogenetics and divergence times of Papilioninae (Lepidoptera) with special reference to the enigmatic genera Teinopalpus and Meandrusa. *Cladistics*, 27, 113–137.

Slatyer, R.A., Nash, M.A., Miller, A.D., Endo, Y., Umbers, K.D.L., and Hoffmann, A.A. (2014). Strong genetic structure corresponds to small-scale geographic breaks in the Australian alpine grasshopper Kosciuscola tristis. *BMC Evolutionary Biology*, 14, 204.

Sloane, T.G. (1905). Australian Carabidae: Check-list; Part i; Subfamily Carabinae. *Supplement to the Proceedings of the Linnean Society of New South Wales (Part I)*, 30, 1–18.

Sloane, T.G. (1915). On the faunal subregions of Australia. *Proceedings of the Royal Society of Victoria*, 28, 139–148.

Sota, T., Takami, Y., Monteith, G.B., and Moore, B.P. (2005). Phylogeny and character evolution of endemic Australian carabid beetles of the genus *Pamborus* based on mitochondrial and nuclear gene sequences. *Molecular Phylogenetics and Evolution*, 36, 391–404.

Spencer, W.B. (1896). *Report on the Work of the Horn Scientific Expedition to Central Australia*. Part 1: Introduction, narrative, summary of results, supplement to zoological report, map. Melville, Mullen and Slade, Melbourne, Australia, 220 p.

Staunton, K.M., Robson, S.K.A., Burwell, C.J., Reside, A.E., and Williams, S.E. (2014). Projected distributions and diversity of flightless ground beetles within the Australian Wet Tropics and their environmental correlates. *PLOS ONE*, 9, 1–16.

Stork, N.E., and S.M. Turton (Eds.). (2008). *Living in a Dynamic Tropical Forest Landscape*. Blackwell, Oxford, UK.

Tate, R. (1889). *On the Influence of Physiological Changes in the Distribution of Life in Australia*. Report of the first meeting of the Australian Association for the Advancement of Science, pp. 312–326.

Tierney, S.M., Smith, J.A., Chenoweth, L., and Schwarz, M.P. (2008). Phylogenetics of allodapine bees: A review of social evolution, parasitism and biogeography. *Apidologie*, 39, 3–15.

Tillyard, R.J. (1924). Origin of the Australian and New Zealand insect faunas. *Report of the Australasian Association for the Advancement of Science*, 16, 407–413.

Torsvik, T.H., and Cocks, L.R.M. (2013). Gondwana from top to base in space and time. *Gondwana Research*, 24, 999–1030.

Toussaint, E.F.A., Condamine, F.L., Hawlitschek, O., Watts, C.H., Porch, N., Hendrich, L., and Balke, M. (2015). Unveiling the diversification dynamics of Australasian predaceous diving beetles in the Cenozoic. *Systematic Biology*, 64, 3–24.

Van Doesburg, P.H., Cassis, G., and Monteith, G.B. (2010). Discovery of a living fossil: A new xylastodorine species from New Caledonia (Heteroptera: Thaumastocoridae) and first record of the subfamily from the Eastern Hemisphere. *Zoologische Mededelingen*, 84(6),93–115.

Ward, P.S. (2001). Taxonomy, phylogeny and biogeography of the ant genus *Tetraponera* (Hymenoptera: Formicidae) in the Oriental and Australian regions. *Invertebrate Taxonomy*, 15, 589–665.

Ward, P.S., and Brady, S.G. (2003). Phylogeny and biogeography of the ant subfamily Myrmeciinae (Hymenoptera: Formicidae). *Invertebrate Systematics*, 17, 361–386.

Ware, J.L., Beatty, C.D., Herrera, M.S., Valley, S., Johnson, J., Kerst, C., May, M.L., and Theischinger, G. (2014). The petaltail dragonflies (Odonata: Petaluridae): Mesozoic habitat specialists that survive to the modern day. *Journal of Biogeography*, 41, 1291–1300.

Watts, C.H.S. (1982). A blind terrestrial water beetles from Australia. *Memoirs of the Queensland Museum*, 20, 527–531.

Wedmann, S., and Yeates, D.K. (2008). Eocene records of bee flies (Insecta, Diptera, Bombyliidae, *Comptosia)*, their paleobiogeographic implications and remarks on the evolutionary history of bombyliids. *Palaeontology*, 51, 231–240.

Weirauch, C., and Schuh, R.T. (2010). Southern Hemisphere distributional patterns in plant bugs (Hemiptera: Miridae: Phylinae): *Xiphoidellus*, gen. nov. from Australia and *Ampimpacoris*, gen. nov. from Argentina, show transantarctic relationships. *Invertebrate Systematics*, 24, 473–508.

Wiegmann, B.M., Trautwein, M.D., Winkler, I.S., Barr, N.B., Kim, J.W., Lambkin, C., *et al.* (2011). Episodic radiations in the fly tree of life. *Proceedings of the National Academy of Science of the USA*, 108, 5690–5695.

Williams, K.J., Ferrier, S., Rosauer, D., Yeates, D., Manion G., Harwood, T., Stein, J., Faith, D.P., Laity, T., and Whalen, A. (2010). *Harnessing Continent-Wide Biodiversity Datasets for Prioritising National Conservation Investment*. A report prepared for the Department of Sustainability, Environment, Water, Population and Communities. CSIRO, Canberra, Australia.

Yeates, D.K., Bouchard, P., and Monteith, G.B. (2002). Patterns and levels of endemism in the Australian Wet Tropics rainforest: Evidence from flightless insects. *Invertebrate Systematics*, 16, 605–619.

Yeates, D.K., Harvey, M.S., and Austin, A.D. (2003). New estimates for terrestrial arthropod species-richness in Australia. *Records of the South Australian Museum Monograph Series*, 7, 231–241.

Yeates, D.K., and Irwin, M.E. (1996). Apioceridae (Insecta: Diptera): Cladistics and biogeography. *Zoological Journal of the Linnean Society*, 116, 247–301.

Yeates, D.K., and Monteith, G.B.M. (2008). The invertebrate fauna of the Wet Tropics: Diversity, endemism and relationships, pp. 178–191 in N.E. Stork and S.M. Turton (Eds.), *Living in a Dynamic Tropical Forest Landscape*. Blackwell Publishing, Oxford UK.

Zwick, P. (1981). Plecoptera, pp. 1169–1182 in A. Keast (Ed.), *Ecological Biogeography of Australia*. Dr. W. Junk, The Hague, the Netherlands.

10

The Biogeography of Australasian Arachnids

**Mark S. Harvey, Michael G. Rix, Danilo Harms,
Gonzalo Giribet, Cor J. Vink and David E. Walter**

CONTENTS

Arachnida

The Arachnida, a class of arthropod animals that includes prominent examples such as spiders, ticks and scorpions (Figure 10.1), comprises some of the most successful biological radiations on Earth. The lineage is extremely ancient and has a fossil record that dates back to the Palaeozoic (Dunlop 2010), but it is also highly diverse, with some 114,000 named species (Zhang 2013). The vast majority of arachnids are terrestrial, but some, such as water mites and marine mites, have independently evolved an aquatic lifestyle (Walter and Proctor 2013).

Arachnids are excellent subjects for biogeographic research, as they are globally distributed and one of the oldest terrestrial lineages, with fossils dating back to the Silurian and most extant orders known from the Palaeozoic (Dunlop 2010). This, combined with the relatively limited dispersal capabilities of many taxa, allows biogeographic hypotheses to be tested at different temporal and spatial scales, from deep intercontinental (vicariance and dispersal) relationships to finer-scale intraspecific phylogeographies. Australasia and the other southern continents have played a significant role in discussions of arachnid biogeography, especially vicariance biogeography, due largely to the presence of many early-branching lineages that retain ancient biogeographic signal. However, the region is also home to a highly diverse suite of more derived and recently evolved taxa. Both groups of organisms are suitable for revealing the evolutionary history of the regional Australasian biota.

FIGURE 10.1 (See colour insert.) Live images of representative arachnid taxa from Australasia. (a–c) Spiders (Araneae): (a) *Missulena* sp. (Actinopodidae); (b) *Malkara loricata* (Malkaridae); (c) *Zephyrarchaea vichickmani* (Archaeidae). (d,e) Scorpions (Scorpiones): (d) *Urodacus planimanus* (Urodacidae); (e) *Cercophonius* sp. (Bothriuridae). (f) Amblypygid, *Charon* sp. (Amblypygi: Charontidae). (g,h) Harvestmen (Opiliones): (g) *Forsteropsalis wattsi* (Neopilionidae), with parasitic mites (order Trombidiformes); (h) *Karripurcellia* sp. (Pettalidae). (i) Schizomid, *Draculoides bramstokeri* (Schizomida: Hubbardiidae). (j,k) Pseudoscorpions (Pseudoscorpiones): (j) unidentified New Zealand chernetid (Chernetidae); (k) *Neopseudogarypus scutellatus* (Pseudogarypidae). (l–o) Mites: (l) unidentified opilioacarid from South Africa (Opilioacarida: Opilioacaridae); (m) *Amblyomma triguttatum* (Ixodida: Ixodidae); (n) unidentified chyzeriid from New Zealand (Trombidiformes: Chyzeriidae); (o) unidentified mesostigmatan mite (Mesostigmata). Images (a) and (b) courtesy of G. Anderson. Other images by G. Giribet (g, h, j, l, n), M. Harvey (d–f, i, m) and M. Rix (c, k).

We examine arachnid biogeography in Australasia by first discussing the composition, diversity and origins of the fauna, followed by a bioregional treatment that aims to broadly summarise known patterns, key data sets and relevant contributions, both past and present. We emphasise that our understanding of arachnid evolution and biogeography in the region remains patchy and inevitably biased towards the more prominent taxa such as spiders, and thus a secondary aim of this review is to highlight gaps in our knowledge and avenues for future research.

Australasian Arachnid Fauna

Arachnids, as in other Chelicerata, can be characterised by the absence of antennae and the presence of an anterior region bearing a pair of preoral chelicerae, a pair of pedipalps and up to four pairs of legs in adults (Snodgrass 1948; Beccaloni 2009). Although the monophyly of the Chelicerata is well supported (e.g. Giribet *et al.* 2002; Coddington *et al.* 2004; Regier *et al.* 2010), a monophyletic Arachnida is still open to contention, even when using large phylogenomic data sets (Sharma *et al.* 2014). The chelicerate outgroups to the Arachnida are the Xiphosura (horseshoe crabs), Eurypterida (extinct sea scorpions) and Pycnogonida (sea spiders) – all marine organisms. The Arachnida is currently divided into 16 recent orders, with 4 additional fossil ones (Haptopoda, Phalangiotarbida, Uraraneida and Trigonotarbida) that became extinct in the Palaeozoic (Dunlop and Selden 2009; Dunlop and Penney 2012). The Australasian fauna (here defined as Australia, mainland New Guinea and other islands as far west as Lydekker' s Line, New Caledonia and New Zealand; Figure 10.2) is highly diverse and comprises 14 orders (Figure 10.1): spiders (Araneae; Figure 10.1a–c), scorpions (Scorpiones; Figure 10.1d,e), whip spiders (Amblypygi; Figure 10.1f), whip scorpions (Uropygi), harvestmen (Opiliones; Figure 10.1g,h), schizomids (Schizomida; Figure 10.1i), pseudoscorpions (Pseudoscorpiones; Figure 10.1j,k), palpigrades (Palpigradi) and six acarine orders (Trombidiformes, Sarcoptiformes, Holothyrida, Ixodida,

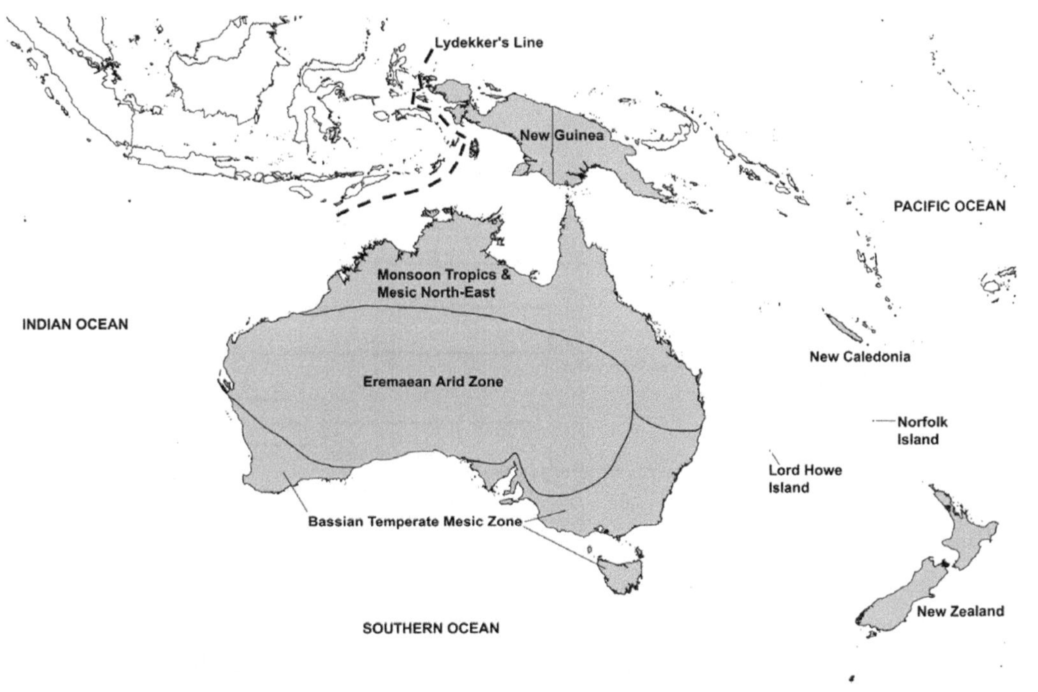

FIGURE 10.2 Map of Australasia, southern Southeast Asia and the western Pacific, showing bioregions discussed in the text. Australasian landmasses, as here defined, are highlighted in grey.

Mesostigmata and Opilioacarida; Figure 10.1l–o). Only two arachnid orders are absent from Australasia: the ricinuleids (Ricinulei) and sun/camel spiders (Solifugae).

Composition, Diversity and Distributions

Australasia has a complex geological and climatic history, which is concomitantly reflected in the composition and diversity of the arachnid fauna. The Austral landmasses that now comprise Australia, New Guinea, New Caledonia and New Zealand were, during the Mesozoic era, part of the supercontinent Gondwana (along with Africa, Madagascar, India, South America and Antarctica) (McLoughlin 2001; Sanmartín and Ronquist 2004). These landmasses started to rift during the mid-Mesozoic ca. 165 mya, culminating in the relatively late separation of Australia from Antarctica sometime during the Early Oligocene ca. 52–35 mya (Sanmartín and Ronquist 2004), while New Zealand became isolated considerably earlier during the Late Cretaceous (ca. 80 mya) (Figure 10.3). The order and timing of Gondwanan continental drift has been the subject of considerable debate over the past few decades (e.g. Nelson and Platnick 1981; Sanmartín and Ronquist 2004), reflected in controversies such as the *drowned Zealandia* or *goodbye Gondwana* hypotheses (e.g. McGlone 2005; Giribet and Boyer 2010) and our increased understanding of the deep and complex history of the subtropical and temperate western Pacific (e.g. McDougall *et al.* 1981; Grandcolas *et al.* 2008). Vicariant biogeographic hypotheses have dominated the Austral arachnological discourse over recent decades, but dispersal and relatively recent in situ radiations have clearly been integral in generating biodiversity and influencing arachnid biogeography throughout the region, especially in the face of major climatic fluctuations since the Miocene (Byrne *et al.* 2008). To make things even more complicated, New Caledonia, the northernmost part of the Zealandia continental fragment – and traditionally considered a continental island – may have suffered a deep drowning event, reemerging during the Eocene 37 mya, and thus most of its biota is now considered to have reached the island by dispersal (Grandcolas *et al.* 2008), mostly from other Austral areas.

The Australasian arachnid fauna is imperfectly known and new taxa, including previously unnamed species and genera, are frequently being described, further improving our knowledge of a distinctive and highly diverse biota. For example, the named Australian fauna currently comprises nearly 7600 species (ABRS 2009) and the New Zealand fauna nearly 2700 species (Sirvid *et al.* 2010). Mites include the most ecologically diverse arachnid lineages, ranging from free-living and parasitic terrestrial species to those that are dependent on marine and freshwater ecosystems. Most mite orders are found all over Australasia,

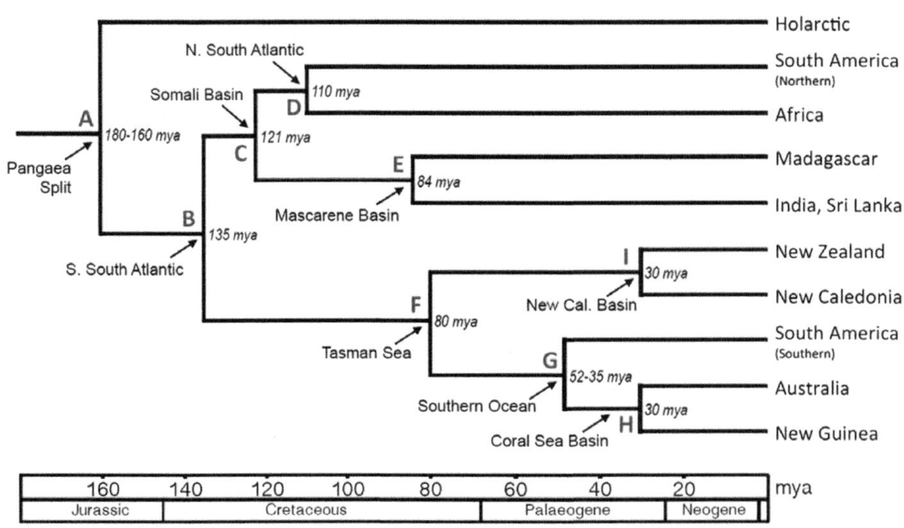

FIGURE 10.3 Dated area cladogram showing the order and timing of different Austral vicariance events, from the mid-Mesozoic to the present (redrawn from Sanmartín, I., and Ronquist, F., *Systematic Biology* 53, 216–243, 2004). See text for references to labelled nodes.

but the Opilioacarida (Figure 10.1l) are restricted to northern Australia and are apparently absent from other regions of Australasia. Spiders are also ubiquitous and have radiated to occupy most terrestrial habitats, from intertidal zones to alpine habitats. Pseudoscorpions and harvestmen are also abundant and widely distributed, but due to their small size or cryptic habits are less well known. Scorpions are absent from New Zealand but occur in Australia, New Caledonia and New Guinea, where they inhabit forested or arid ecosystems. Four other orders (whip spiders, whip scorpions, schizomids and palpigrades) are ecologically specialised and have more confined distributions, resulting in the restriction to tropical and/or subterranean habitats.

Numerous arachnid families are entirely restricted to Australasia, indicating the long-term persistence of at least some lineages on the continental blocks (most of which predate the separation of Gondwana) and/or the likely extinction of related forms elsewhere. While many families of spiders, mites and scorpions are restricted to single Australasian regions, others occur on multiple landmasses, both across Australasia and/or the Southern Hemisphere. For the purposes of this review, we highlight nine such distributions that encompass the major patterns exhibited by Australasian arachnids (Figure 10.4), not including single-landmass endemics (which do little to inform intercontinental biogeography without knowledge of their immediate sister group distributions).

At the broadest scale, there is evidence that a number of lineages display disjunct distributions between Australasia and the Holarctic region (Figure 10.4a). This has been postulated to be the result of vicariant events dating from before the Jurassic when the world's landmasses were connected via the supercontinent Pangaea (Figure 10.3, node a). For these taxa, a vicariant hypothesis has been preferred, as their lifestyle would indicate that transoceanic dispersal events are virtually impossible. For example, pseudoscorpions of the families Pseudotyrannochthoniidae and Pseudogarypidae (Figure 10.1k), the subfamily Syarininae (Syarinidae) and the genus *Oreolpium* (Garypinidae) have species in both temperate Australia and the Holarctic region (Harvey 1998a; Harvey and Šťáhlavský 2010; Harms and Harvey 2013). Likewise, the water mite genera *Tartarothyas* and the *Panisellus* group (Hydryphantidae) and Huitfeldtiinae (Pionidae) have bipolar distributions, which are suspected to be the result of ancient vicariant events rather than dispersal (Harvey 1998b). Colloff (2013) found that both genera and seven of eight species groups of the oribatid mite family Malaconothridae, primarily inhabitants of fresh water, appeared to have evolved as part of the Pangaean fauna, a hypothesis originally proposed by Hammer and Wallwork (1979). A high proportion of mite families in all orders except the Holothyrida have cosmopolitan to semicosmopolitan distributions, suggesting similar ancient origins are common. Two groups of Opiliones with mostly modern Gondwanan distributions have a few species in the Holarctic region. Acropsopilionidae is a small family of Dyspnoi with two species in North America, one shared with Japan, while most species inhabit Australia, New Zealand, South Africa and South America (Groh and Giribet 2014). Triaenonychidae also has one species in North America and a few in East Asia, but most diversity is in the landmasses of the former Gondwana, including Australia, New Zealand, New Caledonia, southern Africa, Madagascar and South America (Giribet and Kury 2007).

The second-most significant scale, and the most obvious pattern for many different arachnid lineages, is the presence of lineages on the landmasses that formed the supercontinent Gondwana. There are very few taxa that are known from all of the former Gondwanan regions (Figure 10.4b; see also Figure 10.3, node b), but one of the best examples is the tiny litter-dwelling mite harvestmen of the family Pettalidae occurring in Australia, New Zealand, South Africa, Sri Lanka and southern South America (Boyer and Giribet 2007) (Figure 10.1h), but not in New Caledonia, New Guinea or Tasmania. Dated phylogenies suggest they originated during the Jurassic (Giribet *et al.* 2010, 2012 and 2016). Another harvestman family, Neopilionidae (Figure 10.1g), shows a similar distribution (Australia, New Zealand, South Africa, southern South America) (Taylor 2011). Similarly, the ancient trapdoor spider family Idiopidae has living representatives on all Austral landmasses except New Caledonia, as well as the Middle East and the Asian subcontinent. The parasitiform mite order Holothyrida has a supposedly relictual Gondwanan distribution, with two of the three known families (Allothyridae, Holothyridae) being found in Australasia and a third family in the Neotropics. Allothyridae appears to be basal and is restricted to Australia and New Zealand. Holothyridae are found only on islands in the Indian (Mauritius, the Seychelles, Sri Lanka) and Pacific Oceans (New Guinea, New Caledonia, Lord Howe Island) associated with ridges and crustal fragments (Krantz and Walter 2009). Holothyrans are large mites (2–7 mm long

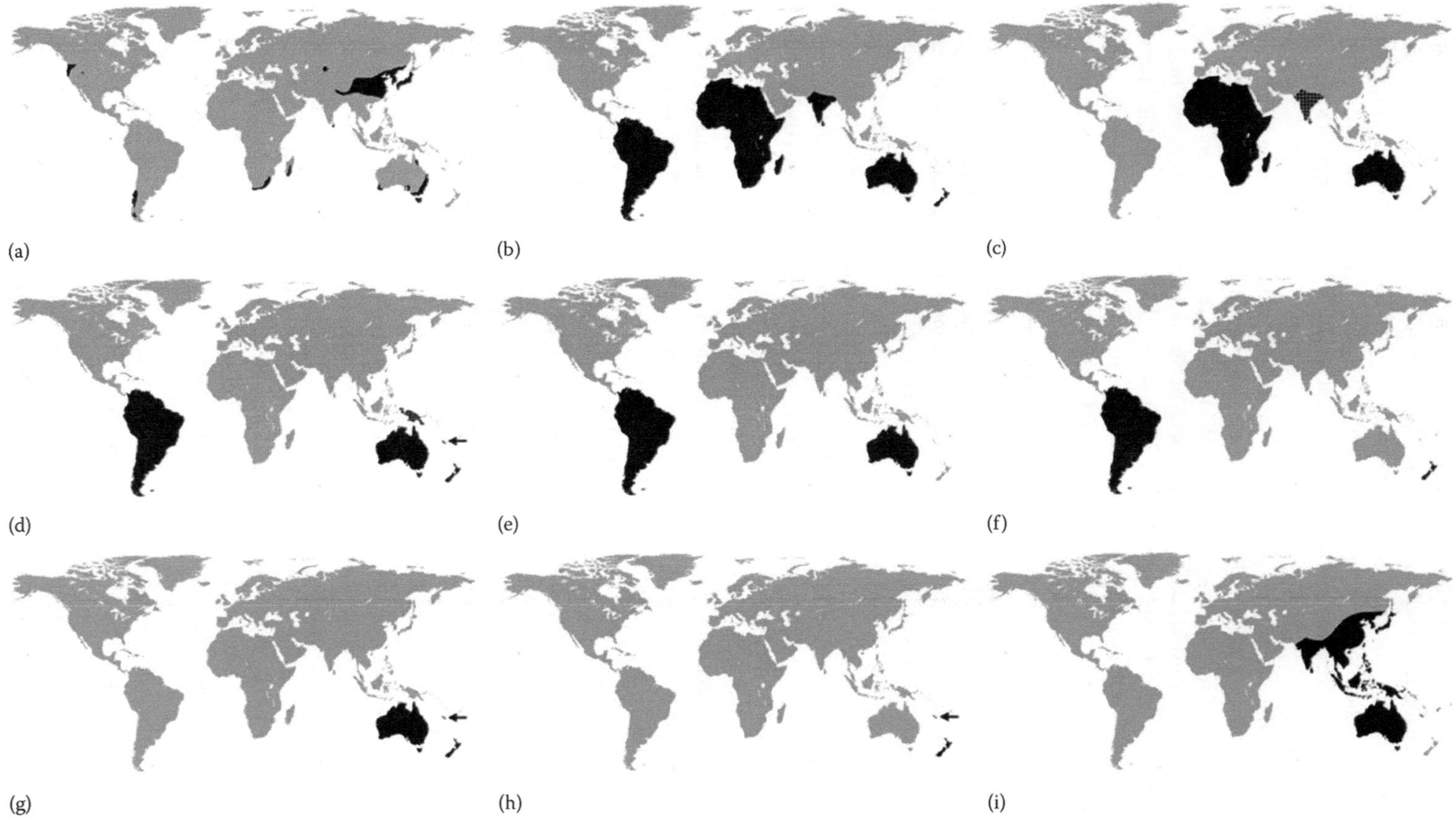

FIGURE 10.4 World maps depicting nine major intercontinental distributions exhibited by Australasian arachnid lineages: (a) relictual (disjunct) Pangaean distribution (of Pseudotyrannochthoniidae); (b) Austral Gondwanan distribution, including the Indian subcontinent; (c) 'western' Gondwanan distribution, including Australia + Africa +/– the Indian subcontinent; (d) 'eastern' Gondwanan distribution, including South America + Australasia; (e) Australian + South American distribution; (f) New Zealand + South American distribution; (g) Australian + New Zealand +/– New Caledonian distribution; (h) Zealandic distribution, including New Zealand + New Caledonia +/– Lord Howe Island; (i) Southeast Asian + Australian distribution. See Section Composition, Diversity and Distributions in this chapter for details.

as adults) with no known dispersal stages, and their current distribution appears to have resulted primarily from tectonic events. Curiously, holothyrans appear to be absent from Africa, a pattern repeated in the mesostigmatan superfamily Fedrizzioidea, associates of passalid beetles (Coleoptera, Passalidae), with one family in Australasia (and north to the Philippines) and one in South America (Seeman 2013), but apparently none in Africa. The passalid-associated mesostigmatan family Diarthrophallidae has species in Australia, New Guinea, the Americas and Africa (Krantz and Walter 2009). The water mite subfamily Notoaturinae (Aturidae) has representatives in southern Africa, temperate South America, New Zealand and Australia (Harvey 1998b), and are unlikely to be able to disperse between continents due to their obligate freshwater lifestyle.

At a third scale, a variety of other Gondwanan patterns can also be discerned. An unusual 'western' Gondwanan signal is evident in some arachnids that are found in Australia, Africa and India, or just in Australia and Africa (Figure 10.4c). These include the assassin spiders of the family Archaeidae (Figure 10.1c), which occur in temperate forested ecosystems (see Section Origins of the Fauna in this chapter) (Rix and Harvey 2012a) and the pseudoscorpion family Garypidae (all genera except the genus *Garypus*), which are relatively widespread across both regions (Harvey 1996). The pseudoscorpion genus *Indohya* (Hyidae) has a highly disjunct distribution with species known only from India, northwestern Australia and Madagascar (Harvey and Volschenk 2007), and the water mite genus *Tiramideopsis* has species in the Indian region, as well as northwestern Australia (Harvey 1998b).

Some taxa show unique distributions and biogeographic patterns that are currently difficult to explain. The New Caledonian endemic harvestman family Troglosironidae is closely related to the Afrotropical–Neotropical families Neogoveidae and Ogoveidae, and not to the temperate Gondwanan family Pettalidae (found in nearby Australia and New Zealand, among other places) (Boyer *et al.* 2007b; Giribet *et al.* 2012) or to the Southeast Asian family Stylocellidae, which gets into the westernmost part of New Guinea (Clouse and Giribet 2007, 2010). Its biogeographic history is, however, a puzzle and difficult to resolve (Sharma and Giribet 2009). Similarly, the terrestrial oribatids in the subfamily Crotoniinae, once thought to be restricted to Gondwanan lands, are now also known from the Oriental region and their current distribution is consistent with complex dynamics around the plate margins of the Pacific Ocean, with older taxa persisting on emerging islands though localised dispersal (Colloff and Cameron 2014).

'Eastern' Gondwanan distributions (i.e. Australasia +/– South America) are more evident among Australasian arachnids, and there are a variety of discrete patterns. These patterns are most likely derived from a time when these landmasses were connected via Antarctica until the Eocene (56–34 mya; Figure 10.3, nodes f–g). Groups such as the spider family Malkaridae (Figure 10.1b) and subfamily Micropholcommatinae are found in temperate South America and Australasia (+/– New Caledonia and montane New Guinea) (Figure 10.4d), and cladistic data for textricellin Micropholcommatinae are consistent with vicariance biogeography, with limited dispersal (Rix and Harvey 2010b). Other discrete patterns may be the result of regional extinctions, including taxa that are only found in Australia and South America (e.g. the spider families Austrochilidae and Actinopodidae, and the scorpion subfamily Bothriurinae) (Figures 10.1a,e and 10.4e) or in New Zealand and South America (e.g. the spider family Mecysmaucheniidae) (Figure 10.4f). In other cases, such distributions have been clearly achieved through long-distance dispersal, as has been evidenced in the harvestman family Zalmoxidae, which colonised the south Pacific from the Neotropics (Sharma and Giribet 2012).

It is surprising that very few higher-level taxa are found solely in Australia and New Zealand, despite the geographical proximity of these landmasses. Notable examples include the spider families Gradungulidae (Forster *et al.* 1987) and Periegopidae (Forster 1995; Vink *et al.* 2013), the mite family Allothyridae (Krantz and Walter 2009) and a number of families of oribatid mites (Colloff and Cameron 2014). The resolution of the harvestman family Pettalidae (Figure 10.1h) is still ambiguous with respect to the monophyly of New Zealand and at least Western Australia (Boyer and Giribet 2007; de Bivort and Giribet 2010; Giribet *et al.* 2012; Giribet *et al.* 2016). However, support for the monophyly of the Australian/New Zealand Enantiobuninae or at least of some Neopilionidae is emerging (Vélez *et al.* 2014). Similarly, few taxa are restricted to Australia, New Zealand and New Caledonia. The spider family Pararchaeidae shows this pattern (Figure 10.4g) but it is possible that the sole New Caledonian species may be the result of a more recent dispersal event (Rix and Harvey 2010a). The ancient continental fragment Zealandia, which comprises the emerged landmasses of New Zealand, New Caledonia and some associated islands such as Lord Howe Island (Lewis *et al.* 2012) has very few endemic arachnid lineages (Figure 10.4h), but the pseudoscorpion subfamily Philomaoriinae and the oribatid mite family Tumerozetidae (Colloff and Cameron 2014) are examples. Within harvestmen, the relationships of the Triaenonychidae of these three regions remain to be tested, but as in the case of Pararchaeidae, the triaenonychids of New Caledonia could be a case of recent dispersal. The fact that so few higher clades are shared between these three landmasses could be due to the purported total submergence of New Caledonia, as discussed later in the chapter (e.g. Grandcolas *et al.* 2008; Espeland and Murienne 2011).

The fourth and final scale is proposed to be much more recent than ancient Gondwanan connections. The northward movement of the Australian plate and its proximity to the Eurasian plate from the mid-Tertiary allowed faunal interchange in both directions. This Eurasian element has been implicated in transcontinental distributions of some arachnid taxa (Figure 10.4i). The best-studied taxa are the scorpion family Hormuridae (Monod and Prendini 2015), the spider genus *Myrmarachne* (Maddison 2008; Bodner and Maddison 2012; Maddison *et al.* 2014) and the harvestman genus *Zalmoxis* (Sharma and Giribet 2012). Older relationships between Southeast Asia and at least New Guinea, in the harvestmen of the family Stylocellidae, are explained by their diversification in the Sibumasu terrane, which became the eastern end of the long, northward-moving Cimmerian palaeocontinent and is now the region extending from northeast India to the southern Thai–Malay Peninsula (see Clouse and Giribet 2010). This terrane is hypothesised to have separated from the northwestern Australian Gondwanan margin in the late Early Permian (Metcalfe 2002, 2013).

Finally, some of the Australasian landmasses host endemic lineages, akin to some famous vertebrate lineages such as monotremes, sphenodontids or leiopelmatid frogs. Notable examples are the harvestman family Synthetonychiidae (Forster 1954), and the spider family Huttoniidae (Forster and Platnick 1984), which are both endemic to New Zealand.

Fossils

There are very few fossil arachnids recorded from Australasia, due in part to a lack of cuticular mineralisation and the poor preservation potential of most taxa (Penney and Selden 2011; Raven et al. 2015). Most Australasian arachnid fossils are amber inclusions, but an impression fossil of a long-legged harvestman was found in Cretaceous deposits in southeastern Australia (Jell and Duncan 1986). Womersley (1956) described an oribatid mite from Cainozoic amber deposits in Victoria. The recent discovery of inclusions within fossilised amber from beaches in northern Queensland include pseudoscorpions and spiders (Hand *et al.* 2010), but the fauna has not been studied in detail and the age and origin of the amber is unknown. The recent discovery of amber invertebrates from Otago and Southland in New Zealand includes pseudoscorpions, mites and spiders (Kaulfuss 2014; Schmidt *et al.* 2014), but these are still unstudied and remain undescribed.

Australasian arachnid fossils have so far not been used for phylogenetic reconstruction – for example, in morphological cladistic studies or for dated phylogenetic analyses. This may partly be due to the relatively recent (i.e. Tertiary) age of the predominantly amber fossils, which are unsuitable for rooting time trees at a deep level.

Biogeographic Synopsis of the Australasian Arachnida

History of Documentation

The biogeographic study of Australasian arachnids has a relatively recent history, with few seminal, explicitly biogeographic contributions prior to the middle of the twentieth century. The earliest faunistic works were facilitated by eighteenth- and nineteenth-century European maritime explorations that returned specimens to European museums, where they were frequently described as new species. The first arachnid to be named from Australasia was the very attractive orb-weaving spider *Aranea fornicata* (now *Gasteracantha fornicata*) by Fabricius (1775), based on a specimen collected by Joseph Banks during

James Cook's voyage to the Southern Ocean in 1768–1771. British settlement and colonisation of Australia and New Zealand saw further specimens being shipped to Europe, mainly London. The documentation of Australasian fauna continued in this piecemeal fashion throughout the nineteenth and early twentieth centuries, but seminal early taxonomic works included Walckenaer (1805), White (1849), Koch and Keyserling (1871–1890), Simon (1908, 1909), Pickard-Cambridge (1879), Urquhart (1885, 1892), Pocock (1888, 1898), Rainbow (1896, 1898, 1906, 1911, 1920), Rainbow and Pulleine (1918) and Hogg (1909), to name but a few.

The modern era of Australasian arachnid biogeography commenced with the postwar contributions of Barbara York Main (1957a,b, 1981a–c), Raymond Forster (1947, 1961, 1975) and Lucien Koch (1977, 1981). These studies explored zoogeographic patterns and distributional similarities or correlations, usually in a taxonomic framework. Many of these early studies were published prior to the acceptance of the theory of continental drift and plate tectonics, which had been first proposed by Wegener (1912, 1915) and revitalised in the 1960s and 1970s (e.g. Dietz and Holden 1970). The evidence for continental drift and the breakup of Pangaea and then Gondwana provided the geological setting for what had been obvious to numerous nineteenth-century natural history explorers, including Alfred Russell Wallace, Charles Darwin and Joseph Hooker: the close similarity of the flora and fauna of the southern continents. Indeed, the transition in biogeographic thinking during this period is exemplified by the works of Forster (1961, 1975). In 1961, Forster was clearly starting to piece together the foundations of a southern temperate biogeography and recognised the strong similarities between the Australian and New Zealand arachnid faunas. However, until the more general acceptance of the continental drift theory, he focused entirely on alternative dispersal hypotheses. In 1975, by contrast, Forster recognised the significance of continental drift and the potential for vicariant Gondwanan elements within the New Zealand fauna.

At about the same time as continental drift was revolutionising biogeographic thinking, the English translation of Hennig's theory of phylogenetic systematics (Hennig 1950, 1966) profoundly transformed evolutionary biology on a global scale (Rosen *et al.*, in Hennig 1966) by introducing cladistic phylogenetic theory as 'the general reference system of biology' (Hennig 1966: 239). This prompted some pioneering contributions in Australasian arachnid systematics that explicitly used cladistic methods to test (and not to narrate, as previously) alternative biogeographic scenarios (e.g. Platnick 1976; Raven 1980, 1984; Gray 1981; Forster and Platnick 1984; Harvey 1985, 1986). Gray (1988) pioneered the use of genetic methods by utilising allozyme data to analyse the relationships of funnel-web spiders (Hexathelidae).

As with many branches of biological research, the advent of molecular biology in the 1990s completely revolutionised our understanding of arachnid phylogeny and biogeography, including within Australasia. Molecular phylogenetic studies have facilitated the quantitative inference of phylogenies for a suite of taxa across numerous lineages and orders, greatly improving our knowledge of their biogeography and evolution. These studies have ranged from modest contributions with a relatively small number of ingroup taxa or loci (e.g. Vink *et al.* 2002; Moulds *et al.* 2007; Harvey *et al.* 2008, 2012; Castalanelli *et al.* 2014) through to more comprehensive analyses using a multilocus approach (e.g. Vink and Paterson 2003; Murphy *et al.* 2006; Boyer and Giribet 2007, 2009; Rix *et al.* 2008; Rix and Harvey 2010a; Sirvid *et al.* 2013; Giribet *et al.* 2014; Harrison *et al.* 2014). A fundamental aspect of modern biogeographic studies is the use of molecular dating approaches, adding a temporal dimension to phylogenetic trees and allowing estimates of the approximate age of cladogenesis. Additional computational methods have also been developed to analyse molecular phylogenetic trees, and used to explicitly test for dispersal versus vicariance (Yu *et al.* 2010; Drummond *et al.* 2012), infer diversification through time (e.g. LTT plots; MEDUSA, LASER) (Pybus and Harvey 2000; Rabosky 2006; Alfaro *et al.* 2009) and infer ancestral distributions of taxa (e.g. Lagrange) (Ree and Smith 2008). These remain infrequently used but are likely to refine further our concept of arachnid evolution in Australasia. Although a dated molecular phylogenetic approach has been applied to only a handful of arachnid taxa in Australasia (e.g. Boyer *et al.* 2007b; Cooper *et al.* 2011; Sharma and Giribet 2011; Giribet *et al.* 2012; Rix and Harvey 2012a; Wood *et al.* 2013), and comprehensive phylogeographic contributions applying aforementioned biogeographic methodology are still rare (e.g. Boyer *et al.* 2007a; Fernández and Giribet 2014), these studies have provided some of the most significant insights into the origins and biogeography of the fauna.

Finally, while the majority of our knowledge on arachnid biogeography is still being deduced from taxonomic revisions, molecular phylogenies and online databases such as the *Atlas of Living Australia* (http://www.ala.org.au/), there have been several regional surveys that have contributed to more detailed concepts of local biogeographic patterns and processes. Three large biotic surveys in Western Australia used analytical techniques to compare the composition of the fauna against a range of variables including climate and soil types to help explain modern distributions: the Carnarvon regional survey (Harvey *et al.* 2000; Main *et al.* 2000; Smith and McKenzie 2000), the southwestern agricultural zone or Wheatbelt survey (McKenzie *et al.* 2003; Harvey *et al.* 2004) and the Pilbara bioregion survey (Durrant *et al.* 2010; Volschenk *et al.* 2010). These remain the only in-depth studies of arachnid faunas using quantitative analytical methods. Although there are a number of comprehensive papers summarising biogeographic patterns in Australasia (e.g. Byrne *et al.* 2008, 2011; Heads 2008; Murienne 2009, 2010; Stelbrink *et al.* 2012; Rix *et al.* 2015), only a few have used arachnid taxa (Boyer and Giribet 2009; Rix *et al.* 2015). This is unfortunate, because arachnids are extremely old, highly diverse and offer enormous biogeographic potential.

Origins of the Fauna

As already discussed in the Section Composition, Diversity and Distributions in this chapter, many elements of the Australasian arachnid fauna have a Gondwanan heritage (referred to throughout as *Gondwanan groups*), meaning that the ancestors of most extant lineages originated before the southern landmasses drifted apart between the Cretaceous and Palaeogene (i.e. they were present on their respective Austral landmasses >35 mya) (Sanmartín and Ronquist 2004). Some lineages may even be older and may have originated well before there was a split between (southern) Gondwana and (northern) Laurasia (Harvey 1998a; Harvey and Šťáhlavský 2010) (Figure 10.4a; see also Figure 10.3, node a).

Forster (1975), Platnick (1976) and Main (1981b, 1981c) were among the first to explicitly document the similarity and the potential shared biogeographic history between New Zealand and Australian arachnids, with other workers expanding on these foundational concepts (e.g. Hopper *et al.* 1996; Boyer *et al.* 2007b; Boyer and Giribet 2009). It is important to note that this Australasian Gondwanan heritage of the fauna does not preclude interlandmass dispersal (e.g. from Australia to New Caledonia) or in situ adaptive radiation during more recent periods, nor does it require explicit intercontinental vicariance to be currently demonstrable; clearly, extinction (both within Australasia and across the Southern Hemisphere, including on Antarctica itself) is to be expected, and some range-restricted taxa (e.g. single-continent endemics) and even some currently dispersal-prone (e.g. Gnaphosidae) or arid-adapted (e.g. Idiopidae) groups are likely to have had their origins in Gondwana. Indeed, the varied distributions of different taxa (Figure 10.4) highlight a potentially complex historical mosaic of vicariance and subsequent extinction throughout former Gondwanan landmasses. Recently, biogeographers have begun investigating such complex biogeographic histories using up-to-date biogeographic techniques (e.g. Boyer *et al.* 2007b; Rix and Harvey 2010b; Wood *et al.* 2013; Frick and Scharff 2014), whereas other studies have inferred an intercontinental dispersal scenario for at least some taxa of Gondwanan origin (e.g. mynoglenine spiders; Frick and Scharff 2014).

In contrast to those taxa of putatively Gondwanan ancestry, some elements of the Australasian arachnid fauna, especially in New Guinea, appear to have tropical affinities with immigrant taxa from Southeast Asia (Figure 10.4i). However, few instances of this have been convincingly demonstrated with phylogenetic data (cf. Clouse and Giribet 2007; Monod and Prendini 2015), and some taxa hypothesised to be northern immigrants (e.g. Barychelidae; Main 1981b) seem less likely to be so now that the world fauna is better known (see Raven 1994 for a taxonomic treatment of the Australasian Barychelidae). Other groups (e.g. spiders of the family Archaeidae) do not currently have extant members in the Northern Hemisphere (restricted today to southern Africa, Madagascar and Australia), but Tertiary amber and even Mesozoic impression fossils show that the family was once widespread across Eurasia (e.g. Selden *et al.* 2008; Wood *et al.* 2013). This demonstrates that other currently Austral or seemingly Australasian lineages may also have been more widely distributed or have a much deeper phylogenetic history, but fossil evidence is currently lacking (Ezcurra and Agnolín 2011).

Bioregional Perspectives

To facilitate discussion we define two major Australasian regions, each with three subregions (Figure 10.2). The first region includes Australia and New Guinea, and comprises (1) the Monsoon Tropics (including the tropical mesic east), (2) the Bassian (temperate) mesic zone and (3) the Eremaean arid zone (Crisp *et al.* 2004; Byrne *et al.* 2008, 2011). The second region is Zealandia, which comprises (4) New Zealand, (5) New Caledonia (including Norfolk Island) and (6) Lord Howe Island (Lewis *et al.* 2012). We are, however, aware that some of these regions may be too large and that short-range endemism (Harvey 2002) is prominent across groups with low vagility, and thus some eastern Australian groups may be more closely related to Zealandian groups than to western Australian ones (Giribet and Edgecombe 2006).

Monsoon Tropics and Mesic Northeast

New Guinea and other islands as far west as Lydekker's Line (see Clouse and Giribet 2007 for discussion) and tropical Australia – including the mesic tropical and subtropical east of Queensland and northern New South Wales, which have their southernmost extensions in the McPherson–Macleay Overlap (Burbidge 1960; Lucky 2011; Rix and Harvey 2012a) (Figure 10.2) – is a region with a large variety of vegetation and habitat types, probably due to its complex geological and climatic history and great topographic diversity (Schneider and Moritz 1998). This region has a diverse and varied arachnid fauna that illustrates a complex biogeographic history. Rainfall in this region, while extremely high in some areas (such as the Wet Tropics of northeastern Queensland), is largely seasonal with a pronounced summer peak, known as the monsoon in areas north of approximately the Tropic of Capricorn. Despite this seasonality, the whole area can be thought of as a tropical mesic biome (extending the more restrictive *ever-wet* definition of Byrne *et al.* 2011), in that rainfall is annually predictable and receives at least 300–500 mm of rain per year. This predictability and volume of rain demarcates the fundamentally more arid and aseasonal Eremaean zone, and has a profound effect on the composition of the biota.

For the purposes of this review, the monsoon tropics and tropical mesic east of Australasia can be simplified into three elements: (1) ever-wet tropical and subtropical rainforests – namely, those of New Guinea and those on and east of the Australian Great Dividing Range, north to approximately Cooktown; (2) other semi-rainforest habitats including the vine thickets of the Kimberley, the Cape York Peninsula, the Northern Territory and New Guinea; and (3) the remaining savannah habitats of northern Australia. Elements (2) and (3) include some of the youngest and clearly the most northerly tropical habitats in Australasia, with orogenesis in New Guinea caused by the collision of the Australian and Southeast Asian (Sundaland) plates since the Oligocene (McLoughlin 2001; Monod and Prendini 2015). Concomitant climatic drying of Australia since the Miocene, combined with the ongoing northern movement of the continent, has had a profound effect on the faunistic composition of these regions, which contains elements of dispersive, arid-adapted and/or putatively Southeast Asian arachnids, as well as 'older' in situ clades. Indeed, it has long been assumed that most arachnid taxa characteristic of the seasonally wet monsoon tropics are either post-Miocene immigrants from Southeast Asia (Main 1981b,c, 1991) or recently arid-adapted crown groups of otherwise Gondwanan lineages. Unfortunately, for the arachnid fauna of this region, there is actually little empirical evidence to support either scenario and, where it does exist, results can be conflicting or ambiguous. For example, epigean schizomids in Australia are entirely restricted to the monsoon tropics and tropical mesic east (Harvey 1992), and given their extensive distribution and diversity throughout Southeast Asia, it has been assumed that they may be relatively recent northern immigrants to Australia (Main 1981a), as must be invoked for the related orders Uropygi and Amblypygi (Figure 10.1f). However, recent phylogeographic analysis of subterranean schizomids (Figure 10.1i) from the arid zone of Western Australia is consistent with a model of Miocene *islandisation* due to progressive aridification and the extirpation of surface habitats (Harvey *et al.* 2008). While this scenario does not necessarily preclude northern immigration (Monod and Prendini 2015), further phylogenetic analysis is required to adequately test an *out of Asia* hypothesis for schizomids. Likewise, the circumtropical (Old World) jumping spider genus *Myrmarachne* has been inferred as belonging to an otherwise Australasian clade (the Astioidea) with a crown group radiation of <15 mya (Maddison 2008; Bodner and Maddison 2012; Maddison *et al.* 2014), implying that Australia/New Guinea may have been the source of some taxa dispersing into Southeast Asia, rather than vice versa. The Astioidea itself is likely older than 30 million years in age (Bodner and Maddison 2012; Maddison *et al.* 2014), concordant with a Gondwanan ancestry for this diverse clade. Similar patterns are evident in the spider genus *Prethopalpus*, which is widespread throughout Asia and northern and eastern Australia forest and savannah ecosystems, but has a distinct subterranean radiation in the Pilbara region (Baehr *et al.* 2012).

However, the lack of a molecular phylogeny of this group must preclude any generalisations of the direction of colonisation.

One tropical lineage that has been studied in detail is the scorpion family Hormuridae, which is found throughout Southeast Asia, tropical eastern Australia and isolated patches of the Australian Monsoon Tropics. In one of the few studies explicitly testing tropical Australian biogeography using an arachnid group, Monod and Prendini (2015) explored alternative biogeographic hypotheses to explain the distribution and possible dispersal of *Liocheles* and *Hormurus* in tropical Australia. They found strong evidence for southern immigration into New Guinea and Australia via Southeast Asia, highlighting the potential for dispersal in the tropical arachnid fauna as postulated by Koch (1977, 1981).

Arid adaptations are omnipresent in the arachnid fauna of northern Australia, and based on the climatic history (Martin 2006) are likely to be post-Oligocene in origin. Groups characteristic of savannah and semiarid monsoonal habitats include various araneid, lycosid, salticid, thomisid and mygalomorph spiders, buthid and urodacid scorpions (Figure 10.1d), and olpiid and garypid pseudoscorpions, to mention a few. The continental origins of most of these taxa are presumed to be Gondwanan (e.g. as is likely for astioid Salticidae, as previously mentioned), but in most cases phylogenetic evidence is lacking. Indeed, many spider groups from savannah habitats are highly dispersive and diverse with worldwide distributions (e.g. Lycosidae), and thus only major dated phylogenies such as those attempted for Salticidae (Bodner and Maddison 2012; Maddison *et al.* 2014) can begin to explain the origins and evolution of the Australian fauna. Murphy *et al.* (2006) explored the phylogeny of the Australian wolf spider (Lycosidae) fauna with molecular data and found evidence for endemic Australian subfamilies that may have Austral rather than Asian origins, but dating was not attempted and worldwide coverage was relatively limited.

The ever-wet tropical and subtropical mesic east of Australia, in stark contrast to the monsoon tropics and savannah regions, contains some of the oldest persistent remnant rainforest habitats in Australia (e.g. Byrne *et al.* 2011) but has had a complex climatic history, with range shifts of rainforest habitats throughout the Neogene and drastic contractions during the Quaternary. Unsurprisingly, many Gondwanan arachnid relicts persist throughout this region, including within the Queensland Wet Tropics, where an otherwise distinctive southern temperate biota persists in cooler montane habitats (e.g. Forster and Platnick 1985; Forster *et al.* 1987; Platnick 1991; Rix and Harvey 2012a). However, a very few outlier southern temperate taxa are currently known from New Guinea in montane habitats (e.g. Harvey 1995; Rix and Harvey 2010b); these are perhaps vicariant relicts, as hypothesised for micropholcommatine spiders (Rix and Harvey 2010b). Clearly, the arachnid fauna characteristic of the tropical and subtropical Great Dividing Range includes a vast array of taxa with varying histories of documentation, taxonomic coverage and phylogenetic analysis. For those studies that have attempted to document patterns of biogeography and speciation in the region, the strongest pattern seen among tropical rainforest endemic lineages is pronounced endemism, with allopatric or parapatric replacement of species between upland rainforest blocks or along altitudinal gradients. This has been convincingly demonstrated in harvestmen (e.g. Popkin-Hall and Boyer 2014), schizomids (e.g. Harvey 1992, 2000) and spiders (e.g. Platnick 1991; Raven 1994; Raven and Stumkat 2005; Baehr *et al.* 2010, 2012; Rix and Harvey 2011, 2012b; Edward and Harvey 2014), and is often characteristic of less-vagile, habitat-restricted taxa more generally (e.g. Yeates *et al.* 2002; Bell *et al.* 2007; Lucky 2011).

Bassian (Temperate) Mesic Zone

The Bassian mesic zone of southeastern and southwestern temperate Australia includes the ever-wet eastern rainforests and wet sclerophyll forests on and east of the Great Dividing Range, the warm- and cool-temperate rainforests and wet sclerophyll forests of Tasmania, and the seasonally winter-wet climates of southern South Australia and southwestern Western Australia (and surrounding habitats; Crisp *et al.* 2004; Byrne *et al.* 2011; Rix *et al.* 2015) (Figure 10.2). The arachnid faunas of these regions are largely dominated by temperate Gondwanan (or Pangaean) groups, many with broader Austral distributions (Figure 10.4) and some of which are known to be explicitly vicariant (Boyer *et al.* 2007b; Rix and Harvey 2010b; Giribet *et al.* 2012; Wood *et al.* 2013). Numerous Bassian taxa have affinities with New Zealand and/or temperate South American taxa, and a few groups are confamilial (e.g. the spiders Austrochilidae, Actinopodidae, Anapidae, Hexathelidae, sternoidine Malkaridae and

Micropholcommatinae; the pseudoscorpions Pseudotyrannochthoniidae) or even congeneric (e.g. the micropholcommatine genus *Gigiella*, the pseudoscorpion genus *Austrochthonius*, the oribatid mite genus *Neotrichozetes*) with taxa from Chile. One genus of Cyphophthalmi, *Karripurcellia*, is also present in this region, but its affinities remain unresolved (Giribet 2003).

The biogeographic history of the Bassian fauna is proposed to be a complex mosaic of continental vicariance plus subsequent extinction – highlighted graphically in Figure 10.4 – overlaid with a more recent signature of in situ radiation, dispersal and speciation in both strictly mesic-adapted and other more generalist taxa – a response to the repeated cycles of expansion/contraction of mesic forest habitats resulting from post-Miocene climatic cycling. Indeed, post-Miocene climate change in Australia has had a profound influence on the biota of the temperate mesic zone, driving extinction, range contraction and allopatric speciation in mesic-adapted taxa, and range expansion in xeric-adapted forms. Thus, while many Bassian lineages can be considered to be niche-conserved with a Gondwanan heritage, meaning that they have broadly retained pre-Oligocene habitat preferences and have almost certainly endured significant range contraction since the Miocene, many have also experienced considerable in situ allopatric/parapatric speciation as a result of these processes, as suggested for the spider families Archaeidae and Migidae (Cooper *et al.* 2011; Rix and Harvey 2012a). Indeed, the distributions of arachnids across southern Australia have contributed to our understanding of some of the major biogeographic breaks in the Bassian region (all of which are congruent with gaps inferred for other animal and/or plant taxa), including but not limited to the Nullarbor divide (i.e. the Nullarbor Plain and its fringing xeric habitats; Rix and Harvey 2012a; Rix *et al.* 2015), the Hunter Valley or Cassilis Gap (Forster *et al.* 1987; Rix and Harvey 2012a), the Australian alpine zone (Rix and Harvey 2012a) and the Mallee Divide or Naracoorte Gap (Rix and Harvey 2012a).

Finally, arachnids are a major component of the distinctive Tasmanian cave fauna (Eberhard *et al.* 1991), which is (along with the subterranean fauna of arid Western Australia; see Section Eremaean Arid Zone in this chapter) undoubtedly one of the most significant in Australasia for troglomorphic arachnids. Significant examples include spiders of the subfamily Micropholcommatinae (Rix and Harvey 2010b) and family Synotaxidae (Forster *et al.* 1990; Dimitrov *et al.* 2016), harvestmen of the genera *Hickmanoxyomma* and *Lohmanella* (Hunt 1990, 1993) and the pseudoscorpion family Pseudotyrannochthoniidae (Dartnall 1970).

Eremaean Arid Zone

The Australian arid zone is the largest of the six defined bioregions in Australasia, and, along with the tropical monsoonal north, is the most recently evolved in terms of climate and vegetation, and the result of the progressive aridification of formerly mesic habitats since the mid-Miocene, possibly starting in the west and extending eastwards only later (Martin 2006). Extending across most of mainland inland Australia (Figure 10.2), the arid zone is characterised by low and usually highly sporadic rainfall (<250–300 mm/year) (Byrne *et al.* 2008). Formed during the massive climatic upheavals of the Miocene, Pliocene and Quaternary, the arid zone is remarkable in that it evolved from highly mesic, warm-wet or cool-temperate rainforests, which today only fringe the region on its eastern and southern boundaries (Martin 2006). Although relatively depauperate and uniform at higher taxonomic levels, the arachnid fauna of the arid zone is surprisingly diverse at the species level, with evidence for major in situ radiations of arid-adapted lineages; this is especially true of spiders (e.g. mygalomorph spiders; Raven 1994; Castalanelli *et al.* 2014), scorpions (e.g. Urodacidae; Koch 1977; Volschenk *et al.* 2012) and pseudoscorpions (e.g. Garypidae Harvey 1987). In the Pilbara, Kimberley and Yilgarn regions of Western Australia there also exists a remarkable assemblage of subterranean arachnid taxa, surviving 'under the desert' in hypogean habitats (e.g. Harvey and Volschenk 2007; Barranco and Harvey 2008; Harvey *et al.* 2008; Platnick 2008; Volschenk and Prendini 2008; Guzik *et al.* 2011; Baehr *et al.* 2012; Harrison *et al.* 2014). These otherwise relictual mesic elements have speciated allopatrically in disconnected habitats across vast areas of the Western Australian arid zone and include representatives of several different orders. In both epigean and subterranean habitats, Miocene or post-Miocene diversification is hypothesised for the arid zone fauna. Several mygalomorph lineages appear to have speciated extensively within the area (e.g. Raven 1994; Harms and Framenau 2013; Castalanelli *et al.*

2014; Miglio *et al.* 2014), and many of these, such as Idiopidae and Actinopodidae, have a former Gondwanan distribution. Contemporary distributions suggest that these lineages may have radiated out of their ancestral niche into the emerging deserts, where spectacular cases of niche differentiation are frequent (Main 2001). Other lineages are clearly not Gondwanan, with the Ctenizidae postulated to have been derived from Asian elements (Main 1981b). Likewise, in a now classic paper uniting cladistics and vicariance biogeography, Platnick (1976) posited that gnaphosid spiders of the genus *Eilica* are similarly Gondwanan, despite occurring across the Australian arid zone. There is no quantitative phylogenetic evidence to suggest that any major arid-zone arachnid clades are northern immigrants from Southeast Asia, although Main (1981b) speculated about the possible Asian affinities of the mygalomorph families Barychelidae, Theraphosidae and Ctenizidae. Contemporary diversity and distributions suggest that the Eremaean fauna comprises both ancient, niche-conserved relict taxa, but also those that have radiated out of the mesic habitats as Australia aridified. Currently, there is only anecdotal evidence to support this claim, and comprehensive studies are clearly needed to piece together the complex history of the Eremaean arachnid fauna.

Zealandia: New Zealand

Modern-day New Zealand constitutes one of the most interesting and difficult to comprehend biogeographical regions. The land area of New Zealand (Figure 10.2) is a small part of Zealandia, a largely submerged continent of nearly 4 million km^2. Zealandia is principally made up of two almost parallel ridges. These are largely underwater and trend northwest through the southern Pacific Ocean. The western ridge includes the Lord Howe Rise and Campbell Plateau. The narrower eastern ridge forms New Caledonia, the Norfolk Ridge, the Northland Peninsula of New Zealand and the Chatham Rise.

The complex geological history of New Zealand is relatively well understood, and two episodes seem to have played a key biogeographic role: (1) the breakup from Gondwana and (2) an Oligocene drowning episode. Late during the breakup of Gondwana, 80 mya, New Zealand began to drift away from West Antarctica, opening the Tasman Sea, and the final terrestrial connections between New Zealand and Australia broke by 60 mya. The idea that the terrestrial biota of New Zealand was inherited from Gondwana during the breakup – the so-called ghosts of Gondwana (Gibbs 2006) – has given New Zealand the name *Moa's Ark*, the title of a popular documentary on the country's natural history (Bellamy *et al.* 1990). The Moa's Ark paradigm is based on charismatic examples of supposed vicariance in ratite birds, chironomid midges, and *Nothofagus* beech trees, or on the presence of ancient relicts such as the tuatara and leiopelmatid frogs, with no close extant relatives outside of New Zealand (Giribet and Boyer 2010). While the Oligocene drowning of New Zealand has been well studied for decades, recent reinterpretations of New Zealand's geology and its fauna have led to very different conclusions, framed as the *goodbye Gondwana* paradigm (McGlone 2005), which questions the antiquity of the land surface and thus of its terrestrial biota, and postulates that virtually every terrestrial group has reached New Zealand through over-water dispersal during the past 22 million years (Landis *et al.* 2008), with subsequent diversification (e.g. Waters and Craw 2006; Trewick *et al.* 2007; Goldberg *et al.* 2008). However, evidence for the total submersion of New Zealand is not supported by a number of recent geological studies (Mildenhall *et al.* 2014) and is also challenged by a few groups of organisms, including some arachnids (Giribet and Boyer 2010; Giribet *et al.* 2012). In addition, recent simulations have challenged previous interpretations of post-Oligocene crown ages for clades endemic to New Zealand (Sharma and Wheeler 2013).

Key players in this debate have been the harvestmen of the family Pettalidae (Boyer *et al.* 2007a; Giribet and Boyer 2010; Giribet *et al.* 2012, 2016; Sharma and Wheeler 2013) (Figure 10.1h), where each of the genera *Aoraki* and *Rakaia* has a pre-Oligocene origin and radiation. Similar patterns are expected (but remain untested) for other Pangean or Gondwanan lineages in the harvestman families Acropsopilionidae, Neopilionidae (Figure 10.1g) and Triaenonychidae, and in numerous families of both mite superorders. For example, about 43 species (10%) and 117 genera (67%) of New Zealand oribatid mites are shared with Australia, closely rivalled by the Neotropics with 43 species (10%) and 107 genera (61%) shared; about half the genera have a Pangaean distribution, indicating a probable more ancient origin (Colloff and Cameron 2014). Diverse New Zealand genera in the Mygalomorphae, such as the trapdoor spiders *Cantuaria*, which have closely related genera in Australia, are also potentially excellent groups with which to further test the drowning hypothesis.

In addition, the study of New Zealand arachnofauna has been used to elucidate more recent evolutionary events. The uplift of the Southern Alps in the last 5 million years (Batt *et al.* 2000; Chamberlain and Poage 2000) and the resulting creation of alluvial habitats appears to have driven the diversification of the endemic lycosid genus *Anoteropsis* (Vink and Paterson 2003). The Southern Alps have also played a part in the divergence of the mite harvestman *Aoraki denticulata* (Boyer *et al.* 2007a; Fernández and Giribet 2014).

The molecular phylogenetics of New Zealand Lycosidae, Pisauridae and Thomisidae have been explored (Vink and Paterson 2003; Vink and Dupérré 2010; Lattimore *et al.* 2011; Sirvid *et al.* 2013), but these families are all known to be efficient dispersers, which appears to have clouded any phylogeographic patterns. The exception is that there are endemic species in Lycosidae and Pisauridae present on the Chatham Islands, an archipelago approximately 850 km to the east of New Zealand that emerged from the sea 4 mya (Heenan *et al.* 2010). The mitochondrial divergence in these species is consistent with their arrival and speciation since the islands' emergence (Vink and Paterson 2003; Vink and Dupérré 2010). Episodes of glaciation have also undoubtedly been important in shaping the distribution of New Zealand arachnids (Forster 1975), but as yet this has not been examined explicitly.

Zealandia: New Caledonia

New Caledonia (Figure 10.2) constitutes the northernmost emerged portion of Zealandia, completely opposite to New Zealand, as discussed in the previous section. New Caledonia may have separated from Australia roughly 66 mya, subsequently drifting in a northeasterly direction, reaching its present position about 50 mya. New Caledonia consists of the main island of Grande Terre, the Isle des Pines (south of the Grande Terre), the Loyalty Islands, the Chesterfield Islands and the Belep archipelago, but some of these island systems are not continental in origin.

The main island is divided in length by a central mountain range whose highest peaks are Mont Panié (1629 m) in the north and Mont Humboldt (1618 m) in the southeast. The east coast is covered by lush vegetation, while the west coast, with its large savannahs and plains, is a drier area more transformed by farming practices. Many ore-rich massifs are found along this coast. The Diahot River is the longest river of New Caledonia, flowing for some 100 km. It has a catchment area of 620 km^2 and opens northwestward into the Baie d'Harcourt, flowing towards the northern point of the island along the western escarpment of Mont Panié. Most of the island is covered by wet evergreen forests, while savannahs dominate the lower elevations (Grandcolas *et al.* 2008).

Although Gondwanan in origin, New Caledonia may be a Darwinian island (e.g. Gillespie and Roderick 2002), based on the evidence for a marine transgression with a deep total submersion during the Palaeocene (see a review of the geology in Grandcolas *et al.* 2008). During the Eocene, collision with the Loyalty Islands arc resulted in a layer of oceanic crust over the submerged continental crust. New Caledonia then emerged during the Oligocene. The present-day mountains are the product of complex orogenesis since the Oligocene. Ephemeral islands in the region may have sustained some Gondwanan elements later transferred to New Caledonia by acting as stepping-stones, favouring groups able to disperse over short distances. Smaller islands such as Norfolk or the Loyalty Islands emerged 3.7 and 2.0 mya, respectively (Grandcolas *et al.* 2008).

The high diversity of New Caledonia has traditionally been seen as a result of its Gondwanan origin, old age and long isolation under stable climatic conditions (the *museum model*, equivalent to New Zealand's Moa's Ark model) (Espeland and Murienne 2011). Alternatively, if New Caledonia was completely submerged after its breakup from Gondwana, as geological evidence indicates, signals should be found in the diversification curves of multiple lineages, with a characteristic slowdown over time according to a diversity-dependent model where species accumulation decreases as space is filled. This pattern is evident in most of the cases examined in a recent study of diversification in New Caledonia (Espeland and Murienne 2011). Evidence for the total submersion is also patent in many recent phylogenetic studies, but these are often based on clades that are too recent to test a 37-million-year-old marine transgression. Biological examples supporting both the museum model (Nattier *et al.* 2011), and the old Darwinian island model (Swenson *et al.* 2014) exist, and a clade of arachnids has played a key role in discussions on the New Caledonia marine transgression (Sharma and Giribet 2009; Giribet *et al.* 2012; cf. Espeland and Murienne 2011). Troglosironidae is a New Caledonian endemic harvestman family that originated during the Permian but diversified much later, during the Cretaceous/Tertiary boundary (Giribet *et al.* 2012). Although this precedes the Eocene marine transgression, the inferred dates cannot entirely reject such a hypothesis, and Espeland and Murienne (2011) postulate that the group presents the same characteristics as other recent diversifications in New Caledonia, suggesting a diversity-dependent process of diversification and recolonisation. However, the simulations of Sharma and Wheeler (2013) and the fact that the sister group of Troglosironidae inhabits West Africa and the Neotropics makes dispersal highly unlikely, and the old age of Cyphophthalmi has also been inferred for other Australasian regions. For example, in a recent study of 27 animal lineages that persisted in Sulawesi through long periods of time, the cyphophthalmid family Stylocellidae was the oldest one, initiating its diversification around the Palaeocene (Stelbrink *et al.* 2012).

Dispersal to New Caledonia has been shown in other groups of Neotropical harvestmen (Zalmoxidae; see Sharma and Giribet 2012), but in this case it has left a long trail of clades colonising all sorts of habitats in the south Pacific, which is not the case for Cyphophthalmi. The shared presence of Triaenonychidae in Australia, New Zealand and New Caledonia (among other landmasses) is exceptional for many arachnids with low vagility, but a phylogenetic framework for the family is still lacking, and dispersal from New Zealand or Australia cannot be ruled out.

Unfortunately, phylogenetic evidence is largely lacking for most other arachnid groups characteristic of New Caledonia, but some myriapod groups show probable recent dispersal between the Australian Wet Tropics and New Caledonia (Edgecombe and Giribet 2009). Cladistic analysis of the Micropholcommatinae (Rix and Harvey 2010b) could not rule out the dispersal of *Micropholcomma* and *Taphiassa* to New Caledonia, and indeed the presence of *Taphiassa* on Lord Howe Island would suggest that this genus is capable of transoceanic dispersal. Similarly, the distribution of oribatid mites in the genus *Crotonia* on Pacific islands, including New Caledonia, suggests the ability to disperse across ocean barriers (Colloff 2010). Anapidae are common and diverse in New Caledonia (Platnick and Forster 1989), although it is unknown how old the crown group radiation is, and the presence of a single species of Pararchaeidae – belonging to an otherwise Australian genus – is consistent with an isolated dispersal event (Rix and Harvey 2010a).

Zealandia: Lord Howe Island

Lord Howe Island, situated in the western Pacific Ocean 580 km east of Australia (Figure 10.2), is a small subtropical Darwinian island formed by volcanic activity 6.4–6.9 mya (McDougall *et al.* 1981; Savolainen *et al.* 2006). It is the larger of two emergent islands situated along the Lord Howe Rise – part of the western ridge of the continent of Zealandia. Despite its small size, Lord Howe possesses a remarkably diverse biota and range of habitats, from coastal dune vegetation to montane cloud forests on the summit of Mount Gower (875 m). The recent terrestrial fauna consists of numerous island endemics, many of which have suffered extinction or near extinction in the last century. The arachnid fauna of Lord Howe Island is a mix of eastern Australian and Zealandic elements (summarised in Gray 1974), all of which are assumed to have arrived by transoceanic dispersal (e.g. *Taphiassa* and *Rayforstia* Micropholcommatinae; Rix and Havery 2010b). However, there are a number of arachnid taxa of biogeographic significance, whose affinities and origins hint at a more complex history. The first is *Patelliella adusta* (Micropholcommatinae), remarkable among members of the subfamily in having no known close relatives. Its presence on Lord Howe Island is therefore puzzling, given the young age of the island. One hypothesis that has been proposed for such enigmatic taxa is that the Lord Howe Rise and the nearby Tasmantid Guyots (McDougall *et al.* 1981) – themselves once composed of other emergent seamounts – have acted as stepping-stone landscapes for shorter-distance dispersals through time, and that the current fauna of Lord Howe Island may in fact be a relictual 'signature' of a biota that was once present on now submerged landmasses. The strongest evidence for this comes from the endemic Lord Howe Island tree lobster (*Dryococelus australis*). This wingless phasmid has relatives in New Guinea, New Caledonia and associated islands, and in a recent dated molecular analysis was found to have diverged from its closest relatives some 22 mya (Buckley *et al.* 2009). Buckley *et al.* (2009) suggest that the tree lobster may have 'evolved on now drowned islands far to the north of Lord Howe and progressively dispersed down the island chain, leaving its ancestral populations to become extinct as their islands eroded away'. Another biogeographically anomalous arachnid on Lord Howe Island is the phrurolithid spider *Dorymetaceus spinnipes*, also of ambiguous affinity to Australasian taxa. However, without dated phylogenies for either *Dorymetaceus* or *Patelliella*, and the absence of a close sister lineage in the latter, the origins of both genera remain a mystery.

Future Research

Despite significant advances made over the last three decades, our understanding of Australasian arachnid biogeography remains frustratingly patchy and limited to just a few examples. This is partly a function of a poorly described fauna, for which perhaps only 20%–30% of species may have been taxonomically documented (Raven 1988; Yeates *et al.* 2004). Similarly, cladograms or molecular phylogenies are unavailable for the vast majority of genera, and only a handful of dated trees have ever been published (e.g. Rix *et al.* 2015). This situation is exemplified by our knowledge of tropical Australian and New Guinean arachnids, for which much has been assumed regarding the origins and modes of vicariance or dispersal, although only three recent cladistic studies have actually explored alternative hypotheses (Rix and Harvey 2010b; Colloff and Cameron 2013; Monod and Prendini 2015). Likewise, our understanding of mite biogeography throughout Australasia is at best woefully inadequate and at worst nonexistent, and indeed the same can be said for most arachnid families.

In short, progressing the study of arachnid biogeography in Australasia requires an increase in the number of comprehensive cladistic and/or molecular phylogenetic studies on the fauna, at all levels and on all taxa, preferably in concert. The power of a combined approach to understanding Australasian biogeography is exemplified by two groups of arachnids for which comprehensive taxonomies have been developed along with dated molecular phylogenies (i.e. cyphophthalmid harvestmen and archaeid spiders; Giribet *et al.* 2012; Rix and Harvey 2012a; Wood *et al.* 2013), both of which have provided insights into the mode and timing of divergence and speciation at continental scales. Other informative studies have likewise used species-level cladistics or undated molecular phylogenies – along with rigorous taxonomic foundations – to provide key insights into the biogeography of the fauna (e.g. Gray 1981; Harvey 1995; Rix 2006; Harvey *et al.* 2008; Harms and Harvey 2009; Rix and Harvey 2010b, 2012a; Frick and Scharff 2014; Monod and Prendini 2015). The foundation for an understanding of any biota must inevitably be collection based and taxonomic, as relevant collections and taxonomic analyses inform biogeographic studies in the first instance (Rix *et al.* 2015). Dateable molecular phylogenies in conjunction with an ever-growing suite of biogeographic software and available data resources will help to refine the potential 'how' of speciation or divergence, in both space and time, but the lack of relevant fossils for most Australasian groups is a serious impediment to this endeavour.

Finally, biogeography is not only essential to understanding the evolutionary history of arachnids in this region, but also assists conservation biology, in that our understanding of the distributions and evolutionary history of organisms can (and should) inform conservation measures where appropriate. Refugial habitats and regions of highest diversity or phylogenetic endemism can inform decision-making in the face of major anthropogenic threatening processes (e.g. land clearing and climate change) (Keppel *et al.* 2012) affecting all parts of the Australasian region. Furthermore, the comparison of dated phylogenies for a single region can also provide evidence for shared histories of the biota, as hypothesised by Nelson and Platnick (1981).

Conclusions

Our review of the biogeography of Australasian arachnids has revealed that this region includes some strikingly clear patterns of both vicariance and dispersal, and that the potential for discovering more examples of these patterns and processes is unparalleled within its extremely rich arachnid fauna and its plethora of both ancient and recently evolved lineages. However, detailed taxonomic, phylogenetic and evolutionary work is mandatory for making progress in the biogeographic arena. The downside is the small number of arachnologists working in this part of the world and the loss of taxonomic expertise in many taxonomic groups to the detriment of other more applied sciences. Progress in this field is badly needed, not only because a lot of this diversity may have been lost since European settlement, but also because the entire region is facing a plethora of current environmental challenges. We hope that this review stimulates all Australasian arachnologists to continue showcasing their unique biota and the patterns they explain.

References

ABRS (Australian Biological Resources Study) (2009). Australian Faunal Directory. http://www.environment.gov.au/biodiversity/abrs/online-resources/fauna/afd/index.html (accessed 10 February 2015).

Alfaro, M.E., Santini, F., Brock, C., Alamillo, H., Dornburg, A., Rabosky, D.L., Carnevale, G., and Harmon, L.J. (2009). Nine exceptional radiations plus high turnover explain species diversity in jawed vertebrates. *Proceedings of the National Academy of Sciences of the USA* 106, 13410–13414.

Baehr, B.C., Harvey, M.S., Burger, M., and Thoma, M. (2012). The new Australasian goblin spider genus *Prethopalpus* (Araneae, Oonopidae). *Bulletin of the American Museum of Natural History* 763, 1–113.

Baehr, B.C., Harvey, M.S., and Smith, H.M. (2010). A review of the new endemic Australian goblin spider genus *Cavisternum* (Araneae: Oonopidae). *American Museum Novitates* 3684, 1–40.

Barranco, P., and Harvey, M.S. (2008). The first indigenous palpigrade from Australia: A new species of *Eukoenenia* (Palpigradi: Eukoeneniidae). *Invertebrate Systematics* 22, 227–233.

Batt, G.E., Braun, J., Kohn, B.P., and McDougall, I. (2000). Thermochronological analysis of the dynamics of the Southern Alps, New Zealand. *Geological Society of America Bulletin* 112, 250–266.

Beccaloni, J. (2009). *Arachnids*. (Natural History Museum: London.)

Bell, K.L., Moritz, C., Moussalli, A., and Yeates, D.K. (2007). Comparative phylogeography and speciation of dung beetles from the Australian Wet Tropics rainforest. *Molecular Ecology* 16, 4984–4998.

Bellamy, D., Springett, B., and Hayden, P. (1990). *Moa's Ark: The Voyage of New Zealand*. (Viking: New York.)

Bodner, M.R., and Maddison, W.P. (2012). The biogeography and age of salticid spider radiations (Araneae: Salticidae). *Molecular Phylogenetics and Evolution* 65, 213–240.

Boyer, S.L., Baker, J.M., and Giribet, G. (2007a). Deep genetic divergence in *Aoraki denticulata* (Arachnida, Opiliones, Cyphophthlami): A widespread 'mite harvestman' defies DNA taxonomy. *Molecular Ecology* 16, 4999–5016.

Boyer, S.L., Clouse, R.M., Benavides, L.R., Sharma, P., Schwendinger, P.J., Karunarathna, I., and Giribet, G. (2007b). Biogeography of the world: A case study from cyphophthalmid Opiliones, a globally distributed group of arachnids. *Journal of Biogeography* 34, 2070–2085.

Boyer, S.L., and Giribet, G. (2007). A new model Gondwanan taxon: Systematics and biogeography of the harvestman family Pettalidae (Arachnida, Opiliones, Cyphophthalmi), with a taxonomic revision of genera from Australia and New Zealand. *Cladistics* 23, 337–361.

Boyer, S.L., and Giribet, G. (2009). Welcome back New Zealand: Regional biogeography and Gondwanan origin of three endemic genera of mite harvestmen (Arachnida, Opiliones, Cyphophthalmi). *Journal of Biogeography* 36, 1084–1099.

Buckley, T.R., Attanayake, D., and Bradler, S. (2009). Extreme convergence in stick insect evolution: Phylogenetic placement of the Lord Howe Island tree lobster. *Proceedings of the Royal Society B: Biological Sciences* 276 (1659), 1055–1062.

Burbidge, N.T. (1960). The phytogeography of the Australian region. *Australian Journal of Botany* 8, 75–221.

Byrne, M., Steane, D.A., Joseph, L., Yeates, D.K., Jordan, G.J., Crayn, D., Aplin, K., *et al.* (2011). Decline of a biome: Evolution, contraction, fragmentation, extinction and invasion of the Australian mesic zone biota. *Journal of Biogeography* 38, 1635–1656.

Byrne, M., Yeates, D.K., Joseph, L., Kearney, M., Bowler, J., Williams, M.A.J., Cooper, S., *et al.* (2008). Birth of a biome: Insights into the assembly and maintenance of the Australian arid zone biota. *Molecular Ecology* 17, 4398–4417.

Castalanelli, M.A., Teale, R., Rix, M.G., Kennington, J.W., and Harvey, M.S. (2014). Barcoding of mygalomorph spiders (Araneae: Mygalomorphae) in the Pilbara bioregion of Western Australia reveals a highly diverse biota. *Invertebrate Systematics* 28, 375–385.

Chamberlain, C.P., and Poage, M.A. (2000). Reconstructing the paleotopography of mountain belts from the isotopic composition of authigenic minerals. *Geology* 28, 115–118.

Clouse, R.M., and Giribet, G. (2007). Across Lydekker's Line: First report of mite harvestmen (Opiliones: Cyphophthalmi: Stylocellidae) from New Guinea. *Invertebrate Systematics* 21, 207–227.

Clouse, R.M., and Giribet, G. (2010). When Thailand was an island: The phylogeny and biogeography of mite harvestmen (Opiliones, Cyphophthalmi, Stylocellidae) in Southeast Asia. *Journal of Biogeography* 37, 1114–1130.

Coddington, J.A., Giribet, G., Harvey, M.S., Prendini, L., and Walter, D.E. (2004). Arachnida. In *Assembling the Tree of Life* (Cracraft, J., and Donoghue, M., Eds.), pp. 296–318. (Oxford University Press: New York.)

Colloff, M.J. (2010). New species of *Crotonia* (Acari: Oribatida: Crotoniidae) from Lord Howe and Norfolk Islands: Further evidence of long-distance dispersal events in the biogeography of a genus of Gondwanan relict oribatid mites. *Zootaxa* 2650, 1–18.

Colloff, M.J. (2013). Species-groups and biogeography of the oribatid mite family Malaconothridae (Oribatida: Malaconothroidea), with new species from the south-western Pacific region. *Zootaxa* 3722, 401–438.

Colloff, M.J., and Cameron, S.L. (2013). A phylogenetic analysis and taxonomic revision of the oribatid mite family Malaconothridae (Acari: Oribatida), with new species of *Tyrphonothrus* and *Malaconothrus* from Australia. *Zootaxa* 3681, 301–346.

Colloff, M.J., and Cameron, S.L. (2014). Beyond Moa's Ark and Wallace's Line: Extralimital distribution of new species of *Austronothrus* (Acari, Oribatida, Crotoniidae) and the endemicity of the New Zealand oribatid mite fauna. *Zootaxa* 3780, 263–281.

Cooper, S.J.B., Harvey, M.S., Saint, K.M., and Main, B.Y. (2011). Deep phylogeographic structuring of populations of the trapdoor spider *Moggridgea tingle* (Migidae) from southwestern Australia: Evidence for long-term refugia within refugia. *Molecular Ecology* 20, 3219–3236.

Crisp, M., Cook, L.G., and Steane, D. (2004). Radiation of the Australian flora: What can comparisons of molecular phylogenies across multiple taxa tell us about the evolution of diversity in present-day communities? *Philosophical Transactions of the Royal Society of London, Series B: Biological Sciences* 359, 1551–1571.

Dartnall, A.J. (1970). Some Tasmanian chthoniid pseudoscorpions. *Papers and Proceedings of the Royal Society of Tasmania* 104, 65–68.

de Bivort, B.L., and Giribet, G. (2010). A systematic revision of the South African Pettalidae (Arachnida: Opiliones: Cyphophthalmi) based on a combined analysis of discrete and continuous morphological characters with the description of seven new species. *Invertebrate Systematics* 24, 371–406.

Dietz, R.S., and Holden, J.C. (1970). The breakup of Pangaea. *Scientific American* 223, 30–41.

Dimitrov, D., Benavides, L.R., Arnedo, M.A., Giribet, G., Griswold, C.E., Scharff, N., and Hormiga, G. (2016). Rounding up the usual suspects: A standard target-gene approach for resolving the interfamilial phylogenetic relationships of ecribellate orb-weaving spiders with a new family-rank classification (Araneae, Araneoidea). *Cladistics*. In press. doi: 10.1111/cla.12165

Drummond, A.J., Suchard, M.A., Xie, D., and Rambaut, A. (2012). Bayesian phylogenetics with BEAUti and the BEAST 1.7. *Molecular Phylogenetics and Evolution* 29, 1969–1973.

Dunlop, J.A. (2010). Geological history and phylogeny of Chelicerata. *Arthropod Structure and Development* 39, 124–142.

Dunlop, J.A., and Penney, D. (2012). *Fossil Arachnids*. Monograph Series (Siri Scientific Press: Manchester, UK.)

Dunlop, J.A., and Selden, P.A. (2009). Calibrating the chelicerate clock: A paleontological reply to Jeyaprakash and Hoy. *Experimental and Applied Acarology* 48, 183–197.

Durrant, B.J., Harvey, M.S., Framenau, V.W., Ott, R., and Waldock, J.M. (2010). Patterns in the composition of ground-dwelling spider communities in the Pilbara bioregion, Western Australia. *Records of the Western Australian Museum*, Supplement 78, 185–204.

Eberhard, S.M., Richardson, A.M.M., and Swain, R. (1991). *The Invertebrate Cave Fauna of Tasmania.* University of Tasmania, Hobart.

Edgecombe, G.D., and Giribet, G. (2009). Phylogenetics of scutigeromorph centipedes (Myriapoda: Chilopoda) with implications for species delimitation and historical biogeography of the Australian and New Caledonian faunas. *Cladistics* 25, 406–427.

Edward, K.L., and Harvey, M.S. (2014). Australian goblin spiders of the genus *Ischnothyreus* (Araneae, Oonopidae). *Bulletin of the American Museum of Natural History* 389, 1–144.

Espeland, M., and Murienne, J. (2011). Diversity dynamics in New Caledonia: Towards the end of the museum model? *BMC Evolutionary Biology* 11(1), 254.

Ezcurra, M.D., and Agnolín, F.L. (2011). A new global palaeobiogeographical model for the Late Mesozoic and Early Tertiary. *Systematic Biology* 61, 553–566.

Fabricius, J.C. (1775). *Systema Entomologiae Sistens Insectorum Classes, Ordines, Genera, Species, Adjectis Synonymis, Locis, Descriptionibus, Observationibus*. (Libraria Kortii: Liepzig, Germany.)

Fernåndez, R., and Giribet, G. (2014). Phylogeography and species delimitation in the New Zealand endemic, genetically hypervariable harvestman species, Aoraki denticulata (Arachnida, Opiliones, Cyphophthalmi). *Invertebrate Systematics* 28, 401–414.

Forster, R.R. (1947). The zoogeographical relationships of the New Zealand Opiliones. *Transactions and Proceedings of the Royal Society of New Zealand* 77: 233–235.

Forster, R.R. (1954). The New Zealand harvestmen (sub-order Laniatores). *Canterbury Museum Bulletin* 2, 1–329.

Forster, R.R. (1961). The New Zealand fauna and its origins. *Proceedings of the Royal Society of New Zealand* 89, 51–55.

Forster, R.R. (1975). The spiders and harvestmen. In *Biogeography and ecology in New Zealand* (Kuschel, G., Ed.), pp. 493–505. (Junk: The Hague, the Netherlands.)

Forster, R.R. (1995). The Australasian spider family Periegopidae Simon, 1893 (Araneae: Sicarioidea). *Records of the Western Australian Museum*, Supplement 52, 91–105.

Forster, R.R., and Platnick, N.I. (1984). A review of the archaeid spiders and their relatives, with notes on the limits of the superfamily Palpimanoidea (Arachnida, Araneae). *Bulletin of the American Museum of Natural History* 178, 1–106.

Forster, R.R., and Platnick, N.I. (1985). A review of the austral spider family Orsolobidae (Arachnida, Araneae), with notes on the superfamily Dysderoidea. *Bulletin of the American Museum of Natural History* 181, 1–230.

Forster, R.R., Platnick, N.I., and Coddington, J. (1990). A proposal and review of the spider family Synotaxidae (Araneae, Araneoidea), with notes on theridiid interrelationships. *Bulletin of the American Museum of Natural History* 193, 1–116.

Forster, R.R., Platnick, N.I., and Gray, M.R. (1987). A review of the spider superfamilies Hypochiloidea and Austrochiloidea (Araneae, Araneomorphae). *Bulletin of the American Museum of Natural History* 185, 1–116.

Frick, H., and Scharff, N. (2014). Phantoms of Gondwana? Phylogeny of the spider subfamily Mynogleninae (Araneae: Linyphiidae). *Cladistics* 30, 67–106.

Gibbs, G. (2006). *Ghosts of Gondwana: The History of Life in New Zealand*. (Craig Potton Publishing: Nelson, New Zealand.)

Gillespie, R.G., and Roderick, G.K. (2002). Arthropods on islands: Colonization, speciation, and conservation. *Annual Review of Entomology* 47, 595–632.

Giribet, G. (2003). *Karripurcellia*, a new pettalid genus (Arachnida: Opiliones: Cyphophthalmi) from Western Australia, with a cladistic analysis of the family Pettalidae. *Invertebrate Systematics* 17, 387–406.

Giribet, G., and Boyer, S.L. (2010). 'Moa's Ark' or 'Goodbye Gondwana': Is the origin of New Zealand's terrestrial invertebrate fauna ancient, recent, or both? *Invertebrate Systematics* 24, 1–8.

Giribet, G., and Edgecombe, G.D. (2006). The importance of looking at small-scale patterns when inferring Gondwanan biogeography: A case study of the centipede *Paralamyctes* (Chilopoda, Lithobiomorpha, Henicopidae). *Biological Journal of the Linnean Society* 89, 65–78.

Giribet, G., Edgecombe, G.D., Wheeler, W.C., and Babbitt, C. (2002). Phylogeny and systematic position of Opiliones: A combined analysis of chelicerate relationships using morphological and molecular data. *Cladistics* 18, 5–70.

Giribet, G., and Kury, A. (2007). Phylogeny and biogeography. In *Harvestmen: The Biology of Opiliones* (Pinto-da-Rocha, R., Machado, G., and Giribet, G., Eds.), pp. 62–87. (Harvard University Press: Cambridge, MA.)

Giribet, G., McIntyre, E., Christian, E., Espinasa, L., Ferreira, R.L., Francke, O.F., Harvey, M.S., *et al.* (2014). The first phylogenetic analysis of Palpigradi (Arachnida): The most enigmatic arthropod order. *Invertebrate Systematics* 28, 350–360.

Giribet, G., Sharma, P., Benavides, L.R., Boyer, S.L., Clouse, R.M., de Bivort, B.L., Dimitrov, D., Kawauchi, G.Y., Murienne, J.Y., and Schwendinger, P.J. (2012). Evolutionary and biogeographic history of an ancient and global group of arachnids (Arachnida, Opiliones, Cyphophthalmi) with a new taxonomic arrangement. *Biological Journal of the Linnean Society* 105, 92–130.

Giribet, G., Vogt, L., González, A.P., Sharma, P., and Kury, A.B. (2010). A multilocus approach to harvestman (Arachnida: Opiliones) phylogeny with emphasis on biogeography and the systematics of Laniatores. *Cladistics* 26, 408–437.

Giribet, G., Boyer, S.L., Baker, C., Fernández, R., Sharma, P.P., de Bivort, B.L., Daniels, S.R., Harvey, M.S., and Griswold, C.E. (2016). A molecular phylogeny of the temperate Gondwanan family Pettalidae (Arachnida, Opiliones, Cyphophthalmi) with biogeographic and taxonomic implications. *Zoological Journal of the Linnean Society.*

Goldberg, J., Trewick, S.A., and Paterson, A.M. (2008). Evolution of New Zealand's terrestrial fauna: A review of molecular evidence. *Philosophical Transactions of the Royal Society B: Biological Sciences* 363(1508), 3319–3334.

Grandcolas, P., Murienne, J., Robillard, T., Desutter-Grandcolas, L., Jourdan, H., Guilbert, E., and Deharveng, L. (2008). New Caledonia: A very old Darwinian island? *Philosophical Transactions of the Royal Society of London, Series B, Biological Sciences* 363, 3309–3317.

Gray, M.R. (1974). Survey of the spider fauna. In *Lord Howe Island: Its Biology and Conservation* (Recher, H.F., and Clark, S.S., Eds.), pp. 50–54. (The Australian Museum: Sydney, Australia.)

Gray, M.R. (1981). A revision of the spider genus *Baiami* Lehtinen (Araneae, Amaurobioidea). *Records of the Australian Museum* 33, 779–802.

Gray, M.R. (1988). Aspects of the systematics of the Australian funnel web spiders (Araneae: Hexathelidae: Atracinae) based upon morphological and electrophoretic data. In *Australian Arachnology* (Austin, A.D., and Heather, N.W., Eds.), pp. 113–125. (Australian Entomological Society: Brisbane, Australia.)

Groh, S., and Giribet, G. (2014). Polyphyly of Caddoidea, reinstatement of the family Acropsopilionidae in Dyspnoi, and a revised classification system of Palpatores (Arachnida, Opiliones). *Cladistics* 31, 277–290.

Guzik, M.T., Austin, A.D., Cooper, S.J.B., Harvey, M.S., Humphreys, W.F., Bradford, T., Eberhard, S.M., *et al.* (2011). Is the Australian subterranean fauna uniquely diverse? *Invertebrate Systematics* 24, 407–418.

Hammer, M., and Wallwork, J.A. (1979). A review of the world distribution of oribatid mites (Acari: Cryptostigmata) in relation to continental drift. *Biologiske Skrifter det Kongelige Dansk Videnskabernes Selskab* 22(4), 1–31.

Hand, S., Archer, M., Bickel, D., Creaser, P., Dettmann, M., Godthelp, H., Jones, A., Norris, B., and Wicks, D. (2010). Australian Cape York amber. In *Biodiversity of Fossils in Amber from the Major World Deposits* (Penney, D., Ed.), pp. 69–79. (Siri Scientific Press: Manchester, UK.)

Harms, D., and Framenau, V.W. (2013). New species of mouse spiders (Araneae: Mygalomorphae: Actinopodidae: *Missulena*) from the Pilbara region, Western Australia. *Zootaxa* 3637, 521–540.

Harms, D., and Harvey, M.S. (2009). Pirates of Australia: Systematics and phylogeny of the Australian mimetid spiders (Araneae: Mimetidae: Mimetinae) with a description of the Western Australian fauna. *Invertebrate Systematics* 23, 231–280.

Harms, D., and Harvey, M.S. (2013). Review of the cave-dwelling species of *Pseudotyrannochthonius* Beier (Arachnida: Pseudoscorpiones: Pseudotyrannochthoniidae) from mainland Australia, with description of two troglobitic species. *Australian Journal of Entomology* 52, 129–143.

Harrison, S.E., Guzik, M.T., Harvey, M.S., and Austin, A.D. (2014). Molecular phylogenetic analysis of Western Australian troglobitic chthoniid pseudoscorpions (Pseudoscorpiones: Chthoniidae) points to multiple independent subterranean clades. *Invertebrate Systematics* 28, 386–400.

Harvey, F.S.B., Framenau, V.W., Wojcieszek, J.M., Rix, M.G., and Harvey, M.S. (2012). Molecular and morphological characterisation of new species in the trapdoor spider genus *Aname* (Araneae: Mygalomorphae: Nemesiidae) from the Pilbara bioregion of Western Australia. *Zootaxa* 3383, 15–38.

Harvey, M.S. (1985). The systematics of the family Sternophoridae (Pseudoscorpionida). *Journal of Arachnology* 13, 141–209.

Harvey, M.S. (1986). The Australian Geogarypidae, new status, with a review of the generic classification (Arachnida: Pseudoscorpionida). *Australian Journal of Zoology* 34, 753–778.

Harvey, M.S. (1987). A revision of the genus *Synsphyronus* Chamberlin (Garypidae: Pseudoscorpionida: Arachnida). *Australian Journal of Zoology*, Supplementary Series 126, 1–99.

Harvey, M.S. (1992). The Schizomida (Chelicerata) of Australia. *Invertebrate Taxonomy* 6, 77–129.

Harvey, M.S. (1995). The systematics of the spider family Nicodamidae (Araneae: Amaurobioidea). *Invertebrate Taxonomy* 9, 279–386.

Harvey, M.S. (1996). Small arachnids and their value in Gondwanan biogeographic studies. In *Gondwanan Heritage: Past, Present and Future of the Western Australian Biota* (Hopper, S.D., Chappill, J.A., Harvey, M.S., and George, A.S., Eds.), pp. 155–162. (Surrey Beatty: Chipping Norton, UK.)

Harvey, M.S. (1998a). Pseudoscorpion groups with bipolar distributions: A new genus from Tasmania related to the Holarctic *Syarinus* (Arachnida, Pseudoscorpiones, Syarinidae). *Journal of Arachnology* 26, 429–441.

Harvey, M.S. (1998b). *The Australian Water Mites: A Guide to the Families and Genera*. (CSIRO Publishing: Melbourne, Australia.)

Harvey, M.S. (2000). *Brignolizomus* and *Attenuizomus*, new schizomid genera from Australia (Arachnida: Schizomida: Hubbardiidae). *Memorie Della Società Entomologica Italiana*, Supplement 78, 329–338.

Harvey, M.S. (2002). Short-range endemism in the Australian fauna: Some examples from non-marine environments. *Invertebrate Systematics* 16, 555–570.

Harvey, M.S., Berry, O., Edward, K.L., and Humphreys, G. (2008). Molecular and morphological systematics of hypogean schizomids (Schizomida: Hubbardiidae) in semi-arid Australia. *Invertebrate Systematics* 22, 167–194.

Harvey, M.S., Sampey, A., West, P.L.J., and Waldock, J.M. (2000). Araneomorph spiders from the southern Carnarvon Basin, Western Australia: A consideration of regional biogeographic relationships. *Records of the Western Australian Museum*, Supplement 61, 295–321.

Harvey, M.S., and Šťáhlavský, F. (2010). A review of the pseudoscorpion genus *Oreolpium* (Pseudoscorpiones: Garypinidae), with remarks on the composition of the Garypinidae and on pseudoscorpions with bipolar distributions. *Journal of Arachnology* 38, 294–308.

Harvey, M.S., and Volschenk, E.S. (2007). The systematics of the Gondwanan pseudoscorpion family Hyidae (Pseudoscorpiones: Neobisioidea): New data and a revised phylogenetic hypothesis. *Invertebrate Systematics* 21, 365–406.

Harvey, M.S., Waldock, J.M., Guthrie, N.A., Durrant, B.J., and McKenzie, N.L. (2004). Patterns in the composition of ground-dwelling araneomorph spider communities in the Western Australian wheatbelt. *Records of the Western Australian Museum*, Supplement 67, 257–291.

Heads, M. (2008). Panbiogeography of New Caledonia, south-west Pacific: Basal angiosperms on basement terranes, ultramafic endemics inherited from volcanic island arcs and old taxa endemic to young islands. *Journal of Biogeography* 35(12), 2153–2175.

Heenan, P.B., Mitchell, A.D., de Lange, P.J., Keeling, J., and Paterson, A.M. (2010). Late-Cenozoic origin and diversification of Chatham Islands endemic plant species revealed by analyses of DNA sequence data. *New Zealand Journal of Botany* 48, 83–136.

Hennig, W. (1950). *Grundzüge einer Theorie der phylogenetischen Systematik*. (Deutscher Zentralverlag: Berlin, Germany.)

Hennig, W. (1966). *Phylogenetic Systematics*. (University of Illinois Press: Urbana.)

Hogg, H.R. (1909). Some New Zealand and Tasmanian Arachnidae. *Transactions of the New Zealand Institute* 42, 273–283.

Hopper, S.D., Harvey, M.S., Chappill, J.A., Main, A.R., and Main, B.Y. (1996). The Western Australian biota as Gondwanan heritage: A review. In *Gondwanan Heritage: Past, Present and Future of the Western Australian Biota* (Hopper, S.D., Chappill, J.A, Harvey, M.S., and George, A.S., Eds.), pp. 1–46. (Surrey Beatty: Chipping Norton, UK.)

Hunt, G.S. (1990). *Hickmanoxyomma*, a new genus of cavernicolous harvestmen from Tasmania (Opiliones: Triaenonychidae). *Records of the Australian Museum* 42, 45–68.

Hunt, G.S. (1993). A revision of the genus *Lomanella* Pocock and its implication for family level classification in the Travunioidea (Arachnida: Opiliones: Triaenonychidae). *Records of the Australian Museum* 45, 81–119.

Jell, P.A., and Duncan, P.M. (1986). Invertebrates, mainly insects, from the freshwater Lower Cretaceous Koonwarra Fossil Bed (Korumburra Group), South Gippsland, Victoria. *Association of Australasian Palaeontologists Memoir* 3, 111–205.

Kaulfuss, U. (2014). Drowned in lakes and trapped in amber: Diverse terrestrial arthropod faunas from Miocene New Zealand (abstract). In *The 63rd New Zealand Entomological Society Conference*, Queenstown, 13. (New Zealand Entomological Society: Queenstown, New Zealand.)

Keppel, G., Van Niel, K.P., Wardell-Johnson, G.W., Yates, C.J., Byrne, M., Mucina, L., Schut, A.G.T., Hopper, S.D., and Franklin, S.E. (2012). Refugia: Identifying and understanding safe havens for biodiversity under climate change. *Global Ecology and Biogeography* 21, 393–404.

Koch, L., and Keyserling, E. (1871–1890). *Die Arachniden Australiens*, 2 vols. (Bauer und Raspe: Nuremberg, Germany.)

Koch, L.E. (1977). The taxonomy, geographic distribution and evolutionary radiation of Australo-Papuan scorpions. *Records of the Western Australian Museum* 5, 83–367.

Koch, L.E. (1981). The scorpions of Australia: Aspects of their ecology and zoogeography. In *Ecological Biogeography of Australia* (Keast, A., Ed.), vol. 2, pp. 875–884. (Junk: The Hague, the Netherlands.)

Krantz, G.W., and Walter, D.E. (Eds.). (2009). *A Manual of Acarology*, 3rd edn. (Texas Tech University Press: Lubbock.)

Landis, C.A., Campbell, H.J., Begg, J.G., Mildenhall, D.C., Paterson, A.M., and Trewick, S.A. (2008). The Waipounamu Erosion Surface: Questioning the antiquity of the New Zealand land surface and terrestrial fauna and flora. *Geological Magazine* 145, 173–197.

Lattimore, V.L., Vink, C.J., Paterson, A.M., and Cruickshank, R.H. (2011). Unidirectional introgression within the genus *Dolomedes* (Araneae: Pisauridae) in southern New Zealand. *Invertebrate Systematics* 25, 70–79.

Lewis, K., Nodder, S.D., and Carter, L. (2012). Sea floor geology: Zealandia; The New Zealand continent. In *Te Ara: The Encyclopedia of New Zealand*. Editor: J. Phillips. (Ministry for Culture and Heritage: Wellington.

Lucky, A. (2011). Molecular phylogeny and biogeography of the spider ants, genus *Leptomyrmex* Mayr (Hymenoptera: Formicidae). *Molecular Phylogenetics and Evolution* 59, 281–292.

Maddison, W., Li, D., Bodner, M., Zhang, J., Xin, X., Liu, Q., and Liu, F. (2014). The deep phylogeny of jumping spiders (Araneae, Salticidae). *ZooKeys* 440, 57–87.

Maddison, W.P. (2008). New cocalodine jumping spiders from Papua New Guinea (Araneae: Salticidae: Cocalodinae). *Zootaxa* 2021, 1–22.

Main, B.Y. (1957a). Adaptive radiation of trapdoor spiders. *Australian Museum Magazine* 12, 160–163.

Main, B.Y. (1957b). Biology of aganippine trapdoor spiders (Mygalomorphae: Ctenizidae). *Australian Journal of Zoology* 5, 402–473.

Main, B.Y. (1981a). A comparative account of the biogeography of terrestrial invertebrates in Australia: Some generalizations. In *Ecological Biogeography of Australia* (Keast, A., Ed.), vol. 2, pp. 1055–1077. (Junk: The Hague, the Netherlands.)

Main, B.Y. (1981b). Eco-evolutionary radiation of mygalomorph spiders in Australia. In *Ecological Biogeography of Australia* (Keast, A., Ed.), vol. 2, pp. 854–872. (Junk: The Hague, the Netherlands.)

Main, B.Y. (1981c). Australian spiders: Diversity, distribution and ecology. In *Ecological Biogeography of Australia* (Keast, A., Ed.), vol. 2, pp. 808–852. (Junk: The Hague, the Netherlands.)

Main, B.Y. (1991). Kimberley spiders: Rainforest strongholds. In *Kimberley Rainforests* (McKenzie, N.L., Johnston, R.B., and Kendrick, P.G., Eds.), pp. 271–293. (Surrey Beatty: Chipping Norton, UK.)

Main, B.Y. (2001). Historical ecology, responses to current ecological changes and conservation of Australian spiders. *Journal of Insect Conservation* 5, 9–25.

Main, B.Y., Sampey, A., and West, P.L.J. (2000). Mygalomorph spiders of the southern Carnarvon Basin, Western Australia. *Records of the Western Australian Museum*, Supplement 61, 281–293.

Martin, H.A. (2006). Cenozoic climatic change and the development of the arid vegetation in Australia. *Journal of Arid Environments* 66, 533–563.

McDougall, I., Embleton, B.J.J., and Stone, D.B. (1981). Origin and evolution of Lord Howe Island, Southwest Pacific Ocean. *Journal of the Geological Society of Australia* 28, 155–176.

McGlone, M.S. (2005). Goodbye Gondwana. *Journal of Biogeography* 32, 739–740.

McKenzie, N.L., Burbidge, A.H., and Rolfe, J.K. (2003). Effect of salinity on small, ground-dwelling animals in the Western Australian wheatbelt. *Australian Journal of Botany* 51, 725–740.

McLoughlin, S. (2001). The breakup history of Gondwana and its impact on pre-Cenozoic floristic provincialism. *Australian Journal of Botany* 49, 271–300.

Metcalfe, I. (2002). Permian tectonic framework and palaeogeography of SE Asia. *Journal of Asian Earth Sciences* 20, 551–566.

Metcalfe, I. (2013). Tectonic evolution of the Malay Peninsula. *Journal of Asian Earth Sciences* 76, 195–213.

Miglio, L.T., Harms, D., Framenau, V.W., and Harvey, M.S. (2014). Four new mouse spider species (Araneae, Mygalomorphae, Actinopodidae, *Missulena*) from Western Australia. *ZooKeys* 410, 121–148.

Mildenhall, D.C., Mortimer, N., Bassett, K.N., and Kennedy, E.M. (2014). Oligocene paleogeography of New Zealand: Maximum marine transgression. *New Zealand Journal of Geology and Geophysics* 57, 107–109.

Monod, L., and Prendini, L. (2015). Evidence for Eurogondwana: The roles of dispersal, extinction and vicariance in the evolution and biogeography of Indo-Pacific Hormuridae (Scorpiones: Scorpionoidea). *Cladistics* 31(1), 71–111.

Moulds, T.A., Murphy, N., Adams, M., Reardon, T., Harvey, M.S., Jennings, J., and Austin, A.D. (2007). Phylogeography of cave pseudoscorpions in southern Australia. *Journal of Biogeography* 34, 951–962.

Murienne, J. (2009). Testing biodiversity hypotheses in New Caledonia using phylogenetics. *Journal of Biogeography* 36(8), 1433–1434.

Murienne, J. (2010). Panbiogeography of New Caledonia: A response to Heads (2008). *Journal of Biogeography* 37(8), 1625–1626.

Murphy, N.P., Framenau, V.W., Donnellan, S.C., Harvey, M.S., Park, Y.-C., and Austin, A.D. (2006). Phylogenetic reconstruction of the wolf spiders (Araneae: Lycosidae) using sequences from the 12S rRNA, 28S rRNA and NADH1 genes: Implications for classification, biogeography and the evolution of web building behaviour. *Molecular Phylogenetics and Evolution* 38, 583–602.

Nattier, R., Robillard, T., Desutter-Grandcolas, L., Couloux, A., and Grandcolas, P. (2011). Older than New Caledonia emergence? A molecular phylogenetic study of the eneopterine crickets (Orthoptera: Grylloidea). *Journal of Biogeography* 38, 2195–2209.

Nelson, G., and Platnick, N.I. (1981). *Systematics and Biogeography: Cladistics and Vicariance.* (Columbia University Press: New York.)

Penney, D., and Selden, P.A. (2011). *Fossil Spiders: The Evolutionary History of a Mega-Diverse Order*. (Siri Scientific Press: Manchester, UK.)

Pickard-Cambridge, O. (1879). On some new and rare spiders from New Zealand, with characters of four new genera. *Proceedings of the Zoological Society of London* 47, 681–703.

Platnick, N.I. (1976). Drifting spiders or continents? Vicariance biogeography of the spider subfamily Laroniinae (Araneae: Gnaphosidae). *Systematic Zoology* 25, 101–109.

Platnick, N.I. (1991). Patterns of biodiversity: Tropical vs. temperate. *Journal of Natural History* 25, 1083–1088.

Platnick, N.I. (2008). A new subterranean ground spider genus from Western Australia (Araneae: Trochanteriidae). *Invertebrate Systematics* 22, 295–299.

Platnick, N.I., and Forster, R.R. (1989). A revision of the temperate South American and Australasian spiders of the family Anapidae (Araneae, Araneoidea). *Bulletin of the American Museum of Natural History* 190, 1–139.

Pocock, R.I. (1888). The species of the genus *Urodacus* contained in the collection of the British (Natural-History) Museum. *Annals and Magazine of Natural History* 2(6), 169–175.

Pocock, R.I. (1898). The Australian scorpions of the genus *Urodacus*, Pet. *Annals and Magazine of Natural History* 2(7),59–67.

Popkin-Hall, Z.R., and Boyer, S.L. (2014). New species of mite harvestmen from southeast Queensland, Australia greatly extend the known distribution of the genus *Austropurcellia* (Arachnida, Opiliones, Cyphophthalmi). *Zootaxa* 3827(1), 517–541.

Pybus, O.G., and Harvey, P.H. (2000). Testing macro-evolutionary models using incomplete molecular phylogenies. *Proceedings of Biological Science* 267(1459), 2267–2272.

Rabosky, D.L. (2006). LASER: A maximum likelihood toolkit for detecting temporal shifts in diversification rates from molecular phylogenies. *Evol Bioinform Online* 2, 273–276.

Rainbow, W.J. (1896). Contributions to a knowledge of the Arachnidan fauna of Australia, no. 1. *Proceedings of the Linnean Society of New South Wales* 21, 634–636.

Rainbow, W.J. (1898). Contribution to a knowledge of the arachnidan fauna of British New Guinea. *Proceedings of the Linnean Society of New South Wales* 23, 328–356.

Rainbow, W.J. (1906). A synopsis of Australian Acarina. *Records of the Australian Museum* 6, 145–196.

Rainbow, W.J. (1911). A census of Australian Araneidae. *Records of the Australian Museum* 9, 107–319.

Rainbow, W.J. (1920). Arachnida from Lord Howe and Norfolk Islands. *Records of the South Australian Museum* 1, 229–272.

Rainbow, W.J., and Pulleine, R.H. (1918). Australian trap-door spiders. *Records of the Australian Museum* 12, 81–169.

Raven, R.J. (1980). The evolution and biogeography of the mygalomorph spider family Hexathelidae (Araneae, Chelicerata). *Journal of Arachnology* 8, 251–266.

Raven, R.J. (1984). Systematics and biogeography of the mygalomorph spider family Migidae (Araneae) in Australia. *Australian Journal of Zoology* 32, 379–390.

Raven, R.J. (1988). The current status of Australian spider systematics. In *Australian Arachnology* (Austin, A.D., and Heather, N.W., Eds.), pp. 37–47. (Australian Entomological Society: Brisbane, Australia.)

Raven, R.J. (1994). Mygalomorph spiders of the Barychelidae in Australia and the western Pacific. *Memoirs of the Queensland Museum* 35, 291–706.

Raven, R.J., Jell, P.A., and Knezour, R.A. (2015). Edwa maryae gen. et sp. nov. in the Norian Blackstone Formation of the Ipswich Basin—the first Triassic spider (Mygalomorphae) from Australia (Alcheringa: An Australasian). *Journal of Palaeontology* 39, 259–263. doi:10.1080/03115518.2015.993300

Raven, R.J., and Stumkat, K.S. (2005). Revisions of Australian ground-hunting spiders: II. Zoropsidae (Lycosoidea: Araneae). *Memoirs of the Queensland Museum* 50, 347–423.

Ree, R.H., and Smith, S.A. (2008). Maximum likelihood inference of geographic range evolution by dispersal, local extinction, and cladogenesis. *Systematic Biology* 57, 4–14.

Regier, J.C., Shultz, J.W., Zwick, A., Hussey, A., Ball, B., Wetzer, R., Martin, J.W., and Cunningham, C.W. (2010). Arthropod relationships revealed by phylogenomic analysis of nuclear protein-coding sequences. *Nature* 463, 1079–1083.

Rix, M.G. (2006). Systematics of the Australasian spider family Pararchaeidae (Arachnida: Araneae). *Invertebrate Systematics* 20, 203–254.

Rix, M.G., Edwards, D.L., Byrne, M., Harvey, M.S., Joseph, L., and Roberts, J.D. (in press). Biogeography and speciation of terrestrial fauna in the south-western Australian biodiversity hotspot. *Biological Reviews.*

Rix, M.G., and Harvey, M.S. (2010a). The first pararchaeid spider (Araneae: Pararchaeidae) from New Caledonia, with a discussion on spinneret spigots and egg sac morphology in *Ozarchaea. Zootaxa* 2414, 27–40.

Rix, M.G., and Harvey, M.S. (2010b). The spider family Micropholcommatidae (Arachnida, Araneae, Araneoidea): A relimitation and revision at the generic level. *ZooKeys* 36, 1–321.

Rix, M.G., and Harvey, M.S. (2011). Australian assassins, part I: A review of the assassin spiders (Araneae, Archaeidae) of mid-eastern Australia. *ZooKeys* 123, 1–100.

Rix, M.G., and Harvey, M.S. (2012a). Phylogeny and historical biogeography of ancient assassin spiders (Araneae: Archaeidae) in the Australian mesic zone: Evidence for Miocene speciation within Tertiary refugia. *Molecular Phylogenetics and Evolution* 62, 375–396.

Rix, M.G., and Harvey, M.S. (2012b). Australian assassins, part III: A review of the assassin spiders (Araneae, Archaeidae) of tropical north-eastern Queensland. *ZooKeys* 218, 1–55.

Rix, M.G., Harvey, M.S., and Roberts, J.D. (2008). Molecular phylogenetics of the spider family Micropholcommatidae (Arachnida: Araneae) using nuclear rRNA genes (18S and 28S). *Molecular Phylogenetics and Evolution* 46, 1031–1048.

Rix, M.G., Edwards, D.L., Byrne, M., Harvey, M.S., Joseph, L., and Roberts, J.D. (2015). Biogeography and speciation of terrestrial fauna in the south-western Australian biodiversity hotspot. *Biological Reviews* 90: 762–793.

Sanmartín, I., and Ronquist, F. (2004). Southern Hemisphere biogeography analyzed with event-based models: plant versus animal patterns. *Systematic Biology* 53, 216–243.

Savolainen, V., Anstett, M.-C., Lexer, C., Hutton, I., Clarkson, J.J., Norup, M.V., Powell, M.P., Springate, D., Salamin, N., and Baker, W.J. (2006). Sympatric speciation in palms on an oceanic island. *Nature* 441(7090), 210–213.

Schmidt, A.R., Kaulfuss, U., Borkent, A., Busch, A., Conran, J.G., D'Haese, C.A., Engel, M.S., *et al.* (2014). Amber inclusions from New Zealand. 9th European Palaeobotany–Palynology Conference, 26–31 August 2014, Padova, Italy, Abstract Book, pp. 245–246.

Schneider, C., and Moritz, C. (1998). Rainforest refugia and evolution in Australia's Wet Tropics. *Proceedings of the Royal Society of London B* 266, 191–196.

Seeman, O. (2013). A review of the *Fedrizzia* mites species (Acari: Mesostigmata: Fedrizzidae) found in association with Australian *Mastachilus* beetles (Coleoptera: Passalidae). *Memoirs of the Queensland Museum: Nature* 58, 33–47.

Selden, P.A., Diying, H., and Dong, R. (2008). Palpimanoid spiders from the Jurassic of China. *Journal of Arachnology* 36, 306–321.

Sharma, P., and Giribet, G. (2009). A relict in New Caledonia: Phylogenetic relationships of the family Troglosironidae (Opiliones: Cyphophthalmi). *Cladistics* 25, 279–294.

Sharma, P.P., and Giribet, G. (2011). The evolutionary and biogeographic history of the armoured harvestmen: Laniatores phylogeny based on ten molecular markers, with the description of two new families of Opiliones (Arachnida). *Invertebrate Systematics* 25, 106–142.

Sharma, P.P., and Giribet, G. (2012). Out of the Neotropics: Late Cretaceous colonization of Australasia by American arthropods. *Proceedings of the Royal Society B: Biological Sciences* 279(1742):3501–3509.

Sharma, P.P., Kaluziak, S.T., Pérez-Porro, A.R., González, V.L., Hormiga, G., Wheeler, W.C., and Giribet, G. (2014). Phylogenomic interrogation of Arachnida reveals systemic conflicts in phylogenetic signal. *Molecular Biology and Evolution* 31(11): 2963–2984.

Sharma, P.P., and Wheeler, W.C. (2013). Revenant clades in historical biogeography: The geology of New Zealand predisposes endemic clades to root age shifts. *Journal of Biogeography* 40, 1609–1618.

Simon, E. (1908). Araneae, part 1. In *Die Fauna Südwest-Australiens* (Michaelsen, W., and Hartmeyer, R., Eds.), vol. 1, pp. 359–446. (Gustav Fischer: Jena, Germany.)

Simon, E. (1909). Araneae, part 2. In *Die Fauna Südwest-Australiens* (Michaelsen, W., and Hartmeyer, R., Eds.), vol. 2, pp. 155–212. (Gustav Fischer: Jena, Germany.)

Sirvid, P.J., Moore, N.E., Chambers, G.K., and Prendergast, K. (2013). A preliminary molecular analysis of phylogenetic and biogeographic relationships of New Zealand Thomisidae (Araneae) using a multilocus approach. *Invertebrate Systematics* 27, 655–672.

Sirvid, P.J., Zhang, Z.-Q., Harvey, M.S., Rhode, B.E., Cook, D.R., Bartsch, I., and Staples, D.A. (2010). Phylum Arthropoda, Chelicerata, horseshoe crabs, arachnids and sea spiders. In *The New Zealand Inventory of Biodiversity: A Species 2000 Symposium Review* (Gordon, D., Ed.), vol. 1, pp. 50–89. (Canterbury University Press: Christchurch, New Zealand.)

Smith, G.T., and McKenzie, N.L. (2000). Biogeography of scorpion communities in the southern Carnarvon Basin, Western Australia. *Records of the Western Australian Museum*, Supplement 61, 269–279.

Snodgrass, R.E. (1948). The feeding organs of Arachnida, including mites and ticks. *Smithsonian Miscellaneous Collections* 110(10), 1–93.

Stelbrink, B., Albrecht, C., Hall, R., and von Rintelen, T. (2012). The biogeography of Sulawesi revisited: Is there evidence for a vicariant origin of taxa on Wallace's 'anomalous island'? *Evolution* 66, 2252–2271.

Swenson, U., Nylinder, S., and Munzinger, J. (2014). Sapotaceae biogeography supports New Caledonia being an old Darwinian island. *Journal of Biogeography* 41, 797–809.

Taylor, C.K. (2011). Revision of the genus *Megalopsalis* (Arachnida: Opiliones: Phalangioidea) in Australia and New Zealand and implications for phalangioid classification. *Zootaxa* 2773, 1–65.

Trewick, S.A., Paterson, A.M., and Campbell, H.J. (2007). Hello New Zealand. *Journal of Biogeography* 34, 1–6.

Urquhart, A.T. (1885). On the spiders of New Zealand. *Transactions of the New Zealand Institute* 17, 31–53.

Urquhart, A.T. (1892). Catalogue of the described species of New Zealand Araneidae. *Transactions of the New Zealand Institute* 24, 220–230.

Vélez, S., Fernández, R., and Giribet, G. (2014). A molecular phylogenetic approach to the New Zealand species of Enantiobuninae (Opiliones: Eupnoi: Neopilionidae). *Invertebrate Systematics* 28, 565–589.

Vink, C.J., and Dupérré, N. (2010). Pisauridae (Arachnida: Araneae). *Fauna of New Zealand* 64, 1–60.

Vink, C.J., Dupérré, N., and Malumbres-Olarte, J. (2013). Periegopidae (Arachnida: Araneae). *Fauna of New Zealand* 70, 1–41.

Vink, C.J., Mitchell, A.D., and Paterson, A.M. (2002). A molecular analysis of phylogenetic relationships of Australasian wolf spider genera (Araneae: Lycosidae). *Journal of Arachnology* 30, 227–237.

Vink, C.J., and Paterson, A.M. (2003). Combined molecular and morphological phylogenetic analyses of the New Zealand wolf spider genus *Anoteropsis* (Araneae: Lycosidae). *Molecular Phylogenetics and Evolution* 28, 576–587.

Volschenk, E.S., Burbidge, A.H., Durrant, B.J., and Harvey, M.S. (2010). Spatial distribution patterns of scorpions (Scorpiones) in the arid Pilbara region of Western Australia. *Records of the* Western Australian *Museum*, Supplement 78, 271–283.

Volschenk, E.S., Harvey, M.S., and Prendini, L. (2012). A new species of *Urodacus* (Scorpiones: Urodacidae) from Western Australia. *American Museum Novitates* 3748, 1–18.

Volschenk, E.S., and Prendini, L. (2008). *Aops oncodactylus*, gen. et sp. nov., the first troglobitic urodacid (Urodacidae: Scorpiones), with a re-assessment of cavernicolous, troglobitic and troglomorphic scorpions. *Invertebrate Systematics* 22, 235–257.

Walckenaer, C.A. (1805). *Tableau des Aranéides ou caractères essentiels des tribus, genres, families et races que renferme le genre Aranea de Linné, avec la désignation des espèces comprises dans chacune de ces divisions*. (Dentu: Paris.)

Walter, D.E., and Proctor, H.C. (2013). *Mites: Ecology, Evolution & Behaviour*, 2nd edn. (Springer: Dordrecht, Heidelberg, New York, London.)

Waters, J.M., and Craw, D. (2006). Goodbye Gondwana? New Zealand biogeography, geology, and the problem of circularity. *Systematic Biology* 55, 351–356.

Wegener, A. (1912). Die Herausbildung der Grossformen der Erdrinde (Kontinente und Ozeane), auf geophysikalischer Grundlage. *Petermanns Geographische Mitteilungen* 63, 185–195, 253–256, 305–309.

Wegener, A. (1915). *Die Entstehung der Kontinente und Ozeane*. (F. Vieweg: Braunschweig, Germany.)

White, A. (1849). Descriptions of apparently new species of Aptera from New Zealand. *Proceedings of the Zoological Society of London* 17, 3–6.

Womersley, H. (1956). A fossil mite (*Acronothrus ramus* n.sp.) from Cainozoic resin at Allendale, Victoria. *Proceedings of the Royal Society of Victoria* 69, 21–23.

Wood, H.M., Matzke, N.J., Gillespie, R.G., and Griswold, C.E. (2013). Treating fossils as terminal taxa in divergence time estimation reveals ancient vicariance patterns in the Palpimanoidea spiders. *Systematic Biology* 62, 264–284.

Yeates, D.K., Bouchard, P., and Monteith, G.B. (2002). Patterns and levels of endemism in the Australian Wet Tropics rainforest: Evidence from flightless insects. *Invertebrate Systematics* 16, 605–619.

Yeates, D.K., Harvey, M.S., and Austin, A.D. (2004). New estimates for terrestrial arthropod species-richness in Australia. *Records of the South Australian Museum*, Monograph Series 7, 231–241.

Yu, Y., Harris, A.J., and He, X. (2010). S-DIVA (Statistical Dispersal-Vicariance Analysis): A tool for inferring biogeographic histories. *Molecular Phylogenetics and Evolution* 56, 848–850.

Zhang, Z.-Q. (Ed.) (2013). Animal biodiversity: An outline of higher-level classification and survey of taxonomic richness (addenda 2013). *Zootaxa* 3703, 1–82.

11

Australasian Subterranean Biogeography

William F. Humphreys

CONTENTS

> They are found … in the dark recesses of caverns and of the waters under the earth, where no storm ruffles the everlasting stillness, no light illuminates the thick darkness, and no sound breaks the eternal silence.
>
> **– Charles Chilton (1894: p. 273), a pioneer student of groundwater fauna**

Introduction

'We did not during our voyage pass a more dull and uninteresting time' (Darwin 1839: 432). Charles Darwin was unimpressed with the natural history of Western Australia, a misconception gained during a brief visit by HMS *Beagle* to Albany in 1836, an area now recognised as a biodiversity hotspot. Nearly 180 years later a similar misconception is being revealed, in that Western Australia in particular, but Australia more broadly, is a biodiversity hotspot for the obligate subterranean species (Guzik *et al.* 2011a). Such species, typified as cave inhabitants, have long been recognised as important subjects in evolutionary studies – including by Darwin (1872) himself – by biospeleologists and increasingly by contemporary evolutionary biologists (Wilkens 2010; Leijs *et al.* 2012; Vergnon *et al.* 2013; Jeffery

2005). This awakening is surprising because Australia was long considered to be a poor prospect for those obligate subterranean species – troglobionts – that inhabit the subterranean voids.

This misconception was on account of the relative scarcity of carbonate karst in Australia and the widespread aridity of its climate compared with the better-studied and biospeleologically diverse areas of the Northern Hemisphere, especially the Dinaric karst. Globally, research on subterranean biology has overwhelmingly been centred on carbonate karst, in which solutional processes create an interconnected hydrogeological network. Such systems may expand to form caves by further solution or by collapse, and changes in hydrology may leave air-filled voids suitable for colonisation by lineages of terrestrial troglobionts, especially insects, myriapods and arachnids, while the flooded voids serve as a habitat for lineages of aquatic troglobionts, which overwhelmingly comprise crustaceans. The aridity of Australia was considered unfavourable for a diverse troglobiont fauna because abundant water was considered necessary both for cave formation and for the maintenance of humidity-dependent troglobionts. Finally, subterranean faunas were thought not to be favoured by small climatic variation – in the north, as one-third of the continent lies in the tropics, and in the south by the absence of widespread phases of glaciation during the Pleistocene – which was considered to be a major driver of troglobiont evolution in the Northern Hemisphere (Holsinger 2000). It is not surprising, therefore, that troglobiont species were initially recorded in the humid, cold south of Australasia, in Tasmania and New Zealand (Hurley 1990). Tasmania, now known to have a troglobiont diversity comparable to some areas in the Northern Hemisphere, is both humid (annual rainfall 1458–2690 mm) and the only area of Australia that was subjected to repeated episodes of glaciation during the Pleistocene. Moderately diverse troglobiont faunas were reported from the more humid areas of New South Wales, such as the Jenolan Caves (mean annual rainfall 968 mm) (Thurgate *et al.* 2001). In the 1980s, diverse troglobiont faunas were found in tropical Queensland (Chillagoe, rainfall 850 mm) (Howarth 1988; Howarth and Stone 1990) and in central Queensland (Undara lava tube, rainfall 789 mm) (Gray 1989; Hoch and Howarth 1989a–c). Since about 1990 it has been recognised that there is a great diversity of subterranean fauna in arid and semiarid parts of Australia (Guzik *et al.* 2011a), occurring in typical Tertiary karst areas such as the Nullarbor karst and Cape Range, as well as in alluvial aquifers (Humphreys 2000a). However, an ever-increasing biodiversity is being discovered in atypical substrates, such as groundwater calcretes in the Yilgarn (Humphreys 2008; Leijs *et al.* 2012) and the Northern Territory (Cho *et al.* 2006; Watts and Humphreys 2009), lacustrine and groundwater calcretes in the Pilbara (Poore and Humphreys 1998; Finston *et al.* 2009), and in the fractured rock and pisolites associated with Precambrian banded iron formations (BIFs) and their derivatives (Halse and Pearson 2014).

It is notable that the great diversity of troglobionts now known in Australia has emerged in the last two decades, in atypical habitats, many of which are in the tropics and/or arid areas, including Cape Range (median rainfall 210 mm), Barrow Island (median rainfall 280 mm) and Christmas Island (median rainfall 2011 mm), Pilbara (spatially averaged median rainfall 298 mm), and in the monsoonal Kimberley region (spatially averaged median rainfall – Northern, Central and Dampierland IBRA bioregions – 515–939 mm) (Humphreys 1991, 1993a, 1995; Humphreys *et al.* 2013; Humphreys and Eberhard 2001; Eberhard *et al.* 2005; reviews: Humphreys 2008, 2012).

Background to Chapter

A series of deep history events have been hypothesised to have significantly influenced the biogeography of the subterranean fauna of Australia. These fall into a number of discrete areas: the tectonic stability of the continent, the absence of widespread glaciation since the Permian, the latitudinal extent of Australia, the spreading aridity since the opening of the Drake Passage allowed the establishment of the Antarctic Circumpolar Current, the scarcity of karst relative to Laurasian continents, past connections with Gondwana and the close connection of the North West Shelf area with Tethys. These are summarised in the schema in Figure 11.1, which is built around our current understanding of the historical biogeography of the subterranean fauna (Humphreys 2012). As knowledge of the subterranean fauna of Australia is so incomplete, I want to set the discussion of the biogeography of the better-known higher taxonomic groups of subterranean fauna against this schema. This may provide a research framework against which more refined hypotheses may emerge, and which may itself be refuted in whole or in part as a consequence of such studies.

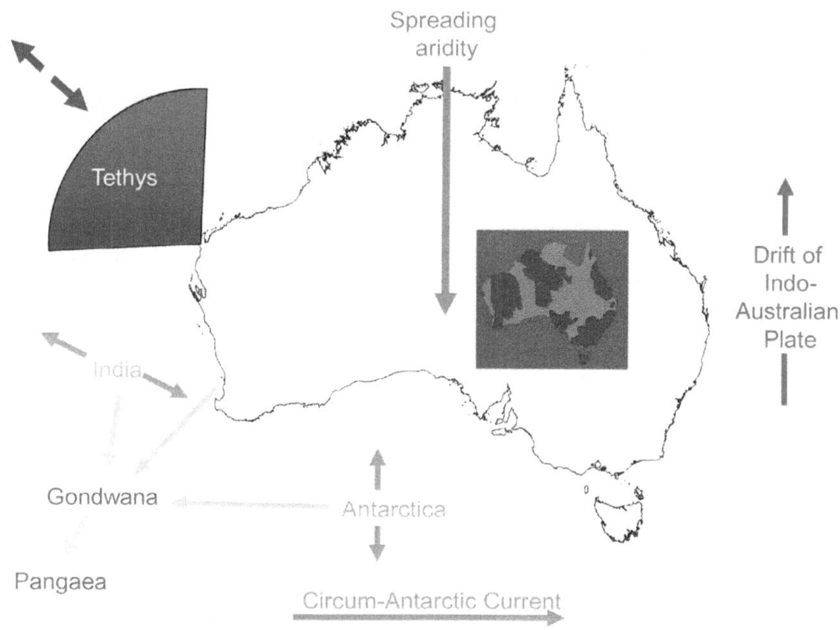

FIGURE 11.1 Schematic diagram depicting various effects resulting from plate tectonic and eustatic events over geological time that are hypothesised to have influenced the biogeography and evolution of the Australian subterranean fauna. The inset map depicts the extent of marine inundation during the Cretaceous. (After Humphreys, W.F., *Encyclopedia of Caves*, Academic Press, San Diego, CA, 2012.)

Characteristics of Obligate Subterranean Animals

Animals have different degrees of affinity with subterranean environments: in this Chapter I am going to deal particularly with obligate subterranean animals that inhabit air-filled or water-filled voids and which are termed *terrestrial troglobionts* and *aquatic troglobionts*, respectively. Owing to their evolutionary adaptations to the subterranean life, such species are incapable of existing in the surface environment in which the lineage evolved, and so they complete their entire life cycle in the subterranean realm. These evolutionary changes include a broad range of morphological, behavioural and physiological adaptations but they are manifest overtly by the characteristic reduction or loss of eyes, the reduction or loss of body pigment, the loss of wings where relevant, the elongation of appendages and the enhancement of nonoptic senses, or, in the case of minute interstitial lineages, by a vermiform body shape, the better to move within sediments (Langecker 2000; Coineau 2000).

Troglobionts were traditionally considered to be the inhabitants of caves – voids large enough for people to enter – largely in carbonate karst terranes where solution processes form a network of voids for colonisation by surface lineages. It was later recognised that subterranean habitats occur also in fissure systems, in lava fields and in nonconsolidated granular sediments (Sket 2008). As foreshadowed in the introduction, in Australia it seems that troglobionts may occur wherever interconnected voids of appropriate size and habitat (aquatic or air) are found, and that inland aquatic troglobionts – elsewhere known from freshwater – may occur in inland waters up to at least typical marine levels of salinity (Humphreys *et al.* 2009).

Troglobionts evolved from surface lineages, and so the integrated study of both the surface and subterranean members of the lineage is essential to understand the nature, degree and rate of the evolutionary changes leading to the troglobiont habit (Leijs and Watts 2008) and may provide a time frame useful for biogeographic analysis. This is commonly not possible in Australia, where surface ancestors are absent and thought often to have been ablated by surface aridity. However, this is not the particular subject of this chapter, which will be restricted to elucidating what is known of the biogeographical affinities of troglobionts that inhabit Australia.

Troglobionts and Biogeography

Troglobionts have evolved, ultimately, from surface lineages and become trapped by their evolution within a given geological context, after which their biogeography is at the whim of geological rather than biological processes and events. This is because troglobionts typically develop a suite of morphological, developmental, behavioural and physiological changes, termed *troglomorphies*, that are considered to be adaptive to their obligate subterranean life – although this is still a major research area and focus of debate – and which make them unable to survive in their ancestral surface habitats. Speciation may occur in situ of the troglobiont habitat through microallopatric vicariance as a result of the changes through time in the continuity of subterranean voids, which may become more integrated, for example, by solution effects in karst or fragmented owing to changing water tables, landscape evolution or infill by siltation and collapse. The net effect is that hypogean lineages commonly have very small geographic extent, what are termed *short-range endemics* – limited to a cave or a part of an aquifer – placing them at high risk in terms of conservation. As they are locked into their geological context they are powerful instruments in historical biogeography, the more so because subterranean habitats can be very persistent, with caves in the European Alps dated at least to the Miocene (Audra *et al.* 2007), while clay from the Jenolan Caves, New South Wales, has been dated to the Early Carboniferous (ca. 340 Ma; Osborne 2010). These processes are, however, dynamic and may not indicate the continuous presence of troglobiont habitat because, for example, the Devonian reef complex in the Kimberley was planed off by the Permian ice sheet, during which the giant grike landscape formed by water flow beneath the ice sheet and was subsequently infilled by sediments. The present cave systems are still being exposed by the erosion of these infill sediments (Playford *et al.* 2009). Consequently, many such habitats may have been present throughout the formation and dissolution of Pangaea and the fragmentation of Gondwana, but each case needs careful scrutiny. Aquifers, similarly, may persist through geological eras, and the Edwards Aquifer, Texas, for example, supports a diverse subterranean community derived from the Cretaceous marine inundation of the area (Holsinger and Longley 1980; Longley 1986).

As subterranean communities may comprise lineages isolated underground in different geological eras, they have the potential to yield information on past geological and climatic events in deep history – that is, to serve as 'living fossils' of certain provenance. This is particularly the case in Australia owing to the widespread tectonic stability and consequent integrity of the geomorphology of the continent, with its extensive shield regions long emergent from the sea. Key lineages are represented in those higher taxa, largely Crustacea, the entire membership of which is confined to subterranean freshwater systems – for example, Bathynellacea and Spelaeogriphacea, which are well represented in Australia (Humphreys 2012). Such taxa, isolated underground and producing no dispersive phase, have the potential to provide the most robust subjects for testing biogeographic hypotheses. Some of the lineages present in Australia's groundwaters are purported to have persisted through geological time (Humphreys 2000b,c; Wilson 2008). There are those whose origins are related to the former extent of Tethys and represented by the anchialine fauna of both northwestern Australia and Christmas Island. Several lineages from the former – remipede-type anchialine fauna (Humphreys and Danielopol 2006) – have congeneric species known elsewhere only from subterranean waters on either side of the North Atlantic, and the latter – procarid-type anchialine fauna – from comparable faunas on Hawaii, Ascension Island and Bermuda. Other lineages have affinities with Pangaea (e.g. Bathynellacea and crangonyctoid amphipods) and Gondwana (e.g. phreatoicidean isopods, Spelaeogriphacea and Candoninae ostracods).

Geological and Climatic Settings

Geologically, large parts of Australia have been stable, especially the *Western Shield* (Hocking *et al.* 1987), and lacking major tectonic uplift save in the eastern highlands and along the western margin with its gently folded anticlines of Tertiary carbonates in the northern Carnarvon Basin, with Cape Range and Barrow Island (Figure 11.2) notable for troglobionts. The Western Shield of Australia comprises the Pilbara and Yilgarn cratons and the related orogens, and lies at the western rim of the western plateau of

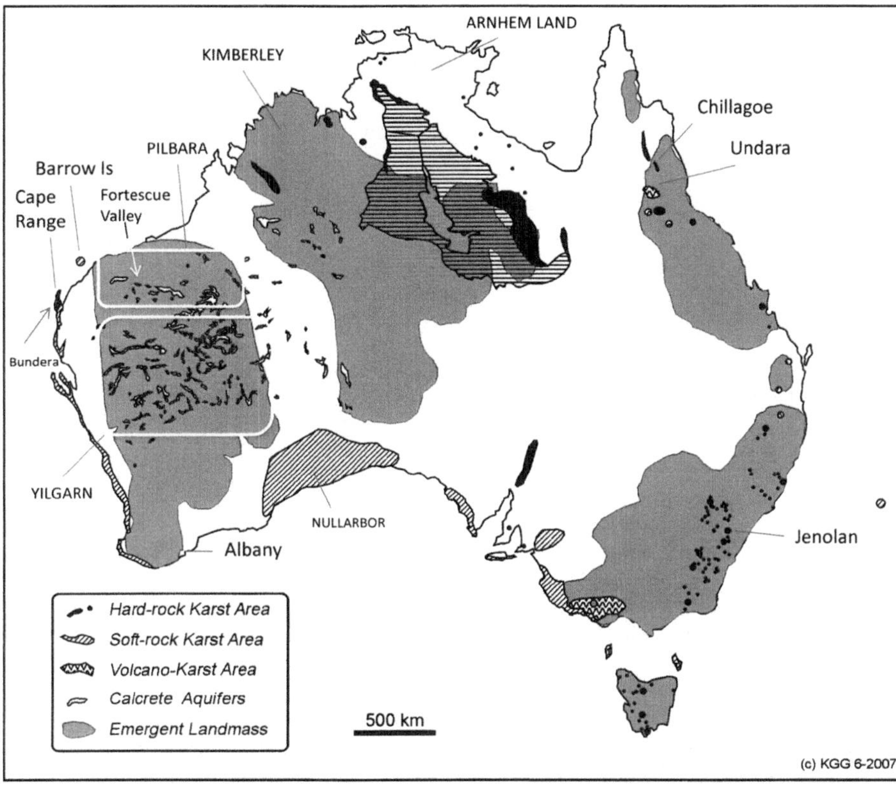

FIGURE 11.2 Australian karst and locations. Black and hashed areas denote karst area of differing types (inset key). The extensive shaded areas denote those plateau areas that were not inundated by the sea during the Cretaceous. The rectangle outlines in white denote the approximate extent of the Pilbara and Yilgarn regions as discussed in the text. (Developed from a base map drawn for me by Ken Grimes.)

Australia. Having been emergent above the sea since at least the Palaeocene, it ranks among the oldest non-marine landmasses on Earth. This must be interpreted carefully because the erosion surface means the landscape is younger (Vasconcelos *et al.* 2008) and may have had a very different geology and geomorphology at the relevant period to be considered. The degree of erosion is examined using the age and distribution of detrital zircons, the volume of deep basin deposits, and palaeothermal and present-day geothermal gradients. Such studies indicate that there has been massive erosion of the cratonic areas, where several kilometres of cover rock was removed from the Yilgarn Craton in the Permian to Early Cretaceous, and ~4.09 km of basement was removed from the Western Shield between the Early Ordovician and the end of the Cretaceous, an average rate of 8.87 m Ma^{-1} (0.009 mm $year^{-1}$). The minimum denudation rate for the Yilgarn Craton was 4.5–5.0 m Ma^{-1} in the Mesozoic and 1.5–2.0 m Ma^{-1} during the Late Cretaceous–Early Tertiary (reviewed by Kohn *et al.* 2002). However, owing to isostatic adjustment these massive bulk losses would have made little difference to the altitude of emergence of the landscape above sea level.

This landscape supports many of the higher taxa considered to be ancient fresh groundwater lineages, such as bathynellaceans, tainisopidean and phreatoicidean isopods, crangonyctoid amphipods and candonine ostracods (Bradbury 1999; Wilson and Johnson 1999; Humphreys 2001; Wilson 2001; I. Karanovic 2003). Denudation, combined with plate tectonic movement and climate change, means that these ancient Gondwanan lineages could have, and in some caves seem to have (Wilson 2008), evolved into troglobionts in a very different zoogeographic and ecophysiological context and in a geological and geomorphological context quite different from that prevailing today. Trapped in their subterranean realm, they may have predominantly moved vertically, up to several kilometres, through the landscape as

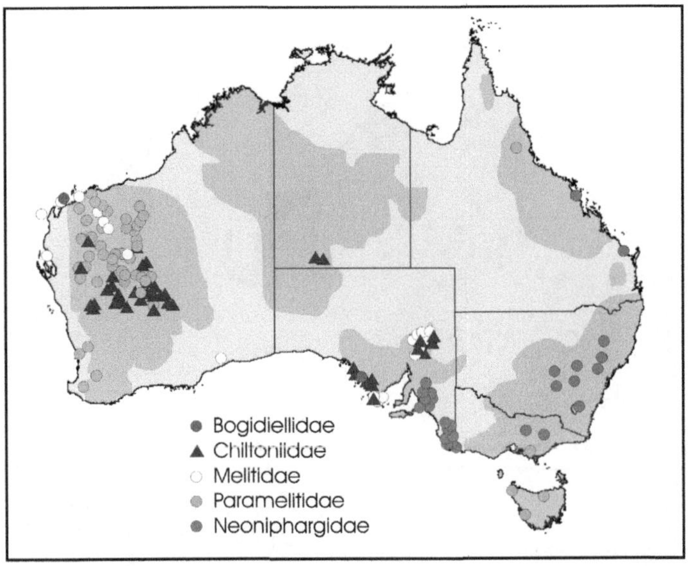

FIGURE 11.3 (See colour insert.) The distribution of five families of troglobiont amphipods overlain on a base map of Australia depicting the distribution of long emergent land areas (deeper shading). Map by Remko Leijs. (From Humphreys, W.F., *Encyclopedia of Caves*, Academic Press, San Diego, CA, 2012.)

the geology changed around them by erosion (surface ablation), solution (karstification; Humphreys and Adams 2001), accretion (groundwater calcretes and pisolites, Spelaeogriphacea; Poore and Humphreys 1998; Schizomida, Harvey *et al.* 2008) and deposition (alluvial interstitial; Boulton 2001), and been progressively isolated by landscape evolution (Humphreys and Adams 2001; Harvey *et al.* 2008).

These long-emergent areas are home to many ancient freshwater lineages that are dominated by subterranean crustaceans, including Bathynellacea, Spelaeogriphacea, phreatoicidean and tainisopidean isopods, crangonyctoid amphipods (Figure 11.3) and candonine ostracods (Bradbury and Williams 1997; Bradbury 1999; Wilson and Johnson 1999; Poore and Humphreys 1998, 2003; Humphreys 2001; Wilson 2001, 2003; I. Karanovic and Marmonier 2003). Species belonging to ancient freshwater lineages have also been recovered from Proterozoic Pentecost sandstone aquifers in the Kimberley (*Crenisopus*, Tainisopidae; Wilson and Keable 1999) and on small continental islands (Kimberley, Koolan Island; G.D.F. Wilson, pers. comm.), including Barrow Island, which has an anchialine system (Humphreys 2002; Cho *et al.* 2006; Humphreys *et al.* 2013). These lineages have clear Gondwanan connections, but the historical biogeography has yet to be supported using molecular phylogeographic analysis.

Intercontinental Connections

The crustacean order Spelaeogriphacea is known from only four species worldwide, all in freshwater, two of which occur in calcrete aquifers in the Fortescue Valley, Pilbara. The type species occurs in fractured sandstone on the top of Table Mountain, South Africa, and the final species occurs in karst in the western Mato Grosso, Brazil (Figure 11.4). The location of one species atop Table Mountain, a site that has not been near sea level since before the separation of Africa and South America, indicates a Gondwanan origin for this fauna (Poore and Humphreys 1998), but no molecular phylogenetic analysis has been conducted that may support this interpretation. The alternative hypothesis, suggested by the South American and Australian locations, could be dispersion through Tethys (Poore and Humphreys 2003), but this is contraindicated by the Table Mountain location.

Phreatoicidean isopod crustaceans live in freshwater, usually associated with groundwater, in India, South Africa, New Zealand and across southern Australia (and in tropical Arnhem Land and the Kimberley). This current Gondwanan distribution belies their origin, as fossils attributable to this order

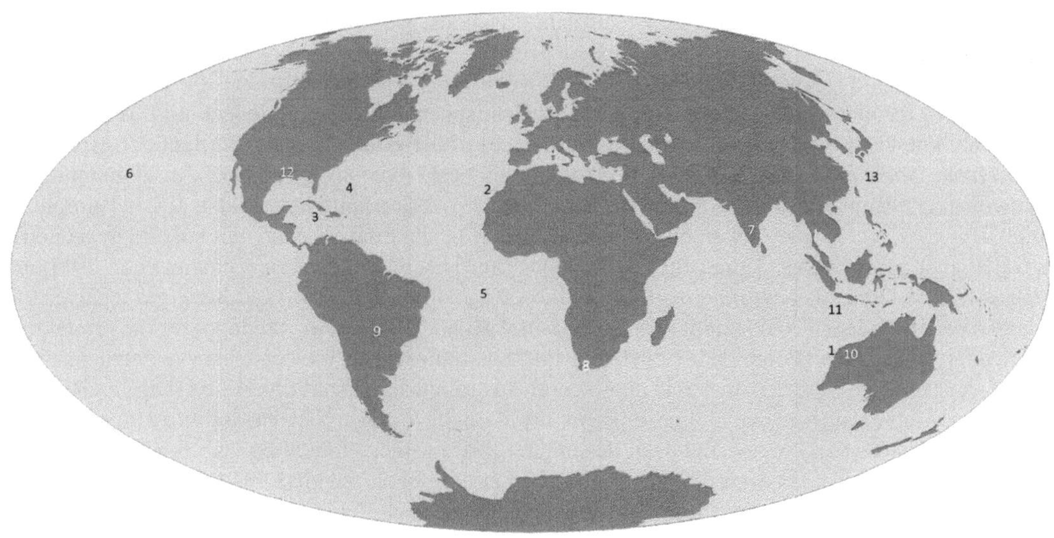

FIGURE 11.4 Worldwide locations mentioned in the text: (a) Cape Range + Bundera, Australia; (b) Canary Islands, Spain; (c) Caribbean; (d) Bermuda; (e) Ascension Island; (f) Hawaii, United States; (g) Andhra Pradesh, India; (h) Table Mountain, South Africa; (i) Mato Grosso del Sul, Brazil; (j) Fortescue Valley, Australia; (k) Christmas Island, Australia; (l) Edwards Aquifer, TX, United States; (m) Minamidaito-jima, Okinawa, Japan. (Base map adapted from http://www.freeworldmaps.net.)

are cosmopolitan. Their Australian distribution is strongly associated with the areas of the continent not submerged by Cretaceous seas, and about 10 species in eight genera are hypogean (cavernicolous or spring emergents). In the Kimberley the stygobiont genus *Crenisopus* is phylogenetically basal to most families in the Phreatoicidea, suggesting divergence after they entered freshwater but prior to the fragmentation of East Gondwana during the Mesozoic era, and so provides a link between African and Australasian lineages of phreatoicideans (Wilson and Keable 1999). Five hypogean species occur on the Precambrian Western Shield, where the family Hypsimetopidae is represented by the genera *Pilbarophreatoicus* in the Pilbara and *Hyperoedesipus* in the Yilgarn regions (separate cratons of the Western Shield). These genera are closely related to the hypogean genus *Nichollsia* found in the Ganges Valley of India and a new genus from caves of Andrah Pradesh (east-central India). Wilson and Keable (2001) contend that as *Nichollsia* is nested phylogenetically within the Australian Hypsimetopodinae, this troglobiont clade has a minimum age of 130 Ma, the timing of the separation of Greater India from Australia. They used this information to propose that a high conservation value should be placed on this group owing to its great phylogenetic age, endemism and localised diversity.

Tethyan Connections

Anchialine systems are inland groundwater near the ocean that are influenced by marine tides but have no surface connection to the ocean – subterranean estuaries – and which exhibit marked hydrogeochemical stratification with marine waters separated from the overlying freshwater by a marked density gradient and layer(s) of hydrogen sulphide. The seawater layer supports a diverse fauna of aquatic troglobionts predominantly comprising higher taxa of crustacean (classes, orders and families) that are endemic to anchialine habitats and typically comprise biogeographic and/or phylogenetic relicts. The general composition of anchialine fauna is predictable however far apart in the world they occur (mainly the Caribbean, the Canary Islands, Cape Range in northwestern Australia; Figure 11.4). The Australian anchialine fauna occurs on the coast fringing the North West Shelf – Cape Range, Barrow Island and the Pilbara coast south of the Fortescue River – but it is fully expressed outside the North Atlantic only in Bundera sinkhole in the Ningaloo Coast World Heritage Area. This fauna comprises atyids, thermosbaenaceans, hadziid amphipods, cirolanid isopods, remipeds, thaumatocypridid ostracods *Humphreysella* (formerly *Danielopolina*) and an array of copepods, including ridgewayid, epacteriscid and pseudocyclopiid calanoids, and speleophriid misophrioids. The fauna includes the only known representative of the class Remipedia in the Southern Hemisphere, and several genera, until reassigned in 2013, were known elsewhere from anchialine systems on either side of the North Atlantic (*Lasionectes*, *Halosbaena*, *Speleophria*, *Danielopolina*). At the larger scale there are a number of troglobiont higher taxa with major disjunctions in their distributions, indicative of vicariant events (Table 11.2 in Humphreys 2008). However, recent study has resolved some higher taxa to be Bundera endemics but closely related to the other anchialine lineages – namely, the remipede family Kumongidae (Hoenemann *et al.* 2013) and genera *Welesina* (thaumatocypridid ostracod; Iglikowska and Boxshall 2013) and *Bunderia* (epacteriscid calanoid copepod; Jaume and Humphreys 2001). *Halosbaena*, however, the predictor for the presence of Remipedia in Australia (Poore and Humphreys 1992), is indicated by molecular phylogenetic analysis, to be a relatively late (Eocene) arrival in continental Australia where it has diversified (Page *et al.* 2016).

Being troglobionts, these anchialine lineages are thought to have poor capability or opportunities for dispersal, particularly across the deep ocean, and this attribute, combined with their distributions, which closely match areas covered by the sea in the Late Mesozoic, suggest that their present distributions may have resulted by vicariance as a result of the movement of tectonic plates. The epicontinental Tethys spread westwards between the fragmenting Pangaean terranes (now the Mediterranean area) into the opening North Atlantic in the Jurassic (Figure 11.5), and it is suggested that anchialine precursors spread through and colonised the shores of these epicontinental seas before being separated by the development of the deep ocean. Thus, the current vicariant distributions have resulted through migration of the tectonic plates. Alternatively, the lineages may have dispersed broadly across Tethys before the closure of the Mediterranean in the Middle Miocene (ca. 14 Ma) constrained distribution. The timing of the diversification of these lineages is crucial to interpreting the current distribution of the anchialine faunas, independently of plate tectonic evidence, using a molecular clock approach. It is possible that both processes have been involved at various temporal and spatial scales, and that different lineages have

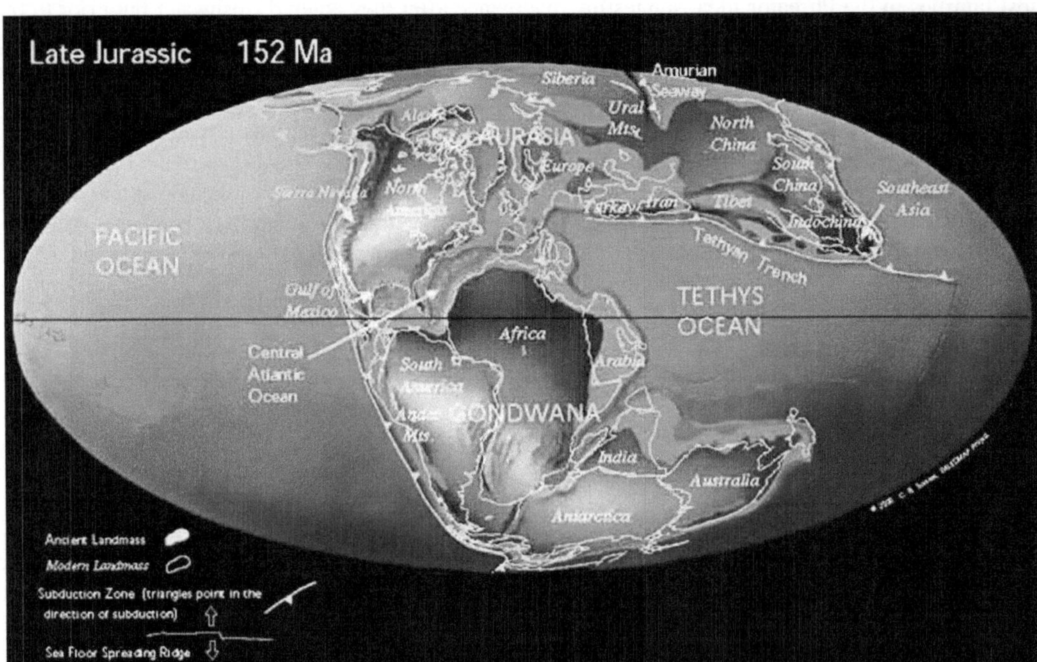

FIGURE 11.5 (See colour insert.) Position of tectonic plates in the Late Jurassic showing the link between northwest Australia, the Canary Islands and Caribbean regions through the westward expansion of Tethys into the developing Atlantic Ocean as Pangaea fragmented. (From Scotese, C.R., PALEOMAP, 2002. http://scotese.com/late1.htm. With permission.)

different origins, although this is belied by the coherent composition and habitat of anchialine faunas. At some stage the fauna became restricted to anchialine habitats, and the presence of a close fossil relative of thaumatocipridid *Welesina* in a crevicular context in Jurassic age rocks in the Czech Republic – then a northern arm of Tethys – suggests an early troglobiont transition in at least one element of the anchialine fauna. Conversely, the atyid shrimp genus *Stygiocaris*, endemic to the Australian anchialine system, is sister to the amphi-Atlantic genus *Typhlatya* (Page *et al.* 2008), from which it appears to have diverged in the Palaeogene (Botello *et al.* 2012) and thus suggests dispersal though the ocean. The present data are inadequate to test hypotheses on the origin of the anchialine fauna using molecular phylogenetics (von Rintelen *et al.* 2012; Botello *et al.* 2012; Hoenemann *et al.* 2013), and testing even the more straightforward hypothesis of trans–ocean basin plate tectonic vicariance (Atlantic, Bauzà-Ribot *et al.* 2012; Indian Ocean, Chakrabarty *et al.* 2012; Parenti and Ebach 2013) has been strongly contested (de Bruyn *et al.* 2013; Phillips *et al.* 2013) due to the claimed use of inappropriate molecular calibration.

While the epicontinental anchialine habitats are characterised by the remipedes, those occurring on isolated seamounts support a distinct anchialine fauna, characterised by the presence of the primitive shrimp *Procaris* (Procarididae) occurring with alpheid, hippolytid and atyid shrimps, an assemblage that is also predictable however far apart they occur. Seamount anchialine faunas are known from all tropical oceans, occurring on Hawaii, Christmas Island, Ascension Island and Bermuda. Recently, two elements of the epicontinental anchialine system have been found on isolated seamounts: the thaumatocypridid *Humphreysella* and the thermosbaenacean *Halosbaena* have been found on Christmas Island, and the latter also on Minamidaito-jima, Okinawa (Shimomura and Fujita 2009; Page *et al.* 2016), which, given the likely age of the islands, suggests the oceanic dispersal (Humphreys and Danielopol 2006) of these two iconic anchialine taxa, although there is no coherent theory to account for this, nor for the occurrence of the anchialine copepod *Speleophria* in the Nullarbor (Karanovic and Eberhard 2009).

Discordant Geological Setting

There are a number of troglobiont lineages that, on the basis of other evidence, inhabit geological settings that are discordant with their phylogenetic age, occurring in geomorphological settings that are too young. Among these are the spelaeogriphacean *Mangkurtu* and the remipedean *Kumonga*. *M. mityula* inhabits a freshwater aquifer in the Millstream Dolomite that probably formed in the Middle-to-Late Tertiary in the Fortescue Valley (Figure 11.4), the latter formed by the Late Jurassic in the Proterozoic Hamersley Group, including the Wittenoom Dolomite. The distribution of the four extant spelaeogriphaceans in South America, Africa and Australia (Figure 11.4) could have resulted from Gondwanan vicariance or Tethyan dispersal (see section Tethyan Connections in this chapter), but either theory would place *M. mityula* in a discordant geological setting. This discordance was explained by the continuity of the habitat with the underlying cavernous Wittenoom Dolomite of Proterozoic age, allowing the troglobiont fauna to migrate up through the developing geology (Poore and Humphreys 1998), but the denudation processes, discussed previously, would provide a means by which the fauna could also have migrated downwards through the landscape as it eroded, coming to occupy the de novo aquifer.

By contrast, *Kumonga*, which belongs to the monotypic crustacean family Kumongidae, is a basal member of the lineage of remipedia found in anchialine systems in parts of the former Tethys – the Canary Islands and some Caribbean islands on the North American Plate, especially the Bahamas – and is hypothesised to have dispersed either on the tectonic plates as Gondwana fragmented, or else through the Tethys sea as it spread into the developing North Atlantic in the Jurassic (Figures 11.4 and 11.5). The current geological setting in Tertiary carbonate karst is discordant with either hypothesis and suggests the lineage (Hoenemann *et al.* 2013) has moved through geological settings as they formed. That the fauna can move through developing landscape is supported by the nature of their habitat – anchialine systems occur where freshwater and seawater form a salinity-stratified subterranean estuary – which will move laterally and vertically to accommodate changing sea levels. The present habitat was only flooded by the sea in the Holocene and for most of the Pleistocene would have been many kilometres seawards and at up to 140 m lower altitude. However, the principle hypothesis of the origin of this fauna – which awaits testing using molecular phylogenetics – may require such migration from the Jurassic onwards.

New Zealand

The subterranean fauna of New Zealand is not well documented, but there are several biogeographic issues that can be usefully addressed using troglobiont lineages. There has been vigorous debate concerning the nature of the *Gondwanan* component of the New Zealand fauna. Waters and Craw (2006) call for rigorous evidence, independent of plate tectonics, that the New Zealand biota survived through the Oligocene to the present. They contend that there is no conclusive evidence for emergent terrestrial landscapes during the Cretaceous–Oligocene submergence of New Zealand, but there is compelling evidence that many purported Gondwanan elements of the biota arrived after the Oligocene, and so they conclude that dispersal may account for the present composition of New Zealand' s biota. Some subterranean lineages provide robust models to test such hypotheses (Humphreys 2008), especially some ancient lineages of obligate freshwater crustaceans such as the stygocarididian syncarids and phreatoicidean isopods, especially if the entire lineage has been troglobiont, like the order bathynellacean syncarids (Schminke 2011). The presence of all these lineages in New Zealand (Scarsbrook and Fenwick 2003), and several other Gondwanan terranes, indicates the presence of at least a freshwater lens within the New Zealand landmass, the lens itself indicative of emergent land allowing the rainfall recharge of the aquifer. The distribution of these lineages across the Gondwanan terranes may permit rigorous molecular phylogeographic interpretation independent of plate tectonic evidence.

Sampling Intensity and Biodiversity Measures

The measurement of biodiversity and the change in biodiversity with area is intricately related to the intensity and thoroughness of the sampling of the fauna (Iknayan *et al.* 2014). Troglobionts are difficult to sample and, where possible, they are difficult to sample thoroughly because most subterranean voids do not open to the surface (Curl 1966). On all other accounts as well, troglobiont fauna is problematic as access is sparse – cave openings or boreholes – and poorly distributed across the landscape. Furthermore, troglobiont species are typically rare, and so numerous sampling occasions are required to find a high proportion of the species at a site (Eberhard *et al.* 2009). In addition, most species are short-range endemics, which presents both taxonomic and sampling intensity issues owing to the high γ- and β-diversity; that is, the species composition of the fauna across the landscape differs over short distances, and the broader taxonomic composition of the troglobiont fauna changes between regions (Finston *et al.* 2007; Humphreys 2008, 2012).

The two best-reported regions are the Yilgarn and Pilbara regions, which are contiguous areas of the Western Shield, but, for several reasons, they are expected to yield considerably greater diversity when they have been adequately sampled. Firstly, the sampling density (no. sites area^{-1}) is low even within these better-sampled regions (about one site per 400–500 km^2, often with poor-quality access [Humphreys 2008]), while in large swathes of the deserts access to groundwater is entirely lacking even where there is a highly prospective troglobiont habitat, such as groundwater calcrete. There are only a few published accounts of intensive sampling of small areas (Guzik *et al.* 2009; Abrams *et al.* 2012; Bradford *et al.* 2013), but there are numerous unpublished site-specific environmental review and management reports, although these typically lack formal taxonomic content, as in the published paper by Halse and Pearson (2014).

Nonetheless, in just over three decades the perception of Australia has changed from having a relative poverty of troglobionts to arguably among the most diverse troglobiont faunas in the world. This is attributed to the accumulation of taxa enabled by the exceptional tectonic stability of the continent, the range of climates and, especially, the range of subterranean habitats present. These include such novelties as groundwater calcretes and pisolites as well as typical Tertiary orogenic karst and anchialine systems (Humphreys 2012), and a range of fresh and saline groundwaters. Guzik *et al.* (2011a) considered that 960 species of noncrustacean terrestrial troglobionts await discovery, mainly in the Western Shield, but this is likely to be a considerable underestimate even for the Pilbara alone, as indicated by a detailed sample of just ~1% of the area of the Pilbara (Halse and Pearson 2014). Similarly, a single calcrete in the Yilgarn, intensively sampled for the proposed development of a uranium mine, yielded 22 morphospecies of copepods, 70% of the previously recorded copepod diversity of the region (Karanovic and Cooper 2012).

Bathynellacea

Bathynellacea are typically minute (<1 mm) obligate freshwater interstitial syncarid crustaceans with a global distribution and considered to have established in the groundwater habitat across Pangaea (Schminke 2011). Some genera have global connections and are thought to have been dispersed by means of plate tectonics, although none of these intercontinental genera has yet been confirmed using molecular phylogenetic methods. The Australian bathynellid fauna is diverse but is still poorly sampled, as is the case everywhere save the Iberian Peninsula (Camacho *et al.* 2014). The order is best known in Australia from the family Parabathynellidae, some species of which have unusual characteristics for this family in that, rather than inhabiting the interstitial habitat, they swim freely in groundwater, are very large (at a body length of 6.3 mm, *Billibathynella humphreysi* Cho 2005 is globally the largest species) or occur near playas in groundwater calcrete aquifers in water with the composition and salinity of seawater (Cho and Humphreys 2010), such as the sympatric *Brevisomabathynella clayi* and *B. uramurdahensis* near Lake Way. Of particular interest is the morphologically well-characterised Gondwanan genus *Atopobathynella* that is widespread in Australasia (Western Australia, Northern Territory, Tasmania and New Zealand), and also found in India, Madagascar and South America. Other genera also have intercontinental distributions (e.g. *Notobathynella*, *Hexabathynella* and *Chilibathynella*), but some genera are endemic to Australia (e.g. *Kimberleybathynella*, *Billibathynella*, *Brevisomabathynella*, *Octobathynella* and other genera are recognised informally from molecular research; Abrams *et al.* 2012). Globally, most parabathynellids are known only from their type locality, and as such they are typically endemic to very small areas (Schminke 2011), although in the Iberian Peninsula, the only region thoroughly sampled shows that many species do have broader ranges (Camacho *et al.* 2014). Parabathynellid species occur in specific bodies of groundwater calcrete in Western Australia (Guzik *et al.* 2008; Abrams *et al.* 2012), and they may have very limited gene flow over short distances, even within aquifers (Asmyhr *et al.* 2014), as reported in other Australian arid-zone lineages of both aquatic and terrestrial troglobionts (diving beetles, amphipods and isopods: Guzik *et al.* 2011b; isopods: Cooper *et al.* 2008; diving beetles: Guzik *et al.* 2009; amphipods: King *et al.* 2012; Bradford *et al.* 2013; millipedes: Humphreys and Adams 2001; Schizomida: Adams and Humphreys 1993; Harvey *et al.* 2008).

Ostracoda

Globally, only about 10% of ostracod species have intercontinental distributions, and 94% of species and 60% of genera are known from only one zoogeographic region (Martens *et al.* 2007). In Australia 92% of ostracod species are endemic, but they remain poorly studied and have the lowest continental diversity (I. Karanovic 2012). In arid Western Australia ostracods are more diverse in subterranean waters than in surface waters and the diversity of Candoninae ostracods in groundwater of the arid Pilbara region is greater than that in Lake Baikal (I. Karanovic 2007), a recognised biodiversity hotspot in the largest freshwater lake. Many species of troglobiont ostracods are endemic to small karst regions and the endemism may occur at the generic level, indicative of an extended period of isolation of these lineages and great phylogenetic age. Suprageneric endemism is rare, however, being restricted to several tribes of Candoninae that occur widely across the Gondwanan terrane (Table 11.1). Remarkably, two of the three tribes of Candoninae (12 of 13 genera) are endemic to the Pilbara, with only one genus being in common with the contiguous Yilgarn region. The Candoniinae, as do the Timiriaseviinae (as *Gomphodella*), indicate the close Tethyan connections of the Australian subterranean fauna (I. Karanovic 2009). Although water chemistry is presumed to be a major factor in ostracod ecology (Radke *et al.* 2003), the distribution of species of candonine ostracods in the Pilbara shows scant relationship with water chemistry (Reeves *et al.* 2007).

TABLE 11.1

Number of Genera in Each of Three Tribes of Candoninae (Candonidae: Podocopida) from the Western Shield of Australia and Their Continental Distribution

Tribe[a]	Australia (Pilbara) (*n*)	South America (*n*)	Africa (*n*)	Pilbara Endemic Genera (%)	Species (*n*)
Candonopsini	2 (1)	2	0	50	42
Danielocandonini	3 (3)	1	1	100	24
Humphreyscandonini	8 (8)	0	0	100	58
Total	13 (12)	3	1	92.3	124

Source: Humphreys, W.F., *Invertebrate Systematics*, 22, 85–101, 2008; Karanovic, I., *Subterranean Biology*, 2, 91–108, 2004; Karanovic, I., *New Zealand Journal of Marine and Freshwater Research*, 39, 29–75, 2005; pers. comm.

Note: Most genera are endemic to the northern part of the Western Shield and these Pilbara endemics are denoted by brackets. The continental distribution of these troglobiont ostracods is indicative of the Gondwanan affinities of these ancient lineages.

[a] The Indian fauna known to date has Holarctic affinities (Karanovic, I., and Ranga Reddy, Y., *Crustaceana*, 81, 861–871, 2008.)

Copepods and the Pulsating Desert Hypothesis

The Western Shield of Australia, which includes the Pilbara and Yilgarn cratons and intervening orogens, is among the oldest emergent landscapes on earth, having been above sea level since the Proterozoic (Figure 11.1) and where the regolith has been in situ since at least the Mesozoic (Bird and Chivas 1988). Consequently, it is surprising that 39% of the species considered by T. Karanovic (2006) belong to genera with clear marine origins, some genera being recent colonisers (*Halicyclops*, *Phyllopodopsyllus* and *Schizopera*), while others have distributions consistent with plate tectonic vicariance associated with the westward expansion of Tethys in the Jurassic (*Parapseudoleptomesochra*, *Archinitocrella*, *Abnitocrella*, *Stygonitocrella* and *Pseudectinosoma*), a hypothesis also suggested in relation to the anchialine fauna of Cape Range (see section Tethyan Connections in this chapter). In addition, the genera *Diacyclops*, *Metacyclops* and *Psammocyclops* represent a strong Eastern Gondwana signal in the Pilbara freshwater copepods.

T. Karanovic (2004, 2006) conducted sequential studies of the groundwater copepods of the Yilgarn and Pilbara regions of the Western Shield and noted the high proportion of species and genera endemic to each region (Table 11.2), with few occurring in both of these contiguous regions (4%, $n = 70$, and 21%, $n = 31$, respectively); those species found in both regions are also found more broadly in Australia or are cosmopolitan. He considered the exceptionally rich subterranean fauna to be the result of a long period of accumulation of species starting after the Permo-Carboniferous glaciation, the last continental glaciation of Australia. He hypothesised that some genera, such as *Parastenocaris* and *Allocyclops*, are likely to have invaded subterranean waters in the Jurassic because they lack both marine relatives and more recent connections with other continents, making it likely that they dispersed through plate tectonic movement. Conversely, some other species, having close surface relatives, are clearly of more recent origin.

T. Karanovic (2010) comments on the strange disjunction in the copepod assemblage between the contiguous Pilbara and Yilgarn regions (repeated in other taxa; see Table 11.3), especially as each of these regions has clear affinities with noncontiguous areas far away to the north (Kimberley), south (southwestern Australia) and east (Pioneer Valley, central Queensland). Like Giribet and Edgecombe (2006), he emphasised the importance of small (presumably subcontinental) scale patterns when inferring Gondwanan biogeography because different regions of Australia have different affinities with Gondwanan faunas, a situation he found unexceptional given the age of the Australian landscape. He is, however, silent on the probable importance of the temporarily distinct separation histories of India and Antarctica from Australia, information that may help to resolve the causes of these, to date, largely inexplicable distributions.

TABLE 11.2

Distribution of Australian Copepod Genera across Gondwanan Terranes

Genus	Family	Species (n)(N = 171)	Africa	Madagascar	South America	Australian Region	India
Kinnecaris	Parastenocarididae	25	●	●		●	●
Attheyella (subgenus *Chappuisiella*)	Canthocamptidae	32			●	●	
Attheyella (subgenus *Delachauxiella*)	Canthocamptidae	42			●	●	
Haplocyclops	Cyclopidae	8	●	●	●	●	●
Diacyclops (*alticola* group)	Cyclopidae	6				●	●
Diacyclops (*michaelseni* group)	Cyclopidae	6	●			●	
Boeckella	Centropagidae	52	●			●	

Source: Data from Tomislav Karanovic, personal communication.
Note: Black-Harpacticoida; dark grey-Cyclopoida; light grey-Calanoida.

Distributional Hypotheses

Pulsating Desert Hypothesis

The hiatus in the distribution of lineages between the Pilbara and Yilgarn on the Western Shield, evident in the Karanovics' studies and formally addressed by T. Karanovic in his *pulsating desert* hypothesis, is presented for a range of troglobiont lineages studied in Table 11.3.

T. Karanovic (2006) proposed his pulsating desert hypothesis to explain the strange affinities and disjunctions recorded in the occurrence of troglobiont copepods between different regions of Australia (T. Karanovic and Cooper 2011). Namely, he advocated that the wide fluctuations in aridity (implied due to the circum-Antarctic ocean flow following the opening of the Tasmanian Gateway in the Late Eocene; Scher and Martin 2006) caused the area of deserts to expand and contract. This process resulted in the extinction of the fauna in the centre during peak aridity and reinvasion from the north and south during more humid times, and explains the diversity and disjunction in the distribution of the subterranean copepods (Box 11.1). This is the only hypothesis that has been proposed to account for the faunistic disjunction between the Pilbara and Yilgarn, but it is inconsistent with the evidence, especially given the almost complete separation of the adjacent faunas. First, Karanovic anchored the meeting place of this climatic convergence on the Tropic of Capricorn, the relative position of which would have moved southwards from about north Kimberley to its present position since the mid-Miocene (calculated from Dyksterhuis *et al.* 2005) as Australia drifted northwards towards the intertropical convergence zone. Thus, there has been no prospect for a consistent response to climate change to remain anchored on the Pilbara through prolonged periods, thus permitting the accumulation of the high species diversity. Second, it is inconsistent with his proposition that the high copepod diversity is due to the stable landscape allowing species to accumulate since the Permian glaciation, a period predominantly before the onset of aridity in the Tertiary. Third, troglobionts generally are not able to disperse widely to take advantage of changing climate as they are entrapped by their evolutionary adaptations, to the extent that their subterranean matrix is interconnected, and by their low capacity for dispersion in the subterranean environment (e.g. Humphreys and Adams 2001; Harvey *et al.* 2008; Guzik *et al.* 2011b; Asmyhr *et al.* 2012; T. Karanovic and Cooper 2012; Bradford *et al.* 2013; cf. Eme *et al.* 2013).

TABLE 11.3

Distribution of Some of the Better-Known Higher Taxa of Troglobiont Invertebrates between the Adjacent Pilbara and Yilgarn Regions on the Australian Western Shield

Aquatic Troglobionts	Taxon	Yilgarn (*n*)	Pilbara (*n*)	Overlap (%)
Copepoda	Species	30	43	4
Copepoda	Genera	15	25	21
Podocopoda: Candoninae: Candonidae	Species	5	58	0
Podocopoda: Candoninae: Candonidae	Genera	1	13	8
Isopoda: Tainisopidea	Species	1[a]	5	17
Coleoptera: Dytiscidae	Species	89	1	1.1
Spelaeogriphacea	Species	0	2	0
Terrestrial Troglobionts				
Blattodea: Nocticolidae	Species	0	9	0
Arachnida: Schizomida	Species	0	26	0
Total Species		125	144	0.7
Total Genera		16	38	14.5

Source: Humphreys, W.F., *Encyclopedia of Caves*, Academic Press, San Diego, CA, 2012; Watts, C.H.S., and McRae, J., *Records of the Western Australian Museum*, 28, 141–143, 2013.

[a] Probably by drainage capture.

BOX 11.1 PULSATING DESERT HYPOTHESIS

This hypothesis assumes that the Tropic of Capricorn was a long-term barrier between the two regions. It is almost a general knowledge now that there have been past episodes of very severe aridity in Australia. During these arid phases the real desert would have spread westwards from the central part of the continent, perhaps wiping out most of the stygofauna and forcing the rest to retreat, first towards the coast and then either northwards or southwards. During the periods of increased humidity the regions would be repopulated and two different faunas would meet around the Tropic of Capricorn. Then another cycle would begin and these faunas really did not have a chance to develop any stronger connections. (T. Karanovic 2006: 230)

Fourth, as seen previously, aridity and troglobiont diversity are not incompatible, and true desertification does not necessarily eliminate troglobionts as both aquatic and terrestrial troglobionts persist in areas covered by Pleistocene red dune systems in Cape Range (Humphreys 2000c). Fifth, there is now considerable evidence from molecular phylogenetic work for several lineages of disparate taxonomic and ecological groups that the development of aridity – biotic implication reviewed by Byrne *et al.* (2008) – has been a driver of the exploitation of subterranean voids and may have facilitated troglobiont adaptation by the elimination of surface ancestors (Leys *et al.* 2003; Leys and Watts 2008). Nonetheless, there is no competing hypothesis to account for the extraordinary troglobiont faunal disjunction between the contiguous Pilbara and Yilgarn regions (Table 11.3) that would be consistent with the geological interpretation that the entire Western Shield has been a continuously emergent landmass since the Proterozoic (Hocking *et al.* 1987), albeit with substantial marine incursions along the coasts and up the palaeovalleys in the Cretaceous and Eocene.

Headwater Isolation and Subterranean Island Models

Based on the environmental conditions found in the Pilbara (Eberhard *et al.* 2005), Humphreys (2001) hypothesised that isolated populations of aquatic troglobionts would occur in headwater tributaries; he reasoned that there would have been reduced flow and consequent upstream progression of salinity as aridity intensified through the Tertiary and confined the fauna in the headwaters, which I will here name the *headwater isolation* model. The model was developed for the Pilbara, where the rivers are still active, if episodic. It has been found that troglobiont amphipods do indeed occur as isolated populations in headwater tributaries, and molecular phylogenies (Finston *et al.* 2007) are consistent with the proposed headwater isolation model, although they are not informative of the processes by which such consistency arose.

The headwater isolation model was, however, first tested, and found wanting, in the northern Yilgarn, the region contiguous with and south of the Pilbara, with a mix of coastal and inland drainages with respectively episodic rivers and palaeodrainages largely lacking flow. In this region the troglobiont habitat comprises isolated groundwater calcretes near salt lakes (playas) spread along the palaeovalleys (Humphreys 2001). The model was appraised against data from the aquatic troglobiont communities that inhabit phreatic calcretes that are deposited from the groundwater flow near salt lakes (Humphreys 2001). Analysis of the molecular phylogeny of all the epigean and diverse troglobiont diving beetle fauna (Dytiscidae) in Australia showed that the distribution of the troglobiont beetles does not map the drainage pattern but rather supports a model of multiple independent invasions of calcretes across a wide region (Leys *et al.* 2003). This model is also applicable to the parabathynellids, amphipods and isopods inhabiting the calcrete aquifers (Cooper *et al.* 2007, 2008; Abrams *et al.* 2012; Guzik *et al.* 2008, 2009; Watts and Humphreys 2009). Thus, although the Yilgarn troglobiont fauna is inconsistent with the headwater isolation model, this research established the *groundwater island* model, whereby each isolated calcrete contains a troglobiont fauna endemic to that calcrete. The model was originally established for aquatic troglobionts (Cooper *et al.* 2002) but is now being extended to terrestrial troglobionts (Javidkar *et al.* 2016.).

Aridity as a Driver of Subterranean Colonisation

A consequence of the northern movement of the Indo-Australian Plate during the Tertiary was the loss of the widespread moist forests that covered Australia and the spreading desertification of the continent reaching peaks in intensity during glacial maxima and fluctuating widely in extent during the Pleistocene glacial–interglacial cycles. This process had profound influence on the assembly of the entire arid-zone biota of Australia (Morton *et al.* 1995; Byrne *et al.* 2008) and the hypothesis that aridity has been a driver for the colonisation of subterranean spaces by surface lineages, and their subsequent isolation and speciation in caves has had a long history in Australia, being invoked early on for cave colonisation in the Nullarbor and even in humid Tasmania (Eberhard and Humphreys 2003). The development of aridity was proposed at a landscape scale to account for the formation in Cape Range of genetic 'provinces' within mostly humid forest lineages of terrestrial troglobionts – belonging to Diplopoda, Schizomida and Crustacea – for which no close relatives of the cave fauna have persisted on the surface. It is hypothesised that the fauna colonised caves as the humid vegetation was lost from the range but retained some continuity though karst conduits in the cavernous Tulki Limestone. As aridity deepened, the humid vegetation was lost from the deep gorges and cut both subterranean and forest dispersal routes because the gorges cut into a noncavernous underlying marly Mandu calcarenite (Humphreys 1993b; Humphreys and Adams 2001).

More recently, consistent evidence that the colonisation of subterranean habitats is associated with the development of aridity has come from studies on aquatic troglobionts in the groundwater calcretes of the Yilgarn region in various lineages of invertebrates (see the previous section; for a review, see Juan *et al.* 2010).

Pleistocene Effects in Periglacial Areas

The Tasmanian cave fauna has distribution patterns in periglacial areas comparable to those found in Europe, North America and New Zealand. In Australia, unlike Northern Hemisphere continents, widespread coverage by ice sheets has not occurred since the Permian. During the Pleistocene in Australia the temperature effects of climate were most severe in Tasmania, where ice cover fluctuated around areas at higher altitude, with some glacier flow towards the lowlands. Several studies have linked the biogeography and diversity of cave faunas of Tasmania to the consequences of the climatic fluctuations during the Pleistocene – namely, the ebb and flow of ice fields in the Tasmanian high country and regional and altitudinal shifts in vegetation zones and humidity zones. Hunt (1990) showed that *Hickmanoxyomma*, a Tasmanian endemic genus of harvestmen (Opiliones), displayed clear morphological separation and regional of endemicity, and hypothesised that the speciation, distribution and expression of troglomorphisms were related to Pleistocene climatic fluctuations, particularly glaciation. While this hypothesis is consistent with the traditional Northern Hemisphere paradigm of Pleistocene-driven subterranean speciation, more recent work, both in Australia and Europe, has placed the roots of many subterranean lineages much earlier, to the late Miocene or before (see section Troglobionts and Biogeography in this chapter).

Although carabid beetles form a major component of subterranean faunas in the Northern Hemisphere (Eberhard and Giachino 2011), they are generally sparse in karst on the Australian mainland save for two highly troglomorphic species in the Nullarbor (Moore 1995), and species of Zuphini and Anilini have recently been found in BIFs in the Pilbara. In contrast, carabids are very diverse in Tasmania, where there are over 300 discrete karst areas, many impounded. There, the Trechinae are represented by 76 species in sixteen genera, of which 17 species in four genera are troglobionts, each endemic to specific karst areas (Eberhard and Giachino 2011). The presence of congeneric sympatric species exhibiting different stages of troglomorphy is proposed as evidence of the heterochronic colonisations of the karsts from adjacent forests by troglophilic lineages during the Pleistocene interglacials, followed by forest retreat isolating the subterranean populations and facilitating troglogenesis (Eberhard and Giachino 2011). Thus, considering the spatial changes in ice cover in Tasmania through the late Pleistocene, and as advocated earlier (Moore 1965), they follow the Northern Hemisphere paradigm, in which the Pleistocene climate changes resulted in the current species diversity and distribution of the cavernicolous Trechini (Figure 11.6).

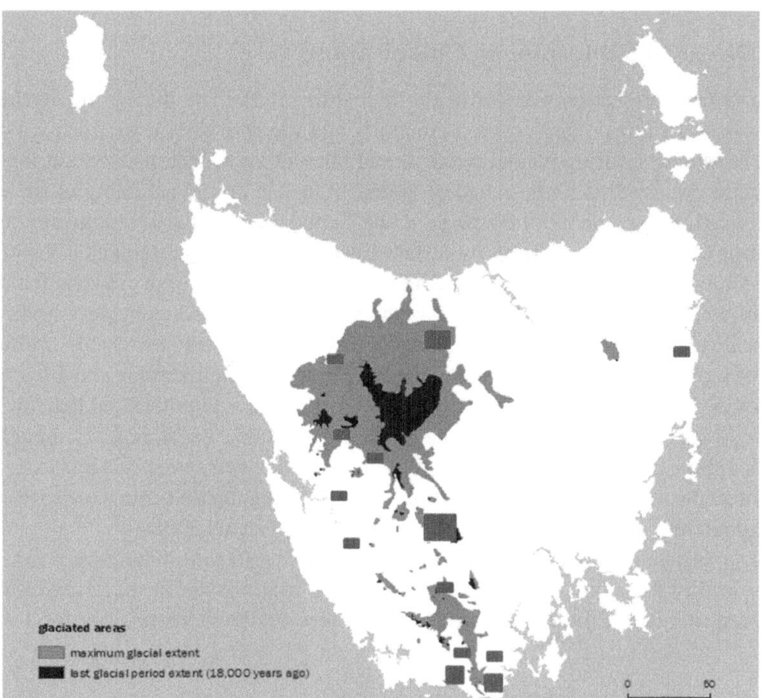

FIGURE 11.6 The extent of glaciation in Tasmania and the distribution of troglobitic Trechini and Zolini carabid beetles of the genera (no. species) *Goedetrechus* (6), *Tasmanotrechus* (4), *Tasmanorites* (1), *Idacarabus* (4) and *Pterocyrtus* (2). The size of squares denotes the number of species at a site (1–4). (Derived from Eberhard, S., and Giachino, P.M., *Subterranean Biology*, 9, 1–72, 2011. Base map of glacial extent from http://soer.justice.tas.gov.au/2009/image/162/index.php.)

The two studies of Tasmanian troglomorphic lineages, while of great interest, lack both phylogenies and any temporal context that would better able the testing of the proposed hypothesis. A molecular phylogeographic study of both these lineages would help unravel the apparent differences between the origins of the cavernicolous faunas of southeastern Australia and the apparent great age of origin of those elsewhere on the continent. The origin of troglogenesis starting in the Quaternary and driven by Pleistocene glacial cycles was a long-held interpretation in the Northern Hemisphere, especially for trechine beetles, which are such a prominent and well-studied component of the troglobiont fauna of the Europe. However, this interpretation is increasingly being shown to be invalid and the morphological interpretation of the relationships between trechines (and other troglobionts, both aquatic and terrestrial) is masked by the convergent morphologies exhibited by troglobionts. Faille *et al.* (2013 and earlier papers therein) have demonstrated, using molecular phylogenies, that most clades of carabid beetles belonging to the tribe Trechini, in the European Alps and the Pyrenean and Dinaric karsts, started their subterranean isolation at various phases since the Oligocene (possibly the Eocene) but that the trogloneogenesis continued into the Plio-Pleistocene. This time frame is in accord with studies in other systems in Australia (Leijs *et al.* 2003) and the resolution of the evolutionary timing of troglogenesis in the Tasmanian trechines must await molecular studies. It is notable that oscillating areas of refugia driven by Pleistocene climatic fluctuations are hypothesised to have resulted in the rapid radiation of a species flock in the epigean genus *Sternopriscus* (Dytiscidae: Hydroporini) in the freshwaters of southeastern Australia (Hawlitschek *et al.* 2012); it would be very interesting if the trogloneogenesis of the Tasmanian carabids and harvestmen proves to be so recent because, given their much greater degree of morphological divergence than that found in *Sternopriscus*, it would suggest a very rapid rate of evolution. However, the divergence of evolutionary rates between surface and troglobiont lineages, or the high rates of evolution in subterranean habitats of both morphological and genetic characters, has been proposed previously on both theoretical and evidential grounds (Rétaux and Casane 2013; Rohner *et al.* 2013).

Concluding Remarks

Patchy sample coverage, incomplete sampling in the areas sampled and an undeveloped taxonomy all impede the development of a robust biogeographical understanding of the subterranean fauna, and some approaches to resolve these limitations follow. Most of the sampling for troglobionts has been conducted on the shield regions where ancient relictual faunas may be expected, but there has been minimal work in the areas inundated by the major flooding events in the Cenomanian (ca. 100 mya) and the Eocene (Figure 11.2). This is surprising as there is a rich body of literature and terminology on the origin of aquatic troglobionts from coastal invasion and, especially, due to stranding inland by eustatic changes (Boutin and Coineau 2000). The latter, particularly, can provide independent time markers to complement those derived from molecular phylogenetic methods, although the methodology requires the selection of the appropriate lineages to derive defensible analyses (Phillips *et al.* 2013). Information is needed on whether these 'lowland' areas contain suitable habitats to support troglobiont fauna and, if so, whether fauna is present and what the affinities of its members are. To delineate regions where troglobiont fauna really could be absent it is important that sampled sites and their characteristics be recorded even if no troglobionts are present.

In addition to eye and pigment loss – the overt troglomorphies commonly mentioned – troglobionts have highly convergent morphological characters which often obscure the relationship between species, even in the best-studied faunas (Faille *et al.* 2013). In practice it has often been found necessary to undertake molecular work on troglobionts as a prelude to identifying suitable taxonomic characters, as well as to establish the phylogenetic relationships (Cooper *et al.* 2008; Guzik *et al.* 2008; Abrams *et al.* 2012). Indeed, molecular methods are commonly used to indicate the presence of both described and undescribed species during the environmental management assessment of major mining projects in the absence of an adequate, even any, taxonomic base for many lineages. Molecular methods could also be used on a regional scale to ascertain the diversity and affinities of troglobionts in the broad areas yet to be sampled and for which a complete taxonomic base for troglobionts is unlikely to be feasible, owing both to convergence and a lack of taxonomic resources.

Finally, investment in the systematics of certain higher taxa that are especially well represented and/or informative, and relatively easy to collect, would facilitate progress on understanding biogeographic relationships, albeit on a narrower path that may mask important information. Among aquatic troglobionts, amphipods and copepods would be good targets on account of both their diversity and ubiquity in subterranean habitats, and the likely wide range of time since they transitioned to subterranean life. Troglobiont copepods have the disadvantage of very small size. Terrestrial troglobionts are more problematic to recommend as the distribution of higher taxa is more varied regionally. Oniscidean isopods are ubiquitous, diverse, narrow-range endemics (M. Javidkar pers. comm., University of Adelaide) and relatively simple to collect, whereas true spiders (Araneae) are ubiquitous but are less easily collected in nontraditional subterranean habitats.

Despite the immaturity of biogeographic studies on Australian troglobionts, where research has been conducted in ample depth and to a sufficient spatial extent, it is clear that troglobiont faunas have considerable untapped potential for the investigation of the timing and origin of broad elements of the fauna of Australia. This arises because of the mix of the attributes of subterranean animals, the geomorphological context in which they are found and the age and stability of the continent. The fauna itself is diverse – arguably the phylogenetically most diverse troglobiont fauna globally – on account of the age of the fauna and its varied intercontinental connections, and yet there are endemic higher taxa (to suborder) restricted to groundwater. Additionally, the continent covers a broad range of climates, ranging from the Wet Tropics, the humid, temperate eastern seaboard and, in places, the alpine south, together with the broad arid centre. Further, there is a wide range of aquatic habitats, with inland groundwater ranging from fresh- to seawater salinity and greater, as well as the salinity-stratified anchialine systems – groundwater estuaries – with actual seawater at depth. Troglobionts are known from all types of subterranean voids ranging from classical karst and coarse alluvial deposits to novel systems such as groundwater calcrete and goethite pisolite formations. Yet the troglobiont fauna of Australia is sparsely known, with by far the greater proportion of the country having been unexplored for troglobionts. Over much of the continent, access to troglobiont communities will be obtained through artificial access (boreholes) to atypical, even novel, habitats, and these are proving to contain the most species-rich troglobiont communities in Australia, possibly globally.

Acknowledgements

I thank Ken Grimes and Remko Leijs for their help with the figures.

References

There are a number of collected works on the cave biology of Australia that include broad introductions (Wilkens *et al.* 2000; Wicks and Humphreys 2011) and edited volumes concerning subterranean fauna (Humphreys 1993, 2012; Humphreys and Harvey 2001; Austin *et al.* 2008). The fauna of epikarst – a perched aquifer lying between surface water and vadose water – an exciting area for ecological, evolutionary and biogeographic research (Pipan 2005; Pipan and Culver 2007), remains largely unpublished in Australia, save in the context of shallow subterranean habitats (Chapter 5 in Culver and Pipan 2014).

Abrams, K.M., Guzik, M.T., Cooper, S.J., Humphreys, W.F., King, R.A., Cho, J.L., and Austin, A.D. 2012. What lies beneath: Molecular phylogenetics and ancestral state reconstruction of the ancient subterranean Australian Parabathynellidae (Syncarida, Crustacea). *Molecular Phylogenetics and Evolution* 64: 130–144.

Adams, M., and Humphreys, W.F. 1993. Patterns of genetic diversity within selected subterranean fauna of the Cape Range peninsula, Western Australia: Systematic and biogeographic implications. In Humphreys, W.F. (ed.), *The Biogeography of Cape Range, Western Australia*, pp. 145–164. *Records of the Western Australian Museum*, Supplement 45. Western Australian Museum, Perth, WA.

Asmyhr, M.G., Hose, G., Graham, P., and Stow, A.J. 2014. Fine-scale genetics of subterranean syncarids. *Freshwater Biology* 59: 1–11.

Audra, P., Bini, A., Gabrovšek, F., Häuselmann, P., Hobléa, F., Jeannin, P.Y., Kunaver, J., *et al.* 2007. Cave and karst evolution in the Alps and their relation to paleoclimate and paleotopography. *Acta Carsologica* 36(1):53–67.

Austin, A.D., Cooper, S.J.B., and Humphreys, W.F. (eds.). 2008. Subterranean connections: Biology and evolution in troglobiont and groundwater ecosystems. *Invertebrate Systematics* 22: 85–310.

Bauzà-Ribot, M.M., Juan, C., Nardi, F., Oromí, P., Pons, J., and Jaume, D. (2012). Mitogenomic phylogenetic analysis supports continental-scale vicariance in subterranean thalassoid crustaceans. *Current Biology* 22: 2069–2074.

Bird, M.I., and Chivas, A.R. 1988. Oxygen isotope dating of the Australian regolith. *Nature* 331: 514–516.

Botello, A., Iliffe, T., Alvarez, F., Juan, C., Pons, J., and Jaume, D. 2012. Historical biogeography and phylogeny of *Typhlatya* cave shrimps (Decapoda: Atyidae) base on mitochondrial and nuclear data. *Journal of Biogeography* 40: 594–607.

Boulton, A.J. 2001. Twixt two worlds: Taxonomic and function biodiversity at the surface water/groundwater interface. *Records of the Western Australian Museum*, Supplement 64: 1–13.

Boutin, C., and Coineau, N. 2000. Evolutionary rates and phylogenetic age in some stygobiontic species. In H. Wilkens, D.C. Culver and W.F. Humphreys (eds.), *Ecosystems of the World*, vol. 30, pp. 433–451. Subterranean Ecosystems. Elsevier, Amsterdam, the Netherlands.

Bradbury, J.H. 1999. The systematics and distribution of Australian freshwater amphipods: A review. In F.R. Schram and J.C. von Vaupel Klein (eds.), *Crustaceans and the Biodiversity Crisis*, pp. 533–540. Proceedings of the Fourth International Crustacean Congress, Amsterdam, the Netherlands, 20–24 July 1998. Brill, Leiden, the Netherlands.

Bradbury, J.H., and Williams, W.D. 1997. Amphipod (Crustacea) diversity in underground waters in Australia: An Aladdin's Cave. *Memoirs of Museum Victoria* 56: 513–519.

Bradford, T., Adams, M., Guzik, M., Humphreys, W., Austin, A., and Cooper, S. 2013. Patterns of population genetic variation in sympatric chiltoniid amphipods within a calcrete aquifer reveal a dynamic subterranean environment. *Heredity* 111: 77–85. Advance online publication, 3 April 2013.

Byrne, M., Yeates, D.K., Joseph, L., Kearney, M., Bowler, J., Williams, A.J., Cooper, S., *et al.* 2008. Birth of a biome: Insights into the assembly and maintenance of the Australian arid zone biota. *Molecular Ecology* 17: 4398–4417.

Camacho, A.I., Dorda, B.A., and Rey, I. 2014. Iberian Peninsula and Balearic Island Bathynellacea (Crustacea, Syncarida) database. *ZooKeys* 386: 1–20.

Chakrabarty, P., Davis, M.P., and Sparks, J.S. 2012. The first record of a trans-oceanic sister-group relationship between obligate vertebrate troglobites. *PLoS ONE* 7(8): e44083.

Chilton, C. 1894. The subterranean Crustacea of New Zealand: With some general remarks on the fauna of caves and wells. *Transactions of the Linnean Society*, London, 2nd series 6(2): 163–584.

Cho, J.L. 2005. A primitive representative of the Parabathynellidae (Bathynellacea, Syncarida) from the Yilgarn Craton of Western Australia. *Journal of Natural History* 39: 3423–3433.

Cho, J.L., and Humphreys, W.F. 2010. Ten new species of the genus *Brevisomabathynella* Cho, Park and Ranga Reddy, 2006 (Malacostraca, Bathynellacea, Parabathynellidae) from Western Australia. *Journal of Natural History* 44: 993–1079.

Cho, J.L., Humphreys, W.F., and Lee, S.D. 2006. Phylogenetic relationships within the genus *Atopobathynella* Schminke, 1973 (Bathynellacea, Parabathynellidae): With the description of six new species from Western Australia. *Invertebrate Systematics* 20: 9–41.

Coineau, N. 2000. Adaptations to interstitial groundwater life. In H. Wilkens, D.C. Culver and W.F. Humphreys (eds.), *Ecosystems of the World*, vol. 30, *Subterranean Ecosystems*, pp. 189–210. Elsevier, Amsterdam, the Netherlands.

Cooper, S.J.B., Bradbury, J.H., Saint, K.M., Leys, R., Austin, A.D., and Humphreys, W.F. 2007. Subterranean archipelago in the Australian arid zone: Mitochondrial DNA phylogeography of amphipods from central Western Australia. *Molecular Ecology* 16: 1533–1544.

Cooper, S.J.B., Hinze, S., Leys, R., Watts, C.H.S., and Humphreys, W.F. 2002. Islands under the desert: Molecular systematics and evolutionary origins of stygobitic water beetles (Coleoptera: Dytiscidae) from central Western Australia. *Invertebrate Systematics* 16: 589–598.

Cooper, S.J.B., Saint, K.M., Taiti, S., Austin, A.D., and Humphreys, W.F. 2008. Subterranean archipelago II: Mitochondrial DNA phylogeography of stygobitic isopods (Oniscidea: *Haloniscus*) from the Yilgarn region of Western Australia. *Invertebrate Systematics* 22: 195–206.

Culver, D.C., and Pipan, T. 2014. *Shallow Subterranean Habitats: Ecology, Evolution and Conservation.* Oxford University Press, Oxford, UK.

Curl, R.L. 1966. Caves as a measure of karst. *Journal of Geology* 74: 798–830.

Darwin, C. 1839. *Journal of Researches into the Geology and Natural History of the Various Countries Visited by H.M.S.* Beagle. H. Colburn, London.

Darwin, C. 1872. *On the Origin of Species by Means of Natural Selection, or the Preservation of Favoured Races in the Struggle for Life*, 6th edn. John Murray, London.

de Bruyn, M., Stelbrink, B., Page, T.J., Phillips, M.J., Lohman, D.J., Albrecht, C., Hall, R., *et al.* 2013. Time and space in biogeography: Response to Parenti and Ebach (2013). *Journal of Biogeography* 40: 2204–2208.

Dyksterhuis, S., Müller, R.D., and Albert, R.A. 2005. Paleostress field evolution of the Australian continent since the Eocene. *Journal of Geophysical Research* 110: B05102, 1–13.

Eberhard, S., and Giachino, P.M. 2011. Tasmanian Trechinae and Psydrinae (Coleoptera, Carabidae): A taxonomic and biogeographic synthesis, with description of new species and evaluation of the impact of Quaternary climate changes on evolution of the subterranean fauna. *Subterranean Biology* 9: 1–72.

Eberhard, S.E., and Humphreys, W.F. 2003. The crawling, creeping and swimming life of caves. In B. Finlayson and E. Hamilton-Smith (eds.), *Beneath the Surface: A Natural History of Australian Caves*, pp. 127–147 and plates. University of New South Wales Press, Sydney, Australia.

Eberhard, S.M., Halse, S.A., and Humphreys, W.F. 2005. Stygofauna in the Pilbara region, north-west Western Australia: A systematic review. *Journal of the Royal Society of Western Australia* 88: 167–176.

Eberhard, S.M., Halse, S.A., Williams, M.R., Scanlon, M.D., Cocking, J., and Barron, H.J. 2009. Exploring the relationship between sampling efficiency and short-range endemism for groundwater fauna in the Pilbara region, Western Australia. *Freshwater Biology* 54: 885–901.

Eme, D., Malard, F., Konecny-Dupré, L., Lefébure, T., and Douady, C.J. (2013). Bayesian phylogeographic inferences reveal contrasted colonization dynamics among European groundwater isopods. *Molecular Ecology* 22: 6685–5699.

Faille, A., Casale, A., Balke, M., and Ribera, I. 2013. A molecular phylogeny of Alpine subterranean Trechini (Coleoptera: Carabidae). *BMC Evolutionary Biology* 13: 248.

Finston, T.L., Francis, C.J., and Johnson, M.S. 2009. Biogeography of the stygobitic isopod *Pygolabis* (Malacostraca: Tainisopidae) in the Pilbara, Western Australia: Evidence for multiple colonisations of the groundwater. *Molecular Phylogenetics and Evolution* 52: 448–460.

Finston, T.L., Johnson, M.S., Humphreys, W.F., Eberhard, S., and Halse, S. 2007. Cryptic speciation in two widespread subterranean amphipod genera reflects historical drainage patterns in an ancient landscape. *Molecular Ecology* 16: 355–365.

Giribet, G., and Edgecombe, G.D. 2006. The importance of looking at small-scale patterns when inferring Gondwanan biogeography: A case study of the centipede *Paralamyctes* (Chilopoda, Lithobiomorpha, Henicopida). *Journal of the Linnaean Society* 89: 65–78.

Gray, M.R. 1989. Cavernicolous spiders (Araneae) from Undara, Queensland and Cape Range, Western Australia. *Helictite* 27(2): 87–89.

Guzik, M.T., Abrams, K.M., Cooper, S.J.B., Humphreys, W.F., Cho, J.L., and Austin, A. 2008. Phylogeography of the ancient Parabathynellidae (Crustacea: Bathynellacea) from the Yilgarn region of Western Australia. *Subterranean Connections Invertebrate Systematics* 22: 205–216.

Guzik, M.T., Austin, A.D., Cooper, S.J.B., Harvey, M.S., Humphreys, W.F., Bradford, T., *et al.* 2011a. Is the Australian subterranean fauna uniquely diverse? *Invertebrate Systematics* 24: 407–418.

Guzik, M.T., Cooper, S.J.B., Humphreys, W.F., and Austin, A.D. 2009. Fine-scale comparative phylogeography of a sympatric sister species triplet of subterranean diving beetles from a single calcrete aquifer in Western Australia. *Molecular Ecology* 18: 3683–3698.

Guzik, M.T., Cooper, S.J.B., Humphreys, W.F., Ong, S., Kawakami, T., and Austin, A.D. 2011b. Evidence for population fragmentation within a subterranean aquatic habitat in the Western Australian desert. *Heredity* 107: 215–230.

Halse, S.A., and Pearson, G.B. 2014. Troglofauna in the vadose zone: Comparison of scraping and trapping results and sampling adequacy. *Subterranean Biology* 13: 17–34.

Harvey, M.S., Berry, O., Edward, K.L., and Humphreys, G. 2008. Molecular and morphological systematics of hypogean schizomids (Schizomida: Hubbardiidae) in semi-arid Australia. *Invertebrate Systematics* 22: 167–194.

Hawlitschek, O., Hendrich, L., Espeland, M., Toussaint, E.F.A., Genner, M.J., and Balke, M. 2012. Pleistocene climate change promoted rapid diversification of aquatic invertebrates in Southeast Australia. *BMC Evolutionary Biology* 12: 142.

Hoch, H., and Howarth, F.G. 1989a. Six new cavernicolous cixiid planthoppers in the genus *Solonaima* from Australia (Homoptera: Fulgoridea). *Systematic Entomology* 14: 377–402.

Hoch, H., and Howarth, F.G. 1989b. Reductive evolutionary trends in two new cavernicolous species of a new Australian cixiid genus (Homoptera: Fulgoroidea). *Systematic Entomology* 14: 179–196.

Hoch, H., and Howarth, F.G. 1989c. The evolution of cave-adapted cixiid planthoppers in volcanic and limestone caves in north Queensland, Australia (Homoptera: Fulgoroidea). *Mèmoires de biospèologie* 16: 17–24.

Hocking, R.M., Moors, H.T., and van de Graaff, W.J.E. 1987. *Geology of the Carnarvon Basin, Western Australia*. Geological Survey of Western Australia, Bulletin 133. Department of Mines, Perth, Australia.

Hoenemann, M., Neiber, M.T., Humphreys, W.F., Iliffe, T.M., Li, D., Schram, F.R., and Koenemann, S. 2013. Phylogenetic analyses and systematic revision of Remipedia (Nectiopoda) from Bayesian analysis of molecular data. *Journal of Crustacean Biology* 33: 603–619.

Holsinger, J.R. 2000. Ecological derivation, colonization, and speciation. In H. Wilkens, D.C. Culver and W.F. Humphreys (eds.), *Ecosystems of the World*, vol. 30, pp. 399–415. Subterranean Ecosystems. Elsevier, Amsterdam, the Netherlands.

Holsinger, J.R., and Longley, G. 1980. The subterranean amphipod crustacean fauna of an artesian well in Texas. *Smithsonian Contributions to Zoology* 308: 1–62.

Howarth, F.G. 1988. Environmental ecology of north Queensland caves: Or why there are so many troglobites in Australia. In L. Pearson (ed.), *17th Biennial Conference, Australian Speleological Federation Tropicon Conference, Lake Tinaroo, Far North Queensland, 27–31 December 1988*, pp. 76–84. Australian Speleological Federation, Cairns, Australia. 139 p.

Howarth, F.G. 1993. High-stress subterranean habitats and evolutionary change in cave-inhabiting arthropods. *American Naturalist* 142: 65–77.

Humphreys, G., Alexander, J., Harvey, M.S., and Humphreys, W.F. 2013. The subterranean fauna of Barrow Island, northwestern Australia: 10 years on. *Records of the Western Australian Museum*, Supplement 83: 145–158.

Humphreys, W.F. 1991. Experimental re-establishment of pulse-driven populations in a terrestrial troglobite community. *Journal of Animal Ecology* 60: 609–623.

Humphreys, W.F. (ed.). 1993a. The biogeography of Cape Range, Western Australia. *Records of the Western Australian Museum*, Supplement 45: 1–248.

Humphreys, W.F. 1993b. The significance of the subterranean fauna in biogeographical reconstruction: Examples from Cape Range peninsula, Western Australia. *Records of the Western Australian Museum*, Supplement 45: 165–192.

Humphreys, W.F. 1995. *Limestone of the East Kimberley, Western Australia: Karst and Cave Fauna*. Report to the Australian Heritage Commission and the Western Australian Heritage Committee. 190 p. + xix. Unpublished report. Western Australian Museum. http://museum.wa.gov.au/sites/default/files/Humphreys%201995_Limestone%20E%20Kimberey_OCR.pdf.

Humphreys, W.F. 2000a. The hypogean fauna of the Cape Range peninsula and Barrow Island, northwestern Australia. In H. Wilkens, D.C. Culver and W.F. Humphreys (eds.), *Ecosystems of the World*, vol. 30, pp. 581–601. Subterranean Ecosystems. Elsevier, Amsterdam, the Netherlands.

Humphreys, W.F. 2000b. Relict faunas and their derivation. In H. Wilkens, D.C. Culver and W.F. Humphreys (eds.), *Ecosystems of the world*, vol. 30, pp. 417–432. Subterranean Ecosystems. Elsevier, Amsterdam, the Netherlands.

Humphreys, W.F. 2000c. Karst wetlands biodiversity and continuity through major climatic change: An example from arid tropical Western Australia. In B. Gopal, W.J. Junk and J.A. Davis (eds.), *Biodiversity in Wetlands: Assessment, Function and Conservation*, vol. 1, pp. 227–258. Backhuys Publishers, Leiden, the Netherlands. 353 p.

Humphreys, W.F. 2001. Groundwater calcrete aquifers in the Australian arid zone: The context to an unfolding plethora of stygal biodiversity. In W.F. Humphreys and M.S. Harvey (eds.), *Subterranean Biology in Australia* 2000, pp. 63–83. *Records of the Western Australian Museum*, Supplement 64. Western Australian Museum, Perth, Australia.

Humphreys, W.F. 2002. The subterranean fauna of Barrow Island, northwestern Australia, and its environment. *Mémoires de biospéologie* (International Journal of Subterranean Biology) 28: 107–127.

Humphreys, W.F. 2008. Rising from Down Under: Developments in subterranean biodiversity in Australia from a groundwater fauna perspective. *Invertebrate Systematics* 22: 85–101.

Humphreys, W.F. 2012. Diversity patterns in Australia. In W.B. White and D.C. Culver (eds.), *Encyclopedia of Caves*, 2nd edn., pp. 203–219. Academic Press, San Diego, CA.

Humphreys, W.F., and Adams, M. 2001. Allozyme variation in the troglobitic millipede *Stygiochiropus communis* (Diplopoda: Paradoxosomatidae) from arid tropical Cape Range, northwestern Australia: Population structure and implications for the management of the region. *Records of the Western Australian Museum*, Supplement 64: 15–36.

Humphreys, W.F., and Danielopol, D.L. 2006. *Danielopolina* (Ostracoda, Thaumatocyprididae) on Christmas Island, Indian Ocean, a sea mount island. *Crustaceana* 78: 1339–1352.

Humphreys, W.F., and Eberhard, S. 2001. Subterranean fauna of Christmas Island, Indian Ocean. *Helictite* 37(2): 59–74.

Humphreys, W.F., and Harvey, M.S. (eds.). 2001. *Subterranean Biology in Australia 2000*, 242 p. *Records of the Western Australian Museum*, Supplement 64. Western Australian Museum, Perth, Australia.

Humphreys, W.F., Watts, C.H.S., Cooper, S.J.B., and Leijs, R. 2009 Groundwater estuaries of salt lakes: Buried pools of endemic biodiversity on the western plateau, Australia. *Hydrobiologia* 626: 79–95. Erratum: *Hydrobiologia* 632: 377.

Hunt, G.S. 1990. *Hickmanoxyomma*, a new genus of cavernicolous harvestmen from Tasmania (Opiliones: Triaenonychidae). *Records of the Australian Museum* 42: 45–68.

Hurley, D.E. 1990. Charles Chilton: The Phreatoicidea and other interests of a phreatic pioneer from down under. *Bijdragen tot de Dierkunde* 60(3/4): 233–238.

Iglikowska, A., and Boxshall, G.A. 2013. *Danielopolina* revised: Phylogenetic relationships of the extant genera of the family Thaumatocyprididae (Ostracoda: Myodocopa). *Zoologischer Anzeiger* 252: 469–485.

Iknayan, K.J., Tingley, M.W., Furnas, B.J., and Beissinger, S.R. 2014. Detecting diversity: Emerging methods to estimate species diversity. *Trends in Ecology and Evolution* 29: 97–106.

Jaume, D., and Humphreys, W.F. 2001. A new genus of epacteriscid calanoid copepod from an anchialine sinkhole in northwestern Australia. *Journal of Crustacean Biology* 21: 157–169.

Javidkar, M., Cooper, S.J.B., King, R., Humphreys, W.F., Bertozzi, T., Stevens, M.I., and Austin, A.D. 2016. Molecular systematics and biodiversity of oniscidean isopods in the groundwater calcretes of central Western Australia. *Molecular Phylogenetics and Evolution* 104: 83–98.

Jeffery, W.R. 2005. Adaptive evolution of eye degeneration in the Mexican blind cavefish. *Journal of Heredity* 96: 185–196.

Juan, C., Guzik, M.T., Jaume, D., and Cooper, S.J.B. 2010. Evolution in caves: Darwin's 'wrecks of ancient life' in the molecular era. *Molecular Ecology* 19: 3865–3880.

Karanovic, I. 2003. Towards a revision of Candoninae (Crustacea, Ostracoda): Description of two new genera from Australian ground-waters. *Species Diversity* 8: 353–383.

Karanovic, I. 2004. Towards a revision of Candoninae (Crustacea, Ostracoda): On the genus *Candonopsis* Vavra, with description of new taxa. *Subterranean Biology* 2: 91–108.

Karanovic, I. 2005. Towards a revision of Candoninae (Crustacea, Ostracoda): Australian representatives of the subfamily, with description of three new genera and seven new species. *New Zealand Journal of Marine and Freshwater Research* 39: 29–75.

Karanovic, I. 2007. Candoninae ostracodes from the Pilbara region in Western Australia. *Crustaceana Monographs* 7: 1–432.

Karanovic, I. 2009. Four new species of *Gomphodella* de Deckker, with a phylogenetic analysis and a key to the living representatives of the subfamily Timiraseviinae (Ostracoda). *Crustaceana* 82: 1133–1176.

Karanovic, I. 2012. *Recent Freshwater Ostracods of the World: Crustacea, Ostracoda, Podocopida*. Springer, New York.

Karanovic, I., and Marmonier, P. 2003. Three new genera and nine new species of the subfamily Candoninae (Crustacea, Ostracoda, Podocopida) from the Pilbara Region (Western Australia). *Beaufortia* 53: 1–51.

Karanovic, I., and Ranga Reddy, Y. 2008. The second representative of the genus *Indocandona* Gupta (Ostracoda) from a well in southeastern India. *Crustaceana* 81: 861–871.

Karanovic, T. 2004. *Subterranean Copepoda from Arid Western Australia*. Crustaceana Monographs, No. 3. Brill, Leiden, the Netherlands.

Karanovic, T. 2006. Subterranean copepods (Crustacea, Copepoda) from the Pilbara region in Western Australia. *Records of the Western Australian Museum*, Supplement 70: 1–239.

Karanovic, T. 2010. First record of the harpacticoid genus *Nitocrellopsis* (Copepoda, Ameiridae) in Australia, with descriptions of three new species. *Annals of Limnology* 46: 249–280.

Karanovic, T., and Cooper, S.J.B. 2011. Third genus of parastenocarid copepods from Australia supported by molecular evidence. In D. Defaye, E. Suárez and J.C. von Vaupel Klein (eds.), *Studies on Freshwater Copepoda*, pp. 293–337. Koninklijke Brill, Leiden, the Netherlands.

Karanovic, T., and Cooper, S.J.B. 2012. Explosive radiation of the genus *Schizopera* on a small subterranean island in Western Australia (Copepoda: Harpacticoida): Unravelling the cases of cryptic speciation, size differentiation and multiple invasions. *Invertebrate Systematics* 26: 115–192.

Karanovic, T., and Eberhard, S.M. 2009. Second representative of the order Misophrioida (Crustacea, Copepoda) from Australia challenges the hypothesis of the Tethyan origin of some anchialine faunas. *Zootaxa* 2059: 51–68.

King, R.A., Bradford, T., Austin, A.D., Humphreys, W.F., Cooper, S.J.B. 2012. Divergent molecular lineages and not-so-cryptic species: The first descriptions of stygobitic chiltoniid amphipods (Talitroidea: Chiltoniidae) from Western Australia. *Journal of Crustacean Biology* 32: 465–488.

Kohn, B.P., Gleadow, A.J.W., Brown, R.W., Gallagher, K., O'Sullivan, P.B., and Foster, D.A. 2002. Shaping the Australian crust over the last 300 million years: Insights from fission track thermotectonic imaging and denudation studies in key terranes. *Australian Journal of Earth Sciences* 49: 697–717.

Langecker, T.G. 2000. The effect of continuous darkness on cave ecology and cavernicolous evolution. In H. Wilkens, D.C. Culver and W.F. Humphreys (eds.), *Ecosystems of the World*, vol. 30, pp. 135–157. Subterranean Ecosystems. Elsevier, Amsterdam, the Netherlands.

Leijs, R., van Nes, E.H., Watts, C.H., Cooper, S.J.B., Humphreys, W.F., and Hogendoorn, K. 2012. Evolution of blind beetles in isolated aquifers: A test of alternative modes of speciation. *PLoS ONE* 7(3):e34260.

Leys, R., and Watts, C.H. 2008. Systematics and evolution of the Australian subterranean hydroporine diving beetles (Dytiscidae), with notes on *Carabhydrus*. *Invertebrate Systematics* 22: 217–225.

Leys, R., Watts C.H.S., Cooper S.J.B., and Humphreys, W.F. 2003. Evolution of subterranean diving beetles (Coleoptera: Dytiscidae: Hydroporini, Bidessini) in the arid zone of Australia. *Evolution* 57: 2819–2834.

Longley, G. 1986. The biota of the Edwards Aquifer and the implications for palaeozoogeography. In P.L. Abbott and C.M. Woodruff, Jr. (eds.), *The Balcones Escarpment*, pp. 51–54. Geological Society of America, San Antonio, TX.

Martens, K., Schön, I., Meisch, C., and Horne, D.J. 2007. Global diversity of ostracods (Ostracoda, Crustacea). *Hydrobiologia* 595: 185–193.

Moore, B.P. 1965. Present-day cave beetle fauna of Australia: A pointer to past climatic change. *Helictite* 3: 3–9.

Moore, B.P. 1995. Two remarkable new genera and species of troglobitic Carabidae (Coleoptera) from Nullarbor caves. *Australian Journal of Entomology* 34: 159–161.

Morton, S.R., Short, J., and Barker, R.D. 1995. *Refugia for Biological Diversity in Arid and Semi-Arid Australia*. Biodiversity Series, Paper No. 4. Biodiversity Unit, Department of Environment, Sport and Territories, Canberra, Australia.

Osborne, A. 2010. Rethinking eastern Australian caves. In P. Bishop and B. Pillans (eds.), *Australian Landscapes*, pp. 289–308. Geological Society, London.

Page, T.J., Humphreys, W.F., and Hughes, J.M. 2008. Shrimps Down Under: Evolutionary relationships of subterranean crustaceans from Western Australia (Decapoda: Atyidae: *Stygiocaris*). *PLoS ONE* 3(2): e1618, 1–12.

Page, T.J., Hughes, J.M., Real, K.M., Stevens, M.I., King, R.A., and Humphreys, W.F. 2016. Allegory of the cave crustacean: Systematic and biogeographic reality of *Halosbaena* (Peracarida: Thermosbaenacea) sought with molecular data at multiple scales. *Marine Biodiversity* doi: 10.1007/s12526-016-0565-3. Accessed 19 September, 2016.

Parenti, L.R., and Ebach, M.C. 2013. Evidence and hypothesis in biogeography. *Journal of Biogeography* 40: 813–820.

Phillips, M.J., Page, T.J., de Bruyn, M., Huey, J.A., Humphreys, W.F., Hughes, J.M., Santos, S.R., Schmidt, D.J., and Waters, J.M. (2013). The linking of plate tectonics and evolutionary divergence. *Current Biology* 23: R603–R605.

Pipan, T. 2005. *Epikarst: A Promising Habitat*. Založba ZRC, Ljubljana, Slovenia, 101 p.

Pipan, T., and Culver, D.C. 2007. Regional species richness in an obligate subterranean dwelling fauna: Epikarst copepods. *Journal of Biogeography* 34: 854–861.

Playford, P.E., Hocking, R.M., and Cockbain, A.E. 2009. Devonian reef complexes of the Canning Basin, Western Australia. *Geological Survey of Western Australia*, Bulletin 145: 1–471.

Poore, G.C.B., and Humphreys, W.F. 1992. First record of Thermosbaenacea (Crustacea) from the Southern Hemisphere: A new species from a cave in tropical Western Australia. *Invertebrate Taxonomy* 6: 719–725.

Poore, G.C.B., and Humphreys, W.F. 1998. First record of Spelaeogriphacea from Australasia: A new genus and species from an aquifer in the arid Pilbara of Western Australia. *Crustaceana* 71: 721–742.

Poore, G.C.B., and Humphreys, W.F. 2003. Second species of *Mangkurtu* (Spelaeogriphacea) from north-western Australia. *Records of the Western Australian Museum* 22: 67–74.

Radke, L., Juggins, S., Halse, S.A., De Deckker, P., and Finston, T. 2003. Chemical diversity in south-eastern Australian saline lakes II: Biotic implications. *Marine and Freshwater Research* 54: 895–912.

Reeves, J.M., De Deckker, P., and Halse, S.A. 2007. Groundwater ostracods from the arid Pilbara region of northwestern Australia: Distribution and water chemistry. *Hydrobiologia* 585: 99–118.

Rétaux, S., and Casane, D. 2013. Evolution of eye development in the darkness of caves: Adaptation, drift, or both? *EvoDevo* 4: 26.

Rohner, N., Jarosz, D.F., Kowalko, J.E., Yoshizawa, M., Jeffery, W.R., Borowsky, R.L., Lindquist, S., and Tabin, C.J. 2013. Cryptic variation in morphological evolution: HSP90 as a capacitor for loss of eyes in cavefish. *Science* 342(6164): 1372–1375.

Scarsbrook, M.R., and Fenwick, G.D. 2003. Preliminary assessment of crustacean distribution patterns in New Zealand groundwater aquifers. *New Zealand Journal of Marine and Freshwater Research* 37: 405–413.

Scher, H.D., and Martin, E.E. 2006. Timing and climatic consequences of the opening of the Drake Passage. *Science* 312: 428–230.

Schminke, H.K. 2011. Arthropoda, Crustacea, Malacostraca, Bathynellacea: Parabathynellidae. *Invertebrate Fauna of the World* 21(1): 1–244. (Fauna and Flora of Korea, National Institute of Biological Resources, Ministry of the Environment.)

Shimomura, M., and Fujita, Y. 2009. First record of the thermosbaenacean genus *Halosbaena* from Asia: *H. daitoensis* sp. nov. (Peracarida: Thermosbaenacea: Halosbaenidae) from an anchialine cave of Minamidaito-jima Is., in Okinawa, southern Japan. *Zootaxa* 1990: 55–64.

Sket, B. 2008. Can we agree on an ecological classification of subterranean animals? *Journal of Natural History* 42: 1549–1563.

Thurgate, M.E., Gough, J.S., Spate, A., and Eberhard, S.M. 2001. Subterranean biodiversity in New South Wales: From rags to riches. *Records of the Western Australian Museum*, Supplement 64: 37–47.

Vasconcelos, P.M., Knesel, K.M., Cohen, B.E., and Heim, J.A. 2008. Geochronology of the Australian Cenozoic: A history of tectonic and igneous activity, weathering, erosion, and sedimentation. *Australian Journal of Earth Sciences* 55: 865–914.

Vergnon, R., Leijs, R., van Nes, E.H., and Scheffer, M. 2013. Repeated parallel evolution reveals limiting similarity in subterranean diving beetles. *The American Naturalist* 182: 67–75.

von Rintelen, K., Page, T.J., Cai, Y., Roe, K., Stelbrink, B., Kuhajda, B.R., Iliffe, T.M., Hughes, J., and von Rintelen, T. 2012. Drawn to the dark side: A molecular phylogeny of freshwater shrimps (Crustacea: Decapoda: Caridea: Atyidae) reveals frequent cave invasions and challenges current taxonomic hypotheses. *Molecular Phylogenetics and Evolution* 63: 82–96.

Waters, J.M., and Craw, D. 2006. Goodbye Gondwana? New Zealand biogeography, geology, and the problem of circularity. *Systematic Biology* 55: 351–356.

Watts, C.H.S., and Humphreys, W.F. 2009. Fourteen new Dytiscidae (Coleoptera) of the genera *Limbodessus* Guignot, *Paroster* Sharp and *Exocelina* Broun, from underground waters in Australia. *Transactions of the Royal Society of South Australia* 133: 62–107.

Watts, C.H.S., and McRae, J. 2013. *Limbodessus bennetti* sp. nov., the first stygobitic Dytiscidae (Coleoptera) from the Pilbara region of Western Australia. *Records of the Western Australian Museum* 28: 141–143.

Wicks, C., and Humphreys, W.F. (eds.). 2011 Anchialine ecosystems: Reflections and prospects. *Hydrobiologia* (Special Issue) 677(1): 1–168.

Wilkens, H. 2010. Genes, modules and the evolution of cave fish. *Heredity* 105: 1–10.

Wilkens, H., Culver, D.C., and Humphreys, W.F. (eds.). 2000. *Ecosystems of the World*, vol. 30. Subterranean Ecosystems. Elsevier, Amsterdam, the Netherlands. 791 p.

Wilson, G.D.F. 2001. Australian groundwater-dependent isopod crustaceans. *Records of the Western Australian Museum*, Supplement 64: 239–240.

Wilson, G.D.F. 2003. A new genus of Tainisopidae fam. nov. (Crustacea: Isopoda) from the Pilbara, Western Australia. *Zootaxa* 245: 1–20.

Wilson, G.D.F. 2008. Gondwanan groundwater: Subterranean connections of Australian phreatoicidean isopods (Crustacea) to India and New Zealand. *Invertebrate Systematics* 22: 301–310.

Wilson, G.D.F., and Johnson, R.T. 1999. Ancient endemism among freshwater isopods (Crustacea, Phreatoicidea). In W. Ponder and D. Lunney (eds.), *The Other 99%: The Conservation and Biodiversity of Invertebrates*, pp. 264–268. Transactions of the Royal Zoological Society of New South Wales, Mosman, Australia.

Wilson, G.D.F., and Keable, S.J. 1999. A new genus of phreatoicidean isopod (Crustacea) from the north Kimberley Region, Western Australia. *Zoological Journal of the Linnean Society* 126: 51–79.

Wilson, G.D.F., and Keable, S.J. 2001. Systematics of the Phreatoicidea. In B. Kensley and R.C. Brusca (eds.), *Isopod Systematics and Evolution*, pp. 175–194. Balkema, Rotterdam, the Netherlands.

12

Molecular Biogeography of Australian and New Zealand Reptiles and Amphibians

Mitzy Pepper, J. Scott Keogh and David G. Chapple

CONTENTS

Introduction

Reptiles and amphibians are a major and ubiquitous component of the vertebrate fauna of Australia and New Zealand. They are found across all biomes and in environments as diverse as rainforests, snow-capped peaks, arid salt pans and desert dune fields. Despite the proximity of Australia and New Zealand (~1500 km), these adjacent landmasses belong to separate continents (New Zealand is part of the largely submerged subcontinent Zealandia). Thus, the taxonomic composition and evolutionary origins of the herpetofauna are markedly different; Australia is a large, ancient continent with subdued topography and a vast, arid centre, while New Zealand is a comparatively young, geologically active group of oceanic islands, with high altitude peaks and plateaus and extensive alpine habitats. Global tectonic processes, such as the breakup of Gondwana, have shaped the higher-level herpetological composition and diversity of Australia and New Zealand, while recent climatic fluctuations and associated vegetation changes have fuelled adaptive radiations in different groups (Morton and James 1988; Byrne *et al.* 2008; Marin *et al.* 2013).

Compared with current world figures, the number of reptile and amphibian families in Australia is low. Australia has only one of the three major groups of living amphibians – the order Anura, comprising frogs and toads (Table 12.1). The number of squamate (snakes and lizards) families in Australia is also low by world standards, with just 7 of approximately 37 lizard families, and 5 or 6 of 26 snake families (Table 12.1). With the exception of the legless geckos (Pygopodidae), these families are all represented elsewhere in the world (Hutchinson and Donnellan 1993). Of the 12 freshwater turtle families, 2 are found in Australia, while crocodylians are the only represented family of

TABLE 12.1

Summary of Molecular Studies of Australian and New Zealand Reptiles and Amphibians since 1990

Region	Taxon	Genera (*n*)	Species (*n*)	Origin	Origin Reference	Phylogeography Reference
	Iguania					
Australia	Agamidae	15	81	Southeast Asia	Macey *et al.* 2000; Honda *et al.* 2000b; Melville *et al.* 2001; Hugall and Lee 2004; Amer and Kumazawa 2005; Hugall *et al.* 2008; Schulte *et al.* 2003	Scott and Keogh 2000 (*Tympanocryptis*); Melville *et al.* 2001 (*Ctenophorus*), 2008 (*Ctenophorus/Rankinia*), 2011 (*Amphibolurus/Lophognathus*); Shoo *et al.* 2008 (*Tympanocryptis*); Edwards and Melville 2011 (*Diporiphora*); Smith *et al.* 2011 (*Diporiphora*); Ng *et al.* 2013 (*Rankinia*); Pepper *et al.* 2014 (*Amphibolurus*)
	Gekkota				Donnellan *et al.* 1999; Han *et al.* 2004; Gamble *et al.* 2008	
Australia	Gekkonidae	7	47	Southeast Asia	Strasburg and Kearney 2005; Jackman *et al.* 2008; Heinicke *et al.* 2010, 2011; Wood *et al.* 2012	Kearney *et al.* 2003 (*Heteronotia*); Heinicke *et al.* 2010 (*Nactus*); Fujita *et al.* 2010 (*Heteronotia*); Pepper *et al.* 2011b, 2013b (*Heteronotia*); Sistrom *et al.* 2013, 2014 (*Gehyra*)
Australia	Diplodactylidae	17	105	Gondwana	Melville *et al.* 2004; Oliver *et al.* 2007, 2012a,b; Oliver and Sanders 2009; Oliver and Bauer 2011	Hoskin *et al.* 2003 (*Orraya/Phyllurus/Saltuarius*); Melville *et al.* 2004 (*Strophurus*); Pepper *et al.* 2006, 2008 (*Diplodactylus*); Oliver *et al.* 2007 (*Diplodactylus*), 2010, 2013a (*Credadactylus*), 2012a, 2014b (*Oedura*), 2013b (*Pseudothecadactylus*), 2014c (*Diplodactylus*); Couper *et al.* 2008 (*Saltuarius*); Pepper *et al.* 2011a (*Rhynchoedura*); Oliver and Bauer 2011 (*Nephrurus*)
Australia	Pygopodidae	7	43	Gondwana	Jennings *et al.* 2003; Heinicke *et al.* 2011; Lee *et al.* 2009; Oliver and Sanders 2009	
	Scincomorpha				Honda *et al.* 2000a; Reeder 2003	
Australia	Scincidae; Egernia	4	49	Southeast Asia	Rabosky *et al.* 2007; Chapple and Keogh 2004; Skinner *et al.* 2011	Donnellan *et al.* 2002 (*Liopholis*); Adams *et al.* 2003 (*Menetia*); Chapple and Keogh 2004 (*Liopholis*); Chapple *et al.* 2005 (*Liopholis*); Gardner *et al.* 2008 (*Egernia* Group); Doughty *et al.* 2011 (*Egernia*)
Australia	Scincidae; Sphenomorphus	17	273	Southeast Asia	Reeder 2003; Skinner *et al.* 2013	O'Connor and Moritz 2003 (*Eulamprus*); Hodges *et al.* 2007 (*Eulamprus*); Rabosky *et al.* 2007, 2009, 2014 (*Ctenotus*); Kay and Keogh 2012 (*Ctenotus*)

Australia	Scincidae; Eugongylus	19	132	Southeast Asia	Smith *et al.* 2007; Stuart-Fox *et al.* 2002	Moussalli *et al.* 2005 (*Saproscincus*); Horner and Adams 2007 (*Cryptoblepharus*); Dolman and Hugall 2008 (*Carlia*); Dubey and Shine 2010 (*Bassiana*); Bell *et al.* 2010 (*Lampropholis*); Chapple *et al.* 2011a,b (*Lampropholis*); Reeder and Reichert 2011 (*Hemiergis*); Haines *et al.* 2014 (*Pseudomoia*)
	Varanoidea				Vidal *et al.* 2012	
Australia	Varanidae	1	29	Southeast Asia	Fuller *et al.* 1998; Ast 2001; Schulte *et al.* 2003; Jennings and Pianka 2004; Fitch *et al.* 2006; Schuett *et al.* 2009; Vidal *et al.* 2012	Fitch *et al.* 2006 (*Varanus*); Smissen *et al.* 2013 (*Varanus*)
	Scolecophidia					
Australia	Typhlopidae	1	43	Gondwana	Rabosky *et al.* 2004 (*Ramphotyphlops*); Marin *et al.* 2013a,b	Marin *et al.* 2013a,b (*Ramphotyphlops*)
	Booidae					
Australia	Pythonidae	4	13	Southeast Asia	Rawlings and Donnellan 2003; Rawlings *et al.* 2004, 2008	Rawlings and Donnellan 2003 (*Morelia*)
	Colubroidea					
Australia	Acrochordidae	1	2	Southeast Asia	Sanders *et al.* 2010	
Australia	Colubridae	5	7	Southeast Asia		
Australia	Elapidae; Hydrophiinae (terrestrial)	26	99	Southeast Asia	Slowinski *et al.* 1997; Keogh 1998; Keogh *et al.* 1998; Scanlon and Lee 2004; Kuch *et al.* 2005; Wuster *et al.* 2005; Williams *et al.* 2008; Kelly *et al.* 2009	Keogh *et al.* 2003 (*Hoplocephalus*); Skinner *et al.* 2005 (*Pseudonaja*); Kuch *et al.* 2005 (*Pseudechis*); Dubey *et al.* 2010 (*Drysdalia*)
Australia	Elapidae; Hydrophiinae (true sea snakes)	12	33	Southeast Asia	Sanders *et al.* 2008, 2013; Sanders and Lee 2008; Rasmussen *et al.* 2014	Lukoschek and Keogh 2006 (sea snakes); Lukoschek *et al.* 2007 (*Aipysurus*)
Australia	Elapidae; Laticaudinae (sea kraits)	1	2	Southeast Asia	Slowinski *et al.* 1997; Keogh 1998; Keogh *et al.* 1998; Scanlon and Lee 2004; Sanders *et al.* 2008, 2013; Sanders and Lee 2008	
Australia	Homalopsidae	4	5	Southeast Asia	Voris *et al.* 2002; Alfaro *et al.* 2004, 2008	

(*Continued*)

TABLE 12.1 (CONTINUED)

Summary of Molecular Studies of Australian and New Zealand Reptiles and Amphibians since 1990

Region	Taxon	Genera (*n*)	Species (*n*)	Origin	Origin Reference	Phylogeography Reference
	Anura					
Australia	Bufonidae	1	1	Introduced	Slade and Moritz 1998	Estoup *et al.* 2004
Australia	Hylidae	2	85	?		McGuigan *et al.* 1998 (*Litoria*); James and Moritz 2000 (*Litoria*); Bell *et al.* 2012 (*Litoria*)
Australia	Microhylidae	2	24	?		Hoskin *et al.* 2011 (*Cophixalus*)
Australia	Myobatrachidae	14	87	Gondwana	Read *et al.* 2001; Edwards 2007	Schauble and Moritz 2001 (*Limnodynastes*); Morgan *et al.* 2007 (*Heleioporus*); Catullo *et al.* 2011, 2014; Catullo and Keogh 2014 (*Uperoleia*)
Australia	Ranidae	1	1	Southeast Asia	Bossuyt *et al.* 2006	
Australia	Limnodynastidae	8	40	Gondwana		
	Testudines				Crawford *et al.* 2014	
Australia	Chelidae	7	24	Gondwana	Le *et al.* 2013; Georges *et al.* 2014; Todd *et al.* 2014a	Todd *et al.* 2013, 2014a (*Elseya*); Todd *et al.* 2014b (*Emydura*); Hodges *et al.* 2014 (*Chelodina*); Georges *et al.* 2014 (*Elseya*)
	Carettochelyidae	1	1	?		
	Crocodilia					
Australia	Crocodylus	1	2	?	Oaks 2011; Brochu and Storrs 2012	

	Anura					
New Zealand	Hylidae	1	3	Introduced (Australia)	Voros *et al.* 2008	
New Zealand	Leiopelmatidae	1	4	Gondwana	Holyoake *et al.* 2001; Roelants and Bossuyt 2005; Irisarri *et al.* 2010; Pyron and Wiens 2011	Fouquet *et al.* 2010
	Rhynchocephalia					
New Zealand	Sphenodontidae	1	1	Gondwana	Rest *et al.* 2003; Hugall *et al.* 2007; Cree 2014	Hay *et al.* 2010
	Gekkota					
New Zealand	Diplodactylidae	7	43	Australia	Nielsen *et al.* 2011	Nielsen *et al.* 2011
	Scincomorpha					
New Zealand	Scincidae; Eugongylus (native)	1	61	New Caledonia	Hickson *et al.* 2000; Smith *et al.* 2007; Chapple *et al.* 2009	Berry and Gleeson 2005; Greaves *et al.* 2007, 2008; Chapple and Patterson 2007; Bell and Patterson 2008; Hare *et al.* 2008; O'Neill *et al.* 2008; Liggins *et al.* 2008a,b; Chapple *et al.* 2008a,b,c, 2011c, 2012; Miller *et al.* 2009; Patterson *et al.* 2013
New Zealand	Scincidae; Eugongylus (introduced)	1	1	Australia (1960s)	Chapple *et al.* 2013	

Note: This table does not include sea turtles or sea snakes that are occasional visitors or vagrants in New Zealand. Nor does it include the red-eared slider. It also does not include new species description papers despite the fact that they often now contain molecular data as part of the description.

crocodiles (Uetz *et al.* 2016). In contrast, at the species level, several widespread groups including skinks (Scincidae), blindsnakes (Typhlopidae), terrestrial elapids and sea snakes (Elapidae), dragons (Agamidae) and goannas (Varanidae) have radiated extensively, with the former three reaching their greatest diversity in Australia (Hutchinson and Donnellan 1993) (Table 12.1). The higher-level herpetological diversity of New Zealand is considerably more depauperate than Australia, comprising just four families. There is a single lineage of frog (with four species), the last remaining species of the reptile order Rhynchocephalia (the tuatara, *Sphenodon punctatus*), one genus of skinks (*Oligosoma*, ~61 species) and seven genera of geckos (~43 species) (Table 12.1). Terrestrial snakes, freshwater turtles and crocodiles are absent. Like Australia, while diversity at higher taxonomic levels is low, species-level diversity in the few lineages of herpetofauna is exceptionally high in New Zealand (Table 12.1).

In the past two decades, the description of new species has increased rapidly worldwide (Bickford *et al.* 2007; Padial *et al.* 2010; Goldstein and DeSalle 2011; Blackwell 2011; Costello *et al.* 2013), fuelled largely by the increasing availability and use of DNA sequence data. While this phenomenon is particularly evident in poorly studied organisms (i.e. fungi; Buee *et al.* 2009) or regions (i.e. New Guinea; Riedel *et al.* 2013), even groups that have received thorough taxonomic assessment continue to herald new discoveries (i.e. mammals, Ceballos and Ehrlich 2009; birds, Lohman *et al.* 2010). In Australia, the number of currently recognised reptile and amphibian species has nearly doubled in the past 40 years, from 664 recognised species in 1975 to more than 1218 species in the most recent catalogue of the Australian herpetofauna (Cogger 2014). While some of this new biodiversity represents morphologically well-differentiated taxa uncovered by fieldwork in unexplored regions (e.g. Hoskin 2013; Hoskin and Couper 2013; Hoskin 2014, from an expedition to the remote Cape Melville), a major consequence of recent molecular assessments is that many previously unrecognised species have been detected within morphologically similar forms that comprise *species complexes* of a larger number of cryptic taxa (Oliver *et al.* 2013a, 2014c). In the Australian herpetofauna, such complexes are particularly well documented in geckos (Oliver *et al.* 2007, 2009, 2010, 2014a,b; Fujita *et al.* 2010; Pepper *et al.* 2011a, 2013a,b; Shea *et al.* 2011; Sistrom *et al.* 2013), but also have been detected in blindsnakes (Marin *et al.* 2013b), skinks (Rabosky *et al.* 2004; Horner and Adams 2007; Smith and Adams 2007; Dolman and Hugall 2008), dragons (Smith *et al.* 2011) and frogs (Catullo *et al.* 2011, 2014). A similar story has unfolded in New Zealand, first with allozyme electrophoresis data and later mitochondrial and nuclear DNA sequence data uncovering cryptic species within several widespread New Zealand skinks (Daugherty *et al.* 1990; Patterson and Daugherty 1990; Greaves *et al.* 2007, 2008; Chapple *et al.* 2008a,b, 2009, 2011; Bell and Patterson 2008; Chapple and Ritchie 2013; Patterson *et al.* 2013) and geckos (Nielsen *et al.* 2011). However, much of the recognised lizard biodiversity (~45%) in New Zealand remains to be formally described (Hitchmough *et al.* 2013).

Data from molecular genetics can play a major role in our understanding of patterns of species richness on global, regional and local scales. The herpetofauna in particular provide an excellent system for exploring evolutionary and biogeographic hypotheses (Vitt *et al.* 2003), due to their age (an overwhelming proportion of the anuran and chelonian faunas are Gondwanan), extreme diversity, near-worldwide distribution and the dispersal ability of many taxa across oceanic basins (Vences *et al.* 2003; Rocha *et al.* 2006). The broad-level biogeography of the Australian and New Zealand herpetofauna has been addressed previously in several important publications (i.e. Storr 1964; Cogger and Heatwole 1981; Cracraft 1991; Hutchinson and Donnellan 1993; Merrick *et al.* 2006 and references within). However, as molecular approaches have driven our recent enhanced understanding of the diversity and distribution of the Australian and New Zealand herpetofauna, it is timely to reconsider their biogeographic patterns. In this chapter, we summarise our current understanding of the biogeography of Australian and New Zealand herpetofauna based on molecular data and new analytical methods, with an emphasis on intracontinental distributional and phylogeographic patterns. We focus on regions rather than biogeographical accounts by families in order to showcase the contribution of reptiles and amphibians to improving our understanding of broader evolutionary patterns and processes emerging in these regions. We begin with an overview of the geophysical and climatic history of Australia and New Zealand to provide important context within which to interpret modern biogeographical patterns of the herpetofauna.

Major Geologic and Climatic Determinants Underlying Patterns in Herpetological Biodiversity

An understanding of earth history is crucial to any discussion of biogeographic patterns and undoubtedly goes a long way towards explaining present species richness and evolutionary history. For example, the large-scale movement of continents over geologic time has been the principal explanation for global distribution patterns of organisms (Craw *et al.* 1999). At the regional scale, it also is well known that geology and geophysical processes play a major role in shaping the evolutionary dynamics of organisms. Tectonic uplift can drive diversification both by creating new habitats and isolating populations on either side of mountain ranges (Hughes and Eastwood 2006). Similarly, vicariance processes involving river barriers (Hall and Harvey 2002) or the intermittent connection of land bridges (Riddle *et al.* 2000) also impact on evolutionary dynamics. At the local scale, the importance of geological heterogeneity in shaping plant diversification and distribution patterns via edaphic specialisation is widely recognised (Kruckeberg 2002; Fine *et al.* 2005), with colour variations in animals also related to variations in geological substrates (Rosenblum and Harmon 2011). As a general rule, geological diversity (and therefore habitat heterogeneity) is positively correlated to species richness (Anderson and Feree 2010).

In addition to geological processes, historical climate change also drives speciation and biotic diversification (Hewitt 1996, 2000). Global climate has fluctuated greatly through the Cenozoic, particularly from the mid-Neogene (23.0–2.5 mya) and throughout the Quaternary (2.5 mya–present), when rapid global cooling led to diminishing precipitation and the instigation of sharply oscillating temperatures of the glacial and interglacial cycles of the Pleistocene. The expansion and contraction of different habitats during these climatic oscillations led to great changes in species distributions (Hewitt 2000), with the persistence and diversification of some lineages in refugia allowing populations to evolve independently with limited gene flow, facilitating biotic diversification and speciation. This phenomenon has been well documented in herpetofauna (and other taxa) across all biomes, including in boreal (i.e. Ursenbacher *et al.* 2006), tropical (i.e. Schauble and Moritz 2001), temperate (i.e. Chapple *et al.* 2011a,b) and arid systems (i.e. Pepper *et al.* 2011b).

Australia and New Zealand both have a long tectonic history following the breakup of Gondwana, and more than 80 Ma of subsequent geographic isolation. However, given they are characterised by strikingly different modern-day environments and palaeogeological histories, they are addressed separately in the following sections.

Australia: An Old and Stable Continent

The Australian continent possesses the oldest known materials on earth, as well as the oldest landforms (Gale 1992). Many features of the landscape have been geologically stable for tens of millions of years (Gale 1992; Twidale 2000), and as a result Australia's 'mountainous' regions are considerably more subdued than those of other continents, and these have been progressively eroding, with little in the way of modern tectonic uplift. Recent tectonic events have been confined largely to the leading edge of the Australia–New Guinea Plate (Hill and Hall 2002).

Modern surface landforms across Australia are directly influenced by the underlying geology. For example, the western two-thirds of Australia is situated on higher-elevation regions associated with the Western Plateau (Mabbutt 1988) and comprises a number of inland ranges, including the Pilbara, Kimberley and central Australian ranges (Wasson 1982). These regions are ancient, uplifted exposures of the underlying Australian craton, characterised by rugged landscapes of razor-back ridges, scarps and scattered mesas. In contrast, much of eastern Australia is situated on the desert floodplains of the Interior or Central Lowlands (Wasson 1982; Mabbutt 1988). This landscape is low-lying with little topographic variation, and is dominated by the extensive Lake Eyre and Murray–Darling Basins. The main topographic feature of eastern Australia is the Great Dividing Range (GDR), stretching more than 3500 km down the entire eastern coast. While the timing of the uplift of these eastern highlands is contentious, with various hypotheses ranging from the Palaeozoic to the Cenozoic (Van der Beek *et al.*

1999), substantial uplift along this mountain chain is almost certainly related to the Cretaceous rifting of the Tasman Sea ~94 mya (O'Sullivan *et al.* 2000).

Despite the apparent antiquity of the landscapes, the Australian continent has had a dynamic tectonic history during the Cenozoic, having migrated more than 3000 km to the north-northeast over the past 45 Ma as part of the Indo-Australian Plate (Quigley *et al.* 2010). This greatly increased convergence between the Australian and Pacific Plates lead to the immigration of lineages with Southeast Asian origins, to an Australian biota that had previously evolved in isolation since separation from Antarctica (discussed in more detail later in the chapter).

The biggest change in more recent Australian history has undoubtedly been the aridification of the continental interior, with vast inland seas and tropical ecosystems replaced over the last 15 Ma by increasingly arid landscapes and ecosystems (Frakes *et al.* 1987; Fujioka and Chappell 2010). The height of arid conditions in Australia appears to correlate with the transition from high-frequency, low-amplitude glaciations (every 40 ka) that characterised the Late Pliocene/Early Pleistocene, to the low-frequency, high-amplitude glaciations (every 100 ka) that became established in the Middle Pleistocene (Huybers 2007). This led to increasingly severe aridification and the development of the vast inland sand deserts and dune systems (Bowler 1976; Frakes *et al.* 1987; Mabbutt 1988; Martin 2006; Fujioka *et al.* 2009; McLaren and Wallace 2010) as recently as 1 mya (Fujioka *et al.* 2009). The mesic fringes of Australia were not immune to the influence of aridification, with palaeoenvironmental evidence suggesting much drier conditions in the tropical north of Australia during Pleistocene glacial cycles, in conjunction with cooler temperatures, especially in lowland regions (Reeves *et al.* 2013a). Evidence of extensive sand dune activity also has been discovered beneath presently forested regions along the humid margins of eastern Australia, indicating substantially drier conditions during the Last Glacial Maximum (LGM) (Thom *et al.* 1994; Hesse *et al.* 2004). While on many continents glacial cycles of the Pleistocene promoted widespread glaciation, in Australia the Koskiuszko Massif in the Snowy Mountains of New South Wales and the Tasmanian Highlands are thought to be the only regions affected by glacial activity during this period (Barrows *et al.* 2002).

New Zealand: A Gondwanan Fragment with a Post-Oligocene Makeover

In contrast to Australia, New Zealand has had a complex and tumultuous geological and climatic history. Due to the quirks of its evolutionary history, New Zealand displays the blended characteristics of both a continental fragment, with Gondwanan heritage, and an isolated oceanic archipelago (Daugherty *et al.* 1993; Gibbs 2006; reviewed in Wallis and Trewick 2009). Zealandia separated from Gondwana ~82 mya, with the formation of the Tasman Sea by ~65 mya resulting in the 1500 km isolation of New Zealand from Australia, which has been maintained through to the present day (Cooper and Millener 1993; Gibbs 2006; Campbell and Hutching 2007; Wallis and Trewick 2009). Over 40 Ma (64–24 mya) Zealandia was slowly stretched and thinned, which resulted in the substantial thinning of the continental crust and the gradual subsidence and marine inundation of New Zealand (Gibbs 2006; Trewick *et al.* 2007; Landis *et al.* 2008; Neall and Trewick 2008; Wallis and Trewick 2009). While there is consensus that this marine inundation reached its peak in the Oligocene with the Oligocene transgression (or *Oligocene drowning*; Suggate *et al.* 1978; Cooper and Millener 1993; Gibbs 2006; Wallis and Trewick 2009), there is considerable debate as to whether New Zealand was completely submerged (e.g. Trewick *et al.* 2007; Landis *et al.* 2008) or simply reduced to a series of low-lying islands (e.g. Cooper and Cooper 1995; Lee *et al.* 2009).

The rebirth of New Zealand following the Oligocene drowning has been driven by tectonic and volcanic activity. New Zealand is situated at the boundary of the Pacific and Indo-Australian Plates, and continual uplift along the Alpine Fault (running southwest–northeast along the majority of the South Island) since the Late Oligocene (~25–23 mya) has led to the formation of modern New Zealand (Gibbs 2006; Trewick *et al.* 2007; Landis *et al.* 2008). Volcanic activity associated with this tectonism occurred until 13 mya in the South Island and until the present day in the North Island (Wallis and Trewick 2009). The Southern Alps that currently characterise the South Island are the result of tectonic activity along the Alpine Fault that commenced during the Miocene and intensified during the Pliocene (Gage 1980; Suggate 1982; Stevens *et al.* 1995; Landis *et al.* 2008). During this time, severe tectonic uplift (up to

16 km, 2–11 mm/year; Wellman 1979) has been countered by high rates of erosion (up to 12 km; Craw 1995), but has still resulted in the formation of prominent, high-elevation regions (>3000 m) within the Southern Alps (Chamberlain *et al.* 1999; Lee *et al.* 2001; Gibbs 2006; Wallis and Trewick 2009). The stable warm-/cool-temperate climate that had prevailed since the Oligocene (Fleming 1975; Cooper and Millener 1993; Lee *et al.* 2001) eventually gave way in the Late Pliocene to rapid cooling that continued through to Pleistocene glacial cycles (Cooper and Millener 1993; Newnham *et al.* 1999). While large areas of the South Island (up to 30%) were covered by glaciers during the Pleistocene, the North Island was not subjected to any extensive glaciation (Newnham *et al.* 1999; Carter 2005). Despite the post-Oligocene topographic makeover of New Zealand, episodes of regional marine inundation have still occurred during the Pliocene (Manawatu Strait [the inundation of the lower North Island]) and Pleistocene (Northland, Cook Strait [separating the North Island and South Island]; Foveaux Strait [separating the South Island and Stewart Island]) (Lewis *et al.* 1994; Worthy and Holdaway 2002).

Recapitulating Biogeographical Origins with Molecular Data

A number of authors have previously reviewed the age, origin and regional affinities of the Australian and New Zealand herpetofauna (e.g. Storr 1964; Cogger and Heatwole 1981; Cracraft 1991; Hutchinson and Donnellan 1993; Table 12.1). These evolutionary hypotheses have been based largely on species distribution/area relationships and fossils, in combination with geological and climate history. More than two decades of subsequent research, in particular ever-increasing amounts of molecular data as well as increased taxonomic effort, has seen many of these ideas tested in a phylogenetic framework (Table 12.1).

For example, it has long been considered that the Australian squamate fauna consists predominantly of Indo-Malay lineages, with major adaptive radiations within Australia of elapid snakes and diplodactyline geckos, as well as major lizard groups of the agamids, varanids and skinks thought to be derived from ancestors that arrived no later than the mid-Tertiary (Cogger and Heatwole 1981; Figure 12.1). Recent genetic data from agamids (Hugall *et al.* 2008) confirmed this using mitochondrial and nuclear loci, suggesting agamids immigrated into Australia from Southeast Asia via northern mesic forest biomes within the last 30 Ma, with diversification initially within mesic habitats followed by radiations of xeric taxa into the emerging arid zone. Scincid lizards in the clade Lygosominae also are thought to have dispersed on multiple occasions to Australia from Southeast Asia within the last 25 Ma, subsequent to the rifting of Australia and Antarctica (Skinner *et al.* 2011). Similarly, varanids appear to have dispersed to Australia from Southeast Asia in the Late Eocene/Oligocene around 32 mya (Vidal *et al.* 2012), while the venomous elapid snakes may have arrived from Southeast Asia as recently as 10 mya, and then radiated extensively in both the terrestrial and marine environments (Keogh 1998; Keogh *et al.* 1998, 2001; Lukoscheck and Keogh 2006; Sanders *et al.* 2008). In contrast, the limbless Pygopodidae have long been thought to have a Gondwanan history, owing to their endemicity in the Australian–Papuan region, and fossil evidence that establishes their presence in Australia well before the time when other biotic elements of Asian origin invaded Australia (Hutchinson 1997; Jennings *et al.* 2003). Recent molecular evidence has not only confirmed that pygopodids and diplodactytline geckos are sister groups (Donnellan *et al.* 1999), but in addition indicates much of the Australian gekkonid fauna (carphodactyalids, diplodactylids and pygopodids) comprises ancient Gondwanan lineages estimated to have arisen 70 mya (Oliver and Sanders 2009). Myobatrachid frogs also are a Gondwanan group (Littlejohn *et al.* 1993; Pyron 2014), while the evolutionary history of the Australian hylid frogs is less clear, with some authors suggesting they arrived via long-distance rafting from South America during the Palaeocene/Eocene (Pyron 2014).

The strong affinity between the herpetofauna of Australia and New Guinea has been emphasised (Cogger and Heatwole 1981), with these regions sharing a diverse array of taxa, especially at the generic level (Allison 2006; Menzies 2006; Todd *et al.* 2014a). More detailed examination using phylogenetic data has shed light on the patterns and directions of dispersal between these regions, highlighting differences between herpetofaunal lineages. For example, freshwater turtles in the family Chelidae appear to have dispersed from Australia to New Guinea on multiple occasions during the Miocene and Pliocene

FIGURE 12.1 (See colour insert.) A recently delineated species complex of gecko, formerly *Diplodactylus conspicillatus*. (From Oliver, P. M., *et al.*, *PLoS One*, 2014.) Species of the *D. conspicillatus* complex in life: (a) *D. conspicillatus* from 10 km north of Barkley Hwy on Ranken to Alexander Station Road, northeastern Northern Territory (image: Ross Sadlier); (b) *D. conspicillatus* Alice Springs, Northern Territory (image: Eric Vanderduys); (c) *D. hillii*, Dorat Road, Northern Territory (Image: Paul Horner); (d) *D. laevis* in life from Morgan Range, Western Australia (Image: Mark Hutchinson); (e) *D. platyurus*, Brooklyn Station, north Queensland (Image: Eric Vanderduys); (f) *D. platyurus* Myendetta Stn., Charleville, Queensland (image: Steve Wilson); (g) *D. bilybara* sp. nov. Onslow, Western Australia (image: Ryan Ellis); (h) *D. custos* sp. nov. Gibb River Road turnoff via Wyndham, Western Australia (image: Steve Wilson).

(Le *et al.* 2013; Todd *et al.* 2014), with Papuan members of rainbow skinks (Stuart-Fox *et al.* 2002), elapid snakes (Wuster *et al.* 2005; Williams *et al.* 2008) and forest dragons (Hugall *et al.* 2008) also appearing to be derived from Australian ancestors. In contrast, genetic examples are emerging that also clearly demonstrate Australian lineages that have been derived from New Guinean taxa, such as in the geckos *Cyrtodactylus* (Oliver *et al.* 2012b) and *Nactus* (Heinicke *et al.* 2010).

The New Zealand herpetofauna have played a pivotal role in our understanding of the biogeographic significance of the Oligocene drowning. In the absence of strong geological evidence for the continuous presence of land throughout the Oligocene (reviewed in Landis *et al.* 2008), some authors have suggested that New Zealand was completely submerged (Waters and Craw 2006; Trewick *et al.* 2007; Goldberg *et al.* 2008; reviewed in Wallis and Trewick 2009). However, the presence of Leiopelmatid frogs and the tuatara in New Zealand, two lineages with strong evidence (including molecular data) for Gondwanan origins (Table 12.1) and a limited capacity for long-distance over-water dispersal (Rest *et al.* 2003; Roelants and Bossuyt 2005; Hugall *et al.* 2007; Cree 2014), has formed a central pillar of the evidence for the continued presence of land throughout the Oligocene (Daugherty *et al.* 1993; Wallis and Trewick 2009; Sharma and Wheeler 2013). In contrast, recent molecular studies have indicated diplodactylid geckos colonised New Zealand from Australia during the Oligocene (between 40.2–24.4 mya; Nielsen *et al.* 2011), with Eugongylus group skinks reaching New Zealand from New Caledonia (via the Lord Howe Rise and Norfolk Ridge) during the Early Miocene (24.4 mya, range 16–22.6 mya; Chapple *et al.* 2009) (Table 12.1).

Regional Biogeography: Species Richness and Endemism

At the continental scale, Australia and New Zealand have markedly contrasting biogeographic patterns; Australia generally has large-scale, biome-level patterns, which may be a reflection of widespread climate-driven diversification. In contrast, patterns in the New Zealand herpetofauna exhibit continental-level biogeographic patterns (e.g. deep genetic splits) over short geographic distances, likely influenced by regional and local tectonic or eustatic processes (e.g. the Alpine Fault, Cook Strait, Taupo Line).

Various authors have used ever-increasing database records to identify geographic patterns of endemism and species richness in the herpetofauna of Australia (Pianka and Schall 1981; Cogger and Heatwole 1981; Gambold and Woinarski 1993; Williams and Pearson 1997; Williams and Hero 2001; Slatyer *et al.* 2007; Rosauer *et al.* 2009; Powney *et al.* 2010) and New Zealand (Gibbs 2006; Chapple *et al.* 2009; Nielsen *et al.* 2011; Di Virgilio *et al.* 2014; Cree 2014). As physiological constraints (e.g. the permeable skin of amphibians) strongly influence the biomes and geographic distributions where different taxa are concentrated, centres of species diversity are different between reptiles and amphibians, and even within major groups (Cogger and Heatwole 1981). For example, the greatest amphibian species richness in Australia exists along the humid east and southeast coasts, with major centres of diversity in the Wet Tropics of Queensland, and the coastal region near the Queensland/New South Wales border (Slatyer *et al.* 2007; Powney *et al.* 2010). In contrast, for reptiles, a group thought to be generally preadapted physiologically and behaviourally to xeric conditions (Pianka and Schall 1981), species diversity across many lineages is especially high in the arid zone, particularly geckos and agamids (Cogger and Heatwole 1981; Powney *et al.* 2010). Indeed, the Great Victoria Desert of Western Australia contains the highest diversity of lizards anywhere on earth (Mittermeier *et al.* 1999). In contrast, elapid snakes and scincid lizards have their greatest species diversity along the mesic east coast, while varanid lizards reach their highest species diversity in the Monsoon Tropics of northern Australia (Cogger and Heatwole 1981; Powney *et al.* 2010). In New Zealand, despite some differences between skinks and geckos, species richness and endemism is highest in Northland, Nelson/Marlborough and Otago/Southland (Chapple *et al.* 2009; Nielsen *et al.* 2011; Di Virgilio *et al.* 2014).

Knowledge on the evolutionary history of the herpetofauna of Australia and New Zealand is patchy and variable within and between taxonomic groups and geographic areas. For example, in Australia, considerable molecular systematic research has been focused on the herpetofauna of the mesic areas of the east coast (e.g. McGuigan *et al.* 1998; Schneider *et al.* 1998; James and Moritz 2000; Schauble and Moritz 2001; Stuart-Fox *et al.* 2001; Moussalli *et al.* 2005; Symula *et al.* 2008; Edwards 2010; Bell *et al.* 2011; Chapple *et al.* 2011a,b; Pepper *et al.* 2014) and in the arid zone (Kearney *et al.* 2003; Chapple and Keogh 2004; Chapple *et al.* 2004; Strasburg and Kearney 2005; Pepper *et al.* 2006, 2008, 2011a,b, 2014; Rabosky *et al.* 2014; Oliver and Bauer 2011; Oliver *et al.* 2007). However, some regions of Australia still remain vastly underexplored in a molecular sense, and it is thought that significant evolutionary diversity is not reflected in current taxonomy (Oliver *et al.* in press). For example, increased taxonomic effort in the remote Monsoon Tropics region of northern Australia is revealing this region to be much more diverse and complex than previously thought, with remarkably high endemism over restricted geographic ranges in a number of lizard genera (reviewed later in this chapter) (Fujita *et al.* 2010; Oliver *et al.* 2010, 2013a, 2014a; Pepper *et al.* 2011b; Smith *et al.* 2011; Marin *et al.* 2013b) as well as in amphibians (Doughty and Anstis 2007; Doughty and Roberts 2008; Doughty *et al.* 2009; Anstis *et al.* 2010; Catullo *et al.* 2014; Oliver and Parkin 2014).

In New Zealand, the biogeographic and phylogeographic patterns have been explored most extensively in skinks (Berry and Gleeson 2005; Chapple and Patterson 2007; Greaves *et al.* 2007, 2008; Hare *et al.* 2008; Liggins *et al.* 2008a,b; O'Neill *et al.* 2008; Chapple *et al.* 2008a,b,c, 2009, 2011c, 2012; Miller *et al.* 2009; Patterson *et al.* 2013; Di Virgilio *et al.* 2014) and the tuatara (Hay *et al.* 2010; Cree 2014), with relatively less attention on the frogs (Holyoake *et al.* 2001; Fouquet *et al.* 2010) and geckos (Towns *et al.* 1985; Nielsen *et al.* 2011; Di Virgilio *et al.* 2014). However, many regions of New Zealand are yet to be explored in detail to assess their herpetological diversity. Recent expeditions to some remote regions (e.g. Jewell and Tocher 2005; Jewell 2007) have uncovered new lizard species (Chapple and Patterson 2007; Bell and Patterson 2008; Patterson and Bell 2009; Chapple *et al.* 2011c), and thus the true herpetological diversity of New Zealand is likely higher than currently recognised.

Patterns within Local Regions of Endemism

Because fossil and geological records are often poor, and palaeoclimate reconstructions for one area are regularly extrapolated and applied to distant geographic regions, phylogenetic data provide an important means to develop and refine hypotheses about when, and how, organisms adapted to and diversified within biomes or regions of endemism. For example, despite the sedimentary hiatus that typically characterises the geological records of xeric landforms, dated molecular studies of arid-zone taxa across the globe show deep divergences in line with the postulated onset and development of aridification, with intensifying aridity, persistence in localised mesic refugia and the movement of mobile sand deserts thought to drive phylogenetic divergences and phylogeographic structuring. Similarly, biogeographic studies in New Zealand have been used in concert with geological evidence to gain a better understanding of the Oligocene drowning (reviewed in Wallis and Trewick 2009), the uplift of the Southern Alps (Liggins *et al.* 2008a; Wallis and Trewick 2009) and the formation of the Chatham Islands (Liggins *et al.* 2008b; Wallis and Trewick 2009). Largely due to their extreme diversity, distribution across multiple biomes and habitats and low vagility, the herpetofauna are an ideal system in which to explore phylogeographic patterns, and as such they are well represented in the phylogeographic literature. In particular, the regular use of common molecular markers (e.g. the mitochondrial loci ND2, ND4 and 16SrRNA, as well as the nuclear loci RAG1 and Cmos) have made it possible to compare the levels and timing of divergences across lineages. In the following sections, we consider the major biomes within both Australia and New Zealand, and use molecular studies of the herpetofauna to characterise shared patterns in order to better understand biome origins, assembly and maintenance.

Australia

Broadly speaking, the Australian continent can be classified into a number of different biogeographical units corresponding to major biome types (i.e. Crisp and Cook 2007). For simplicity, we consider the following four broad regions: the vast central arid zone, the northern Monsoon Tropics, the Eastern Mesic Biome (EMB) and the Southwestern Mesic Biome (SMB) (Catullo and Keogh 2014; Figure 12.2a). Within these broadly defined regions, several authors have identified areas of endemism based on congruent biogeographic patterns of flora and fauna (e.g. Cracraft 1991; Crisp and Cook 2007). However, the relationships between areas of endemism within a given biome have been rather obscure, particularly

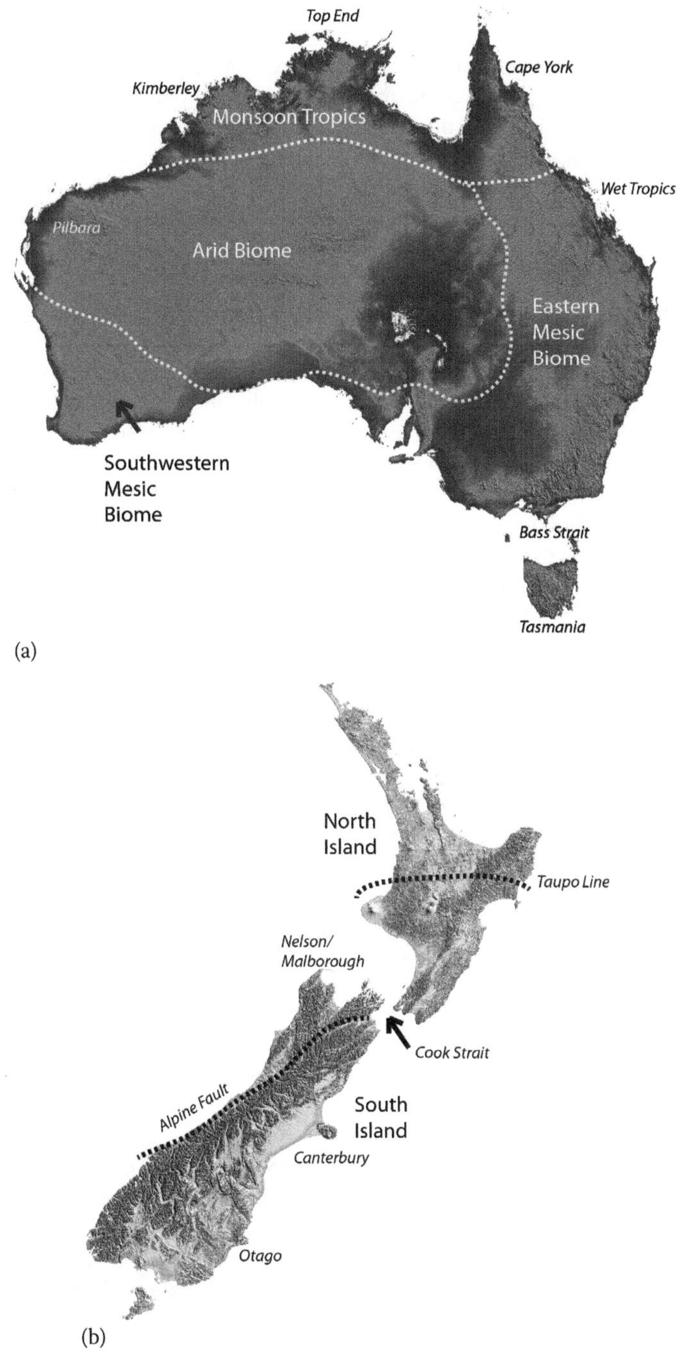

FIGURE 12.2 The location of areas of endemism and biogeographic barriers in (a) Australia and (b) New Zealand mentioned in the text.

for vast areas such as the arid zone (Cracraft 1991). Molecular data based on the sampling of widespread taxa are revealing that biogeographic barriers identifiable at higher taxonomic levels also act at the intraspecific level to restrict gene flow, leading to congruent phylogeographic patterns at finer genetic scales. As such, we are beginning to better understand how genetic diversity is structured within biomes, and the relationships among areas of endemism.

Arid Zone

The Australian arid zone comprises more than 70% of the continent and is one of the largest arid systems in the world. While often portrayed in the literature as a homogeneous region in the centre of Australia, the landscape is considerably more complex and comprises a matrix of different desert systems. While the topography of the region is generally subdued, ancient and isolated rocky uplands occur in the Pilbara, the central Australian ranges and the Flinders ranges. In stark contrast, the vast intervening lowlands comprise stony and sandy deserts, the latter of which are thought to have formed less than a million years ago at the height of the Pleistocene (Fujioka and Chappell 2010). A major geomorphic division separates the western desert systems (the Great Sandy, Little Sandy, Great Victoria, Gibson and Tanami Deserts) formed on the tablelands of the Great Western Plateau, from the eastern desert systems (the Strzelecki, Simpson and Sturts Stony Deserts) and floodplains of the Central-Eastern Lowlands. Importantly for desert organisms, watercourses in the western deserts are thought to have ceased flowing in the Late Miocene (Quilty 1994), while palaeochannels (and indeed current riparian systems and floodplains) in the eastern arid zone indicate the large volume of water that continued to drain from the uplands of the Monsoon Tropics during wet periods.

Cracraft (1991) identified four centres of endemism in the arid zone: the Pilbara, the Western Desert, the Eastern Desert and the Northern Desert, noting that the Western Desert and the Northern Desert were particularly ill-defined. Molecular data from a number of reptile species generally support this cohesion, with distinct lineages in the Pilbara, as well as the topographically differentiated Western and Eastern Deserts (Pepper *et al.* 2011a; Oliver *et al.* 2014c). While sampling is particularly sparse in the Northern Desert, this region also is emerging genetically as a distinct area of endemism (Smith *et al.* 2011; Catullo *et al.* 2014). In addition, genetic data sets accumulating for arid-zone reptiles suggest that range shifts due to climate fluctuations had diverse impacts on genetic structure, with complex phylogeographic patterns indicative of habitat specialisation (Chapple and Keogh 2004; Pepper *et al.* 2006, 2008; Couper and Hoskin 2008; Shoo *et al.* 2008), persistence and diversification within mesic refugia (Fujita *et al.* 2010; Pepper *et al.* 2011a,b), and recent range expansion into the deserts (Jennings *et al.* 2003; Fujita *et al.* 2010; Oliver and Bauer 2011; Pepper *et al.* 2011a,b; Oliver *et al.* 2014c).

Divergence dates estimated for numerous arid-zone herpetofauna suggest the majority of taxa are much older than the age of the desert systems they now live in. Recent research has focused on the role of the rocky uplands as ancient centres of persistence and diversification for mesic-adapted reptiles as the surrounding landscapes dried out (Pepper *et al.*, 2008, 2011a,b, 2013b, 2014; Fujita *et al.* 2010). These mesic refugia typically are characterised by elevated levels of genetic diversity, with Miocene-aged crown nodes reflecting long-term persistence and the repeated movement in and out of refugia (Pepper *et al.* 2011b). However, the desert lowlands also are home to a remarkably diverse assemblage of arid-adapted lineages that have evolved in the absence of topographic and other physical barriers to dispersal. In stark contrast to saxicolous taxa, the genetic signatures of a number of desert lizards are characterised by exceptionally low genetic diversity over large geographic areas (Pepper *et al.* 2011c). While deeper divergences within these lineages are inferred to be Miocene in age, this pattern is indicative of the rapid expansion of arid-adapted lineages into large areas of suitable territory (Castoe *et al.* 2007), concomitant with the development of the widespread sand deserts less than 1 mya.

To date, few data sets include comprehensive sampling across the arid zone (but see Fujita *et al.* 2010; Pepper *et al.* 2011a,c; Oliver *et al.* 2014c), and our understanding of processes that may be responsible for diversification within true deserts is limited. Identifying whether common patterns can be found across the arid zone and whether they relate to different desert systems (i.e. stony vs. sandy, eastern vs. western), and whether and where arid refugia (i.e. centres of persistence for xeric-adapted taxa) existed, will be a fruitful area of future research in an area still very much unknown.

Eastern Mesic Biome

The eastern coastline of Australia encompasses a steep climatic gradient, ranging from tropical in the north, warm temperate in the central coast, to cool temperate in the southeast and Tasmania. Spanning more than 23° in latitude, the GDR mountain system runs the entire length of the coast from northern

Queensland to southern Victoria. Cracraft (1991) considered four areas of endemism along the east coast: the Atherton Plateau, eastern Queensland, the Southeastern Forest and Tasmania, with boundaries between the mainland areas corresponding to major breaks in the uplands (e.g. the Burdekin Gap, Hunter Valley, etc.). In a study on *Lampropholis* skinks, Chapple *et al.* (2011a) summarised major biogeographic barriers in eastern Australia, as well as a number of additional barriers repeatedly identified in the genetic signatures of numerous vertebrate and plant taxa.

The formation of the GDR during the Cretaceous undoubtedly had a profound influence on the evolution of taxa on the east coast of Australia (Dubey *et al.* 2010). In addition, the nonuniform relief along its great length of more than 3500 km, and width of over 300 km in some sections, would have provided multiple microclimatic refuges during Plio-Pleistocene cycles of aridification, dividing formerly continuous distributions and allowing the development of genetically distinct local populations. Indeed, genetic studies of numerous herpetological taxa along the GDR show congruent phylogenetic patterns indicative of persistence in higher-elevation mesic refugia during arid phases, with deep genetic breaks across lowland areas inferred to be historically dry biogeographic barriers to dispersal (e.g. Horton 1972; Schäuble and Moritz 2001; Chapple *et al.* 2005, 2011a,b; Symula *et al.* 2008; Bell *et al.* 2010; Pepper *et al.* 2014). The genetic impacts of repeated forest expansion and contraction has been particularly well studied in the herpetofauna of the Queensland Wet Tropics (Schneider *et al.* 1998; McGuigan *et al.* 1998; Schauble and Moritz 2001; Stuart-Fox *et al.* 2001; Couper and Hoskin 2008; Moussalli *et al.* 2009; Moritz *et al.* 2009; Edwards and Melville 2010; Bell *et al.* 2010). Using extensive data based on palaeoclimate surfaces to predict rainforest distribution since the last glacial cycle, the size and location of refugia have been reconstructed and compared with species distribution models and phylogeographic patterns of extant populations. As well as evaluating refugia, detailed phylogeographic data sets for diverse herpetofauna in the Wet Tropics have been used to explore a broad range of evolutionary questions, including the location of suture zones (Moritz *et al.* 2009), reproductive isolation at contact zones (Singhal and Moritz 2013), differences in thermal physiological parameters (Moritz *et al.* 2012) and divergent selection in phenotypic traits (Hoskin *et al.* 2011).

In addition to the expansion and contraction of forest habitats, topographic barriers such as mountain ranges, river valleys and lava fields also are thought to have played an important role in shaping the evolutionary history of modern herpetofauna in eastern Australia (Chapple *et al.* 2011a,b; Haines *et al.* 2014; Smissen *et al.* 2013; Ng *et al.* 2014). Furthermore, while significant glaciation during the Plio-Pleistocene only occurred in Tasmania and had little impact on mainland taxa (Barrows *et al.* 2001), the intermittent land connection (and current isolation of Tasmania) across the Bass Strait during times of low sea level was also important in historical lineage divergence within a number of reptile and amphibian species (reviewed in Ng *et al.* 2014; cf. Chapple *et al.* 2005; Symula *et al.* 2008; Dubey *et al.* 2010).

Monsoon Tropics

The Monsoon Tropics biome of northern Australia lies within the seasonally dry tropics, and presently has a summer (November–April) rainfall regime originating from tropical depressions, thunderstorms and the northern Australian monsoon trough (Wende 1997). Temperatures are high year round, with monthly averages between 25°C and 35°C (Waples 2007). The landscapes are dominated by ranges and dissected escarpments of Proterozoic sandstone, with extensive savannah woodland habitats as well as rainforest, heath and shrublands, mangroves and salt marsh flats (Woinarski *et al.* 2007; Bowman *et al.* 2010).

The Monsoon Tropics can be broadly divided into three areas of endemism, each comprising major sandstone regions: the Kimberley Plateau, Arnhem Land and the Cape York Peninsula (Cracraft 1991). A number of well-known biogeographic barriers have been identified across these regions (reviewed in Bowman *et al.* 2010; Eldridge *et al.* 2011; Catullo *et al.* 2014), largely associated with breaks in the ranges and arid lowlands surrounding major ephemeral river systems. Recently, the Monsoon Tropics have become the focus of increased phylogeographic research (Bowman *et al.* 2010; Fujita *et al.* 2010; Toon *et al.* 2010; Melville *et al.* 2011; Potter *et al.* 2012; Moritz *et al.* 2013; Catullo *et al.* 2014; Catullo and Keogh 2014), with fine-scale molecular data shedding light on the taxonomic composition and distribution of cryptic species complexes, as well as landform attributes that may have contributed to these genetic patterns (i.e. Pepper and Keogh 2014). In addition, these molecular data are helping to unravel the complex evolutionary relationships between the broader arid zone to the south, and New Guinea and other islands to the north. In particular, the herpetofauna are playing a central role in understanding the complex biogeographic history of the region, with large research programs focused on comprehensive sampling and generating large molecular data sets for a number of lizard species complexes such as *Carlia*, *Cryptoblepharus*, *Gehyra* and *Heteronotia*.

The most striking pattern to come out of recent studies of the herpetofauna of the Monsoon Tropics is the extreme level of highly localised endemism (Fujita *et al.* 2010; Pepper *et al.* 2011b; Smith *et al.* 2011; Oliver *et al.* 2010, in press; Catullo *et al.* 2014; C. Moritz *et al.*, unpublished data). Sampling from Arnhem Land and Cape York remains sparse due to inaccessibility and restricted access to private indigenous land; however, concentrated effort in the Kimberley region in the northwest continues to reveal cryptic diversity in the herpetofauna at a scale not seen anywhere else on the continent (e.g. Pepper *et al.* 2011b; Pepper and Keogh 2014; Oliver *et al.* 2010, 2013a,b; C. Moritz *et al.*, unpublished data). For example, the gecko species *Heteronotia planiceps* comprises lineage diversity in the Kimberley on par with the diversity found across the entire continental range of *H. binoei* (Fujita *et al.* 2010; Pepper *et al.* 2011b), with an additional nine deeply diverged lineages uncovered in the Kimberley since the publication of the latter (C. Moritz *et al.*, unpublished data). Similarly, another gecko in the genus *Crenadactylus* currently consisting of two subspecies in the Kimberley was recently found to comprise multiple highly divergent and allopatric lineages, including seven candidate new taxa (Oliver *et al.* 2010, 2013a). This kind of extreme short-range endemism may in part be explained by the particularly heterogeneous nature of the Kimberley landscape, with plateaus, gorges, rivers and numerous islands providing ample barriers to dispersal for nonvagile organisms (Phillips *et al.* 2009). The uplands of the Kimberley Plateau would also provide a more thermally buffered environment than that of the surrounding lowlands, and species diversity would therefore be expected to be high in this region. Indeed, the rugged and deeply dissected uplands in the high-rainfall area of northwest Kimberley has particularly high species diversity and levels of endemism compared with more topographically subdued parts of the region (Slatyer *et al.* 2007; Doughty 2011; Gonzalez-Orozco *et al.* 2011; Maslin *et al.* 2013).

Southwestern Mesic Biome

The SMB has been identified as a global 'biodiversity hotspot'; based largely on plant species richness and endemism, it is the only Australian region in the top 25 (Myers *et al.* 2000). The SMB is surrounded by xeric regions to the north and east, and oceans to the south and west, and is geographically isolated from other temperate mesic zones in Australia. The landscapes are ancient and highly weathered, with little in the way of mountainous topography, and the region has been tectonically stable since the Carboniferous/Permian (Hopper and Gioia 2004).

Considerable research attention has been focused on the high diversity and endemism of southwestern plant taxa (reviewed in Hopper and Gioia 2004; Rix *et al.* 2014), with molecular phylogenies suggesting multiple dispersal events both into, out of and within the SMB throughout the Cretaceous and the Cenozoic (Hopper and Gioia 2004). A significant number of the herpetofauna species also are endemic, including 26% of the reptiles and over 80% of the amphibians (Myers *et al.* 2000; Morgan *et al.* 2007). Two alternative hypotheses have featured prominently in trying to understand the origin of southwestern Australian herpetological diversity (Morgan *et al.* 2007); a historical relationship between the SMB and other regions of Australia has long been suggested to explain the number of genera or related genera found in both the SMB and the EMB (the *multiple invasion hypothesis*), with this hypothesis extensively tested using higher-level frog systematics and distributional data (i.e. Main *et al.* 1958; Lee 1967; Main 1968; Littlejohn 1981). However, in situ diversification within the SMB (the *endemic speciation hypothesis*) has also been proposed, with a number of recent phylogenetic studies supporting high intraspecific genetic divergence between populations in the southwest, rather than repeated migrations across Australia (Read *et al.* 2001; Jennings *et al.* 2003; Morgan *et al.* 2007; Rix *et al.* 2014).

Within the SMB, major climatic gradients and physical features across the landscape are thought to have been important in shaping phylogeographic patterns (Kay and Keogh 2012). For example, strong rainfall and temperature gradients between the High Rainfall Zone of the southwest coast and the inland

Transitional Rainfall Zone (Hopper 1979) appear to have been significant barriers to frogs (Wardell-Johnson and Roberts 1993; Edwards *et al.* 2007, 2008) and skinks (Kay and Keogh 2012). In addition, physical barriers such as the Darling Scarp (Kay and Keogh 2012) and the waterlogged Pingerup Plains (Wardell-Johnson and Roberts 1993; Edwards *et al.* 2008) as well as sea-level fluctuations in coastal areas (Edwards *et al.* 2008; Kay and Keogh 2012) also appear to have been important in the regional diversification of herpetofauna in the SMB.

New Zealand

The introduction of a diverse suite of mammals into New Zealand, which was previously largely devoid of native terrestrial mammals (Worthy and Holdaway 2002; Worthy *et al.* 2006), has resulted in widespread range contractions of many native herpetofauna species (Towns *et al.* 1985; Hitchmough *et al.* 2013; Tingley *et al.* 2013; Nelson *et al.* 2015). This has acted to disrupt and obscure many biogeographic patterns within New Zealand, with molecular or fossil data sometimes required to reconstruct the patterns and processes that have driven the diversity of the native herpetofauna (Worthy and Holdaway 2002; Chapple *et al.* 2008a; Lee *et al.* 2009a). In addition, the New Zealand herpetofauna includes two relictual lineages (Leiopelmatid frogs, tuatara) that are the last remnants of ancient, previously more diverse groups (Holyoake *et al.* 2001; Cree 2014; Table 12.1). This leaves their biogeographic history to be pieced together using the fossil record and the current distribution of the remaining members of the lineages (e.g. *Leiopelma hochstetteri*; Fouquet *et al.* 2010). For instance, the tuatara was previously distributed throughout the North Island, South Island and Stewart Island, but now has a relictual distribution on several Cook Strait Islands and on several island groups off the northeast of the North Island (Cree 2014). The only substantial genetic breaks evident within the tuatara are between the Cook Strait and northeastern island regions (Hay *et al.* 2010).

Previous detailed investigations of the biogeography of New Zealand skinks and geckos (e.g. Towns *et al.* 1985) are now considerably out of date due to the extensive taxonomic revisions that have occurred in both lineages over the last two decades (e.g. Chapple *et al.* 2009; Nielsen *et al.* 2011; Chapple and Ritchie 2013). Recent molecular phylogenetic studies for both skinks (Chapple *et al.* 2009) and geckos (Nielsen *et al.* 2011) have shed light on interspecific biogeographic patterns, while extensive phylogeographic studies of New Zealand skinks also enable the examination of intraspecific patterns. Here we outline six main biogeographic patterns that are present in New Zealand lizards (Figure 12.2b).

1. *Northland diversification*: The extent of landmass in the Northland region, and the degree of connection between the North Island and offshore islands, has been in continual flux since the Pliocene, particularly due to sea-level changes associated with Pleistocene glacial cycles (Rogers 1989; King 2000; Worthy and Holdaway 2002). In particular, some island groups (e.g. Poor Knights Islands, Three Kings Islands) have been isolated from the North Island mainland for 1–2 Ma (Hayward 1986, 1991). While the repeated connection and separation of landmasses and islands in the Northland region appears to have driven substantial diversification in *Oligosoma* skinks (Chapple *et al.* 2008a,b,c, 2009; Hare *et al.* 2008; Patterson *et al.* 2013) and geckos (*Dactylocnemis* and *Naultinus*; Nielsen *et al.* 2011), no consistent patterns have emerged regarding their placement or timing (Wallis and Trewick 2009).
2. *Taupo Line biogeographic barrier*: The Manawatu Strait, which inundated the lower North Island during the Pliocene (Bull and Whitaker 1975; Rogers 1989; Worthy and Holdaway 2002), appears to have had a long-lasting influence on the distribution of lizards (Chapple *et al.* 2009). The Taupo Line, which runs between the present-day Hawkes Bay and Taranaki regions, corresponds to the northern boundary of the Manawatu Strait and represents a major species turnover zone in *Oligosoma* skinks (Hare *et al.* 2008; Liggins *et al.* 2008a; Chapple *et al.* 2009) and geckos (McCann 1955, 1956; Nielsen *et al.* 2011).
3. *Cook Strait*: The North and South Islands of New Zealand are separated by the Cook Strait, a narrow and shallow waterway (Lewis *et al.* 1994). Land bridges formed intermittently during glacial maxima during the Pleistocene that provided the potential for the faunal interchange of taxa between the two islands (Lewis *et al.* 1994; Worthy and Holdaway 2002). Genetic

studies inferred recent dispersal or connection across the Cook Strait in several skinks species (*Oligosoma maccanni*, *O. infrapunctatum*, *O. lineoocellatum*, *O. polychroma*; Greaves *et al.* 2007, 2008; O'Neill *et al.* 2008; Liggins *et al.* 2008a). In addition, several gecko species have distributions that span the Cook Strait (e.g. *Mokopirirakau granulatus*, *Woodworthia maculatus*; Nielsen *et al.* 2011). However, the Cook Strait represents a distributional barrier for several *Oligosoma* skinks (e.g. *O. aeneum*, *O. ornatum*, *O. whitakeri*; Chapple *et al.* 2008a, 2009; Miller *et al.* 2009) and geckos (Nielsen *et al.* 2011). Interestingly, lizard species whose distributions fail to span the Cook Strait are more likely to be continuously distributed across the Taupo Line (reviewed in Chapple *et al.* 2009).

4. *Alpine Fault*: The Alpine Fault has been responsible for the uplift of the Southern Alps since the Miocene. The presence of the Southern Alps is believed to be responsible for the east–west biogeographic breaks (of Miocene–Pliocene origin), or distributional limits, of several South Island lizard species, including *O. polychroma* (Liggins *et al.* 2008a), *O. lineoocellatum* (Greaves *et al.* 2007), *O. infrapunctatum* (Greaves *et al.* 2008) and several gecko species (*Dactylocnemis*, *Mokopirirakau*, *Naultinus*; Nielsen *et al.* 2011). As a result, many lizard species have distributions that are restricted entirely to either the east or west of the Southern Alps (Chapple *et al.* 2009; Nielsen *et al.* 2011).
5. *North–south splits in the South Island*: On the eastern side of the Southern Alps, a combination of tectonic activity and climatic process (mostly Pleistocene glacial cycles) have resulted in alternating extremes of endemism from north to south: Nelson–Marlborough (high), Canterbury (low), Otago-Southland (high) (Wardle *et al.* 1988, Wardle 1991; Craw 1989; Gibbs 2006). As a result, north–south biogeographic breaks in the South Island are common at both the intra- and interspecific level in both skinks (e.g. *O. polychroma*, *O. maccanni*, *O. chloronoton-lineoocellatum*, *O. otagense-waimatense*; Greaves *et al.* 2007; O'Neill *et al.* 2008; Liggins *et al.* 2008a; Chapple *et al.* 2012) and geckos (*Naultinus*, *Woodworthia*; Nielsen *et al.* 2011).
6. *East–west splits in Otago*: The Nevis–Cardrona fault system, which is delineated by the Cardrona and Nevis rivers, has been active since the Late Miocene and constitutes the boundary between eastern and western Otago (reviewed in Waters *et al.* 2001). Several skink species (*O. grande*, *O. maccanni*, *O. otagense*; Berry and Gleeson 2005; O'Neill *et al.* 2008; Chapple *et al.* 2012) exhibit deep genetic breaks across this region, with species breaks also evident in geckos (*Woodworthia*; Nielsen *et al.* 2011). Similarly, tectonic activity in the southern South Island appears to have resulted in several isolated or disjunct species of Miocene–Pliocene origin in mountainous regions (Bell and Patterson 2008; Patterson and Bell 2009; Chapple *et al.* 2011c; Nielsen *et al.* 2011).

Future Research

Our ability to understand the complex biogeographic history of the Australian and New Zealand herpetofauna is rapidly gaining momentum. The application of next-generation sequencing technologies and the availability of phylogenomic data, combined with powerful inferential tools of statistical phylogeography, have made it possible to robustly estimate population connectivity and gene flow, test shared biogeographic barriers and detect underlying mechanisms of diversification in a statistical framework. In particular, emerging model-based analytical methods to infer parameters and compare models (reviewed in Hickerson *et al.* 2010) provide a powerful means for statistically testing complex and competing biogeographical hypotheses, including vicariance versus ecological scenarios, while methods for model-based comparative phylogeographical inference such as approximate Bayesian computation (ABC) can test for simultaneous divergence times (Leache *et al.* 2007) or congruence in biogeographical scenarios across co-distributed taxa (Carnaval *et al.* 2009).

In addition, the increasing availability of online species distribution databases (e.g. the *Atlas of Living Australia* [ALA] website, www.ala.org.au, and BioWeb Herpetofauna Database 2016) allows digital access to point- and museum-based records, expert range maps and checklists, with ever-improving model-based infrastructure (Jetz *et al.* 2012), including the ability to integrate biodiversity data with phylogenies (i.e. Phylojive). While Australian-based museums now routinely collect and store tissue for genetic analyses, this is not common practice in New Zealand. However, new methods for extracting genetic material from formalin-preserved material or dried skins are overcoming the challenges associated with working with highly degraded DNA from historic samples (Bi *et al.* 2013). Access to all these types of data, along with ever-improving analytical techniques, are allowing researchers to get a better handle on true species diversity and distributions, which in turn improves our understanding of biogeographic patterns and processes. Targeted sampling in poorly studied regions (such as the Australian Monsoon Tropics) will be particularly fruitful in this regard.

The delimitation and description of new species, especially those belonging to cryptic species complexes, is being improved through an increasingly utilised *integrative taxonomy* framework (Padial *et al.* 2010; Miralles and Vences 2013). With an emphasis on the demographic and evolutionary processes responsible for lineage diversification, the advantage of coalescent-based approaches is that they have clear and objective underpinnings. When such approaches are combined with more traditional phylogenetic inference methods, as well as with detailed morphological, geographical and ecological data, they provide more complete and robust information on species distributions and boundaries.

Finally, advances in GIS, remote sensing and biodiversity spatial modelling are revolutionising our ability to visualise and analyse biogeography. Not only is access to quality global environmental data layers continuously improving, but the increase in data-sampling points combined with sophisticated ecological niche modelling (e.g. Laffan *et al.* 2010; Rosauer *et al.* 2013) is enhancing our ability to predict not only current species distributions but also to estimate palaeodistributions to understand biogeographic history. The next decade of molecular biogeography and phylogenomics will undoubtedly continue to see reptiles and amphibians playing a central role in the exploration of questions related to a broad range of evolutionary applications in biogeography and ecology.

References

Adams, M., Foster, R., Hutchinson, M. N., Hutchinson, R. G., and Donnellan, S. C. (2003). The Australian scincid lizard *Menetia greyii*: A new instance of widespread vertebrate parthenogenesis. *Evolution*, *57*(11), 2619–2627.

Alfaro, M. E., Karns, D. R., Voris, H. K., Abernathy, E., and Sellins, S. L. (2004). Phylogeny of Cerberus (Serpentes: Homalopsinae) and phylogeography of Cerberus rynchops: Diversification of a coastal marine snake in Southeast Asia. *Journal of Biogeography*, *31*(8), 1277–1292.

Alfaro, M. E., Karns, D. R., Voris, H. K., Brock, C. D., and Stuart, B. L. (2008). Phylogeny, evolutionary history, and biogeography of Oriental–Australian rear-fanged water snakes (Colubroidea: Homalopsidae) inferred from mitochondrial and nuclear DNA sequences. *Molecular Phylogenetics and Evolution*, *46*(2), 576–593.

Allison, A. (2006). *Reptiles and Amphibians of the Trans-Fly Region, New Guinea*. Contribution No. 2006–039 (Pacific Biological Survey: Madang, Papua New Guinea), 50 p.

Amer, S. A. M., and Kumazawa, Y. (2005). Mitochondrial genome of Pogona vitticepes (Reptilia; Agamidae): Control region duplication and the origin of Australasian agamids. *Gene*, *346*, 249–256.

Anderson, M. G., and Ferree, C. E. (2010). Conserving the stage: Climate change and the geophysical underpinnings of species diversity. *PLoS ONE*, *5*(7), e11554.

Ast, J. (2001). Mitochondrial DNA evidence and evolution in Varanoidea (Squamata). *Cladistics*, *17*(3), 211–226.

Barrows, T. T., Stone, J. O., Fifield, L. K., and Cresswell, R. G. (2001). Late Plesitocene glaciation of the Kosciusko Massif, Snowy Mountains, Australia. *Quaternary Research*, *55*, 179–189.

Barrows, T. T., Stone, J. O., Fifield, L. K., and Cresswell, R. G. (2002). The timing of the last glacial maximum in Australia. *Quaternary Science Reviews*, *21*(1), 159–173.

Bell, R.C., MacKenzie, J.B., Hickerson, M.J., *et al.* (2011). Comparative multi-locus phylogeography confirms multiple vicariance events in co-distributed rainforest frogs. *Proceedings of the Royal Society of London Series B: Biological Sciences*, 279, 991–999.

Bell, R. C., MacKenzie, J. B., Hickerson, M. J., Chavarria, K. L., Cunningham, M., Williams, S., and Moritz, C. (2012). Comparative multi-locus phylogeography confirms multiple vicariance events in co-distributed rainforest frogs. *Proceedings of the Royal Society B: Biological Sciences*, *279*(1730), 991–999.

Bell, R. C., Parra, J. L., Tonione, M., Hoskin, C. J., MacKenzie, J. B., Williams, S. E., and Moritz, C. (2010). Patterns of persistence and isolation indicate resilience to climate change in montane rainforest lizards. *Molecular Ecology*, *19*, 2531–2544.

Bell, T. P., and Patterson, G. B. (2008). A rare alpine skink *Oligosoma pikitanga* n. sp. (Reptilia: Scincidae) from Llawrenny Peaks, Fiordland, New Zealand. *Zootaxa*, *1882*, 57–68.

Berry, O., and Gleeson, D. M. (2005). Distinguishing historical fragmentation from a recent population decline: Shrinking or pre-shrunk skink from New Zealand? *Biological Conservation*, *123*(2), 197–210.

Bi, K., Linderoth, T., Vanderpool, D., Good, J. M., Nielsen, R., and Moritz, C. (2013). Unlocking the vault: Next-generation museum population genomics. *Molecular Ecology*, *22*(24), 6018–6032.

Bickford, D., Lohman, D. J., Sodhi, N. S., Ng, P. K. L., Meier, R., Winker, K., *et al.* (2007). Cryptic species as a window on diversity and conservation. *Trends in Ecology and Evolution*, *22*(3), 148–155.

Blackwell, M. (2011). The Fungi: 1, 2, 3 … 5.1 million species? *American Journal of Botany*, *98*(3), 426–438.

Bossuyt, F., Brown, R., Hillis, D., Cannatella, D., and Milinkovitch, M. (2006). Phylogeny and biogeography of a cosmopolitan frog radiation: Late Cretaceous diversification resulted in continent-scale endemism in the family ranidae. *Systematic Biology*, *55*(4), 579–594.

Bowler, J. M. (1976). Aridity in Australia: Age, origins and expression in Aeolian landforms and sediments. *Earth Science Reviews*, *12*, 279–310.

Bowman, D. M. J. S., Brown, G. K., Braby, M. F., Brown, J. R., Cook, L. G., Crisp, M. D., *et al.* (2010). Biogeography of the Australian monsoon tropics. *Journal of Biogeography*, *37*(2), 201–216.

Brochu, C. A., and Storrs, G. W. (2012). A giant crocodile from the Plio-Pleistocene of Kenya, the phylogenetic relationships of Neogene African crocodylines, and the antiquity of Crocodylus in Africa. *Journal of Vertebrate Palaeontology*, *32*(3), 587–602.

Bueé, M., Reich, M., Murat, C., Morin, E., Nilsson, R. H., Uroz, S., and Martin, F. (2009). 454 Pyrosequencing analyses of forest soils reveal an unexpectedly high fungal diversity. *New Phytologist*, *184*(2), 449–456.

Bull, P. C., and Whitaker, A. H. (1975). The amphibians, reptiles, birds and mammals. In *Biogeography and Ecology in New Zealand* (Ed. G Kuschel), pp. 231–276. (Dr. W. Junk: The Hague, the Netherlands).

Byrne, M., Yeates, D. K., Joseph, L., Kearney, M., Bowler, J., Williams, M. A. J., *et al.* (2008). Birth of a biome: Insights into the assembly and maintenance of the Australian arid zone biota. *Molecular Ecology*, *17*(20), 4398–4417.

Campbell, H., and Hutching, G. (2007). *In Search of Ancient New Zealand* (Penguin: Auckland, New Zealand).

Carnaval, A. C., Hickerson, M. J., Haddad, C. F. B., Rodrigues, M. T., and Moritz, C. (2009). Stability predicts genetic diversity in the Brazilian atlantic forest hotspot. *Science*, *323*, 785–789.

Carter, R. M. (2005). A New Zealand climatic template back to c. 3.9 Ma: ODP Site 1119, Canterbury Bight, south-west Pacific Ocean, and its relationship to onland successions. *Journal of the Royal Society of New Zealand*, *35*, 9–42.

Castoe, T. A., Spencer, C. L., and Parkinson, C. L. (2007). Phylogeographic structure and historical demography of the western diamondback rattlesnake (Crotalus atrox): A perspective on North American desert biogeography. *Molecular Phylogenetics and Evolution*, *42*(1), 193–212.

Catullo, R. A., Doughty, P., Roberts, J. D., and Keogh, J. S. (2011). Multi-locus phylogeny and taxonomic revision of *Uperoleia* toadlets (Anura: Myobatrachidae) from the western arid zone of Australia, with a description of a new species. *Zootaxa*, *2902*, 1–43.

Catullo, R. A., and Keogh, J. S. (2014). Aridification drove repeated episodes of diversification between Australian biomes: Evidence from a multi-locus phylogeny of Australian toadlets (*Uperoleia*: Myobatrachidae). *Molecular Phylogenetics and Evolution*, *79*, 106–117.

Catullo, R. A., Lanfear, R., Doughty, P., and Keogh, J. S. (2014). The biogeographical boundaries of northern Australia: Evidence from ecological niche models and a multi-locus phylogeny of *Uperoleia* toadlets (Anura: Myobatrachidae). *Journal of Biogeography*, *41*(4), 659–672.

Ceballos, G., and Ehrlich, P. R. (2009). Discoveries of new mammal species and their implications for conservation and ecosystem services. *Proceedings of the National Academy of Sciences of the United States of America*, *106*(10), 3841–3846.

Chamberlain, C. P., Poage, M. A., Craw, D., and Reynolds, R. C. (1999). Topographic development of the Southern Alps recorded by the isotopic composition of authigenic clay minerals, South Island, New Zealand. *Chemical Geology (including Isotope Geoscience)*, *155*, 279–294.

Chapple, D. G., Keogh, J. S., and Hutchinson, M. N. (2004). Molecular phylogeography and systematics of the arid-zone members of the Egernia whitii (Lacertilia: Scincidae) species group. *Molecular Phylogenetics and Evolution*, *33*, 549–561.

Chapple, D. G., Keogh, J. S., and Hutchinson, M. N. (2004). Molecular phylogeography and systematics of the arid-zone members of the Egernia whitii (Lacertilia: Scincidae) species group. *Molecular Phylogenetics and Evolution*, *33*, 5497–5561.

Chapple, D. G., Bell, T. P., Chapple, S. N. J., Miller, K. A., Daugherty, C. H., and Patterson, G. B. (2011c). Phylogeography and taxonomic revision of the New Zealand cryptic skink (*Oligosoma inconspicuum*; Reptilia: Scincidae) species complex. *Zootaxa*, *2782*, 1–33.

Chapple, D. G., Birkett, A., Miller, K. A., Daugherty, C. H., and Gleeson, D. M. (2012). Phylogeography of the endangered Otago skink, *Oligosoma otagense*: Population structure, hybridisation and genetic diversity in captive populations. *PLoS ONE*, *7*, e34599.

Chapple, D. G., Chapple, S. N. J., and Thompson, M. B. (2011b). Biogeographic barriers in south-eastern Australia drive phylogeographic divergence in the garden skink, *Lampropholis guichenoti*. *Journal of Biogeography*, *38*(9), 1761–1775.

Chapple, D. G., Daugherty, C. H., and Ritchie, P. A. (2008a). Comparative phylogeography reveals pre-decline population structure of New Zealand *Cyclodina* (Reptilia: Scincidae) species. *Biological Journal of the Linnean Society*, *95*, 388–408.

Chapple, D. G., Hoskin C. J., Chapple, S. N. J., and Thompson, M. B. (2011a). Phylogeographic divergence in the widespread delicate skink (Lampropholis delicata) corresponds to dry habitat barriers in eastern Australia. *BMC Evolutionary Biology*, *11*, 191.

Chapple, D. G., and Keogh, J. S. (2004). Parallel adaptive radiations in arid and temperate Australia: Molecular phylogeography and systematics of the *Egernia whitii* (Lacertilia: Scincidae) species group. *Biological Journal of the Linnean Society*, *83*(2), 157–173.

Chapple, D. G., Keogh, J. S., and Hutchinson, M. N. (2005). Substantial genetic substructuring in southeastern and alpine Australia revealed by molecular phylogeography of the *Egernia whitii* (Lacertilia: Scincidae) species group. *Molecular Ecology*, *14*(5), 1279–1292.

Chapple, D. G., and Patterson, G. B. (2007). A new skink species (*Oligosoma taumakae* sp. nov.; Reptilia: Scincidae) from the Open Bay Islands, New Zealand. *New Zealand Journal of Zoology*, *34*, 347–357.

Chapple, D. G., Patterson, G. B., Bell, T., and Daugherty, C. H. (2008b) Taxonomic revision of the New Zealand Copper Skink (*Cyclodina aenea*; Squamata: Scincidae) species complex, with description of two new species. *Journal of Herpetology*, *42*(3), 437–452.

Chapple, D. G., Patterson, G. B., Gleeson, D. M., Daugherty, C. H., and Ritchie, P. A. (2008c). Taxonomic revision of the marbled skink (*Cyclodina oliveri*, Reptilia: Scincidae) species complex, with a description of a new species. *New Zealand Journal of Zoology*, *35*, 129–146.

Chapple, D. G., and Ritchie, P. A. (2013). A retrospective approach to testing the DNA barcoding method. *PLoS ONE*, *8*, e77882.

Chapple, D. G., Ritchie, P. A., and Daugherty, C. H. (2009). Origin, diversification, and systematics of the New Zealand skink fauna (Reptilia: Scincidae). *Molecular Phylogenetics and Evolution*, *52*(2), 470–487.

Cogger, H. (2014). *Reptiles and Amphibians of Australia*, 7th edn. (CSIRO Publishing: Collingwood, Australia).

Cogger, H., and H. Heatwole. (1981). The Australian reptiles: Origins, biogeography, distribution patterns and island evolution. In *Ecological Biogeography of Australia* (Ed. A. Keast), pp. 1333–1373 (Dr. W. Junk: The Hague, the Netherlands).

Cooper, A., and Cooper, R. A. (1995). The Oligocene bottleneck and New Zealand biota: Genetic record of a past environmental crisis. *Proceedings of the Royal Society of London, Series B*, *261*, 293–302.

Cooper, R. A., and Millener, P. R. (1993). The New Zealand biota: Historical background and new research. *Trends in Ecology* and *Evolution*, *8*(12), 429–433.

Costello, M. J., May, R. M., and Stork, N. E. (2013). Can we name earth's species before they go extinct? *Science*, *339*(6118), 413–416.

Couper, P. J., and Hoskin, C. J. (2008). Litho-refugia: The importance of rock landscapes for the long-term persistence of Australian rainforest fauna. *Australian Zoologist*, *34*(4), 554–560.

Couper, P. J., Sadlier, R. A., Shea, G. M., and Wilmer, J. W. (2008). A reassessment of *Saltuarius swaini* (Lacertilia: Diplodactylidae) in southeastern Queensland and New South Wales: Two new taxa, phylogeny, biogeography and conservation. *Records of the Australian Museum*, *60*(1), 87–118.

Cracraft, J. (1991). Patterns of diversification within continental biotas: Hierarchical congruence among the areas of endemism of Australian vertebrates. *Australian Systematic Biology*, *4*, 211–227.

Craw, D. (1995). Reinterpretation of the erosion profile across the southern portion of the Southern Alps, Mount Aspiring area, Otago, New Zealand. *New Zealand Journal of Geology and Geophysics*, *38*, 501–507.

Craw, R. (1989). New Zealand biogeography: A panbiogeography approach. *New Zealand Journal of Zoology*, *16*, 527–547.

Craw, R., Grehan J. R., Heads M. J. (1999). *Panbiogeography: Tracking the History of Life* (Oxford University Press: New York).

Crawford, N. G., Parham, J. F., Sellas, A. B., Faircloth, B. C., Glenn, T. C., Papenfuss, T. J., *et al.* (2014). A phylogenomic analysis of turtles. *Molecular Phylogenetics and Evolution*, *83*, 250–257. doi:10.1016/j.ympev.2014.10.021.

Cree, A. (2014). *Tuatara: Biology and Conservation of a Venerable Survivor* (Canterbury University Press: Christchurch, New Zealand), 583 p.

Crisp, M. D., and Cook, L. G. (2007). A congruent molecular signature of vicariance across multiple plant lineages. *Molecular Phylogenetics and Evolution*, *43*(3), 1106–1117.

Daugherty, C. H., Gibbs, G. W., and Hitchmough, R. A. (1993). Mega-island or micro-continent? New Zealand and its fauna. *Trends in Ecology and Evolution*, *8*, 437–442.

Daugherty, C. H., Patterson, G. B., Thorn, C. J., and French, D. C. (1990). Differentiation of the members of the New Zealand *Leiolopisma nigriplantare* species complex (Lacertilia: Scincidae). *Herpetological Monographs*, *4*, 61–76.

Di Virgilio, G., Laffan, S. W., Ebach, M. C., and Chapple, D. G. (2014). Spatial variation in the climatic predictors of species composition turnover and endemism. *Ecology and Evolution*, *4*, 3264–3278.

Dolman, G., and Hugall, A. F. (2008). Combined mitochondrial and nuclear data enhance resolution of a rapid radiation of Australian rainbow skinks (Scincidae: Carlia). *Molecular Phylogenetics and Evolution*, *49*(3), 92–104.

Donnellan, S. C, Hutchinson, M. N., Dempsey, P., and Osborne, W. S. (2002). Systematics of the *Egernia whitii* species group (Lacertilia: Scincidae) in south-eastern Australia. *Australian Journal of Zoology*, *50*(5), 439.

Donnellan, S. C., Hutchinson, M. N., and Saint, K. M. (1999). Molecular evidence for the phylogeny of Australian gekkonoid lizards. *Biological Journal of the Linnean Society*, *67*, 97–118.

Doughty, P., and Anstis, M. (2007). A new species of rock-dwelling hylid frog (Anura: Hylidae) from the eastern Kimberley region of Western Australia. *Records of the Western Australian Museum*, *23*, 241–257.

Doughty, P., and Roberts, J. D. (2008). A new species of Uperoleia (Anura: Myobatrachidae) from the northwest Kimberley, Western Australia. *Zootaxa*, *1939*, 10–18.

Doughty, P., Anstis, M., and Price, P. (2009). A new species of *Crinia* (Anura: Myobatrachidae) from the high rainfall zone of the northwest Kimberley, Western Australia. *Records of the Western Australian Museum*, *25*, 125–144.

Doughty, P. (2011). An emerging frog diversity hotspot in the northwest Kimberley of Western Australia: Another new frog species from the high rainfall zone. *Records of the Western Australian Museum*, *26*, 209–216.

Doughty, P., Kealley, L., and Donnellan, S. C. (2011). Revision of the pygmy spiny-tailed skinks (*Egernia depressa* species-group) from Western Australia, with descriptions of three new species. *Records of the Western Australian Museum*, *26*(2), 115–127.

Dubey, S., and Shine, R. (2010). Evolutionary diversification of the lizard genus *Bassiana* (Scincidae) across Southern Australia. *PLoS ONE*, 5, e112982.

Dubey, S., Keogh, J. S., and Shine, R. (2010). Plio-Pleistocene diversification and connectivity between mainland and Tasmanian populations of Australian snakes (Drysdalia, Elapidae, Serpentes). *Molecular Phylogenetics and Evolution*, *56*(3), 1119–1125.

Edwards, D. L. (2007). Biogeography and speciation of a direct developing frog from the coastal arid zone of Western Australia. *Molecular Phylogenetics and Evolution*, *45*(2), 494–505.

Edwards, D. L., Roberts, J. D., and Keogh, J. S. (2007). Impact of Plio-Pleistocene arid cycling on the population history of a south- western Australian frog. *Molecular Ecology*, *16*, 2782–2796.

Edwards, D. L., Roberts, J. D., and Keogh, J. S. (2008). Climatic fluctuations shape the phylogeography of a mesic direct-developing frog from the south-western Australian biodiversity hotspot. *Journal of Biogeography*, *35*, 1803–1815.

Edwards, D. L., and Melville, J. (2010). Phylogeographic analysis detects congruent biogeographic patterns between a woodland agamid and Australian wet tropics taxa despite disparate evolutionary trajectories. *Journal of Biogeography*, *37*, 1543–1556.

Edwards, D., and Melville, J. (2011). Extensive phylogeographic and morphological diversity in *Diporiphora nobbi* (Agamidae) leads to a taxonomic review and a new species description. *Journal of Herpetology*, *45*, 530–546.

Eldridge, M. D. B., Potter, S., and Cooper, S. J. B. (2011). Biogeographic barriers in north-western Australia: An overview and standardisation of nomenclature. *Australian Journal of Zoology*, *59*(4), 270.

Estoup, A., Beaumont, M., Sennedot, F., Moritz, C., and Cornuet, J.-M. (2004). Genetic analysis of complex demographic scenarios: Spatially expanding populations of the cane toad, *Bufo marinus*. *Evolution*, *58*(9), 2021–2036.

Fine, P., Daly, D. C., and Cameron, K. M. (2005). The contribution of edaphic heterogeneity to the evolution *and diversity of burseracear trees in the western amazon. Evolution, 59*(7), 1464–1478.

Fitch, A. J., Goodman, A. E., and Donnellan, S. C. (2006). A molecular phylogeny of the Australian monitor lizards (Squamata: Varanidae) inferred from mitochondrial DNA sequences. *Australian Journal of Zoology*, *54*(4), 253.

Fleming, C. A. (1975). The geological history of New Zealand and its biota. In *Biogeography and Ecology in New Zealand* (Ed. G. Kuschel), pp. 1–81 (Dr. W. Junk: Auckland, New Zealand).

Fouquet, A., Green, D. M., Waldman, B., Bowsher, J. H., McBride, K. P., and Gemmell, N. J. (2010). Phylogeography of *Leiopelma hochstetteri* reveals strong genetic structure and suggests new conservation priorities. *Conservation Genetics*, *11*, 907–919.

Frakes, L. A., McGowran, B., and Bowler, J. M. (1987). Evolution of the Australian environments. In *Fauna of Australia*, Vol. 1a (Eds. G. R. Dyne and D. W. Walton), pp. 1–16 (Australian Government Publishing Service: Canberra, Australia).

Fujioka, T., and Chappell, J. (2010). History of Australian aridity: Chronology in the evolution of arid landscapes. In *Australian landscapes* (Eds. P. Bishop and B. Pillans), pp. 121–139. Special Publication *346* (Geological Society: London).

Fujioka, T., Chappell, J., Fifield, L. K., and Rhodes, E. J. (2009). Australian desert dune fields initiated with Pliocene–Pleistocene global climatic shift. *Geology*, *37*, 51–54.

Fujita, M. K., McGuire, J. A., Donnellan, S. C., and Moritz, C. M. (2010). Diversification at the arid-monsoonal interface: Australia-wide biogeography of the Bynoe's gecko (*Heteronotia binoei*; Gekkonidae). *Evolution*, *64*, 2293–2314.

Fuller, S., Baverstock, P., and King, D. (1998). Biogeographic origins of goannas (Varanidae): A molecular perspective. *Molecular Phylogenetics and Evolution*, *9*(2), 294–307.

Gage, M. (1980). *Legends in the Rocks: An Outline of New Zealand Geology*. (Whitcoulls: Christchurch, New Zealand).

Gale, S. J. (1992). Long-term landscape evolution in Australia. *Earth Surface Processes and Landforms*, *17*, 323–343.

Gamble, T., Bauer, A. M., Greenbaum, E., and Jackman, T. R. (2008). Evidence for Gondwanan vicariance in an ancient clade of gecko lizards. *Journal of Biogeography*, *35*, 88–104.

Gambold, N., and Woinarski, J. (1993). Distributional patterns of herpetofauna in monsoon rainforests of the Northern Territory, Australia. *Australian Journal of Ecology*, *18*(4), 431–449.

Gardner, M. G., Hugall, A. F., Donnellan, S. C., Hutchinson, M. N., and Foster, R. (2008). Molecular systematics of social skinks: Phylogeny and taxonomy of the *Egernia* group (Reptilia: Scincidae). *Zoological Journal of the Linnean Society*, *154*(4), 781–794.

Georges, A., Zhang, X., Unmack, P., Reid, B. N., Le, M., and McCord, W. P. (2014). Contemporary genetic structure of an endemic freshwater turtle reflects Miocene orogenesis of New Guinea. *Biological Journal of the Linnean Society*, *111*(1), 192–208.

Gibbs, G. (2006). *Ghosts of Gondwana: The History of Life in New Zealand* (Craig Potton Publishing: Nelson, New Zealand), 232 p.

Goldberg, J., Trewick, S. A., and Paterson, A. M. (2008). Evolution of New Zealand's terrestrial fauna: A review of molecular evidence. *Philosophical Transactions of the Royal Society of London B*, *363*, 3319–3334.

Goldstein, P. Z., and DeSalle, R. (2010). Integrating DNA barcode data and taxonomic practice: Determination, discovery, and description. *BioEssays*, *33*(2), 135–147.

González-Orozco, C. E., Laffan, S. W., and Miller, J. T. (2011). Spatial distribution of species richness and endemism of the genus *Acacia* in Australia. *Australian Journal of Botany*, *59*(7), 601.

Greaves, S. N. J., Chapple, D. G., Daugherty, C. H., Gleeson, D. M., and Ritchie, P. A. (2008). Genetic divergences pre-date Pleistocene glacial cycles in the New Zealand speckled skink, *Oligosoma infrapunctatum*. *Journal of Biogeography*, *35*, 853–864.

Greaves, S. N. J., Chapple, D. G., Gleeson, D. M., Daugherty, C. H., and Ritchie, P. A. (2007). Phylogeography of the spotted skink (*Oligosoma lineoocellatum*) and green skink (*O. chloronoton*) species complex (Lacertilia: Scincidae) in New Zealand reveals pre-Pleistocene divergence. *Molecular Phylogenetics and Evolution*, *45*, 729–739.

Haines, M. L., Moussalli, A., Stuart-Fox, D., Clemann, N., and Melville, J. (2014). Phylogenetic evidence of historic mitochondrial introgression and cryptic diversity in the genus *Pseudemoia* (Squamata: Scincidae). *Molecular Phylogenetics and Evolution*, *81*, 86–95.

Hall, J. P. W., and Harvey, D. J. (2002). The phylogeography of Amazonia revisited: New evidence from riodinid butterflies. *Evolution*, *56*(7), 1489–1497.

Han, D., Zhou, K., and Bauer, A. M. (2004). Phylogenetic relationships among gekkotan lizards inferred from C-mos nuclear DNA sequences and a new classification of the Gekkota. *Biological Journal of the Linnean Society*, *83*(3), 353–368.

Hare, K. M., Daugherty, C. H., and Chapple, D. G. (2008). Comparative phylogeography of three skink species (*Oligosoma moco, O. smithi* and *O. suteri*; Reptilia: Scincidae) in northeastern New Zealand. *Molecular Phylogenetics and Evolution*, *46*, 303–315.

Hay, J. M., Sarre, S. D., Lambert, D. M., Allendorf, F. W., and Daugherty, C. H. (2010). Genetic diversity and taxonomy: A reassessment of species designation in tuatara (*Sphenodon*: Reptilia). *Conservation Genetics*, *11*, 1063–1081.

Hayward, B. W. (1986). Origin of the offshore islands of northern New Zealand and their landform development. *The Offshore Islands Of Northern New Zealand*, New Zealand Department of Lands and Survey Information Series, *16*, 129–138.

Hayward, B. W. (1991). Geology and geomorphology of the Poor Knights Islands, northern New Zealand. *Tane*, *33*, 23–37.

Heinicke, M. P., Greenbaum, E., Jackman, T. R., and Bauer, A. M. (2010). Molecular phylogenetics of Pacific *Nactus* (Squamata: Gekkota: Gekkonidae) and the diphyly of Australian species. *Proceedings of the California Academy of Sciences*, *61*(7), 633.

Heinicke, M. P., Greenbaum, E., Jackman, T. R., and Bauer, A. M. (2011). Phylogeny of a trans-Wallacean radiation (Squamata, Gekkonidae, *Gehyra*) supports a single early colonization of Australia. *Zoologica Scripta*, *40*(6), 584–602.

Hesse, P. P., Magee, J. W., and van der Kaars, S. (2004). Late Quaternary climates of the Australian arid zone: A review. *Quaternary International, 118–119*, 87–102.

Hewitt, G. (2000). The genetic legacy of the Quaternary ice ages. *Nature*, *405*(6789), 907–913.

Hewitt, G. M. (1996). Some genetic consequences of ice ages, and their role in divergence and speciation. *Biological Journal of the Linnean Society*, *58*(3), 247–276.

Hickerson, M. J., Carstens, B. C., Cavender-Bares, J., Crandall, K. A., Graham, C. H., Johnson, J. B., *et al.* (2010). Phylogeography's past, present, and future: 10 years after Avise, 2000. *Molecular Phylogenetics and Evolution*, *54*(1), 291–301.

Hickson, R. E., Slack, K. E., and Lockhart, P. (2000). Phylogeny recapitulates geography, or why New Zealand has so many species of skinks. *Biological Journal of the Linnean Society*, *70*, 415–433.

Hill, K. C., and Hall, R. (2003). Mesozoic–Cenozoic evolution of Australia's New Guinea margin in a west Pacific context (22nd ed.). *Geological Society of America*, *372*, 265–290.

Hitchmough, R., Anderson, P., Barr, B., Monks, J., Lettink, M., Reardon, J., Tocher, M., and Whitaker, T. (2013). *Conservation Status of New Zealand Reptiles, 2012*. New Zealand Threat Classification Series 2 (Department of Conservation: Wellington, New Zealand).

Hodges, K., Donnellan, S., and Georges, A. (2014). Phylogeography of the Australian freshwater turtle *Chelodina expansa* reveals complex relationships among inland and coastal bioregions. *Biological Journal of the Linnean Society*, *111*(4), 789–805.

Hodges, K. M., Rowell, D. M., and Keogh, J. S. (2007). Remarkably different phylogeographic structure in two closely related lizard species in a zone of sympatry in south-eastern Australia. *Journal of Zoology*, *272*(1), 64–72.

Holyoake, A., Waldman, B., and Gemmell, N. J. (2001). Determining the species status of one of the world's rarest frogs: A conservation dilemma. *Animal Conservation*, *4*, 29–35.

Honda, M., Ota, H., Kobayashi, M., Nabhitabhata, J., Yong, H.-S., and Hikida, T. (2000a). Phylogenetic relationships, character evolution, and biogeography of the subfamily Lygosominae (Reptilia: Scincidae) inferred from mitochondrial DNA sequences. *Molecular Phylogenetics and Evolution*, *15*(3), 452–461.

Honda, M., Ota, H., Kobayashi, M., Nabhitabhata, J., Yong, H.-S., Sengoku, S., and Hikida, T. (2000b). Phylogenetic relationships of the family Agamidae (Reptilia: Iguania) inferred from mitochondrial DNA sequences. *Zoological Science*, *17*(4), 527–537.

Hopper, S. D. (1979). Biogeographical aspects of speciation in the southwest Australian flora. *Annual Review of Ecology and Systematics*, *10*, 399–422.

Hopper, S. D., and Gioia, P. (2004). The southwest Australian floristic region: Evolution and conservation of a global hot spot of biodiversity. *Annual Review of Ecology, Evolution, and Systematics*, 623–650.

Horner, P., and Adams, M. (2007). A molecular systematic assessment of species boundaries in Australian Cryptoblepharus (Reptilia: Squamata: Scincidae): A case study for the combined use of allozymes and morphology to explore cryptic biodiversity. *The Beagle*, Rec. Mus. Art. Gall. North. Terr., Supp. *3*, 1–19.

Horton, D. R. 1972. Evolution in the genus *Egernia* (Lacertilia: Scincidae). *Journal of Herpetology*, *6*, 101–109.

Hoskin, C. J. (2013). A new frog species (Microhylidae: *Cophixalus*) from boulder-pile habitat of Cape Melville, north-east Australia. *Zootaxa*, *3722*(1), 61.

Hoskin, C. J. (2014). A new skink (Scincidae: *Carlia*) from the rainforest uplands of Cape Melville, north-east Australia. *Zootaxa*, *3869*(3), 224.

Hoskin, C. J., and Couper, P. (2013). A spectacular new leaf-tailed gecko (Carphodactylidae: *Saltuarius*) from the Melville Range, north-east Australia. *Zootaxa*, *3717*(4), 543.

Hoskin, C. J., Couper, P. J., and Schneider, C. J. (2003). A new species of *Phyllurus* (Lacertilia: Gekkonidae) and a revised phylogeny and key for the Australian leaf-tailed geckos. *Australian Journal of Zoology*, *51*(2), 153.

Hoskin, C. J., Tonione, M., Higgie, M., MacKenzie, J. B., Williams, S. E., VanDerWal, J., and Moritz, C. (2011). Persistence in peripheral refugia promotes phenotypic divergence and speciation in a rainforest frog. *The American Naturalist*, *178*(5), 561–578.

Hugall, A. F., Foster, R., Hutchinson, M., and Lee, M. S. (2008). Phylogeny of Australasian agamid lizards based on nuclear and mitochondrial genes: Implications for morphological evolution and biogeography. *Biological Journal of the Linnean Society*, *93*(2), 343–358.

Hugall, A. F., Foster, R., and Lee, M. S. Y. (2007). Calibration choice, rate smoothing, and the pattern of tetrapod diversification according to long nuclear gene RAG-1. *Systematic Biology*, *56*, 543–563.

Hugall, A. F., and Lee M. S. (2004). Molecular claims of Gondwanan age for Australian agamid lizards are untenable. *Molecular Biology and Evolution*, *21*(11), 2102–2110.

Hughes, C., and Eastwood, R. (2006). Island radiation on a continental scale: Exceptional rates of plant diversification after uplift of the Andes. *Proceedings of the National Academy of Sciences of the United States of America*, *103*(27), 10334–10339.

Hutchinson, M. N. 1997. The first fossil pygopod (Squamata, Gekkota), and a review of mandibular variation in living species. *Memoirs Queensland Museum*, *41*, 355–366.

Hutchinson, M. N., and S. C. Donnellan. (1993). Biogeography and phylogeny of the Squamata. In *Fauna of Australia*, Vol. 2A: *Amphibia and Reptilia* (Eds. C. J. Glasby, G. J. B. Ross and P. L. Beesley), pp. 210–220 (AGPS Press: Canberra, Australia).

Huybers, P. (2007). Early Pleistocene glacial cycles and the integrated summer insolation forcing. *Science*, *313*, 508–511.

Irissari, I., San Mauro, D., Green, D. M., and Zardoya, R. (2010). The complete mitochondrial genome of the relict frog *Leiopelma archeyi*: Insights into the root of the frog tree of life. *Mitochondrial DNA 20*, 173–182.

Jackman, T. R., Bauer, A. M., and Greenbaum, E. (2008). Phylogenetic relationships of geckos of the genus *Nactus* and their relatives (Squamata: Gekkonidae). *Acta Herpetologica*, *3*(1), 1–18.

James, C. H., and Moritz, C. (2000). Intraspecific phylogeography in the sedge frog *Litoria fallax* (Hylidae) indicates pre-Pleistocene vicariance of an open forest species from eastern Australia. *Molecular Ecology*, *9*(3), 349–358.

Jennings, W. B., and Pianka E. R. (2004). Tempo and timing of the Australian *Varanus* radiation. In *Varanoid Lizards of the World* (Eds. E. R. Pianka, D. R. King and R. A. King), pp. 77–87 (Indiana University Press: Bloomington, IL).

Jennings, W. B., Pianka, E. R., and Donnellan, S. (2003). Systematics of the lizard family pygopodidae with implications for the diversification of Australian temperate biotas. *Systematic Biology*, *52*(6), 757–780.

Jetz, W., McPherson, J. M., and Guralnick, R. P. (2012). Integrating biodiversity distribution knowledge: Toward a global map of life. *Trends Ecology Evolution, 27,* 151–159.

Kay, G., and Keogh, J. S. (2012). Molecular phylogeny and morphological revision of the *Ctenotus labillardieri* (Reptilia: Squamata: Scincidae) species group and a new species of immediate conservation concern in the southwestern Australian biodiversity hotspot. *Zootaxa*, *3390*, 1–18.

Kearney, M., Moussalli, A., Strasburg, J., Lindenmayer, D., and Moritz, C. (2003). Geographic parthenogenesis in the Australian arid zone: I. A climatic analysis of the *Heteronotia binoei* complex (Gekkonidae). *Evolutionary Ecology Research*, *5*(7), 953–976.

Kelly, C. M. R., Barker, N. P., Villet, M. H., and Broadley, D. G. (2009). Phylogeny, biogeography and classification of the snake superfamily Elapoidea: A rapid radiation in the Late Eocene. *Cladistics*, *25*(1), 38–63.

Keogh, J. S. (1998). Molecular phylogeny of elapid snakes and a consideration of their biogeographic history. *Biological Journal of the Linnean Society*, *63*(2), 177–203.

Keogh, J. S., Shine, R., and Donnellan, S. (1998). Phylogenetic relationships of terrestrial Australo-Papuan elapid snakes (subfamily hydrophiinae) based on cytochrome b and 16S rRNA sequences. *Molecular Phylogenetics and Evolution*, *10*(1), 67–81.

Keogh, S. J., Scott, I. A., Fitzgerald, M., and Shine, R. (2003). Molecular phylogeny of the Australian venomous snake genus *Hoplocephalus* (Serpentes, Elapidae) and conservation genetics of the threatened *H. stephensii*. *Conservation Genetics*, *4*(1), 57–65.

King, P. R. (2000). Tectonic reconstructions of New Zealand: 40 Ma to the present. *New Zealand Journal of Geology and Geophysics 43*, 611–638.

Kruckeberg, A. R. 2002. *Geology and Plant Life* (University of Washington Press: Seattle).

Kuch, U., Keogh, J. S., Weigel, J., Smith, L. A., and Mebs, D. (2005). Phylogeography of Australia's king brown snake (*Pseudechis australis*) reveals Pliocene divergence and Pleistocene dispersal of a top predator. *Naturwissenschaften*, *92*(3), 121–127.

Laffan, S. W., Lubarsky, E., and Rosauer, D. F. (2010). Biodiverse, a tool for the spatial analysis of biological and related diversity. *Ecography*, *33*(4), 643–647.

Landis, C. A., Campbell, H. J., Begg, J. G., Mildenhall, D. C., Paterson, A. M., and Trewick, S. A. (2008). The Waipounamu Erosion Surface: Questioning the antiquity of the New Zealand land surface and terrestrial fauna and flora. *Geological Magazine*, *145*(2), 173–197.

Le, M., Reid, B. N., McCord, W. P., Naro-Maciel, E., Raxworthy, C. J., Amato, G., and Georges, A. (2013). Resolving the phylogenetic history of the short-necked turtles, genera *Elseya* and *Myuchelys* (Testudines: Chelidae) from Australia and New Guinea. *Molecular Phylogenetics and Evolution*, *68*(2), 251–258.

Leache, A. D., Crews, S. C., and Hickerson, M. J. (2007). Two waves of diversification in mammals and reptiles of Baja California revealed by hierarchical Bayesian analysis. *Biology Letters*, *3*(6),646–650.

Lee, A. K. (1967). Studies in Australian amphibian: II. Taxonomy, ecology and evolution of the genus *Heleioporus* Gray (Anura: Leptodactylidae). *Australian Journal of Zoology*, *15*(2), 367–439.

Lee, D. E., Lee, W. G., and Mortimer, N. (2001). Where and why have all the flowers gone? Depletion and turnover in the New Zealand Cenozoic angiosperm flora in relation to palaeography and climate. *Australian Journal of Botany*, *49*, 341–356.

Lee, M. S. Y., Hutchinson, M. N., Worthy, T. H., Archer, M., Tennyson, A. J. D., Worthy, J. P., and Scofield, R. P. (2009a). Miocene skinks and geckos reveal long-term conservatisms of New Zealand's lizard fauna. *Biology Letters*, *5*, 833–837.

Lewis, K. B., Carter, L., and Davey, F. J. (1994). The opening of Cook Strait: Interglacial tidal scour and aligning basins at a subduction to transform plate edge. *Marine Geology*, *116*, 293–312.

Liggins, L., Chapple, D. G., Daugherty, C. H., and Ritchie, P. A. (2008a). A SINE of restricted gene flow across the Alpine Fault: Phylogeography of the New Zealand common skink (*Oligosoma nigriplantare polychroma*). *Molecular Ecology*, *17,* 3668–3683.

Liggins, L., Chapple, D. G., Daugherty, C. H., and Ritchie, P. A. (2008b). Origin and post-colonization evolution of the Chatham Islands skink (*Oligosoma nigriplantare nigriplantare*). *Molecular Ecology*, *17*, 3290–3305.

Littlejohn, M. J. (1981). The Amphibia of the mesic southern Australia: A zoogeographical perspective. A. Keast (Ed.), *Ecological Biogeography of Australia. Monographiae Biologicae*, *41*, 1301–1330.

Littlejohn, M. J., Roberts, J. D., Watson, G. F., and Davies, M. (1993). Family Myobatrachidae. In *Fauna of Australia*, Vol. 2A: *Amphibia and Reptilia* (Eds. C. J. Glasby, G. J. B. Ross and P. L. Beesley), pp. 41–57 (Australian Government Printing Service: Canberra, Australia).

Lohman, D. J., Ingram, K. K., Prawiradilaga, D. M., Winker, K., Sheldon, F. H., Moyle, R. G., *et al.* (2010). Biological Conservation. *Biological Conservation*, *143*(8), 1885–1890.

Lukoschek, V., and Keogh, J. S. (2006). Molecular phylogeny of sea snakes reveals a rapidly diverged adaptive radiation. *Biological Journal of the Linnean Society*, *89*(3), 523–539.

Lukoschek, V., Waycott, M., and Marsh, H. (2007). Phylogeography of the olive sea snake, *Aipysurus laevis* (Hydrophiinae) indicates Pleistocene range expansion around northern Australia but low contemporary gene flow. *Molecular Ecology*, *16*(16), 3406–3422.

Mabbutt, J. A. 1988. Australian desert landscapes. *Geojournal*, *16*(4), 355–369.

Macey, J. R., Schulte, J. A., Larson, A., Ananjeva, N. B., Wang, Y., Pethiyagoda, R., *et al.* (2000). Evaluating trans-Tethys migration: An example using acrodont lizard phylogenetics. *Systematic Biology*, *49*(2), 233–256.

Main, A. R., Lee, A. K., and Littlejohn, M. J. (1958). Evolution in three genera of Australian frogs. *Evolution*, *12*, 224–233.

Main, A. R. (1968). Ecology, systematics and evolution of Australian frogs. *Advances in Ecological Research* *5*, 37–86.

Marin, J., Donnellan, S. C., Blair Hedges, S. B., Doughty, P., Hutchinson, M. N., Cruaud, C., and Vidal, N. (2013a). Tracing the history and biogeography of the Australian blindsnake radiation. *Journal of Biogeography*, *40*, 928–937.

Marin, J., Donnellan, S. C., Hedges, S. B., Puillandre, N., Aplin, K. P., Doughty, P., *et al.* (2013b). Hidden species diversity of Australian burrowing snakes (Ramphotyphlops). *Biological Journal of the Linnean Society*, *110*(2), 427–441.

Martin, H. A. (2006). Cenozoic climatic changes and the development of the arid vegetation of Australia. *Journal of Arid Environments*, *66*, 533–563.

Maslin, B. R., Barrett, M. D., and Barrett, R. L. (2013). A baker's dozen of new wattles highlights significant *Acacia* (Fabaceae: Mimosoideae) diversity and endemism in the north-west Kimberley region of Western Australia. *Nuytsia*, *23*, 543–587.

McCann, C. (1955). The lizards of New Zealand: Gekkonidae and Scincidae. *Dominion Museum Bulletin*, *17*, 1–127.

McCann, C. (1956). The distribution of New Zealand reptiles. *Proceedings of the New Zealand Ecological Society*, *4*, 15–16.

McGuigan, K., McDonald, K., Parris, K., and Moritz, C. (1998). Mitochondrial DNA diversity and historical biogeography of a wet forest-restricted frog (*Litoria pearsoniana*) from mid-east Australia. *Molecular Ecology*, *7*(2), 175–186.

McLaren, S., and Wallace, M. W. (2010). Plio-Pleistocene climate change and the onset of aridity in southeastern Australia. *Global and Planetary Change*, *71*, 55–72.

Melville, J., Ritchie, E. G., Chapple, S. N. J., Glor, R. E., and Schulte, J. A., II. (2011). Evolutionary origins and diversification of dragon lizards in Australia's tropical savannas. *Molecular Phylogenetics and Evolution*, *58*(2), 257–270.

Melville, J., Schulte, J. A., and Larson, A. (2001). A molecular phylogenetic study of ecological diversification in the Australian lizard genus *Ctenophorus*. *Journal of Experimental Zoology*, *291*(4), 339–353.

Melville, J., Schulte, J. A., II, and Larson, A. (2004). A molecular study of phylogenetic relationships and evolution of antipredator strategies in Australian Diplodactylus geckos, subgenus *Strophurus*. *Biological Journal of the Linnean Society*, *82*, 123–138.

Melville, J., Shoo, L. P., and Doughty, P. (2008). Phylogenetic relationships of the heath dragons (*Rankinia adelaidensis* and *R. parviceps*) from the south-western Australian biodiversity hotspot. *Australian Journal of Zoology*, *56*(3), 159.

Menzies, J. I. (2006). *The Frogs of New Guinea and the Soloman Islands* (Pensoft: Sofia, Bulgaria).

Merrick, J. R., Archer, M., Hickey, G. M., and Lee, M. S. Y. (2006). *Evolution and Biogeography of Australasian Vertebrates* (Australian Scientific Publishing: Oatlands, Australia).

Miller, K. A., Chapple, D. G., Towns, D. R., Ritchie, P. A., and Nelson, N. J. (2009). Assessing genetic diversity for conservation management: A case study of a threatened reptile. *Animal Conservation*, *12*, 163–171.

Miralles, A., and Vences, M. (2013). New metrics for comparison of taxonomies reveal striking discrepancies among species delimitation methods in *Madascincus* lizards. *PLoS ONE*, *8*(7), e68242.

Mittermeier, R. A., Myers, N., Gil, P. R., and Mittermeier, C. G. (1999). *Hotspots: Earth's Biologically Richest and Most Endangered Terrestrial Ecoregions* (Cemex, Conservation International and Agrupacion Sierra Madre: Monterrey, Mexico).

Morgan, M. J., Roberts, J. D., and Keogh, J. S. (2007). Molecular phylogenetic dating supports an ancient endemic speciation model in Australia's biodiversity hotspot. *Molecular Phylogenetics and Evolution*, *44*(1), 371–385.

Moritz, C., Ens, E. J., Potter, S., and Catullo, R. (2013). The Australian monsoonal tropics: An opportunity to protect unique biodiversity and secure benefits for Aboriginal communities. *Pacific Conservation Biology*, *19*, 343–355.

Moritz, C., Hoskin, C. J., MacKenzie, J. B., Phillips, B. L., Tonione, M., Silva, N., *et al.* (2009). Identification and dynamics of a cryptic suture zone in tropical rainforest. *Proceedings of the Royal Society B: Biological Sciences*, *276*(1660), 1235–1244.

Moritz, C., Langham, G., Kearney, M., Krockenberger, A., VanDerWal, J., and Williams, S. (2012). Integrating phylogeography and physiology reveals divergence of thermal traits between central and peripheral lineages of tropical rainforest lizards. *Philosophical Transactions of the Royal Society B: Biological Sciences*, *367*(1596), 1680–1687.

Morton, S. R., and James, C. D. (1988). The diversity and abundance of lizards in arid Australia: A new hypothesis. *American Naturalist*, *132*(2), 237–256.

Moussalli, A., Hugall, A. F., and Moritz, C. (2005). A mitochondrial phylogeny of the rainforest skink genus *Saproscincus*, Wells and Wellington (1984). *Molecular Phylogenetics and Evolution*, *34*(1), 190–202.

Moussalli, A., Moritz, C., Williams, S. E., and Carnaval, A. C. (2009). Variable responses of skinks to a common history of rainforest fluctuation: Concordance between phylogeography and palaeo-distribution models. *Molecular Ecology*, *18*(3), 483–499.

Myers, N., Mittermeier, R. A., Mittermeier, C. G., da Fonseca, G. A., and Kent, J. (2000). Biodiversity hotspots for conservation priorities. *Nature*, *403*(6772), 853–858.

Neall, V. E., and Trewick, S. A. (2008). The age and origin of the Pacific islands: A geological overview. *Philosophical Transactions of the Royal Society of London B*, *363*, 3293–3308.

Nelson, N. J., Hitchmough, R., and Monks, J. M. (2015). New Zealand reptiles and their conservation. In *Austral Ark: The State of Wildlife in Australia and New Zealand* (Eds. A. Stow, N. Maclean and G. I. Holwell), pp. 382–404. (Cambridge University Press: Cambridge, UK).

Newnham, R. M., Lowe, D. J., and Willaims, P. W. (1999). Quaternary environmental change in New Zealand: A review. *Progress in Physical Geography*, *23*, 567–610.

Ng, J., Clemann, N., Chapple, S. N. J., and Melville, J. (2013). Phylogeographic evidence links the threatened 'Grampians' Mountain Dragon (*Rankinia diemensis* Grampians) with Tasmanian populations: Conservation implications in south-eastern Australia. *Conservation Genetics*, *15*(2), 363–373.

Ng, J., Clemann, N., Chapple, S. N., and Melville, J. (2014). Phylogeographic evidence links the threatened 'Grampians' Mountain Dragon (*Rankinia diemensis* Grampians) with Tasmanian populations: Conservation implications in south-eastern Australia. *Conservation Genetics*, 15, 363–373.

Nielsen, S. V., Bauer, A. M., Jackman, T. R., Hitchmough, R. A., and Daugherty, C. H. (2011). New Zealand geckos (Diplodactylidae): Cryptic diversity in a post-Gondwanan lineage with trans-Tasman affinities. *Molecular Phylogenetics and Evolution*, *59*(1), 1–22.

O'Connor, D., and Moritz, C. (2003). A molecular phylogeny of the Australian skink genera *Eulamprus, Gnypetoscincus* and *Nangura*. *Australian Journal of Zoology*, *51*(4), 317.

O'Neill, S. B., Chapple, D. G., Daugherty, C. H., and Ritchie, P. A. (2008). Phylogeography of two New Zealand lizards: McCann's skink (*Oligosoma maccanni*) and the brown skink (*O. zelandicum*). *Molecular Phylogenetics and Evolution*, *48*, 1168–1177.

O'Sullivan, P. B., Mitchell, M. M., O'Sullivan, A. J., Kohn, B. P., and Gleadow, A. (2000). Thermotectonic history of the Bassian Rise, Australia: Implications for the breakup of eastern Gondwana along Australia's southeastern margins. *Earth and Planetary Science Letters*, *182*(1), 31–47.

Oaks, J. R. (2011). A time-calibrated species tree of crocodylia reveals a recent radiation of the true crocodiles. *Evolution*, *65*(11), 3285–3297.

Oliver, P., Hugall, A., Adams, M., Cooper, S. J. B., and Hutchinson, M. (2007). Genetic elucidation of cryptic and ancient diversity in a group of Australian diplodactyline geckos: The *Diplodactylus vittatus* complex. *Molecular Phylogenetics and Evolution*, *44*(1), 77–88.

Oliver, P. M., Adams, M., and Doughty, P. (2010). Molecular evidence for ten species and Oligo-Miocene vicariance within a nominal Australian gecko species (*Crenadactylus ocellatus*, Diplodactylidae). *BMC Evolutionary Biology*, *10*(1), 386.

Oliver, P. M., Adams, M., Lee, M. S. Y., Hutchinson, M. N., and Doughty, P. (2009). Cryptic diversity in vertebrates: Molecular data double estimates of species diversity in a radiation of Australian lizards (*Diplodactylus*, Gekkota). *Proceedings of the Royal Society B: Biological Sciences*, *276*(1664), 2001–2007.

Oliver, P. M., and Bauer, A. M. (2011). Systematics and evolution of the Australian knob-tail geckos (*Nephrurus*, Carphodactylidae, Gekkota): Plesiomorphic grades and biome shifts through the Miocene. *Molecular Phylogenetics and Evolution*, *59*(3), 664–674.

Oliver, P. M., Bauer, A. M., Greenbaum, E., Jackman, T., and Hobbie, T. (2012a). Molecular phylogenetic evidence for the paraphyly of the arboreal Australian gecko genus *Oedura* Gray 1842 (Gekkota: Diplodactylidae): Yet another plesiomorphic grade. *Molecular Phylogenetics and Evolution*, *63*(2), 255–264.

Oliver, P. M., Doughty, P., and Palmer, R. (2013a). Hidden biodiversity in rare northern Australian vertebrates: The case of the clawless geckos (*Crenadactylus*, Diplodactylidae) of the Kimberley. *Wildlife Research*, *39*(5), 429.

Oliver, P. M., Laver, R. J., Smith, K. L., and Bauer, A. M. (2013b). Long-term persistence and vicariance within the Australian Monsoonal Tropics: The case of the giant cave and tree geckos (*Pseudothecadactylus*). *Australian Journal of Zoology*, *61*(6), 462.

Oliver, P. M., and Parkin, T. (2014). A new phasmid gecko (Squamata: Diplodactylidae: Strophurus) from the Arnhem PlateMore new diversity in rare vertebrates from northern Australia. *Zootaxa*, *3878*(1), 37.

Oliver, P. M., Pepper, M., and Couper, P. (2014c). Independent transitions between monsoonal and arid biomes revealed by systematic revision of a complex of Australian geckos (*Diplodactylus*; Diplodactylidae). *PLoS One*. *9*(12), e111895.

Oliver, P. M., Richards, S. J., and Sistrom, M. (2012b). Phylogeny and systematics of Melanesia's most diverse gecko lineage (Cyrtodactylus, Gekkonidae, Squamata). *Zoologica Scripta*, *41*(5), 437–454.

Oliver, P. M., and Sanders, K. L. (2009). Molecular evidence for Gondwanan origins of multiple lineages within a diverse Australasian gecko radiation. *Journal of Biogeography*, *36*(11), 2044–2055.

Oliver, P. M., Smith, K. L., Laver, R. J., Doughty, P., and Adams, M. (2014b). Contrasting patterns of persistence and diversification in vicars of a widespread Australian lizard lineage (the *Oedura marmorata* complex). *Journal of Biogeography*, *41*(11), 2068–2079.

Padial, J. M., Miralles, A., De la Riva, I., and Vences, M. (2010). The integrative future of taxonomy. *Frontiers in Zoology*, *7*(1), 16.

Patterson, G. B., and Bell, T. P. (2009). The Barrier skink *Oligosoma judgei* n. sp. (Reptilia: Scincidae) from the Darran and Takitimu Mountains, South Island, New Zealand. *Zootaxa*, *2271*, 43–56.

Patterson, G. B., and Daugherty, C. H. (1990). Four new species and one new subspecies of skinks, genus *Leiolopisma* (Reptilia: Lacertilia: Scincidae) from New Zealand. *Journal of the Royal Society of New Zealand*, *20*, 65–84.

Patterson, G. B., Hitchmough, R. A., and Chapple, D. G. (2013). Taxonomic revision of the ornate skink (*Oligosoma ornatum*; Reptilia: Scincidae) species complex from northern New Zealand. *Zootaxa*, *3736*, 54–68.

Pepper, M., Barquero, M. D., Whiting, M. J., and Keogh, J. S. (2014). A multi-locus molecular phylogeny for Australia's iconic Jacky dragon (Agamidae: *Amphibolurus muricatus*): Phylogeographic structure along the Great Dividing Range of south-eastern Australia. *Molecular Phylogenetics and Evolution, 71*, 149–156.

Pepper, M., Doughty, P., Arculus, R., and Keogh, J. S. (2008). Landforms predict phylogenetic structure on one of the world's most ancient surfaces. *BMC Evolutionary Biology*, *8*(1), 152.

Pepper, M., Doughty, P., Fujita, M. K., Moritz, C., and Keogh, J. S. (2013b). Speciation on the rocks: Integrated systematics of the *Heteronotia spelea* species complex (Gekkota; Reptilia) from western and central Australia. *PLoS ONE*, *8*(11), e78110.

Pepper, M., Doughty, P., Hutchinson, M. N., and Keogh, J. S. (2011a). Ancient drainages divide cryptic species in Australia's arid zone: Morphological and multi-gene evidence for four new species of beaked geckos (*Rhynchoedura*) *Molecular Phylogenetics and Evolution*, *61*(3), 810–822.

Pepper, M., Doughty, P., and Keogh, J. S. (2006). Molecular phylogeny and phylogeography of the Australian *Diplodactylus stenodactylus* (Gekkota; Reptilia) species-group based on mitochondrial and nuclear genes reveals an ancient split between Pilbara and non-Pilbara *D. stenodactylus*. *Molecular Phylogenetics and Evolution*, *41*(3), 539–555.

Pepper, M., Doughty, P., and Keogh, J. S. (2013a). Geodiversity and endemism in the iconic Australian Pilbara region: A review of landscape evolution and biotic response in an ancient refugium. *Journal of Biogeography*, *40*(7), 1225–1239.

Pepper, M., Fujita, M. K., Moritz, C., and Keogh, J. S. (2011b). Palaeoclimate change drove diversification among isolated mountain refugia in the Australian arid zone. *Molecular Ecology*, *20*(7), 1529–1545.

Pepper, M., Ho, S. Y. W., Fujita, M. K., and Keogh, J. S. (2011c). The genetic legacy of aridification: Climate cycling fostered lizard diversification in Australian montane refugia and left low-lying deserts genetically depauperate *Molecular Phylogenetics and Evolution*, *61*(3), 750–759.

Pepper, M., and Scott Keogh, J. (2014). Biogeography of the Kimberley, Western Australia: A review of landscape evolution and biotic response in an ancient refugium. *Journal of Biogeography*, *41*(8), 1443–1455.

Phillips, R. D., Storey, A. W., and Johnson, M. S. (2009). Genetic structure of *Melanotaenia australisat* local and regional scales in the east Kimberley, Western Australia. *Journal of Fish Biology*, *74*(2), 437–451.

Pianka, E. R., and Schall, J. J. (1981). 59 Species densities of Australian vertebrates. *Ecological Biogeography of Australia*, *3*, 1675.

Potter, S., Eldridge, M. D. B., Taggart, D. A., and Cooper, S. J. B. (2012). Multiple biogeographical barriers identified across the monsoon tropics of northern Australia: Phylogeographic analysis of the *brachyotis* group of rock-wallabies. *Molecular Ecology*, *21*(9), 2254–2269.

Powney, G. D., Grenyer, R., Orme, C. D. L., Owens, I. P. F., and Meiri, S. (2010). Hot, dry and different: Australian lizard richness is unlike that of mammals, amphibians and birds. *Global Ecology and Biogeography*, *19*, 386–396.

Pyron, R. A., and Wiens, J. J. (2011). A large-scale phylogeny of Amphibia including over 2,800 species, and a revised classification of extant frogs, salamanders, and caecilians. *Mol Phylogenet and Evolution 61*, 543–583.

Pyron, R. A. (2014). Biogeographic analysis reveals ancient continental vicariance and recent oceanic dispersal in amphibians. *Systematic Biology*, *63*(5), 779–797.

Quigley, M. C., Clark, D., and Sandiford, M. (2010). Tectonic geomorphology of Australia. *Geological Society, London, Special Publications*, *346*(1), 243–265.

Quilty, P. G. 1994 The background: 144 million years of Australian palaeoclimate and palaeogeography. In *History of the Australian Vegetation: Cretaceous to Recent* (Ed. R. S. Hill), pp. 14–43 (Cambridge University Press: Cambridge).

Rabosky, D. L., Aplin, K. P., Donnellan, S. C., and Hedges, S. B. (2004). Molecular phylogeny of blindsnakes (Ramphotyphlops) from Western Australia and resurrection of *Ramphotyphlops bicolor* (Peters, 1857). *Australian Journal of Zoology*, *52*(5), 531.

Rabosky, D. L., Donnellan, S. C., Talaba, A. L., and Lovette, I. J. (2007). Exceptional among-lineage variation in diversification rates during the radiation of Australia's most diverse vertebrate clade. *Proceedings of the Royal Society B: Biological Sciences*, *274*(1628), 2915–2923.

Rabosky, D. L., Talaba, A. L., Donnellan, S. C., and Lovette, I. J. (2009). Molecular evidence for hybridization between two Australian desert skinks, *Ctenotus leonhardii* and *Ctenotus quattuordecimlineatus* (Scinidae: Squamata). *Molecular Phylogenetics and Evolution*, *53*, 368–377.

Rabosky, D. L., Hutchinson, M. N., Donnellan, S. C., Talaba, A. L., and Lovette, I. J. (2014). Phylogenetic disassembly of species boundaries in a widespread group of Australian skinks (Scincidae: *Ctenotus*). *Molecular Phylogenetics and Evolution*, *77*, 71–82.

Rasmussen, A. R., Sanders, K. L., Guinea, M. L., and Amey, A. P. (2014). Sea snakes in Australian waters (Serpentes: Subfamilies Hydrophiinae and Laticaudinae): A review with an updated identification key. *Zootaxa*, *3869*(4), 351.

Rawlings, L. H., Barker, D., and Donnellan, S. C. (2004). Phylogenetic relationships of the Australo-Papuan *Liasis* pythons (Reptilia: Macrostomata), based on mitochondrial DNA. *Australian Journal of Zoology*, *52*(2), 215.

Rawlings, L. H., and Donnellan, S. C. (2003). Phylogeographic analysis of the green python, *Morelia viridis*, reveals cryptic diversity. *Molecular Phylogenetics and Evolution*, *27*(1), 36–44.

Rawlings, L. H., Rabosky, D. L., Donnellan, S. C., and Hutchinson, M. N. (2008). Python phylogenetics: Inference from morphology and mitochondrial DNA. *Biological Journal of the Linnean Society*, *93*(3), 603–619.

Read, K., Keogh, J. S., Scott, I. A. W., Roberts, J. D., and Doughty, P. (2001). Molecular phylogeny of the Australian frog genera Crinia, Geocrinia, and allied taxa (Anura: Myobatrachidae). *Molecular Phylogenetics and Evolution*, *21*(2), 294–308.

Reeder, T. W. (2003). A phylogeny of the Australian Sphenomorphus group (Scincidae: Squamata) and the phylogenetic placement of the crocodile skinks (Tribolonotus): Bayesian approaches to assessing congruence and obtaining confidence in maximum likelihood inferred relationships. *Molecular Phylogenetics and Evolution*, *27*(3), 384–397.

Reeder, T. W., and Reichert, J. D. (2011). Phylogenetic relationships within the Australian limb-reduced lizard genus *Hemiergis* (Scincidae: Squamata) as inferred from the Bayesian analysis of mitochondrial rRNA gene sequences. *Copeia*, 113–120.

Reeves, J. M., Barrows, T. T., Cohen, T. J., Kiem, A. S., Bostock, H. C., Fitzsimmons, K. E., Jansen, J. D., Kemp, J., Krause, C., Petherick, L., and Phipps, S. J. (2013b). Climate variability over the last 35,000 years recorded in marine and terrestrial archives in the Australian region: An OZ-INTIMATE compilation. *Quaternary Science Reviews*, *74*, 21–34.

Rest, J. S., Ast, J. C., Austin, C. C., Waddell, P. J., Tibbetts, E. A., Hay, J. M., and Mindell, D. P. (2003). Molecular systematics of primary reptilian lineages and the tuatara mitochondrial genome. *Molecular Phylogenetics and Evolution*, *29*, 289–297.

Riddle, B. R., Hafner, D. J., Alexander, L. F., and Jaeger, J. R. (2000). Cryptic vicariance in the historical assembly of a Baja California peninsular desert biota. *Proceedings of the National Academy of Sciences of the United States of America*, *97*(26), 14438–14443.

Riedel, A., Sagata, K., Suhardjono, Y. R., Tänzler, R., and Balke, M. (2013). Integrative taxonomy on the fast track – towards more sustainability in biodiversity research. *Frontiers in Zoology*, *10*, 15.

Rix, M. G., Edwards, D. L., Byrne, M., Harvey, M. S., Joseph, L., and Roberts, J. D. (2015). Biogeography and speciation of terrestrial fauna in the south-western Australian biodiversity hotspot. *Biological Reviews*. *90*(3), 762–793. doi:10.1111/brv.12132

Rocha, S., Carretero, M. A., Vences, M., Glaw, F., and James Harris, D. (2006). Deciphering patterns of trans-oceanic dispersal: The evolutionary origin and biogeography of coastal lizards (*Cryptoblepharus*) in the western Indian Ocean region. *Journal of Biogeography*, *33*(1), 13–22.

Roelants, K., and Bossuyt, F. (2005). Archaeobatrachian paraphyly and Pangean diversification of crown group frogs. *Systematic Biology*, *54*, 111–126.

Rogers, G. M. (1989). The nature of the lower North Island floristic gap. *New Zealand Journal of Botany*, *27*, 221–241.

Rosauer, D., Ferrier, S., Williams, K. J., Manion, G., Keogh, J. S., and Laffan, S. W. (2013). Phylogenetic generalised dissimilarity modelling: A new approach to analysing and predicting spatial turnover in the phylogenetic composition of communities. *Ecography*, 37, 21–32. doi:10.1111/j.16000587.2013.00466.x.

Rosauer, D., Laffan, S. W., Crisp, M. D., Donnellan, S. C., and Cook, L. G. (2009). Phylogenetic endemism: A new approach for identifying geographical concentrations of evolutionary history. *Molecular Ecology*, *18*, 4061–4072.

Rosenblum, E. B., and Harmon, L. J. (2011). 'Same same but different': Replicated ecological speciation at white sands. *Evolution*, *65*(4), 946–960.

Sanders, K. L., and Lee, M. S. Y. (2008). Molecular evidence for a rapid late-Miocene radiation of Australasian venomous snakes (Elapidae, Colubroidea). *Molecular Phylogenetics and Evolution 46*: 1165–1173. doi:10.1016/j.ympev.2007.11.013.

Sanders, K. L., Lee, M. S. Y., Leys, R., Foster, R., and Scott Keogh, J. (2008). Molecular phylogeny and divergence dates for Australasian elapids and sea snakes (hydrophiinae): Evidence from seven genes for rapid evolutionary radiations. *Journal of Evolutionary Biology*, *21*(3), 682–695.

Sanders, K. L., Lee, M. S. Y., Mumpuni, Bertozzi, T., and Rasmussen, A. R. (2013). Multilocus phylogeny and recent rapid radiation of the viviparous sea snakes (Elapidae: Hydrophiinae). *Molecular Phylogenetics and Evolution*, *66*(3), 575–591.

Sanders, K. L., Mumpuni, Hamidy, A., Head, J. J., and Gower, D. J. (2010). Phylogeny and divergence times of filesnakes (Acrochordus): Inferences from morphology, fossils and three molecular loci. *Molecular Phylogenetics and Evolution*, *56*(3), 857–867.

Scanlon, J. D., and Lee, M. (2004). Phylogeny of Australasian venomous snakes (Colubroidea, Elapidae, Hydrophiinae) based on phenotypic and molecular evidence. *Zoologica Scripta*, *33*(4), 335–366.

Schäuble, C. S., and Moritz, C. (2001). Comparative phylogeography of two open forest frogs from eastern Australia. *Biological Journal of the Linnean Society*, *74*(2), 157–170.

Schneider, C. J., Cunningham, M., and Moritz, C. (1998). Comparative phylogeography and the history of endemic vertebrates in the Wet Tropics rainforests of Australia. *Molecular Ecology*, *7*(4), 487–498.

Schuett, G. W., Reiserer, R. S., and Earley, R. L. (2009). The evolution of bipedal postures in varanoid lizards. *Biological Journal of the Linnean Society*, *97*(3), 652–663.

Schulte, J. A., Melville, J., and Larson, A. (2003). Molecular phylogenetic evidence for ancient divergence of lizard taxa on either side of Wallace's Line. *Proceedings of the Royal Society B: Biological Sciences*, *270*(1515), 597–603.

Scott, I. A., and Keogh, J. S. (2000). Conservation genetics of the endangered grassland earless dragon *Tympanocryptis pinguicolla* (Reptilia: Agamidae) in Southeastern Australia. *Conservation Genetics*, *1*(4), 357–363.

Sharma, P. P., and Wheeler, W. C. (2013). Revenant clades in historical biogeography: The geology of New Zealand predisposes endemic clades to root shifts. *Journal of Biogeography*, *40*, 1609–1618.

Shea, G., Couper, P., Worthington Wilmer, J., Amey, A. (2011). Revision of the genus *Cyrtodactylus* Gray, 1827 (Squamata: Gekkonidae) in Australia. *Zootaxa*, *3146*, 1–63.

Shoo, L. P., Rose, R., Doughty, P., Austin, J. J., and Melville, J. (2008). Diversification patterns of pebble-mimic dragons are consistent with historical disruption of important habitat corridors in arid Australia. *Molecular Phylogenetics and Evolution*, *48*(2), 528–542.

Singhal, S., and Moritz, C. (2013). Reproductive isolation between phylogeographic lineages scales with divergence. *Proceedings of the Royal Society B: Biological Sciences*, *280*(1772), 20132246–20132246.

Sistrom, M., Donnellan, S. C., and Hutchinson, M. N. (2013). Delimiting species in recent radiations with low levels of morphological divergence: A case study in Australian *Gehyra* geckos. *Molecular Phylogenetics and Evolution*, *68*(1), 135–143.

Sistrom, M., Hutchinson, M., Bertozzi, T., and Donnellan, S. (2014). Evaluating evolutionary history in the face of high gene tree discordance in Australian *Gehyra* (Reptilia: Gekkonidae). *Heredity*, *113*(1), 52–63.

Skinner, A., Donnellan, S. C., Hutchinson, M. N., and Hutchinson, R. G. (2005). A phylogenetic analysis of *Pseudonaja* (Hydrophiinae, Elapidae, Serpentes) based on mitochondrial DNA sequences. *Molecular Phylogenetics and Evolution*, *37*(2), 558–571.

Skinner, A., Hugall, A. F., and Hutchinson, M. N. (2011). Lygosomine phylogeny and the origins of Australian scincid lizards. *Journal of Biogeography*, *38*(6), 1044–1058.

Skinner, A., Hutchinson, M. N., and Lee, M. S. Y. (2013). Phylogeny and divergence times of Australian Sphenomorphus group skinks (Scincidae, Squamata). *Molecular Phylogenetics and Evolution*, *69*(3), 906–918.

Slade, R. W., and Moritz, C. (1998). Phylogeography of *Bufo marinus* from its natural and introduced ranges. *Proceedings: Biological Sciences/The Royal Society*, *265*(1398), 769–777.

Slatyer, C., Rosauer, D., and Lemckert, F. (2007). An assessment of endemism and species richness patterns in the Australian Anura. *Journal of Biogeography*, *34*(4), 583–596.

Slowinski, J. B., Knight, A., and Rooney, A. P. (1997). Inferring species trees from gene trees: A phylogenetic analysis of the Elapidae (Serpentes) based on the amino acid sequences of venom proteins. *Molecular Phylogenetics and Evolution*, *8*(3), 349–362.

Smissen, P. J., Melville, J., Sumner, J., and Jessop, T. S. (2013). Mountain barriers and river conduits: Phylogeographical structure in a large, mobile lizard (Varanidae: *Varanus varius*) from eastern Australia. *Journal of Biogeography*, *40*(9), 1729–1740.

Smith, K. L., Harmon, L. J., Shoo, L. P., and Melville, J. (2011). Evidence of constrained phenotypic evolution *in a cryptic species complex of agamid lizards. Evolution, 65*(4), 976–992.

Smith, L. A., and Adams, M. (2007). Revision of the *Lerista muelleri* species-group (Lacertilia: Scincidae) in Western Australia, with a redescription of *L. muelleri* (Fischer, 1881) and the description of nine new species. *Records Western Australian Museum*, *23*, 309–358.

Smith, S. A., Sadlier, R. A., Bauer, A. M., Austin, C. C., and Jackman, T. (2007). Molecular phylogeny of the scincid lizards of New Caledonia and adjacent areas: Evidence for a single origin of the endemic skinks of Tasmantis. *Molecular Phylogenetics and Evolution*, *43*(3), 1151–1166.

Symula, R., Keogh, J. S., and Cannatella, D. (2008). Ancient phylogeographic divergence in southeastern Australia among populations of the widespread common froglet, *Crinia signifera*. *Molecular Phylogenetics and Evolution*, *47*, 569–580.

Stevens, G., McGlone, M., and McCulloch, B. (1995). *Prehistoric New Zealand* (Reed Books: Auckland, New Zealand).

Storr, G. M. (1964). Some aspects of the geography of Australian reptiles. *Senckenbergiana Biologica, 45*, 577–589.

Strasburg, J. L., and Kearney, M. (2005). Phylogeography of sexual *Heteronotia binoei* (Gekkonidae) in the Australian arid zone: Climatic cycling and repetitive hybridization. *Molecular Ecology*, *14*(9), 2755–2772.

Stuart-Fox, D. M., Hugall, A. F., and Moritz, C. (2002). A molecular phylogeny of rainbow skinks (Scincidae: *Carlia*): Taxonomic and biogeographic implications. *Australian Journal of Zoology*, *50*(1), 39.

Stuart-Fox, D. M., Schneider, C. J., Moritz, C., and Couper, P. J. (2001). Comparative phylogeography of three rainforest-restricted lizards from mid-east Queensland. *Australian Journal of Zoology*, *49*(2), 119.

Suggate, R. P. (1982). The geological perspective. In *Landforms of New Zealand* (Eds. J. M. Soons and M. J. Selby), pp. 1–13 (Longman Paul: Auckland, New Zealand).

Suggate, R. P., Stevens, G. R., and Te Punga, M. T. (1978). *The Geology of New Zealand* (Government Printer: Wellington, New Zealand).

Thom, B. G., Hesp, P. A., Bryant, T., 1994. Last glacial 'coastal dunes' in Eastern Australia and implications for landscape stability during the Last Glacial Maximum. *Palaeogeography Palaeoclimatology Palaeoecology*, *111*, 229–248.

Tingley, R., Hitchmough, R. A., and Chapple, D. G. (2013). Life-history traits and extrinsic threats determine extinction risk in New Zealand lizards. *Biological Conservation*, *165*, 62–68.

Todd, E. V., Blair, D., Farley, S., Farrington, L., FitzSimmons, N. N., Georges, A., *et al.* (2013). Contemporary genetic structure reflects historical drainage isolation in an Australian snapping turtle, Elseya albagula. *Zoological Journal of the Linnean Society*, *169*(1), 200–214.

Todd, E. V., Blair, D., Georges, A., Lukoschek, V., and Jerry, D. R. (2014a). A biogeographical history and timeline for the evolution of Australian snapping turtles (*Elseya*: Chelidae) in Australia and New Guinea. *Journal of Biogeography*, *41*(5), 905–918.

Todd, E. V., Blair, D., and Jerry, D. R. (2014b). Influence of drainage divides versus arid corridors on genetic structure and demography of a widespread freshwater turtle, Emydura macquarii krefftii, from Australia. *Ecology and Evolution*, *4*(5), 606–622.

Toon, A., Hughes, J. M., and Joseph, L. (2010). Multilocus analysis of honeyeaters (Aves: Meliphagidae) highlights spatio-temporal heterogeneity in the influence of biogeographic barriers in the Australian monsoonal zone. *Molecular Ecology*, *19*(14), 2980–2994.

Towns, D. R., Daugherty, C. H., and Newman, D. G. (1985). An overview of the ecological biogeography of the New Zealand lizards (Gekkonidae, Scincidae). In *Biology of Australasian Frogs and Reptiles* (Eds. G. Grigg, R. Shine and H. Ehmann), pp. 107–115 (Royal Zoological Society of New South Wales: Sydney, Australia).

Trewick, S. A., Paterson, A. M., and Campbell, H. J. (2007). Hello New Zealand. *Journal of Biogeography*, *34*, 1–6.

Twidale, C. R. (1994). Gondwanan (Late Jurassic and Cretaceous) palaeosurfaces of the Australian craton. *Palaeogeography, Palaeoclimatology, Palaeoecology*, *112(1–2)*, 157–186.

Ursenbacher, S., Carlsson, M., Helfer, V., Tegelstrom, H., and Fumagalli, L. (2006). Phylogeography and Pleistocene refugia of the adder (*Vipera berus*) as inferred from mitochondrial DNA sequence data. *Molecular Ecology*, *15*(11), 3425–3437.

Van der Beek, P. A., Braun, J., and Lambeck, K. (1999). Post-Palaeozoic uplift history of southeastern Australia revisited: Results from a process-based model of landscape evolution. *Australian Journal of Earth Sciences*, *46*(2), 157–172.

Vences, M., Vieites, D. R. Glaw, F., Brinkmann, H., Kosuch, J., Veith, M., and Meyer, A. (2003). Multiple overseas dispersal in amphibians. *Proceedings of the Royal Society of London*, *270B*, 2435–2442.

Vidal, N., Marin, J., Sassi, J., Battistuzzi, F. U., Donnellan, S., Fitch, A. J., *et al.* (2012). Molecular evidence for an Asian origin of monitor lizards followed by Tertiary dispersals to Africa and Australasia. *Biology Letters*, *8*(5), 853–855.

Vitt, L. J., Pianka, E. R., Cooper, W. E., Jr., and Schwenk, K. (2003). History and the global ecology of squamate reptiles. *The American Naturalist*, *162*(1), 44–60.

Voris, H. K., Alfaro, M. E., Karns, D. R., Starns, G. L., Thompson, E., and Murphy, J. M. (2002). Phylogenetic relationships of the Oriental–Australian rear-fanged water snakes (colubridae: Homalopsinae) based on mitochondrial DNA sequences. *Copeia*, 4, 906–915.

Voros, J., Mitchell, A., Waldman, B., Goldstein, S., and Gemmell, M. J. (2008). Crossing the Tasman Sea: Inferring the introduction history of *Litoria aurea* and *Litoria raniformis* (Anura: Hylidae) from Australia into New Zealand. *Austral Ecology*, *33*, 623–629.

Wallis, G. P., and Trewick, S. A. (2009). New Zealand phylogeography: Evolution on a small continent. *Molecular Ecology*, *18*(17), 3548–3580.

Waples, K. (2007). *Kimberley Biodiversity Review*. Report prepared for the EPA Services Unit and the Environmental Management Branch (Department of Environment and Conservation: Perth, Australia).

Wardell-Johnson, G., and Roberts, J. D. (1993). Biogeographic barriers in a subdued landscape: The distribution of the *Geocrinia rosea* (Anura: Myobatrachidae) complex in south-western Australia. *Journal of Biogeography*, *20*(1), 95–108.

Wardle, P. (1991). *Vegetation of New Zealand* (Cambridge University Press: Cambridge, UK).

Wardle, P., Harris, W., and Buxton, R. P. (1988). Effects of glacial climates on floristic distribution in New Zealand: 2. The role of long-distance hybridisation in disjunct distributions. *New Zealand Journal of Botany*, *26*, 557–564.

Wasson, R. J. (1982). Landform Development in Australia. In *Evolution of the Flora and Fauna of Arid Australia* (Eds. W. R. Barker and P. J. M. Greenslade), pp. 23–33. (Peacock Publications: Adelaide, Australia).

Waters, J. M., and Craw, D. (2006). Goodbye Gondwana? New Zealand Biogeography, Geology, and the Problem of Circularity. *Systematic Biology*, *55*(2), 351–356.

Waters, J. M., Craw, D., Youngson, J. H., and Wallis, G. P. (2001). Genes meet geology: Fish phylogeographic pattern reflects ancient, rather than modern, drainage connectivity. *Evolution*, *55*(9), 1844–1851.

Wellman, H. W. (1979). An uplift map for the South Island of New Zealand and a model for uplift of the Southern Alps. In *The Origin of the Southern Alps* (Eds. R. I. Walcott and M. M. Cresswell), pp. 13–20 (The Royal Society of New Zealand Bulletin: Wellington, New Zealand).

Wende, R. (1997). Aspects of the fluvial geomorphology of the Eastern Kimberley Plateau, Western Australia. Doctor of Philosophy thesis, School of Geosciences, University of Wollongong. http://ro.uow.edu.au/theses/1965.

Williams, D. J. O'Shea, M., Daguerre, R. L., Pook, C. E., Wüster, W., Hayden, C. J., and McVay, J. D. *et al.* (2008). Origin of the eastern brownsnake, *Pseudonaja textilis* (Dumé ril, Bibron and Dumé ril) (Serpentes: Elapidae: Hydrophiinae) in New Guinea: Evidence of multiple dispersals from Australia, and comments on the status of *Pseudonaja textilis* pughi Hoser 2003. *Zootaxa*, *1703*, 47–61.

Williams, S. E., and Hero, J.-M. (2001). Multiple determinants of Australian tropical frog biodiversity. *Biological Conservation*, *98*(1), 1–10.

Williams, S. E., and Pearson, R. G. (1997). Historical rainforest contractions, localized extinctions and patterns of vertebrate endemism in the rainforests of Australia's wet tropics. *Proceedings of the Royal Society B: Biological Sciences*, *264*(1382), 709–716.

Woinarski, J. C. Z., Mackey, B., Nix, H. A., and Traill, B. (2007). *The Nature of Northern Australia: Its Natural Values, Ecological Processes and Future Prospects* (ANU ePress: Canberra, Australia).

Wood, P. L., Jr., Heinicke, M. P., Jackman, T. R., and Bauer, A. M. (2012). Phylogeny of bent-toed geckos (*Cyrtodactylus*) reveals a west to east pattern of diversification. *Molecular Phylogenetics and Evolution*, *65*(3), 992–1003.

Worthy, T. H., and Holdaway, R. N. (2002). *The Lost World of the Moa: Prehistoric Life of New Zealand* (Indiana University Press: Bloomington, IN).

Worthy, T. H., Tennyson, A. J. D., Archer, M., Musser, A. M., Hand, S. J., Jones, C., Douglas, B. J., McNamara, J. A., and Beck, R. M. D. (2006). Miocene mammal reveals a Mesozoic ghost lineage on insular New Zealand, southwest Pacific. *PNAS*, *103*(51), 19419–19423.

Wüster, W., Dumbrell, A. J., Hay, C., Pook, C. E., Williams, D. J., and Fry, B. G. (2005). Snakes across the Strait: Trans-Torresian phylogeographic relationships in three genera of Australasian snakes (Serpentes: Elapidae: Acanthophis, Oxyuranus, and Pseudechis). *Molecular Phylogenetics and Evolution*, *34*(1), 1–14.

13

The Biogeographical History of Non-Marine Mammaliaforms in the Sahul Region

Robin M.D. Beck

CONTENTS

Introduction

Today, the biogeographical region comprising Australia, New Guinea and the adjacent islands is the only part of the globe where representatives of all three major extant mammalian clades occur together – namely, monotremes, marsupials and placentals (Flannery 1995b; Van Dyck and Strahan 2008; Wilson and Reeder 2005). Monotremes (five species) are currently found nowhere else, while >240 described marsupial species comprise ~40% of the total terrestrial mammal diversity in the region. Placentals dominate the mammal faunas of most continental landmasses, but in Australia and New Guinea only two placental clades have achieved moderate diversity: murine rodents (>160 species; ~25% of the total) and bats (>130 species; ~20% of the total). Most other non-marine placental clades seem to have been entirely absent from Australia and New Guinea prior to human-related introductions during the Holocene. This unique overall pattern of mammalian biodiversity, so different from that seen elsewhere in the world, has fascinated a long line of researchers, including Darwin and Wallace themselves.

Evidence from the fossil record reveals further striking patterns and complexities. For example, Metatheria (the clade which includes modern marsupials) probably originated in the Northern Hemisphere by the middle Cretaceous at the latest (ca. 125 mya; Luo *et al.* 2003), but metatherians appear to have been absent from the southern continents, including Australia, until the latest Cretaceous or earliest Palaeocene, some 40–60 Ma later (Case *et al.* 2005; Goin *et al.* 2012b; Kielan-Jaworowska *et al.* 2004; Pascual and Ortiz-Jaureguizar 2007; Rougier *et al.* 2011b; Woodburne *et al.* 2014). Prior to the late Oligocene, the fossil record of mammals and their extinct relatives (collectively, Mammaliformes) in Australia is very poor, but it includes high-level taxa that are known nowhere else, such as the middle

Cretaceous ausktribosphenids and *Kollikodon*, and biogeographical enigmas such as the putative eutherian 'condylarth' *Tingamarra* from the early Eocene (Archer *et al.* 1999a; Long *et al.* 2002). The richer Oligo-Miocene record of Australia is dominated by marsupials, including members of most living families, with monotremes and representatives of multiple bat families also present (Archer *et al.* 1999a; Black *et al.* 2012; Long *et al.* 2002). However, the oldest murine rodent fossils from Australia are only ca. 4 Ma old and the oldest from New Guinea only 3.0–3.5 Ma old (Aplin 2006; Aplin and Ford 2014; Godthelp 1999), several Ma younger than the origin of Murinae as a whole (Fabre *et al.* 2013; Jacobs and Flynn 2005; Schenk *et al.* 2013).

This fossil evidence, when combined with phylogenies, divergence dates estimated from molecular and stratigraphic data, and geological information, gives insight into the biogeographical history of Australian and New Guinean Mammaliformes, and provides clues as to how the current mammal fauna of the region developed. At finer taxonomic scales, phylogeographic studies of molecular data are beginning to reveal the roles that environmental change and putative barriers to gene flow have played in shaping the biogeography of modern species and populations. Nevertheless, major gaps in our current knowledge – due to factors such as the highly incomplete fossil record of Mammaliformes in the region, uncertainty regarding the phylogeny and even alpha taxonomy of many taxa, and a lack of detailed, quantitative biogeographic analyses – mean that numerous uncertainties remain.

This chapter represents an attempted synthesis of our current knowledge of terrestrial mammaliaform biogeography in Australia, New Guinea and the adjacent islands (specifically, the region east of Lydekker's Line; Figure 13.1d), integrating the available fossil, phylogenetic and geological data. My major focus is above the species level, with extensive discussion of the fossil record, but I also briefly discuss phylogeographic studies of modern species. I end with a short summary of the biggest lacunae in our current knowledge, and the prospects for improving our understanding of the biogeographical history of Mammaliformes in what is undoubtedly one of the most fascinating and remarkable regions on earth.

Phylogenetic Definitions and Scope

Table 13.1 lists the formal phylogenetic definitions of selected clades discussed in this chapter. I follow most recent studies (e.g. Bi *et al.* 2014; Meredith *et al.* 2011; O'Leary *et al.* 2013) in restricting the name Mammalia to the crown clade only. I use Mammaliformes to refer to the more inclusive synapsid clade corresponding to 'traditional' definitions of Mammalia (e.g. Kielan-Jaworowska *et al.* 2004), following the stem-based definition of Sereno (2006: Table 10.1). Most fossil Mammaliformes found in Sahul to date appear to be members of Mammalia (Long *et al.* 2002), but a few (e.g. *Kollikodon*) may fall outside the crown clade.

Within Mammalia, I follow a crown clade definition of Theria – that is to say, the clade circumscribed by Marsupialia and Placentalia plus all other fossil taxa within that clade (O'Leary *et al.* 2013; Sereno 2006). As in most recent studies (e.g. O'Leary *et al.* 2013; Sereno 2006; Vullo *et al.* 2009), I restrict Marsupialia and Placentalia to the crown clades, with Metatheria and Eutheria referring to their respective total clades. Vullo *et al.* (2009) proposed the name Marsupialiformes to correspond to the 'traditional', more inclusive definition of Marsupialia (e.g. Kielan-Jaworowska *et al.* 2004). Vullo *et al.* (2009: 19910) gave an approximate definition of Marsupialiformes: 'crown group Marsupialia (extant marsupials and related extinct fossil taxa) plus all stem marsupialiform taxa that are more closely related to them, as their sister taxa, than to Deltatheroida and basal Metatheria'. However, I propose a new, less ambiguous definition here (Table 13.1). It is unclear whether only (crown clade) marsupials reached Sahul or whether non-marsupial marsupialiforms were also present (Beck 2014; Sigé *et al.* 2009). Fossil metatherians from Sahul and elsewhere in Gondwana that cannot be definitively placed within the crown clade will therefore be referred to as marsupialiforms here.

Several clades within Marsupialia are of particular biogeographical relevance, and it is appropriate to discuss their phylogenetic definitions. The clade Australidelphia was originally named by Szalay (1982) to include the four Sahulian marsupial orders – Dasyuromorphia (predominantly carnivorous forms such as quolls, dunnarts, the numbat and the thylacine), Diprotodontia ('possums', kangaroos, wombats

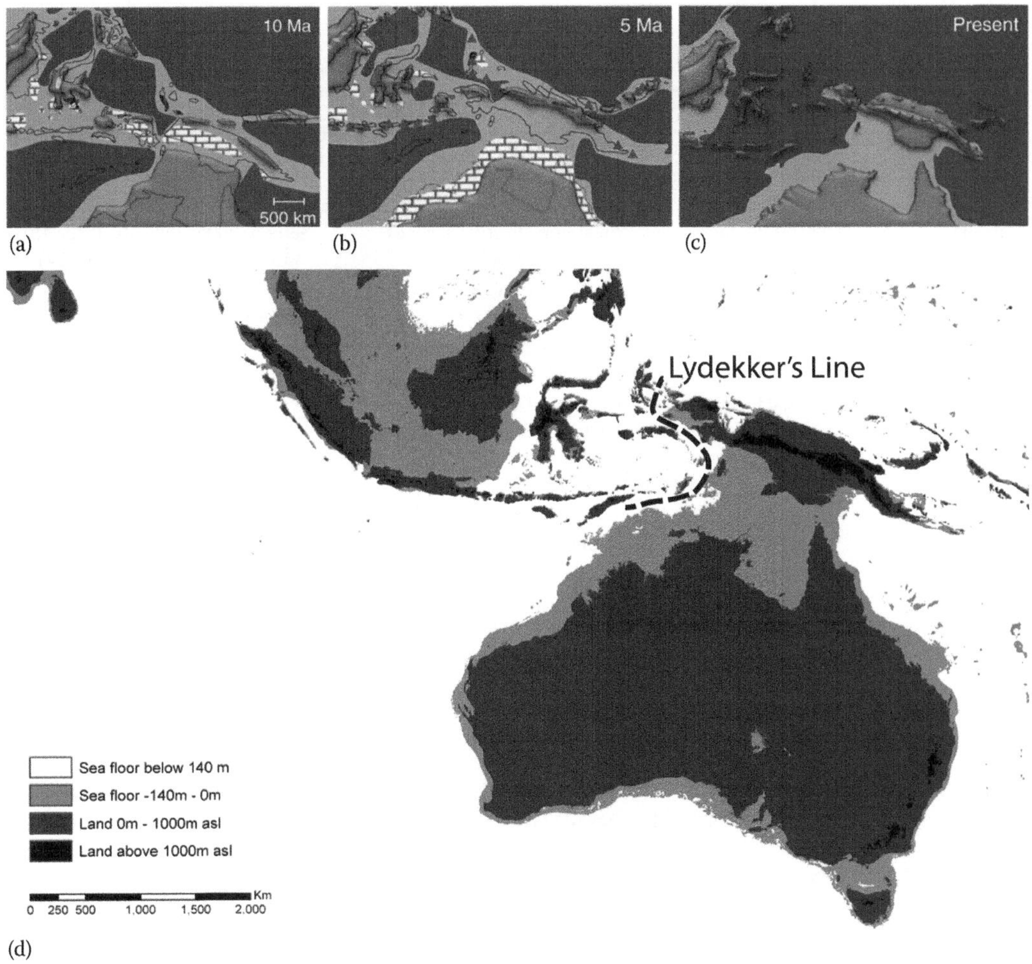

FIGURE 13.1 (See colour insert.) (a–c) Palaeotectonic evolution and emergence of New Guinea over the last 10 Ma (green, land; dark blue, deep sea; lighter blue, shallow sea; red white brick, calcareous plateaus possibly exposed at times; orange, highland; grey, high altitude above 2800 m). (Reprinted from Toussaint *et al.*, *Nature Communications* 5, copyright 2014, with permission from Macmillan.) (d) Geography of Sahul region showing maximum land extent during Pleistocene glacial maxima. (Modified from Aplin, K., and F. Ford, in: H.H.T. Prins and I.J. Gordon (eds). Invasion Biology and Ecological Theory: Insights from a Continent in Transformation, Cambridge University Press, Cambridge, UK, 2014: Figure 10.1). Lydekker's Line is indicated.

and the koala), Notoryctemorphia (marsupial moles) and Peramelemorphia (bandicoots and bilbies) – plus Microbiotheria (represented today by a single species, the South American 'monito del monte', *Dromiciops gliroides*). However, here I follow the apomorphy-based definition for Australidelphia that I proposed previously (Beck 2012: 717; Table 13.1). I use the name crown clade Australidelphia to refer to the clade circumscribed by the five extant australidelphian orders, and propose a formal, node-based phylogenetic definition for this clade (Table 13.1). Within crown clade Australidelphia, the four Sahulian orders appear to form a clade to the exclusion of Microbiotheria (Beck 2008a, 2011; Meredith *et al.* 2009a,b, Nilsson *et al.* 2010), which Archer (1984) named Eomarsupialia. I follow Beck *et al.*'s (2014: 132) crown-based definition for Eomarsupialia.

Marine mammals will not be discussed in this review, because the factors influencing their biogeographical distributions are obviously very different to those affecting non-marine forms. Recent reviews discussing the biogeography of marine mammals that occur in the waters around Sahul include Deméré *et al.* (2003) on pinnipedimorphs, Fordyce (2006) on cetaceans and de Iongh and Domning (2014) on

TABLE 13.1

Summary of Formal Phylogenetic Definitions Assumed Here for Selected Mammaliaform Clades

Clade	Definition	Reference(s)
Mammaliformes	'The most inclusive clade containing *Mus musculus* but not *Tritylodon longaevus* or *Pachygenelus monus*' (branch based)	Sereno (2006: Table 10.1; see also O'Leary *et al.* 2013: Table S4)
Mammalia	'The least inclusive clade containing *Ornithorhynchus anatinus* and *Mus musculus*' (node based)	Sereno (2006: Table 10.1; see also O'Leary *et al.* 2013: Table S4)
Theria	'The least inclusive clade containing *Mus musculus* and *Didelphis marsupialis*' (node based)	Sereno (2006: Table 10.1; see also O'Leary *et al.* 2013: Table S4)
Eutheria	'The most inclusive clade containing *Mus musculus* but not *Didelphis marsupialis*' (branch based)	Sereno (2006: Table 10.1; see also O'Leary *et al.* 2013: Table S4)
Placentalia	'The least inclusive clade containing *Dasypus novemcinctus*, *Elephas maximus*, *Erinaceus europaeus* and *Mus musculus*' (node based)	Sereno (2006: Table 10.1; see also O'Leary *et al.* 2013: Table S4)
Metatheria	'The most inclusive clade containing *Didelphis marsupialis* but not *Mus musculus*' (branch based)	Sereno (2006: Table 10.1; see also O'Leary *et al.* 2013: Table S4)
Marsupialiformes	'The most inclusive clade containing *Didelphis marsupialis* but not *Deltatheridium pretrituberculare*' (node based)	new
Marsupialia	'The least inclusive clade containing *Didelphis marsupialis*, *Caenolestes fuliginosus* and *Phalanger orientalis*' (node based)	Beck *et al.* (2014: 131)
Australidelphia	'The most inclusive clade exhibiting a "continuous lower ankle joint pattern" (CLAJP; that is, confluent ectal and sustentacular facets) and a triple-faceted, or "tripartite", calcaneocuboid (CaCu) joint synapomorphic with that in *Dromiciops*' (apomorphy based)	Beck (2012: 717)
Crown clade Australidelphia	'The least inclusive clade containing *Dromiciops gliroides*, *Phalanger orientalis*, *Perameles nasuta*, *Notoryctes typhlops* and *Dasyurus maculatus*' (node based)	New
Eomarsupialia	'The least inclusive clade containing *Phalanger orientalis*, *Perameles nasuta*, *Notoryctes typhlops*, and *Dasyurus maculatus*' (node based)	Beck *et al.* (2014: 132)

sirenians. I also do not cover the arrival of humans in the region, nor the various other placental species they introduced; recent discussions of these topics can be found in, among others, Johnson (2006), Helgen (2007), Davidson (2014), Dennell and Porr (2014) and Prins and Gordon (2014).

Geographical and Geological Context

From a geographical perspective, this review focuses on the continental and oceanic landmasses of the Sahul shelf, delimited by Lydekker's Line to the west and the Pacific oceanic plate to the east, which includes the mainland and adjacent islands of Australia and New Guinea (Figure 13.1d; Lydekker 1896). However, it should be recognised that this boundary is (like other faunal lines in the region; Simpson 1977) somewhat arbitrary and that numerous typically 'Sahulian' mammals (e.g. marsupials, Old Endemic and New Endemic murine rodents) occur on islands west of Lydekker's Line (Flannery 1995a). I refer to the Australian mainland plus Tasmania as Australia, and Australia plus New Guinea and adjacent islands as Sahul. As will be discussed, the major period of uplift of New Guinea does not appear to have occurred until the late Miocene; thus, I refer to Australia only when discussing biogeographical and geological events prior to this.

An understanding of the geological history of Sahul is key to interpreting the biogeographical history of the species inhabiting it. The following is a brief review of major global tectonic events occurring over the known timeframe of mammaliaform evolution (from the Late Triassic onwards; Kielan-Jaworowska *et al.* 2004), focusing on those directly affecting the geological evolution of the region.

The major phase of the breakup of Pangaea commenced with the opening of the Central Atlantic Ocean ca. 180–195 mya, although complete separation of Laurasia and Gondwana (with the development of a continuous Tethyan Seaway between the two supercontinents) may not have occurred until the Early Cretaceous (Lomolino *et al.* 2010; Seton *et al.* 2012; Torsvik and Cocks 2013). West Gondwana (what would become South America and Africa) and East Gondwana (what would become Antarctica, India, Madagascar, Australia and New Zealand) began to separate ca. 140–170 mya but remained in contact at their southern ends (Chatterjee *et al.* 2013; Lomolino *et al.* 2010; Seton *et al.* 2012; Torsvik and Cocks 2013); the southern tip of South America and the Antarctic Peninsula did not separate fully, via the opening of the Drake Passage, until well into the Cenozoic (Figure 13.2). The first major landmass to break away from Gondwana appears to have been Indo-Madagascar, ca. 130 mya (Chatterjee *et al.* 2013; Lomolino *et al.* 2010; Seton *et al.* 2012). Africa and South America began to separate ca. 120 mya, with the separation largely complete by ca. 100 mya (Lomolino *et al.* 2010; Seton *et al.* 2012). New Zealand began to break away from Australia ca. 80–90 mya (Ericson *et al.* 2014; Lomolino *et al.* 2010; Seton *et al.* 2012), but complete separation may not have been achieved until ca. 52 mya (Ericson *et al.* 2014).

Isotopic evidence indicates an influx of Pacific seawater into the Atlantic, across the Drake Passage between the southern tip of South America and the Antarctic Peninsula, ca. 41 mya (Scher and Martin 2006), although the Drake Passage may already have been open, but shallow (<1000 m), as early as 50 mya (Eagles and Jokat 2014; Lawver *et al.* 2011; Livermore *et al.* 2005). The deepwater opening of the Drake Passage may not have occurred until ca. 30 mya (Eagles and Jokat 2014). Evidence from dinoflagellates and organic geological records indicates a flow of water through the Tasmanian Gateway between Antarctica and Australia ca. 49–50 mya (Bijl *et al.* 2013). This was followed by the deepening of the Tasmanian Gateway ca. 35.5 mya (Stickley *et al.* 2004), leading to the establishment of the Antarctic Circumpolar Current. Terrestrial vertebrate dispersal between Australia and Antarctica would therefore seem highly unlikely beyond 35.5 mya, and so Australia can be reasonably considered an 'island continent' from this date onwards (Figure 13.2).

The geological history of New Guinea is complex and as yet incompletely understood. However, it is now generally accepted that only small areas of land, if any, were emergent north of the Australian continent until at least the middle–late Miocene (Figure 13.1a–c), with the major period of enlargement of the New Guinean landmass occurring over the last 5 Ma (Baldwin *et al.* 2012; Cloos *et al.* 2005; Hall 2002; Hill and Hall 2003; Hocknull 2009; Quarles van Ufford and Cloos 2005; Toussaint *et al.* 2014; Westerman *et al.* 2012). Particularly significant is that there is no geological evidence for a dry-land connection between Australia and New Guinea during the Eocene and/or Oligocene, contra Flannery (1988: Figure 13.2; 1995b: Maps 3–4).

Large fluctuations in eustatic sea level over the last ca. 10 Ma, and particularly the last ca. 2.5 Ma (Haq *et al.* 1987; Miller *et al.* 2005), have resulted in major changes in the extent of exposed land on the Sahul Shelf (Figure 13.1d). This has led to the repeated formation and severing of dry-land connections between New Guinea, mainland Australia and the adjacent islands over the last 3.5 Ma (Coller 2009; Hocknull 2009). The approximate extent and duration of these connections can be visualised using Monash University's SahulTime Web page (http://sahultime.monash.edu.au; Coller 2009). Of particular importance is that over the last ca. 1 Ma, Australia and New Guinea have more often formed a single landmass than they have been separated by sea (Bintanja *et al.* 2005; Coller 2009; Hocknull 2009). However, it is also crucial to note that, even during the lowest periods of sea level (such as during the Last Glacial Maximum), major marine barriers remained in place between Sahul and the land masses west of Lydekker's Line (Figure 13.1d); there has never been a dry-land connection between Sahul and Wallacea, or between Wallacea and the landmasses of the Sunda Shelf (Sundaland), further to the west (Lohman *et al.* 2011). Thus, dispersal to and from Sahul has required the crossing of marine barriers since the deepwater opening of the Tasman Gateway between Antarctica and Australia ca. 35.5 mya. This presumably explains why, among mammals, only murine rodents and bats appear to have successfully dispersed to Sahul from Southeast Asia, despite the existence of highly diverse placental faunas in Sundaland and the presence of a number of terrestrial placentals besides murines in Wallacea (Dennell *et al.* 2014; Flannery 1995a). Marine barriers are obviously less formidable to volant

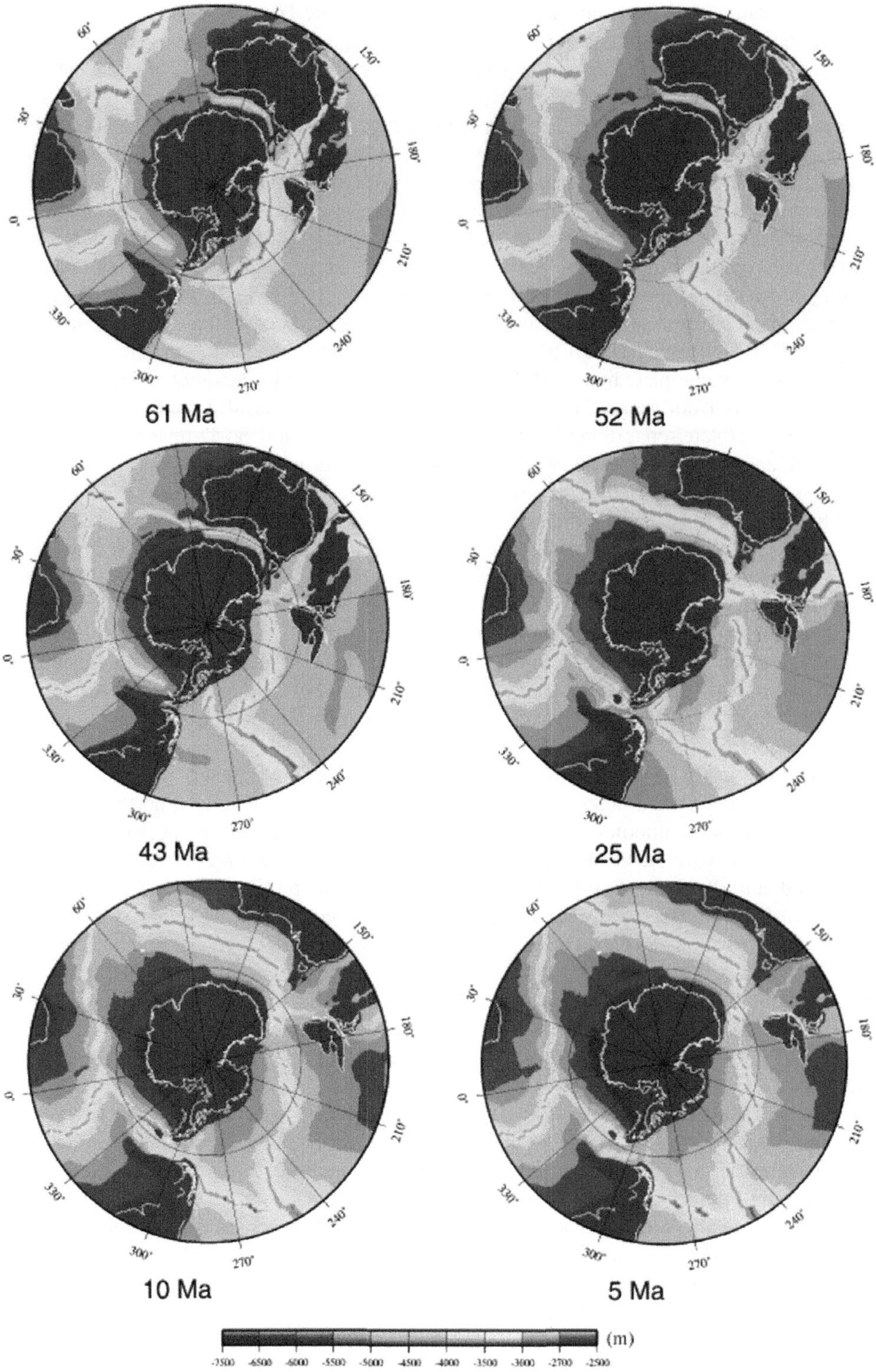

FIGURE 13.2 (See colour insert.) Circum-Antarctic palaeodepth models for the seafloor 61, 52, 43, 25, 10 and 5 mya, illustrating the opening and development of the Drake Passage (between South America and Antarctica) and the Tasmanian Gateway (between Antarctica and Australia). (Reprinted from Brown *et al.*, *Palaeogeography, Palaeoclimatology, Palaeoecology* 231, 158–168, Copyright 2006, with permission from Elsevier.)

bats than to terrestrial mammals, and murines appear to be particularly adept at over-water dispersal (Achmadi *et al.* 2013; van der Geer *et al.* 2010).

Biogeographical History of Non-marine Mammaliaform Clades

?Multituberculata

Multituberculates were conspicuous members of the Late Jurassic, Cretaceous and early Palaeogene mammal faunas of Laurasia (Kielan-Jaworowska *et al.* 2004; Rose 2006). Putative Gondwanan multituberculates, by contrast, are very rare and fragmentary, and their biogeographical interpretation correspondingly controversial. To date, a single probable multituberculate has been described from Australia (Rich *et al.* 2009a): *Corriebaatar marywaltersae* (Figure 13.3f), represented by a partial dentary preserving a plagiaulacoid premolar from the early–middle Aptian (ca. 115–125 Ma old) Flat Rocks site in the Eumeralla (or 'Wonthaggi') Formation, Strzelecki Group, southern Victoria (Figure 13.3a). Rich *et al.* (2009a) tentatively referred *C. marywaltersae* to the multituberculate clade Cimolodonta, but suggested that it might in fact represent a previously unknown mammaliaform lineage. A third possibility not considered by Rich *et al.* (2009a) is that *Corriebaatar* is a representative of the gondwanatherian family Ferugliotheriidae (otherwise known only from the Late Cretaceous and possibly early Palaeogene of South America; Goin *et al.* 2012a), as this group has also been argued to be characterised by the presence of a plagiaulacoid lower premolar (Gurovich and Beck 2009; but see Pascual *et al.* 1999; Pascual and Ortiz-Jaureguizar 2007). Nevertheless, pending further analysis and/or the discovery of additional material, I will follow Rich *et al.*'s (2009a) preferred interpretation here – namely, that *Corriebaatar* is a cimolodontan multituberculate.

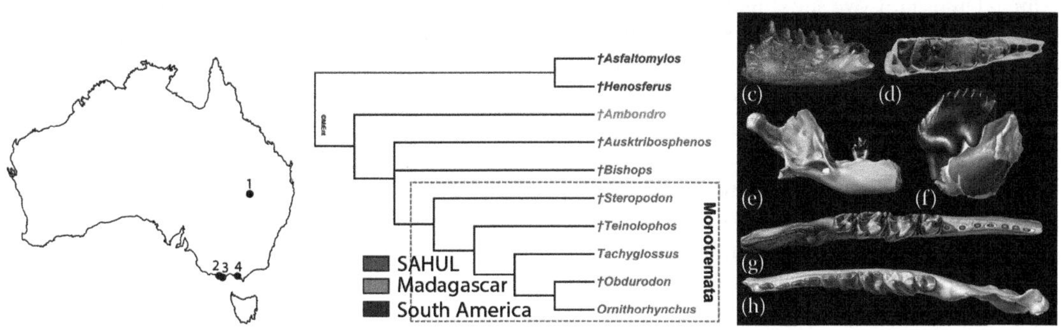

(a) (b)

FIGURE 13.3 (See colour insert.) (a) Locations of known Mesozoic mammaliaform-bearing fossil sites in Australia: (1) Lightning Ridge, northern New South Wales (Albian); (2) Dinosaur Cove, southern Victoria (early–middle Aptian); (3) Eric the Red West, southern Victoria (early–middle Aptian); (4) Flat Rocks, southern Victoria (early–middle Aptian). (b) Composite phylogeny of Australosphenida (based on Rougier, G. W., *et al.*, *American Museum Novitates*, 3566, 1–54, 2007; Bi, S., *et al.*, *Nature*, 514 (7524), 579–584, 2014.) Terminals are colour-coded according to their known biogeographic distributions. Extinct taxa are identified by daggers. Note that the Sahulian terminals (in red) form a single clade to the exclusion of non-Sahulian terminals (see main text). (c) Holotype of the monotreme *Steropodon galmani* (AM F66763, partial right dentary preserving m1–3) from Lightning Ridge (Archer, M., *et al.*, *Nature*, 318, 363–366, 1985.) (d) Holotype of *Kollikodon ritchiei* (AM F96602, partial right dentary preserving m1–3) from Lightning Ridge (Flannery *et al.* 1995). (e) Holotype of monotreme *Teinolophos trusleri* (NMV P208231, partial left dentary preserving m?2) from Flat Rocks (Rich *et al.*, *Records of the Queen Victoria Museum* 106, 1–35, 1999; Rich *et al.*, *Records of the Queen Victoria Museum* 110, 1–9, 2001b; 2005). Illustration: P. Trusler. (f) Holotype of ?cimolodontan multituberculate *Corriebaatar marywaltersae* (NMV P216655; partial left dentary preserving p4 and anterior root of m1) from Flat Rocks (Rich *et al.*, *Acta Palaeontologica Polonica* 54 (1), 1–6, 2009a). Illustration: P. Trusler. (g) Holotype of ausktribosphenid *Ausktribosphenos nyktos* (NMV P208090, partial right dentary preserving p6 m1–3) from Flat Rocks (see Rich, T. H., *et al.*, *Science*, 278, 1438–42, 1997). Illustration: P. Trusler. (h) Holotype of aukstribosphenid *Bishops whitmorei* (NMV P210075, partial left dentary preserving p1–6 m1–3) (Rich *et al.*, *Records of the Queen Victoria Museum* 110, 1–9, 2001a). Illustration: P. Trusler.

The oldest well-dated multituberculates are from the late Bathonian (ca. 166 mya) of England (Butler and Hooker 2005). However, Butler and Hooker (2005) argued for a much earlier origin of Multituberculata, in the Early Jurassic or even earlier. If this is the case, then the apparent presence of multituberculates in Gondwana (based on *Corriebaatar* and other putative Gondwanan records; Gurovich and Beck 2009; Kielan-Jaworowska *et al.* 2007; Krause 2013; Parmar *et al.* 2013; Rich *et al.* 2009a) could reflect an initial trans-Pangaean distribution for Multituberculata and subsequent vicariance (see ‘Geographical and Geological Context’ in this chapter). Of particular relevance is *Indobaatar zofiae* from the Kota Formation in India, which has been described as a probable eobataarid multituberculate (Parmar *et al.* 2013). The age of the Kota Formation remains controversial, with estimates ranging from the Early Jurassic to the Early Cretaceous (Parmar *et al.* 2013; Prasad and Manhas 2007). However, if the Kota Formation is Early Jurassic, then *Indobaatar* suggests that Multituberculata originated prior to the Pangaean breakup, and so the presence of *Corriebaatar* in the Aptian of Australia is plausibly the result of vicariance.

Monotremata

Two definitive fossil monotremes have been described from middle Cretaceous sites in Australia: *Steropodon galmani* (Figure 13.3c) from the Albian Griman Creek Formation at Lightning Ridge (Figure 13.3a) in northern New South Wales (Archer *et al.* 1985), and *Teinolophos trusleri* (Figure 13.3e) from the early–middle Aptian Flat Rocks site (Figure 13.3a) in southern Victoria (Rich *et al.* 1999, 2001b). Other putative records of monotremes from the Cretaceous of Australia are more uncertain. The dentally bizarre *Kollikodon ritchiei* (Figure 13.3d), also from Lightning Ridge, was originally identified as a probable monotreme (Flannery *et al.* 1995) but subsequently suggested to be a nonmammalian mammaliaform (Musser 2006, 2013). However, the recent paper by Pian *et al.* (2016) concluded that *Kollikodon* is indeed a monotreme or close relative. A ‘tachyglossid-like’ partial right humerus from the Albian Dinosaur Cove site (Figure 13.3a) in southern Victoria has been named *Kryoryctes cadburyi* and tentatively identified as a monotreme (Pridmore *et al.* 2005). However, given the lack of data regarding the postcranial morphology of most Mesozoic mammal groups, this identification should be viewed with caution. Additional specimens from Lightning Ridge, including a number of edentulous dentaries, have been suggested to represent monotremes (Musser 2013; Smith 2009), but these have yet to be described in detail. A fossil from Lightning Ridge, which Rich *et al.* (1989) suggested might be a maxilla of *Steropodon galmani*, has more recently been identified as turtle (Smith 2009).

There is currently no Mesozoic record of monotremes outside Australia (Kielan-Jaworowska *et al.* 2004; Rich 2008), and it is tempting to interpret this as evidence of a restricted biogeographical distribution for the group at this time, perhaps as part of a larger eastern Gondwanan radiation within Australosphenida (Figure 13.3b; see ‘Ausktribosphenidae’ in this chapter). Additional evidence that the terrestrial vertebrate fauna of Australia (or eastern Gondwana more generally) may have been biogeographically isolated or relictual during at least the middle Cretaceous include the temnospondyl *Koolasuchus cleelandi* from the early–middle Aptian Eumeralla Formation of Victoria (Warren *et al.* 1997), and a putative dicynodont that appears to be from the Aptian Allaru Formation in central Queensland (Thulborn and Turner 2003), although the exact geological provenance of the latter specimen is uncertain and its identification as a dicynodont has been questioned (Agnolin *et al.* 2010). Outside Australia, the youngest known temnospondyls are from the Jurassic (Schoch 2014), and the youngest known dicynodonts are from the Late Triassic (Fröbisch 2009). Rich *et al.* (2009a) also summarised evidence suggesting a clear distinction between the Early Cretaceous palaeofloras of South America on the one hand and those of Antarctica and Australia on the other, perhaps reflecting a climate-related filter acting through the Antarctic Peninsula. However, the exact biogeographical significance of other terrestrial vertebrates – such as turtles, crocodylomorphs and dinosaurs – from Australian Mesozoic sites is contentious, in part because most are known from highly fragmentary specimens (see the summary by Poropat *et al.* 2014). Some authors have recognised clear evidence of climate-driven provinciality in the middle Cretaceous terrestrial vertebrate faunas of Australia (e.g. Benson *et al.* 2012), whereas others have argued that they show close links to those from elsewhere in Gondwana, particularly South America, implying extensive faunal exchange during the middle Cretaceous (e.g. Agnolin *et al.* 2010).

Focusing specifically on Mammaliformes, direct comparison of the Australian record with the record elsewhere in Gondwana faces two major difficulties. Firstly, very few Mesozoic mammaliaform-bearing sites are currently known from Gondwana, and only a small number of specimens have been obtained from them to date (Kielan-Jaworowska *et al.* 2004; Rich 2008). Secondly, Gondwanan sites exhibit a disjunct temporal distribution (Kielan-Jaworowska *et al.* 2004; Rich 2008): the Australian sites (Figure 13.3a) are all Aptian–Albian in age, whereas those from elsewhere in Gondwana are either much older, much younger or of uncertain age. Thus, it is difficult to determine whether differences between the Mesozoic mammaliaform faunas of Australia and those of other Gondwanan landmasses reflect biogeographical factors or, alternatively, Gondwana-wide changes in faunal composition through time. Bearing these difficulties in mind, perhaps of greatest significance is the apparent absence of monotremes in the rich Late Cretaceous Allenian (i.e. Alamitian South American Land Mammal Age [SALMA]) faunas of Patagonia (Bonaparte 1990; Rougier *et al.* 2009a,b, 2011b). Rich *et al.* (2009a) interpreted this as evidence that terrestrial vertebrate dispersal between South America and Australia was unlikely during the middle Cretaceous but more probable during the Late Cretaceous, when falling global temperatures allowed the cool-adapted high-latitude fauna and flora of Australia–Antarctica to spread to lower latitudes, including into South America.

Remains of a fossil monotreme, *Monotrematum sudamericanum*, are known from the early Palaeocene (Peligran SALMA, ca. 65.7–63.5 Ma old; Clyde *et al.* 2014) 'Banco Negro Inferior' of the Salamanca Formation of Patagonia (Forasiepi and Martinelli 2003; Pascual *et al.* 1992, 2002). There seem two plausible interpretations for this. Firstly, monotremes may have been common to Australia, Antarctica and South America at least during the Cretaceous, in which case *M. sudamericanum* is a South American post-Cretaceous survivor. If so, then the apparent absence of monotremes in the Late Cretaceous Allenian faunas of South America is an artefact, which is possible given that monotreme fossils are rare even in Australia (R. Pian, pers. comm.). Alternatively, *M. sudamericanum* is the result of a dispersal event from Australia to South America (presumably via Antarctica), after the Allenian but before the Peligran – that is, during the latest Cretaceous or earliest Palaeocene. This is congruent with Rich *et al.*'s (2009a) preferred biogeographical hypothesis, and coincides approximately with the likely timing of the dispersal of marsupials from South America to Australia (Beck 2012; Beck *et al.* 2008; 'Marsupialiformes' in this chapter.). In either case, monotremes are expected to have been present in Antarctica, but have yet to be found; however, the only known Antarctic terrestrial mammal faunas, from the Early–Middle Eocene La Meseta Formation on Seymour Island off the coast of the Antarctic Peninsula, are as yet poorly known (Gelfo *et al.* 2014; Reguero *et al.* 2013). Besides *Monotrematum*, definitive monotremes are also unknown from South America.

Focusing now on the Cenozoic record of monotremes in Sahul, it is interesting that members of this clade have to date not been found in the early Eocene (ca. 54.6 mya; Godthelp *et al.* 1992) Tingamarra Local Fauna (Figure 13.4a) in northeastern Australia (pers. obs.). Again, this may be an artefact of sampling. However, a plausible alternative explanation is that average temperatures at Tingamarra during the early Eocene were too high for monotremes. Today, the modern platypus, *Ornithorhynchus anatinus*, does not occur north of the southern end of the Cape York Peninsula (~15°S), apparently because it is intolerant of the higher temperatures farther north (Grant 2007; Nicol 2013). If the platypus ecomorphotype is ancestral for crown clade monotremes (Mirceta *et al.* 2013; Phillips *et al.* 2009, 2010; but see Ashwell 2013; Camens 2010; Musser 2013), and if echidnas (family Tachyglossidae) did not originate until after the early Eocene (Phillips *et al.* 2009), then it is plausible that early Palaeogene monotremes were semiaquatic (Mirceta *et al.* 2013) platypus-like forms. If so, they may have had similar thermal physiologies to *Ornithorhynchus anatinus*. Although Australia was ~20° farther south during the early Eocene, the Tingamarra Local Fauna (Figure 13.4a) only slightly postdates the Palaeocene–Eocene Thermal Maximum, when global temperatures were as much as 12°C warmer than today (Zachos *et al.* 2001). Monotremes may therefore have been restricted to more southerly latitudes at this time, in southern Australia or perhaps in Antarctica (see also Musser 2013); however, this hypothesis remains speculative in the absence of additional early Palaeogene mammaliaform-bearing sites and better sampling.

The oldest Cenozoic monotremes currently known from Australia are *Obdurodon insignis* and a second, currently unnamed *Obdurodon* species from the latest Oligocene Etadunna and Namba Formations in central Australia (Woodburne and Tedford 1975; Woodburne *et al.* 1994). The discovery of a nearly

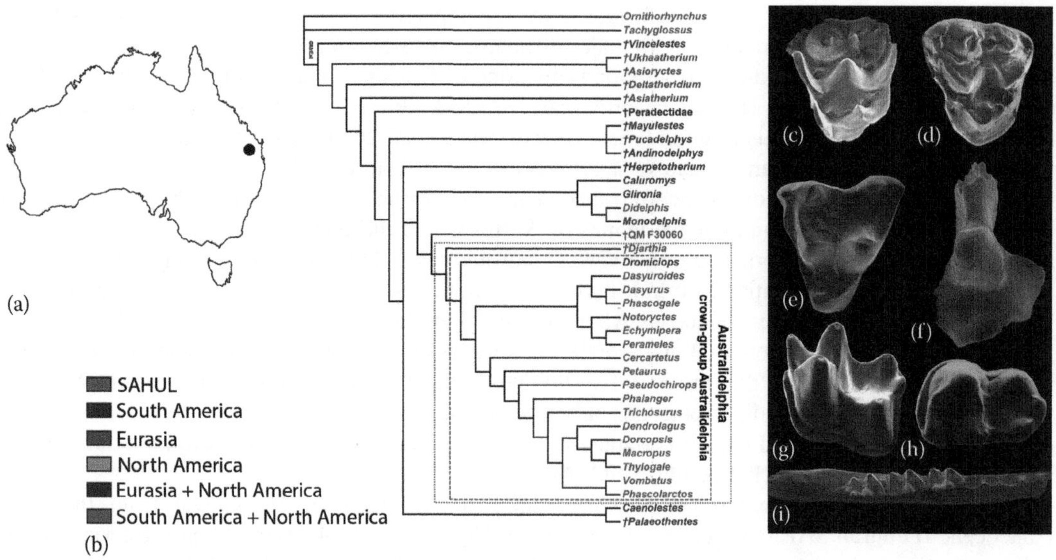

FIGURE 13.4 (See colour insert.) (a) Location of Australia's only early Palaeogene mammal-bearing fossil site, the early Eocene Tingamarra Local Fauna, near Murgon in southeastern Queensland. (b) Phylogeny of Metatheria based on maximum parsimony analysis of a 260 morphological character matrix with relationships constrained using a molecular scaffold (modified from Beck, R. M. D., *Naturwissenschaften*, 99(9), 715–29, 2012: Figure 6a). Terminals are colour-coded according to their known biogeographic distributions. Extinct taxa are identified by daggers. Note that the Sahulian terminals (in red) do not form a single clade to the exclusion of non-Sahulian terminals, and hence the single-dispersal model for the presence of marsupialiforms in Sahul can be rejected (see Beck, R. M. D., *Naturwissenschaften*, 99[9], 715–29, 2012 for full details). (c) Holotype of the bunodont marsupialiform *Chulpasia jimthorselli* (QM F50411, left M1 or 2) from Tingamarra (Sigé, B., *et al.*, *Geobios*, 42[6], 813–23, 2009). (d) Holotype of the bunodont marsupialiform *Chulpasia mattaueri* (CHU 30, left M1 or 2) from the ?late Palaeocene Chulpas locality, Laguna Umayo red mudstone unit, Peru (Sigé, B., *et al.*, *Geobios*, 42[6], 813–23, 2009). (e) Holotype of the faunivorous marsupialiform *Archaeonothos henkgodthelpi* (QM F53825, left M2 or 3) from Tingamarra (Beck, R. M. D., *Acta Palaeontologica Polonica*, 2014). (f) Calcaneus of an 'ameridelphian' marsupial (QM F30060) from Tingamarra (Beck, R. M. D., *Naturwissenschaften*, 99[9], 715–29, 2012). (g) Holotype of the bat *Australonycteris clarkae* (QM F19147, left m?2) from Tingamarra (Hand, S., *et al.*, *Journal of Vertebrate Paleontology*, 14[3], 375–81, 1994). (h) Holotype of the ?condylarth eutherian *Tingamarra porterorum* (QM F20564, right lower molar) from Tingamarra (Godthelp, H., *et al.*, *Nature*, 356, 514–16, 1992). (i) Holotype of the australidelphian marsupial *Djarthia murgonensis* (QM F31458, left partial dentary preserving m2–4) from Tingamarra (Godthelp *et al.* 1999; Beck, R. M. D., *et al.*, *PLoS ONE*, 3[3], e1858, 2008).

complete skull of a third *Obdurodon* species, the early Miocene *Ob. dicksoni*, confirms that *Obdurodon* was platypus-like in cranial morphology (Musser and Archer 1998). The recent description of a fourth species (Pian *et al.* 2013), the middle Miocene or Pliocene *Ob. tharalkooschild*, largely fills the temporal gap between the older *Obdurodon* species and the earliest remains of *Ornithorhynchus*, which are reportedly Pliocene in age (Musser 2013; Rich 1991). Thus, there is evidence for the continued presence of monotremes in mainland Australia since the late Oligocene.

There are five living species of living monotreme: the platypus *Ornithorhynchus anatinus*, the short-beaked echidna *Tachyglossus aculeatus*, and three species of long-beaked echidna: *Zaglossus attenboroughi*, *Z. bartoni* and *Z. bruijni* (Augee *et al.* 2006; Flannery 1995b; Flannery and Groves 1998; Van Dyck and Strahan 2008; Wilson and Reeder 2005). Recent point estimates for the divergence between *Ornithorhynchus* and the tachyglossids *Tachyglossus* and *Zaglossus* based on molecular data are 27.7–37.8 mya (Meredith *et al.* 2011; Phillips *et al.* 2009), but confidence intervals (CIs) on this divergence vary markedly depending on the analysis (composite CI = 13.3–103.1 mya). Thus, it is unclear whether or not the *Ornithorhynchus*–tachyglossid split occurred before or after the opening of the Tasman Gateway between Australia and Antarctica; however, it probably occurred before the emergence of the major New Guinean landmass (see 'Geographical and Geological Context' in this chapter).

The split between *Tachyglossus* and *Zaglossus* was estimated by Phillips *et al.* (2009) to be 5.5 mya (95% highest posterior distribution = 1.8–10.6 mya).

Today, *Ornithorhynchus anatinus* occurs only along the eastern side of mainland Australia, at latitudes >15°S, and in Tasmania (Gongora *et al.* 2012; Grant 2007; Kolomyjec *et al.* 2013). It has undergone range reduction since European colonisation, and may now be extinct in South Australia (apart from an introduced population on Kangaroo Island) and throughout much of the Murray–Darling Basin (Grant 2007). However, it appears never to have occurred naturally in Western Australia. *O. anatinus* is known from the southern end of the Cape York Peninsula, but (as discussed previously) is not found farther north, apparently because the ambient temperature is too high (Grant 2007; Nicol 2013). If so, this may explain why *O. anatinus* seems never to have been present in New Guinea.

Mitochondrial sequence data indicate the existence of two major clades within *O. anatinus*: an Australian mainland clade and a second clade comprising individuals from Tasmania and King Island (Gongora *et al.* 2012). The divergence time between these two clades was estimated by Gongora *et al.* (2012) at 0.7–0.94 mya, which is long before the last dry-land connector between Tasmania and the mainland was severed (ca. 14 ka ago; Lambeck and Chappell 2001). Gongora *et al.* (2012) therefore suggested that these clades diverged in mainland Australia, with subsequent extinction of the Tasmanian mitochondrial haplotype on the mainland. Recovery of ancient DNA from subfossil *O. anatinus* specimens should allow the testing of this hypothesis. The northeastern Queensland population is also genetically quite divergent (Gongora *et al.* 2012), and mitochondrial sequences and microsatellites indicate additional geographic structuring within populations, with different clusters coinciding with major river drainages (Gongora *et al.* 2012; Kolomyjec *et al.* 2013). *O. anatinus* exhibits Bergmann's rule across its latitudinal range as a whole; however, at smaller spatial scales (e.g. within a single river basin), an inverse relationship between temperature and body size (i.e. the opposite of Bergmann's rule) is observed (Furlan *et al.* 2011). In southeastern Australia, both lower rainfall and higher temperatures were found to be associated with larger-sized individuals by Furlan *et al.* (2011).

Tachyglossus aculeatus occurs in both Australia and New Guinea, with five subspecies currently recognised (Augee *et al.* 2006; Griffiths 1978). These subspecies appear to show clear biogeographic structuring (Augee *et al.* 2006; Griffiths 1978): *T. a. acanthion* in the Northern Territory, northern Queensland, inland Australia and Western Australia; *T. a. aculeatus* in eastern New South Wales, Victoria and southern Queensland; *T. a. lawesii* in the lowlands of New Guinea and possibly also the rainforests of northeastern Queensland; *T. a. multiaculeatus* in mainland South Australia and Kangaroo Island; and *T. a. setosus* in Tasmania. Intriguingly, body size in *T. aculeatus* corresponds to the reverse of Bergmann's rule, with the subspecies occurring at the highest latitudes (the Tasmanian *T. a. setosus*) being the smallest (Augee *et al.* 2006). The presence of *T. a. lawesii* in New Guinea is probably best explained as a relatively recent dispersal from Australia, given that the same subspecies is reportedly present in northeastern Queensland (Griffiths 1978). Dry-land connectors have formed repeatedly between Australia and New Guinea since the major emergence of New Guinea ca. 5 mya, and have been the rule rather than the exception over the last ca. 1 Ma (see 'Geographical and Geological Context' in this chapter), which presumably facilitated the dispersal of *T. aculeatus*, perhaps as part of a shared Austral–East Torresian mammal fauna (Lavery *et al.* 2013).

The three living species of long-beaked echidna occur only in New Guinea, with *Z. bruijnii* occurring in the west, *Z. bartoni* in central and eastern regions, typically at high elevations, and *Z. attenboroughi* in the Cyclops Mountains in the island's north (Baillie *et al.* 2009; Flannery 1995b; Flannery and Groves 1998; Helgen *et al.* 2012). Flannery and Groves (1998) identified four subspecies within *Z. bartoni*, with three of them forming a longitudinal cline in body size along the New Guinean central cordillera: the smallest, *Z. b. smeenki*, in the east, the largest, *Z. b. diamondi*, in the west, and the intermediate-sized *Z. b. bartoni* between them. A fourth subspecies, *Z. b. clunius*, is known from the Huon Peninsula in the east of the island. Recognisable long-beaked echidnas (most likely *Zaglossus* sp.) are depicted in Aboriginal rock art, possibly Late Pleistocene in age, from Arnhem Land in the Northern Territory of Australia (Helgen *et al.* 2012). In addition, Helgen *et al.* (2012) discussed a long-beaked echidna specimen in the Natural History Museum, London, that was apparently collected in 1901 in the West Kimberley region of northern Western Australia, and which they identified as *Z. bruijnii* (currently endemic to western New Guinea). Helgen *et al.* (2012) also discussed accounts by Aboriginal people living in the East Kimberley that may refer to *Zaglossus*, and suggest that it may still occur in the Kimberley region. Depending on the exact relationship between the putative Kimberley *Zaglossus* specimen and New Guinean populations, this raises the possibility of a complex biogeographical history for the genus, particularly given the existence of putative *Zaglossus* specimens from multiple Plio-Pleistocene sites in Australia.

Turning now to the fossil record of tachyglossids, possibly the oldest known remains are from a goldmine in New South Wales (Dun 1895), which has been suggested to date to the middle Miocene, but which may in fact be Pleistocene in age (Augee *et al.* 2006; Musser 2006, 2013). This material has been referred to the modern genus *Zaglossus* or to the extinct genus *Megalibgwilia*, as *Z.* or *M. robusta*, by different authors (Augee *et al.* 2006; Griffiths *et al.* 1991; Long *et al.* 2002; Musser 2003, 2006, 2013). The next-oldest tachyglossids are from Plio-Pleistocene deposits in mainland Australia. Of these, the short-beaked forms have been referred to the extant species *Tachyglossus aculeatus* (although these putative *T. aculeatus* specimens are markedly larger than modern individuals; Augee *et al.* 2006; Pledge 1980). The long-beaked forms have been referred to the extant genus *Zaglossus* and/or the fossil genus *Megalibgwilia* (Augee *et al.* 2006; Musser 2006; Murray 1978a,b). None of the long-beaked material from Australia has been referred to a living *Zaglossus* species, apart from postcranial remains from the Pleistocene Henschke's Quarry Cave at Naracoorte in South Australia, which Murray (1978b) provisionally referred to the living *Z. bruijnii*; however, Murray (1978a) subsequently referred this material to *Zaglossus* sp., and Pledge (1980) suggested it probably represents *Tachyglossus*. Hocknull (2009) also reported a manual ungual of an indeterminate tachyglossid from the Middle Pleistocene (at least 330 ka old; Hocknull *et al.* 2007) of Mt. Etna in northern Queensland. Tachyglossids (and other monotremes) are not known from the pre-Pleistocene of New Guinea, although the fossil record is poor, with only a single mammal-bearing deposit known – namely, the middle Pliocene Otibanda Formation (Flannery *et al.* 1993; Long *et al.* 2002; Plane 1967).

A better understanding of the biogeographical history of tachyglossids is prevented by current uncertainties regarding their taxonomy and phylogeny. The alpha taxonomy of the living genera *Tachyglossus* and *Zaglossus* is in need of revision and testing with molecular data. Broad-scale phylogeographic analyses of molecular sequence data and associated estimates of divergence times are also required to clarify when *T. aculeatus* dispersed to New Guinea, and whether there was a single or multiple dispersals. Likewise, the biogeographical history of *Zaglossus* will remain unclear until the relationships and divergence times between the New Guinean *Zaglossus* species, the Plio-Pleistocene long-beaked echidnas of mainland Australia (*Zaglossus* and *Megalibgwilia* species) and the Kimberley *Zaglossus* specimen (Helgen *et al.* 2012) are clarified. To this end, a thorough revision and formal phylogenetic analysis of living and fossil long-beaked echidnas is sorely needed; in particular, the validity of *Megalibgwilia*, distinct from *Zaglossus*, needs to be tested. Ideally, such an analysis should incorporate DNA sequence data from multiple representatives of extant New Guinean *Zaglossus* species, and (if DNA can be obtained) from the Kimberley specimen and subfossil material of long-beaked echidnas from Australia.

Ausktribosphenidae

Ausktribosphenos nyktos and *Bishops whitmorei* are tribosphenic mammals known from multiple partial lower jaws from the early–middle Aptian Flat Rocks site (Rich *et al.* 1997, 2001a), with a species of *Bishops* (possibly *B. whitmorei*) also known from the similarly aged Eric the Red West site (Rich *et al.* 2009b). *Ausktribosphenos* and *Bishops* are currently classified as the only named representatives of the family Ausktribosphenidae (Rich *et al.* 1997, 2001a; Kielan-Jaworowska *et al.* 2004). It has been proposed that ausktribosphenids are eutherians (Rich *et al.* 1997, 2001a; Woodburne *et al.* 2003). If so, this would pose something of a biogeographic conundrum: the majority of fossil evidence supports a Laurasian origin for Eutheria (Kielan-Jaworowska *et al.* 2004; Ji *et al.* 2002; Luo *et al.* 2011), and there is no evidence of either eutherians or metatherians in the Late Cretaceous Allenian faunas of southern South America. Both eutherians and metatherians were widespread and diverse throughout Laurasia during the Cretaceous (Kielan-Jaworowska *et al.* 2004), and were also highly diverse in early-to-middle Palaeocene faunas known from South America (Bonaparte *et al.* 1993; Gelfo *et al.* 2007;

Goin *et al.* 1992; Marshall and Muizon 1988; Muizon 1991; Muizon and Cifelli 2001), suggesting that therian mammals in fact first entered Gondwana during the latest Cretaceous or early Palaeocene (see 'Marsupialiformes' in this chapter).

However, most studies have concluded that *Ausktribosphenos* and *Bishops* fall outside Theria, and are members of a group that independently evolved a tribosphenic dentition (Davis 2011; Kielan-Jaworowska *et al.* 2004; Luo *et al.* 2001, 2002). In support of this conclusion, most published phylogenetic analyses (e.g. Bi *et al.* 2014; Kielan-Jaworowska *et al.* 2004; Luo *et al.* 2001, 2002; Rougier *et al.* 2007) place *Ausktribosphenos* and *Bishops* in a clade with monotremes and three other tribosphenic mammals from the Jurassic of Gondwana: *Henosferus molus* and *Asfaltomylos patagonicus* from the Toarcian of South America (Martin and Rauhut 2005; Rauhut *et al.* 2002; Rougier *et al.* 2007), and *Ambondro mahabo* from the Bathonian of Madagascar (Flynn *et al.* 1999). This clade was named Australosphenida by Luo *et al.* (2001).

Relationships within Australosphenida have yet to be fully resolved and are likely to remain so pending the discovery of more complete material of this enigmatic group; as such, the biogeography of the group is also uncertain. However, based on the Early Jurassic age of *Henosferus* and *Asfaltomylos* (see Cúneo *et al.* 2013), Australosphenida originated prior to the breakup of Gondwana. Recent phylogenetic analyses suggest that ausktribosphenids and monotremes form a clade to the exclusion of *Asfaltomylos* and *Ambondro* (Bi *et al.* 2014; Rougier *et al.* 2007), raising the possibility that Ausktribosphenidae + Monotremata represents a localised eastern Gondwanan or Australian radiation of australosphenidans. Certainly, it is interesting that australosphenidans are unknown from Cretaceous deposits in South America (notably the diverse Allenian faunas), and so it is possible that the group had become extinct in western Gondwana by the Late Cretaceous but survived in the east, along with other possibly relictual taxa such as the temnospondyl *Koolasuchus* (see the discussion in 'Monotremata' in this chapter).

Marsupialiformes

The oldest generally accepted metatherian is *Sinodelphys szalayi* from the 125 Ma old Yixian Formation in northeastern China (Luo *et al.* 2003). The subsequent description of the apparent eutherian *Juramaia sinensis* from the Upper Jurassic Daohugou Beds in the same region by Luo *et al.* (2011) appears to push the age of the Metatheria–Eutheria split back to a minimum of 160 mya. However, doubts as to both the affinities and age of *Juramaia* have been raised (Jansa *et al.* 2014: supporting information; Sullivan *et al.* 2014). Recent molecular estimates for this divergence are broad, spanning 140.5–215.3 Ma when confidence intervals are taken into account (dos Reis *et al.* 2012, 2014; Meredith *et al.* 2011).

Critically, whereas metatherians and eutherians are common elements of middle-to-late Cretaceous faunas in Laurasia, they have not been found in similarly aged faunas in the Southern Hemisphere (Kielan-Jaworowska *et al.* 2004), with the exception of a questionable marsupialiform known from a single partial molar from the Maastrichtian of Madagascar (Krause 2001; but see Averianov *et al.* 2003), and a few eutherians from the Maastrichtian of India (Khosla and Verma 2014; Prasad *et al.* 1994, 2007; Rana and Wilson 2003). The entire published Mesozoic mammaliaform record in Australia is restricted to four Aptian–Albian localities (Figure 13.3a) – Lightning Ridge in New South Wales, and the Dinosaur Cove, Flat Rocks and Eric the Red West sites in southern Victoria – and only seven taxa have been named from these. Thus, the apparent absence of therian mammals from the Mesozoic of Australia could plausibly be an artefact of incomplete sampling. Of greater significance is the lack of therians from the much richer Mesozoic mammal record of South America. Prior to the Late Cretaceous, the South American record is relatively sparse (Rougier *et al.* 2011b). However, the comparatively rich Late Cretaceous Allenian faunas of southern South America lack any trace of therians, preserving instead a diverse, apparently highly endemic mammalian fauna dominated by non-therian dryolestoids and gondwanatherians (Bonaparte 1990; Rougier *et al.* 2009a,b, 2011b). The simplest explanation for the lack of therians in these diverse Allenian faunas, and one assumed by numerous authors (e.g. Beck 2008b; Goin *et al.* 2012b; Pascual 2006; Pascual and Ortiz-Jaureguizar 2007; Szalay 1994), is that they were genuinely absent from South America at this time. Given that the most likely point of entry of therians into Gondwana was from North America to South America (Case *et al.* 2005), their apparent absence from South America during the Allenian is likely an indication that they were also absent from at least those parts of Gondwana in direct contact with South America at this time – namely, Antarctica and Australia.

The precise age of the Allenian faunas is unknown, but they are most likely late Campanian–early Maastrichtian in age (Pascual and Ortiz-Jaureguizar 2007; Rougier *et al.* 2009a,b). A few mammaliaform fossils are known from a younger South American Cretaceous site, the middle Maastrichtian Pajcha Pata locality in Bolivia (Gayet *et al.* 2001); they include at least one 'dryolestoid', but definitive therians have not been found there (Rougier *et al.* 2011b; pers. obs.). However, the Pajcha Pata fauna is still too poorly known to confidently rule out the presence of therians. A conservative maximum bound on the age of entry of therians (i.e. marsupialiforms and eutherians) into South America is therefore the maximum age of the Campanian, 83.6 Ma, based on a conservative maximum age of the Allenian. The diverse mammal fauna from the Banco Negro Inferior of the Salamanca Formation in southern Argentina includes marsupialiforms and eutherians, as well as non-therian taxa (Bonaparte *et al.* 1993; Gelfo *et al.* 2007; Goin *et al.* 1992). This fauna forms the basis of the Peligran SALMA, and has recently been dated as early-to-middle Danian (ca. 65.7–63.5 mya; Clyde *et al.* 2014). This date provides a minimum age for the dispersal of therians from North America to South America. Thus, the probable time of dispersal of both marsupialiforms and eutherians from North to South America can be constrained to between 83.6 and 63.5 Ma.

A combination of geological, fossil and phylogenetic evidence suggests that the presence of marsupialiforms in Australia is the result of dispersal from South America, across Antarctica, prior to the deepwater opening of the Drake Passage and Tasman Gateway (Beck 2008b, 2012; Beck *et al.* 2008; Kemp 2005; Lawver *et al.* 2011; Woodburne and Case 1996). The oldest marsupialiforms known from Australia are from the Tingamarra Local Fauna in southeastern Queensland (Figure 13.4a; Beck 2012, 2014; Beck *et al.* 2008; Godthelp 1999; Sigé *et al.* 2009), which has been radiometrically dated as 54.6 Ma old (Godthelp *et al.* 1992). This provides a minimum date for the dispersal of marsupialiforms to Australia. The maximum likely date for dispersal can be set at 83.6 mya, based on the maximum likely date for the entry of marsupialiforms into South America.

As discussed by Kemp (2005: 218–221) and Beck (2008b, 2012), a key issue is whether there was (1) a single dispersal by marsupialiforms (restricted to crown clade marsupials) from South America to Australia only, which would imply that major dispersal barriers were already in place by the Late Cretaceous–early Palaeogene (the single-dispersal model); (2) multiple independent dispersals between South America and Australia, implying less severe dispersal barriers (the multiple-dispersals model); or (3) a single, relatively continuous marsupialiform fauna stretching from at least the southern part of South America, across Antarctica to Australia (collectively, the Austral Kingdom; Aragón *et al.* 2011; Goin *et al.* 2007, in press; Morrone 2002) during at least the early part of the Palaeogene, implying weak or absent dispersal barriers (the continuous fauna model). Under the continuous fauna model, faunal differences between South America and Australia would have developed subsequently as a result of vicariance – most plausibly the deepwater opening of the Drake Passage and Tasman Gateway – and differential extinction (Beck 2012).

Recent molecular analyses of modern marsupials suggest that the four modern Sahulian marsupial orders form a clade, Eomarsupialia (see Table 13.1), with respect to the three extant South American marsupial orders (Didelphimorphia, Paucituberculata and Microbiotheria), which form a paraphyletic assemblage outside Eomarsupialia (Beck 2008a; Meredith *et al.* 2009a,b, 2011, 2014; Nilsson *et al.* 2010). An obvious interpretation of this pattern is that the presence of marsupials in Sahul is the result of a single dispersal event, from South America, with an ancestor giving rise to the entire Sahulian marsupial radiation – that is, the single-dispersal model (Meredith *et al.* 2009a,b; Nilsson *et al.* 2010). However, for this to be true, all fossil marsupialiforms from Sahul must also be part of the same radiation, and recent studies indicate that this is not the case. Specifically, material from the early Eocene Tingamarra Local Fauna (Figure 13.4a) – namely, specimens referable to *Djarthia murgonensis* (Figure 13.4i) and also an isolated calcaneus (QM F30060; Figure 13.4f) that represents a second taxon – falls within Marsupialia but outside crown clade Australidelphia (which includes the South American *Dromiciops*) in published phylogenetic analyses (Figure 13.4b), suggesting that the single-dispersal model can be rejected (Beck 2012; Beck *et al.* 2008). Although yet to be included in formal phylogenetic analyses, the Tingamarran species *Thylacotinga bartholomaii*, *Chulpasia jimthorselli* (Figure 13.4c) and *Archaeonothos henkgodthelpi* (Figure 13.4e) also show no clear evidence of belonging to Eomarsupialia; *T. bartholomaii* and *C. jimthorselli* show the closest similarities to *C. mattaueri* (Figure 13.4d) from the late Palaeocene or early Eocene of Peru (Sigé *et al.* 2009), while *A. henkgodthelpi* resembles *Kasserinotherium tunisiense* from the early Eocene of Tunisia and *Wirunodon chanku* from the ? middle–late Eocene of Peru (Beck 2014). Thus, monophyly of the modern Sahulian marsupial radiation relative to modern South American marsupials appears to be the result of the extinction of non-eomarsupialian lineages in Australia, some time between the early Eocene and late Oligocene.

As I have noted previously (Beck 2012), distinguishing between the other two possibilities – namely, the multiple dispersals and continuous fauna models – is difficult given the deficiencies in the available fossil record, particularly from the early Palaeogene of Australia and Antarctica. However, the apparent absence from Tingamarra and younger Australian sites of typically 'South American' marsupialiforms, such as the dentally distinctive polydolopids (which are also common in the middle Eocene La Meseta Fauna of Antarctica; Chornogubsky *et al.* 2009) and also eutherian groups such as 'meridiungulates' and xenarthrans, argues against the continuous fauna model (Beck 2012). Similarly, recent molecular estimates of divergence times indicate that the modern Sahulian marsupial orders had diverged from each other by the middle Eocene at the latest (Beck 2008a; Meredith *et al.* 2009a,b, 2011; Mitchell *et al.* 2014), but unequivocal members of these orders have not been found in any South American site or the La Meseta Fauna. Thus, on present evidence, the multiple dispersals model appears most likely. However, the number of dispersals is unclear, and it is uncertain whether they were all from South America to Australia or whether dispersal(s) in the reverse direction also occurred. For example, it is possible that the presence of microbiotherians in western Antarctica and South America is the result of a back-dispersal from Australia (Beck 2012; Beck *et al.* 2008).

The multiple dispersals model implies the presence of dispersal barriers between South America and Australia during the Late Cretaceous–early Palaeogene, preventing the formation of a single continuous mammalian fauna (Beck 2008b, 2012). I have previously speculated (Beck 2008b) that Antarctica may have posed such a barrier due to (1) the narrow, mountainous, high-latitude connection between western and eastern Antarctica; (2) low temperatures (particularly at the highest latitudes), even during the 'greenhouse' conditions of the Late Cretaceous and early Palaeogene (Francis and Poole 2002; Kemp *et al.* 2014; Poole *et al.* 2005); and possibly (3) the Valdivian-type flora that extended from southern South America across Antarctica and into Australia (Case *et al.* 1988). Based on this, I suggested that trans-Antarctic dispersal by terrestrial mammals during the latest Cretaceous and early Palaeogene would have been more likely for those taxa characterised by a high basal metabolic rate, cold tolerance, the ability to hibernate, small body size and arboreal adaptations (Beck 2008b).

Turning now to the modern Sahulian radiation, a key issue has been to clarify the precise biogeographical relationships between the Australian and New Guinean marsupial faunas. Flannery (1988) proposed that the development of a seaway between the two landmasses led to the marsupial faunas of the two landmasses becoming distinct by the early Miocene at the latest, and that a dry-land connection was not re-established until the Pleistocene. However, this scenario conflicts with current evidence (summarised in 'Geographical and Geological Context' in this chapter) that (1) the majority of the New Guinean landmass has become emergent only in the last 5 Ma (Figure 13.1a–c) and (2) that dry-land connectors have formed repeatedly between Australia and New Guinea since at least 3.5 mya (Figure 13.1d). Several studies have used molecular estimates of divergence dates between modern Australian and New Guinean taxa to infer the timing of putative dispersal events between the two landmasses, and determine whether they conform to Flannery's (1988) model (e.g. Aplin *et al.* 1993; Kirsch and Springer 1993; Meredith *et al.* 2010; Mitchell *et al.* 2014; Raterman *et al.* 2006; Westerman *et al.* 2012). The most recent and comprehensive of these is that of Mitchell *et al.* (2014), who observed that most of their inferred dispersals between Australia and New Guinea can be dated as having occurred within the last 5 Ma, after the major emergence of the New Guinean landmass. This is congruent with Murray's (1992) conclusion that Plio-Pleistocene zygomaturine diprotodontids known from New Guinea are the result of dispersal from Australia during the middle-to-late Pliocene.

However, Mitchell *et al.* (2014) argued that the divergence times of three predominantly New Guinean marsupial clades – namely, (1) a clade within Phalangeridae comprising the genera *Ailurops*, *Phalanger*,

Spilocuscus and *Strigocuscus*; (2) the *Murexia sensu lato* dasyurid clade (which encompasses the genera *Murexia*, *Micromurexia*, *Murexechinus*, *Paramurexia* and *Phascomurexia*; Groves 2005a; Krajewski *et al.* 2007; Van Dyck 2002); and (3) peroryctid bandicoots (*Echymipera*, *Microperoryctes*, *Peroryctes* and *Rhynchomeles*) – suggest early dispersals from Australia to New Guinea, probably sometime between 9 and 11 mya. Given the apparent synchronicity of these dispersals, Mitchell *et al.* (2014) considered that they are unlikely to have been over marine barriers, *contra* current geological evidence that there was probably no land connection between New Guinea and Australia until after the major period of uplift of New Guinea ca. 5 mya (see 'Geographical and Geological Context' in this chapter).

There are, however, at least two alternative explanations for the presence of these three 'old' clades in New Guinea. Firstly, and perhaps more prosaically, it may reflect problems with the molecular clock analyses implemented by Mitchell *et al.* (2014), resulting in overestimated divergence dates; many current molecular clock models appear unable to fully account for rapid changes in the rate of molecular evolution, and have likely overestimated the ages of certain divergences in published studies (Dornburg *et al.* 2012, 2014; Kitazoe *et al.* 2007; Steiper and Seiffert 2012; Waddell 2008). Indeed, such issues should be borne in mind when considering all of the molecular divergence dates presented in this chapter.

Perhaps a more interesting alternative explanation is that the apparent antiquity of these three clades is due to the extinction of Australian relatives. Support for this interpretation comes from the rich Middle Pleistocene vertebrate fauna from Mount Etna in northern Queensland, which has revealed the presence of multiple typically 'New Guinean' taxa that have gone extinct in Australia within the last 280 ka (Cramb *et al.* 2009; Hocknull 2005, 2009; Hocknull *et al.* 2007). These include probable representatives of the three 'old' New Guinean clades identified by Mitchell *et al.* (2014) – most notably, (1) *Phalanger gymnotis* (Hocknull 2009), (2) a dasyurid that Cramb *et al.* (2009) identified as cf. *Micromurexia habbema* and (3) three peramelemorphians that appear most similar among living bandicoots to *Peroryctes* and *Microperoryctes* (Hocknull 2005, 2009). Thus, it seems that the current lack of representatives of these three clades in Australia – with the exception of the peroryctid *Echymipera rufescens* and the phalangerids *Phalanger mimicus* and *Spilocuscus maculatus*, all of which appear to represent recent divergences from New Guinean conspecifics – is the result of extinction, rather than prolonged biogeographical isolation from New Guinea. Congruent with this interpretation, several other marsupials from Middle Pleistocene deposits at Mt. Etna have also been argued to be more closely related to New Guinean than to Australian species among living taxa, including species of *Dendrolagus*, *Dactylopsila* and *Pseudochirops* (Hocknull 2005, 2009). Also of relevance is the presence at Mt. Etna of several highly distinctive taxa that appear to represent major lineages that are now entirely extinct. These include the phascolarctid *Invictokoala monticola* (Price and Hocknull 2011) and representatives of three as yet undescribed dasyurid genera (Cramb *et al.* 2009). This further demonstrates that the modern fauna of Australia has been shaped by the complete loss of several major marsupial lineages within the last 280 ka.

In fact, it is perhaps more appropriate to consider northeastern Australia and New Guinea, as well as islands such as the Moluccas, as essentially having a shared fauna throughout at least the Pliocene and Early Pleistocene, intermittently divided during interglacial periods; this corresponds to Lavery *et al.*'s (2013) concept of the Austral–East Torresian zoogeographic province, spanning the northeast of Australia and southwest of New Guinea (see also Flannery 1995b; Helgen 2007). If members of this shared fauna then became extinct in Australia, perhaps with the onset of more xeric conditions 280–205 ka ago (Hocknull *et al.* 2007), this might explain the relatively ancient divergences between predominantly New Guinean and predominantly Australian sister taxa found in molecular studies of modern species, without recourse to early (and geologically implausible) dispersals to New Guinea of the kind proposed by Mitchell *et al.* (2014).

Focusing specifically on the modern New Guinean marsupial fauna, Helgen (2007) represents a recent, comprehensive overview of the taxonomy and distribution of marsupials and other mammals on the island. However, there is a need for detailed phylogeographic studies, both to clarify species boundaries and to determine the biogeographical factors that have shaped their distributions. For example, they would allow further testing and refinement of the Oceanic, Tumbanan and Austral zoogeographic provinces recognised by Lavery *et al.* (2013; see also Flannery 1995b; Helgen 2007), and help us to understand the role that uplift and altitudinal gradients have played shaping marsupial diversity on the island (Helgen 2007; Macqueen *et al.* 2011; Meredith *et al.* 2010). One of the few such studies published to date is Macqueen *et al.*'s (2011) phylogeographic analysis of New Guinean pademelons (*Thylogale* spp.). Macqueen *et al.* (2011) found evidence for the existence of 'western' and 'eastern' *Thylogale* clades that do not correspond to the current species taxonomy (Flannery 1995b; Groves 2005b) or to Flannery's (1992) recognition of northern, central and southern groups; instead, the major split coincides with the Ramu–Markham and Watut–Tauri valleys in the east of New Guinea. Additional genetic structuring within the western and eastern *Thylogale* clades is complex, and likely reflects both geological (e.g. recent uplift) and climatic (e.g. vegetational changes between glacial and interglacial periods) factors (Macqueen *et al.* 2011). It remains to be seen whether other New Guinean marsupial species show congruent or contrasting biogeographical patterns.

Published phylogeographic studies of Australian marsupials have been far more numerous (e.g. Brown *et al.* 2006; Cooper *et al.* 2000; Eldridge *et al.* 2011; Firestone *et al.* 1999; Hazlitt et al. 2014; Macqueen *et al.* 2012; Malekian *et al.* 2010; Neaves *et al.* 2009; Pavlova *et al.* 2010; Pope *et al.* 2000, 2001; Potter *et al.* 2012; Spencer *et al.* 2001), and cannot be fully summarised here. These studies have uncovered complex biogeographical patterns, which have been interpreted to be the result of numerous different factors, including the existence of past and present biogeographical barriers to gene flow, climate change and habitat contraction and expansion. Perhaps most striking is that biogeographic patterns found in these studies appear to be highly individualistic, with little evidence of commonalities between different marsupial species. For example, the Burdekin Gap (an area of dry woodland separating the Wet Tropics of Queensland from mesic forest habitats farther south) appears to have formed a major barrier to gene flow in the yellow-bellied glider (*Petaurus australis*; Brown *et al.* 2006) but not in the red-legged pademelon (*Thylogale stigmatica*; Eldridge et al. 2011), even though both are forest-adapted species. Similarly, Potter et al. (2012) found evidence of a deep divergence between populations of *brachyotis* group rock wallabies either side of the East–West Kimberley Divide, whereas similar phylogeographic structuring was not observed in the scaly tailed possum (*Wyulda squamicaudata*; Potter *et al.* 2014), despite the fact that both are rock-dwelling species.

These individualistic biogeographic patterns are perhaps unsurprising, given species' differing dispersal abilities, niche requirements and other aspects of basic biology. Nevertheless, they may not be easily predictable: for example, Eldridge *et al.* (2014) found evidence of strong genetic structuring across the Carpentarian Barrier (an area of semiarid grassland in northern Australia separating more mesic savannah woodland habitats farther west and east) in the common wallaroo (*Macropus robustus*), even though the species is a habitat generalist that is continuously distributed throughout the region. Conversely, the antilopine wallaroo (*M. antilopinus*) showed little genetic differentiation across the Carpentarian Barrier, despite the fact that it is a tropical woodland specialist that does not occur in the region of the barrier (Eldridge *et al.* 2014). In fact, the phylogeographic structuring observed in *M. robustus* and *M. antilopinus* in northern Australia is the reverse of *a priori* predictions based on the known biology and distributions of the two species (Eldridge *et al.* 2014). Similar studies of other marsupial species may reveal further surprises.

Two factors are likely to complicate future phylogeographic studies of modern Sahulian marsupials. Firstly, it is clear that during the Late Pleistocene–Holocene, numerous New Guinean (Aplin and Pasveer 2006; Aplin *et al.* 1999b; Helgen 2007) and Australian (Burbidge *et al.* 2008; Johnson 2006; McKenzie *et al.* 2007) species have gone entirely extinct. Many others have radically reduced distributions, with the loss of entire populations, at least some of which likely represent major, evolutionarily distinct lineages. Conversely, there is also evidence that current distributions of some New Guinean marsupials have been influenced by human-mediated introductions (Macqueen *et al.* 2011; Heinsohn 2010). As a result, studies that use molecular and/or distributional data from current populations only may be misled as to the likely biogeographical history of the species in question; in fact, this might explain the lack of evidence for common biogeographical patterns in phylogeographic studies published to date. Once again, evidence from the fossil and subfossil record, and also from museum specimens collected during the early years of European colonisation (before the extinction or extreme range reduction of many small- and medium-sized marsupial species in Australia) will be critical; this might be simply distributional data (as used, for example, in the biogeographic analyses of Burbidge *et al.* 2008 and Lavery *et al.* 2013) or, if obtainable, ancient DNA.

?Condylartha

Godthelp *et al.* (1992) described an isolated tribosphenic lower molar (Figure 13.4h) from the early Eocene Tingamarra Local Fauna (Figure 13.4a) that they named *Tingamarra porterorum*. They tentatively referred *Tingamarra* to 'Condylarthra', a non-monophyletic assemblage of eutherians, some of which likely gave rise to the various 'ungulate' groups (Archibald 1998; Rose 2006). 'Condylarths' are first known in South America from the early Palaeocene (Peligran SALMA; Clyde *et al.* 2014; Gelfo 2007; Gelfo *et al.* 2007), probably the result of dispersal from North America, at around the same time that marsupialiforms seem to have taken the same route (Muizon and Cifelli 2001; see 'Marsupialiformes' in this chapter). Thus, the dispersal of this group to Australia, again presumably along the same route taken by marsupialiforms (namely, from South America via Antarctica), seems plausible.

However, the single known tooth of *T. porterorum* does not show close similarities to, and is also markedly smaller than, any known South American 'condylarth' (Gelfo 2007; Gelfo and Sigé 2011; Gelfo *et al.* 2007; Muizon and Cifelli 2000). Furthermore, eutherians besides bats are otherwise entirely absent from Australian fossil deposits until the first appearance of murine rodents ca. 4 mya (Archer *et al.* 1999b; Long et al. 2002); if *T. porterorum* is indeed a eutherian, this lineage appears to have gone extinct in Australia before the late Oligocene. Another lower molar that shows a very similar overall morphology to *T. porterorum* but is distinctly larger has been found at Tingamarra but remains undescribed (Godthelp *et al.* 2001). The study of this material, together with an isolated petrosal from Tingamarra that apparently represents a small nonvolant eutherian, should shed further light on this biogeographically puzzling taxon (Beck *et al.* in prep.).

Chiroptera

Besides the putative condylarth *Tingamarra* (see previous section) and murine rodents (see 'Murinae' in this chapter), bats (order Chiroptera) are the only eutherians known to have reached Australia without human assistance. This is perhaps unsurprising: as the only truly volant mammals, bats are excellent dispersers, and they are often the only native mammals on otherwise isolated landmasses such as oceanic islands (Flannery 1995a; Simmons 2005; van der Geer *et al.* 2010). Hand (2006) provided a recent and comprehensive overview of Sahulian bat diversity and biogeography, and the following is largely a summary of this work.

Recent phylogenetic analyses of placental mammals place bats within the superordinal clade Laurasiatheria, together with artiodactyls, perissodactyls, carnivorans, pangolins and eulipotyphlan 'insectivores' (Meredith *et al.* 2011; O'Leary *et al.* 2013). As the name suggests, current evidence suggests that this superorder probably originated in Laurasia (Springer *et al.* 2011). However, the oldest record of bats in Australia is also one of the earliest globally – namely, *Australonycteris clarkae* from the early Eocene Tingamarra Local Fauna (Figure 13.4g; Hand *et al.* 1994); only a few European bats from the earliest Eocene may be slightly older (Tabuce *et al.* 2009). Indeed, bats appear to have achieved an essentially global distribution early in the Eocene (Smith *et al.* 2012).

Based on unpublished postcranial material, *A. clarkae* appears to have been fully volant (S. J. Hand, pers. comm.). Its phylogenetic relationships have yet to be fully assessed, but preliminary analyses indicate that it lies outside crown clade Chiroptera (S.J. Hand, pers comm.). If so, it represents an early dispersal to Australia unrelated to later dispersals by crown clade forms (Hand 2006). It is possible that *Australonycteris* followed the same dispersal route into Australia as did marsupialiforms and the condylarth *Tingamarra* (and possibly, in the opposite direction, the monotreme *Monotrematum*), namely from South America, via Antarctica. However, the oldest known bats from South America are younger (ca. 48–52 Ma) than *A. clarkae*, and do not appear to be closely related to it (Tejedor *et al.* 2005). Furthermore, given that bats are clearly capable of dispersing over large distances over water, there seems no reason why *Australonycteris* need necessarily have taken a terrestrial route to Australia. Further study of early bats from other landmasses, such as those from the early Eocene Vastan fauna of India (Smith *et al.* 2007), and their incorporation into broad-scale phylogenetic analyses, should help clarify their biogeographical relationships. Until then, the closest relative(s) and likely biogeographical origin of *Australonycteris* will remain obscure.

After *Australonycteris*, the fossil record of bats (and, indeed, other mammals) in Sahul is blank until the late Oligocene. However, representatives of at least eight crown clade families are known from Sahul: Mystacinidae, Hipposideridae, Megadermatidae, Molossidae, Emballonuridae, Vespertilionidae, Rhinolophidae and Pteropodidae (Hand 2006). Of these, most are known from Oligo-Miocene sites in Australia (Hand 2006). However, emballonurids are first known from the Rackham's Roost site at Riversleigh World Heritage Area in northwestern Queensland (Hand 2006), which was originally suggested to be Pliocene in age based on biocorrelation, but on the basis of radiometric dates now appears to be Early Pleistocene (Woodhead *et al.* 2014). Pteropodids also have no pre-Pleistocene Australian record (Hand 2006).

Based on their known fossil and modern distributions, it seems plausible that the presence of these families in Sahul is the result of dispersal from Sundaland, via Wallacea (Hand 2006), with the notable exception of Mystacinidae. However, uncertainty remains regarding the phylogeny and even the alpha taxonomy of many of these families (e.g. Almeida *et al.* 2014; Lavery *et al.* 2014; Reardon *et al.* 2014), limiting the biogeographical inferences that can be drawn at present. In addition, it is clear that the fossil record of bats in Sahul exhibits extreme sampling biases, with the vast majority of specimens known only from the Riversleigh World Heritage Area (Hand 2006). Resolving the biogeographical relationships of these families – for example, determining the extent to which Sahul itself has acted as a centre of diversification (e.g. for Hipposideridae; Hand and Kirsch 1998) – will require more comprehensive phylogenetic analyses that integrate molecular and fossil evidence.

Unlike the other bat families discussed previously, the presence of mystacinids in Australia is unlikely to be the result of dispersal from Sundaland (Hand 2006). Mystacinidae is a member of the superfamily Noctilionoidea (Meredith *et al.* 2011; Teeling *et al.* 2005), the other members of which are largely restricted to the Southern Hemisphere (Czaplewski and Morgan 2012; Gunnell *et al.* 2014; Simmons 2005). Noctilionoids are entirely unknown from Sundaland or Wallacea (Czaplewski and Morgan 2012; Gunnell *et al.* 2014; Simmons 2005). Today, Mystacinidae is represented by a single extant species, *Mystacina tuberculata*, which is endemic to New Zealand; a second New Zealand species, *M. robusta* appears to have gone extinct within the last 50 years (Simmons 2005). At least two mystacinid species were present in the early Miocene St. Bathans Fauna, Central Otago, on New Zealand's South Island, demonstrating that the group has been present in New Zealand for at least the last 16 Ma (Hand *et al.* 2013). However, at least four mystacinid species are known from late Oligocene to middle Miocene deposits in Australia (Hand *et al.* 2005). Recent molecular estimates suggest that Mystacinidae diverged from other noctilionoids 41.0–52.9 mya (Meredith *et al.* 2011; Miller-Butterworth *et al.* 2007); this largely postdates recent estimates for the separation of New Zealand from Antarctica (Ericson *et al.* 2014; see 'Geographical and Geological Context' in this chapter), suggesting that, regardless of the precise origin of mystacinids, their dispersal either into or out of New Zealand was probably over water. However, the fossil record and phylogeny of Mystacinidae are still too poorly known to determine whether the family originated in Australia and dispersed to New Zealand, vice versa, or whether it originated on another landmass, such as Antarctica (Hand *et al.* 1998, 2013; Kirsch *et al.* 1998; Teeling *et al.* 2003).

Murinae

Today, murine rodents comprise >160 species in Sahul, ~25% of the total native mammal fauna (Aplin 2006; Aplin and Ford 2014; Breed and Ford 2007; Flannery 1995b; Musser and Carleton 2005; Van Dyck and Strahan 2008). This number is likely to rise considerably as a result of ongoing fieldwork, taxonomic revisions and DNA sequencing of collected specimens, uncovering cryptic diversity. At least five species (*Mus musculus/domesticus* and at least four species of *Rattus*) present in Sahul are the result of human-mediated introductions (Aplin and Ford 2014: Table 10.3) and will not be discussed here. The remaining extant species are recognised as comprising two distinct groups: the Old Endemics, numbering at least 140 species in at least 35 genera within the tribe Hydromyini, and the New Endemics, comprising at least 20 species all within the genus *Rattus* (Figure 13.5; Aplin 2006; Aplin and Ford 2014; Breed and Ford 2007; Flannery 1995b; Musser and Carleton 2005; Van Dyck and Strahan 2008).

Based on current fossil and molecular evidence, Murinae probably originated in Southeast Asia during the Middle Miocene (Schenk *et al.* 2013; Jacobs and Flynn 2005; Fabre *et al.* 2013). Recent large-scale

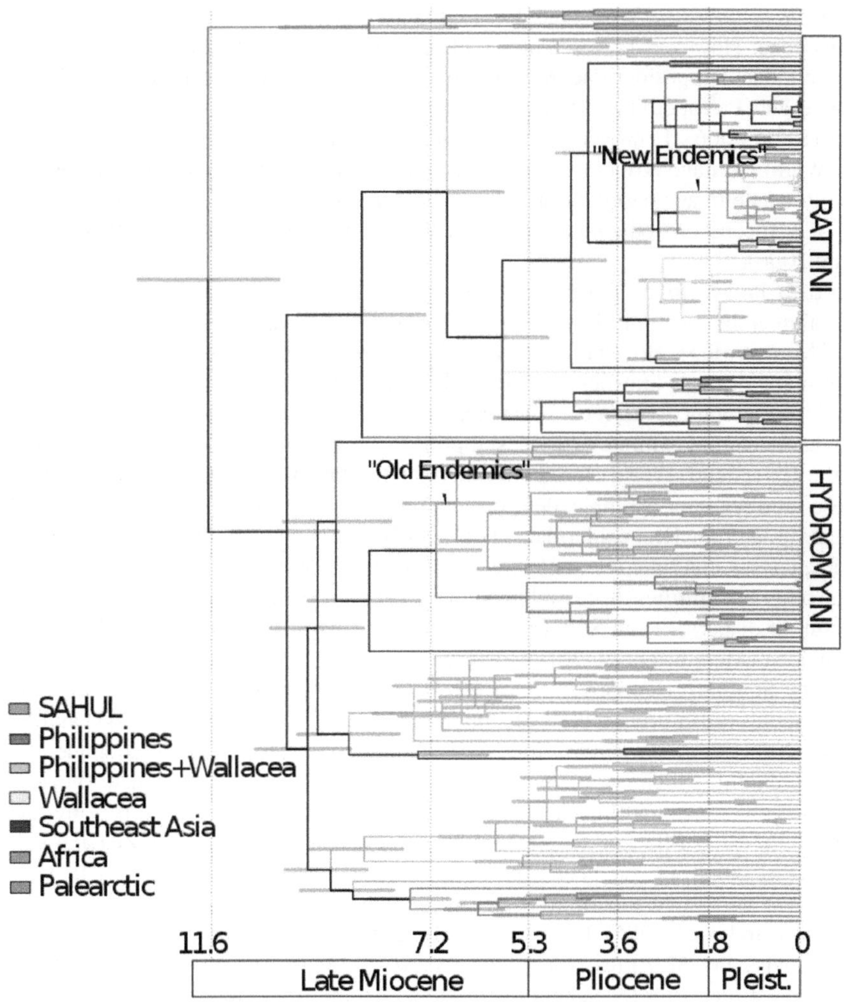

FIGURE 13.5 (See colour insert.) Phylogeny of murid rodents based on a Bayesian relaxed molecular clock analysis of one mitochondrial (cytochrome b) and two nuclear (growth hormone receptor exon 10; interphotoreceptor retinoid binding protein exon 1) protein-coding genes using *BEAST* (modified from Fabre, P. H., *et al.*, *Zoological Journal of the Linnean Society*, 169[2], 408–47, 2013: Figure 13.3). Branches are colour-coded according to their known biogeographic distributions, and the predominantly Sahulian Old Endemics and New Endemics are indicated (see Fabre, P. H., *et al.*, *Zoological Journal of the Linnean Society*, 169[2], 408–47, 2013 for full details).

molecular phylogenies indicate that the Old Endemics and the New Endemics are each nested within otherwise predominantly Southeast Asian clades (Figure 13.5; Rowe *et al.* 2008; Schenk *et al.* 2013). This topology implies one dispersal by the ancestor of the Old Endemics and another, later dispersal by the ancestor of the New Endemics into Sahul, most likely from Sundaland via Wallacea (Aplin and Ford 2014; Fabre *et al.* 2013; Rowe *et al.* 2008; Schenk *et al.* 2013). Both dispersal events must have involved the crossing of marine barriers (Figure 13.1d; see 'Geographical and Geological Context' in this chapter). The presence of both Old Endemics (e.g. species of *Melomys*; Flannery 1995a) and New Endemics (e.g. *Rattus morotaiensis* from Halmahera in the north Moluccas; Fabre *et al.* 2013) in Wallacea suggests back-dispersals from Sahul by both groups. It seems likely that further collection and analysis of modern, subfossil and fossil murines from the region will complicate this story still further (Aplin and Ford 2014; Aplin and Helgen 2010). Godthelp's (2001) report of a '*Potwarmus*-like "dendromurine"' and a second taxon that 'seems to have affinities with the widespread, primitive *Chiropodomys* group' from Southeast Asia, from the Rackham's Roost deposit (probably Early Pleistocene in age; Woodhead *et al.*

2014) at the Riversleigh World Heritage Area, is particularly intriguing. Today, dendromurines occur only in Africa, whereas *Chiropodomys* species are found in mainland Asia and Sundaland (Musser and Carleton 2005), and there is otherwise no recent or fossil record of either group from Sahul. However, these potentially highly significant taxa have yet to be formally described.

The likely times of arrival of the Old Endemics and the New Endemics in Sahul can be broadly constrained based on the known fossil record and molecular estimates of divergence times. As noted by Aplin and Ford (2014), the apparent absence of murines in the diverse Hamilton Fauna (currently estimated as 4.46 Ma old; Turnbull *et al.* 2003) in western Victoria suggests that they had probably failed to reach Australia by this time. The oldest well-dated evidence of murines in Sahul is indeterminate material from the Bluff Downs Local Fauna in northeastern Queensland (Aplin and Ford 2014: Table 10.2), which has a minimum radiometric date of 3.6 Ma (Mackness *et al.* 2000), with material from a slightly younger (3.6–2.8 Ma) deposit on Barrow Island including representatives of modern Australian Old Endemic genera (*Pseudomys* and *Zyzomys*; Aplin and Ford 2014: Table 10.2). Thus, the likely time of the first entry of murines into Australia can be constrained to between 4.46 and 3.6 mya. Fabre *et al.*'s (2013; Figure 13.5) and Schenk *et al.*'s (2013) pooled molecular estimates suggest entry into Sahul by the ancestor of the Old Endemics after 9.6 mya (the maximum age of divergence from their nearest non-Sahulian sister taxon) but before 5.5 mya (the minimum age of divergence between the Sahulian lineages). The fossil record and molecular dating are both subject to potentially misleading biases; however, assuming that both sources of evidence are accurate in this case, this implies that the Old Endemics were present in New Guinea for at least 1 Ma before reaching Australia (Aplin and Ford 2014; Rowe *et al.* 2008: 97).

Schenk *et al.* (2013: 852) found some evidence for an increase in diversification in a subclade comprising the predominantly Australian *Pseudomys*, *Mesembriomys* and *Uromys* groups (see Aplin and Ford 2014: Table 10.1), and hypothesised that this may reflect the first colonisation of Australia by the Old Endemics. If so, Schenk *et al.*'s (2013) estimated divergence dates suggest that this occurred 4.8–4.2 mya, which is nicely congruent with the fossil evidence discussed previously. Given that murines must have crossed water barriers to enter New Guinea, a dry-land connector may have been unnecessary for at least some lineages to reach Australia. However, the dispersal of others (e.g. the arboreal rainforest specialist *Pogonomys*) from New Guinea to Australia may have been facilitated by the repeated development of land connections from the latest Miocene onwards (Figure 13.1d; see 'Geographical and Geological Context' in this chapter).

The arrival of the New Endemics in Sahul was clearly a more recent event (Figure 13.5). The oldest securely dated and identified *Rattus* fossils from Australia are from Speaking Tube Cave at Mt. Etna in eastern-central Queensland (Cramb 2012; Cramb and Hocknull 2010), which is estimated to be ca. 280 ka old (Hocknull *et al.* 2007); *Rattus* is notable by its absence from older Mt. Etna sites, despite their rich mammal faunas that include multiple Old Endemic murine species (Cramb 2012). However, molecular estimates for the first arrival of *Rattus* in Sahul are considerably older than this (Fabre *et al.* 2013; Rowe *et al.* 2011a; Schenk *et al.* 2013): after 3 mya (the maximum age of divergence from their nearest non-Sahulian sister taxon) but before 0.85 mya (the minimum age of divergence between the Sahulian lineages). Again, making the questionable assumption that both the fossil record and the molecular divergence dates are accurate, this implies that the New Endemics also experienced a prolonged period (>500 ka) in New Guinea before dispersing to Australia.

The biogeographical relationships between Australian and New Guinean murine species appear highly complex. Among the Old Endemics, the molecular phylogeny of Rowe *et al.* (2008) implies at least nine dispersals between Australia and New Guinea (five from New Guinea to Australia, two from Australia to New Guinea and two equivocal), while Aplin and Ford (2014: Table 10.3) identified at least 14 (11 from New Guinea to Australia, and three from Australia to New Guinea). Among the New Endemics, Aplin and Ford (2014: Table 10.3) identified three dispersals, two from New Guinea to Australia and one in the reverse direction. This pattern is perhaps unsurprising given the repeated formation and severing of dry-land connections between Australia and New Guinea over the last 5 Ma. However, as for marsupials (see 'Marsupialiformes' in this chapter), it may be that northern Australia and southern New Guinea (corresponding to the Austral–East Torresian zoogeographic province recognised by Lavery *et al.* 2013) had an at least partially shared murine fauna for much of the Plio-Pleistocene, intermittently divided during interglacial periods, and that faunal differences developed as a result of differential extinction with the onset of more xeric conditions in northern Australia ca. 280 ka ago. The reported presence in Middle Pleistocene (>280 ka) deposits at Mt. Etna of *Abeomelomys* (currently a New Guinean endemic; Cramb 2012) and species of *Pogonomys* and *Uromys* that appear more closely related to New Guinean than Australian species among living taxa (Hocknull 2005, 2009; Hocknull *et al.* 2007) is congruent with this interpretation. However, adequate testing of this hypothesis will require considerable improvements in the Plio-Pleistocene fossil records of northern Australia and New Guinea.

Assuming a prolonged period in New Guinea before dispersing to Australia, both the Old Endemics and New Endemics presumably radiated initially in predominantly mesic environments. However, in Australia, multiple Old Endemic lineages have evolved to occupy the xeric habitats, most within the *Pseudomys* group (including the species-rich *Pseudomys* and *Notomys*), but also a member of the *Mesembriomys* group, *Leporillus*. By contrast, most New Endemic species are restricted to mesic, coastal environments mainly in eastern Australia; however, *Rattus villosissimus* and (at least until recently) *R. tunneyi* are both known from wetter areas of central Australia.

To date, only Australian murines have been subjected to phylogeographic studies. Bryant and Fuller's (2014) DNA sequence and microsatellite analysis of the Old Endemic *Melomys cervinipes*, which occurs along the east coast of Australia, found evidence that a number of recognised potential biogeographical barriers (the Brisbane Valley Barrier, the St. Lawrence Gap and the Burdekin Gap) have influenced the phylogeographic structure of the species, as have habitat fragmentation and contraction, local extinctions and later re-expansions. Rowe *et al.*'s (2011b) study of another Old Endemic, the Hastings River mouse (*Pseudomys oralis*), found evidence of two major mitochondrial lineages that appear to have diverged 300–900 ka ago, with the area of lineage overlap corresponding to the northern limit of the Macleay–McPherson Overlap (Burbidge 1960; Ebach *et al.* 2013). Strikingly, however, there is no indication from Rowe *et al.*'s (2011b) results that the Brisbane Valley Barrier farther north has posed a barrier to gene flow in *P. oralis*, *contra* Bryant and Fuller's (2014) findings regarding *M. cervinipes*. Although many more phylogeographic studies are required, this hints at the likelihood that biogeographical patterns within Australian (and probably also New Guinean) murines are likely to be highly species specific, as they appear to be in marsupials.

Like small- and medium-sized marsupials, the native murines of Australia have suffered severely in post-European colonisation, with the introduction of invasive placental species (particularly cats and foxes) and widespread habitat modification (Burbidge *et al.* 2008; Johnson 2006; McKenzie *et al.* 2007). As a result, several species have gone extinct and many others have seen enormous range reductions. New Guinean species also appear to have experienced extinctions and range reductions, particularly on islands off the New Guinean mainland (Aplin *et al.* 1999a). As already discussed, the loss of entire populations (at least some of which likely represent evolutionarily distinct lineages) is likely to mislead biogeographical studies that are restricted to extant representatives of a particular species. Again, the inclusion of distributional data (as in Burbidge *et al.* 2008 and Lavery *et al.* 2013) and (if possible) ancient DNA from subfossil and fossil specimens will be critical for an accurate understanding of the biogeographical history of Sahulian murines.

Questionable Records

Clemens *et al.* (2003) described a fossilised vertebrate tooth (AMF 118621) from the Albian Griman Creek Formation at Lightning Ridge in New South Wales as a probable upper molar of a 'dryolestoid' mammal. The oldest 'dryolestoid' from South America is *Cronopio dentiacutus* from the Cenomanian La Buitera locality in southern Argentina (Rougier *et al.* 2011a), and 'dryolestoids' were highly diverse in the Late Cretaceous Allenian faunas of southern South America (Bonaparte 1990; Rougier *et al.* 2009a,b, 2011b). Thus, the apparent presence of at least one 'dryolestoid' in Australia during the middle Cretaceous would not be surprising, and would provide the closest link yet between the Mesozoic mammaliaform faunas of Australia and South America, which otherwise appear to show high-level endemism (Rich 2008; Rich *et al.* 2009a). However, it is unclear whether AMF 118621 is indeed a 'dryolestoid', or even a synapsid. Perhaps most concerning is its very large size: at 12.2 mm long by 14.0 mm wide, it is far bigger than the upper molars of any definitive Mesozoic 'dryolestoid' (Clemens *et al.* 2003).

Furthermore, as noted by Clemens *et al.* (2003), some crocodyliforms are known to have evolved complex, superficially mammal-like cheek teeth (including Cretaceous forms from Gondwana; O'Connor *et al.* 2010). It is entirely possible that AMF 118621 does not represent a mammaliaform.

Mastodon australis and *Notelephas australis* were described and identified as proboscideans by Owen (1845, 1882) based on specimens apparently collected from Australia: *M. australis* from 'further in the interior than [the caves] of Wellington Valley' (presumably inland New South Wales) and *N. australis* from the Darling Downs in southeastern Queensland. Among mammals, proboscideans have excellent over-water dispersal capabilities (Johnson 1980; van der Geer *et al.* 2010): in 1856, an elephant was reported to have swum 48 km to land after falling overboard in the Atlantic (Johnson 1980), and remains of the fossil proboscidean *Stegodon* are known from multiple islands in Wallacea (Dennell *et al.* 2014; van der Geer *et al.* 2010), undoubtedly the result of over-water dispersal(s). However, *Stegodon* fossils have not been found east of Lydekker's Line, and their apparent absence from New Guinea is perhaps the biggest argument against their reaching Australia, at least as live individuals. The decaying, gas-filled carcass of a large-bodied, thick-skinned mammal such as a proboscidean might be capable of being transported prolonged distances over water (Archer 1999), but this would not explain why *M. australis* at least was reportedly found far inland. Owen's (1845, 1882) identification of these specimens as proboscideans has not been questioned (Archer *et al.* 1999b), but doubts have been raised as to whether they were genuinely collected in Australia (Longman 1916). Pending careful reassessment of the known material and reported provenance of *M. australis* (known from a single molar, apparently now lost) and *N. australis* (known from a fragmentary tusk held at the Natural History Museum, London), or the discovery of additional specimens, they will remain biogeographical enigmas.

Some Concluding Thoughts: Problems and Prospects

As this chapter should make abundantly clear, there are numerous areas of uncertainty in our understanding of the biogeography of Mammaliformes in Sahul, most of which are due to a sheer lack of evidence. Inevitably, this leads to the all-too-familiar cry of 'more data are required'. But what precisely is needed, and how likely is this to be obtained? Our understanding of the complex geological history of Sahul, as well as related aspects such as sea level and palaeoclimate, will undoubtedly increase with further study. Ever-improving sequencing technology will allow researchers to obtain sequence data from multiple individuals of extant taxa increasingly quickly and cheaply. Remarkable advances in methods for extracting and sequencing ancient DNA allow sequence data to be acquired increasingly effectively from subfossil and fossil remains, with the latest methods successfully obtaining genuine sequences from specimens >400 ka old (Dabney *et al.* 2013; Meyer *et al.* 2014; Orlando *et al.* 2013). If such methods can be successfully applied to subfossil and fossil remains from Sahul, they will open up entire new vistas of research, enabling extinct species and populations to be incorporated directly into phylogenetic and phylogeographic analyses based on molecular data (e.g. Austin *et al.* 2013; Sheng *et al.* 2014; Stiller *et al.* 2014). If ancient DNA proves unobtainable from particular specimens, methods that obtain amino acid sequences from proteins such as collagen may still allow phylogenetically useful molecular data to be acquired (Buckley 2013). Ongoing collecting is likely to result in the discovery of new mammalian species and also additional populations of currently known species, particularly in New Guinea and the adjacent islands (Helgen 2007), which is likely to lead to the revision and refinement of current biogeographical hypotheses.

Improvements in the Sahulian fossil record, however, may prove more elusive, particularly for the Mesozoic and early Palaeogene. Despite continuing exploration, the entire published Mesozoic record of Mammaliformes in Australia is from four middle Cretaceous localities (Kielan-Jaworowska *et al.* 2004; Long *et al.* 2002; Rich and Vickers-Rich 2004; Rich *et al.* 2009b): the Aptian Dinosaur Cove, Flat Rocks and Eric the Red West sites in southern Victoria, and the Albian Griman Creek Formation at Lightning Ridge in northern New South Wales (Figure 13.3a). Similarly, there is only a single early Palaeogene mammal-bearing fossil site known in Australia, the early Eocene Tingamarra Local Fauna (Figure 13.4a; Godthelp *et al.* 1992; Long *et al.* 2002), to fill the 85 Ma old gap between the four Aptian–Albian sites and the richer fossil record of mammals known from the late Oligocene onwards. Given that these pre-Oligocene sites are so few in number and that only a handful of mammaliaform taxa have been described from them to date, it is unclear whether the apparent absence of particular groups from these sites (e.g. the apparent lack of monotremes at Tingamarra) is an accurate reflection of the mammaliaform fauna of the time or simply an artefact of the very limited sampling to date. Thus, biogeographical inferences drawn from the pre-Oligocene mammaliaform record of Australia should be treated with particular caution.

Progress is also likely to come from the development and use of new, quantitative methods of phylogenetic and biogeographic analysis. Particularly promising are methods that allow the phylogenetic relationships and divergence times of fossil and extant taxa to be calculated simultaneously (Pyron 2011; Ronquist *et al.* 2012), although the accuracy of these methods remains to be fully established (Beck and Lee 2014). Also of interest are new parametric methods for analysing biogeography within an explicitly statistical, model-based framework (Matzke 2013, 2014; Ree and Sanmartin 2009; Ree and Smith 2008; Ree *et al.* 2005). Such methods (1) permit both vicariance and dispersal, (2) take into account divergence times/temporal branch lengths and (if specified) changes in connectivity between areas through time, and (3) allow the best-fitting model(s) to be identified using explicit model selection criteria. These methods are starting to be applied to Sahulian mammaliaform clades (Fabre *et al.* 2013; Mitchell *et al.* 2014; Westerman *et al.* 2012).

As noted repeatedly throughout this chapter, the incorporation of evidence from the fossil, subfossil and historical record – whether simply as distributional data or via the inclusion of extinct taxa in phylogenetic and phylogeographic analyses, ideally in the form of ancient DNA sequences – is likely to be critical for the accurate reconstruction of biogeographical histories (Buerki *et al.* 2013; Lieberman 2002). Specifically, the complete extinction of species or populations may account for biogeographical patterns that are otherwise hard to explain, such as the 'old' New Guinean marsupial clades found by Mitchell *et al.* (2014) and the Tasmanian platypus clade found by Gongora *et al.* (2012). Thus, active collaboration between palaeontologists and researchers working on living taxa is likely to prove at least as important as simple improvements in data and methods.

The prospects are therefore good for major improvements in our understanding of at least some areas of Sahulian mammaliaform biogeography. At the same time, however, the modern mammal fauna of Sahul is under threat, with numerous species experiencing severe and ongoing declines (George 1979; Woinarski *et al.* 2010, 2011). Let us hope that current and future biogeographical studies do not become monuments to yet more vanished populations and species.

Acknowledgements

My thanks to the editor, Malte Ebach, for inviting me to contribute this chapter. I thank Guillermo Rougier, Erich Fitzgerald, Tom Rich, Kris Helgen and John Schenk for their discussion and for supplying me with references. Pierre-Henri Fabre, Mike Archer, Peter Trusler, Ken Aplin and Fred Ford generously provided me with images that have been used in some of the figures. I am particularly grateful to Scott Hocknull for his thorough and constructive review. My thanks also to Sue Hand, Rebecca Pian, Julien Louys, Karen Black and Mike Woodburne, all of whom read earlier drafts of this chapter and gave extremely helpful comments. Financial support for my research on Australian mammaliaform biogeography has been provided by the Leverhulme Trust (via Study Abroad Studentship SAS/30110), Phil Creaser and the CREATE fund at the University of New South Wales (via a CREATE scholarship), the National Science Foundation (via grant DEB-0743039, in collaboration with Rob Voss at the American Museum of Natural History) and the Australian Research Council (via Discovery Early Career Researcher Award DE120100957).

References

Achmadi, A. S., J. A. Esselstyn, K. C. Rowe, I. Maryanto and M. T. Abdullah. 2013. Phylogeny, diversity, and biogeography of Southeast Asian spiny rats (*Maxomys*). *Journal of Mammalogy* 94 (6):1412–23.

Agnolin, F. L., M. D. Ezcurra, D. F. Pais and S. W. Salisbury. 2010. A reappraisal of the Cretaceous non-avian dinosaur faunas from Australia and New Zealand: Evidence for their Gondwanan affinities. *Journal of Systematic Palaeontology* 8:257–300.

Almeida, F. C., N. P. Giannini, N. B. Simmons and K. M. Helgen. 2014. Each flying fox on its own branch: A phylogenetic tree for *Pteropus* and related genera (Chiroptera: Pteropodidae). *Molecular Phylogenetics and Evolution* 77:83–95.

Aplin, K., and F. Ford. 2014. Murine rodents: Late but highly successful invaders. In *Invasion Biology and Ecological Theory: Insights from a Continent in Transformation*, edited by Herbert H. T. Prins and Iain J. Gordon, 196–240. Cambridge, UK: Cambridge University Press.

Aplin, K. P. 2006. Ten million years of rodent evolution in Australasia: Phylogenetic evidence and a speculative historical biogeography. In *Evolution and biogeography of Australasian vertebrates,* edited by J. R. Merrick, M. Archer, G. M. Hickey and M. S. Y. Lee, 707–44. Sydney, Australia: Auscipub.

Aplin, K., and J. Pasveer. 2006. Mammals and other vertebrates from Late Quaternary archaeological sites on Pulau Kobroor, Aru Islands, Eastern Indonesia. In *The Archaeology of the Aru Islands, Eastern Indonesia*, edited by S. O'Connor, M. Spriggs and P. Veth, 41–62. Canberra, Australia: ANU E Press.

Aplin, K. P., P. R. Baverstock and S. C. Donnellan. 1993. Albumin immunological evidence for the time and mode of origin of the New Guinean terrestrial mammal fauna. *Science in New Guinea* 19 (3):131–45.

Aplin, K. P., and K. M. Helgen. 2010. Quaternary murid rodents of Timor, part I: New material of *Coryphomys buehleri* Schaub, 1937, and description of a second species of the genus. *Bulletin of the American Museum of Natural History* 341:1–80.

Aplin, K. P., J. M. Pasveer and W. E. Boles. 1999a. Quaternary vertebrates from the Bird's Head Peninsula, Irian Jaya, Indonesia, including descriptions of two previously unknown marsupial species. *Records of the Western Australian Museum*, Supplement 57:351–87.

Aplin, K. P., J. M. Pasveer and W. E. Boles. 1999b. Late Quaternary deposits from the Bird's Head Peninsula, Irian Jaya, Indonesia, including descriptions of two previously unknown marsupial species. *Records of the Western Australian Museum*, Supplement 57:351–87.

Aragón, E., F. J. Goin, Y. E. Aguilera, M. O. Woodburne, A. A. Carlini and M. F. Roggiero. 2011. Palaeogeography and palaeoenvironments of northern Patagonia from the Late Cretaceous to the Miocene: The Palaeogene Andean gap and the rise of the North Patagonian High Plateau. *Biological Journal of the Linnean Society* 103:305–15.

Archer, M. 1984. The Australian marsupial radiation. In *Vertebrate Zoogeography and Evolution in Australasia*, edited by Michael Archer and Georgina Clayton, 633–808. Perth, Australia: Hesperian Press.

Archer, M. 1999. A legacy of boats, bloats and floaters. *Nature Australia* 26:70–1.

Archer, M., R. Arena, M. Bassarova, K. Black, J. Brammall, B. Cooke, P. Creaser *et al.* 1999a. The evolutionary history and diversity of Australian mammals. *Australian Mammalogy* 21:1–45.

Archer, M., T. F. Flannery, A. Ritchie and R. E. Molnar. 1985. First Mesozoic mammal from Australia: An Early Cretaceous monotreme. *Nature* 318:363–6.

Archer, M., H. Godthelp, M. Gott, Y. Wang and A. Musser. 1999b. The evolutionary history of notoryctids, yingabalanarids, yalkaparidontids and other enigmatic groups of Australian mammals. *Australian Mammalogy* 21:13–15.

Archibald, J. D. 1998. Archaic ungulates ('Condylarthra'). In *Evolution of Tertiary Mammals of North America*, Vol. 1: *Terrestrial Carnivores, Ungulates, and Ungulate-Like Mammals,* edited by Christine M. Janis, Kathleen M. Scott and Louis L. Jacobs, 292–311. Cambridge, UK: Cambridge University Press.

Ashwell, K. W. S. 2013. Reflections: Monotreme neurobiology in context. In *Neurobiology of Monotremes: Brain Evolution in Our Distant Mammalian Cousins*, edited by Ken Ashwell, 285–98. Collingwood, Australia: CSIRO Publishing.

Augee, M. L., B. Gooden and A. Musser. 2006. *Echidna: Extraordinary Egg-Laying Mammal*. Collingwood, Australia: CSIRO Publishing.

Austin, J. J., J. Soubrier, F. J. Prevosti, L. Prates, V. Trejo, F. Mena and A. Cooper. 2013. The origins of the enigmatic Falkland Islands wolf. *Nat Commun* 4:1552.

Averianov, A. O., J. D. Archibald and T. Martin. 2003. Placental nature of the alleged marsupial from the Cretaceous of Madagascar. *Acta Palaeontologica Polonica* 48 (1):149–51.

Baillie, J. E. M., S. T. Turvey and C. Waterman. 2009. Survival of Attenborough's long-beaked echidna *Zaglossus attenboroughi* in New Guinea. *Oryx* 43 (1):146–8.

Baldwin, S. L., P. G. Fitzgerald and L. E. Webb. 2012. Tectonics of the New Guinea region. *Annual Review of Earth and Planetary Sciences* 40:495–520.

Beck, R. M. D. 2008a. A dated phylogeny of marsupials using a molecular supermatrix and multiple fossil constraints. *Journal of Mammalogy* 89 (1):175–89.

Beck, R. M. D. 2008b. Form, function, phylogeny and biogeography of enigmatic Australian metatherians. Unpublished PhD thesis, School of Biological, Earth and Environmental Sciences, University of New South Wales, Sydney, Australia.

Beck, R. M. D. 2012. An 'ameridelphian' marsupial from the Early Eocene of Australia supports a complex model of Southern Hemisphere marsupial biogeography. *Naturwissenschaften* 99 (9):715–29.

Beck, R. M. D. 2015. A peculiar faunivorous metatherian from the Early Eocene of Australia. *Acta Palaeontologica Polonica*. 60:1237–129.

Beck, R. M. D., H. Godthelp, V. Weisbecker, M. Archer and S. J. Hand. 2008. Australia's oldest marsupial fossils and their biogeographical implications. *PLoS ONE* 3 (3):e1858.

Beck, R. M. D., and M. S. Y. Lee. 2014. Ancient dates or accelerated rates? Morphological clocks and the antiquity of placental mammals. *Proceedings of the Royal Society B: Biological Sciences* 281:20141278.

Beck, R. M. D., K. J. Travouillon, K. P. Aplin, H. Godthelp and M. Archer. 2014. The osteology and systematics of the enigmatic Australian Oligo-Miocene metatherian *Yalkaparidon* (Yalkaparidontidae; Yalkaparidontia; ?Australidelphia; Marsupialia). *Journal of Mammalian Evolution* 21 (2):127–72.

Benson, R. B. J., T. H. Rich, P. Vickers-Rich and M. Hall. 2012. Theropod fauna from southern Australia indicates high polar diversity and climate-driven dinosaur provinciality. *PLoS ONE* 7 (5):e37122.

Bi, S., Y. Wang, J. Guan, X. Sheng and J. Meng. 2014. Three new Jurassic euharamiyidan species reinforce early divergence of mammals. *Nature* 514 (7524):579–584.

Bijl, P. K., J. A. Bendle, S. M. Bohaty, J. Pross, S. Schouten, L. Tauxe, C. E. Stickley *et al.* 2013. Eocene cooling linked to early flow across the Tasmanian Gateway. *Proceedings of the National Academy of Sciences of the United States of America* 110 (24):9645–50.

Bintanja, R., R. S. van de Wal and J. Oerlemans. 2005. Modelled atmospheric temperatures and global sea levels over the past million years. *Nature* 437 (7055):125–8.

Black, K. H., M. Archer, S. J. Hand and H. Godthelp. 2012. The rise of Australian marsupials: A synopsis of biostratigraphic, phylogenetic, palaeoecologic and palaeobiogeographic understanding. In *Earth and Life: Global Biodiversity, Extinction Intervals and Biogeographic Perturbations through Time*, edited by J. A. Talent, 983–1078. Dordrecht, the Netherlands: Springer.

Bonaparte, J. F. 1990. New Late Cretaceous mammals from the Los Alamitos Formation, northern Patagonia. *National Geographic Research* 6 (1):63–93.

Bonaparte, J. F., L. Van Valen and A. Kramarz. 1993. La fauna local de Punta Peligro: Paleoceno inferior de la Provincia de Chubut, Patagonia, Argentina. *Evolutionary Monographs* 14:1–61.

Breed, B., and F. Ford. 2007. *Native Mice and Rats*. Collingwood, Australia: CSIRO Publishing.

Brown, M., H. Cooksley, S. M. Carthew and S. J. B. Cooper. 2006. Conservation units and phylogeographic structure of an arboreal marsupial, the yellow-bellied glider (*Petaurus australis*). *Australian Journal of Zoology* 54 (5):305–17.

Buckley, M. 2013. A molecular phylogeny of *Plesiorycteropus* reassigns the extinct mammalian order 'Bibymalagasia'. *PLoS ONE* 8 (3):e59614.

Buerki, S., D. S. Devey, M. W. Callmander, P. B. Phillipson and F. Forest. 2013. Spatio-temporal history of the endemic genera of Madagascar. *Botanical Journal of the Linnean Society* 171 (2):304–29.

Burbidge, A. A., N. L. McKenzie, K. E. C. Brennan, J. C. Z. Woinarski, C. R. Dickman, A. Baynes, G. Gordon, P. W. Menkhorst and A. C. Robinson. 2008. Conservation status and biogeography of Australia's terrestrial mammals. *Australian Journal of Zoology* 56:411–22.

Burbidge, N. T. 1960. Phytogeography of the Australian region. *Australian Journal of Botany* 8:75–211.

Butler, P. M., and J. J. Hooker. 2005. New teeth of allotherian mammals from the English Bathonian, including the earliest multituberculates. *Acta Palaeontologica Polonica* 50 (2):185–207.

Bryant, L. M., and Fuller, S. J. 2014. Pleistocene climate fluctuations influence phylogeographical patterns in *Melomys cervinipes* across the mesic forests of eastern Australia. *Journal of Biogeography* 41:1923–1935.

Camens, A. B. 2010. Were early Tertiary monotremes really all aquatic? Inferring paleobiology and phylogeny from a depauperate fossil record. *Proceedings of the National Academy of Sciences of the United States of America* 107 (4):E12.

Case, J. A., F. J. Goin and M. O. Woodburne. 2005. 'South American' marsupials from the Late Cretaceous of North America and the origin of marsupial cohorts. *Journal of Mammalian Evolution* 12 (3–4):461–94.

Case, J. A., M. O. Woodburne and D. S. Chaney. 1988. A new genus and species of polydolopid marsupial from the La Meseta Formation, Late Eocene, Seymour Island, Antarctic Peninsula. In *Geology and Paleontology of Seymour Island*, edited by R. M. Feldmann and Michael O. Woodburne, 505–21. Boulder, CO: Geological Society of America.

Chatterjee, S., A. Goswami and C. R. Scotese. 2013. The longest voyage: Tectonic, magmatic, and paleoclimatic evolution of the Indian plate during its northward flight from Gondwana to Asia. *Gondwana Research* 23 (1):238–67.

Chornogubsky, L., F. J. Goin and M. Reguero. 2009. A reassessment of Antarctic polydolopid marsupials (Middle Eocene, La Meseta Formation). *Antarctic Science* 21 (03):285.

Clemens, W. A., G. P. Wilson and R. E. Molnar. 2003. An enigmatic (synapsid?) tooth from the Early Cretaceous of New South Wales, Australia. *Journal of Vertebrate Paleontology* 23 (1):232–7.

Cloos, M., B. Sapiie, A. Q. van Ufford, R. J. Weiland, P. Q. Warren and T. P. McMahon. 2005. Collisional delamination in New Guinea: The geotectonics of subducting slab breakoff. *GSA Special Papers* 400:1–51.

Clyde, W. C., P. Wilf, A. Iglesias, R. L. Slingerland, T. Barnum, P. K. Bijl, T. J. Bralower *et al.* 2014. New age constraints for the Salamanca Formation and lower Río Chico Group in the western San Jorge Basin, Patagonia, Argentina: Implications for Cretaceous–Paleogene extinction recovery and land mammal age correlations. *Geological Society of America Bulletin* 126 (3–4):289–306.

Coller, M. 2009. SahulTime: Rethinking archaeological representation in the digital age. *Archaeologies: Journal of the World Archaeological Congress* 5 (1):110–23.

Cooper, S. J. B., M. Adams and A. Labrinidis. 2000. Phylogeography of the Australian dunnart Sminthopsis crassicaudata (Marsupialia: Dasyuridae). *Australian Journal of Zoology* 48 (5):461–73.

Cramb, J. 2012. Taxonomy and palaeoecology of Quaternary faunas from caves in eastern tropical Queensland: A record of broad-scale environmental change. Unpublished PhD thesis, Queensland University of Technology, Brisbane, Australia.

Cramb, J., and S. Hocknull. 2010. Fossil *Rattus*: The arrival and diversification of new endemic rodents. In *Australian Mammal Society 56th Meeting Conference*. Australian Academy of Science, Canberra, Australia.

Cramb, J., S. Hocknull and G. E. Webb. 2009. High diversity Pleistocene rainforest dasyurid assemblages with implications for the radiation of the Dasyuridae. *Austral Ecology* 34:663–9.

Cúneo, R., J. Ramezani, R. Scasso, D. Pol, I. Escapa, A. M. Zavattieri and S. A. Bowring. 2013. High-precision U–Pb geochronology and a new chronostratigraphy for the Cañadón Asfalto Basin, Chubut, central Patagonia: Implications for terrestrial faunal and floral evolution in Jurassic. *Gondwana Research* 24 (3–4):1267–75.

Czaplewski, N. J., and G. S. Morgan. 2012. New basal noctilionoid bats (Mammalia: Chiroptera) from the Oligocene of subtropical North America. In *Evolutionary History of Bats: Fossils, Molecules and Morphology*, edited by G. F. Gunnell and N. B. Simmons, 162–209. Cambridge, UK: Cambridge University Press.

Dabney, J., M. Knapp, I. Glocke, M. T. Gansauge, A. Weihmann, B. Nickel, C. Valdiosera *et al.* 2013. Complete mitochondrial genome sequence of a Middle Pleistocene cave bear reconstructed from ultrashort DNA fragments. *Proceedings of the National Academy of Sciences of the United States of America* 110 (39):15758–63.

Davidson, I. 2014. Peopling the last new worlds: The first colonisation of Sahul and the Americas. *Quaternary International* 285:1–29.

Davis, B. M. 2011. Evolution of the tribosphenic molar pattern in early mammals, with comments on the 'dual-origin' hypothesis. *Journal of Mammalian Evolution* 18:227–44.

de Iongh, H. H., and D. Domning. 2014. The biological invasion of Sirenia into Australasia. In *Invasion Biology and Ecological Theory: Insights from a Continent in Transformation*, edited by H. H. T. Prins and I. J. Gordon, 118–37. Cambridge, UK: Cambridge University Press.

Deméré, T. A., A. Berta and P. J. Adam. 2003. Pinnipedimorph evolutionary biogeography. *Bulletin of the American Museum of Natural History* 279:32–76.

Dennell, R., and M. Porr. 2014. *Southern Asia, Australia and the Search for Human Origins*. New York: Cambridge University Press.

Dennell, R. W., J. Louys, H. J. O'Regan and D. M. Wilkinson. 2014. The origins and persistence of *Homo floresiensis* on Flores: Biogeographical and ecological perspectives. *Quaternary Science Reviews* 96:98–107.

Dornburg, A., M. C. Brandley, M. R. McGowen and T. J. Near. 2012. Relaxed clocks and inferences of heterogeneous patterns of nucleotide substitution and divergence time estimates across whales and dolphins (Mammalia: Cetacea). *Molecular Biology and Evolution* 29 (2):721–36.

Dornburg, A., J. P. Townsend, M. Friedman and T. J. Near. 2014. Phylogenetic informativeness reconciles ray-finned fish molecular divergence times. *BMC Evolutionary Biology* 14 (1):169.

dos Reis, M., P. C. J. Donoghue and Z. Yang. 2014. Neither phylogenomic nor palaeontological data support a Palaeogene origin of placental mammals. *Biology Letters* 10 (1):20131003.

dos Reis, M., J. Inoue, M. Hasegawa, R. J. Asher, P. C. Donoghue and Z. Yang. 2012. Phylogenomic datasets provide both precision and accuracy in estimating the timescale of placental mammal phylogeny. *Proceedings Biological sciences* 279 (1742):3491–500.

Dun, W. S. 1895. Notes on the occurrence of monotreme remains in the Pliocene of New South Wales. *Records of the Geological Survey of New South Wales* 4:118–26.

Eagles, G., and W. Jokat. 2014. Tectonic reconstructions for paleobathymetry in Drake Passage. *Tectonophysics* 611:28–50.

Ebach, M. C., A. C. Gill, A. Kwan, S. T. Ahyong, D. J. Murphy and G. Cassis. 2013. Towards an Australian Bioregionalisation Atlas: A provisional area taxonomy of Australia's biogeographical regions. *Zootaxa* 3619 (3):315–42.

Eldridge, M. D. B., K. Heckenberg, L. E. Neaves, C. J. Metcalfe, S. Hamilton, P. M. Johnson and R. L. Close. 2011. Genetic differentiation and introgression amongst *Thylogale* (pademelons) taxa in eastern Australia. *Australian Journal of Zoology* 59 (2):103–17.

Eldridge, M. D. B., S. Potter, C. N. Johnson and E. G. Ritchie. 2014. Differing impact of a major biogeographic barrier on genetic structure in two large kangaroos from the monsoon tropics of Northern Australia. *Ecology and Evolution* 4 (5):554–67.

Ericson, P. G. P., S. Klopfstein, M. Irestedt, J. M. T. Nguyen and J. A. A. Nylander. 2014. Dating the diversification of the major lineages of Passeriformes (Aves). *BMC Evolutionary Biology* 14:8.

Fabre, P. H., M. Pages, G. G. Musser, Y. S. Fitriana, J. Fjeldsa, A. Jennings, K. A. Jonsson *et al.* 2013. A new genus of rodent from Wallacea (Rodentia: Muridae: Murinae: Rattini), and its implication for biogeography and Indo-Pacific Rattini systematics. *Zoological Journal of the Linnean Society* 169 (2):408–47.

Firestone, K. B., M. S. Elphinstone, W. B. Sherwin and B. A. Houlden. 1999. Phylogeographical population structure of tiger quolls *Dasyurus maculatus* (Dasyuridae: Marsupialia), an endangered carnivorous marsupial. *Molecular Ecology* 8 (10):1613–25.

Flannery, T. F. 1988. Origins of the Australo-Pacific land mammal fauna. *Australian Zoological Reviews* 1:15–24.

Flannery, T. F. 1992. Taxonomic revision of the *Thylogale brunii* complex (Macropodidae: Marsupialia) in Melanesia with description of a new species. *Australian Mammalogy* 15:7–23.

Flannery, T. F. 1995a. *Mammals of the South-West Pacific and Moluccan Islands*. Chatswood, Australia: Reed Books.

Flannery, T. F. 1995b. *Mammals of New Guinea*. Chatswood, Australia: Reed Books.

Flannery, T. F., M. Archer, T. H. Rich and R. Jones. 1995. A new family of monotremes from the Cretaceous of Australia. *Nature* 377:418–20.

Flannery, T. F., and C. P. Groves. 1998. A revision of the genus *Zaglossus* (Monotremata, Tachyglossidae), with description of new species and subspecies. *Mammalia* 62:367–96.

Flannery, T. F., E. Hoch and K. Aplin. 1993. Macropodines from the Pliocene Otibanda Formation, Papua New Guinea. *Alcheringa: An Australasian Journal of Palaeontology* 13 (2):145–52.

Flynn, J. J., J. M. Parrish, B. Rakotosamimanana, W. F. Simpson and A. R. Wyss. 1999. A Middle Jurassic mammal from Madagascar. *Nature* 401:57–60.

Forasiepi, A. M., and A. G. Martinelli. 2003. Femur of a monotreme (Mammalia, Monotremata) from the Early Paleocene Salamanca Formation of Patagonia, Argentina. *Ameghiniana* 40 (4):625–30.

Fordyce, R. E. 2006. A southern perspective on cetacean evolution and zoogeography. In *Evolution and Biogeography of Australasian Vertebrates*, edited by J. R. Merrick, M. Archer, G. M. Hickey and M. S. Y. Lee, 755–78. Oatlands, Australia: Auscipub.

Francis, J. E., and I. Poole. 2002. Cretaceous and Early Tertiary climates of Antarctica: Evidence from fossil wood. *Palaeogeography, Palaeoclimatology, Palaeoecology* 182:47–64.

Fröbisch, J. 2009. Composition and similarity of global anomodont-bearing tetrapod faunas. *Earth-Science Reviews* 95 (3–4):119–57.

Furlan, E., J. Griffiths, N. Gust, R. Armistead, P. Mitrovski, K. A. Handasyde, M. Serena, A. A. Hoffmann and A. R. Weeks. 2011. Is body size variation in the platypus (*Ornithorhynchus anatinus*) associated with environmental variables? *Australian Journal of Zoology* 59 (4):201–15.

Gayet, M., L. G. Marshall, T. Sempere, F. J. Meunier, H. Cappetta and J. C. Rage. 2001. Middle Maastrichtian vertebrates (fishes, amphibians, dinosaurs and other reptiles, mammals) from Pajcha Pata (Bolivia). Biostratigraphic, palaeoecologic and palaeobiogeographic implications. *Palaeogeography, Palaeoclimatology, Palaeoecology* 169:39–68.

Gelfo, J. N. 2007. The 'condylarth' *Raulvaccia peligrensis* (Mammalia: Didolodontidae) from the Paleocene of Patagonia, Argentina. *Journal of Vertebrate Paleontology* 27 (3):651–60.

Gelfo, J. N., T. Mörs, M. Lorente, G. M. López and M. Reguero. 2014. The oldest mammals from Antarctica, Early Eocene of the La Meseta Formation, Seymour Island. *Palaeontology* 58:101–110.

Gelfo, J. N., E. Ortiz-Jaureguizar and G. W. Rougier. 2007. New remains and species of the 'condylarth' genus *Escribania* (Mammalia: Didolodontidae) from the Palaeocene of Patagonia, Argentina. *Earth and Environmental Science Transactions of the Royal Society of Edinburgh* 98:127–38.

Gelfo, J. N., and B. Sigé. 2011. A new didolodontid mammal from the Late Paleocene–earliest Eocene of Laguna Umayo, Peru. *Acta Palaeontologica Polonica* 56 (4):665–78.

George, G. G. 1979. The status of endangered Papua New Guinea mammals. In *The Status of Endangered Australasian Wildlife*, edited by M. J. Tyler, 93–100. Adelaide, Australia: Royal Zoological Society of South Australia.

Godthelp, H. 1999. Diversity, relationships and origins of the Tertiary and Quaternary rodents of Australia. *Australian Mammalogy* 21:32–5.

Godthelp, H., Wroe, S., and Archer, M. 1999. A new marsupial from the Early Eocene Tingamarra Local Fauna of Murgon, southeastern Queensland: A prototypical Australian marsupial? *Journal of Mammalian Evolution* 6: 289–313.

Godthelp, H. 2001. The Australian rodent fauna, flotilla's flotsam or just fleet footed? In *Faunal and Floral Migrations and Evolution in S.E. Asia–Australasia*, edited by I. Metcalfe, J. M. B. Smith, M. Morwood and I. Davidson, 319–22. Lisse, the Netherlands: A. A. Balkema.

Godthelp, H., M. Archer, R. L. Cifelli, S. J. Hand and C. F. Gilkeson. 1992. Earliest known Australian Tertiary mammal fauna. *Nature* 356:514–16.

Godthelp, H. J., M. Archer, S. J. Hand and K. Aplin. 2001. New species of the enigmatic mammal genus *Tingamarra*: A placental mammal in the court of king kangaroo. *Journal and Proceedings of the Royal Society of New South Wales* 134:105.

Goin, F. J., A. M. Forasiepi, A. M. Candela, E. O. Jaureguizar, R. Pascual, M. Archer, H. Godthelp *et al.* 1992. Earliest Paleocene marsupials from Patagonia. *Abstracts of the I International Palaeontological Congress*, Sydney: 68.

Goin, F. J., J. N. Gelfo, L. Chornogubsky, M. O. Woodburne and T. Martin. 2012b. Origins, radiations, and distribution of South American mammals: From greenhouse to icehouse worlds. In *Bones, Clones, and Biomes: An 80-Million Year History of Modern Neotropical Mammals*, edited by Bruce D. Patterson and Leonora P. Costa, 20–50. Chicago, IL: University of Chicago Press.

Goin, F. J., M. F. Tejedor, L. Chornogubsky, G. M. Lopez, J. N. Gelfo, M. Bond, M. O. Woodburne, Y. Gurovich and M. Reguero. 2012a. Persistence of a Mesozoic, non-therian mammalian lineage (Gondwanatheria) in the mid-Paleogene of Patagonia. *Naturwissenschaften* 99 (6):449–63.

Goin, F. J., A. N. Zimicz, A. M. Forasiepi, L. C. Chornogubsky and M. A. Abello. In press. The rise and fall of South American metatherians: Contexts, adaptations, radiations, and extinctions. In *Origins and Evolution of Cenozoic South American Mammals*, edited by A. L. Rosenberger and M. F. Tejedor. New York: Springer.

Goin, F. J., N. Zimicz, M. A. Reguero, S. N. Santillana, S. A. Marenssi and J. J. Moly. 2007. New marsupial (Mammalia) from the Eocene of Antarctica, and the origins and affinities of the Microbiotheria. *Revista de la Asociacion Geologica Argentina* 62 (4):597–603.

Gongora, J., A. B. Swan, A. Y. Chong, S. Y. W. Ho, C. S. Damayanti, S. Kolomyjec, T. Grant *et al.* 2012. Genetic structure and phylogeography of platypuses revealed by mitochondrial DNA. *Journal of Zoology* 286 (2):110–19.

Grant, T. 2007. *Platypus*. Collingwood, Australia: CSIRO Publishing.

Griffiths, M. 1978. *The Biology of the Monotremes*. New York: Academic Press.

Griffiths, M., R. T. Wells and D. J. Barrie. 1991. Observations on the skulls of fossil and extant echidnas (Monotremata: Tachyglossidae). *Australian Mammalogy* 14:87–101.

Groves, C. P. 2005a. Order Dasyuromorphia. In *Mammal Species of the World*, 3rd edn., edited by D. E. Wilson and D. M. Reeder, 23–37. Baltimore, MA: Johns Hopkins University Press.

Groves, C. P. 2005b. Order Diprotodontia. In *Mammal Species of the World*, 3rd edn., edited by D. E. Wilson and D. M. Reeder, 43–70. Baltimore: Johns Hopkins University Press.

Gunnell, G. F., N. B. Simmons and E. R. Seiffert. 2014. New Myzopodidae (Chiroptera) from the Late Paleogene of Egypt: Emended family diagnosis and biogeographic origins of Noctilionoidea. *PLoS ONE* 9 (2):e86712.

Gurovich, Y., and R. M. D. Beck. 2009. The higher-level relationships of the enigmatic mammalian clade Gondwanatheria. *Journal of Mammalian Evolution* 16 (1):25–49.

Hall, R. 2002. Cenozoic geological and plate tectonic evolution of SE Asia and the SW Pacific: Computer-based reconstructions, model and animations. *Journal of Asian Earth Sciences* 20 (4):353–431.

Hand, S., M. Archer and H. Godthelp. 2005. Australian Oligo-Miocene mystacinids (Microchiroptera): Upper dentition, new taxa and divergence of New Zealand species. *Geobios* 38 (3):339–52.

Hand, S., and J. Kirsch. 1998. A southern origin for the Hipposideridae (Microchiroptera)? Evidence from the Australian fossil record. In *Bat Biology and Conservation*, edited by T. Kunz and P. Racey, 72–90. Washington, DC: Smithsonian Institution Press.

Hand, S., M. Novacek, H. Godthelp and M. Archer. 1994. First Eocene bat from Australia. *Journal of Vertebrate Paleontology* 14 (3):375–81.

Hand, S. J. 2006. Bat beginnings and biogeography: The Australasian record. In *Evolution and Biogeography of Australasian Vertebrates*, edited by J. R. Merrick, M. Archer, G. M. Hickey and M. S. Y. Lee, 673–706. Oatlands, Australia: Auscipub.

Hand, S. J., P. Murray, D. Megirian, M. Archer and H. Godthelp. 1998. Mystacinid bats (Microchiroptera) from the Australian Tertiary. *Journal of Paleontology* 72 (3):538–45.

Hand, S. J., T. H. Worthy, M. Archer, J. P. Worthy, A. J. D. Tennyson and R. P. Scofield. 2013. Miocene mystacinids (Chiroptera, Noctilionoidea) indicate a long history for endemic bats in New Zealand. *Journal of Vertebrate Paleontology* 33 (6):1442–8.

Haq, B. U., J. Hardenbol and P. R. Vail. 1987. Chronology of fluctuating sea levels since the Triassic. *Science* 235:1156–67.

Hazlitt, S. L., A. W. Goldizen, J. A. Nicholls and M. D. B. Eldridge. 2014. Three divergent lineages within an Australian marsupial (*Petrogale penicillata*) suggest multiple major refugia for mesic taxa in southeast Australia. *Ecology and Evolution* 4 (7):1102–16.

Heinsohn, T. E. 2010. Marsupials as introduced species: Long-term anthropogenic expansion of the marsupial frontier and its implications for zoogeographic interpretation. In *Altered Ecologies: Fire, Climate and Human Influence on Terrestrial Landscapes*, edited by S. Haberle, J. Stevenson and M. Prebble, 133–76. Canberra, Australia: ANU E Press.

Helgen, K. M. 2007. A taxonomic and geographic overview of the mammals of Papua. In *The Ecology of Indonesia Series*, Vol. 6: *The Ecology of Papua*, part one, edited by A. Marshall and B. M. Beehler, 689–749. Singapore: Periplus Editions.

Helgen, K. M., R. P. Miguez, J. L. Kohen and L. E. Helgen. 2012. Twentieth century occurrence of the long-beaked echidna *Zaglossus bruijnii* in the Kimberley region of Australia. *Zookeys* (255):103–32.

Hill, K. C., and R. Hall. 2003. Mesozoic–Cenozoic evolution of Australia's New Guinea margin in a west Pacific context. In *Evolution and Dynamics of the Australian Plate*. Geological Society of Australia Special Publication No. 22:265–89.

Hocknull, S. A. 2005. Ecological succession during the Late Cainozoic of central eastern Queensland: Extinction of a diverse rainforest community. *Memoirs of the Queensland Museum* 51 (1):39–122.

Hocknull, S. A. 2009. Late Cainozoic rainforest vertebrates from Australopapua: Evolution, biogeography and extinction. PhD thesis, University of New South Wales, Sydney, Australia.

Hocknull, S. A., J.-X. Zhao, Y.-X. Feng and G. E. Webb. 2007. Responses of Quaternary rainforest vertebrates to climate change in Australia. *Earth and Planetary Science Letters* 264:317–31.

Jacobs, L. L., and L. J. Flynn. 2005. Of mice … again: The Siwalik rodent record, murine distribution, and molecular clocks. In *Interpreting the Past: Essays on Human, Primate, and Mammal Evolution in Honor of David Pilbeam*, edited by D. E. Lieberman, R. J. Smith and J. Kelley, 63–80. Boston, MA: Brill Academic.

Jansa, S. A., F. K. Barker and R. S. Voss. 2014. The early diversification history of didelphid marsupials: A window into South America's 'splendid isolation'. *Evolution* 68 (3):684–95.

Ji, Q., Z. X. Luo, C. X. Yuan, J. R. Wible, J. P. Zhang and J. A. Georgi. 2002. The earliest known eutherian mammal. *Nature* 416:816–22.

Johnson, C. 2006. *Australia's Mammal Extinctions: A 50,000 Year History*. Melbourne, Australia: Cambridge University Press.

Johnson, D. L. 1980. Problems in the land vertebrate zoogeography of certain islands and the swimming powers of elephants. *Journal of Biogeography* 7 (4):383–98.

Kemp, D. B., S. A. Robinson, J. A. Crame, J. E. Francis, J. Ineson, R. J. Whittle, V. Bowman and C. O'Brien. 2014. A cool temperate climate on the Antarctic Peninsula through the latest Cretaceous to Early Paleogene. *Geology* 42 (7):583–6.

Kemp, T. S. 2005. *The Origin and Evolution of Mammals*. Oxford, UK: Oxford University Press.

Khosla, A., and O. Verma. 2014. Paleobiota from the Deccan volcano-sedimentary sequences of India: Paleoenvironments, age and paleobiogeographic implications. *Historical Biology* 27 (7):898–914.

Kielan-Jaworowska, Z., R. L. Cifelli and Z.-X. Luo. 2004. *Mammals from the Age of Dinosaurs: Origins, Evolution, and Structure*. New York: Columbia University Press.

Kielan-Jaworowska, Z., E. Ortiz-Jaureguizar, C. Vieytes, R. Pascual and F. J. Goin. 2007. First ?cimolodontan multituberculate mammal from south America. *Acta Palaeontologica Polonica* 52 (2):257–62.

Kirsch, J. A. W., J. M. Hutcheon, D. G. P. Byrnes and B. D. Lloyd. 1998. Affinities and historical zoogeography of the New Zealand short-tailed bat, *Mystacina tuberculata* Gray 1843, inferred from DNA-hybridization comparisons. *Journal of Mammalian Evolution* 5 (1):33–64.

Kirsch, J. A. W., and M. S. Springer. 1993. Timing of the molecular evolution of New Guinean marsupials. *Science in New Guinea* 19 (3):147–156.

Kitazoe, Y., H. Kishino, P. J. Waddell, N. Nakajima, T. Okabayashi, T. Watabe and Y. Okuhara. 2007. Robust time estimation reconciles views of the antiquity of placental mammals. *PLoS ONE* 4:e384.

Kolomyjec, S. H., T. R. Grant, C. N. Johnson and D. Blair. 2013. Regional population structuring and conservation units in the platypus (*Ornithorhynchus anatinus*). *Australian Journal of Zoology* 61 (5):378–85.

Krajewski, C., R. Torunsky, J. T. Sipiorski and M. Westerman. 2007. Phylogenetic relationships of the dasyurid marsupial genus *Murexia*. *Journal of Mammalogy* 88 (3):696–705.

Krause, D. W. 2001. Fossil molar from a Madagascan marsupial. *Nature* 412:497–8.

Krause, D. W. 2013. Gondwanatheria and ?Multituberculata (Mammalia) from the Late Cretaceous of Madagascar. *Canadian Journal of Earth Sciences* 50 (3):324–40.

Lambeck, K., and J. Chappell. 2001. Sea level change through the last glacial cycle. *Science* 292 (5517):679–86.

Lavery, T. H., D. O. Fisher, T. F. Flannery and L. K. P. Leung. 2013. Higher extinction rates of dasyurids on Australo-Papuan continental shelf islands and the zoogeography of New Guinea mammals. *Journal of Biogeography* 40 (4):747–58.

Lavery, T. H., L. K.-P. Leung and J. M. Seddon. 2014. Molecular phylogeny of hipposiderid bats (Chiroptera: Hipposideridae) from Solomon Islands and Cape York Peninsula, Australia. *Zoologica Scripta* 43 (5):429–442.

Lawver, L. A., L. M. Gahagan and I. W. D. Dalziel. 2011. A different look at gateways: Drake Passage and Australia/Antarctica. In *Tectonic, Climatic, and Cryospheric Evolution of the Antarctic Peninsula*, edited by J. B. Anderson and J. S. Wellner, 5–33. Washington, DC: American Geophysical Union.

Lieberman, B. S. 2002. Phylogenetic biogeography with and without the fossil record: Gauging the effects of extinction and paleontological incompleteness. *Palaeogeography, Palaeoclimatology, Palaeoecology* 178 (1–2):39–52.

Livermore, R., A. Nankivell, G. Eagles and P. Morris. 2005. Paleogene opening of Drake Passage. *Earth and Planetary Science Letters* 236:459–70.

Lohman, D. J., M. de Bruyn, T. Page, K. von Rintelen, R. Hal, P. K. L. Ng, H. T. Shih, G. R. Carvalho and T. von Rintelen. 2011. Biogeography of the Indo-Australian Archipelago. *Annual Review of Ecology, Evolution, and Systematics* 42:205–26.

Lomolino, M. V., B. R. Riddle, R. J. Whittaker and J. H. Brown. 2010. *Biogeography*. Fourth edition. Sunderland: Sinauer Associates.

Long, J. A., M. Archer, T. F. Flannery and S. J. Hand. 2002. *Prehistoric Mammals of Australia and New Guinea: One Hundred Million Years of Evolution*. Sydney, Australia: UNSW Press.

Longman, H. A. 1916. The supposed artiodactyle Queensland fossils. *Proceedings of the Royal Society of Queensland* 28:83–7.
Luo, Z.-X., R. L. Cifelli and Z. Kielan-Jaworowska. 2001. Dual origin of tribosphenic mammals. *Nature* 409:53–7.
Luo, Z.-X., Q. Ji, J. R. Wible and C.-X. Yuan. 2003. An Early Cretaceous tribosphenic mammal and metatherian evolution. *Science* 302:1934–40.
Luo, Z.-X., Z. Kielan-Jaworowska and R. L. Cifelli. 2002. In quest for a phylogeny of Mesozoic mammals. *Acta Palaeontologica Polonica* 47 (1):1–78.
Luo, Z.-X., C. X. Yuan, Q. J. Meng and Q. Ji. 2011. A Jurassic eutherian mammal and divergence of marsupials and placentals. *Nature* 476 (7361):442–5.
Lydekker, R. 1896. *A Geographical History of Mammals*. Cambridge, UK: Cambridge University Press.
Mackness, Brian S., P. W. Whitehead and G. C. McNamara. 2000. New potassium-argon basalt date in relation to the Pliocene Bluff Downs Local Fauna, Northern Australia. *Australian Journal of Earth Sciences* 47 (4):807–11.
Macqueen, P., A. W. Goldizen, J. J. Austin and J. M. Seddon. 2011. Phylogeography of the pademelons (Marsupialia: Macropodidae: *Thylogale*) in New Guinea reflects both geological and climatic events during the Plio-Pleistocene. *Journal of Biogeography* 38:1732–47.
Macqueen, P., J. M. Seddon and A. W. Goldizen. 2012. Effects of historical forest contraction on the phylogeographic structure of Australo-Papuan populations of the red-legged pademelon (Macropodidae: *Thylogale stigmatica*). *Austral Ecology* 37 (4):479–90.
Malekian, M., S. J. B. Cooper and S. M. Carthew. 2010. Phylogeography of the Australian sugar glider (*Petaurus breviceps*): Evidence for a new divergent lineage in eastern Australia. *Australian Journal of Zoology* 58 (3):165–81.
Marshall, L. G., and C. de Muizon. 1988. The dawn of the age of mammals in South America. *National Geographic Research* 4 (1):23–55.
Martin, T., and O. W. M. Rauhut. 2005. Mandible and dentition of *Asfaltomylos patagonicus* (Australosphenida, Mammalia) and the evolution of tribosphenic teeth. *Journal of Vertebrate Paleontology* 25 (2):414–25.
Matzke, N. J. 2013. BioGeoBEARS: BioGeography with Bayesian (and Likelihood) Evolutionary Analysis in R Scripts. University of California, Berkeley, Berkeley, CA.
Matzke, N. J. 2014. Model selection in historical biogeography reveals that founder-event speciation is a crucial process in island clades. *Systematic Biology* 63 (6):951–70.
McKenzie, N. L., A. A. Burbidge, A. Baynes, R. N. Brereton, C. R. Dickman, G. Gordon, L. A. Gibson *et al.* 2007. Analysis of factors implicated in the recent decline of Australia's mammal fauna. *Journal of Biogeography* 34 (4):597–611.
Meredith, R. W., J. E. Janecka, J. Gatesy, O. A. Ryder, C. A. Fisher, E. C. Teeling, A. Goodbla *et al.* 2011. Impacts of the Cretaceous Terrestrial Revolution and KPg extinction on mammal diversification. *Science* 334 (6055):521–4.
Meredith, R. W., C. Krajewski, M. Westerman and M. S. Springer. 2009a. Relationships and divergence times among the orders and families of Marsupialia. *Museum of Northern Arizona Bulletin* 65:383–406.
Meredith, R. W., M. A. Mendoza, K. K. Roberts, M. Westerman and M. S. Springer. 2010. A phylogeny and timescale for the evolution of Pseudocheiridae (Marsupialia: Diprotodontia) in Australia and New Guinea. *Journal of Mammalian Evolution* 17 (2):75–99.
Meredith, R. W., M. Westerman and M. S. Springer. 2009b. A phylogeny of Diprotodontia (Marsupialia) based on sequences for five nuclear genes. *Molecular Phylogenetics and Evolution* 51 (3):554–71.
Meyer, M., Q. M. Fu, A. Aximu-Petri, I. Glocke, B. Nickel, J. L. Arsuaga, I. Martinez *et al.* 2014. A mitochondrial genome sequence of a hominin from Sima de los Huesos. *Nature* 505 (7483):403–6.
Miller, K. G., M. A. Kominz, J. V. Browning, J. D. Wright, G. S. Mountain, M. E. Katz, P. J. Sugarman, B. S. Cramer, N. Christie-Blick and S. F. Pekar. 2005. The Phanerozoic record of global sea-level change. *Science* 310:1293–8.
Miller-Butterworth, C. M., W. J. Murphy, S. J. O'Brien, D. S. Jacobs, M. S. Springer and E. C. Teeling. 2007. A family matter: Conclusive resolution of the taxonomic position of the long-fingered bats, *Miniopterus*. *Molecular Biology and Evolution* 24 (7):1553–61.
Mirceta, S., A. V. Signore, J. M. Burns, A. R. Cossins, K. L. Campbell and M. Berenbrink. 2013. Evolution of mammalian diving capacity traced by myoglobin net surface charge. *Science* 340 (6138):1234192.

Mitchell, K. J., R. C. Pratt, L. N. Watson, G. C. Gibb, B. Llamas, M. Kasper, J. Edson *et al.* 2014. Molecular phylogeny, biogeography, and habitat preference evolution of marsupials. *Molecular Biology and Evolution* 31 (9):2322–30.

Morrone, J. J. 2002. Biogeographical regions under track and cladistic scrutiny. *Journal of Biogeography* 29:149–52.

Muizon, C. de. 1991. La fauna de mamiferos de Tiupampa (Paleoceno Inferior, Formacion Santa Lucia), Bolivia. In *Fossils y Facies de Bolivia*, Vol. 1: *Vertebrados,* edited by R. Suarez-Soruco, 575–624. Santa Cruz, Bolivia: Revista Technica de Yacimientos Petroliferos Fiscales Bolivianos.

Muizon, C. de, and Richard L. Cifelli. 2000. The 'condylarths' (archaic Ungulata, Mammalia) from the Early Palaeocene of Tiupampa (Bolivia): Implications on the origin of the South American ungulates. *Geodiversitas* 22 (1):47–150.

Muizon, C. de, and Richard L. Cifelli. 2001. A new basal 'didelphoid' (Marsupialia, Mammalia) from the Early Paleocene of Tiupampa (Bolivia). *Journal of Vertebrate Paleontology* 21 (1):87–97.

Murray, P. F. 1978a. Late Cenozoic monotreme anteaters. *Australian Zoologist* 20 (1):29–55.

Murray, P. F. 1978b. A Pleistocene spiny anteater from Tasmania (Monotremata, Tachyglossidae, Zaglossus). *Papers and Proceedings of the Royal Society of Tasmania* 112:39–67.

Murray, P. F. 1992. The smallest New Guinea zygomaturines-derived dwarfs or relict plesiomorphs? *The Beagle: Records of the Northern Territory Museum of Arts and Sciences* 9 (1):89–110.

Musser, A. M. 2003. Review of the monotreme fossil record and comparison of palaeontological and molecular data. *Comparative Biochemistry and Physiology Part A* 136:927–42.

Musser, A. M. 2006. Furry egg-layers: monotreme relationships and radiations. In *Evolution and Biogeography of Australasian Vertebrates*, edited by J. R. Merrick, M. Archer, G. M. Hickey and M. S. Y. Lee, 523–47. Oatlands, Australia: Auscipub.

Musser, A. M. 2013. Classification and evolution of the monotremes. In *Neurobiology of Monotremes: Brain Evolution in Our Distant Mammalian Cousins*, edited by Ken Ashwell, 1–16. Collingwood, Australia: CSIRO Publishing.

Musser, A. M., and M. Archer. 1998. New information about the skull and dentary of the Miocene platypus *Obdurodon dicksoni* and a discussion of ornithorhynchid relationships. *Philosophical Transactions of the Royal Society of London B* 353:1063–79.

Musser, G. G., and M. D. Carleton. 2005. Superfamily Muroidea. In *Mammal Species of the World,* Third Edition, edited by Wilson D. E. and D. M. Reeder, 894–1531. Baltimore, USA: The Johns Hopkins University Press.

Neaves, L. E., K. R. Zenger, R. I. T. Prince, M. D. B. Eldridge and D. W. Cooper. 2009. Landscape discontinuities influence gene flow and genetic structure in a large, vagile Australian mammal, Macropus fuliginosus. *Molecular Ecology* 18 (16):3363–78.

Nicol, S. C. 2013. Behaviour and ecology of monotremes. In *Neurobiology of Monotremes: Brain Evolution in Our Distant Mammalian Couins*, edited by Ken Ashwell, 17–30. Collingwood, Australia: CSIRO Publishing.

Nilsson, M. A., G. Churakov, M. Sommer, N. V. Tran, A. Zemann, J. Brosius and J. Schmitz. 2010. Tracking marsupial evolution using archaic genomic retroposon insertions. *PLoS Biology* 8 (7):e1000436.

O'Connor, P. M., J. J. W. Sertich, N. J. Stevens, E. M. Roberts, M. D. Gottfried, T. L. Hieronymus, Z. A. Jinnah, R. Ridgely, S. E. Ngasala and J. Temba. 2010. The evolution of mammal-like crocodyliforms in the Cretaceous Period of Gondwana. *Nature* 466 (7307):748–51.

O'Leary, M. A., J. I. Bloch, J. J. Flynn, T. J. Gaudin, A. Giallombardo, N. P. Giannini, S. L. Goldberg *et al.* 2013. The placental mammal ancestor and the post-K-Pg radiation of placentals. *Science* 339 (6120):662–7.

Orlando, L., A. Ginolhac, G. J. Zhang, D. Froese, A. Albrechtsen, M. Stiller, M. Schubert *et al.* 2013. Recalibrating Equus evolution using the genome sequence of an early Middle Pleistocene horse. *Nature* 499 (7456):74–8.

Owen, R. 1845. Description of a fossil molar tooth of a *Mastodon* discovered by Count Strzlecki in Australia. *The Annals and Magazine of Natural History* 14:268–71.

Owen, R. 1882. Description of portions of a tusk of a proboscidian mammal from Australian Pleistocene deposits. *Philosophical Transactions of the Royal Society of London* 173:777–81.

Parmar, V., G. V. R. Prasad and D. Kumar. 2013. The first multituberculate mammal from India. *Naturwissenschaften* 100 (6):515–23.

Pascual, R. 2006. Evolution and geography: the biogeographic history of South American land mammals. *Annals of the Missouri Botanical Garden* 93 (2):209–30.

Pascual, R., M. Archer, E. Ortiz-Jaureguizar, J. L. Prado, H. Godthelp and S. J. Hand. 1992. First discovery of monotremes in South America. *Nature* 356:704–5.

Pascual, R., F. J. Goin, L. Balarino and D. E. Udrizar Sauthier. 2002. New data on the Paleocene monotreme *Monotrematum sudamericanum*, and the convergent evolution of triangulate molars. *Acta Palaeontologica Polonica* 47 (3):487–92.

Pascual, R., F. J. Goin, D. W. Krause, E. Ortiz-Jaureguizar and A. A. Carlini. 1999. The first gnathic remains of *Sudamerica*: Implications for gondwanathere relationships. *Journal of Vertebrate Paleontology* 19 (2):373–82.

Pascual, R., and E. Ortiz-Jaureguizar. 2007. The Gondwanan and South American episodes: Two major and unrelated moments in the history of the South American mammals. *Journal of Mammalian Evolution* 14 (2):75–137.

Pavlova, A., F. M. Walker, R. van der Ree, S. Cesarini and A. C. Taylor. 2010. Threatened populations of the Australian squirrel glider (*Petaurus norfolcensis*) show evidence of evolutionary distinctiveness on a Late Pleistocene timescale. *Conservation Genetics* 11 (6):2393–407.

Phillips, M. J., T. H. Bennett and M. S. Lee. 2009. Molecules, morphology, and ecology indicate a recent, amphibious ancestry for echidnas. *Proceedings of the National Academy of Sciences of the United States of America* 106 (40):17089–94.

Phillips, M. J., T. H. Bennett and M. S. Y. Lee. 2010. Reply to Camens: How recently did modern monotremes diversify? *Proceedings of the National Academy of Sciences of the United States of America* 107 (4):E13

Pian, R., M. Archer and S. J. Hand. 2013. A new, giant platypus, *Obdurodon tharalkooschild*, sp. nov. (Monotremata, Ornithorhynchidae), from the Riversleigh World Heritage Area, Australia. *Journal of Vertebrate Paleontology* 33 (6):1255–9.

Pian, R., Archer, M., Hand, S. J., Beck, R. M. D., and Cody, A. (2016). The upper dentition and relationships of the enigmatic Australian Cretaceous mammal *Kollikodon ritchiei*. *Memoirs of Museum Victoria* 74: 97–105.

Plane, M. D. 1967. Stratigraphy and vertebrate fauna of the Otibanda Formation, New Guinea. *Australian Bureau of Mineral Resources, Geology and Geophysics Bulletin* 86:1–64.

Pledge, N. S. 1980. Giant echidnas in South Australia. *South Australian Naturalist* 55:27–30.

Poole, I., D. Cantrill and T. Utescher. 2005. A multi-proxy approach to determine Antarctic terrestrial palaeoclimate during the Late Cretaceous and Early Tertiary. *Palaeogeography, Palaeoclimatology, Palaeoecology* 222:95–121.

Pope, L. C., A. Estoup and C. Moritz. 2000. Phylogeography and population structure of an ecotonal marsupial, Bettongia tropica, determined using mtDNA and microsatellites. *Molecular Ecology* 9 (12):2041–53.

Pope, L., D. Storch, M. Adams, C. Moritz and G. Gordon. 2001. A phylogeny for the genus *Isoodon* and a range extension for *I. obesulus peninsulae* based on mtDNA control region and morphology. *Australian Journal of Zoology* 49 (4):411–34.

Poropat, S. F., P. Upchurch, P. D. Mannion, S. A. Hocknull, B. P. Kear, T. Sloan, G. H. K. Sinapius and D. A. Elliott. 2014. Revision of the sauropod dinosaur *Diamantinasaurus matildae* Hocknull *et al.* 2009 from the mid-Cretaceous of Australia: implications for Gondwanan titanosauriform dispersal. *Gondwana Research* 27 (3):995–1033.

Potter, S., M. D. B. Eldridge, D. A. Taggart and S. J. B. Cooper. 2012. Multiple biogeographical barriers identified across the monsoon tropics of northern Australia: Phylogeographic analysis of the *brachyotis* group of rock-wallabies. *Molecular Ecology* 21 (9):2254–69.

Potter, S., D. Rosauer, J. S. Doody, M. J. Webb and M. D. B. Eldridge. 2014. Persistence of a potentially rare mammalian genus (*Wyulda*) provides evidence for areas of evolutionary refugia within the Kimberley, Australia. *Conservation Genetics* 15 (5):1084–94.

Prasad, G. V. R., J. J. Jaeger, A. Sahni, E. Gheerbrant and C. K. Khajuria. 1994. Eutherian mammals from the Upper Cretaceous (Maastrichtian) Intertrappean Beds of Naskal, Andhra Pradesh, India. *Journal of Vertebrate Paleontology* 14 (2):260–77.

Prasad, G. V. R., and B. K. Manhas. 2007. A new docodont mammal from the Jurassic Kota Formation of India. *Palaeontologia Electronica* 10 (2):7A.

Prasad, G. V., O. Verma, A. Sahni, V. Parmar and A. Khosla. 2007. A Cretaceous hoofed mammal from India. *Science* 318 (5852):937.

Price, G. J., and S. A. Hocknull. 2011. *Invictokoala monticola* gen. et sp. nov. (Phascolarctidae, Marsupialia), a Pleistocene plesiomorphic koala holdover from Oligocene ancestors. *Journal of Systematic Palaeontology* 9 (2):327–35.

Pridmore, P. A., T. H. Rich, P. Vickers-Rich and P. P. Gambaryan. 2005. A tachyglossid-like humerus from the Early Cretaceous of south-eastern Australia. *Journal of Mammalian Evolution* 12:359–78.

Prins, H. H. T., and I. J. Gordon. 2014. *Invasion Biology and Ecological Theory: Insights from a Continent in Transformation*. Cambridge, UK: Cambridge University Press.

Pyron, R. A. 2011. Divergence time estimation using fossils as terminal taxa and the origins of Lissamphibia. *Systematic Biology* 60 (4):466–81.

Quarles van Ufford, A., and M. Cloos. 2005. Cenozoic tectonics of New Guinea. *AAPG Bulletin* 89 (1):119–40.

Rana, R. S., and G. P. Wilson. 2003. New Late Cretaceous mammals from the Intertrappean beds of Rangapur, India and paleobiogeographic framework. *Acta Palaeontologica Polonica* 48 (3):331–48.

Raterman, D., R. W. Meredith, L. A. Ruedas and M. S. Springer. 2006. Phylogenetic relationships of the cuscuses and brushtail possums (Marsupialia: Phalangeridae) using the nuclear gene BRCA1. *Australian Journal of Zoology* 54:353–61.

Rauhut, O. W. M., T. Martin, E. Ortiz-Jaureguizar and P. Puerta. 2002. A Jurassic mammal from South America. *Nature* 416:165–8.

Reardon, T. B., N. L. McKenzie, S. J. B. Cooper, B. Appleton, S. Carthew and M. Adams. 2014. A molecular and morphological investigation of species boundaries and phylogenetic relationships in Australian free-tailed bats Mormopterus (Chiroptera: Molossidae). *Australian Journal of Zoology* 62 (2):109–36.

Ree, R. H., B. R. Moore, C. O. Webb and M. J. Donoghue. 2005. A likelihood framework for inferring the evolution of geographic range on phylogenetic trees. *Evolution* 59 (11):2299–311.

Ree, R. H., and I. Sanmartin. 2009. Prospects and challenges for parametric models in historical biogeographical inference. *Journal of Biogeography* 36 (7):1211–20.

Ree, R. H., and S. A. Smith. 2008. Maximum likelihood inference of geographic range evolution by dispersal, local extinction, and cladogenesis. *Systematic Biology* 57 (1):4–14.

Reguero, M., F. Goin, C. A. Hospitaleche, T. Dutra and S. Marenssi. 2013. *Late Cretaceous/Paleogene West Antarctica Terrestrial Biota and Its Intercontinental Affinities*. Dordrecht, the Netherlands: Springer.

Rich, T. H. 1991. Monotremes, placentals, and marsupials: Their record in Australia and its biases. In *Vertebrate Palaeontology of Australasia*, edited by P. Vickers-Rich, J. M. Monaghan, R. F. Baird and T. H. Rich, 893–1070. Melbourne, Australia: Pioneer Design Studio and Monash University Publications Committee.

Rich, T. H. 2008. The palaeobiogeography of Mesozoic mammals: A review. *Arquivos do Museu Nacional, Rio de Janeiro* 66:231–49.

Rich, T. H., T. F. Flannery, P. Trusler, L. Kool, N. van Klaveren and P. Vickers-Rich. 2001a. A second placental mammal from the Early Cretaceous Flat Rocks site, Victoria, Australia. *Records of the Queen Victoria Museum* 110:1–9.

Rich, T. H., Hopson, J. A.,, Musser, A. M., Flannery, T. F., Vickers-Rich, P. 2005. Independent origins of middle ear bones in monotremes and therians. *Science* 307 (5711):910–914. doi:10.1126/science.1105717.

Rich T. H., Vickers-Rich, P., Trusler, P., Flannery, T. F., Cifelli, R. L., Constantine, A., Kool, L., van Klaveren, N. 2001b. Monotreme nature of the Australian Early Cretaceous mammal *Teinolophos trusleri*. *Acta Palaeontol Pol* 46:113–118.

Rich, T. H., and P. Vickers-Rich. 2004. Diversity of Early Cretaceous mammals from Victoria, Australia. *Bulletin of the American Museum of Natural History* 285:36–53.

Rich, T. H., P. Vickers-Rich, A. Constantine, T. F. Flannery, L. Kool and N. van Klaveren. 1997. A tribosphenic mammal from the Mesozoic of Australia. *Science* 278:1438–42.

Rich, T. H., P. Vickers-Rich, A. Constantine, T. F. Flannery, L. Kool and N. van Klaveren. 1999. Early Cretaceous mammals from Flat Rocks, Victoria, Australia. *Records of the Queen Victoria Museum* 106:1–35.

Rich, T. H., P. Vickers-Rich, T. F. Flannery, B. P. Kear, D. J. Cantrill, P. Komarower, L. Kool *et al.* 2009a. An Australian multituberculate and its palaeobiogeographic implications. *Acta Palaeontologica Polonica* 54 (1):1–6.

Rich, T. H., P. Vickers-Rich, T. F. Flannery, D. Pickering, L. Kool, A. M. Tait and E. M. G. Fitzgerald. 2009b. A fourth Australian Mesozoic mammal locality. *Museum of Northern Arizona Bulletin* 65:677–81.

Rich, T. H., P. Vickers-Rich, P. Trusler, T. F. Flannery, R. L. Cifelli, A. Constantine, L. Kool and N. van Klaveren. 2001b. Monotreme nature of the Australian Early Cretaceous mammal *Teinolophos trusleri*. *Acta Palaeontologica Polonica* 46:113–18.

Rich, T. H. V., T. F. Flannery and M. Archer. 1989. A second Cretaceous mammalian specimen from Lightning Ridge, N.S.W., Australia. *Alcheringa* 13:85–8.

Ronquist, F., S. Klopfstein, L. Vilhelmsen, S. Schulmeister, D. L. Murray and A. P. Rasnitsyn. 2012. A total-evidence approach to dating with fossils, applied to the early radiation of the Hymenoptera. *Systematic Biology* 61 (6):973–99.

Rose, K. D. 2006. *The Beginning of the Age of Mammals*. Baltimore, MA: Johns Hopkins University Press.

Rougier, G. W., S. Apesteguia and L. C. Gaetano. 2011a. Highly specialized mammalian skulls from the Late Cretaceous of South America. *Nature* 479 (7371):98–102.

Rougier, G. W., L. Chornogubsky, S. Casadio, N. P. Arango and A. Giallombardo. 2009a. Mammals from the Allen Formation, Late Cretaceous, Argentina. *Cretaceous Research* 30 (1):223–38.

Rougier, G. W., A. M. Forasiepi, R. V. Hill and M. Novacek. 2009b. New mammalian remains from the Late Cretaceous La Colonia Formation, Patagonia, Argentina. *Acta Palaeontologica Polonica* 54 (2):195–212.

Rougier, G. W., L. C. Gaetano, B. Drury, N. Paéz Arango and R. Colella. 2011b. A review of the Mesozoic mammalian record of South America. In *Paleontología y Dinosaurios Desde América Latina*, edited by J. Calvo, J. Porfiri, B. Gonzales Riga and D. Dos Santos, 195–214. Mendoza, Argentina: Editorial de la Universidad Nacional de Cuyo e EDIUNC.

Rougier, G. W., A. G. Martinelli, A. M. Forasiepi and M. J. Novacek. 2007. New Jurassic mammals from Patagonia, Argentina: A reappraisal of australosphenidan morphology and interrelationships. *American Museum Novitates* 3566:1–54.

Rowe, K. C., K. P. Aplin, P. R. Baverstock and C. Moritz. 2011a. Recent and rapid speciation with limited morphological disparity in the genus *Rattus*. *Systematic Biology* 60 (2):188–203.

Rowe, K. C., M. L. Reno, D. M. Richmond, R. M. Adkins and S. J. Steppan. 2008. Pliocene colonization and adaptive radiations in Australia and New Guinea (Sahul): Multilocus systematics of the old endemic rodents (Muroidea: Murinae). *Molecular Phylogenetics and Evolution* 47 (1):84–101.

Rowe, K. M. C., K. C. Rowe, M. S. Elphinstone and P. R. Baverstock. 2011b. Population structure, timing of divergence and contact between lineages in the endangered Hastings River mouse (*Pseudomys oralis*). *Australian Journal of Zoology* 59:186–200.

Schenk, J. J., K. C. Rowe and S. J. Steppan. 2013. Ecological opportunity and incumbency in the diversification of repeated continental colonizations by muroid rodents. *Systematic Biology* 62 (6):837–64.

Scher, H. D., and E. E. Martin. 2006. Timing and climatic consequences of the opening of Drake Passage. *Science* 312:428–30.

Schoch, R. R. 2014. *Amphibian Evolution: The Life of Early Land Vertebrates*. Chichester, UK: Wiley.

Sereno, Paul C. 2006. Shoulder girdle and forelimb in multituberculates: Evolution of parasagittal forelimb posture in mammals. In *Amniote Paleobiology: Perspectives on the Evolution of Mammals, Birds, and Reptiles*, edited by M. T. Carrano, T. J. Gaudin, R. W. Blob and J. R. Wible, 315–66. Chicago, IL: University of Chicago Press.

Seton, M., R. D. Muller, S. Zahirovic, C. Gaina, T. H. Torsvik, G. Shephard, A. Talsma *et al.* 2012. Global continental and ocean basin reconstructions since 200 Ma. *Earth-Science Reviews* 113 (3–4):212–70.

Sheng, G. L., J. Soubrier, J. Y. Liu, L. Werdelin, B. Llamas, V. A. Thomson, J. Tuke *et al.* 2014. Pleistocene Chinese cave hyenas and the recent Eurasian history of the spotted hyena, *Crocuta crocuta*. *Molecular Ecology* 23 (3):522–33.

Sigé, B., M. Archer, J.-Y. Crochet, H. Godthelp, S. Hand and R. M. D. Beck. 2009. *Chulpasia* and *Thylacotinga*, Late Paleocene–earliest Eocene trans-Antarctic Gondwanan bunodont marsupials: New data from Australia. *Geobios* 42 (6):813–23.

Simmons, N. B. 2005. Order Chiroptera. In *Mammal Species of the World*, 3rd edn., edited by D. E. Wilson and D. M. Reeder, 312–529. Baltimore, MA: The Johns Hopkins University Press.

Simpson, G. G. 1977. Too many lines: Limits of Oriental and Australian zoogeographic regions. *Proceedings of the American Philosophical Society* 121 (2):107–20.

Smith, E. T. 2009. Terrestrial and freshwater turtles of Early Cretaceous Australia. Unpublished PhD thesis, University of New South Wales, Sydney, Australia.

Smith, T., J. Habersetzer, N. B. Simmons and G. F. Gunnell. 2012. Systematics and paleobiogeography of early bats. In *Evolutionary History of Bats: Fossils, Molecules and Morphology*, edited by G. F. Gunnell and N. B. Simmons, 23–66. Cambridge, UK: Cambridge University Press.

Smith, T., R. S. Rana, P. Missiaen, K. D. Rose, A. Sahni, H. Singh and L. Singh. 2007. High bat (Chiroptera) diversity in the Early Eocene of India. *Naturwissenschaften* 94:1003–9.

Spencer, P. B. S., S. G. Rhind and M. D. B. Eldridge. 2001. Phylogeographic structure within *Phascogale* (Marsupialia: Dasyuridae) based on partial cytochrome b sequence. *Australian Journal of Zoology* 49 (4):369–77.

Springer, M. S., R. W. Meredith, J. E. Janecka and W. J. Murphy. 2011. The historical biogeography of Mammalia. *Philosophical Transactions of the Royal Society B: Biological Sciences* 366 (1577):2478–502.

Steiper, M. E., and E. R. Seiffert. 2012. Evidence for a convergent slowdown in primate molecular rates and its implications for the timing of early primate evolution. *Proceedings of the National Academy of Sciences of the United States of America* 109 (16):6006–11.

Stickley, C. E., H. Brinkhuis, S. A. Schellenberg, A. Sluijs, U. Rohl, M. Fuller, M. Grauert, M. Huber, J. Warnaar and G. L. Williams. 2004. Timing and nature of the deepening of the Tasmanian Gateway. *Paleoceanography* 19 (4):1–18. doi:10.1029/2004pa001022.

Stiller, M., M. Molak, S. Prost, G. Rabeder, G. Baryshnikov, W. Rosendahl, S. Munzel *et al.* 2014. Mitochondrial DNA diversity and evolution of the Pleistocene cave bear complex. *Quaternary International* 339–340:224–31.

Sullivan, C., Y. Wang, D. W. E. Hone, Y. Wang, X. Xu and F. Zhang. 2014. The vertebrates of the Jurassic Daohugou Biota of northeastern China. *Journal of Vertebrate Paleontology* 34 (2):243–80.

Szalay, F. S. 1982. A new appraisal of marsupial phylogeny and classification. In *Carnivorous Marsupials*, edited by Michael Archer, 621–40. Mosman, Australia: Royal Zoological Society of New South Wales.

Szalay, F. S. 1994. *Evolutionary History of the Marsupials and an Analysis of Osteological Characters*. Cambridge, UK: Cambridge University Press.

Tabuce, R., M. T. Antunes and B. Sige. 2009. A new primitive bat from the earliest Eocene of Europe. *Journal of Vertebrate Paleontology* 29 (2):627–30.

Teeling, E. C., O. Madsen, W. J. Murphy, M. S. Springer and S. J. O'Brien. 2003. Nuclear gene sequences confirm an ancient link between New Zealand's short-tailed bat and South American noctilionoid bats. *Molecular Phylogenetics and Evolution* 28 (2):308–19.

Teeling, E. C., M. S. Springer, O. Madsen, P. Bates, S. J. O'Brien and W. J. Murphy. 2005. A molecular phylogeny for bats illuminates biogeography and the fossil record. *Science* 307 (5709):580–84.

Tejedor, M. F., N. J. Czaplewski, F. J. Goin and E. Aragón. 2005. The oldest record of South American bats. *Journal of Vertebrate Paleontology* 25 (4):990–93.

Thulborn, T., and S. Turner. 2003. The last dicynodont: An Australian Cretaceous relict. *Proceedings of the Royal Society B: Biological Sciences* 270:985–93.

Torsvik, T. H., and L. R. M. Cocks. 2013. Gondwana from top to base in space and time. *Gondwana Research* 24 (3–4):999–1030.

Toussaint, E. F. A., R. Hall, M. T. Monaghan, K. Sagata, S. Ibalim, H. V. Shaverdo, A. P. Vogler, J. Pons and M. Balke. 2014. The towering orogeny of New Guinea as a trigger for arthropod megadiversity. *Nat Commun* 5. doi:10.1038/Ncomms5001.

Turnbull, W. D., E. L. Lundelius, Jr. and M. Archer. 2003. Dasyurids, perameloids, phalangeroids, and vombatoids from the Early Pliocene Hamilton Fauna, Victoria, Australia. *Bulletin of the American Museum of Natural History* 279:513–40.

van der Geer, A., G. Lyras, J. de Vos and M. Dermitzakis. 2010. *Evolution of Island Mammals: Adaptation and Extinction of Placental Mammals on Islands*. Chichester, UK: Wiley-Blackwell.

Van Dyck, S. 2002. Morphology-based revision of *Murexia* and *Antechinus* (Marsupialia: Dasyuridae). *Memoirs of the Queensland Museum* 48:239–330.

Van Dyck, S., and R. Strahan. 2008. *The Mammals of Australia*, 3rd edn. Sydney, Australia: New Holland Publishers.

Vullo, R., E. Gheerbrant, C. de Muizon and D. Neraudeau. 2009. The oldest modern therian mammal from Europe and its bearing on stem marsupial paleobiogeography. *Proceedings of the National Academy of Sciences of the United States of America* 106 (47):19910–15.

Waddell, P. J. 2008. Fit of fossils and mammalian molecular trees: Dating inconsistencies revisited. arXiv:0812.5114.

Warren, A., T. H. Rich and P. Vickers-Rich. 1997. The last labyrinthodonts? *Palaeontographica Abteilung A* 247:1–24.

Westerman, M., B. P. Kear, K. Aplin, R. W. Meredith, C. Emerling and M. S. Springer. 2012. Phylogenetic relationships of living and recently extinct bandicoots based on nuclear and mitochondrial DNA sequences. *Molecular Phylogenetics and Evolution* 62 (1):97–108.

Wilson, D. E., and D. M. Reeder. 2005. *Mammal Species of the World*, 3rd edn. Baltimore, MA: John Hopkins University Press.

Woinarski, J. C. Z., M. Armstrong, K. Brennan, A. Fisher, A. D. Griffiths, B. Hill, D. J. Milne *et al.* 2010. Monitoring indicates rapid and severe decline of native small mammals in Kakadu National Park, northern Australia. *Wildlife Research* 37 (2):116–26.

Woinarski, J. C. Z., S. Legge, J. A. Fitzsimons, B. J. Traill, A. A. Burbidge, A. Fisher, R. S. C. Firth *et al.* 2011. The disappearing mammal fauna of northern Australia: Context, cause, and response. *Conservation Letters* 4 (3):192–201.

Woodburne, M. O., and J. A. Case. 1996. Dispersal, vicariance, and the Late Cretaceous to Early Tertiary land mammal biogeography from South America to Australia. *Journal of Mammalian Evolution* 3:121–61.

Woodburne, M. O., F. J. Goin, M. Bond, A. A. Carlini, J. N. Gelfo, G. M. López, A. Iglesias and A. N. Zimicz. 2014. Paleogene land mammal faunas of South America: A response to global climatic changes and indigenous floral diversity. *Journal of Mammalian Evolution* 21 (1):1–73.

Woodburne, M. O., B. J. MacFadden, J. A. Case, M. S. Springer, N. S. Pledge, J. D. Power, J. M. Woodburne and K. B. Springer. 1994. Land mammal biostratigraphy and magnetostratigraphy of the Etadunna Formation (Late Oligocene) of South Australia. *Journal of Vertebrate Paleontology* 13:483–515.

Woodburne, M. O., T. H. Rich and M. S. Springer. 2003. The evolution of tribospheny and the antiquity of mammalian clades. *Molecular Phylogenetics and Evolution* 28:360–85.

Woodburne, M. O., and R. H. Tedford. 1975. The first Tertiary monotreme from Australia. *American Museum Novitates* 2588:1–11.

Woodhead, J., S. J. Hand, M. Archer, I. Graham, K. Sniderman, D. A. Arena, K. H. Black, H. Godthelp, P. Creaser and E. Price. 2014. Developing a radiometrically-dated chronologic sequence for Neogene biotic change in Australia, from the Riversleigh World Heritage Area of Queensland. *Gondwana Research* 29 (1):153–67.

Zachos, J., M. Pagani, L. Sloan, E. Thomas and K. Billups. 2001. Trends, rhythms, and aberrations in global climate 65 Ma to present. *Science* 292:686–93.

Index

An environmentally friendly book printed and bound in England by www.printondemand-worldwide.com

PEFC Certified

This product is from sustainably managed forests and controlled sources

www.pefc.org

PEFC/16-33-415

This book is made of chain-of-custody materials; FSC materials for the cover and PEFC materials for the text pages.